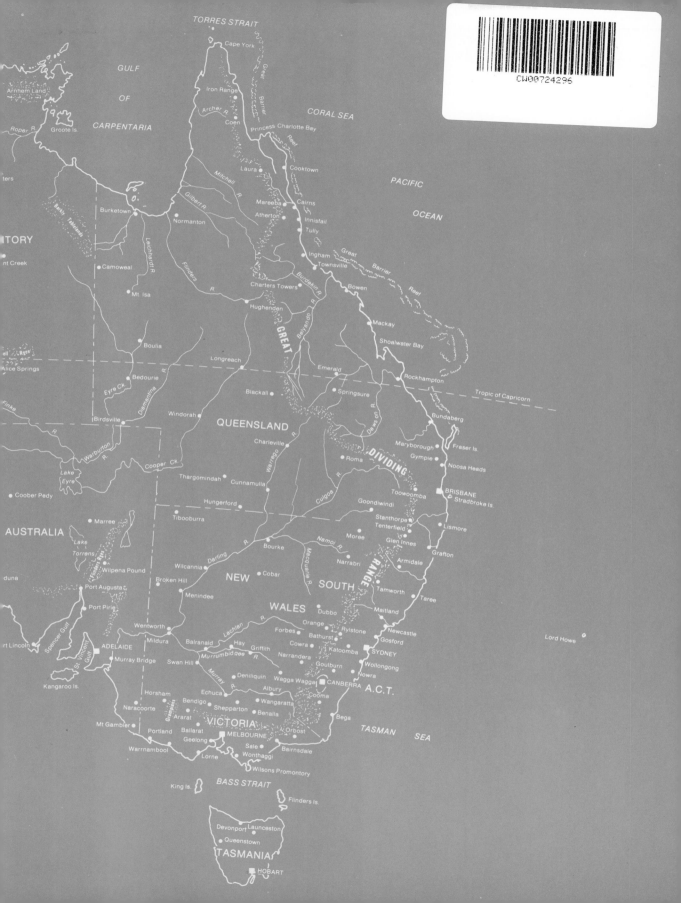

ENCYCLOPAEDIA OF AUSTRALIAN PLANTS

suitable for cultivation

VOLUME THREE

ENCYCLOPAEDIA OF AUSTRALIAN PLANTS

suitable for cultivation

VOLUME THREE

W. Rodger Elliot
David L. Jones

B.Ag.Sc. Dip.Hort.

Line drawings by Trevor L. Blake

LOTHIAN PUBLISHING COMPANY PTY. LTD.

MELBOURNE SYDNEY AUCKLAND

LOTHIAN BOOKS
Lothian Publishing Company Pty Ltd
11 Munro Street, Port Melbourne, Victoria 3027

First published 1984
Reprinted 1989
Copyright © W. Rodger Elliot and David L. Jones 1984

National Library of Australia
Cataloguing-in-publication data:

Elliot, W. Rodger (Winston Rodger), 1941 —
Encyclopaedia of Australian plants suitable for
cultivation. Volume 3.

Bibliography.
Includes index.
ISBN 0 85091 167 2

1. Wild flower gardening — Australia.
I. Jones, D.L. (David Lloyd), 1944 — II. Title

635.9'676'0994

Designed by Arthur Stokes
The text has been photo-set in Baskerville
by Acton Graphic Arts, Melbourne
Printed in Singapore

Frontispiece — Grampians in western Vic. Looking west from Mt. William,
the tallest peak. The Grampians have an extremely rich
native flora.   W.R. Elliot

Contents

Preface

The introduction to Volume Two remains applicable to this volume, with some minor additions and alterations.

We have continued to gather information from an extremely wide range of sources, and further travel has been undertaken both within Australia and overseas.

Approximately 305 genera and 1700 species are covered in this volume.

One important aspect that continues to be apparent is the need for Australian plant growers to keep records of information on cultivation. In 1982-83 the eastern part of Australia experienced one of the most severe droughts known since settlement by white man, and the south-eastern area suffered some of the heaviest frosts recorded this century. Parts were also burnt by the worst bushfires in forty-four years. Much has been learnt from these natural phenomena, and this knowledge should lead us to a better understanding of plant tolerances of climatic extremes and fire.

We have become increasingly aware of the large number of species that are considered rare or endangered in their natural habitat. There is the need for a nationally co-ordinated programme to ensure the conservation of species threatened with extinction. Continuing pressure to develop bushland for agricultural purposes poses the greatest threat to many species, particularly in Queensland and Western Australia.

We welcome the publication of Volumes 1, 8 and 29 of 'Flora of Australia'. This venture needs the support of all those interested in Australian plants. The active botanical work involved with the 'Flora of Australia' project has resulted in many taxonomic changes. While we have endeavoured to keep abreast of current nomenclature, some changes were impossible to include. The proposed appendix volume of this project will bring up to date all such changes.

For the author citations of genera and species in this and subsequent volumes, we have adopted the same system as that used in 'Flora of Australia'; namely, the 'Draft Index of Author Abbreviations' compiled by the Herbarium, Royal Botanic Gardens, Kew, England.

We reiterate that we would be very pleased to receive any additional information and corrections to the text. We would also be grateful to learn of any species being cultivated which may have been omitted from any volume.

Explanation of Text

Examples of layout used for genus and species in this book follow, together with explanations of relevant terms and headings.

Example of Genus Layout

Generic name and author citation
(derivation)
Plant family
Simplified botanical description, including:
> form of plant — shrub, tree etc,
> leaves — shape and form,
> flowers — arrangement, size, colour, often including calyx, petals, stamens,
> fruit — type, size and colour.

Information regarding the distribution of the genus and the number of species.
Cultivation information, plus propagation notes.

N.B. In the case of a genus being monotypic (ie. with only one species in the genus), no generic description is included.

Example of Species Layout

Name and author citation
(derivation)

Distribution within Australia	Vernacular name
Growth dimensions, height x width	Flowering time

Simplified botanical description including:
> form of plant — herb, shrub, tree etc,
> bark — texture, colour,
> branches — form, colour and other characteristics,
> branchlets — as above,
> leaves — type, dimensions, shape, colour and other characteristics (eg. margins, apex, nerves etc),
> inflorescence — type and size,
> flowers — size, colour, arrangement and other differentiating characteristics,
> fruit — type, dimensions, shape, colour and other characteristics.

Specific information on natural habitat, when available.
Requirements regarding soil and climatic conditions, and tolerances.
Specific uses. Propagation notes.

Reference to related species and their differences.

Information regarding previous names, confusion of names and allied facts.

In some cases further subspecies and varieties are included, with differentiating characteristics. When this is done, the species description always refers to the type form, eg. *Acacia pulchella* var. *pulchella*. The other varieties, eg. *A. pulchella* var. *reflexa* appear below.

Cultivars
Cultivars are included with relevant information, the name being presented in quotation marks, eg. *Callistemon* 'Burgundy'.

Synonym Entries
Major and recent entries are included to explain recent name changes, and to facilitate the use of the book.

Group Entries
Group entries are included to cover plant families, major plant groups such as ferns or orchids, interesting features of plants such as aromatic flowers or foliage, and specialized features or groups such as bog plants, cushion plants, annuals etc.

Terms Used in Text
1   For any botanical terms, see glossary.

2   Distribution

   States are given only when records are listed by the various herbaria, or are known by personal observation. There is little emphasis given to overseas distribution, since such information is difficult to find, and often inaccurate.

3   Vernacular name (common name)

   Only those names that are in common use have been included, as there is no Australia-wide convention for such names.

4   Growth dimensions

   These are given as height x width. They should be taken as a guide only, because plant growth can be variable due to differing soil and climatic conditions. In cultivation some plants may be vigorous with dense foliage, whereas in nature the same species will be of open habit, and leggy, straggly or misshapen. In some cases the latter habits of growth add to their beauty.

   HEIGHT GROUPINGS

   Dwarf shrub — 0-1 m
   Small shrub — 1-2 m
   Medium shrub — 2-4 m
   Tall shrub — 4-6 m

   Small tree — 6-12 m
   Medium tree — 12-25 m
   Tall tree — over 25 m.

5   Flowering time

   This is a most variable aspect of plants. The periods given are from records gained from herbaria, publications, and from observation by ourselves and others. As far as possible they relate to all areas of Australia.

Cultivation often results in plants flowering at times which differ from those recorded in the natural habitat. This is often due to artificial watering, and applications of fertilizer which stimulate new growth, resulting in the production of flowers outside the normal season.

In many cases the times given, such as *Sept-Feb*, do not mean that the flowers are displayed for the full length of that period, but rather that flowering occurs during that time.

When a flowering time such as *Sept-Feb; also sporadic* is given, this indicates that limited flowering often occurs outside the period of *Sept-Feb*.

Sometimes the phrase *throughout the year* is used, and this refers to a more or less continual flowering period.

6 Soils are generally described under three headings:
   a) light — including sands, gravels,
   b) medium — including loams,
   c) heavy — including clay-loams and clay soil types.
   For further information, see Soils, page 53, Volume One.

7 Drainage
   The terms related to drainage are as follows:
   a) well drained; free water does not accumulate and has a free passage through the soil,
   b) relatively well drained; free water is retained in the soil for very short periods,
   c) poorly drained; refers to soils in which water is retained for varying periods,
   d) periodic inundation; soils retain water above ground level for short periods, ie. 1-30 days,
   e) waterlogged; heavy soils where the retention of water is maintained at a maximum level for extended periods. This can also result in inundation.

8 Degree of light
   a) full shade; receives no sunlight at all,
   b) semi-shade; receives very little sunlight,
   c) dappled shade or filtered sun; refers to areas having a total overhead canopy of trees or shrubs, allowing some sunlight to penetrate throughout the day,
   d) partial sun; can be an open situation, but only receiving full sunlight for less than half of the day,
   e) full sun; an open position, with virtually no protection from direct sunlight throughout the day.

9 Cultivation
   For more detailed information, see page 47, Volume One.

10 Maintenance
   For details regarding terms used and other relevant information, refer to page 79, Volume One.

11 Propagation
   Procedures in relation to propagation methods are dealt with in detail on page 187, Volume One.

12 Illustrations and colour plates

We have endeavoured to include a representation of as many genera as possible. The ideal would be an illustration of every species, including photographs showing growth habit, close-ups of flower and foliage, and other diagnostic features. This is, however, not possible because of practical and economic restraints, and a selection has had to be made. The black-and-white line drawings can be used as an aid to identification, and such illustrations are often more useful than colour photographs. Many of the species illustrated, especially some from tropical zones, are plants of which illustrations have not been previously published.

Acknowledgements

Once again we would like to express our sincere thanks to many people who have assisted us in all kinds of ways.

As with Volume Two, we are extremely grateful that so many Australian plant enthusiasts have freely shared their expertise, thus making this publication more valuable.

Bruce Gray of Atherton, Qld, continues to be an excellent source of information on elusive tropical species and gives his help willingly. He has made available some very fine photographs.

Brian Crafter from SA has kindly lent photographs, and has also sent many specimens for study and illustration.

Hazel Dempster of WA supplied information and specimens of cultivated *Dampiera* species which were invaluable.

Judy West of CSIRO Herbarium Australiense was always willing to help with information on *Dodonaea*.

David Cannon provided much valuable information on *Dendrobium* hybrids, plus six photographs.

Tony Cavanagh, past leader of the SGAP Dryandra Study Group, gave us access to many *Dryandra* specimens, and read the manuscript relating to this genus.

In our search for information on *Eremophila*, Bob Chinnock of Adelaide Herbarium was ever willing to answer a barrage of questions, even though he was extremely busy with his revision of that genus. Ken Warnes critically read the manuscript, while Neil Marriott, Geoff Needham and Russell Wait provided specimens and cultivation information for many *Eremophila* species.

Roger Hnatiuk from the Bureau of Flora and Fauna, Canberra, supplied detailed information on the genus *Eremaea*.

Thanks to Beryl Blake for compiling the Common Name Index, a task she claims to enjoy.

Once again we are indebted to many people for their help and support in allowing us to photograph plants in their gardens and in the supply of plant material for preparation of the line drawings. Some have provided friendship and the sharing of knowledge on field trips, and also hospitality in their homes. Many people have been always willing to share their cultivation experiences, glorious or otherwise.

There are so many we wish to thank that there is almost the danger that this section could occupy the major portion of the volume. We therefore reiterate our very sincere thanks to all those listed in Volumes One and Two, and we would like to make particular mention of the following people who have provided assistance for this volume. Keith and Sue Alcock, George and Peter Althofer, John and Beth Armstrong, Paul Armstrong, David Beardsell, Marian Beek, Don Bellairs, Lloyd Bird, Bill and Kay

Bond, Bill Cane, Jeannette Closs, John Colwill, Bob Coveny, Eileen Croxford, Alistair and Rosemary Davidson, Kath Deery, John Fanning, Tom and Betty Fawcett, Andrew Garnham, Peter and Wilma Garnham, David Gordon, Chris Goudey, the late Alf Gray, Charlie and Joan Haden, David Hanger, J.F. Hartshorne, Neville and Freda Hatten, Leo and Joyce Hodge, Merv Hodge, Rob Horler, Peter Jones, John and Sue Knight, Alan Lacey, Greg Lamont, John Leaver, Alb and Irene Lindner, Craig Luscombe, Ross and Gwenda Macdonald, Joy Martin, Grant Mattingley, Jack and Peg McAllister, Doug McKenzie, David Nichols, Peter Olde, Royce and Jeanne Raleigh, Helen Richards, Fred and June Rogers, Dulcie Rowley, Alf and Esma Salkin, John and Marion Simmons, Brian and Diana Snape, Ken Stuckey, Lyn Symons, Paul and Pam Thompson, Merv Turner, Barry Walker, Brian Walters, Jenny West, Pauline Wicksteed, and Glen Wilson.

Many institutions have given us support in this venture.

Directors of botanic gardens, chief botanists and staff of the herbaria of Queensland, New South Wales, Victoria, Tasmania, South Australia and Western Australia have provided valuable information on nomenclature and have identified specimens. For this we are most grateful. We thank the directors of the botanic gardens for permission to use maps of botanical regions of each state, on which the map in this volume is based.

Helen Cohn of the Melbourne Herbarium library has always been ready to locate elusive information. Herbarium staff who have helped in various ways include Gordon Guymer, Rod Henderson, Laurie Jessup, Bill McDonald and Estelle Ross (Qld), Ben Wallace and Peter Hind (NSW), Helen Aston, Margaret Corrick, Stephen Forbes and Peter Lumley (Vic), A.E. Orchard (Tas), D. Whibley (SA) and T.E.H. Aplin, R. Cranfield, Kevin Kenneally and N.S. Lander (WA).

The staff of the National Botanic Gardens, Canberra, has once again supported us in various ways, especially through the Herbarium and the Australian Cultivar Registration Authority. We thank the past curator John Wrigley, and his successor Jim Armstrong. A special word of appreciation is due to Geoff Butler, ACRA Registrar, who supplied detailed information on many cultivars.

Alex George, the Executive Editor of 'Flora of Australia', has assisted in many ways.

As with the earlier volumes of the encyclopaedia, we have tried to produce an accurate and up-to-date publication. Such an aim is not achieved without some frustrations. We have consulted many botanists specializing in particular genera, or groups of plants, and they have responded generously. We gratefully acknowledge their support. In particular, we thank Munir Abid (*Dicrastylis*), Bob Anderson (*Correa*), Bruce Andrews (Ferns), Jim Armstrong (*Rutaceae*), Don Blaxell (Orchidaceae and Rainforest Plants), Roger Carolin (*Dampiera*), Clyde Dunlop (Plants of NT), Gordon Guymer (Sterculiaceae and Corynocarpaceae), Gwen Harden (NSW Rainforest Plants), Tom Hartley (Rutaceae), Rod Henderson (*Dianella*), Stephen Hopper (*Conostylis*), Bernie Hyland (Rainforest Plants, especially Lauraceae), Tony Irvine (Palms), Betsy Jackes (Vitaceae), Laurie Jessup (Celastraceae), Jean Jarman (*Epacris*), Greg Keighery (Myrtaceae and other genera), John Maconochie (Cycadaceae), Neville Marchant (Myrtaceae), M.T.M. Rajput (*Dampiera*), Sally Reynolds (Sapindaceae), John Williams (Davidsoniaceae), Karen Wilson (Cyperaceae) and Paul Wilson (Chenopodiaceae, Rutaceae).

Overseas assistance has also been forthcoming for this volume. The Arboretum of the University of California, Santa Cruz, has a large representation of Australian plants in cultivation. The Director Ray Collett, and Brett Hall, Ginny Hunt and Tim Ledwith

Acknowledgements

have been most helpful. Other Californian residents to provide information and assistance include Jeff and Mary Asher, Steve Brigham, Fred Lang, Mildred Mathias, Dean and Jane McHenry, Gilbert Voss and Ray Williams. We sincerely thank George and Betty Marshall of London for their help. Information on the performance of plants in England was supplied by John May of Dorset, a nurseryman who specializes in Australian plants.

We would particularly like to thank those who have supported us greatly from the beginning of this venture: Neville and Elizabeth Bonney of Millicent, SA, Allan Balhorn and Arthur Gulliver (retired), both of 'Your Garden' magazine, Ken Newbey from Ongerup, WA, and that ever helpful person Jim Willis.

We also especially thank Louis Lothian for his understanding and encouraging support, and Judith Brindley for her assistance in many ways during publication.

Finally, it is difficult to adequately praise the efforts of our wives and families during the compilation of this volume. Our wives Gwen and Barbara have typed the manuscript, and their contributions through criticism and editing have been invaluable. This volume owes as much to them as to ourselves.

Abbreviations

aff.	affinity		sp.	species (singular)
alt.	altitude		sp. aff.	species with affinity
av.	average		spp.	species (plural)
°C	degrees Celsius		ssp.	subspecies
cf.	compare		subvar.	subvariety
cm	centimetre		var.	variety
cv.	cultivar		Qld	Queensland
fig.	figure		NSW	New South Wales
gen.	genus		Vic	Victoria
gm	gram		Tas	Tasmania
ht	height		SA	South Australia
K	potassium		WA	Western Australia
kg	kilogram		NT	Northern Territory
km	kilometre		UK	United Kingdom
l	litre		USA	United States of America
m	metre		Jan	January
m²	square metre		Feb	February
max.	maximum		Apr	April
min.	minimum		Aug	August
ml	millilitre		Sept	September
mm	millimetre		Oct	October
N	nitrogen		Nov	November
P	phosphorus		Dec	December
ppm	parts per million			
%	per cent			
q.v.	which see			

Ce

Cedrela australis F. Muell. =
Toona australis (F. Muell.) Harms

Celastraceae R. Br.

A family of dicotyledons consisting of 55 genera and about 850 species widely distributed around the world. About 12 genera and 24 species are native to Australia where they mainly occur in tropical and subtropical regions. Main Australian genera: *Celastrus, Denhamia, Elaeodendron, Hysophila, Maytenus, Psammomoya, Salacia* and *Siphonodon.*

CELASTRUS L.

(from the Greek *celastros,* an evergreen tree)
Celastraceae

Straggly **shrubs** or tall **climbers**; **leaves** alternate, simple; **flowers** small, dioecious, in terminal or axillary panicles; **fruit** a globular capsule; **seeds** enveloped in a fleshy aril.

Species about 30 with 2 found in eastern Australia, one of which is endemic. Neither species is commonly grown. Propagate from seed which must be sown fresh or from cuttings which may be slow to strike.

Celastrus australis Harvey & F. Muell.
(southern)

Qld, NSW, Vic	Staff Climber
5-20 m tall	Sept-June

Bushy or tall woody **climber**; **leaves** 3-8 cm x 2-4 cm, shiny, dark green above, pale beneath, lanceolate, acuminate, often falcate towards the tips, with entire or toothed margins; **panicles** terminal, fairly compact and dense, clothed with dense grey hairs; **flowers** small, greenish, numerous, fragrant; **capsules** 0.2-0.6 cm long, ovoid, leathery, light brown, the inside surface with scattered red dots.

Widespread along stream banks and in rainforest from south-eastern Qld to eastern Vic, and endemic. Bushy to ground level when growing in open areas. A dense, vigorous climber suited only to large gardens where it will climb along fences or over trees. Very hardy and not affected by frost. A useful screening plant to cover unsightly features. Requires well drained soil and grows well in shade. Propagate from seed or cuttings.

Celastrus subspicata Hook.
(somewhat spike-like)

Qld, NSW	
5-20 m tall	Nov-March

Bushy or tall woody **climber**; **leaves** 5-14 cm x 2-7 cm, narrow-lanceolate to elliptical, leathery, dark green above, pale beneath, acute or blunt, often with minutely toothed margins; **panicles** terminal, somewhat loose; **flowers** small, yellowish green, numerous, fragrant; **capsules** 0.7-1 cm long, ovoid, leathery, yellow to orange.

Widespread in rainforests of eastern Australia from the Atherton Tableland to the central coast of NSW. The original plant from which the species was named was cultivated in Kew Gardens during the early 19th century. It is a dense climber suited to covering fences and unsightly features. Requires well drained soil and will grow in a sunny or shady situation. Propagate from seed or cuttings.

CELMISIA Cass.

(after Celmisios, son of the Greek nymph Alciope)
Asteraceae

Perennial **herbs**; **leaves** mainly radical, narrow, entire, hairy; **scapes** more or less leafless, bearing a solitary flower-head; **inflorescence** daisy-like; **outer ray florets** female, spreading; **disc florets** male and female; **fruit** an achene, slightly flattened with pappus.

This genus comprises over 60 species. The main representation is in New Zealand. Australia has only 4, and 3 of these are endemic. In NSW and Vic, the main concentration is in the alpine areas of the NSW-Vic border. One species, *C. asteliifolia,* is a common plant of the alpine herbfields. It often forms large colonies, which are a visual delight during the months of Dec-March.

Celmisias have had limited cultivation. Although they are usually slow-growing, they flower well. They prefer moist but well drained soils, with full sun but will tolerate partial sun. Withstand frost and snow. They like to grow in association with other plants, such as grasses, lilies and prostrate to low shrubs.

Propagation

Plants can be grown from fresh seed, collected as it is beginning to disperse from the spent flower-heads. Alternatively, clumps can be divided, potted into containers and planted out when re-established.

Celmisia asteliifolia Hook.f.
(leaves similar to those of the genus *Astelia*)

NSW, Vic, Tas	Silver Daisy; Snow Daisy
0.1-0.2 m x 0.2-0.5 m	Dec-March

A tufting, perennial **herb**; **leaves** to about 30 cm

Celmisia longifolia

Celmisia asteliifolia   T.L. Blake

long, radical, linear, initially hairy on both surfaces, becoming glabrous above with crowded, silvery white hairs below; **scape** to about 35 cm tall; **flower-heads** daisy-like, to 5 cm diameter, disc florets yellow, ray florets white or rarely pale pink; **fruit** an achene, to 1 cm long.

This species is widespread in the subalpine and alpine regions. It occurs in situations where the soil is more or less continually moist. In cultivation it has often been found to be short-lived, but this is possibly due to inadequate drainage. For best results a situation with partial to full sun in light to medium, very well drained soils is recommended. Well suited to container cultivation.

The var. *latifolia* F. Muell. ex Benth. from the Victorian alps differs in its longer and broader leaves. It is a most ornamental variety and is becoming more widely cultivated. Propagate both varieties from seed or by division.

Celmisia longifolia Cass.
(long leaves)
NSW   Snow Daisy
0.1-0.3 m x 0.2-0.5 m   Dec-March
Dwarf, tufting, perennial **herb**; **leaves** to 25 cm long, linear to linear-lanceolate, radical, entire, margins revolute, hairy to glabrous above, dense white tomentum below; **scape** to 30 cm tall, simple; **flower-heads** daisy-like, to 5 cm diameter, disc florets yellow, ray florets white or pink; **fruit** an achene, to 0.6 cm long.

Whereas the closely allied *C. asteliifolia* is widespread in alpine areas, this celmisia is restricted to moist soils of the high regions in the Blue Mountains. It requires similar cultivation conditions to *C. asteliifolia*. Propagate from seed or by division.

This species is similar in most characteristics to *C. asteliifolia*, which differs in having longer fruits.

Celmisia saxifraga (Benth.) W. M. Curtis
(similar to the genus *Saxifraga*)
Tas
prostrate x 0.5-1 m   Nov-Feb
Mat-like perennial **herb**; **rootstock** erect; **leaves** 1.5-3 cm x about 0.4 cm, lanceolate to narrow-elliptical,

more or less rigid, acute apex, upper surface yellowish to grey-green, covered in hairs, lower surface white or yellowish, with dense hairs, form rosettes at ends of rootstock or branches; **scapes** to 15 cm long, usually solitary; **flower-heads** daisy-like, to about 3 cm diameter.

Not commonly cultivated, this Tas endemic species inhabits mountain slopes and depressions where it can be covered by snow for many months of the year. It requires moist, well drained soils, with partial or full sun. An ideal container plant. Propagate from seed or by division.

Celmisia sericophylla J. H. Willis
(silky-leaved)
Vic   Silky Daisy
0.1-0.2 m x 0.2-0.5 m   Dec-Feb
Perennial alpine **herb**; **leaves** to 25 cm x 1-2.5 cm, linear, silky, drooping; **scapes** to about 30 cm long, usually solitary; **flower-heads** daisy-like, about 3-6 cm diameter, white; **achene** about 0.5 cm long.

Endemic to the Bogong High Plains, this species grows in colonies, inhabiting the rocky stream banks, and can often be seen overhanging the water. It is easily distinguished from *C. asteliifolia* by its drooping foliage. Not well known in cultivation, it requires very well drained soils which are moist for most of the year. Partial or full sun is recommended. Propagate from seed or by division.

Celmisia saxifraga   T.L. Blake

CELTIS L.
(a classical Greek name for the *Lotus* or the Greek name
for another tree)
Ulmaceae

Shrubs or **trees**; **leaves** entire or with toothed
margins, alternate, more or less 3-nerved; **flowers**
bisexual or unisexual, small, in axillary cymes; **fruit** an
ovoid or rounded drupe.

Species about 80 with 3 widespread ones extending
to Australia. A few exotic species are also fairly commonly grown. They are trees of graceful habit, with
small flowers, and fruit relished by birds. They are also
the hosts for some interesting and colourful butterflies.
Propagation is from seed which is best sown while
fresh.

Celtis amblyphylla F. Muell.
(blunt leaves)
NSW (Lord Howe Island)   Cottonwood
4-7 m x 3-4 m   Sept

Tall **shrub** or small bushy **tree**; **branches** willowy;
leaves 5-8 cm x 2-4 cm, ovate-lanceolate, dull green,
with a prominent midrib; **cymes** dense, few-flowered;
flowers about 0.2 cm across, green; **drupe** about
0.8 cm long, black.

A little-known plant endemic to Lord Howe Island.
Not known to be in cultivation but is an attractive,
bushy species that should have potential for coastal
districts in temperate regions. Requires well drained
soil and may need some protection when young.
Propagate from seed which must be sown when fresh.

Celtis paniculata (Endl.) Planchon
(flowers in panicles)
Qld, NSW
3-6 m x 3-4 m   Aug-Sept

Medium **shrub** to small **tree**; **leaves** 5-9 cm x 2-
3 cm, ovate-lanceolate, acuminate, leathery, smooth,
dark green, 2 veins prominent; **cymes** dense, few-
flowered; **flowers** 0.1-0.2 cm across, greenish; **drupe**
0.6-0.9 cm long, ovoid, black.

Widespread in rainforest and coastal scrubs between
Cooktown in north-eastern Qld and Kiama in
southern NSW. When cared for it becomes a graceful
tree with a spreading crown and can be used for a
variety of purposes including home gardens and
municipal parks and gardens. It will tolerate coastal
conditions but the leaves are burnt by severe salt spray.
It needs regular watering during dry periods and will
grow in cool areas of southern Australia. Badly drained
soils are not tolerated. Propagate from seed which
must be sown when fresh.

Celtis philippinensis Blanco
(of the Philippines)
Qld, WA, NT
3-7 m x 2-5 m   Aug-Nov

Tall **shrub** or small **tree**; **leaves** 3-12 cm x 2-6 cm,
ovate, shortly pointed, entire, rounded at the base,
rigid, leathery, dark green, with 3 prominent veins;
cymes loose or crowded, axillary, about 2 cm across;
flowers about 0.2 cm across, cream; **drupe** 0.6-0.8 cm
long, ovoid, yellow.

Celtis amblyphylla   D.L. Jones

Widespread through areas of northern Australia, often in rocky situations, and becoming stunted. An attractive small tree when well grown, with a spreading,
shady canopy. Best suited to tropical regions or inland
areas with a warm climate. May shed its leaves during
dry spells but this can be offset by watering. Needs well
drained soils. Propagate from seed which must be
sown when fresh.

CENARRHENES Labill.
(from the Greek *cenos*, empty; *rrhenos*, anther; refers to
empty or hollow anthers)
Proteaceae

Shrubs or **trees**; **leaves** alternate, toothed; **flowers** in
spikes, axillary or terminal, each one sessile, within a
small bract; **fruit** a drupe, succulent.

A genus of 2 species, one being endemic to Tas, the
other endemic to New Caledonia.

Cenarrhenes nitida Labill.
(shining)
Tas   Native Plum
3-10 m x 3-6 m   Nov-Jan

Tall **shrub** or small **tree**; **branchlets** often pink;
leaves to about 15 cm x 1-3 cm, narrowed at base, apex
blunt, coarsely toothed margins, thick, glabrous;
flower-head a spike, axillary, shorter than the leaves;

5

Centaurium

flowers small, sessile, within small bract, white; **drupe** succulent, to 1.5 cm diameter, purplish black, inedible.

This is generally a slow-growing species, found in rainforests of the central plateau, the south-west coast, and west coast. It has had only limited cultivation, which has indicated a need for moist, well drained soils, with full shade or dappled shade. It may tolerate partial sun, provided the root system is cool and moist. Frost tolerant. Plants can be difficult to propagate from both seed and cuttings.

CENTAURIUM Hill
(an early Greek name)
Gentianaceae

Ephemeral, annual or perennial **herbs**; **leaves** simple, opposite, radical and cauline; **inflorescence** a cyme; **flowers** 5-partite, tubular at the base, then expanding into showy corolla lobes; **calyx** united; **corolla** free, yellow, red or pink; **anthers** twisting spirally after flowering; **fruit** a capsule.

A cosmopolitan genus of about 50 species with a solitary widespread species extending to Australia, and another 3 commonly naturalized. The pink-flowered species are extremely variable and difficult to identify, probably because of hybridization. Propagation is from seed which is fine and must be sown shallowly.

Centaurium spicatum (L.) Fritsch
(flowers in spikes)
Qld, NSW, Vic, SA, WA, NT   Australian Centaury;
                                    Spike Centaury
0.1-0.5 m tall   Aug-Feb; also sporadic

Small, slender, ephemeral to annual **herb**; **leaves** 1-3 cm x 0.2-0.8 cm, elliptic to broadly lanceolate, blunt or pointed, bright green, the basal ones developed into a loose rosette, the rest cauline and reducing in size up the stem; **cymes** spike-like, with few branches; **flowers** about 0.8 cm across, bright pink, numerous, showy; **capsule** about 0.6 cm long, linear.

This very widespread small herb grows in open, harsh situations such as claypans, and may even grow in saline soils. Its flowers are bright and cheery, and the species can be readily induced to naturalize in gardens. Requires a sunny aspect. Propagate from seed which is sown on the soil surface. Previously known as *Erythraea australis* R. Br. Early settlers used this species as a herbal remedy for diarrhoea and dysentery. It is also reputed to help with digestion.

CENTELLA L.
(derivation obscure; may be from the Latin *cento*, a covering of rags or patchwork)
Apiaceae

Erect annual or prostrate perennial **herbs**; **leaves** either orbicular, peltate or deeply cordate and entire or divided, or cuneate at the base and divided; **flowers** small, sometimes unisexual in simple umbels, or also verticillate on the peduncle below the terminal umbel; **petals** entire, acute, imbricate; **fruit** laterally compressed, with several prominent ribs.

A genus comprising about 40 species distributed in

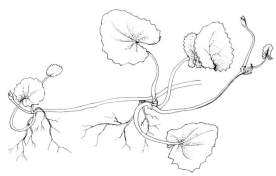

Centella cordifolia, × ·5

Africa, America and New Zealand, with only one representative in Australia.

Centella is very closely related to *Hydrocotyle*. The main difference is that this genus has imbricate petals, while *Hydrocotyle* has valvate petals.

Centella asiatica (L.) Urban =
                        C. cordifolia (Hook.f.) Nannf.

Centella cordifolia (Hook.f.) Nannf.
(heart-shaped leaves)
Qld, NSW, Vic, Tas, SA, WA   Centella
prostrate x 1-2 m   Aug-Jan

Dwarf, creeping, perennial **herb**; **stems** rooting at nodes, glabrous; **leaves** to 4 cm wide, broadly cordate or orbicular to reniform, with very long petioles, margins entire, sinuate or crenate; **flowers** very small, white to pink, in small, loose heads or umbels.

Naturally occurs in margins of swamps, and wet places. Not commonly encountered as a cultivated plant, it is often treated as a weed. It is, however, a most useful groundcover for wet, boggy conditions. A good poolside plant. Will grow in shaded to open locations and can cover a large area. Tolerant of most frosts. Propagate from cuttings or by division of stems with adventitious roots.

Previously known as *C. cordifolia* (L.) Urban and *Hydrocotyle asiatica* L.

CENTIPEDA Lour.
(from the Latin for a centipede; in reference to the creeping habit of the original species described)
Asteraceae

Annual **herbs** or perennial **shrubs**; **leaves** alternate, coarsely toothed, glabrous or hairy; **flowers** in heads, axillary or terminal, usually sessile; **florets** all tubular, the outer ones female, inner ones bisexual; **fruit** an achene.

A small genus of about 6 species with 4 found in Australia. The various species have very limited horticultural appeal and are mainly to be found in research collections or those of enthusiasts. Propagation is from seed or cuttings which strike easily.

6

Some species were used medicinally by the Aborigines, eg. *C. cunninghamii* and *C. thespidioides*, for treatment of colds and infected eyes and throats. During early settlement of Australia *C. minima* was highly regarded as effective in the treatment of sore eyes, while Baron von Mueller recommended the crushed leaves and seeds as an excellent snuff due to its sneeze-inducing properties.

Centipeda cunninghamii (DC.) R. Br. & Asch.
(after Alan Cunningham, early Australian botanist)
Qld, NSW, Vic, SA, WA, NT Common Sneezeweed
0.5-2 m x 0.5-1.5 m   Nov-March
    Dwarf to small **shrub**; **stems** ascending, glabrous or woolly; **leaves** 1-3 cm x 0.1-0.3 cm, oblanceolate, coarsely toothed, glabrous or hairy; **flower-heads** 0.5-1 cm across, globular, greenish yellow; **achenes** 0.15-0.25 cm long, furrowed.
    A widespread shrub most commonly observed in inland regions, growing beside creek beds and in areas subject to periodic inundation. Its ornamental appeal is limited; however, it is a very hardy shrub well suited to a hot situation as in inland regions. Woolly plants are more attractive. In cultivation they respond to water during dry periods and should be regularly pruned to stop them becoming leggy. Propagate from seed or cuttings which strike easily.

Centipeda minima (L.) A. Brown & Asch.
(least; smallest)
all states   Spreading Sneezeweed
0.1-0.5 m x 0.1-0.3 m   Nov-March
    Prostrate, spreading annual **herb**; **stems** procumbent or prostrate, glabrous to slightly woolly; **leaves** 0.3-1.5 cm x 0.2-0.3 cm, obovate, toothed, glabrous to woolly; **flower-heads** about 0.5 cm across, yellowish to greenish white, sessile to shortly pedunculate; **achenes** furrowed.
    A widespread weedy species that extends to Australia, where it is often common and grows in a variety of habitats and soil types. In nature it is usually an ephemeral, but if watered in cultivation survives much longer. It has little appeal and is mainly grown by enthusiasts. Could have some interest as a rockery plant or colonizer of bare soils for inland regions. It likes a sunny situation and responds to additional watering during dry periods. Propagate from seed or cuttings which strike easily.
    The var. *lanuginosa* (DC.) Domin differs in having densely woolly stems and leaves.

CENTRANTHERA R. Br.

(from the Greek *centron*, spike; *anthera*, anther; the anthers have awn-like points)
Scrophulariaceae
    Annual scabrous **herbs** or small **shrubs**; **branches** slender, roughened; **leaves** opposite or alternate, entire; **flowers** tubular, showy, with 5 lobes, almost sessile, axillary or in terminal spikes; **calyx** split; **fruit** a capsule which splits at maturity.
    A small genus of about 9 species with only one being found in tropical Australia. Propagation is from seed which is fine and must be sown shallowly.

Cephalaralia cephalobotrys, × ·45

Centranthera hispida R. Br.
(roughened)
Qld, NT
0.1-0.3 m tall   Nov-May
    Stiff annual **herb**; **branches** roughened, with hispid hairs, simple or with spreading branches; **leaves** 2.5-3 cm x 0.5-1 cm, linear, entire, the basal ones longest; **flowers** 2-2.5 cm long, white, pink, yellow or purple, short-lived; **capsule** about 0.6 cm long, ovoid.
    A common plant of swampy areas or moist grassy depressions, extending north from near Brisbane and across northern Australia. The plants are very showy and have considerable potential as annuals for tropical and subtropical regions. They should succeed best in a moist, sunny situation such as around the margins of a dam soak or spring, or in a bog garden. Propagate from seed. The plants may become self regenerating.

Centrolepidaceae Desv.
    A family of monocotyledons consisting of about 35 species in 5 genera, mainly restricted to Australia and South-East Asia. They are tufted annual or perennial sedge-like herbs, and few have any horticultural appeal. Main Australian genera: *Centrolepis, Hydatella* and *Trithuria*.

Centropappus brunonis Hook.f. =
    Senecio brunonis (Hook.f.) J. H. Willis

7

Cephalipterum drummondii   W.R. Elliot

CEPHALARALIA Harms
(from the Greek *cephale*, head; literally with heads like
the genus *Aralia*)
Araliaceae
A monotypic genus endemic in eastern Australia.

Cephalaralia cephalobotrys (F. Muell.) Harms
(head like a bunch of grapes)
Qld, NSW   Climbing Panax
10-15 m tall   Jan-April

Tall woody **climber**; young shoots covered with
pale hairs; **leaves** trifoliolate, on petioles about 5 cm
long; **leaflets** 5-9 cm x 2-4 cm, lanceolate, dark green
above, paler beneath; **flowers** small, green to purple, in
slender, terminal panicles; **capsules** globular, black.

Widespread in rainforests. Not well known but occasionally cultivated by enthusiasts. Prefers a cool,
semi-shaded position with abundant moisture during
dry periods. The fruits are quite decorative. Propagate
from seed which must be sown while fresh or stem cuttings.

CEPHALIPTERUM A. Gray
(from the Greek *cephale*, head; *pteron*, wing; alluding to
the numerous conspicuous laminae of the compound
head)
Asteraceae
A monotypic, endemic genus commonly found
growing on red loam, amongst *Acacia* spp. in the semi-
arid areas of SA and WA.

Cephalipterum drummondii A. Gray
(after James Drummond, 1st Government Botanist,
WA)
SA, WA
0.2-0.5 m x 0.2-0.75 m   most of year

Slender, erect **annual**; **leaves** 1-5 cm long, alternate,
lower ones broad, upper ones linear-lanceolate, sparsely
hairy; **inflorescence** globular, terminal, about 2.5 cm
diameter, composed of 10-15 sessile, partial flower-
heads, borne on hairy stems; **partial flower-heads** with
about 20 flowers, white, yellow, yellow-green or rarely
pink; **fruit** an achene, with dense hooked hairs, shed
when ripe.

Not commonly cultivated, this species requires well
drained soils, in an open, sunny location. Seed germinates readily and can be sown directly into soil,
which will create a massed display of flowers (see Annuals, Volume 2, page 202). Seed can also be sown in
containers, and transplanted when easy to handle. It is
important to keep seed moist until germination, and
until plants become established. Plants can also be
grown from cuttings. They withstand frost, and light
pruning will prolong their life.

CEPHALOMANES C. Presl
(from the Greek *cephale*, head; *manos*, soft)
Hymenophyllaceae
Small **ferns** growing in clumps; **fronds** erect, thin-

textured, translucent green, crowded; **segments** overlapping; **sori** tubular.

Species about 10 with one being found in Australia.

Cephalomanes atrovirens Presl
(dark green)
Qld, NSW
5-25 cm tall

Roots coarse, stilt-like; **fronds** in a rosette, semi-erect, once-divided, pale bluish green; **segments** crowded, overlapping, harsh, the veins protruding beyond the margins.

A rarely collected fern found in moist positions in dark rainforest. Resents disturbance and is difficult to establish in cultivation. Requires cool, constantly humid conditions. Best in a bottle, terrarium or aquarium. See Filmy Ferns.

Cephalotaceae Dumort.
A monotypic family of dicotyledons endemic to WA. The solitary species, *Cephalotus follicularis*, is a popular subject for pot culture.

CEPHALOTUS Labill.
(from the Greek *cephalotos*, having a head; in reference to the staminal filaments)
Cephalotaceae
A monotypic genus endemic in south-western WA.

Cephalotus follicularis Labill.
(like a small bag or sack)
WA   Western Australian Pitcher Plant
10-60 cm tall   Jan-March

Small perennial **herb**; **rootstock** thick, branching; **leaves** of 2 types; **normal leaves** 5-7 cm long, oval, shiny green, fleshy; **pitcher-leaves** 4-8 cm long; **pitcher** jug-like, up to 5 cm long, with a hinged lid, green, bronze or brilliant red; **flowers** 0.3-0.4 cm across, without petals, sepals greenish white, on stems up to 60 cm tall.

Found on small elevated tussocks in reedy swamps of black peaty soil, or along small stream banks. Leaves become colourful on exposure to sun. An interesting

plant, popular in cultivation because of its unique development and biology. Insects are attracted to the pitchers, where they are trapped by slippery sides and digested by fluids in the bottom of the pitcher. This digestion aids the plant's nutrition but is not essential for its survival in cultivation. It has been cultivated in glasshouses of Europe for at least 50 years. It grows easily in a pot of sandy loam and peat moss or sphagnum, with the base of the pot standing in a tray of water. Best treated as a glasshouse plant. Propagate from seed which should be sown on the surface of a pot of moist peat moss, the pot standing in water and covered by a piece of glass. Also leaf node cuttings are successful.

CERATOPETALUM J. Smith
(from the Greek *ceras*, horn; *petalon*, petal; one species has petals resembling stag's horns)
Cunoniaceae

Shrubs or **trees**; **leaves** opposite, simple or tri-foliolate, dark green; **inflorescences** much-branched, axillary or terminal; **flowers** small, numerous, petals often absent; **calyx** enlarging and subtending the fruit.

A small genus of 5 species, all endemic in Australia and New Guinea. They grow in moist forests and rainforests, sometimes in large stands. All are handsome trees or shrubs worthy of cultivation. The flowers are not generally showy, but after fertilization the sepals enlarge and become colourful, lasting for many weeks. One species is grown commercially for these colourful calyces which are sold in bunches for indoor decoration. All species are easy to grow in well drained soil in a sunny or semi-shady situation. Propagation is usually from seed which is ripe when the calyx with the fruit attached drops from the plant. These are covered lightly and kept moist to ensure germination. At least one species can be propagated from cuttings.

Ceratopetalum apetalum D. Don
(without petals)
Qld, NSW   Coachwood
10-20 m x 5-8 m   Oct-Dec

Small to medium **tree**; young shoots with 4 conspicuous ridges; **leaves** 3-25 cm x 1-7 cm, oblong to lanceolate, dark green, the margins shallowly toothed; **inflorescences** terminal or axillary; **flowers** (sepals) 0.6-0.8 cm across, greenish white; **sepals** enlarge in fruit and become bright red.

Common in rainforests on the eastern side of the dividing range, extending from the McPherson Ranges to Batemans Bay. Frequently cut as a timber tree, the pink wood being extensively used for cabinet making. Develops into a very attractive small tree, ideal for larger gardens or parks. Plants flower when still small. Very showy when the bracts develop pink or red shades. Requires well drained soil and responds to regular watering and the use of fertilizers. Likes a sunny position but is hardy enough to grow under established eucalypts. Also a good coastal plant. Will grow successfully in southern Australia. Propagate from seed.

Cephalotus follicularis in its natural habitat   T.L. Blake

9

Ceratopetalum gummiferum

Ceratopetalum apetalum

D.L. Jones

Ceratopetalum gummiferum J. Smith
(yielding gum)
NSW   NSW Christmas Bush
3-10 m x 2-6 m   Sept-Nov

Tall **shrub** or small bushy **tree**; young shoots rounded; **leaves** trifoliolate; **leaflets** 3-7 cm x 0.6-1.4 cm, narrow-oblong, the apex blunt, the margins shallowly toothed, dark green above, paler beneath; **inflorescences** terminal; **flowers** about 0.6 cm across, white, numerous; **sepals** enlarge in fruit and become bright red, rarely white.

Endemic in NSW where it is common in rainforest and open forest. Widely cultivated in gardens because of its massed display of red sepals after flowering. Also planted commercially by cut flower growers who market the foliage and coloured sepals in bunches. Will regenerate and grow vigorously after heavy pruning. Very hardy and ideally suited to gardens, where it remains compact and bushy. Responds to water during dry periods and the use of fertilizers. Not affected by frost. Best colouring is obtained by planting in a sunny situation. Two interesting cultivars are registered.

A variegated form (cv. Christmas Snow) has recently been introduced into cultivation. It has striking green and white variegated foliage and is an excellent plant for contrast. It is less vigorous and makes an attractive container plant. The cultivar 'White Christmas' develops white calyces instead of the normal red form, and provides a pleasant contrast when planted with the typical form. A form with attractive deep burgundy red leaves is occasionally grown.

Propagate from seed. Selected forms are best propagated from cuttings which may take several months to strike.

Ceratopetalum macrophyllum Hoogl.
(large leaves)
Qld
10-20 m x 5-10 m   Oct-Dec

Small to medium **tree**; young shoots initially with 4 conspicuous ridges, becoming terete; **leaves** unifoliolate, 5-25 cm x 2-10 cm, elliptic to oblong, distinct petiole, prominent nerves, cuspidate apex with long tip, margins entire, base rounded or with small notch; **inflorescence** terminal or axillary; **flowers** (sepals) about 1-2 cm across, greenish or white; **sepals** enlarge in fruit.

A recently named species that is confined to rainforest in north-eastern Qld, where it occurs to elevations of 500 m. It is fairly restricted and only found on northern slopes, along ridges, in swampy areas and along creeks. It has close affinities to *C. apetalum*, which has smaller sepals. Evidently not in cultivation, but certainly has potential for tropical and subtropical regions. Propagate from seed.

Ceratopetalum succirubrum C. White
(producing red sap)
Qld   Satin Sycamore; Blood-in-the-Bark
10-25 m x 5-15 m   Sept-Nov

Small to medium **tree**; young shoots slightly flattened; **leaves** trifoliolate; **leaflets** 3-14 cm x 1-5 cm, oblong to lanceolate, glossy on both sides, the margins slightly denticulate; **inflorescences** terminal; **flowers** (sepals) about 0.6 cm across, greenish white; **sepals** enlarge in fruit.

Confined to rainforest above 1000 m alt. on the Atherton Tableland, usually on rich soils. Cut for timber which is used for indoor building. Red sap oozes

Ceratopetalum gummiferum in flower

D.L. Jones

Ceratopetalum gummiferum with coloured calyces   B.W. Crafter

from damaged bark, hence the common name. Not widely cultivated but has many good features, and should prove an outstanding tree for large gardens or parks. Requires deep soil and plenty of water during dry periods. Propagate from seed.

Ceratopetalum virchowii F. Muell.
(after Prof. R. Virchow)
Qld   Dogwood
10-20 m x 5-12 m   March
    Small to medium **tree**; **leaves** trifoliolate; **leaflets** 4-15 cm x 1-5 cm, oblong, the apex drawn out into a long point, the margins slightly toothed, the surfaces slightly glossy; **inflorescences** terminal, leafless; **flowers** (sepals) about 0.6 cm across, greenish white; **sepals** enlarge in fruit.
    Confined to rainforest above 800 m alt. on the Atherton Tableland, usually in poorer soils than *C. succirubrum*. It also produces red sap from damaged bark, but not in large quantities. The species has many features indicating that it is worthy of cultivation. See comments under *C. succirubrum*. Propagate from seed.

Ceratophyllaceae Gray
    A small family of dicotyledons comprising about 10 species in the solitary genus *Ceratophyllum*. They are rootless aquatics that are widespread throughout the world with a solitary species in Australia.

CERATOPHYLLUM L.
(from the Greek *ceras*, horn; *phyllon*, leaf)
Ceratophyllaceae
    Slender **aquatics**; **stems** rootless; **leaves** forked, in

whorls; male and female **flowers** borne at separate nodes; **fruit** 1-seeded, black.
    Species about 10 with a single widespread species extending to Australia.

Ceratophyllum demersum L.
(submerged)
all mainland states   Hornwort
                                                       irregular
    Submerged **aquatic**; **stems** slender, brittle, rootless, flexuose, much-branched; **leaves** 1.5-4 cm long, slender, dark green, forked, 7-12 per whorl; **flowers** minute, solitary at the nodes.
    Widespread in still or slow-flowing fresh water, forming tangled colonies. Useful in ponds or aquaria as a shelter or egg bed for fish, but may spread and take over unless the pond is thickly planted with other aquatics. Easily controlled by pulling or raking out. Propagate from pieces.

CERATOPTERIS Brongn.
(from the Greek *ceras*, horn; *pteris*, fern)
Parkeriaceae
    Small aquatic **ferns**, floating or rooting in mud; **stems** spongy; **fronds** 1-3 times divided, variable in shape and breadth of segments, light green; **sori** borne on the inside of the fronds, covered by curled margins.
    A small genus of 2 species, both extending to northern Australia.

Ceratopteris cornuta (Beauv.) Lepr.
(horned or spurred)
Qld, ?WA, ?NT   Water Fern
10-30 cm tall
    Aquatic **fern** rooting in mud, either submerged or floating free; **fronds** twice-divided, bearing numerous plants; **segments** broad, blunt, rounded.
    Found in billabongs and swamps of tropical areas, but the range is uncertain due to confusion with *C. thalictroides*. Spreads by plantlets which float free until they take root. Easily grown in tropical areas, but cold sensitive and in temperate regions can only be grown in a heated glasshouse or tropical aquarium. Makes an ideal aquarium plant. Propagate by plantlets. The spores germinate on wet mud.

Ceratopteris thalictroides (L.) Brongn.
(like the genus *Thalictrum*)
WA, Qld, NT   Water Fern
10-35 cm tall
    Aquatic **fern** rooting in mud, either submerged, emergent or floating free; **fronds** 2-3 times divided, bearing plantlets; **segments** linear, blunt or pointed.
    A widespread species which extends into swamps and billabongs of tropical areas of northern Australia, usually in colonies. The species spreads by plantlets which float free and develop until sufficiently large to take root. An easy species to grow in tropical areas, where it makes a useful plant for fish ponds, dams etc. It is also widely sold as a plant for tropical aquaria. Very frost tender. The fronds were eaten as a vegetable by the Aborigines. Propagate by plantlets. Spores germinate on wet mud.

11

CERBERA L.
(from the dog *Cerberus*; an apparent reference to poisonous properties)
Apocynaceae

Shrubs or **trees**; **leaves** alternate, entire, crowded; **cymes** or **panicles** terminal; **flowers** large, tubular, fragrant; **fruit** a large drupe.

A small genus of about 6 species with 4 found in north-eastern Australia, one of which is endemic. The fruits apparently float, and are a familiar sight on tropical shores. The plants grow in rainforest, frequently along stream banks. They are handsome and worthy of garden culture for the leaves and fragrant flowers. They require protection from hot sun and strong winds, and grow best in loams rich in organic matter. They make attractive pot plants when young and are well suited to indoor decoration. Propagate from seed which must be sown fresh.

Cerbera floribunda Schumann
(numerous flowers)
Qld   Cassowary Plum
10-20 m x 3-8 m   July-Sept
Small to medium **tree**; **sap** milky; **leaves** 5-20 cm x 2-5 cm, narrow-lanceolate, on long petioles; **cymes** sparse or dense, held erect on terminal peduncles 4-12 cm long; **flowers** about 3 cm long, white with a red centre, very fragrant; **drupe** 6-10 cm x 4-5 cm, bright blue, mature Aug-Sept.

Confined to rainforests of north-eastern Qld, in both highland and coastal districts. The blue fruits are avidly eaten by cassowaries, with the poisonous seed passing through the bird. A showy species, little-known in cultivation. Best suited to tropical and subtropical areas. Needs protection from hot sun when young. Young plants have attractive leaves and make excellent indoor plants. Propagate from seed which must be sown while fresh.

Cerbera inflata S. T. Blake
(swollen; inflated; in reference to the corolla tube)
Qld   Grey Milkwood
10-20 m x 5-10 m   March-Sept
Small to medium **tree**; **sap** milky; **leaves** 10-20 cm x 1.5-2.5 cm, narrowly elliptical, shiny green, bluntly pointed, crowded towards the ends of branchlets; **panicle** 10-15 cm long, terminal; **flowers** 3 cm long, white or cream, very fragrant, funnel-shaped; **drupe** 5-7 cm long, ovoid, purple, mature Nov.

Restricted to highland rainforests of the Atherton Tableland but there widespread. An attractive tree with large leaves and masses of showy white flowers which have a delightful fragrance. Sensitive to over-exposure to sun when young. Needs well drained soil rich in organic matter. An attractive pot plant suitable for indoor decoration. Propagate from seed which is best sown when fresh.

Cerbera manghas L.
(the Portuguese name for the mango)
Qld
2-7 m x 2-5 m   Jan-April; also sporadic
Tall **shrub** or small **tree**; **sap** milky; **leaves** 8-20 cm x 4-6 cm, narrow-obovate, dark green, very glossy, on long petioles, the apex somewhat drawn out; **cymes** dense, terminal; **flowers** 2.5-5 cm across, white with a pink throat, very fragrant, with coloured bracts; **drupe** 5-7 cm, ovoid, often paired, black to purplish.

A widespread Asian species extending to coastal districts of north-eastern Qld where it grows on rocky and sandy sea coasts near the beachfront. An excellent ornamental species for coastal subtropical or tropical conditions. Has attractive glossy foliage and showy flowers which release an extremely pleasant fragrance. Cold sensitive but excellent for coastal conditions, tolerating considerable exposure. Needs well drained soil. Propagate from seed.

Cerbera odollam Gaertner
(an Indian plant name)
Qld
2-7 m x 2-5 m   Jan-April
Tall **shrub** or small **tree**; **branches** thick, sappy; **leaves** 10-30 cm x 4-8 cm, oblong or lanceolate, dark glossy green, with a long drawn out point, on long petioles; **cymes** dense, terminal; **flowers** 5-8 cm across, white with a yellow throat, sweetly scented, with coloured bracts; **drupe** about 5 cm long, flattened, globular, purplish.

A widespread Asian species extending to coastal districts of north-eastern Qld where it grows along the banks of tidal rivers, sometimes extending a fair distance inland. An excellent ornamental species for coastal subtropical or tropical conditions. Showy when in flower, with a delightful fragrance. Very cold sensitive. Needs well drained soil, in a sunny position. Seedlings are useful for indoor decoration. Propagate from seed. Stem cuttings would be worth trying.

CERIOPS Arn.
(from the Greek *ceras*, horn; *opsis*, appearance; in reference to the horn-like fruit)
Rhizophoraceae

Shrubs or small **trees**; **leaves** small, bright green; **flowers** white or brownish, small, in terminal clusters; **fruit** ovoid.

Species about 2, widespread through the Pacific region and extending to Australia.

Ceriops decandra (Griffith) Ding Hou
(10 stamens)
Qld   Flat-leaved Spurred Mangrove
1-4 m x 1-3 m   Aug-Sept
Small, spreading **shrub** or straggly **tree**; **pneumatophores** short, knobby; young branches brown; **leaves** 2-3 cm long, obovate, blunt, erect; **flowers** in axillary clusters; **fruit** brown, wrinkled.

Found on tropical coastal mud flats between Cairns and Ingham. The seeds germinate on the bush before the fruit falls. Important for coastal areas. Seeds can be planted direct from the bush. Seedlings transplant easily.

Ceriops tagal (Perr.) C. Robinson
(Indian name for the plant)
Qld, WA, NT   Spurred Mangrove
2-6 m x 2-4 m   Oct-Dec
Medium to tall **shrub** or small, spreading **tree**; **bark**

smooth, grey; **pneumatophores** short, knobby; **leaves** 5-11 cm x 2-7 cm, obovate, blunt, thick, yellowish, with entire or wavy margins; **flowers** 0.5-0.7 cm across, white, in pairs or in terminal clusters; **fruit** up to 2.5 cm long, pear-shaped, brown.

Widespread around the coast of tropical Australia. The seeds germinate on the bush before the fruit falls. The typical form has flowers in clusters, while var. *australis* C. T. White bears the flowers in pairs in the leaf axils. Important for coastal stabilization and reclamation work, and may be used by beach protection authorities. Seeds can be planted direct from the bush. Seedlings transplant easily.

CEROPEGIA L.

(from the Greek *ceros*, wax; *pege*, fountain; in allusion to the waxen appearance of the flowers)
Asclepiadaceae

Weak **climbing** or **trailing plants**; **stems** arising from a tuberous rhizome; **leaves** thin-textured or fleshy; **flowers** odd-shaped, tubular or jug-like, veined or patterned, hairy, in axillary cymes or umbels; **fruit** a follicle.

A large genus of about 160 species with one species indigenous in Australia. All species are unusual plants, and a number of exotic species are commonly cultivated in glasshouses for their interesting leaves and flowers. They are very drought resistant and make excellent pot plants, and the unusual flowers provide an interesting conversation piece. Propagation is from cuttings which strike easily or from seed which must be sown fresh.

Ceropegia papuana
(from Papua)
Qld

20-60 cm tall   Feb-April

Weak slender **climber**; **rootstock** thickened, tuberous; **leaves** 6-10 cm long, lanceolate, thin-textured; **cymes** shortly branched, 1-sided; **flowers** 2-2.5 cm long, tubular, flared to about 1 cm at the tip, hairy, brownish, patterned; **follicles** up to 10 cm long, pointed, often in pairs.

Rare in Australia, being found in rocky areas of north-eastern Qld, adjacent to rainforest. The stems die back to a tuberous rootstock during the hot dry season and reappear with the onset of rains. Not well known in cultivation but being grown by enthusiasts, and will become a popular glasshouse plant. Prefers an open, well drained mixture. Propagate from seed which is produced freely. Stem cuttings strike readily.

CHAETANTHUS R. Br.

(from the Greek *chaete*, mane; *anthos*, a flower)
Restionaceae

An endemic, monotypic genus confined to south-western WA.

Chaetanthus leptocarpoides R. Br.
(similar to the genus *Leptocarpus*)
WA

0.3 m x 0.3 m   Sept-Dec

Dwarf, clumping perennial **herb**; **stems** 0.3 m, slender, simple, persistent, with closely appressed, sheathing scales; **flowers** small, borne in spikelets forming a terminal cluster, male and female on separate plants; **male flowers** brown and pendulous; **female flowers** red and erect; **fruit** an indehiscent nut.

This tufting herb occurs near the coast north of Perth and in the Busselton-Augusta region. Little is known of its performance in cultivation, but it is recorded as being horticulturally desirable. It should grow in most soil types, with a situation that is in dappled shade or receives partial to full sun. Probably will need summer moisture. Propagate from seed and possibly by division.

CHAETOSPORA R. Br.

(from the Greek *chaete*, mane; *spora*, seed; in allusion to the bristly scales on the seed)
Cyperaceae

A monotypic genus endemic in Australia.

Chaetospora sphaerocephala R. Br.
(spherical heads)
all states except WA   Button Grass
1-2 m tall   all year

Stout perennial **herb**; **tussocks** coarse, 1-2 m tall; **leaves** 0.6-1.3 m tall, basal, tough, flat, brown at the base, fringed with white hairs; **culms** erect, 1-2 m tall, rigid, flattened, smooth, polished; **heads** 1-2 cm across, globular, brown or black, shiny, subtended by 3-5 fringed bracts; **nuts** 0.4 cm long, obovoid, shiny.

A tough perennial found growing in colonies in moist depressions and along stream banks, sometimes forming extensive stands. A useful rush for a wet position or a bog garden. Quite decorative, especially when in flower. Adaptable to full shade or sun. Frost hardy. Previously known as *Gymnoschoenus sphaerocephalus*. Propagate from seed or by division. Seeds must be sown on moist peat moss in a pot standing in water.

CHAMAESCILLA F. Muell.

(from the Greek *chamai*, on the ground; *scilla*, squill or sea onion; a Liliaceous plant of the Mediterranean region)
Liliaceae

Herbaceous **perennials**; **roots** tuberous; **leaves** basal, grass-like, channelled, glabrous; **flowers** 6 petals, spreading, bright blue, twisted after flowering; **stamens** 6; **fruit** a 3-celled capsule.

Comprising 2 species, this genus is endemic to Australia. One, *C. spiralis*, is restricted to WA, while the other, *C. corymbosa*, is widespread over southern Australia. They commonly grow in grassland and beneath trees and shrubs, on a wide variety of soils. Cultivation has, at this stage, been more or less negligible. Chamaescilla have potential for gardens and as container plants, where their bright blue flowers would be shown to advantage. Propagation can be carried out from seed, although there are virtually no records of results that have been achieved by this method. Probably best to sow fresh seed at the base of an existing plant, but conventional methods can also be used.

13

Chamaescilla corymbosa, × ·5

Chamaescilla corymbosa (R. Br.) F. Muell. ex Benth.
(corymbose)
all states except Qld   Blue Stars; Blue Squill
0.1-0.15 m x 0.1-0.2 m   Aug-Nov

Herbaceous **perennial**; **leaves** to about 15 cm x up to
1.5 cm, basal; **stem** erect, to 15 cm high, branched;
flowers to 1.5 cm diameter, borne in a loose terminal
cluster, bright blue, petals spirally twisted after
flowering; **capsule** 0.5-1 cm long, 3-lobed.

This species will adapt to most well drained soils,
but is not easy to keep growing as it has a tendency to
die out. It is best suited to dappled shade or partial sun,
but will tolerate full sun. Frost tolerant, and withstands
limited waterlogging. For gardens and containers.
Propagate from seed.

Chamaescilla spiralis (Endl.) F. Muell. ex Benth.
(spiral)
From WA, is very similar to the above species, but
differs in the flowers being very compact.

CHAMAESYCE Gray
(from the Greek *chamai*, on the ground; *syce*, a fig)
Euphorbiaceae

Annual and perennial **herbs**; **perennials** with a
thickened rootstock; **branches** erect or spreading,
woody or fleshy; **leaves** simple, entire, opposite, often
fleshy, glabrous or hairy; **flowers** small, unisexual, the
glands bearing an enlarged petal-like appendage; **fruit**
a capsule.

A widespread genus of about 250 species with about
17 known from Australia. They are often included as a
section within the genus *Euphorbia*. They are mostly annual or perennial herbs of little floristic value but a
couple of tropical species have horticultural qualities.
Propagation is from seed or cuttings. Those species
which die back to a perennial rootstock transplant
readily.

Chamaesyce schizolepis (F. Muell. ex Boiss.) Hassall
(referring to the lobed glands)
WA, NT
prostrate-0.3 m x 0.3-0.4 m   Nov-Feb

Dwarf perennial **herb** dying back to a thick, carrot-
like rootstock; **branches** thick and fleshy, obliquely
ascending to form a mat of foliage; **leaves** 1-2.5 cm x
0.8-1 cm, ovate, fleshy, dark green, with a few short
hairs; **flowers** small, in terminal, leafy cymes, surrounded by white-lobed bracts.

A small perennial herb which grows and flowers
rapidly with the onset of rains in the wet season and
dies down to the thick rootstock during the dry. It
grows in clay soil in open forest country. It forms an attractive mat of leaves and flowers and could be suitable
for a rockery in tropical or inland districts. Propagate
by transplants, or from stem cuttings or seed.

Chamaesyce vachellii (Hook. & Arn.) Hara
(after Rev. George Harvey Vachell)
NT
prostrate-0.3 m x 0.3-0.4 m   Nov-Jan

Dwarf perennial **herb** dying back to a thick rootstock; **branches** prostrate, ground hugging, thick and
fleshy; **foliage** in a mat; **leaves** 1-2 cm x 0.5-0.8 cm,
elliptical, dark green with a prominent pale midrib,
glabrous, somewhat fleshy; **flowers** small, in dense terminal cymes, surrounded by large, showy, lobed white
bracts.

An interesting perennial which has a dormant
period during the dry conditions of winter and spring,
and which grows and flowers during the wet season. Its
stems tend to be straggly but end in dense clusters of
flowers, of which the white bracts are most prominent
and showy. It grows in clay soil in open forest country
and could be an interesting garden plant for the
tropics. Propagate by transplants, or from stem cuttings or seed.

Chamaesyce vachellii   D.L. Jones

Geraldton wax plantation for cut flower production, near
Toowoomba, Qld   D.L. Jones

CHAMAEXEROS Benth.

(from the Greek *chamai*, on the ground; *xeros*, dry)
Xanthorrhoeaceae

Tufted perennial **herbs**; **leaves** radical, rigid, young
ones with dry, thin, lacerated margins; **inflorescence** a
panicle or umbel-like cluster; **flowers** bisexual; **perianth** segments 6, the outer 3 more rigid than the inner
3; **stamens** 6; **capsule** globular, smooth.

This genus of 3 species is endemic to the south-west
of WA. Propagation is from seed and possibly by
division.

Chamaexeros fimbriata (F. Muell.) Benth.

(fringed)
WA
0.3-0.5 m x 0.3-0.5 m   July-Sept

Tufted **perennial**; **leaves** to 50 cm long, sheathing
stem-base in 2 rows, terete to slightly flattened, erect,
often curved, rigid; **scape** axillary; **flowers** about 0.3 cm
long, yellow, borne in loose panicle about 30 cm long.

This interesting small species opens its flowers at
night. It is ideal for adding to a garden to provide a
natural appearance. Will grow in a wide range of well
drained soils, with dappled shade or partial sun, and
will possibly tolerate full sun. Can also be grown as a
container plant. Propagate from seed.

Chamaexeros macranthera Kuchel

(large anthers)
WA
0.2-0.3 m x 0.2-0.3 m   Aug-Oct

Tufted **perennial**; **leaves** to 30 cm long, sheathing
stem-base in 2 rows, flattened, 0.3 cm wide, erect, rigid,
sometimes falcate; **scape** filiform, axillary; **flowers** to
0.4-0.5 cm long, yellow to pale orange, borne in loose
panicle to 30 cm long.

A recently named species. It has been confused with
C. fimbriata, which has terete leaves. It is most ornamental and deserves to be cultivated. In nature it occurs over a very wide range of soil and climatic conditions, which means it should be adaptable in

cultivation, and suitable for gardens or containers.
Propagate from seed or by division.

Chamaexeros serra (Endl.) Benth.

(toothed)
WA
0.2-0.3 m x 0.2-0.3 m   Aug-Sept; also sporadic.

Tufted **perennial**; **leaves** to 30 cm long, sheathing
stem-base in 2 rows, flattened, 0.4 cm wide, rigid, often
falcate; **scape** axillary, to 10 cm long; **flowers** to 0.5 cm
long, yellow to pale orange, borne in umbel-like cluster
of up to 12 stalks to 1.5 cm long.

A decorative species that will grow best in well
drained, light to medium soils, with dappled shade or
partial sun. Suited for containers in addition to garden
use. Propagate from seed. *Acanthocarpus serra* (Endl.) F.
Muell. is synonymous.

CHAMELAUCIUM Desf.

(derivation is not clear; apparently from the Greek
chamai, dwarf or on the ground; and probably *leucos*,
white; or *leuce*, white poplar)
Myrtaceae

Shrubs; **leaves** opposite or rarely scattered, small,
narrow, sessile, usually glabrous, aromatic; **flowers**
sessile or on short stalks, in upper axils or terminal
clusters; **petals** 5, orbicular, spreading, white, pink or
red; **stamens** 10; **bracteoles** thin, enclosing young bud,
usually deciduous; **fruit** a 1-celled capsule, formed by
the hardened base of the persistent calyx.

An endemic genus of about 21 species, restricted to
south-western WA. They grow in a wide range of soil
and climatic conditions, occurring most commonly in
sandheath plant communities, either near the coast or
further inland. Some species are found in granite out-
crop areas, while many grow amongst shrubs and taller
trees. Some chamelauciums grow in semi-arid areas,
where temperatures can be very high, and rainfall
negligible during the summer months.

This is a genus in which most species have a high
horticultural potential, due to their floriferous capacity
and their ability to withstand harsh conditions. Some
are useful as cut flowers. *C. uncinatum* is grown commercially for this purpose. The foliage of most species
has a pleasant, sweet aroma when crushed, which is
typical of many Myrtaceae. They have gained the common name of Wax Flower, because of the waxy appearance and feel of the petals.

Generally their cultivation requirements are for well
drained soils which can be light to heavy-textured, and
a warm location with partial or full sun. Most species
will tolerate frost and many are drought tolerant. They
respond beneficially to pruning after flowering. They
do not seem to be attacked by many pests or diseases;
however, they are sensitive to root-rotting fungi. Root
damage can occur if the ground is too wet for extended
periods. The occurrence of scale can be controlled with
applications of white oil.

Propagation is generally carried out from cuttings of
firm new growth, although *C. uncinatum* often strikes
well using semi-hardwood cuttings. Many commercial
growers use very soft wood under mist.

This genus is currently undergoing botanical revision which should clarify certain relationships within
the genus.

Chamelaucium axillare

Chamelaucium confertiflorum   T.L. Blake

Chamelaucium axillare (F. Muell.) ex Benth.
(axillary)
WA   Esperance Wax
1-2 m x 1-2 m   Sept-Feb
   Small **shrub**; **leaves** 1-2.5 cm x 0.2 cm, terete to
3-sided, acute apex; **flowers** to 2 cm diameter, white,
solitary in upper axils.
   A most ornamental species that occurs along the
southern coastline, extending eastwards from Esperance to Pt. Dover, where it grows in sand. In cultivation it
requires well drained soils, and grows best in partial or
full sun. Propagate from cuttings.
   For many years *C. floriferum* has been confused with
this species and wrongly known as Esperance Wax, as
it occurs in the Walpole district west of Albany.

Chamelaucium brevifolium Benth.
(short leaves)
WA
0.7-1.5 m x 0.5-1 m   Aug-Nov
   An upright, small **shrub**; **branchlets** short, many;
leaves to about 0.5 cm long, narrow, opposite, almost
terete, or concave, erect, often in small clusters,
glabrous; **flowers** open-petalled, about 1 cm diameter,
borne on short stalks in upper axils, or in terminal
clusters, white ageing to pink.
   Not well known in cultivation, this ornamental
species will grow best amongst other plants, in very
well drained soils, with partial or full sun. Withstands
light to moderate frosts. Responds well to light
pruning. Propagate from cuttings.

Chamelaucium ciliatum Desf.
(fringed with fine hairs)
WA
0.5-1 m x 0.5-1 m   Sept-Nov
   Dwarf **shrub**; **leaves** to 1 cm long, linear-lanceolate,
acute apex; **flowers** about 1 cm diameter,
white ageing to red, borne in large leafy clusters at or
near ends of branchlets.
   This very variable species grows in sand, sandy loam
or loam, over a large area of the south-west. An inland
form with massed flowering shows excellent potential.
It is best cultivated in moist, well drained soils, with

full sun. Does well as a container plant. Propagate
from seed or cuttings.
   Plants of *C.* species (Avon, Irwin and Roe) have been
available commercially under the name of *C. ciliatum*
for many years in eastern Australia.

Chamelaucium confertiflorum Domin
(crowded flowers)
WA
0.5-2 m x 1-2 m   Aug-Oct
   Dwarf to small **shrub**; **branches** slender, sparse,
glabrous; **leaves** to 1 cm long, opposite, narrow-linear,
erect, initially flat, becoming concave to almost terete,
apex acute, crowded; **flowers** about 0.5 cm diameter,
white to pink, numerous, in crowded terminal heads.
   An ornamental species which occurs in the Stirling
Range. It is not at all well known in cultivation.
Requires well drained, light to medium soils, with dappled shade or partial sun, and probably will tolerate
full sun. Plants could become leggy and need pruning.
Propagate from cuttings.

Chamelaucium drummondii Meisner
(after James Drummond, 1st Government Botanist, WA)
WA
0.5-2 m x 1-2 m   Aug-Jan
   Dwarf to small **shrub**; **branches** twiggy; **leaves** to
1 cm long, scattered or crowded, not opposite, linear,
blunt or slightly pointed, long hairs on margins, grey-
green to bronze; **flowers** about 0.5 cm diameter, nearly

Chamelaucium megalopetalum. × ·5

16

sessile, borne in terminal heads or clusters, white or pink.

Not well known in cultivation, it occurs over a wide area of southern WA, growing mainly in sandheath plant communities. Requires well drained soils, with partial or full sun. Tolerates light to moderate frost, and withstands light pruning. Suitable for containers or gardens. Propagate from cuttings.

Chamelaucium megalopetalum (F. Muell.) ex Benth.
(large petals)
WA
1-2 m x 0.5-1 m   Aug-Nov

An upright, rigid, small **shrub**; **branches** few, erect; **leaves** to about 1 cm x 0.3 cm, opposite, oblong, flat or concave above, convex below, thick, erect, glabrous, acute or blunt apex; **flowers** about 1.5 cm diameter, solitary, borne on stalks 1 cm long in upper axils to form a cluster, petals white or cream ageing to red, greenish centre.

An outstanding ornamental species from inland. It occurs in coarse sands and gravels, growing amongst other shrubs. In cultivation it requires very well drained soils, with partial or full sun. Withstands frosts and extended dry periods. Pruning will produce dense growth if required. Suitable as a container plant. Propagate from cuttings.

Chamelaucium micranthum Turcz.
(small flowers)
WA
1-3 m x 1-2 m   Aug-Nov

A small to medium **shrub**; **branches** slender, erect or pendulous, much-branched; **leaves** 1-3.5 cm long, opposite, linear, almost terete, blunt or short point at apex; **flowers** about 0.5 cm diameter, borne in loose axillary racemes, profuse, white to pink.

A species from low-rainfall areas, where it occurs in sandy soils that can be saline. There is little knowledge of its reaction to cultivation, but it should grow in very well drained soils, with partial or full sun. Withstands

Chamelaucium uncinatum white flower form   W.R. Elliot

light to moderate frosts, and extended dry periods. It should also be suitable for containers. Propagate from cuttings.

Chamelaucium pauciflorum (Turcz.) Benth.
(few-flowered)
WA
0.5-2 m x 0.5-1.5 m   Aug-Nov

An erect dwarf to small **shrub**; **branches** twiggy; **leaves** to 0.6 cm, opposite, linear to oblong, blunt, erect or spreading, thick, concave, glabrous; **flowers** to 1.5 cm diameter, nearly sessile, borne in upper axils or loose terminal heads, white ageing to deep pink.

The specific name does not do justice to this ornamental species. It usually inhabits sandheath plant communities, and for cultivation will require well drained soils, with partial or full sun. Tolerates light frosts and extended dry periods. Responds to light pruning if required. Suitable for gardens and as a container plant. Propagate from cuttings.

Chamelaucium uncinatum Schauer
(hooked)
WA   Geraldton Wax
2-5 m x 2-6 m   Aug-Jan

A medium to tall **shrub**, open to dense; **branches** many; **leaves** to 4 cm long, opposite, linear, almost

Chamelaucium megalopetalum   T.L. Blake

Chamelaucium virgatum

terete, hooked apex, spreading to reflexed, glabrous; **flowers** open-petalled, 1.25-2.5 cm diameter, borne on stalks in small terminal clusters of 2-4, dull red, purple, mauve or pink to white.

Undoubtedly this is the most well known chamelaucium. It has been cultivated to a great extent, both in Australia and overseas. It promises to be one of Australia's most important cut flowers both locally and overseas. Large plantations are being established in various parts of Australia including south-eastern Qld. Countries such as Israel are also in commercial production. Although known as Geraldton Wax, it grows over a wide area in the sandplains north of Perth. It requires relatively well drained soils, with partial or full sun. Ideally suited to semi-arid conditions, as it withstands extended dry periods, but it will also grow in many other areas including subtropical regions. Hardy to most frosts, and will tolerate some alkalinity in the soil. It relishes pruning during or after flowering, and is an excellent cut flower. Useful as an informal hedge or screen plant, and in soil erosion control. Propagate from cuttings or seed.

There are many different forms which are sold as named cultivars and some are listed below. These must be propagated from cuttings to retain their characteristics.

'Album' — white flowers.
'Bundara Excelsior' — large pink flowers, grows well in Qld.
'Bundara Mystic Pearl' — pink-mauve to pale pink flowers.
'Bundara Supreme White' — white flowers.
'Dowell' — purple flowers.
'Giganteum' — large flowers of over 2.5 cm diameter; pale pink with deep red centre.
'Grandiflorum' — large deep pink flowers.
'Munns' — pale pinkish purple petals edged with reddish purple.
'Newmarracarra' — pale pink petals that change to reddish purple with maturity.
'Purple Pride' — reddish purple flowers in dense clusters.
'University' — not as vigorous as most other forms. Initially flowers are purple, becoming deep purple with maturity. Sometimes available as 'University Rubrum' or 'University Red'.

Chamelaucium virgatum Endl.
(twiggy)
WA
1.5-3 m x 1-3 m   Sept-Dec
A small to medium **shrub**; **branches** rigid, twiggy; **leaves** to about 2.5 cm long, opposite, linear, almost terete, blunt, glabrous, crowded; **flowers** about 1 cm diameter, borne on short stalks in the upper axils, in clusters of 2-4, white to pink, bracteoles persistent.

Virtually unknown in cultivation. This ornamental species occurs naturally in deep yellow sand, and will require very well drained soils, with partial or full sun. Pruning should provide a dense bush if required. Propagate from cuttings.

Chamelaucium species (Avon)
WA
0.5-1 m x 0.5-2 m   Sept-Nov
Dwarf **shrub**; **leaves** to 1.5 cm long, narrow-linear, terete, acute apex; **flowers** about 0.5 cm diameter, white, borne in terminal racemes.

This unnamed species occurs in saline, sandy soils in the Avon District. It has adapted well to cultivation, and grows best in well drained soils, with a full sun situation. It will also tolerate partial sun. Hardy to light frosts. Propagate from cuttings.

Chamelaucium species (Avon and Irwin A)
WA
0.5-1 m x 1-3 m   Nov-Feb
Dwarf **shrub**, spreading; **leaves** to 1 cm long, linear, terete, acute apex; **flowers** about 0.5 cm diameter, pale purple, solitary or borne in axillary racemes.

This species occurs on acid, sandy loam in the Irwin and Avon Districts. It is not at all well known in cultivation, and requires very well drained soils in a warm to hot situation. Propagate from cuttings.

Chamelaucium species (Avon and Irwin B)
WA
0.25-0.5 m x 0.5-1 m   Sept-Feb
Dwarf **shrub**, spreading; **leaves** to 1 cm long, linear,

Chamelaucium uncinatum 'Purple Pride'   B.W. Crafter

Chamelaucium uncinatum pale pink flower form

W.R. Elliot

3-sided, blunt, often in bundles on older wood; **flowers** about 0.5 cm diameter, purple, borne in axillary racemes.

An inhabitant of saline, sandy soils. It is not at all well known in cultivation. Should do best in well drained, light soils, with partial or full sun. Worth trying in saline situations. Propagate from cuttings.

Chamelaucium species (Avon, Irwin and Roe)
WA
1.5-3 m x 1.5-2 m   Sept-Nov; also sporadic
A dense, small to medium **shrub**; **branches** many; **leaves** to 1 cm long, narrow, opposite, almost terete, hooked tip, glabrous, crowded; **flowers** about 0.5 cm diameter, borne on very short stalks in upper axils, profuse, white deepening to pink.

This species occurs naturally along protected areas of the south coast, and inland. It prefers relatively well drained soils, with partial or full sun, but also tolerates dappled shade in hot areas. Withstands most frosts, and responds well to light pruning. Suitable as an ornamental garden shrub and as a screen or low windbreak. Propagate from cuttings. Has been commonly available as *C. ciliatum*.

Chamelaucium species (Eyre A)
WA
0.5-1 m x 0.5-1 m   Sept-Nov
Dwarf **shrub**; **branches** erect; **leaves** to 0.7 cm long,

linear, 3-sided; **flowers** to 1 cm diameter, white ageing to red, solitary in upper axils, forming leafy racemes.

Not well known in cultivation, this species occurs in sand with a clay subsoil in the Eyre District. It should therefore prove adaptable to a wide range of well drained soils. Should be suited to partial or full sun. Tolerates light frosts. Propagate from cuttings.

Chamelaucium species (Eyre B)
WA
1 m x 0.5-1 m   Sept-Nov
Small **shrub**; **branches** slender; **leaves** to 1 cm long, linear, 3-sided, mucronate; **flowers** to 1 cm diameter, pink, solitary, borne in axils near ends of branchlets.

This species is evidently not in cultivation. It will require very well drained soils as it occurs on leached white sands of the Eyre District. Prefers partial or full sun. Tolerates light frosts. Propagate from cuttings.

Chamelaucium species (Eyre and Roe)
WA
1-1.5 m x 1-1.5 m   Sept-Nov
Small **shrub**; **branches** twiggy; **leaves** 0.5-1 cm long, linear to obovate, blunt apex; **flowers** to 1 cm diameter, white ageing to red.

This inhabitant of sandy or lateritic soils has yet to be described. It should adapt to most well drained soils, with partial or full sun. Tolerates light frosts. Propagate from cuttings.

Chamelaucium

Chamelaucium species (Irwin)
WA
1-2 m x 1-2 m   Oct-Nov
Small **shrub**; **leaves** 1-2 cm long, linear, 3-sided, apex hooked; **flowers** about 1 cm diameter, purple-red, borne in clusters in upper axils.

A most ornamental, as yet undescribed species, worthy of cultivation. It occurs in sandy soils of the Irwin District. Will require very well drained soils in a hot situation to grow at its best. Tolerates light frosts. Should do well as a container plant, especially for south-eastern Australia. Propagate from cuttings.

Chamelaucium species (Irwin North)
WA
1-3 m x 1-2 m   Oct-Dec
Small to medium **shrub**; **leaves** to 2 cm long, narrow-linear, acute apex; **flowers** to 1 cm diameter, yellow-green, borne in clusters in the upper axils.

An unnamed species that has adapted well to cultivation. It comes from the northern Irwin District, where it grows on sandstone and sandy soils. It needs well drained soils, and is best suited to full sun. Appreciates moist soils. Readily propagated from seed or cuttings.

Chamelaucium species (Walpole)
WA   Walpole Wax
1.5-3 m x 1.5-3 m   Aug-Nov
A dense, small to medium **shrub**; **branches** many; **leaves** to 2.5 cm long, narrow, almost terete, opposite, pointed, channelled above, glandular, glabrous; **flowers** open-petalled, about 1.5 cm diameter, borne on stalks in upper axils, white ageing to pink or purple.

A decorative species in foliage and flower, which is restricted naturally to the Walpole region west of Albany. It will grow in most well drained soils, with partial or full sun, but also tolerates shade. In nature it often grows in association with other shrubs, and benefits from similar protection in cultivation. Withstands light to moderate frosts, and responds well to pruning after flowering. Appreciates summer watering. Suitable as an ornamental shrub, or may be used as a low screen. Is sensitive to cinnamon fungus. It is excellent as a cut flower and is grown commercially. Propagate from cuttings.

Has been confused for many years with *C. axillare*, the Esperance wax, but it does not occur in that area.

An excellent hybrid cultivar, with this species and *C. uncinatum* as the parents, has recently been introduced into cultivation. It has clusters of pale pink flowers, and is a first class cut flower.

CHEESEMANIA O. Schulz
(after T. F. Cheeseman, 20th-century New Zealand botanist)
Brassicaceae
Perennial **herbs**; **rootstock** thick, often branching; **leaves** in rosettes at ends of branches or on flowering stems; **sepals** 4, almost equal; **petals** 4, spreading, white; **fruit** a compressed capsule, formed by 2 joined carpels more than 3 times as long as broad.

A genus of 6 species, with 1 endemic species occurring in Australia, plus a further 5 in New Zealand.

Chamelaucium uncinatum × species (Walpole), × ·66

Cheesemania radicata (Hook.f.) O. Schulz
(arising from the root)
Tas
0.1-0.3 m x 0.1-0.5 m   Dec-Feb
A perennial **herb**; **rootstock** thick, often branching; **branches** procumbent; **leaves** to 10 cm long, tapering to base, glabrous, thick, margins toothed, arranged in rosettes at ends of branches; **flowers** about 1 cm diameter, borne in terminal raceme, petals 4, spreading, white.

A dweller of rock crevices or mountains at above 1200 m alt. This species requires moist, well drained soils. It will grow in partial or full sun, but must have protection to provide a cool root system. It is evidently not well known in cultivation, and is ideally suited to growing between boulders in rocky outcrops. Propagate from seed or cuttings.

CHEILANTHES Sw.
(from the Greek *cheilos*, lip; *anthos*, a flower; in reference to the lip-like indusium)
Sinopteridaceae
Small **ferns** forming clumps; **fronds** erect, glabrous or hairy, narrow or broad, 2-3 times divided; **sori** marginal on the segments, protected by folded leaf-margins.

Species about 180 with 11 found in Australia. They are mostly attractive little ferns, many of which are difficult to grow. They resent interference with their root system, prefer a sunny, well drained position and must not be overwatered.

Cheilanthes caudata R. Br.
(tailed)
Qld
10-25 cm tall
Small **fern** forming discrete clumps; **fronds** glabrous,

20

Chamelaucium species (Walpole)   B.W. Crafter

slender, sparse, 1-2 times divided; **segments** linear, slender, with recurved margins.

Grows in heavy soils along the margins of streams. Rarely encountered because of its inconspicuous nature. Very difficult to maintain in cultivation. Probably best suited to tropical areas.

Cheilanthes distans (R. Br.) Mett.
(far apart)
all states except Tas   Bristly Cloak Fern
5-15 cm tall

Small **fern** forming discrete clumps; **fronds** very slender, compact, hairy, 1-2 times divided, narrow; **segments** blunt, becoming smaller towards the base of the frond.

A widespread species usually found in drier situations. One of the easiest to grow, requiring a warm, well drained position such as in a sunny rockery. Resents shade and must not be overwatered. Propagate from spore or by division.

Cheilanthes fragillima F. Muell.
(very brittle)
Qld, NT
10-20 cm tall

Small **fern** forming discrete clumps; **fronds** triangular, erect, finely divided, dark green, somewhat hairy, very brittle, on black stipes; **segments** blunt, sparsely hairy.

A small tropical fern found on rocky slopes. It dies back to the rhizome during the dry winter and grows with the onset of rains in the summer. The clumps are small and usually consist of less than 5 fronds. Cultivation as for *C. vellea*.

Cheilanthes hirsuta (Poir.) Mett.
(hairy)
Qld, WA, NT
10-20 cm tall

Small **fern** forming discrete clumps; **fronds** erect, variously hairy from nearly glabrous to densely covered with rusty hairs, twice-divided; **segments** blunt.

Widespread in rocky areas and sometimes in sandy soils under low scrub. Resents disturbance and difficult to re-establish. Best in a sheltered, sunny rockery. Propagate from spore.

Cheilanthes lasiophylla Pichi-Serm.
(hairy leaves)
all states except Tas   Woolly Cloak Fern
5-20 cm tall

Small **fern** forming scattered clumps; **fronds** erect, broad to narrow, bearing numerous hairs and scales, brownish, 1-2 times divided; **segments** largest towards the middle of the frond, blunt.

Found in hot dry areas. Very drought resistant. Cultivation as for *C. vellea*.

Cheilanthes nudiuscula (R. Br.) T. Moore
(almost leafless)
Qld, NT
10-20 cm tall

Small **fern** forming spreading clumps; **fronds** deltoid-triangular, erect, bright green, dull, glabrous, 1-2 times divided; **segments** very broad, blunt; **stipes** black.

A drought resistant fern that grows in sandy clay-loams, usually in the lee of boulders or in crevices. It is very attractive but difficult to maintain in cultivation. Best suited to warm, dry areas of low humidity. Propagate by division or from spore.

Cheilanthes prenticei Luerssen
(after Dr. C. Prentice)
Qld, NT
5-20 cm tall

Small **fern** forming discrete clumps; **fronds** erect, narrow, sparse, often stunted, covered with white or brown hairs; **segments** largest towards the base of the frond, blunt.

Found in dry sandy areas, usually in open forest or under light scrub. Cultivation as for *C. vellea*.

Cheilanthes lasiophylla   T.L. Blake

Cheilanthes pumilo (R. Br.) F. Muell.
(dwarfed)
Qld, WA, NT
10-20 cm tall

Small **fern** growing in clumps; **fronds** erect, slender, glabrous, thin-textured, pale green; **segments** lanceolate, blunt.

Widespread in rocky areas of northern Australia. Very difficult to grow. Resents disturbance but may be successful in a warm, sheltered rockery. Propagate from spore or by division.

Cheilanthes sieberi Kunze
(after F. W. Sieber, Austrian botanist)
Qld, NSW, Vic, Tas, NT   Mulga Fern
20-40 cm tall

Dwarf **fern** forming spreading clumps; **fronds** very narrow, erect, bright green, glabrous, 2-3 times divided; **segments** narrow, blunt.

A common fern in a variety of habitats, but always in sunny situations, often amongst rocks. Readily confused with *C. tenuifolia* but distinguished by the much narrower fronds. Poisonous to stock. One of the best to grow, requiring a warm, well drained position such as in a rockery. Resents shade but will respond to water during dry periods. Propagate from spore or by division.

Cheilanthes tenuifolia (Burm.f.) Sw.
(slender leaves)
all states   Rock Fern
10-80 cm tall

Dwarf **fern** forming spreading clumps; **fronds** broadly triangular, erect, bright green or shiny, glabrous, 2-3 times divided; **segments** narrow, blunt.

Widespread in a variety of habitats, but often amongst rocks, and usually not extending very far inland. Fairly popular in cultivation as it grows well in a rockery in a semi-protected position. Forms a nice clump of bright green foliage. Resents shade and poor drainage. Propagate from spore or by division.

Cheilanthes vellea (R. Br.) F. Muell.
(fleece-like)
Qld, SA, WA, NT   Woolly Cloak Fern
5-15 cm tall

Dwarf **fern** forming discrete clumps; **fronds** narrow, erect, dull green or whitish from a dense covering of hairs, 1-2 times divided; **segments** blunt, woolly, hairy.

A drought resistant fern found in hot, dry areas, almost invariably amongst rocks. Very difficult to grow. Resents humidity, cold and excess moisture. May be best suited to warm, dry inland areas. Intolerant of bad drainage. Propagate from spore.

CHEIRANTHERA Cunn. ex Brongn.
(from the Greek *cheir*, hand; *anthera*, anthers; alluding to anthers being spread out like fingers of a hand)
Pittosporaceae

Small **shrubs** or light **climbers**; **branches** flexuose or twining; **leaves** narrow; **flowers** borne in terminal corymbs or cymes, or pendulous from terminal solitary

Cheiranthera cyanea   W.R. Elliot

stalks; **petals** 5, spreading; **stamens** 5 arranged on one side of flower; **fruit** an oblong capsule.

An endemic genus of 4 species restricted to southern Australia. They often occur in poor, stony soils that receive a low rainfall. At this stage only *C. cyanea* and *C. filifolia* have been grown to any extent. They have proved to be adaptable to a relatively wide range of soils and climatic conditions, and are also ideal as container plants. As all species are blue-flowered, undoubtedly they will become better known, and sought after for cultivation.

Propagation

Plants can be grown readily from seed, or from cuttings of firm new growth.

Cheiranthera cyanea Brongn.
(blue)
NSW, Vic, SA   Finger Flower
0.5-1 m x 0.5-1 m   Oct-Dec

An upright dwarf **shrub**; **branches** slender, glabrous, can have suckering stems; **leaves** 2-5 cm x 0.2 cm, erect, acute or obtuse, entire or minutely toothed, or margins incurved, sometimes small leaves clustered in axils, glabrous; **flowers** 3-4 cm diameter, deep blue, borne on stalks beyond foliage, prominent golden anthers.

A most decorative species that is best grown in well drained soils that are not wet for extended periods. Prefers dappled shade or partial sun, and will also tolerate full sun. Frost tolerant, and will withstand dry periods. Ideal as a garden or container plant. Propagate from seed or cuttings.

This species was previously known as *C. linearis* Cunn. ex Lindl.

Cheiranthera filifolia Turcz.
(thread-like leaves)
WA
1.5-4 m x 1-3 m   Aug-Jan

A shrubby **twiner**; **branches** slender, often flexuose or almost twining; **leaves** 0.8-2.5 cm long, very narrow, thick or almost terete, more or less blunt-tipped;

flowers to about 3 cm diameter, deep blue, prominent golden anthers, borne on branching stalks above foliage.

An outstanding ornamental species that will grow in well drained soils (including gravels), with dappled shade or partial sun. It could prove adaptable to a much wider range of soil and climatic conditions. Propagate from seed or cuttings.

Cheiranthera linearis Cunn. ex Lindl. =
C. cyanea Brongn.

Cheiranthera preissiana Putterl.
(after Ludwig Preiss, 19th-century German botanist)
WA
1-1.5 m x 1-2 m   Oct-March
A shrubby **twiner**; **branches** slender, can be short and flexuose, or long and twining, glabrous or hairy; **leaves** to about 4 cm long, more or less sessile, linear to lanceolate, glabrous or hairy, margins revolute; **flowers** to about 2 cm diameter, borne on long, slender terminal stalks, blue or blue-mauve.

Evidently not in cultivation, but should prove to be a useful ornamental plant. It requires well drained soils, with dappled shade or partial sun, and will tolerate full sun, provided it has a protected root system. Propagate from seed or cuttings.

Cheiranthera volubilis Benth.
(twining)
SA
1-2 m tall   Oct-Dec
A light **twiner**; **branches** glabrous; **leaves** 2-4 cm long, deeply channelled, appearing almost terete, usually ending in a short recurved point; **flowers** about 2.5 cm diameter, solitary, borne on a long, slender terminal stalk, blue.

Cheiranthera filifolia   B.W. Crafter

Endemic to Kangaroo Island, this is a useful light twiner for cultivation. It requires fairly well drained soils, with dappled shade or partial sun. Ideal for growing through other shrubs. Propagate from seed or cuttings.

CHEIROSTYLIS Blume
(from the Greek *cheir*, hand; *stylos*, style)
Orchidaceae
Small terrestrial **orchids**; **rhizome** creeping, thicker than the short stems; **leaves** in a scattered rosette; **inflorescences** terminal, few-flowered; **flowers** small; **sepals** joined at the base to form a tube; **fruit** a capsule.

Delicate orchids with patterned leaves. The genus of about 22 species is widespread throughout the Pacific region with a single species extending to Australia.

Cheirostylis ovata (Bailey) Schltr.
(referring to the leaf shape)
Qld, NSW
10-25 cm tall   June-Sept
Rhizome creeping, constricted at the nodes, flattened if in contact with rocks; **stem** short, erect; **leaves** 2-7, about 3 cm long, ovate, pointed, pale green, with conspicuous venation; **inflorescence** erect, glabrous; **flowers** 1-6, about 0.8 cm across, white.

Found in rainforest, usually growing in colonies on rocks. The leaves wither during the dry season and the stems and rhizomes become dormant. They recommence growth during the wet season and grow up through leaf litter which has fallen during the dry. The plants feed by numerous root hairs which become directly attached to the decaying leaves. Hard to grow and must be given a dormant period each year, and topped up with crushed leaf mould. Must be maintained under cool, humid situations while growing.

Chenopodiaceae Vent.
A family of dicotyledons consisting of about 1500 species in 100 genera, with a sporadic distribution in saline soils around the world. They are mostly herbs or shrubs, and are a very well developed group in Australia (24 genera, 200 species). The shrubby species are very important fodder plants and are commonly known as bluebushes or saltbushes. Not many species are commonly grown in gardens; however, their foliage provides a pleasant contrast with other shrubs. Most species also have fire retardant properties by virtue of the high salt content of their sap, and are useful as borders around highways, parks and gardens in fire-prone areas. Main Australian genera: *Atriplex*, *Babbagia*, *Chenopodium*, *Dissocarpus*, *Dysphania*, *Enchylaena*, *Halosarcia*, *Maireana*, *Rhagodia*, *Salicornia*, *Sclerolaena*, *Sclerostegia* and *Suaeda*.

CHENOPODIUM L.
(from the Greek *chen*, goose; *podion*, little foot; referring to shape of leaves)
Chenopodiaceae
Herbs or **shrubs**, annual or perennial; **leaves** alternate, flat, entire, toothed or divided; **flowers** small,

23

sessile, in clusters, either axillary or in interrupted terminal spikes or panicles; **fruit** partially or completely concealed by persistent perianth segments.

Distributed world-wide, this genus contains about 250 species, of which there are about 15-20 indigenous to Australia. Approximately 6 exotic species have become naturalized here as weeds. Their main representation is in the dry and arid regions. They generally have limited application to horticulture, but some species are highly regarded as browse shrubs for grazing animals. Some species are, however, cyanogenetic if animals browse continually without fodder alternatives. Some are fire retardant, and therefore can be useful for planting around buildings or along roadsides in fire-prone areas.

Plants are readily propagated from seed or cuttings.

Chenopodium auricomum Lindl.
(golden hairs)
Qld, NSW, SA, NT   Golden Goosefoot
1-2 m x 1-2 m   June-Oct

Small perennial **shrub**; **branches** glaucous, mealy; **leaves** 2-5 cm long, variable in shape, glaucous, mealy; **flowers** small, borne in dense clusters forming a long terminal panicle, golden; **perianth** mealy and hairy, concealing the fruit.

This species is ornamental, and also very highly regarded as fodder for stock. It is usually found on clays in inland drainage areas. For cultivation it will grow in most soils, with full sun. Withstands frost and drought. Recommended for semi-arid to arid regions. Propagate from seed or cuttings.

Chenopodium cristatum (F. Muell.) F. Muell.
(crested)
Qld, NSW, Vic, SA, WA, NT   Crested Goosefoot
0.1-0.3 m x 0.5-1 m   June-Oct

Dwarf, spreading, annual **herb**; **stems** many, arising from the base, hairy; **leaves** 1-2 cm long, elliptic to obovate, with slender petioles, entire to sinuate, with short hairs; **flowers** in sessile axillary clusters; **fruiting perianth** small, with 5 crested segments.

Occurs naturally in a wide range of conditions, in areas of relatively low rainfall. Rarely cultivated, but has potential, especially for soil erosion control in semi-arid regions. Suited to well drained soils, with partial or full sun. Propagate from seed or cuttings.

Aborigines used this species for treating bad skin infections.

Chenopodium nitrariaceum (F. Muell.)
F. Muell. ex Benth.
(similar to the genus *Nitraria*)
all mainland states   Nitre Goosefoot
2-3 m x 2-4 m   most of year

Spreading medium **shrub**; **branches** slender, rigid; **branchlets** hoary and often spiny; **leaves** to about 2.5 cm long, tapering to base, blunt, pale green or greyish; **flowers** small dense spikes, at ends of branchlets, or in terminal panicles, yellow or brown; **fruit** enclosed in mealy perianth.

A common plant of dry areas, occurring on many soil types, but most common on heavy clays. It is valued as a fodder plant. It will grow in most soils, with

full sun, but will also tolerate partial sun. Propagate from seed or cuttings.

Chenopodium rhadinostachyum F. Muell. =
    Dysphania rhadinostachya (F. Muell.) A. J. Scott

CHILOGLOTTIS R. Br.
(from the Greek *cheilos*, lip; *glottis*, mouth of the windpipe; an allusion to the flowers resembling a young bird waiting to be fed)
Orchidaceae

Small terrestrial **orchids**; **tubers** ovoid or oblong, subterranean; **leaves** basal, in pairs, flat; **flower-stem** erect; **flowers** bird-like, dull coloured; **labellum** hinged, bearing numerous calli.

Species about 8, most of which are endemic in Australia. They are small orchids, usually found in colonies in sheltered areas, and are popularly known as bird orchids from the shape of the flowers. They are a popular group with orchid enthusiasts and are fairly easy to grow in bushhouses or cool glasshouses. They have a period of active growth from autumn to spring and then become dormant over summer. They should be watered regularly while in active growth and allowed to dry out while dormant. Potting mixtures should be well drained and contain some well rotted organic material. Most species increase their tuber number by vegetative reproduction and need repotting when they become crowded. When in growth they respond to regular applications of weak liquid fertilizers. They are severely attacked by slugs and snails, and must be protected continually by baits.

Chiloglottis cornuta Hook.f.
(horned or spurred)
NSW, Vic, Tas   Green Bird Orchid
        Oct-Jan

Leaves 3-7 cm x 1.5-4 cm, ovate-lanceolate, bright green; **flower-stem** 2-4 cm tall; **flower** 1.5-2 cm long, greenish all over; **labellum** ovate, green, with purplish calli.

Usually found in cool, sheltered places, often in scattered colonies. Unspectacular when in flower. Difficult to grow as it requires continual moisture and high humidity, and even when dormant should never be allowed to dry out completely. It does best in a large pot of well drained mixture containing plenty of partially rotted organic material and should be grown in a cool, protected position. It can also be grown in a pot of fresh sphagnum moss or the chopped up fibres of soft tree-fern, *Dicksonia antarctica*.

Chiloglottis formicifera Fitzg.
(ant-like)
NSW   Ant Orchid
        Sept-Nov

Leaves 3-5 cm x 1.5-2 cm, ovate or oblong, dark green, the margins irregularly crinkled; **flower-stem** 6-10 cm long; **flower** 1-1.5 cm long, green with brown or purplish markings; **labellum** trapeziform, bearing a mass of calli which resemble the shape of an ant.

Confined to NSW, where it is not uncommon in central coastal districts. It grows easily in cultivation,

Chiloglottis reflexa   J. Fanning

reproducing rapidly, and quickly building up in numbers. It flowers freely in cultivation, and when in a pot in flower, it is an intriguing sight. Needs regular watering while in active growth and responds to the use of liquid fertilizers.

Chiloglottis gunnii Lindl.
(after Ronald C. Gunn, 19th-century collector, Tas)
NSW, Vic, Tas   Common Bird Orchid
                                         Sept-Jan
Leaves 5-10 cm x 2-4 cm, ovate, dark green; **flower-stem** 5-8 cm tall; **flower** 2-4 cm long, greenish to dark purple, usually large and conspicuous; **labellum** broadly ovate, with a variable number of large, stalked glands.
Common within its range, usually forming colonies in cool, moist places. Grows easily if given cool, humid conditions. Slow to increase. Requires an abundance of water while in active growth, and even when dormant should never be allowed to dry out completely. It does best in a large pot of well drained mixture containing plenty of well rotted organic material. Can also be grown in a pot of fresh sphagnum moss. In cool districts it can be successfully established as a garden plant, provided it is protected from slugs and snails.

Chiloglottis × pescottiana R. Rogers
(after E. E. Pescott, 20th-century orchidologist, Vic)
NSW, Vic
                                         Sept-Dec
Leaves 4-6 cm x 1-3 cm, oblong or lanceolate, dark green; **flower-stem** 8-20 cm tall; **flower** 2-3 cm long, greenish or dark purplish; **labellum** oblong, rounded at the apex, with large and small calli scattered on the surface.

Chiloschista phyllorhiza

A rare species which is a natural hybrid between *C. gunnii* and *C. trapeziformis*. It grows quite easily in cultivation, increasing in number regularly and flowering easily. It is not quite as vigorous as *C. trapeziformis* but is certainly more so than *C. gunnii*. Requires plenty of water while in active growth and responds to the use of liquid fertilizers.

Chiloglottis reflexa (Labill.) Druce
(reflexed petals)
Qld, NSW, Vic, Tas   Autumn Bird Orchid
                                         March-May
Leaves 2-5 cm x 1-1.5 cm, narrow-ovate, light or dark green, with conspicuously crisped margins; **flower-stem** 8-15 cm tall; **flower** 1.5-2 cm long, green or purplish; **labellum** obovate, blunt, with a large central mass of black calli.
A widespread species usually found in coastal areas and extending from northern Qld to southern Tas. It grows and flowers well in cultivation, but the plants are very susceptible to rotting as the leaves are unfurling. Watering should be carried out carefully at this stage. It reproduces fairly freely and has a short dormant period, often beginning growth in January. The flowers appear with the leaves, and may be produced at any time of the year.

Chiloglottis trapeziformis Fitzg.
(trapeziform-shaped labellum)
Qld, NSW, Vic   Broad Lip Bird Orchid
                                         Sept-Nov
Leaves 6-8 cm x 1-1.5 cm, narrow-lanceolate, with irregular margins; **flower-stem** 5-12 cm long; **flower** 1.5-2 cm long, green, purplish or bronze coloured; **labellum** trapeziform, with a congested group of calli near the centre.
Widespread from south-eastern Qld to eastern Vic, usually in colonies in open forest country. The easiest of the genus to grow, thriving in cultivation. It reproduces rapidly, and quickly builds up in numbers to form a potful. It is hardy and will tolerate neglect, but responds to regular watering and feeding with liquid fertilizers when in active growth. It flowers freely in cultivation. The potting mixture seems relatively unimportant as long as it is well drained and with some organic material. Needs repotting every couple of years.

CHILOSCHISTA Lindl.
(from the Greek *cheilos*, lip; *schistos*, cleft; divided; in reference to the shape of the labellum)
Orchidaceae
Small, leafless epiphytic **orchids**; **roots** flat, roughened, green; **racemes** slender; **flowers** short-lived, opening wide; **fruit** a capsule.
A small genus of 3 species with a widespread one extending to north-eastern Australia.

Chiloschista phyllorhiza (F. Muell.) Schltr.
(leaf-like roots)
Qld
                                         Nov-Dec
Roots 5-50 cm x 0.5-1 cm, flattened, dark green,

25

roughened; **racemes** 4-15 cm long, bearing 5-20 flowers; **flowers** 1-1.5 cm across, white, fragrant, short-lived, opening in batches along the racemes.

Found on small trees in coastal scrubs, frequently in open areas. Although it is leafless, the roots have taken on the function of the leaves. The species is easy to grow on a slab of hardwood or cork or very hard tree-fern. It should be tied direct to the slab without any moss underneath. It resents disturbance and is slow to re-establish but once the roots are growing it is all right. It needs regular water throughout the year and responds to liquid fertilizers. Requires a heated glass-house in southern Australia, with a minimum temperature of 10° C and humid conditions. Propagate from seed.

CHIONOCHLOA Zotov
(from the Greek *chion*, snow; *chloe*, young grass)
Poaceae

Perennial **grasses** forming large tussocks; **culms** glabrous or hairy, often shiny; **leaves** flat or folded; **inflorescence** an erect panicle; **spikelets** flattened.

A genus of 19 species, all but one of which are endemic to New Zealand. They are snow grasses that form dense tussocks. Propagate by division.

Chionochloa frigida (Vick.) Conert
(cold)
NSW   Ribbony Grass
0.5-1.2 m tall   Jan-March

Robust perennial forming dense **tussocks**; **culms** smooth and shining; **leaves** flat or folded, greyish or straw coloured, about 1 cm wide at the base, tapered, drawn out into long filiform apex; **panicles** 10-15 cm long, sparse, nodding; **spikelets** 5-7 flowered.

A distinctive grass endemic to the Kosciusko region, where it grows in communities in the subalpine zones. It is little-known in cultivation but is a striking grass for a rockery or ornamental border. Best suited to southern Australia in high-rainfall areas. Requires well drained soil and responds to regular watering and mulching. Propagate by division.

CHIONOHEBE Briggs & Ehrend.
(from the Greek *chion*, snow; *Hebe*, an allied genus)
Scrophulariaceae

Dwarf to small spreading **shrubs** or **cushion plants**; **stems** prostrate to decumbent, much-branched, rooting at the nodes; **leaves** simple, opposite, sessile, shortly connate at the base, usually with ciliate margins; **flowers** solitary in the leaf axils, each subtended by a pair of connate bracts; **calyx** 6-lobed; **corolla** 5-6 lobed, the basal part tubular; **fruit** a capsule.

A small genus of 6 species mainly developed in New Zealand, with 2 species extending to the subalpine regions of Australia. They were previously placed in the genus *Pygmaea* Hook. They are colourful plants with excellent horticultural potential for temperate areas with cold winters. Propagate from seed or cuttings. The cushion plants could probably be propagated by division.

Chionohebe ciliolata (Hook.f.) Briggs & Ehrend.
var. **fiordensis** Ashwin
(bearing a few long cilia; from fiords)
Tas   Cushion Plant
                                                 Dec-Feb

Cushion plant forming dense, compact, round cushions; **branchlets** spreading, much-branched; **leaves** 0.2-0.4 cm x 0.15-0.2 cm, narrow-ovate to broadly spathulate, stiff, overlapping and rosette-like near the apex of the branchlets, greyish green, the apex nearly acute, with a few long hairs near the tip; **flowers** about 1 cm across, white, open-petalled, tubular at the base; **capsule** about 0.2 cm long.

This New Zealand cushion plant has recently been discovered on Hamilton Crags in south-west Tas. It grows in bare rocky places in subalpine areas. Its response in cultivation is virtually unknown; however, it could prove to be an excellent container plant or for a rockery in areas with cool to cold climates. Would require a well structured, organically rich soil. Propagate from cuttings or by division.

Chionohebe densifolia (F. Muell.) Briggs & Ehrend.
(dense leaves)
NSW
0.1-0.3 m x 0.3-0.6 m   Nov-Jan

Prostrate **shrub** forming dense mats; **stems** prostrate, rooting at the nodes; **branchlets** erect, to 10 cm high, hairy; **leaves** 0.4-0.7 cm x 0.2-0.3 cm, obovate to oblong, thick and leathery, sessile, opposite, overlapping, the margins thickened and fringed with hairs; **flowers** 1.5-2.5 cm across, white, pink or violet, in the upper axils; **capsule** 0.4 cm x 0.3 cm, glabrous or hairy.

A beautiful little subapline plant found in New Zealand and also Mount Kosciusko, where it forms dense mats in high areas. It is a very colourful and showy plant that has excellent potential for containers or as a rockery plant in cold districts (it needs cold winters to flower freely). It likes an organically rich, well structured soil in a sunny position and must be kept moist over winter. Propagate from cuttings which strike easily.

CHISOCHETON Blume
(derivation obscure)
Meliaceae

Shrubs or trees; **leaves** pinnate, alternate, glabrous or hairy; **inflorescence** a panicle; **flowers** small, numerous, with 4-5 petals; **fruit** a capsule.

A large genus of about 100 species with one endemic in Qld.

Chisocheton longistipitatus (F. M. Bail.) L. S. Smith
(long stalks)
Qld
8-20 m x 3-8 m   Jan-March

Small to medium **tree**; **leaves** pinnate; **leaflets** 4-18, 15-22 cm x 3-6 cm, oblong or lanceolate, with prominent nerves; **panicles** 15-22 cm long, in the axils of the upper leaves; **flowers** with 5 petals, bearing stellate hairs; **capsule** 2-3 cm across, globular, covered with velvety hairs, on long stalks.

A widespread species that in Australia is confined to

lowland rainforests of north-eastern Qld. An attractive shelter tree for tropical regions. May shed its leaves during dry periods but this can be offset by watering. Needs well drained soils in a sunny or semi-shady situation. Propagate from seed which should be sown while fresh.

CHLAENOSCIADIUM Norman

(from the Greek *chlaina*, cloak; *sciaderon*, umbrella; in reference to the sizeable bracts around the flower-heads)

Apiaceae

An endemic, monotypic genus restricted to southern WA.

Chlaenosciadium gardneri Norman

(after Charles Gardner, 20th-century WA botanist)

WA

0.1-0.2 m x 0.3 m   Oct-Jan

Dwarf, spreading, perennial **herb**; **stems** slightly hairy, several emerging from tufted basal leaves; **basal leaves** to 3 cm x 1 cm, linear-lanceolate-dentate; **stem-leaves** smaller and more or less entire, often hairy; **flower-heads** to 2 cm across, white to pale yellow, with prominent bracts.

A little-known plant from the Avon, Coolgardie and Roe Districts. It is grown mainly by enthusiasts. Should grow in most well drained, light to medium soils, with partial to full sun. Has potential for cultivation in containers.

Propagate from seed or cuttings.

Chloanthaceae Hutch.

A family of dicotyledons consisting of about 7 genera and 20 species, apparently restricted to Australia. The exact delimitation of the family is unclear at present and the Australian species may be referred to as Verbenaceae or Dicrastylidaceae by various authors. Many species have horticultural merit but few are widely grown as yet. Chief genera: *Chloanthes*, *Cyanostegia*, *Dicrastylis*, *Lachnostachys*, *Newcastelia*, *Pityrodia* and *Spartothamnella*.

CHLOANTHES R. Br.

(from the Greek *chloros*, green; *anthos*, a flower; alluding to the greenish yellow flowers)

Chloanthaceae

Perennial **shrubs**; **stems** erect, branched, tomentose; **leaves** simple, sessile, decussate or in whorls of 3, upper surface scabrous, overlapping; **flowers** axillary, solitary, a corolla of 5 fused petals with 5 spreading lobes, hairy; **calyx** 5 fused sepals; **stamens** 4; **fruit** a dry 4-celled drupe.

An endemic genus of 4 species, with 3 in eastern Australia and a solitary species in south-western WA. None of the species is commonly cultivated. *C. coccinea* is the most outstanding for its scarlet flowers, but all species have attractive foliage. They are generally slow-growing. Regular tip pruning promotes bushy growth. Plants can be propagated from seed, but this process is slow and unreliable. Cuttings of firm young

Chloanthes parviflora, × ·66   flowers, × 1·3

growth have proved successful, with a varying time range before roots are produced. The application of a rooting hormone can be beneficial.

Chloanthes coccinea Bartling

(scarlet)

WA

0.3-0.8 m x 0.3-1 m   July-Dec

Dwarf **shrub**; **branches** many, upright; **leaves** 1-3 cm x 0.2-0.4 m, narrow-linear, appearing terete due to revolute margins, rough surfaces, light to dark green; **flowers** tubular, to 3.5 cm long, scarlet, hairy and glandular outside, in spike-like clusters at ends of branchlets; **fruit** more or less globular, about 0.4 cm diameter.

An outstanding species from inland and some coastal areas of south-western WA. It occurs on a wide range of deep and shallow soils with clay subsoil. Needs very well drained conditions, with partial to full sun. Withstands dappled shade in warm climates. Tolerates extended dry periods and light to medium frosts. Suitable for semi-arid regions. Excellent potential as a container plant. Propagate from seed or cuttings.

Chloanthes glandulosa R. Br.

(gland-bearing)

NSW

0.3-1 m x 0.5-1 m   July-Jan

Dwarf **shrub**; **leaves** 3.5-8 cm x 0.4-1.1 cm, lanceolate or near-lanceolate, margins scarcely recurved, rough upper surface, not woolly below; **flowers** tubular, to 4.5 cm long, greenish yellow, slightly glandular and hairy outside, 2 upper lobes, 2 middle lobes and 1 lower lobe; **fruit** to 0.6 cm x 0.5 cm.

This species occurs mainly in the Blue Mountains. Evidently not in cultivation, it will require well drained, light to medium soil, with partial or full sun. Should tolerate dappled shade. Withstands light frosts. Propagate from seed or cuttings.

Allied to *C. stoechadis*, which has narrower leaves with revolute margins.

Chloanthes parviflora Walp.
(small flowers)
Qld, NSW
0.3-1 m x 0.5-1 m   June-Nov; also sporadic
Dwarf **shrub**; often several **branches** near ground level; **leaves** 1-4 cm x 0.2-0.5 cm, linear or appearing terete due to revolute margins, pale green and rough above, densely woolly below but often concealed by revolute margins; **flowers** tubular, to 3.2 cm long, pale mauve with purple-spotted throat, hairy and glandular outside and inside, lower lobe much larger than other 4, axillary, solitary; **fruit** oblong, to 0.5 cm long.

The most widespread of the genus. It occurs in a wide variety of soils, from north-eastern Qld to south-eastern NSW. Not common in cultivation. Its general requirements are for relatively well drained, light to medium soils. It does well in partial to full sun, but also tolerates dappled shade. Frost tolerant. Propagate from seed or cuttings.

Chloanthes stoechadis R. Br.
(refers to similarity to French lavender)
Qld, NSW, ? WA
0.3-1 m x 0.5-1 m   July-Dec
Dwarf **shrub**; **branches** many; **leaves** 1-5 cm x 0.1-0.5 cm, narrow-linear, linear-lanceolate, or appearing terete due to revolute margins, rough upper surface, densely woolly below; **flowers** tubular, 2-4.5 cm long, greenish yellow to greenish blue, glandular and densely hairy on outside, upper lip comprising 2 lobes, lower with 3 lobes, more or less sessile; **fruit** about 0.5 cm diameter.

Its main representation is in central-eastern NSW, such as in the Blue Mountains, with scattered localities in other regions. Has had only very limited cultivation. Requires well drained soils, with partial or full sun. Tolerates light to medium frosts. Propagate from seed or cuttings.

The var. *parviflora* Benth. is now considered conspecific. *C. stoechadis* is closely allied to *C. parviflora*, which can be readily distinguished by its pale mauve flowers with the purple-spotted throat.

CHLORIS Sw.
(from the Greek *chloros*, green)
Poaceae
Small to medium sized annual or perennial **grasses** forming spreading tussocks; **stems** prostrate or erect, ending in branches which spread like the ribs of an umbrella; **spikelets** bearing awns.

Species about 40 with 8 indigenous plus a couple of exotic pasture grasses that have become widely naturalized. They are tussock grasses with interesting and decorative flower-heads. Propagation is by division or from seed which germinates best after dry storage.

Chloris acicularis Lindl. =
          Enteropogon acicularis (Lindl.) Lazarides

Chloris truncata R. Br.
(cut off abruptly)
all states except Tas   Windmill Grass
0.3-0.5 m tall   Oct-Jan
Small to medium sized perennial **grass** forming compact tufts; **stems** oblique or erect; **leaves** basal, dull green, roughened; **panicle** of 5-9 branches spreading like a windmill; **spikelets** bearing long, roughened awns.

Widespread on plains, frequently on heavy clay soil. Conspicuous when in flower. An attractive grass for a rockery. The flower branches are bright purple at flowering. Cultivated as an ornamental grass in the USA. Propagate from seed.

Chloris ventricosa R. Br.
(swollen or inflated)
Qld, NSW, Vic   Tall Windmill Grass
0.3-0.6 m tall   Oct-Jan
Medium sized perennial **grass** forming spreading tufts; **stems** spreading or prostrate, then erect; **leaves** dull green, flat, drawn out into a long point; **panicle** of 3-6 branches spreading like a windmill; **spikelets** closely packed along the branches, each bearing a short awn.

An attractive grass suitable for cultivation in a rockery or amongst low shrubs. Requires good drainage and prefers a sunny position. Very hardy but responds to regular watering. Cultivated as an ornamental grass in the USA. Propagate from seed.

CHLOROPHYTUM Ker Gawler
(from the Greek *chloros*, green; *phyton*, a leaf)
Liliaceae
Perennial, stemless **herbs**; **rhizome** short; **roots** fibrous or tuberous; **leaves** in basal rosette or tussock, linear to lanceolate, with a fimbriate margin; **inflorescence** a raceme or panicle; **flowers** 6-partite, the perianth free, spreading or reflexed; **fruit** a capsule.

A genus of 100-300 species, mainly tropical, with 2 found naturally in Australia. A couple of exotic species are widely cultivated in gardens and sold in nurseries but the native species are rarely cultivated. One of the exotic species, *C. comosum* (Thumb.) Jacques, sometimes becomes naturalized in moist, shady places. Propagate from seed or by division of the clumps. Seed germinates readily when fresh.

Chlorophytum alpinum (Hook.f.) Benth. ex Baker
                    see *Caesia alpina* Hook.f.

Chlorophytum laxum R. Br.
(loose; open)
Qld, NT
10-60 cm tall   Jan-Dec
Perennial **herb** forming tussocks; **roots** often bearing tubers 1-5 cm long; **leaves** 10-50 cm x 4-12 cm, 4-12 per clump, linear to lanceolate, dark green, channelled, with a prominent midrib, in a basal rosette, the outer ones arched, inner ones erect; **peduncle** 10-60 cm tall,

Choricarpia subargentea

Choricarpia leptopetala, × ·5

erect, usually unbranched, straight or flexuose; **bracts** 0.5-1.2 cm long, scarious; **flowers** 0.6-1 cm across, greenish white, with linear segments; **capsule** 0.5-1 cm long, globose, 3-lobed.

A little-known lily which grows in shady, rocky areas, sometimes in sparse colonies. In the Top End it is frequently found in bamboo thickets. Although not showy, it is an interesting plant suitable for growing with ferns or in a shaded rockery. Best suited to tropical and subtropical regions. Propagate from seed or by division of the clumps.

CHORETRUM R. Br.

(from the Greek *chloris*, separate; *etron*, abdomen; in reference to the torus being separate from the perianth)
Santalaceae

Parasitic or semi-parasitic **shrubs**; **leaves** alternate, scale-like; **flowers** minute, bisexual, subtended by several bracts, borne in axillary pedunculate heads; **perianth lobes** 5, thick, fleshy, incurved; **stamens** 5; **style** short, with 5-lobed stigma; **fruit** a fleshy drupe with persistent perianth segments.

An endemic genus with 6 spp. Their main representation is in arid and semi-arid regions. Only *C. glomeratum* seems to have been tried in cultivation. Due to the parasitic or semi-parasitic characteristics of this genus, successful cultivation for any extended period is extremely doubtful. Further experimentation using different host plants is required, which may provide positive results.

Choretrum glomeratum R. Br.

(pressed together)

Qld, NSW, Vic, SA, WA — Common Sour-bush; Berry Broombush

1-3 m x 0.6-2 m — Sept-Jan

Small to medium **shrub**; **branches** erect, slender, angular, broom-like; **leaves** when present scale-like; **flowers** small, white, in clusters, forming loose racemes 2-5 cm long; **drupe** ovoid to globular, about 0.5 cm long, yellow.

Widespread over temperate Australia and in the outer parts of semi-arid regions. This species is not well known in cultivation (see comments under genus entry). As its common name implies, it has characteristics of the non-indigenous broom plants. Should do best in well drained, light to medium soils, with a warm to hot situation. Tolerant of most frosts and extended dry periods. Propagate from seed which can take a few months to germinate.

CHORICARPIA Domin

(from the Greek *chorizo*, to divide; *carpos*, fruit; a reference to the divided fruit)
Myrtaceae

Trees; **leaves** opposite, entire, bearing oil dots; **flowers** in globular heads, arising in the leaf axils; **flowers** lacking petals; **calyx** tubular; **fruit** a small capsule opening by 2-3 valves.

A small genus of 2 species, both of which are endemic in eastern Australia. Propagation is from seed which is very fine and must be covered lightly. The bog method would be worth trying (see Volume 1, page 204).

Choricarpia leptopetala (F. Muell.) Domin

(slender petals)

Qld, NSW — Brush Turpentine

5-12 m x 3-5 m — Nov-Jan

Tall **shrub** or small **tree**; **leaves** 6-12 cm x 2-4 cm, lanceolate, drawn out into a long point at the apex, dark green above, paler beneath; **flowers** crowded in dense globular heads, 1.5-2 cm across, white or cream, on stalks 1-3 cm long; **capsules** dark brown.

Widespread on moist soils from Brisbane to just south of Sydney. Highly ornamental as it flowers while very young and the decorative fruits hang on the bush for several months. Easily grown in a variety of positions. Prefers deep soil but will grow in clays. Responds to water during dry periods. Needs good drainage. Frost hardy and will grow in southern Australia. Needs a sunny or semi-shady position. Propagate from seed or cuttings which are slow to strike.

Choricarpia subargentea (C. T. White) L. Johnson

(silvery beneath)

Qld — Giant Ironwood

10-30 m x 10-15 m — Dec-Feb

Small to large **tree**; young shoots rusty; **leaves** 5-8 cm x 2-4 cm, ovate or elliptical, green above, white or silvery beneath; **flower-heads** about 1 cm across, globular, cream, densely covered with rusty hairs, on stalks 0.6-1 cm long, borne in clusters at the ends of branchlets.

Common in the rainforest of south-eastern Qld, where it is cut as a timber tree. Not commonly grown as it develops into a tree too large for the average home garden, but could be very useful in parks or large gardens or for acreage planting. It is a handsome tree with foliage that contrasts pleasantly as the wind blows. Seedlings are attractive and would probably make excellent indoor plants. Requires well drained soil in a protected situation. Propagate from seed.

CHORICERAS Baillon
(from the Greek *chorizo*, to divide; *ceras*, a horn)
Euphorbiaceae

Shrubs or **trees**; young shoots hairy; **leaves** opposite, entire; **flowers** unisexual, in axillary clusters; **tepals** 5-6; **stamens** 5-6; **fruit** a capsule.

A small genus of 2 species found in Australia with one extending to New Guinea. They are generally drab shrubs with little to recommend them for cultivation. Propagation is from cuttings which strike readily.

Choriceras majus Airy-Shaw
(large; big)
Qld
3-8 m x 1-2 m   May-June; also sporadic

Tall **shrub** or small **tree**; young shoots hairy, with bright red young leaves; **leaves** 7-14 cm x 3-7 cm, ovate to broad-elliptic, leathery, dark green and shiny above, pale beneath, minutely toothed margins; **flowers** in axillary clusters, white; **male flowers** in many-flowered, congested cymes, on slender pedicels; **female flowers** in pairs, on thick pedicels.

Restricted to coastal rainforests of north-eastern Qld. A fairly drab species with attractive new growth and dark shiny leaves. Best suited to tropical regions. Requires well drained soils and some protection when young. Propagate from seed or cuttings.

Choriceras tricorne (Benth.) Airy-Shaw
(3-horned)
Qld, NT
1-3 m x 0.5-1.5 m   all year

Medium to tall **shrub**; young shoots hairy; **leaves** 2.5-4 cm x 1-1.5 cm, elliptical to lanceolate, blunt, leathery, shiny above, pale beneath, with crenulate margins; **flowers** in sessile clusters, green, inconspicuous, the male and females borne separately, the **males** on slender pedicels 0.6 cm long, the **females** on thick pedicels 0.2 cm long; **capsule** 0.6-0.8 cm across, 3-cornered.

Confined to heathlands of northern Australia but there often locally common. A hardy shrub for tropical areas, with little to recommend it except the shiny green foliage. Grown mainly by enthusiasts. Requires well drained soils in a sunny position. Very frost tender. Propagate from seed or cuttings which strike very easily.

CHORILAENA Endl.
(from the Greek *chorizo*, to divide; *chlaina*, cloak; refers to divided bracts around flower-heads)
Rutaceae

A monotypic genus endemic to WA.

Chorilaena hirsuta Benth. = C. quercifolia Endl.

Chorilaena quercifolia Endl.
(leaves similar to those of the genus *Quercus*)
WA   Chorilaena
1-4 m x 1-3 m   Sept-Dec

A small to medium **shrub**; **branches** densely hairy; **leaves** 2-4 cm x 1-3 cm, simple, alternate, lobed mar-

Chorilaena quercifolia, × ·5

gins, often slightly hairy above, undersurface always densely hairy; **flower-heads** pendulous, about 1 cm x 1.5 cm, borne on axillary stalks, composed of many flowers surrounded by many narrow bracts, cream, yellow-green or reddish.

A decorative foliage species that occurs as an understorey shrub in the karri forests. It requires moist, well drained soils, preferably with organic matter. Appreciates dappled shade or partial sun. It grows well under existing eucalypts. Useful as a screening plant for gardens, and does well in containers. Responds well to organic fertilizers. Attractive to nectar-eating birds. Tolerates most frosts. Propagate from seed or cuttings.

CHORISTEMON H. B. Williamson
(from the Greek *chorizo*, to divide; *stemon*, stamen)
Epacridaceae

A monotypic genus endemic in Vic.

Choristemon humilis H. B. Williamson
(low)
Vic   Choristemon
0.1-0.2 m x 0.5-1 m   Sept-Nov

A dwarf **shrub**; **branches** twiggy, glabrous; **leaves** about 0.7 cm long, lanceolate, stiff, acute, concave, minutely hairy, 3-ribbed undersurface; **flowers** tubular, very small, hairy inside, 1-3 at ends of branches, or rarely axillary and solitary, white to cream.

This species is found only in the Brisbane Ranges, west of Melbourne. It is very closely allied to *Leucopogon virgatus*, which is distinguished by its very hairy corolla lobes. In cultivation it would require very well drained soils, with dappled shade, partial sun or possibly full sun. Withstands light to moderate frosts. Propagate from cuttings.

CHORIZANDRA R. Br.

(from the Greek *chorizo*, to divide; *andros*, a man; in reference to the separate male flowers)
Cyperaceae

Perennial **herbs**; **stems** subterranean, creeping; **leaves** scale-like or basal; **stems** erect, rush-like, usually septate; **flower-heads** globose, usually subtended by a sheath; **fruit** a nut.

A small genus of 4 species, all endemic to Australia. They are useful rush-like plants, rarely cultivated although providing an interesting contrast in the garden. Propagation is by division of the clumps or from seed.

Chorizandra cymbaria R. Br.

(boot-like)
Qld, NSW, Vic, Tas, WA, NT   Heron Bristle Rush
0.6-1.5 m tall   Sept-Nov

Culms 0.2-0.6 cm thick, erect, bright green, fleshy, septate; **flower-heads** globular, sheathed by a bract, brownish; **spikelets** about 0.5 cm long; **nut** 0.3 cm wide, globular.

Common in marshy or wet sandstone areas, usually in colonies. An unusual rush-like plant, useful for breaking up lines of foliage or shrubbery. Best suited to a moist position but will also survive in a well drained soil if regularly watered. Propagate by division.

Chorizandra enodis Nees

(without nodes)
NSW, Vic, Tas, SA, WA, NT   Black Bristle Rush
0.3-0.45 m tall   July-Nov

Culms 0.1-0.15 cm thick, erect, greyish green, wiry, without septa; **flower-heads** globular, brownish; **spikelets** 0.35 cm long; **nut** 0.25 cm long, globular.

Widespread in moist, sandy heathlands, damp places and along stream banks. A useful greyish rush-like plant for growing in a rockery or amongst small shrubs. Can be grown as a bog plant or in a moist position but also hardy enough to grow in a dry position, provided extra water is supplied. Propagate by division.

Chorizandra sphaerocephala R. Br.

(globose heads)
Qld, NSW
0.3-0.6 m tall   Aug-Oct

Culms fleshy, septate, greenish; **flower-heads** globular, not enclosed by a bract; **spikelets** globose, about 1 cm across, black; **nut** ovate, with longitudinal ribs.

Found in wet patches. Not commonly grown. Remarks and requirements as for other species. Propagate by division.

CHORIZEMA Labill.

(from the Greek *choros*, a dance; *zema*, a food or drinking vessel; allegedly because Labillardiere and his party were almost exhausted when they found water and this plant at the same location.)

Chorizandra enodis, × ·5

Fabaceae

Small **shrubs** or **twiners**; **leaves** usually alternate, simple, entire or prickly-toothed, stipules small; **flowers** pea-shaped, in terminal racemes, rarely axillary, usually orange, red or pink with yellow; **calyx lobes** more or less equal; **standard** broad, longer than wings and keel; **stamens** free; **ovary** 1-celled, ovules 4 or more; **fruit** an inflated pod.

An endemic genus of about 18 species, restricted to the South-West Botanical Province of WA, except for one species found in south-east Qld and eastern NSW. In nature they grow in varied conditions, but usually in association with other shrubs and often with an overhead canopy of eucalypts. The soils are mainly sands, but can be gravels.

Chorizemas are very floriferous, and renowned for their brightness of colour, which has given rise to the common name of Flame Peas. There is often a combination of colours in each flower, and plants can be used to advantage for creating areas of colour in the landscape. Some species have been cultivated since the early 19th century.

Their requirements are usually for moist but well drained soils, with dappled shade or partial sun. Some species will tolerate full sun if the root system has some protection. Their roots do not penetrate very deeply into the soil, and if they dry out it usually means the death of the plant.

It is recommended that some form of mulching material be used over the root area if plants are exposed in summer months. Some species will tolerate

Chorizema aciculare

W.R. Elliot

light frost, but others are frost tender. Nearly all chorizemas respond well to light pruning after flowering, with some being ideal as cut flowers. The main pest attack is usually from leaf-eating caterpillars, aphids and other sap-sucking insects. Propagation is readily carried out from seed. Being a legume, the seeds have a hard outer coating, and need treatment before sowing (see Treatments and Techniques to Germinate Difficult Species, Volume 1, page 206). Most species have been grown successfully from cuttings of firm new growth.

Chorizema aciculare (DC.) A. W. Hill
(needle-like)
WA   Needle-leaved Chorizema
0.5-1 m x 0.5-1 m   July-Nov
   A small, sometimes suckering **shrub**; **branches** erect to spreading; **leaves** to 2 cm long, tapering to a pungent point, thick, spreading, prominent central nerve above, broad nerve below, reticulate; **flowers** pea-shaped, about 1 cm x 1 cm, solitary, borne on stalks up to 1 cm long, axillary, or terminal forming spikes at end of branches; **keel** pinkish red; **standard** pink and yellow.
   An outstanding ornamental species, occurring naturally in sands and gravels. It will grow in a wide range of well drained soils. It prefers partial or full sun, but tolerates dappled shade. Withstands light frosts. Responds well to light pruning. Suitable for gardens or containers. Propagate from seed or cuttings.

Chorizema glycinifolium, × ·33

32

Chorizema dicksonii   W.R. Elliot

Chorizema angustifolium Benth. =
                C. glycinifolium (Smith) Druce

Chorizema cordatum Lindl.
(cordate)
WA   Heart-leaved Flame Pea
1-2 m x 1-2 m   July-Dec; also sporadic

A dense **shrub** or **semi-climber**; **branches** slender,
many; **leaves** to 5 cm long, cordate at base, tapering to
soft point, margins sometimes toothed or lobed; **flowers**
pea-shaped, about 1.5 cm x 1.5 cm, borne in terminal
racemes to about 15 cm long; **standard** orange, red and
yellow; **keel** purple-pink.

The most widely known and grown species. It can

Chorizema cordatum   B.W. Crafter

act as a light climber if it comes in contact with other
plants. Best grown in moist but well drained soils, with
dappled shade or partial sun. Pruning after flowering is
recommended, as this prolongs the life of a plant. Can
have sporadic bursts of flowering throughout the year.
Various colour forms including one with a yellow standard have been introduced to cultivation. Suitable as a
garden or container plant. Propagate from seed or cuttings.

Chorizema cytisoides Turcz.
(similar to the genus *Cytisus*)
WA
0.3-0.5 m x 0.5-1 m   July-Nov

A spreading dwarf **shrub**; **stems** many; **branches**
few; **leaves** to 2.5 cm x 0.3 cm, scattered, tapering to a
small sharp point, margins revolute, glabrous above,
silky hairs below; **flowers** pea-shaped, about 1 cm x
1 cm, borne in terminal racemes to about 8 cm long,
profuse; **standard** dull orange and yellow; **keel** pink.

An ornamental species that undoubtedly will be
more widely grown in the future. In nature it usually
grows amongst other shrubs in sandy or gravelly soils.
For cultivation it should adapt to most well drained
soils, with dappled shade or partial sun. Light pruning
after flowering will produce a dense plant. Suitable for
gardens or containers including hanging baskets.
Propagate from seed or cuttings.

Chorizema dicksonii R. A. Graham
(after James Dickson & Sons, 19th-century Scottish
nurserymen)
WA   Yellow-eyed Flame Pea
1-1.5 m x 1-2 m   Aug-Dec

Small **shrub**; **branches** many; **leaves** to about 2 cm
long, narrow, curved, tapering to a pungent point,
prominent midrib and veins, glabrous; **flowers** pea-
shaped, about 2 cm across, borne in loose terminal
racemes; **standard** orange-red and yellow; **keel** orange-
red.

A very ornamental species, well-known in cultivation. It grows naturally in lateritic gravels and is
usually found in the jarrah forests. Growing requirements are for well drained, medium to heavy soils, with
dappled shade or partial sun. It will tolerate light soils
and full sun, provided there is protection for the root
system. Pruning after flowering is recommended.
Withstands light frosts. There can be colour variation
in the flowers and some selected forms are grown by
enthusiasts. Propagate from seed or cuttings.

Chorizema diversifolium A. DC.
(variable leaves)
WA
1-2 m x 1-2 m   Sept-Dec

A light, twining **shrub**; **branches** slender; **leaves** to
2.5 cm x about 0.5 cm, linear to oval, tapering both
ends, blunt, glabrous above, glabrous or hairy below;
flowers pea-shaped, about 1 cm across, borne in loose
racemes; **standard** orange and yellow; **keel** pink to purple.

A very useful, light twiner or climber. It prefers to
grow in the association of other plants, which it uses
for the support of its slender branches. Will grow in

33

Chorizema ericifolium

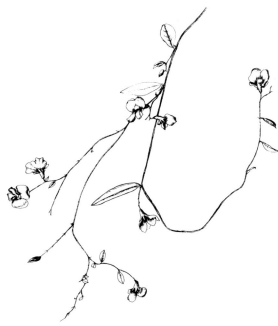

Chorizema diversifolium, × ·4

most relatively well drained soils, and prefers dappled shade or partial sun. Frost tender. Withstands hard pruning if required. Suitable for gardens or containers including hanging baskets. Propagate from seed or cuttings.

Chorizema ericifolium Meisner
(leaves similar to those of the genus *Erica*)
WA
0.5-1.5 m x 0.5-1.5 m   July-Oct
A small **shrub**; **branches** slender, many, smaller ones often spiny; **leaves** about 0.5 cm long, narrow, alternate, irregularly opposite or verticillate, not crowded, margins revolute; **flowers** pea-shaped, small, borne in loose terminal racemes, profuse, red and yellow.
Evidently not in cultivation. This species usually inhabits grey sandy soils, north of Perth, which suggests its growing requirements would be for well drained soils, with dappled shade or partial sun. Propagate from seed or cuttings.

Chorizema genistoides (Meisner) C. Gardner
(similar to the genus *Genista*)
WA
0.2-0.7 m x 0.5-1.5 m   Sept-Oct
A dwarf **shrub**; **branches** few, hairy; **leaves** to 5 cm x 1 cm, prominent nerves, margins revolute, rigid; **flowers** pea-shaped, about 1 cm x 1 cm, borne in loose terminal racemes to 15 cm long, profuse; **standard** orange-red with yellow; **keel** pink.
An ornamental species from the Irwin District, where it grows in sandy or rocky soil, amongst other shrubs. Its cultivation requirements are for well

drained soils, with dappled shade or partial sun, but it will tolerate full sun if the root zone is protected. Withstands light frosts and limited dry periods. Suitable for gardens or containers. Propagate from seed or cuttings.

Chorizema glycinifolium (Smith) Druce
(leaves similar to those of the genus *Glycine*)
WA
0.3-1 m x 0.5-1.5 m   July-Oct
A dwarf **shrub** or light **twiner**; **branches** ascending; **leaves** to 5 cm long, alternate, lower ones oval to lanceolate, pungent, rigid, sometimes with toothed margins, often absent, upper ones usually linear, margins recurved, glabrous and reticulate above, silky hairs below; **flowers** pea-shaped, about 1 cm across, borne in loose terminal racemes, profuse; **standard** orange and yellow; **keel** pink.
An ornamental species that is not widely cultivated. It grows naturally south of Perth, and along the southern coast. It prefers well drained soils, with dappled shade or partial sun, and appreciates the protection of other plants. Propagate from seed or cuttings.
This species was previously known as *C. angustifolium* Benth.

Chorizema ilicifolium Labill.
(leaves similar to those of the genus *Ilex*)
WA   Holly Flame Pea
1-3 m x 1-3 m   Aug-Jan
A small to medium **shrub**; **branches** slender, slightly hairy when young; **leaves** to 8 cm long, narrow or broad, margins undulate, prickly-toothed, or lobed, reticulate, glabrous; **flowers** pea-shaped, about 1 cm across, borne in loose axillary or terminal racemes; **standard** broad, orange-red with yellow; **keel** purple-pink.
This species is similar in many ways to *C. cordatum*, including its habit of acting as a semi-climber if in contact with other shrubs. It has been widely cultivated, and will grow in most relatively well drained soils, with dappled shade or partial sun. It also likes to grow in association with other shrubs. Pruning after flowering is beneficial. Frost tender. Grown successfully under glass in England. Propagate from seed or cuttings. It seems likely that *C. varium* is a form of this species.

Chorizema nervosum T. Moore
(prominent nerves)
WA
0.5-1.5 m x 1-1.5 m   June-Oct
A dwarf to small **shrub**; **branches** hairy; **leaves** about 1.5 cm long, broad, oval, pungent point, undulate, not toothed, reticulate, leathery, usually glabrous; **flowers** pea-shaped, about 1 cm across, borne in loose, few-flowered terminal racemes, orange and yellow; **pods** about 1 cm long.
Evidently not well known in cultivation, but it is a showy species. It requires relatively well drained soils, with dappled shade to partial sun. It usually occurs in gravelly soils near the south coast. Suitable for gardens or containers. Propagate from seed or cuttings.

Chorizema glycinifolium   B.W. Crafter

Chorizema parviflorum Benth.
(small flowers)
Qld, NSW
0.3-0.5 m x 0.5-1.5 m   Aug-Sept; also sporadic

A dwarf **shrub**; **rootstock** stout; **stems** angular, ascending, slightly hairy; **leaves** to 4 cm long, variable on same branch, alternate, linear to oblong, pungent-pointed, recurved margins, lower surface hairy; **flowers** pea-shaped, under 1 cm across, borne in loose terminal racemes, yellow with red; **pods** 0.6-0.8 cm long, turgid.

This is the only species from eastern Australia, and occurs mainly on or near the coast of southern Qld and northern NSW. It is not widely grown, except by enthusiasts. Requirements are for well drained soils, with dappled shade or partial sun. Grows well amongst other small shrubs or as a container plant. Withstands light to medium frosts, but heavy frosts may cause damage. Propagate from seed or cuttings.

Chorizema reticulatum Meisner
(netted veins)
WA   Showy Flame Pea
0.3-0.75 m x 0.5-1 m   Aug-Oct

A dwarf **shrub**; **branches** erect or often slightly ascending, slightly hairy; **leaves** to 3 cm long, lanceolate, rigid, acute, reticulate, margins flat or recurved, crowded; **flowers** pea-shaped, about 1 cm across, borne in erect terminal racemes to 15 cm long; **standard** pink and yellow; **keel** pink.

A most ornamental species, which occurs naturally in gravelly sands of the Stirling district. It requires well drained soils, with dappled shade or partial sun. Should be able to withstand some dry periods. Suitable for gardens or containers including hanging baskets. Propagate from seed or cuttings.

Chorizema rhombeum R. Br.
(rhomboid)
WA
0.2-0.3 m x 0.5-1 m   Aug-Jan

A prostrate to dwarf **shrub**; **branches** slender, spreading, often hairy; **leaves** to about 3 cm long, lower

ones oval to rhomboid, upper ones lanceolate, flat or recurved margins, glabrous above, sometimes hairy below; **flowers** pea-shaped, about 1 cm across, borne on long terminal stalks in loose racemes; **standard** orange and yellow; **keel** pink.

This species is certainly worthy of cultivation. It occurs in the south-west corner of WA, where it is commonly growing on lateritic soils and gravels. It needs well drained soils, with dappled shade or partial sun. Light pruning after flowering is recommended. For gardens or containers including hanging baskets. Propagate from seed or cuttings.

Chorizema trigonum Turcz.
(3-cornered)
WA
0.5-1 m x 0.5-1.5 m   Aug-Nov

A dwarf, spreading **shrub**; **leaves** to 7 cm x 0.3 cm, folded lengthwise, ending in a recurved point, rigid, reticulate, leathery; **flowers** pea-shaped, about 1 cm across, on short stalks, borne in loose terminal racemes to 10 cm long; **standard** dull orange-red with yellow; **keel** dull orange-red.

A species of coast and inland areas, where it grows in gravelly and rocky, sandy soils, amongst low shrubs. In cultivation it requires well drained soils, with dappled shade or partial sun. Withstands light frost and limited dry periods. An ornamental plant for gardens or containers. Propagate from seed or cuttings.

Chorizema uncinatum C. R. P. Andrews
(hooked)
WA
0.3-0.5 m x 0.5-1 m   Aug-Nov

A dwarf, spreading, open **shrub**; **branches** few; **leaves** to 1 cm x 0.3 cm, margins revolute, reticulate;

Chorizema reticulatum, × ·5

flowers pea-shaped, about 1 cm across, borne in terminal racemes to about 15 cm long, profuse; **standard** dull orange and red; **keel** light pink.

This decorative species requires well drained soils, with dappled shade or partial sun. It grows naturally in gravelly and coarse sandy loam soils, with large shrubs. Tolerates light frosts and limited periods of dryness. Suitable for gardens or containers. Propagate from seed or cuttings.

Chorizema varium Benth. ex Lindl. — is most likely a form of *C. ilicifolia* Labill.

CHRISTELLA Léveillé
(derivation unknown)
Thelypteridaceae

Medium to large **ferns** with an erect or creeping rootstock; **fronds** scattered or radiating in a tussock, bipinnatifid; **pinnae** spreading, the lower 1-5 pairs gradually reduced in size, the lowest not less than 2 cm long; **sori** rounded, with a kidney-shaped cover.

A small genus of about 60 species with 5 in Australia. They grow in a variety of situations from rainforest to exposed creek banks and swamps. In general they are easy to grow, being hardy in a variety of situations.

Christella arida (D. Don) Holttum
(dry)
Qld
0.3-1.2 m

Medium sized **fern** forming spreading patches; **rootstock** creeping; **fronds** erect, sparse, narrow, pale green, thin-textured; **segments** lanceolate, lobed, reduced in size towards the base of the fronds.

Found in moist soaks and depressions, usually in open forest country. Recently introduced into cultivation and not well known as yet. Grows readily in a moist position but resents total shade and needs exposure to some sun. Very frost tender and requires protection in southern Australia. Propagate by division of the rhizome or from spore.

Christella dentata (Forsskal) Brownsey & Jermy
(toothed)
Qld, NSW, Vic, SA   Binung
0.5-1 m tall

Medium sized **fern** forming clumps; **fronds** erect or arching, light or dark green; **segments** linear, shallowly lobed, gradually reduced in size towards the base of the frond.

Found in a range of habitats but most common along stream banks in forest country, often in scattered colonies. Popular in cultivation as it forms a neat, radiating tussock. Fast growing. May be damaged by frost, but shoots again in spring. Resents total shade and needs exposure to some sun. Responds to regular watering during dry periods. Propagate from spore which germinates easily.

Christella hispidula (Decne.) Holttum
(small hairs)
Qld
20-60 cm tall

Medium sized **fern** forming clumps; **fronds** erect, in a radiating tussock, light green, very narrow, thin-textured; **segments** linear, slightly curved, pointed, gradually reduced in size towards the base of the frond.

Recently introduced into cultivation, where it has proved easy to grow. Prefers a semi-shady position with an abundance of water during dry periods. Frost tender. Suited to cultivation in milder areas such as around Sydney. Propagate from spore.

Christella parasitica (L.) Léveillé
(parasitic)
Qld, NSW
20-50 cm tall

Medium sized **fern** forming clumps; **fronds** erect, in a radiating tussock, yellowish, narrow, thin-textured; **segments** deeply lobed, pointed, not greatly reduced in size towards the base of the frond.

Previously confused with *C. dentata*, from which it can be distinguished by the basal segments which end abruptly without great reduction in size. Fast growing and suited to cultivation in milder districts. Frost tender. Needs some exposure to sun but protection from winds. Propagate from spore.

Christella subpubescens (Blume) Holttum
(somewhat hairy)
Qld
0.5-1 m tall

Medium sized **fern** forming clumps; **fronds** erect, in a radiating tussock, dark green, thin-textured; **segments** broad, spreading, shallowly lobed, the lower ones gradually reduced in size and deflexed.

A widespread fern that is relatively uncommon in Australia, where it is confined to a few localities in north-eastern Qld along watercourses and the margins of swamps. Easily grown in the tropics or subtropics but cold sensitive and difficult to grow in temperate areas. Well suited to moist soils in a semi-shady situation.

Chrysobalanaceae R. Br.

A family of dicotyledons consisting of about 17 genera and 400 species mainly distributed through the tropics and subtropics. They are trees or shrubs with simple, alternate leaves and mainly bisexual zygomorphic flowers carried on panicles or racemes. The family is represented in Australia by *Parinari* and *Stylobasium*, although the latter is sometimes placed in its own family, Stylobasiaceae.

CHRYSOGONUM L.
(from the Greek *chryso*, golden; *gony*, a knee; in reference to flowers borne at the nodes)
Asteraceae

Perennial **herbs** to small **shrubs**; **leaves** opposite, entire or toothed; **flower-heads** yellow, borne on long peduncles, receptacle with chaffy bracts; **ray flowers** female; **disc flowers** bisexual but sterile; **fruit** a flat-

Christella dentata   D.L. Jones

tened achene, without wings.

In Australia there are 2-3 endemic species with further representation of about 2 species in eastern North America.

Chrysogonum trichodesmoides (F. Muell.) F. Muell.
(similar to the genus *Trichodesma*)
WA
0.6-1.5 m x 0.5-1 m   May-Oct; also sporadic
    Dwarf to small **shrub**; **branches** rigid, usually erect, rough; **leaves** 2-6 cm long, lanceolate, nearly sessile, opposite, 3-nerved, entire, hairy; **flower-heads** small, yellow, borne in a terminal panicle.

    A little-known species with interesting flowers. It occurs mainly in coastal regions of the Eremaean and Northern Botanical Provinces. Has had limited cultivation, and is probably best suited to tropical and subtropical regions. Possibly would grow well as a container plant in temperate climates. Needs very well drained, light to medium soil, with plenty of sunshine. May suffer frost damage. Propagate from seed which germinates readily or from cuttings.

CHRYSOPHYLLUM L.
(from the Greek *chryso*, golden; *phyllon*, a leaf; in reference to the prominent golden hairs on the underside of the leaves)
Sapotaceae
    Trees; young shoots hairy, the hairs usually golden; **leaves** entire, alternate, bearing golden hairs on the underside; **flowers** small, axillary, in some species the buds develop over many years; **fruit** a fleshy drupe, with 1-2 large seeds.
    A large genus of about 150 species mainly found in tropical America. They are large, handsome trees, some of which are grown in the tropics for their edible fruit. The genus has been revised and 3 of the 4 Australian members have been transferred to other genera (*Amorphospermum* and *Niemeyera*). The remaining Australian species is rarely encountered in cultivation.

Chrysophyllum antilogum (F. Muell.) Vink =
                    Amorphospermum antilogum F. Muell.

Chrysophyllum chartaceum (Bailey) Vink =
                    Niemeyera chartacea (Bailey) C. T. White

Chrysophyllum lanceolatum (Blume) DC.
(lanceolate leaves)
Qld
8-15 m x 5-8 m   Nov-Jan
    Small to medium **tree**; **leaves** 6-15 cm x 2-3.5 cm, oblong, glabrous, dark green and glossy above, dull beneath; **flowers** small (about 0.4 cm long), white or greenish, in clusters of up to 20; **fruit** 2-3 cm long, globose, with 3-5 ribs; **seeds** 2-5, large.
    A little-known species from rainforests of northeastern Qld. It is a handsome tree, well suited to culture in parks or large gardens in tropical or subtropical regions. May require protection from direct sun when young. Requires well drained soils. Propagate from seed which must be sown while fresh.

Chrysophyllum pruniferum F. Muell. =
                    Niemeyera prunifera (F. Muell.) F. Muell.

CHRYSOPOGON Trin.
(from the Greek *chryso*, golden; *pogon*, a beard; a probable reference to the golden hairs)
Poaceae
    Tufted perennial **grasses**; **leaves** spreading, basal, hairy; **stems** erect or flexuose, hairy; **inflorescence** a panicle; **spikelets** slender, with a bent awn.
    Species about 25 with 6 being found in northern Australia.

Chrysopogon fallax S. T. Blake
(deceptive; false)
Qld, NSW, SA, WA, NT   Golden Beard Grass
0.3-1.2 m tall   Dec-April
    Tall tufted **grass**; **leaves** 10-45 cm long, in a basal tussock, smooth, hairy; **stems** erect, slightly bent; **inflorescence** a loose panicle, about 20 cm long; **spikelets** narrow, on long, slender stalks.
    A useful slender grass which develops a dense tussock and tall flower-heads. Mixes well with low shrubs. Very hardy and drought resistant. Suited to warm inland areas. Propagate by division of the rhizome or from seed.

CIENFUGOSIA Cav.
    The Australian species previously included in this genus have now been transferred to *ALYOGYNE* Alef. and *GOSSYPIUM* L.

CINNAMOMUM J. Schaeffer
(from the Greek *Cinnamomon*, the classical name for cinnamon)
Lauraceae
    Shrubs or trees; **leaves** opposite or alternate, entire, glabrous; **inflorescences** of axillary or terminal pan-

Cinnamomum oliveri new growth   D.L. Jones

icles; **flowers** unisexual, the females fewer and larger in the panicle, bell-shaped, small; **perianth segments** 6; **fruit** a drupe, developing on the enlarged perianth tube.

A large genus of 250 species with about 4 in Australia, 3 of which are endemic. Another species (Camphor Laurel — *C. camphora*) is commonly cultivated as an ornamental, and is widely naturalized on roadsides and cleared land in NSW and Qld, and is also invading rainforests as a serious weed. The native species are dense trees, often with aromatic bark and foliage. They are rarely encountered in cultivation but generally grow easily. The fruit of all species is avidly eaten by fruit pigeons. Propagation of all species is from seed which must be sown fresh for best results.

Cinnamomum laubatii F. Muell.
(after Marchionis de Chasseloup-Laubat)
Qld   Camphorwood
10-20 m x 5-10 m   Jan-March
Small to medium **tree**; **bark** light brown, wrinkled or scaly; young shoots silvery, clothed with silky hairs; **leaves** 10-15 cm x 2-6 cm, lanceolate, dark green and shiny above, pale beneath, the apex drawn out and blunt; **panicles** in the axils of the upper leaves; **flowers** about 1 cm diameter, greenish cream; **drupe** about 1.8 cm x 1.5 cm, oval, black, mature Nov.

Widespread in tropical rainforests between Eungella and Cape York Peninsula, often at high elevations. An attractive, dense tree with fragrant bark and timber. Best suited to a large garden or parkland situation. Easily grown in well drained soil rich in organic matter, but requires protection from direct sun and strong winds when young. Seedlings make decorative and adaptable indoor plants. Propagate from seed which must be sown fresh.

Cinnamomum oliveri Bailey
(after Prof. D. Oliver, Keeper of Kew Herbarium)
Qld, NSW   Oliver's Sassafras
15-30 m x 10-15 m   Oct-Nov
Medium to tall **tree**; **bark** brown, rough, fragrant; **leaves** 7-18 cm x 2-3 cm, narrow-lanceolate, glossy green above, pale or glaucous beneath, with minute oil dots; **panicles** borne at the ends of branchlets, the

stems bearing white, downy hairs; **flowers** 0.5 cm long, creamy white, tubular, with soft hairs; **drupe** 0.8 cm long, oval, mature March-April.

Found in dense rainforests between the Eungella Range and the Illawarra District. Cut as a timber tree, the pale fragrant wood being used for a variety of purposes. The tree is handsome with a wide, shady canopy but is rarely cultivated because of its size. Ideally suited as a park tree, for stock shelter or for adornment and shelter in large gardens. Requires well drained soils and responds to water during dry periods. The foliage has a pleasant sassafras odour when crushed, and young plants are ideal for indoor decoration. Hardy in southern areas. The leaves were used in treatment of diarrhoea up to the late 19th century. Propagate from seed which must be sown fresh.

Cinnamomum propinquum Bailey
(related to or resembling another)
Qld
10-15 m x 4-8 m   Sept-Oct
Small to medium **tree**; young shoots 4-angled; **leaves** 3-10 cm x 2-4 cm, ovate-lanceolate, thick and leathery, glossy green above, ashy or bluish grey beneath; **panicles** borne in pairs near the ends of the branches; **flowers** about 1 cm across, greenish white or cream; **drupe** about 1.5 cm x 1 cm, ovoid, purplish black, mature March.

Found in highland rainforests of northern Qld. A handsome, dense tree with glossy leaves, grey beneath, which produces a pleasant alternating contrast when blown by the wind. Best suited to parks or large gardens but should be sufficiently hardy to grow in southern Australia, where its size is reduced by the climate. Needs well drained soils and protection from sun and wind when young. Should make an excellent indoor plant. Propagate from seed which must be sown fresh.

Cinnamomum virens R. Baker
(green)
Qld, NSW   Red-barked Sassafras
15-25 m x 8-12 m   Feb-July
Medium **tree**; **bark** reddish brown, scaly; **leaves** 5-12 cm x 2-3 cm, elliptical, green on both surfaces, glossy green above; **panicles** borne at the ends of branchlets; **flowers** about 0.4 cm long, greenish; **drupe** 1-1.2 cm long, oval, smooth, black, succulent, mature Feb-July.

A little-known species found in dense rainforest of the Macpherson Ranges in south-eastern Qld to the Comboyne area of NSW. A handsome tree suited to parks or acreage planting. Requires well drained soil and may need some protection when young. Propagate from seed which must be sown fresh.

CINNAMON FUNGUS (*Phytophthora cinnamomi*)
see Volume 1, page 168

CISSUS L.
(from *cissos*, the Greek name for ivy)
Vitaceae
Small to large woody **climbers**; young shoots glab-

rous or covered with hairs; **leaves** alternate, entire or divided into leaflets, glabrous, shiny or dull; **inflorescences** of axillary cymes; **flowers** small, numerous, greenish or brown; **fruit** a leathery, fleshy berry.

Species about 350 with about 16 found in Australia, 6 of which are undescribed. They are interesting climbers which grow in a variety of situations. They are a rather difficult group to distinguish botanically and the identity of some species is subject to much confusion. A number of exotic species are cultivated for their decorative leaves, either as glasshouse or indoor plants. Propagation is from seed which must be sown while fresh or from stem cuttings which generally strike easily.

Cissus adnata (Wall.) Roxb.
(adnate)

Qld, WA, NT	Endeavour River Vine
5-15 m tall	Nov-Jan

Tall, woody, deciduous **climber**; young shoots covered with brown hairs; **leaves** 6-12 cm x 4-8 cm, simple, broad-orbicular or heart-shaped, dull green or brownish, with shallowly toothed margins; **corymbs** borne in the axils near the ends of shoots; **flowers** 0.1-0.2 cm across, brownish; **berry** 0.6-0.8 cm across, globular, black.

Found in scrubs along stream banks in open forest or monsoonal rainforest. A drab vine hardly worth cultivating, but included in some collections such as botanical gardens. Introduced into cultivation in the early 1900s from the Endeavour River, hence the common name. Very frost tender. Requires a sunny, well drained position. Retains its leaves longer if watered. Used for medicinal purposes in India. Propagate from stem cuttings or seed.

Cissus antarctica Vent.
(southern)

Qld, NSW	Kangaroo Vine
5-15 m tall	Dec-March

Tall woody **climber**; young shoots flattened, bearing rusty hairs; **leaves** 8-15 cm x 3-5 cm, simple, broadly ovate, dark green, shiny, entire or with dentate margins; **cymes** axillary, dense; **flowers** 0.4-0.6 cm across, greenish; **berry** about 1 cm across, globular, black.

Widespread from northern Qld to southern NSW, in or near rainforest. The fruits are edible but rather tasteless or acrid. A handsome climber well worth growing for the curtain of shiny green foliage that it forms. Prefers a protected position but will grow in full sun if given ample water. Not over-vigorous, and easily contained by pruning. Makes a very useful pot plant or basket plant for indoors as it tolerates neglect and dark situations. Also grows well outdoors as a groundcover. Is damaged by heavy frosts. Propagate from stem cuttings which strike easily or from seed which must be sown while fresh.

Cissus brachypoda (F. Muell.) Planchon
(a short petiole)

Qld	
2-4 m tall	Oct-Jan

Scrambling **climber**; **branches** terete; **leaves** compound, of 3 leaflets; **leaflets** 5-10 cm x 2-4 cm, ovate,

leathery, dark shiny green above, pale beneath, on thick, roughened petioles; **flowers** small, yellowish green, in complex panicles; **berry** globular, blackish.

Found in coastal scrubs of north-eastern Qld. Develops into an attractive climber when cultivated and regularly watered and fertilized. Useful for trailing over fences and covering unsightly buildings etc. Forms a curtain of lustrous foliage. Best suited to tropical and subtropical regions. Propagate from seed or cuttings.

Cissus hypoglauca A. Gray
(bluish green beneath)

Qld, NSW, Vic	Water Vine
10-25 m tall	Oct-Dec

Tall, vigorous woody **climber**; young shoots bearing rusty hairs; **leaves** palmately divided into 3-5 spreading leaflets; **leaflets** 5-8 cm long, dark bluish green above, glaucous beneath, ovate to oblong, entire or with toothed margins; **cymes** leaf-opposed; **flowers** about 0.4 cm across, yellowish green; **berry** 1-2 cm across, globular, black.

A common vine of rainforests and moist areas between eastern Vic and north-eastern Qld. Vigorous in cultivation and generally too strong-growing for home gardens, although it is an attractive climber. Occasionally grown in specialized collections such as botanical gardens. Prefers a sheltered position and requires water during dry periods. Will grow in full shade. Makes an adaptable indoor plant in a pot or a basket, and withstands neglect. Propagate from seed or cuttings.

Cissus antarctica, × ·45

Cissus oblonga (Benth.) Planchon
(oblong leaves)
Qld, NSW
5-10 m tall   Jan-March
  Medium sized, woody, deciduous **vine**; young shoots
covered with rusty hairs; **leaves** 4-7 cm x 2-4 cm, simple, ovate or broadly oblong, dark green, dull, blunt,
with entire or slightly toothed margins; **cymes** small,
borne near the ends of shoots; **berry** 0.6-0.8 cm across,
obovoid, black.
  Found in open forest, usually in moist areas such as
along stream banks. Sheds its leaves during long dry
periods. A drab vine hardly worth cultivating, but occasionally grown in botanical collections. Very frost
tender. Requires a sunny, well drained position.
Retains its leaves longer if watered. New growth can be
attractive. Has proved to be an adaptable indoor plant.
  Propagate from seed or stem cuttings.

Cissus adnata   D.L. Jones

Cissus opaca F. Muell.
(with thick growth)
Qld, NSW   Pepper Vine
5-10 m tall   Dec-March
  Small or weak **climber**; young shoots glabrous; **roots**
bearing numerous fleshy tubers up to 25 cm long;
leaves palmately divided into 3-5 leaflets; **leaflets** 2-
5 cm long, elliptical, with entire or slightly toothed
margins, bright green, often purplish beneath; **cymes**
leaf-opposed, sparse; **flowers** 0.2 cm across, yellowish;
berry 0.6-0.8 cm across, globular, bluish.
  A common species found in open country in both
inland and coastal areas, frequently in dry rainforest.
The tubers on the roots are edible and were collected
by the Aborigines. The berries are also edible but are
rather tasteless. A useful small vine occasionally grown
for its interest value and attractive foliage. Withstands
salt-laden winds and is very good for coastal conditions. Can be established by transplanting the tubers.
Requires a sunny position in well drained soil. Propagate from stem cuttings or seed.

Cissus penninervis (F. Muell.) Planchon
(feather-veined)
Qld
3-5 m tall   Jan-April
  Scrambling **climber**; **stems** glabrous, terete; **leaves**
compound, of 3-5 leaflets; **leaflets** 4-8 cm x 2-3 cm,
obovate, papery, blunt, dull green; **flowers** small,
brownish green, in cymes or umbels; **berry** 1-1.5 cm
long, obovate, dull black.
  Found in rainforests of coastal and adjacent inland
ranges of north-eastern Qld, and extending to New
Guinea. Not known to be in cultivation. A fairly drab
climber, probably best suited to tropical and subtropical regions. Propagate from seed or cuttings.

Cissus reniformis Domin
(kidney-shaped)
NT
2-5 m tall   Nov-Jan
  Tall, woody, deciduous **climber**; young shoots glabrous; **tendrils** up to 15 cm long, robust; **leaves** 6-10 cm
x 4-8 cm, broad-orbicular or reniform, glabrous, bright

green on both sides, with crenate margins; **cymes**
dense, leaf-opposed; **flowers** about 0.2 cm across,
brownish, on thick peduncles; **berry** about 1 cm
across, globose, black.
  A little-known species found along stream banks of
near-coastal districts of the Gulf country. Not known to
be in cultivation but could be a useful climber for
tropical regions. It is similar in many respects to *C. adnata* but is readily distinguished by the glabrous leaves.
Propagate from seed or cuttings.

Cissus repens Lam.
(creeping)
Qld, NT
5-15 m tall   Dec-March
  Medium sized **climber**; young shoots glabrous, dark
purple; **leaves** 4-15 cm x 3-8 cm, heart-shaped, dull
green when mature, but handsome when juvenile,
being bluish green and white above, bright purple
beneath; **cymes** leaf-opposed; **flowers** 0.2-0.4 cm
across, brownish; **berry** about 1 cm across, globular,
flattened, black.
  Found in highland and cooler rainforests, often
spreading over the ground as a cover. A very useful
light climber. Prefers a cool, protected position with
abundant water during dry periods. Frost tender and
needs some protection such as planting under trees.
The juvenile leaves are very attractive, and the species
makes a very decorative pot plant or basket plant for
indoor or glasshouse use.
  Propagate from stem cuttings which strike readily or
seed which germinates easily while fresh.

Cissus sterculiifolia (F. Muell. ex Benth.) Planchon
(leaves like those of the genus *Sterculia*)
Qld, NSW   Yaroong
10-20 m tall   Oct-Dec
  Medium to tall **climber**; **leaves** palmately divided
into 3 spreading leaflets; **leaflets** 7-16 cm long, narrowed at the base, oblong, entire, green on both surfaces;
cymes leaf-opposed; **flowers** about 0.4 cm across,
greenish; **berry** 1.5-2 cm across, ovoid, black.
  Found in moist areas from south-eastern Qld to near
Sydney. Similar to *C. hypoglauca* but with leaflets

40

Cissus reniformis   D.L. Jones

narrowed at the base. Vigorous, but occasionally grown in botanical collections. Requires a sunny, well drained position. Could be useful as an indoor plant as the leaves are attractive.

Propagate from seed or stem cuttings.

CITRIOBATUS Cunn. ex Loud.

(from the Greek *citria*, citron; *batos*, a prickly bush; from the common name of orange thorn)
Pittosporaceae

Small **shrubs** to small **trees**; **stems** bearing spines or the branches ending in stiff points; **leaves** entire, sessile, bright green; **flowers** axillary, small; **fruit** a berry.

Species about 5 with 4 endemic in Australia. They bear 2 distinct types of flowers which are probably male and female. Not commonly cultivated. The fruits are reported edible and were collected and eaten by the Aborigines. The leaves have a characteristic taste of carrots, which is a useful diagnostic feature. Propagation is from seed or cuttings. Seed should be sown while fresh and may take several months to germinate.

Citriobatus lancifolius Bailey
(lance-leaved)
Qld, NSW
3-6 m x 2-4 m   Sept-Nov

Medium to tall **shrub** or small bushy **tree**; **bark** grey or white; small **branchlets** terminating in spines; **leaves** 2-6 cm x 1-2 cm, lanceolate, shiny, dark green, dotted with orange oil-glands, entire; **flowers** about 0.8 cm across, greenish, solitary or in pairs in the leaf axils, sessile; **berry** about 1 cm across, rounded, orange, thin-shelled.

Confined to rainforests and rainforest margins in the Qld-NSW border region. A handsome, dense shrub with attractive shiny foliage. Bushy to ground level and an excellent plant for screening or as a hedge. Hardy in southern Australia. Requires a semi-protected position in a well drained, acid soil. Responds to regular watering during dry periods and the use of slow-release fertilizers. Propagate from seed or cuttings.

Citriobatus linearis (Bailey) C. T. White
(linear)
Qld, NSW
1-3 m x 1.5 m   Sept-Nov

Small to medium **shrub**; **branches** stiff or spiny; **leaves** 1-3 cm x 0.4-0.8 cm, linear or linear-lanceolate, dark green, entire; **flowers** about 0.5 cm across, white, on pedicels 0.2-0.5 cm long, solitary in the leaf axils; **berry** about 1 cm long, orange, thin-shelled.

Common in moist forests, on stream banks and along drier rainforest margins. A dense bush with shiny dark green leaves. It grows well in a shady, protected position in well drained loamy soils. Will grow successfully in southern Australia. The flowers are very sweetly scented. Propagate from seed or cuttings.

Citriobatus multiflorus Cunn. ex Loud. =
C. spinescens (F. Muell.) Druce

Citriobatus pauciflorus Cunn. ex Loud. =
C. spinescens (F. Muell.) Druce

Citriobatus spinescens (F. Muell.) Druce
(spiny)
Qld, NSW, WA, NT   Native Orange
2-5 m x 1-1.5 m   Sept-Nov

Small to erect, tall straggly **shrub**; **branches** hairy, bearing numerous thorns; **leaves** 1-3 cm long, obovate, sessile, green and shiny, entire; **flowers** about 0.6 cm across, white, solitary in the leaf axils; **berry** 1.5-2.5 cm across, orange, thick-shelled.

Found in moist scrubs of north-eastern NSW and

Citriobatus lancifolius, × ·45

south-eastern Qld. Previously known as *C. pauciflorus* Cunn. ex Loud.; also as *C. multiflorus* Cunn. ex Loud. A useful groundcover or shrub for protected or shady areas. Needs regular watering during dry periods. Propagate from seed.

CITRONELLA D. Don
(containing the oil of *citronella*)
Icacinaceae

Shrubs or **trees**; **leaves** alternate, entire, simple, glabrous or hairy; **inflorescence** a narrow panicle; **flowers** small, with 5 overlapping petals; **fruit** a drupe.

A small genus of 7 species with 2 species endemic in Australia. Propagation is from seed which must be sown fresh.

Citronella moorei (F. Muell. ex Benth.) Howard
(after C. Moore)

Qld, NSW   Churnwood
8-15 m x 5-8 m   Oct-Dec

Small to medium **tree**; **bark** grey, corky; young shoots reddish, glabrous; **leaves** 7-10 cm x 3-5 cm, ovate-lanceolate, leathery, shiny; **panicles** narrow, about 10 cm long, arising opposite the leaves; **flowers** about 0.5 cm across, sessile, white; **drupe** about 2.5 cm long, globular, roughened, blackened.

Widespread in rainforest but usually of scattered occurrence. A handsome tree suited to parks or large gardens in tropical, subtropical or mild areas of southern Australia. Needs well drained soils and protection from frost while young. Not drought resistant and needs regular watering during dry periods. Propagate from seed.

Citronella smythii (F. Muell.) Howard
(after R. B. Smyth)

Qld   White Oak; Soap Box
3-12 m x 2-8 m   Nov-Dec

Tall **shrub** or small **tree**; **bark** grey, corky; young shoots densely clothed with hairs; **leaves** 8-17 cm x 3-7 cm, ovate, hairy beneath; **panicles** narrow, 5-10 cm long, arising opposite the leaves; **flowers** about 0.8 cm across, white; **fruit** about 2 cm long, oval, black, succulent.

Found in rainforests of north-eastern Qld. An attractive plant with decorative foliage and flowers. Best suited to tropical or subtropical regions, where it requires well drained soil in a semi-shady position. Very frost tender. Propagate from seed.

CLADIUM R. Br.
(from the Greek *cladion*, branchlet; referring to the branched inflorescence)
Cyperaceae

Perennial **herbs**; **rhizome** subterranean, creeping; **leaves** spirally arranged, with a distinct upper and lower surface, revolute when dry; **culms** erect, 3-angled; **inflorescence** a panicle, loose or dense; **flowers** in spikelets; **fruit** a nut.

A small genus of 3 species, 1 of which is found in Australia. Sections of the family Cyperaceae have recently been revised and most species of *Cladium* trans-

ferred to other genera, particularly *Baumea*. Propagation is by division of the rhizome or from seed which should be sown on moist peat and covered with a sheet of glass.

Cladium procerum S. T. Blake
(tall and slender)

Qld, NSW, Vic, SA, WA, NT   Leafy Twig Rush
1-2 m tall   Nov-Jan

Tall perennial **herb**; **leaves** flat, grass-like, roughened, keeled, with sharp minutely toothed margins; **culms** erect, with several nodes; **panicle** 20-30 cm long, dense, leafy; **nut** pale brown, shiny.

Widespread in moist areas along stream banks, particularly in coastal areas, sometimes in saline conditions. A useful tussock species for softening lines of shrubbery. Suited to moist or wet soils, either sandy or clays. Frost hardy. Needs exposure to sun. Propagate from seed or by division.

CLAOXYLON Juss.f.
(with brittle wood)
Euphorbiaceae

Shrubs or small **trees**; **leaves** entire or toothed, alternate, often large, on petioles; male and female **flowers** borne on separate racemes on the same plant or on separate plants; **fruit** a 2-celled capsule.

Species about 80 with 3 endemic in Australia. Propagation is from seed which is best sown while fresh.

Claoxylon australe Baillon
(southern)

Qld, NSW
3-8 m x 1-4 m   Nov-Jan; also sporadic

Tall **shrub** or small straggly **tree**; **leaves** 4-12 cm x 2-5 cm, ovate or oblong, blunt or pointed, with toothed margins, pale green or translucent, often reddish beneath; **male flowers** in racemes 5-8 cm long, the flowers about 2.5 cm long, white, fluffy; **female flowers** reddish, on racemes 3-4 cm long; **capsule** 1-2 cm across, red, 3-lobed.

Widespread in cool, sheltered places from south-eastern Qld to south-eastern NSW. Worth growing for the decorative fluffy catkins of male flowers, although not common in cultivation. Inclined to become straggly unless pruned. Hardy in southern Australia. Needs well drained soil, and prefers a cool, sheltered position with abundant water during dry periods. Propagate from seed or cuttings.

Claoxylon tenerifolium (Baillon) F. Muell.
(soft, tender leaves)

Qld, NT
4-9 m x 4-8 m   Feb-April

Tall **shrub** or small **tree**; young shoots purplish, hairy; **leaves** 15-20 cm x 7-10 cm, ovate, dark shiny green, the margins toothed, on petioles 2.5-5 cm long; **racemes** axillary; **flowers** greenish yellow, in clusters; **male racemes** 5-10 cm long, the **female racemes** less than 5 cm long, the flowers sparse; **capsule** about 1 cm across, 3-lobed.

Claytonia australasica   T L Blake

Widespread in rainforests of north-eastern Qld from the coast to the tablelands. A little-known species with interesting shiny leaves and colourful new growth. It is rarely encountered in cultivation. Requires well drained soil and some protection when young. Would be worth trying as an indoor plant. Propagate from seed and possibly cuttings.

CLAUSENA Burm.f.
(after P. Clauson, a Danish botanist)
Rutaceae
Shrubs or small **trees**; **leaves** alternate, pinnate; **leaflets** alternate, bearing oil dots; **flowers** bearing 4-5 petals, white, in large, loose panicles; **fruit** a small berry.

Species about 30 with one endemic species found in Australia. Propagate from seed which is best sown while fresh.

Clausena brevistyla Oliver
(short style)
Qld
2-3 m x 1-1.5 m
Medium sized **shrub**; young shoots bearing hairs; **leaflets** 10-15, 5-10 cm long, ovate, bright green, glabrous, bearing minute oil dots; **panicles** terminal, loose, large; **flowers** about 0.8 cm across, white; **berry** about 1 cm across, dry.

Restricted to Hope Island off the coast of northern Qld. Not known to be in cultivation. An attractive shrub with aromatic leaves and fragrant flowers, suited to cultivation in tropical areas. Propagate from seed.

CLAY SOILS — see Volume 1, page 53

CLAYTONIA L.
(after John Clayton, an early American botanist)
Portulacaceae
Annual or perennial **herbs**; **branches** usually succulent and glabrous; **leaves** radical, petiolate; **stem-leaves** alternate or opposite; **stipules** none; **flowers** in terminal racemes or cymes, rarely solitary; **petals** 5; **stamens** 5, opposite petals and attached to their base;

capsule globular or ovoid, opening in 3 valves.

This genus is mainly represented in the temperate zones of the northern hemisphere and the Arctic regions. Only one species is endemic to the southern hemisphere, occurring in New Zealand and Australia. *Claytonia* is closely allied to *Calandrinia*, which has a variable number of stamens.

Claytonia australasica Hook.f.
(Australian)
NSW, Vic, Tas, SA, WA   White Purslane
prostrate x 1-2 m   Aug-April
Perennial; **stems** creeping, rooting at nodes, glabrous; **leaves** 4-10 cm x 0.2-1 cm, linear-lanceolate, blunt apex, winged at base, surrounding stem, glabrous; **inflorescence** a loose terminal cyme of 1-6 flowers, borne on stalks to 3.5 cm long; **flowers** about 1.5-2 cm wide, spreading petals, white, fragrant; **capsule** about 0.3 cm long.

A widely distributed species that occurs at all altitudes, usually found on moist soils, and on occasions as an aquatic. It has adapted very well to cultivation, tolerating most soils, but with a preference for moisture throughout the year. It has the capacity to be dormant during dry periods and regrow when conditions are suitable. Grows best in partial or full sun. Frost tolerant. Propagate from seed, cuttings or by division.

There is some disagreement amongst botanists regarding the genus to which this species belongs, some including it in *Montia* and others in *Neopaxia*.

CLEISTANTHUS Hook.f. ex Planch.
(from the Greek *cleistos*, closed; shut; *anthos*, a flower; in reference to the inconspicuous flowers)
Euphorbiaceae
Shrubs or **trees**; **leaves** alternate, entire, often leathery; **flowers** small, greenish, in clusters or spikes, male and female flowers separate on the same plant; **fruit** a capsule.

A large genus of about 140 species with 9 species endemic in Australia. They have little horticultural potential and are rarely encountered in cultivation. Propagation is from seed which is best sown while fresh or from cuttings which may be slow to strike.

Cleistanthus cunninghamii Muell. Arg.
(after A. Cunningham, 19th-century botanist)
Qld, NSW
3-7 m x 1-3 m   all year
Tall **shrub** or small straggly **tree**; **leaves** 5-8 cm long, ovate, lanceolate or elliptical, dull green above, glaucous beneath; **flowers** small, greenish, in dense axillary clusters; **capsule** about 0.6 cm across, globular, 3-furrowed.

Widespread in dry rainforests and along rainforest margins and stream banks. Occasionally found in botanical collections, but the species has few particular merits and is hardly worthy of cultivation. New leaves are an attractive pinkish red. Needs well drained soil in a protected situation. Propagate from seed. Cuttings would probably strike easily.

Cleistanthus dallachyanus

Cleistanthus dallachyanus (Baillon) Baillon ex Benth.
(after J. Dallachy, past Superintendent of Royal
Botanic Gardens, Melbourne)
Qld
4-8 m x 3-5 m   July-Sept

Tall **shrub** or small **tree**; young shoots and inflorescence covered with rusty hairs; **leaves** 5-10 cm x 2-4 cm, ovate, stiff, blunt, leathery, dark glossy green above, green or glaucous beneath; **spikes** 3-8 cm long, leafy; **flowers** small, greenish yellow, starry, in sessile clusters; **capsule** about 0.8 cm across, sessile.

Widespread in coastal districts of north-eastern Qld, often along the shores of large rivers and estuaries. Makes a handsome plant, with its lustrous foliage and rusty brown new growth but is rarely encountered in cultivation. Best suited to tropical and subtropical areas. Needs well drained soil in a sunny situation, and should be tested in coastal areas. Propagate from seed or cuttings.

Cleistanthus hylandii Airy-Shaw
(after B. P. M. Hyland, contemporary botanist, CSIRO
Division of Forest Research, Atherton, Qld)
Qld
6-8 m x 1-3 m   Sept-Nov

Tall **shrub** or small **tree**; **branchlets** glabrous, terete; **leaves** 3-16 cm x 2-8 cm, ovate to obovate, dull green, the margins entire or shortly reflexed, the apex with a slender point 1-2 cm long; **flowers** about 0.4 cm across, greenish cream, in axillary clusters of about 8 unisexual flowers; **capsule** about 0.8 cm across, glabrous, red.

A little-known species of dry rainforests and stream banks of the McIlwraith and Iron Range areas of Cape York Peninsula. It has some horticultural appeal, being quite decorative in fruit, and may be a useful screening plant for tropical and subtropical regions. Propagate from seed or cuttings.

Cleistanthus xerophilus Domin
(growing in dry places)
Qld
5-8 m x 1-2 m   May-July

Medium to tall **shrub** or small shrubby **tree**; **branchlets** bearing rusty hairs; **leaves** 4-11 cm x 1-4 cm, elliptical to narrowly elliptical, dull green, glabrous; **flowers** about 0.4 cm across, pink outside, white within, in dense, sessile axillary bundles; **capsule** about 0.5 cm across, trilocular, reddish.

A rather drab species usually found in rocky areas along stream banks and extending from Cape York Peninsula to near Townsville. It could perhaps be a hardy screening shrub for tropical and subtropical gardens. Requires an organically rich soil and free drainage, and a sunny to partially protected position. Propagate from seed or cuttings.

CLEISTOCALYX Blume
(from the Greek *cleistos*, closed; shut; *calyx*, the sepals; an apparent reference to the fused calyx)
Myrtaceae

Trees; **leaves** entire, simple, opposite, bearing numerous oil-glands; **inflorescence** a cyme or panicle; **flowers** numerous, when in bud the sepals shaped like

44

Clematicissus angustissima   W.R. Elliot

a beak; **stamens** numerous, conspicuous; **fruit** a berry; **seed** solitary.

A small genus of about 20 species, although the exact number is uncertain due to confusion with the closely related genus *Syzygium*. *Cleistocalyx* is distinguished by the beak-like point on the buds, formed by the fused calyx, although some botanists regard the distinguishing features as insignificant and prefer to include them with *Syzygium*. The mainland Australian species have been included with *Syzygium*; however, the status of the species from Lord Howe Island is uncertain and awaits further study. Propagate from seed which must be sown fresh.

Cleistocalyx fullageri (F. Muell.) Merr. & Perry
(after M. Fullager, 1st collector of the species)
NSW (Lord Howe Island)   Scalybark
15-30 m x 8-15 m   Jan-March

Medium to tall **tree**; **bark** brown, scaly; **leaves** 5-7.5 cm x 2-3 cm, obovate, leathery, dark green and shiny above, dull beneath, blunt; **cymes** simple, 5-20 cm long, in the upper axils; **buds** with a large, broad, leathery operculum; **flowers** large, yellowish brown; **berry** about 3 cm long, pear-shaped, glabrous, shiny, yellowish.

A little-known species endemic to the rainforests of Lord Howe Island where it is common and was widely cut as a timber tree. Although scarcely known in cultivation, the species is a handsome tree well suited to parks or large gardens in temperate or subtropical regions. Needs deep, well drained, acid soil. Propagate from seed which must be sown when fresh. Cuttings would probably strike readily.

CLELANDIA J. Black
(after Prof. J. B. Cleland, its discoverer)
Violaceae
    A monotypic genus restricted to SA.

Clelandia convallis J. Black
(with grooves)
SA
0.5-1 m x 0.5-1 m   May-Oct
    A dwarf **shrub**; **branches** glabrous; **leaves** to about
1.5 cm x 0.2 cm, stiff, prominent midrib below, margins
recurved, recurved point; **flowers** open-petalled, about
0.7 cm diameter, borne on solitary axillary stalks to
about 1.2 cm long; **sepals** prominent; **petals** 5, with one
minute, white; **fruit** globular, about 0.5 cm diameter,
leathery.
    This species occurs in the Wilpena Pound of the
Flinders Ranges. It is not well known in cultivation. It
requires fairly well drained soils, with partial or full
sun. Tolerates light to moderate frosts and extended
dry periods. Suitable for gardens or containers. Propagate from seed or cuttings.
    Similar to *Hybanthus floribundus*, but differs in the
petal arrangement.

CLEMATICISSUS Planchon
(from the Greek *clema*, a vine shoot; *cissos*, ivy; in
reference to its appearance)
Vitaceae
    Monotypic, endemic genus confined to WA.

Clematicissus angustissima (F. Muell.) Planchon
(very narrow)
WA
prostrate-3 m x 2-4 m   Aug-April
    Light **climber** or scrambling prostrate **shrub**; **bran-
chlets** slender, glabrous; **leaves** can be deeply lobed,
but usually with 5 spreading leaflets to 10 cm long,
narrowed at base, entire to coarsely toothed or lobed,
or can have only 3 coarsely toothed leaflets; **flowers**
small, cream to yellow, in compact cymes on long
peduncles; **petals** 5; **berries** small, globular.
    This species has ornamental foliage. It occurs in well
drained soils that can be alkaline. Often observed in
exposed coastal situations, and is common in the
Kalbarri-Murchison River area. Evidently not well
known in cultivation, but certainly has potential. It
should adapt to a wide range of well drained soils, but
will need a warm, sunny location to grow well. Propagate from seed or cuttings.

CLEMATIS L.
(from the Greek *clema*, a vine shoot; *clematis* being a
general name for a climber)
Ranunculaceae
    Climbers; young stems glabrous or hairy, slender;
leaves simple or 1-3 times divided; **leaflets** ovate or lanceolate, glabrous or hairy, shiny or dull, with entire or
toothed margins; **inflorescences** of axillary panicles;
male and **female flowers** borne on separate plants;
flowers starry, of 4-8 segments; **fruit** an achene, with a
long tail bearing silky plumes.

Clematis aristata, × ·5

    Species about 250 with 6 or 7 endemic in Australia.
A number of exotic species and hybrids are commonly
cultivated in southern Australia. Propagation is from
seed or stem cuttings. Seed is often difficult to germinate, and the results are improved if it is sown fresh.
Stem cuttings are cumbersome because of the long in-
ternodes, and may consist of 2 nodes only, one at the
base for rooting and one near the top for growth. They
may be very slow to strike.

Clematis aristata R. Br. ex DC.
(awned)
Qld, NSW, Vic, Tas   Austral Clematis; Goat's Beard
5-15 m tall   Aug-March
    Small to medium sized **climber**; **stems** glabrous or
hairy; **leaves** simple or more frequently trifoliolate;
leaflets 3-8 cm x 1.5-2.5 cm, ovate, shiny green, entire
or with toothed margins; **panicles** short, in the upper
leaf axils; **flowers** 3-6 cm across, white, starry, anther
appendages long, pointed; **achene** ovate, with a tail
2.5-4 cm long.
    A widespread and variable climber found in a
variety of situations from dry scrubs to wet sclerophyll forests. It extends as far west as the Grampians in Vic.
Flowering is usually confined to the spring and summer
months but odd flowers can be produced at intervals
throughout the year. The species is very popular and
widely cultivated in eastern Australia. It is a versatile
climber, useful for mingling amongst shrubbery or
growing over trellises, walls etc. Plants are very showy
when in flower, sometimes becoming smothered
beneath masses of starry blossoms. The fruit of the
female plants persists on the vines for a long period
and is an additional decorative feature. Foliage of
suckers and seedlings is colourful, being purplish
beneath, with silver markings on the upper surface.
    The species is variable and a number of forms are
available for selection. Flower size is variable and some
forms from moist gullies have large flowers up to 8 cm
across. A form with broad, very shiny dark green

45

leaflets (about 8 cm long x 4 cm across) and long narrow sepals occurs near Lorne, Vic, while plants from Mount Elizabeth in eastern Vic are very hairy. The following varieties have been recognized:

C. aristata var. *blanda* (Hook.) Benth. from Tas and Vic is a coastal form with small, often twice-divided leaflets and small flowers.

C. aristata var. *dennisae* W. R. Guilfoyle from highland forests near Healesville, Vic, has long, very coarsely toothed leaves and striking red staminal filaments of the flowers.

C. aristata var. *longiseta* Bailey from Nerang in southeastern Qld has ovate-lanceolate leaflets with margins bordered by bristle-like teeth and yellowish, hairy flowers in short racemes during Nov.

An easy species to grow, tolerating a variety of soil types (but not bad drainage) and aspects varying from shade to full sun. It is hardy enough to survive and grow in dry spots under established eucalypts. Responds to extra water during dry summer months. Propagate from seed or stem cuttings.

Clematis fawcettii F. Muell.
(after C. Fawcett)
Qld, NSW
5-15 m tall   April; Sept-Oct

Small to medium sized **climber**; **leaves** up to 20 cm long, 2-3 times divided; **leaflets** 2-4 cm long, ovate to lanceolate, deeply and irregularly toothed; **panicles** short, in the upper leaf axils; **flowers** about 2 cm across, white, the anthers without any appendage.

Confined to a few highland areas of central northern NSW where it is endemic. Not known to be in cultivation; however, it should be an attractive climber for subtropical or temperate areas. Propagate from seed or stem cuttings.

Clematis gentianoides DC.
(like the genus *Gentiana*)
Tas   Bushy Clematis
0.1-0.5 m x 0.5-1.5 m   Sept-Nov

Small, spreading **shrub**; **stems** erect, in tufts, from underground rhizomes; **leaves** 2-8 cm long, simple, ovate-lanceolate, thick-textured, with entire or toothed margins; **flowers** 3-4 cm across, white, solitary on the ends of branchlets; **achene** linear-ovate, with a tail 2-3 cm long.

Endemic in Tas, where it is widespread in open forest, often on rocky slopes. A useful small shrub for a sunny aspect in well drained soils. Hardy but responds to extra water during dry periods. Not widely cultivated but grown mainly by enthusiasts. Propagate from seed or stem cuttings.

Clematis glycinoides DC.
(like the genus *Glycine*)
Qld, NSW, Vic   Forest Clematis
5-20 m tall   Sept-Nov

Medium to large **climber**; **stems** glabrous; **leaves** simple or trifoliolate; **leaflets** 3-6 cm x 1-2 cm, ovate or lanceolate, shiny green, glabrous, with entire margins; **panicles** short, in the upper leaf axils; **flowers** 3-4 cm across, white or greenish, starry; **anther** appendages

short, blunt; **achene** narrow-lanceolate, with a tail 5-6 cm long.

Widespread in moist forests and rainforests from Cape York Peninsula to eastern Vic. In Qld it is called headache vine as the crushed leaves are supposed to alleviate headaches.

Very attractive when in flower, but not as widely grown as *C. aristata*. It seems less adaptable than that species and requires cool, moist conditions with an abundance of water during dry periods. It is a useful climber for trailing over the framework of a fernery. Plants are very showy when in flower and the fruit of female plants is also decorative.

The species is variable in leaf size, shape and colour, and flower size and colour, but little selection has been carried out. Some forms may be frost tender.

Propagate from seed or cuttings.

The var. *submutica* Benth. has leaves loosely hairy beneath, shorter, broader, very hairy sepals and the anthers may be entire or tipped with a small gland. It is found in rainforests of south-eastern Qld.

Clematis microphylla DC.
(small leaves)
Qld, NSW, Vic, Tas, SA, WA   Small-leafed Clematis
5-10 m tall   Aug-Nov

Small to medium sized **climber**; **stems** glabrous, slender; **leaves** 2-3 times divided; **leaflets** 1-3 cm x 0.3-0.6 cm long, oblong, slender, blunt, dull green or greyish; **panicles** short, in the upper leaf axils; **flowers** 2-4 cm across, greenish cream; **achenes** wrinkled or warty, with a tail 4-6 cm long.

The most widespread clematis in Australia, usually found in dry scrubs or heathland, and extending into coastal dune vegetation. Despite its wide distribution it is fairly uniform throughout its range. The var. *leptophylla* F. Muell. ex Benth. has small, narrow leaflets (0.5-1 cm x 0.1-0.3 cm) but all variations exist and it is probably not a good variety. It is an attractive light climber, popular in cultivation. Grows equally well on

Clematis microphylla, × ·5

Clematis microphylla in fruit

D.L. Jones

light, sandy or heavy clay soils but is intolerant of bad drainage and prefers a sunny aspect. A very useful climber or groundcover for coastal districts. The leaves have been used to relieve skin irritations but excessive use can cause blistering. Propagate from seed or stem cuttings.

Clematis pickeringii A. Gray
(after a collector by the name of Pickering)
Qld, WA, NT
5-20 m tall   Dec-Jan
   Medium to tall **climber**; **stems** and petioles finely ribbed; **leaves** up to 20 cm long, ternate; **leaflets** 6-12 cm x 3-5 cm, ovate to lanceolate, entire, dark green; **panicles** to 30 cm long, terminal, much-branched and floriferous; **flowers** 1.5-3 cm across, white, the sepals with tomentose margins; **anthers** of 35-55 male flowers, with a small appendage; **fruit** bearing plumose styles 4-6 cm long.
   This species is widely distributed in islands to the north of Australia such as Java, Timor and New Guinea. In Australia it is confined to monsoonal rainforests of the north, where it grows as a vigorous vine flowering in the canopies of trees. Although flowering plants have not been observed in the wild in Australia, cultivated specimens have flowered. The species should prove to be a showy, if vigorous, climber for gardens in the tropics. Plants may become deciduous during the long, dry winter period. Well drained soil and a support for the plant to climb on are essential.

Propagate from seed or stem cuttings. Dormant plants transplant fairly easily.

Clematis pubescens Huegel
(shortly hairy)
WA   Western Clematis
5-10 m tall   Aug-Nov
   Medium sized **climber**; young stems very hairy; **leaves** trifoliolate; **leaflets** 3-6 cm x 1-1.5 cm, ovate, dull green, hairy, with toothed margins; **panicles** short, in the upper leaf axils; **flowers** 2-4 cm across, white; **achene** ovate, with a tail 3-4 cm long.
   Endemic in WA, where it is widespread in the jarrah and karri forests of the south-west. Distinguished by the hairy stems and young leaves; however, some forms of *C. aristata* are also hairy, and a study may show that the western clematis is but a variant of that widespread species. Versatile and hardy in cultivation, although the species is not widely grown. Seed is frequently available from merchants. Quite hardy in southern Australia. Propagate from seed or cuttings.

Cleomaceae (Pax) Airy-Shaw
   A family of dicotyledons consisting of 12 genera and 275 species widely distributed throughout tropical and subtropical regions of the world. It is poorly represented in Australia by 2 genera and about 7 species. The family is included in Capparaceae by some authors. Australian genera: *Cleome* and *Justago*.

47

Cleome

CLEOME L.
(from an old Latin name)
Cleomaceae

Glabrous or more usually hairy **herbs**; **leaves** entire or divided; **flowers** solitary or in terminal racemes, showy, with 4 spreading petals and a bunch of protruding stamens; **fruit** a capsule.

Species about 150 with 3 small annuals endemic in northern Australia. A couple of exotic species are commonly cultivated as annuals or bedding plants.

Cleome oxalidea F. Muell.
(like the genus *Oxalis*)
Qld, WA, NT
0.1-0.15 m x 0.05 m   April-Oct

Dwarf annual **herb**; **leaves** glabrous, in a basal rosette, each of 1-3 leaflets, on a fleshy petiole 2-10 cm long; **leaflets** 0.5-1 cm long, fleshy, orbicular, dark green; **flowers** on a slender stalk 3-8 cm long, about 1 cm across, each of four bright pink spreading petals and a bunch of yellow stamens.

A very hardy little plant of central and northern Australia, growing on bare sand ridges and ironstone outcrops. The plant is small, although spectacular when seen in colonies. Should be a useful annual for warm inland areas. Could possibly be grown as a summer annual in a sunny position in southern Australia. Propagate from seed.

Cleome tetrandra Banks
(4 stamens)
Qld, WA, NT
0.1-0.3 m x 0.1 m   Jan-Aug

Dwarf annual **herb**; **stems** erect, slender, with a few sticky hairs; **leaves** chiefly basal, divided; **leaflets** 3-5, about 2 cm long, linear-lanceolate; **flowers** 1-1.2 cm across, yellow with reddish markings, in long racemes; **capsule** 2-3 cm long, slender.

Found in inland and northern areas, usually along the banks and flood plains of inland rivers. Widespread and locally common in good seasons. Showy when seen in mass. Should be a useful annual for warm inland or northern areas. Could possibly be grown as a summer annual in a sunny position in southern Australia. Propagate from seed.

Cleome viscosa L.
(sticky)
Qld, NSW, SA, WA, NT   Sticky Cleome
0.3-0.6 m x 0.3 m   April-Oct

Erect **herb** or dwarf **shrub**; **stems** covered with sticky, glandular hairs; **leaves** divided; **leaflets** 3-7, 2-4 cm long, obovate or oblong, light green, hairy; **flowers** 2-2.5 cm across, yellow, each with a bunch of spreading stamens, borne in long racemes; **capsule** about 2.5 cm long.

Common in inland and northern areas, and abundant in years of good rains. Very showy when flowering in colonies. A weedy-type species but nevertheless a very useful annual for warm inland or tropical areas. Germinates readily from seed and may naturalize itself. Can be grown as a summer annual in a sunny position in southern Australia. Both Aborigines and Asian people have used this species for various

Clerodendrum cunninghamii   B. Gray

medicinal purposes, such as relieving headaches and dressing wounds.

CLERODENDRUM L.
(from the Greek *cleros*, a drawing of lots; *dendron*, a tree; in reference to uncertain medicinal properties)
Verbenaceae

Shrubs, trees or **climbers**; **leaves** simple, opposite; **flowers** in axillary cymes, terminal panicles or corymbs, numerous, white, tubular, with long spreading stamens; **fruit** a drupe, frequently surrounded by a fleshy, enlarged calyx.

Species about 400 with about 8 in Australia, 6 of which are endemic. They are mainly tropical and grow in open forests, often along stream banks and frequently in disturbed areas. They are very floriferous, and in good years become smothered with white, tubular flowers. These are relatively delicate and short-lived but are fragrant during the day and in early evening, and attract butterflies and moths. The fruit of some species is also very decorative. All species have excellent horticultural potential but only a couple of species are seen with any regularity. They are generally easy to grow in any sunny situation, but require well drained soil and resent cold. They have potential as glasshouse plants in southern Australia. Propagate from seed which must be sown fresh or from cuttings.

Clerodendrum cunninghamii Benth.
(after Allan Cunningham, 19th-century botanist)
Qld, NT
4-6 m x 1-2 m   Nov-Feb

Tall **shrub**; young shoots hairy; **leaves** 4-15 cm x 3-5 cm, ovate, narrowed at the base, membranous, bright

48

green, with conspicuous veins, shortly pointed; **flowers** white, with a long slender tube, up to 8 cm long, in dense terminal corymbs; **stamens** protruding about 3 cm; **drupe** about 1 cm across, black, surrounded by a purplish, funnel-shaped calyx.

Restricted to tropical areas, where it grows in rainforests and along stream banks. May become deciduous during dry periods. A very floriferous and showy shrub for tropical or subtropical areas. Requirements are for a well drained soil in a sunny situation. Grows well in coastal districts. Could be an attractive glasshouse plant. The flowers are highly fragrant and attract moths and butterflies. Propagate from seed and cuttings.

Clerodendrum floribundum R. Br.
(free-flowering)

Qld, NSW, NT	Thurkoo; Lolly Bush
3-5 m x 1-3 m	Aug-Oct

Tall **shrub** or small slender **tree**; young shoots hairy, rapidly becoming glabrous; **leaves** 4-7 cm x 2-3 cm, broadly ovate, blunt or pointed, rounded at the base, glabrous; **cymes** loose, many-flowered, in the upper leaf axils; **flowers** white, about 0.8 cm across, the tube about 3 cm long; **stamens** protruding about 2.5 cm; **drupe** about 1 cm across, narrowed at the base, black, shiny, the surrounding calyx scarlet.

Widespread along stream banks and in moist areas of open forest, extending south as far as Taree in NSW. A variable species usually with a fairly open growth habit. The form from NT appears quite distinct. An outstanding horticultural subject, showy when in flower and with spectacular clusters of jet black fruit surrounded by a fiery red calyx. Young plants often have attractive purple leaves. Very easily grown in tropical, subtropical and even temperate regions although somewhat cold sensitive. Inclined to become leggy, and responds to hard pruning after fruiting. Also responds to the application of complete fertilizers and manures. Requirements are for a well drained soil in a sunny or semi-shady situation. Resents summer dryness and may need regular watering. Could have potential in southern Australia as a glasshouse plant. Aborigines used the extract obtained after boiling wood, for relief of aches and pains. Propagate from seed, cuttings or root suckers.

Clerodendrum hemiderma F. Muell.
(half-skinned)

Qld	
3-10 m tall	Aug-Nov

Tall bushy **climber**; young shoots hairy; **leaves** 5-8 cm x 2-3 cm, broadly ovate, blunt, bright green; **flowers** white, numerous, in dense terminal corymbs, about 0.5 cm across, the tube about 0.6 cm long; **stamens** protruding about 1 cm; **drupe** about 0.8 cm long, oblong, hairy, winged at the base.

Found in shrubs and open forest along streams, sometimes climbing very high into trees. A useful climber for tropical areas. Can be kept in check by regular pruning and in a sunny, open position may become bushy. The flowers are fragrant and attract butterflies. Very frost tender. Propagate from seed or cuttings.

Clerodendrum floribundum, (NT form)   D.L. Jones

Clerodendrum inerme (L.) Gaertner
(unarmed)

Qld, NSW, WA, NT	
3-5 m tall	May-Aug

Straggly **shrub** or bushy **climber**; young shoots slightly hairy; **leaves** 5-12 cm x 1-5 cm, oblong or ovate, light green, on long petioles; **cymes** axillary, usually clustered at the ends of the new shoots; **flowers** about 2 cm across, the tube 4-5 cm long, the petals white; **stamens** prominent, purplish, about 3 cm long; **drupe** 1-2 cm across, rounded.

Found in coastal scrubs and along rainforest margins, occasionally growing with mangroves, and tolerating saline inundation. The flowers give off a sweet fragrance on warm days and attract butterflies. An attractive climber for tropical regions. Can be kept bushy by regular pruning. Very useful in coastal areas. Prefers a sunny position. May have potential as a glasshouse plant in southern Australia. Propagate from seed or cuttings.

Clerodendrum lanceolatum F. Muell.
(lanceolate leaves)

Qld, WA, NT	
1-4 m x 1-2.5 m	Jan-April

Tall **shrub**; young shoots bearing soft hairs; **leaves**

Clerodendrum inerme   D.L. Jones

Clianthus formosus   T.L. Blake

4-7.5 cm x 1-2 cm, ovate-lanceolate, bright green, blunt or pointed; **cymes** axillary, short, few-flowered; **flowers** white, about 0.6 cm across, the tube 1.5-2 cm long; **stamens** about 2.5 cm long; **drupe** about 1 cm across, black.

Confined to northern localities, usually in scrubs along stream banks. A handsome shrub suitable for cultivation in tropical areas such as Darwin. The flowers are fragrant and attract butterflies. Prefers a sunny position in well drained soil. Propagate from seed or cuttings.

Clerodendrum tomentosum R. Br.
(bearing woolly hairs)
Qld, NSW, WA, NT
3-6 m x 1-3 m   Sept-Jan

Medium **shrub** or small slender **tree**; young shoots covered with velvety hairs; **leaves** 5-10 cm x 2-5 cm, elliptical or lanceolate, hairy, on petioles about 2 cm long; **flowers** white, in dense terminal corymbs, about 1 cm across, the tube about 2 cm long; **stamens** protruding 2.5 cm; **drupe** about 2 cm across, black, shiny.

Widespread and common in sheltered gullies and rainforests. Often becomes dominant after clearing and then puts on a dazzling display of flowers. The flowers are fragrant in warm weather. After flowering, the calyx enlarges and becomes bright red. This contrasts with the shiny black fruits and is an additional decorative feature of the species. Prefers a semi-shaded position in an acid, well drained soil. Slightly frost tender, especially while young. Requires regular water during dry periods and responds to side dressings of slow-release fertilizer. Makes a very attractive pot plant. Propagate from seed or cuttings.

Clerodendrum tracyanum F. Muell.
(after Dr. R. T. Tracy)
Qld
5-10 m x 3-5 m   Oct-March

Tall **shrub** or small, sparse **tree**; young shoots covered with velvety hairs; **leaves** 10-20 cm x 8-10 cm, broadly ovate, rounded at the base, bright green, shortly pointed, on petioles up to 10 cm long; **flowers** white, about 1 cm long, in dense terminal cymes about 10 cm across; **stamens** just protruding; **drupe** about 1 cm across, black, the calyx enlarged and purplish.

An attractive species restricted to tropical areas, where it grows along stream banks and in moist scrubs. The plants tend to be sparse, but in good years flower profusely. The species is best suited to tropical or subtropical areas, in a warm, sunny situation. It is drought resistant; however, the appearance of the plants is improved by regular watering. Needs a soil of unimpeded drainage. The flowers are fragrant and freely attract butterflies. Propagate from seed.

CLIANTHUS Solander ex Lindl.
(from the Greek *cleos*, glory; *anthos*, a flower; an obvious reference to the large showy flowers)
Fabaceae

Herbaceous **perennials** or small **shrubs**; **leaves** pinnate, hairy; **flowers** pea-shaped, large, showy, in umbel-like racemes, on erect axillary stalks; **fruit** a legume.

A genus of 2 species, 1 endemic to Australia, the other to New Zealand.

Clianthus formosus (G. Don) Ford & Vick.
(beautiful)

Qld, NSW, SA, WA, NT	Sturt's Desert Pea
prostrate-0.6 m x 1-4 m	May-March

Prostrate annual or perennial **herb**; **branches** densely covered with soft hairs; **leaves** pinnate; **leaflets** 9-21, to 3 cm long, oval, grey-green, usually only slightly hairy above, densely hairy below; **flowers** pea-shaped, to about 7.5 cm long, borne in axillary racemes of about 5-6, red-and-black, all red, or rarely red-and-white; **pods** to about 6 cm long, leathery, with soft hairs.

This is one of Australia's most admired wildflowers. After rains, the inland areas can become a magnificent and colourful sight, with Sturt's Pea one of the predominant species, covering large areas and displaying its outstanding flowers. It is widely cultivated, with varied degrees of success. It will grow well in conditions similar to those of its natural habitat, with well drained, deep sand or stony ground in exposed locations. It resents cold, wet conditions as occur in southern winters and very humid conditions as can occur in summer. Plants need a sunny situation exposed to adequate air movement. They are successfully grown in neutral to alkaline soils. Plants are very sensitive to root-rotting fungi and the leaves may be attacked by grey mould. Is probably best treated as an annual in cultivation. Because of the desire to cultivate this ornamental in cooler areas, many methods have been used.

Germination of seed is usually very rapid, taking about 7-10 days if pre-sowing treatment is used and conditions are warm. A popular method is to place the seed in a container, cover with warm water and allow to soak for 6-48 hours. Alternatively, the seed can be rubbed sparingly between 2 sheets of fine sandpaper, placed in a container of cool water until swollen, then removed and sown. A suitable time for the sowing of seed is Aug-Dec.

Some difficulty has been encountered in growing strong plants after transplanting seedlings. It is therefore recommended that plants be grown in situ from seed, or that seed be germinated in peat pots which are then placed into the soil with the seedling.

It is also possible to propagate plants from cuttings.

Propagation by grafting was practised in England and Germany during the 19th century, with good success. At present *Clianthus formosus* is successfully grafted on to *Colutea arborescens* as a cotyledon graft.

In Australia, an increasing amount of experimentation is being undertaken, and regular success is achieved with cotyledon grafts using the New Zealand species *Clianthus puniceus* as the stock. When 2-3 days

Clianthus formosus with a red boss   B.W. Crafter

old, seedlings of *Clianthus formosus* are grafted on to seedlings of *C. puniceus* which are about 2 weeks old. The grafts are kept under glass jars for two weeks, then hardened off by gradual removal of the jars.

One major problem, other than the taking of the graft, is the likelihood of 'damping off'. Necessary precautions should be taken to avoid the development of any fungi.

For further information on cotyledon grafts see Volume 1, page 247.

Some successful cultivation methods:
1. In containers. Use a well drained mixture with a mulch of coarse sand about 3 cm deep. Keep in a sunny position and apply regular fortnightly applications of a water-soluble fertilizer.
2. In areas of raised soil, in a sunny location. Soil level should be 15-30 cm above existing level, to provide excellent drainage. Area must be well prepared, and can contain well rotted compost, blood-and-bone at 60 gm/m², or superphosphate at 30 gm/m². Place 2-4 seeds at required intervals if a massed display is desired. Plants can be thinned out later if germination is excellent. Soil should be kept moist, not wet, as stem rot can occur. It is best to water by flood irrigation rather than from overhead. Once established, plants should not need much artificial water.
3. Excellent plants have also been grown hydroponically in artificially prepared solutions containing the elements essential for growth.

Flowers and leaves of *C. formosus* can be used as a dyeing material.

CLIMBERS

Climbers are a group of plants that need some form of support to enable them to grow and reach their potential. In nature this support is usually provided by plants, which may be grasses, dwarf shrubs, ferns, or anything up to tall trees. In cultivation the range of supports becomes considerably wider, and includes buildings, fences or specially constructed frameworks. There are only a very limited number of Australian climbing species that can attach themselves to supports without twining or using tendrils (eg. *Raphidophora pachyphylla*); therefore, additional support is required for plants climbing against a structure such as a solid wall or fence.

There are over 250 species of climbers in Australia, and their main representation is in coastal or near-coastal regions, with a great variety of species in tropical locations such as temperate or monsoonal rainforest. South-west Western Australia has a large diversity of climbers, especially in the genera *Billardiera* and *Kennedia*. A few species are restricted to the drier regions of inland Australia, eg. *Leichhardtia australis* and *Parsonsia eucalyptophylla*.

The vigour of Australian climbers varies greatly, with some very strong-growing species from the rain-forests, eg. *Calamus muelleri*, Lawyer Cane; *Merremia peltata* and *Mucuma gigantea*, Burny Bean; and weak, light climbers such as *Thysanotus patersonii*, Climbing Fringe-lily.

Many climbers have ornamental foliage, flowers and/or fruits. Those renowned for decorative foliage include *Calamus* spp., *Cissus* spp., *Faradaya splendida*, *Pandorea* spp., *Passiflora* spp., *Piper* spp., *Raphidophora pinnata* and *Stemona australiana*. Amongst the ferns there is an interesting group of ornamental-foliaged climbers belonging to the genus *Lygodium*.

There is an amazing variation in flowers of the climbers. The shape of flowers can vary from species with broad, spreading petals to others with narrow tubular bells. There are also many species with pea-shaped flowers. Size can be minute, eg. *Cayratia clematidea* and *Glycine clandestina*, to generous as in *Abelmoschus moschatus* var. *tuberosus*, Climbing Hibiscus, *Hibbertia scandens* and *Ipomoea* spp. All colours are represented, with some species having striking reds and pinks, while others have combinations of colouring, eg. *Billardiera ringens* (yellow, orange and red), *Chorizema diversifolium* (orange, apricot, magenta or purple) and *Pandorea* spp. (often a mixture of cream, yellow, browns, reds or purple).

Australian climbers are richly endowed with fragrant flowers, eg. *Hardenbergia comptoniana*, *Hoya* spp., *Jasminum* spp., *Kennedia glabrata*, *K. stirlingii* and *Milletia megasperma*. Some species such as *Hibbertia scandens* can have an off-putting odour.

Fruits of climbers can be small and insignificant, but there are many which have very ornamental fruits. Some are edible and were readily partaken by the Aborigines and early settlers. A very large-fruited species is *Entada phaseoloides*, Matchbox Bean, which can have pods of over 1 m in length, with a width of about 10 cm. *Cardiospermum halicacabum*, 'Balloon Vine', produces inflated green fruits that hang in clusters. Climbers that belong to the family Asclepiadaceae have elongated fruits, some of which are most decorative, eg. *Hoya* spp., *Leichhardtia australis* and *Marsdenia rostrata*. Plants of *Cissus* spp. and *Morinda* spp. produce more or less globular fruits that can be quite colourful. *Billardiera longiflora* has shiny purple-blue, oblong berries that contain many small black seeds.

Many climbers are widely cultivated, and have proved to be very adaptable, eg. *Billardiera* spp., *Clematis* spp., *Hardenbergia* spp., *Kennedia* spp. and *Pandorea* spp. There are many other species that have horticultural potential but are as yet virtually untried. This applies especially to species from the north of the continent, eg. *Hoya* spp., *Morinda* spp. and *Piper* spp.

The most common method of growing climbers is on some form of artificial support. The alternative way is to allow light climbers to scramble through and along other plants as they do in their natural habitat. Species such as *Billardiera bignoniacea*, *B. scandens*, *Kennedia coccinea* and *Pronaya fraseri* are ideally grown in this way. Vigorous climbers such as *Kennedia nigricans* and *K. rubicunda* are not recommended for such situations unless they are on sturdy, mature trees. These and similar vigorous climbers have choked young trees, and because of their excessive weight have broken branches.

For detailed information on propagation see genera and species descriptions.

A selection of genera that contain climbing species which are presently grown, or have horticultural potential:

Abrus	*Jasminum*
Adenia	*Kennedia*
Agapetes	*Kunstleria*
Aristolochia	*Luffa*
Asparagus	*Lygodium*
Austrodolichos	*Merremia*
Billardiera	*Milletia*
Brachyscome	*Morinda*
Calamus	*Mucuna*
Canavalia	*Neosepicaea*
Cardiospermum	*Nepenthes*
Cayratia	*Pandorea*
Ceropegia	*Parsonsia*
Cheiranthera	*Passiflora*
Chorizema	*Piper*
Cissus	*Pothos*
Clematicissus	*Prionotes*
Clematis	*Pronaya*
Convolvulus	*Raphidophora*
Derris	*Rhyssopterys*
Dioscorea	*Salacia*
Dischidia	*Scaevola*
Eustrephus	*Scindapsus*
Faradaya	*Sollya*
Geitonoplesium	*Stemona*
Glycine	*Stephania*
Gompholobium	*Strongylodon*
Gymnanthera	*Tecomanthe*
Hardenbergia	*Tetracera*
Hibbertia	*Tetrastigma*
Hoya	*Trichosanthes*
Ipomoea	*Tylophora*

CLITORIA L.

(from supposed resemblance of the flower to a clitoris)
Fabaceae

Shrubs with erect or twining **stems**; **leaves** simple, trifoliolate or pinnate, alternate; **stipules** persistent, striate; **flowers** pea-shaped, inverted, large, solitary or in clusters in the leaf axils or in short racemes; **fruit** a pod.

A genus of about 40 species widespread in tropical and subtropical regions. Many are showy when in flower and are cultivated for ornamental purposes. One species is native to northern Australia and another, *C. ternatea* L., is widely naturalized in eastern Qld.

Clitoria australis Benth. = *C. laurifolia* Poiret

Clitoria laurifolia Poiret
(leaves like those of the genus *Laurus*)
Qld, NT
30-60 cm x 0.5-1 m   Feb-May
Rootstock **perennial**, subterranean; **branches** herbaceous, dying back during the dry, hairy; **leaves** simple or more usually trifoliolate; **leaflets** 3-6 cm x 1-3 cm, ovate, blunt, glabrous above, covered with silky hairs beneath; **flowers** 3-3.5 cm long, pea-shaped, in short clusters in the leaf axils.
A little-known plant from northern areas, which seems to have good horticultural potential for tropical regions. It is very hardy to dryness and grows rapidly with the onset of the wet season. It is suitable for inclusion in a general shrubbery or in a large rockery in a sunny position. Propagate from seed which is hard and must be treated to germinate.

CLUB MOSSES
Species from the genera *Lycopodium*, *Phylloglossum* and *Selaginella* are commonly referred to as club mosses. They are closely allied to ferns, but differ in having true leaves that are arranged to give the appearance of moss, and they have spore cases in a spike or are club-like (hence the common name). The spore cases are formed in the leaf axils.
Of the club mosses, *Lycopodium* with about 15 species, has the largest representation in Australia. They grow as epiphytes and terrestrials. The epiphytes are commonly called tassel ferns, and these are more or less restricted to the tropical regions. The foliage of these species is pendulous and most decorative. They are highly regarded for hanging basket culture. For further detailed information, including cultivation notes, see Tassel Ferns. A limited number of terrestrial species can be grown. In general they resent disturbance, and are best suited to cultivation in containers. The potting mixture should be friable, able to retain moisture, yet well drained.
Phylloglossum is a monotypic genus confined to Australia. *P. drummondii* is a very small species that has proved extremely difficult to cultivate.
Selaginella is a large cosmopolitan genus, yet there are only about 10 species represented in Australia, from the total of about 600 species. The genus differs from *Lycopodium* and *Phylloglossum* in having two kinds of spore cases. They are low-spreading plants that occur in moist, often peaty locations. Some are prostrate, others have erect branchlets. Most species are easily grown in gardens, where they prefer the protection of other plants. They are also suitable for pot culture, and some species are ideal for terraria. *Selaginella* are readily propagated by rhizome divisions.
For more details on cultivation and propagation of these related plants, see the genus and species descriptions.

Cochlospermum gillivraei   B. Gray

Clusiaceae Lindl. (alternative *Guttiferae*)
A large family of dicotyledons consisting of 40 genera and about 1000 species widely distributed throughout the tropical and subtropical regions of the world. They are mainly trees or shrubs with entire leaves that often contain prominent oil dots. The family is represented in Australia by only about 5 genera and 15 species, and of these a few species of *Calophyllum* and *Garcinia* are grown on a limited scale.

Cochlospermaceae Planchon
A small family of dicotyledons consisting of about 2 genera and 30 species distributed through the drier parts of the tropical regions of the world. About 5 species of the genus *Cochlospermum* are native to northern Australia.

COCHLOSPERMUM Kunth
(from the Greek *cochlos*, a spiral shell; *sperma*, a seed; a reference to the coiled seeds)
Cochlospermaceae
Small spindly **trees** or sparse **shrubs**; **leaves** palmately lobed or divided; **inflorescences** are few-flowered, either in terminal racemes or panicles; **flowers** large, yellow, showy; **fruits** are large globular capsules; **seeds** embedded in kapok.
Species about 20 with 4 extending to Australia.

53

Cochlospermum fraseri

Found in hot, open forest. All Australian species are deciduous, dropping their leaves in the dry season and flowering while still leafless. Most species have a yellow sap. The kapok in the seed pods has given all of the species the common name of kapok bush. They are showy shrubs, well suited to drier tropical, subtropical and inland areas with a hot climate. Propagate from seed. Young seedlings may be slow and tricky to handle. They are very susceptible to attacks by root-rotting fungi and damping off.

Cochlospermum fraseri Planchon
(after C. Fraser)
Qld, WA, NT   Kapok Bush
3-5 m x 1-2 m   Aug-Oct
    Tall slender **shrub** or small **tree**; **leaves** lobed, 7-8 cm x 10-12 cm, dark green, with about five lobes; **inflorescence** a loose hairy panicle; **flowers** 6-8 cm across, pale yellow; **sepals** densely covered in soft hair; **capsule** 8-12 cm long, oblong.
    Common in open forests of northern Australia. Very showy and worthy of cultivation but best suited to tropical, subtropical or inland areas. Needs well drained soil in sunny positions. For the best floral display, water should be withheld so that the leaves fall naturally before flowering. Small plants can be transplanted in the wet. Propagate from seed.

Cochlospermum gillivraei Benth.
(after J. McGillivray)
Qld, NT   Kapok Bush
3-12 m x 2-4 m   Aug-Oct
    Medium slender **shrub** to small **tree**; **leaves** 5-7 cm x 8-10 cm, deeply lobed with 5-7 ovate-lanceolate lobes, dark green, glabrous; **panicle** short, loose, glabrous; **flowers** about 10 cm across, bright yellow; **sepals** glabrous; **capsule** oblong, about 8 cm long.
    Widespread in northern coastal and inland districts, often in rocky situations. Very showy and well worthy of cultivation. Best suited to tropical, subtropical or warm inland areas. Other requirements as for *C. fraseri*. Propagate from seed.

Cochlospermum gregorii F. Muell.
(after A. C. Gregory, an early Australian explorer)
Qld, NT   Kapok Bush
3-5 m x 2-4 m   Aug-Oct
    Medium to tall **shrub**; **leaves** deeply divided into 5-7 lanceolate, entire segments 5-8 cm long, glabrous; **flowers** borne in short panicles, about 8 cm across, bright yellow; **sepals** glabrous; **capsule** oblong, about 7 cm long.
    Found in hot, dry areas. Showy and worthy of cultivation in tropical or subtropical areas. Requirements as for *C. fraseri*. Propagate from seed. This species is sometimes included as a subspecies of *C. gillivraei* — (ssp. *gregorii* [F. Muell.] Poppendieck).

Cochlospermum heteroneurum F. Muell.
(with more than one type of vein)
WA, NT   Kapok Bush; Cotton Tree
3-5 m x 2-3 m   May-Aug
    Medium to tall **shrub**; young shoots hairy; **leaves** 8-12 cm x 10-15 cm, orbicular, dark green, shallowly

Fruit of *Cochlospermum fraseri*   D.L. Jones

lobed with 5-9 broad, rounded, blunt, crenate lobes; **panicle** loose, glabrous; **flowers** 5-6 cm across, bright yellow; **sepals** glabrous; **capsule** 8-12 cm long, ovoid or globular.
    Confined to sandstone escarpments of north-western WA. A straggly plant, well suited to cultivation in tropical, subtropical and inland areas with a hot climate. Showy when in flower. Requires a sunny situation in well drained soil. Propagate from seed. This species is sometimes included as a subspecies of *C. fraseri* — (ssp. *heteroneurum* [F. Muell.] Poppendieck).

COCOS L.
(from the resemblance of the nut to a monkey's head)
Arecaceae
    A monotypic genus which is widespread around the tropics and grows on the northern coastline of Australia.

Cocos nucifera L.
(nut-bearing)
Qld, NT   Coconut
30-35 m tall   all year
    Trunk slender, grey, prominently ringed, always curved; **leaves** in a terminal crown, 3-6 m long, pinnate; **inflorescence** a dense panicle, bearing separate male and female flowers; **fruit** 20-30 cm long, enclosed in a thick, fibrous sheath.
    A familiar tropical palm which is doubtfully indigenous to Australia. It is very widespread along the tropical coastland, often in isolated areas. It is doubtful if any of the trees is more than one hundred years old,

and they were probably introduced for their valuable fruit. They are widely cultivated in tropical Australia for decorative purposes but trees outside the tropics rarely produce fruit. Very cold sensitive and will rarely survive south of the Qld border. Not very drought tolerant and requires access to underground water or regular watering during dry periods. Fruit are distributed by coastal currents. Plants are very easy to transplant in the summer. Fruit can be germinated by half burying in a pot of sandy, coarse propagating medium in a glasshouse, held at 15°C or above, and watering regularly. They also germinate if hung in a glasshouse in plastic bags containing moist peat or sphagnum moss. The nuts must be potted soon after emergence of the root.

CODIAEUM Blume
(from the Malayan name Codebo)
Euphorbiaceae

Shrubs or **trees**; **leaves** alternate, petiolate, entire, glabrous; **flowers** small, unisexual, the males usually in clusters, the female flowers in racemes; **petals** 5-6; **stamens** numerous; **fruit** a 2-valved capsule.

A small genus of 15 species widely distributed throughout the Pacific region with 2 widespread species extending to north-eastern Qld. The various species are popularly known as crotons, and are widely cultivated for their colourful leaves in tropical and subtropical regions around the world. They are very easy to grow but are sensitive to excessive cold and waterlogged soils. They make excellent pot plants for glasshouse and indoor decoration. Propagate from seed or cuttings.

Codiaeum membranaceum S. Moore
(thin-textured; membranous)
Qld
3-6 m x 2-4 m   April-June

Tall shrub; young shoots glabrous; **leaves** 10-16 cm x 3-6 cm, obovate to oblong, dark green, thin-textured, the apex blunt or shortly toothed, on petioles 3-4 cm long; **racemes** 15-20 cm long, axillary; **male flowers** about 0.2 cm across, bearing numerous stamens, in clusters; **female flowers** about 0.5 cm across, solitary, on thick peduncles; **capsule** about 1 cm across.

A little-known species that in Australia is confined to scrubs of Cape York Peninsula. It is virtually unknown in cultivation but has merit because of its hardiness, dense growth and dark green leaves. Best suited to tropical and subtropical areas. Requires well drained soil in a sunny situation. Propagate from seed or cuttings.

Codiaeum variegatum (L.) Blume
var. **moluccanum** (Decne.) Muell. Arg.
(irregularly coloured; from the Moluccas Islands)
Qld   Croton
3-8 m x 2-4 m   March-May

Medium **shrub** to small **tree**, glabrous in all parts; **leaves** 10-20 cm x 4-8 cm, variable in shape, obovate to elliptical to spathulate, thick and leathery, dark green on both surfaces or more usually blotched and variegated with cream and white; **racemes** 8-15 cm long, axillary, the male racemes longer; **male flowers** in clusters of 2-6; **female flowers** solitary, on thick pedicels; **capsule** 0.6-0.8 cm across, smooth.

A widespread species that in Australia is restricted to coastal ranges north of Townsville, where it grows in rainforest. It is a handsome, dense shrub or small tree with attractive foliage, and is suited to cultivation in tropical and subtropical areas. Forms with variegated leaves are decorative and the colour is accentuated by planting in positions exposed to full sunlight. Will also grow in complete shade. Needs well drained soil and is cold sensitive, especially when small. Makes an excellent hedge plant. Grows well in a pot or tub and is ideal for indoor decoration.

A large range of colourful croton cultivars is grown in Australia. These have arisen by selection and hybridization, mainly from within the typical form of *Codiaeum variegatum* (L.) Blume. This form does not occur naturally in Australia, and most of the cultivars have been imported or have been bred within Australia by crossing various overseas selections and cultivars. It is not known if any selections have been made from Australian material of *Codiaeum variegatum* var. *moluccanum*. Crotons are easily raised from cuttings which strike readily or from seed which should be sown soon after collection.

CODONOCARPUS Cunn. ex Endl.
(from the Greek *codon*, a bell; *carpos*, a fruit; alluding to the shape of the fruit)
Gyrostemonaceae

Tall **shrubs** or small **trees**; **leaves** linear or broad; **flowers** dioecious or monoecious, in leafless racemes, axillary or terminal, female borne on the leafless base of current year's growth; **fruit** composed of numerous closely connected carpels, bell-shaped.

An endemic genus comprising 3 species, 2 of which are widespread in the arid interior with the other occurring in the rainforests of eastern Australia.

Codonocarpus attenuatus (Hook.) H. Walter
(slender; drawn out)
Qld, NSW   Bell-fruit Tree
3-8 m x 2-4 m   Dec-Feb

Medium **shrub** to small **tree**; **bark** corky; **leaves** 5-8 cm x 1-2 cm, lanceolate, entire, bright green, somewhat succulent, drawn into long attenuated tips; **flowers** unisexual; **male flowers** sessile, in axillary spikes 3-8 cm long; **female flowers** solitary; **fruit** about 1.8 cm across, bell-shaped, on stalks about 4 cm long.

Widespread between central Qld and northern NSW, where it is common as regrowth along roadsides and in cleared rainforest. It occurs mainly in coastal or near-coastal districts but is also occasionally found in western areas. The species has a symmetrical growth habit and is useful as a quick-growing screen or for protecting slow-growing plants. It is frost resistant and grows in most well drained soils. Propagate from seed which germinates rapidly when fresh.

Codonocarpus pyramidalis   W R Elliot

Codonocarpus continifolius (Desf.) F. Muell.
(leaves similar to those of the genus *Cotinus*)
Qld, NSW, Vic, SA, WA, NT   Bell-fruit Tree;
                                    Native Poplar
3-12 m x 1.5-4 m   Sept-Feb
    A medium **shrub** to small **tree**; **bark** smooth,
pinkish; **leaves** 2-5 cm long, broadly lanceolate,
tapering to base, glabrous, grey-green; **flowers** small,
unisexual, borne in separate racemes in upper axils or
on separate trees, often forming terminal panicle,
males on very short stalks, females on long stalks; **fruit**
bell-shaped, about 1 cm long, borne in pendulous
clusters, lime green.
    A very fast growing species, usually with a life span
of about 10-15 years. It will grow in most well drained
soils, with full sun. Frost and drought resistant. Useful
for quick temporary windbreaks, and is ornamental.
Recommended for semi-arid regions. Will transplant
readily, even when 1-2 m high. Pruning could increase
life span. The bark, leaves and twigs have a distinctive
'horse-radish' taste. Propagates readily from fresh
seed.

Codonocarpus pyramidalis (F. Muell.) F. Muell.
(pyramid-shaped)
NSW, SA
3-12 m x 1.5-4 m   Sept-Nov
    Medium **shrub** to small **tree**; **bark** smooth, pinkish;

leaves 5-12 cm long, linear, tapering to a pointed apex,
glabrous, greyish green; **flowers** small, unisexual, males
on short stalks, females on long stalks; **fruit** bell-
shaped, to 1.5 cm long.
    Although similar to *C. cotinifolius* in many ways, this
species is readily distinguished by its long, narrow
leaves and larger fruit. Requires similar conditions in
cultivation. Propagate from fresh seed.

COELACHNE R. Br.
(from the Greek *coelos*, hollow; *achne*, a glume; in
reference to the hollow glumes)
Poaceae
    Tufted or creeping perennial **grasses**; **leaves** flat,
erect or flaccid, usually small; **stems** slender, erect;
panicles slender, loose; **spikelets** 2-flowered.
    A small genus of about 10 species widely spread over
tropical Asia with one extending to northern Australia.

Coelachne pulchella R. Br.
(pretty)
Qld, NT
10-15 cm tall   Jan-April
    Perennial **grass** forming spreading mats; **leaves** 2-
4 cm long, linear, flat, flaccid or decumbent along the
ground; **panicle** 2-7 cm long, narrow, wispy; **spikelets**
pinkish, hairy.
    An attractive grass that is apparently confined to
northern areas and there growing in moist patches of
soil amongst rocks and along stream banks. It is a
pretty grass when in flower and would make an ideal
rockery plant or mat plant for tropical and subtropical
areas. Propagate by division of the clumps.

COELORHACHIS Brongn.
(from the Greek *coelos*, hollow; *rhachis*, backbone; in
reference to the hollow rhachis of the racemes)
Poaceae
    Perennial **grasses** forming large, spreading clumps;
leaves erect; **inflorescence** an erect, racemose panicle;
spikelets in pairs.
    A genus of 12 species of tropical grasses with a soli-
tary species widespread in northern Australia.

Coelorhachis rottboellioides (R. Br.) Stapf.
(resembling the genus *Rottboelia*)
WA, NT   Cane Grass
1.5-3 m tall   Jan-March
    Perennial **grass** forming tall, spreading clumps;
leaves 1-2 m x 0.5-1 cm, coarse, dark green, flat; **culms**
1.5-3 m tall, erect, stout, cane-like, with many nodes;
panicle 10-30 cm long, terminal, leafy, the individual
branches 5-10 cm long, slender; **spikelets** 0.3-0.5 cm
long.
    A tall grass found naturally in moist areas along the
banks of streams of northern Australia, often forming
thickets. It could have excellent potential as a windbreak and shelter for water-birds if planted on the perimeter of ponds or dams. Frost tender and suited only to
the tropics. Propagate by division of the clumps.

Coelospermum paniculatum   D.L. Jones

COELOSPERMUM Blume
(from the Greek *coelos*, hollow; *sperma*, a seed)
Rubiaceae

Straggly **shrubs** or **climbers**; **stipules** acuminate, sometimes conspicuous; **leaves** entire; **inflorescence** an axillary umbel, cyme or terminal panicle; **flowers** tubular; **lobes** 4-5; **fruit** a drupe containing 4 hard 1-seeded segments.

A genus of about 15 species with 2 found in northern Australia. Neither species is commonly grown in cultivation, although they do have some horticultural merit.

Coelospermum paniculatum F. Muell.
(flowers in panicles)
Qld, NSW
2-5 m tall   Oct-Jan

Woody, glabrous **climber**; **leaves** 5-10 cm x 2-5 cm, ovate to oblong, leathery, glossy dark green; **inflorescence** of axillary cymes forming a dense terminal panicle; **flowers** 0.8 cm long, white, tubular, fragrant; **corolla lobes** 5, somewhat spreading; **drupe** about 1 cm across, globular, purplish black.

An interesting climber widespread in coastal regions throughout Qld and north-eastern NSW. Quite showy when in flower. Very hardy and useful for coastal districts, tolerating some exposure to salt-laden winds. Needs well drained soil in a sunny situation. Propagate from seed which must be sown fresh. Cuttings have proved to be slow and difficult to strike.

Coelospermum reticulatum Benth.
(with reticulate venation)
Qld, NT
1-3 m tall   Nov-Feb

Straggly **shrub** or **climber**; **branches** glabrous, often twining; **leaves** 3-6 cm x 1-2 cm, obovate, rigid, pointed, somewhat prickly, dull green with prominent reticulate venation; **umbels** axillary or terminal, sometimes in clusters; **flowers** about 0.8 cm long, tubular, hairy inside, fragrant; **corolla lobes** 4-5; **drupe** about 0.8 cm across, globular, blackish.

Widespread in coastal scrubs, sometimes extending inland along large rivers and also on offshore islands. Very straggly but decorative when in flower. Could perhaps be improved in cultivation by pruning, fertilizing and regular watering. Best suited to a sunny position in tropical and subtropical regions. Propagate from seed which must be sown fresh.

COLEANTHERA Stschegl.
(from the Greek *coleos*, a sheath; *anthera*, anther; refers to the sheathing arrangement of the anthers)
Epacridaceae

Small **shrubs**; **branches** many; **leaves** alternate, flat or concave, small; **flowers** tubular, bearded inside throat, small, 2 or 3 together on short axillary stalk, or solitary, lobes rolled under to expose anthers; **fruit** a drupe.

An endemic genus of 3 species, rare in nature, and confined to the South-West Botanical Province of WA. They usually occur in sandy soils, with dappled shade or partial sun. They have potential for cultivation, but to date there is very little information available on their growth performance. Propagation from cuttings is suggested, although difficulty of propagation is undoubtedly the reason that plants have not been available commercially to date.

Coleanthera coelophylla (Cunn. ex DC.) Benth.
(hollow leaves)
WA
0.3-0.5 m x 0.5-1 m   Oct-Nov

A dwarf **shrub**; **branches** with soft hairs; **leaves** to about 1 cm long, ovate to lanceolate, blunt with a small pungent point, strongly veined beneath, margins and base hairy; **flowers** tubular, about 0.7 cm long, solitary or rarely 2, axillary, white to deep pink; **calyx margins** densely hairy.

This species occurs east of Albany, on and near the south coast. It grows in well drained soils, with dappled shade or partial sun. As a garden plant it should be useful for sheltered coastal situations, or for containers. Propagate from cuttings.

Coleanthera myrtoides Stschegl.
(similar to the genus *Myrtus*)
WA
0.5-2 m x 0.5-1.5 m   Sept-Jan

An erect dwarf to small **shrub**; **branches** glabrous or slightly hairy; **leaves** up to 1 cm long, ovate to orbicular, concave or nearly flat, often clustered at ends of branchlets, veins not very conspicuous; **flowers** tubular, about 0.7 cm long, white or purple-pink, lobes

57

longer than the tube, strongly reflexed, bearded; **calyx margins** slightly hairy.

This species has a wider range than *C. coleophylla*, occuring further inland. The requirements and applications of both species would be very similar. Propagate from cuttings.

C. virgata Stschegl., the other species in this genus, is a dwarf shrub. It has lanceolate leaves, white flowers, and is very similar to *C. myrtoides*. There is limited information on this species in its natural habitat, and it is evidently not in cultivation.

COLEUS Lour.
(from the Greek *coleos*, a sheath; an allusion to the sheathed stamens)
Lamiaceae

Small **herbs** or **shrubs**; **stems** square in cross-section; **leaves** opposite, decussate, fragrant; **inflorescences** of erect racemes, panicles or cymes; **flowers** tubular, lobed, blue or white.

Species about 150 with a solitary widespread species scattered through tropical Australia. Many exotic species, forms and hybrids are cultivated in the tropics and as glasshouse or indoor plants for their very decorative leaves.

Coleus scutellarioides Benth.
(like the genus *Scutellaria*)
Qld, WA, NT
0.1-0.45 m x 0.3 m   all year

Tall **herb** or dwarf **shrub**; **leaves** 3-7 cm x 1.5-5 cm, ovate or lanceolate, with coarsely toothed margins, dull green, fragrant when bruised; **racemes** erect, terminal on the shoots; **flowers** about 0.5 cm long, purple or bluish.

Widespread in a variety of habitats but frequently near water, such as along the margins of streams and swamps. Hairy and glabrous forms are known. A weedy-type species but nevertheless very useful as a groundcover or rockery plant for tropical, subtropical or warm inland areas. Requires a glasshouse in southern Australia. Germinates readily from seed and may naturalize itself. Strikes very quickly and easily from stem cuttings.

COLOBANTHUS Bartling
(from the Greek *colobos*, mutilated; stunted; *anthos*, a flower)
Caryophyllaceae

Small perennial **herbs**; **leaves** opposite, narrow, linear to subulate, often stiff, crowded in tufts; **flowers** solitary, on short or long stalks; **petals** none; **sepals** 5; **fruit** a capsule.

A genus of about 14 species distributed mainly in Antarctic regions, or mountains of South America and New Zealand. Australia has 3 or 4 indigenous species.

Colobanthus affinis (Hook.) Hook.f.
(affinity)
NSW, Vic, Tas   Alpine Colobanth
0.5 m x 0.1-0.3 m   Nov-Feb

Densely tufted perennial **herb**; **leaves** about 5 cm long, narrow, linear, more or less limp, pointed but not ending with a mucro; **flowers** small, no petals, borne on stalks slightly longer than leaves, white to cream; **sepals** usually 5.

This species is not at all common in cultivation, but could prove to be useful, especially in gardens which have been planned with a natural concept. It could be used amongst boulders or other low-growing plants. In nature it occurs in mountainous areas, and will require moist, well drained soils, rich in well composted organic matter. Best suited to partial or full sun. For further information on cultivation, see Cushion Plants, page 135. Propagate from seed or cuttings.

Colobanthus apetalus (Labill.) Druce
(without petals)
NSW, Vic, Tas   Tufted Colobanth
0.02-0.08 m x 0.1-0.3 m   Nov-Feb

Dwarf, tufted perennial **herb**; **leaves** to 2.5 cm long, linear to subulate, ending in short mucro; **flowers** small, cream, solitary; **sepals** usually 5; **petals** none.

Whereas the other *Colobanthus* species are restricted to alpine areas, this one is found mainly in sandy coastal regions, although it does occur also in subalpine areas. Should be suited to moist, well drained soils, with partial to full sun. Propagate from seed, cuttings or by division.

Differs from *C. affinis* (Hook.) Hook.f., which has softer and longer leaves.

Colobanthus nivicola M. Gray
(snow-dweller)
NSW   Soft Cushion Plant
0.05 m x 0.1-0.3 m   Dec-March

Dwarf cushion **plant**; **leaves** to 0.5 cm x up to 0.1 cm, subulate, imbricate, more or less held erect, forming a dense, moss-like mat; **flowers** small, cream, solitary; **sepals** usually 5; **petals** none; **fruiting peduncles** about 1 cm long.

A recently described cushion plant from Mount Kosciusko area. It grows in situations where it is covered by deep snow during the winter. Not well known in cultivation but has potential for growing amongst boulders and as a groundcover plant. Should do best in moist, well drained soils with a high organic content. Worth trying as a container plant. See Cushion Plants, page 135 for further information on cultivation. Propagate from seed or cuttings and possibly by division.

Differs from *C. pulvinatus* F. Muell., which has harsh, broad leaves and shorter fruiting peduncles.

Colobanthus pulvinatus F. Muell.
(cushion-like)
NSW   Hard Cushion Plant
0.05 m x 0.1-0.2 m   Dec-March

Dwarf cushion **plant**; **leaves** to 0.5 cm long, subulate, rigid, spreading, ending in sharp point, forming a dense mat; **flowers** small, cream, solitary; **sepals** usually 5; **petals** none; **fruiting** peduncles about 0.5 cm long.

A cushion plant from Mount Kosciusko, where it usually occurs at lower elevations than *C. nivicola*. It is important in its natural habitat for colonizing bare rocky slopes. Also not well known in cultivation, this species has similar needs to those of *C. nivicola*. See Cushion Plants, page 135 for cultivation information. Propagate from seed or cuttings and possibly by division.

Differs from *C. nivicola* M. Gray, which has softer and more erect leaves.

COLOCASIA Schott
(Greek name for the rhizome of *C. antiquorum*)
Araceae

Perennial **herbs**; **rhizomes** prostrate, fleshy; **leaves** erect, arising alternately along the rhizomes, each on a long, fleshy stalk topped by an expanded lamina, the stalk attached to the underside of the leaves; **flowers** on a spadix, subtended by a spathe; **berries** enclosed by the persistent base of the spathe.

A small genus of about 8 species with one native (or naturalized) in northern Australia.

Colocasia esculenta (L.) Schott
var. **antiquorum** Schott
(edible; ancient)
Qld, NT   Taro; Elephant Ears
1-1.5 m x 1-5 m   Jan-March

Fleshy perennial **herb**; **rhizomes** slender, long-creeping and forming colonies; **leaves** 0.5-1.5 m, erect, fleshy, bluish green; **leaf-stalk** thick, channelled at the base; **leaf-blade** 20-30 cm wide, heart-shaped, blunt, with triangular basal lobes; **flowers** greenish, tiny, crowded on to a spadix, subtended by a hood-shaped spathe; **berries** small, containing numerous seeds.

Widespread and cultivated as an important food plant in the Pacific region. Grows in colonies along creeks and swamps in coastal areas of north Qld, and is common in NT. It is uncertain whether these populations are indigenous or naturalized. Taro must be thoroughly prepared and cooked before eating; otherwise severe irritation and poisoning can result. Makes an attractive plant for a moist situation in tropical or subtropical regions. Very frost tender. Propagate by division of the rhizomes or from seed.

COLUBRINA Rich. ex Brongn.
(from the Latin *colubrinus*, shaped like or resembling a snake; in reference to the sinuous branches of some species)
Rhamnaceae

Shrubs, trees or shrubby **climbers**; **leaves** alternate, simple, entire or toothed, 3-nerved at the base; **inflorescence** an axillary cyme; **flowers** small, bisexual; **calyx** 5-lobed; **petals** 5, cucullate; **fruit** a drupe.

A small genus of about 16 species with a solitary widespread Asian species extending to tropical Australia. The genus has little to recommend it for horticulture. Propagate from seed or cuttings of semi-ripened wood.

Colubrina asiatica (L.) Brongn.
(Asian)
Qld, WA, NT
3-7 m x 1-3 m   Aug-Feb
Medium to tall **shrub** or small **tree**; **branches** slen-

der, flexuose, glabrous; **leaves** 5-8 cm x 2-3 cm, ovate, cordate at the base, the margins toothed, dark green and shiny, with 3 prominent nerves; **cymes** axillary, dense; **flowers** about 0.4 cm across, greenish; **drupe** about 0 8 cm across, furrowed, somewhat sunken at the top.

An unusual plant which is a shrubby climber. The slender branches weave through surrounding vegetation. It has very little to recommend it for cultivation except perhaps its hardiness. Plants are available from specialist nurseries in Brisbane. In Asia the people crush the leaves in water to obtain a concoction for washing their hair. A sunny situation in well drained soil is suitable for its culture. Propagate from seed or cuttings.

COLYSIS C. Presl
Polypodiaceae

Medium sized **ferns** forming creeping patches on trees; **fronds** erect or weeping, thin-textured, lobed, with conspicuous venation; **sori** elongated, narrow.

Species about 30 with 2 found in Australia.

Colysis ampla (F. Muell. ex Benth.) Copel.
(derivation unknown)
Qld
20-60 cm long

Rootstock thick, long-creeping, clothed with brown, papery scales; **fronds** semi-weeping, entire or with a few lobes, dark green.

Confined to coastal and highland rainforests of northeastern Qld. Resents disturbance and is slow to reestablish. Easy to maintain once growing. Needs humid and moist conditions. May be slow-growing in cold areas. An excellent basket plant. Hardy in unheated glasshouses in southern Australia. Propagate by division of the rhizome or from spore.

Colysis sayeri (F. Muell. & Bak.) Copel.
(after W. Sayer)
Qld
20-60 cm long

Rootstock thin or thick, clothed with brown scales; **fronds** erect or semi-creeping, with numerous narrow lobes, greyish green.

Found on trees in rainforest at high elevations. Makes an excellent basket fern. Needs cool, humid conditions and is intolerant of dryness. Easier to establish than *C. ampla*. Hardy in a bushhouse in southern Australia. Propagate by division of the rhizome or from spore.

Combretaceae R. Br.
A family of dicotyledons comprising about 500 species in 20 genera, widely distributed in tropical regions of the world. About 35 species are found in Australia, mostly endemic, with the majority being in the genus *Terminalia*. Other Australian genera are *Lumnitzera* and *Macropteranthes*.

COMESPERMA Labill.

(from the Greek *come*, hair; *sperma*, a seed; alluding to tufts of hair on the seeds)
Polygalaceae

Herbs, **shrubs** or **climbers**; **leaves** alternate, entire, or reduced to minute scales; **flowers** irregular, similar to pea-flowers, borne in racemes; **sepals** 5, resembling petals; **petals** 3; **fruit** a capsule, containing 2 seeds, covered in hairs.

An endemic genus of about 24 species. The majority are represented in the South-West Botanical Province of WA, with further species mainly in the southern half of Australia, including Tas.

They are mostly small shrubs, and occur over a wide range of soil and climatic conditions, growing in light to heavy soils, from coastal to mountain areas.

Flower colours are in shades of purple, pink, mauve, blue or occasionally yellow. Some species are very showy in flower, eg. *C. volubile* and *C. ericinum*, and these are probably the two most commonly cultivated.

Others such as *C. retusum* should soon find their way into cultivation. In general, comespermas prefer well drained soils, with dappled shade or partial sun.

Most species can be readily propagated from seed or cuttings.

Comesperma calymega Labill.

(large cover)

Vic, Tas, SA, WA	Blue-spike Milkwort
0.3-0.5 m x 0.3-1 m	Oct-Jan

Dwarf **shrub**; **branches** erect, rigid, glabrous; **leaves** 1-2 cm long, elliptic-oblong to lanceolate, thick, green; **flowers** small, borne in dense terminal racemes about 5 cm long, usually blue, rarely pink or white; **capsule** about 1 cm long, wedge-shaped.

Usually found growing naturally on sandy soils. This small ornamental species will grow in most well drained soils, with dappled shade or partial sun. It tolerates full sun if root area is protected. Withstands light to moderate frosts, but not extended dry periods. Propagate from seed or cuttings.

Comesperma ciliatum Steetz see *C. volubile* Labill.

Comesperma confertum Labill.

(crowded)

WA	
0.5-1 m x 1 m	Sept-Dec

Dwarf **shrub**; **branches** erect; **leaves** about 2.5 cm x 0.2 cm, narrow-linear, thick, acute, crowded, margins revolute; **flowers** small, bluish pink, borne in slender, dense racemes 5-10 cm long; **capsule** about 0.6 cm long, more or less rounded.

An ornamental species that occurs mainly in sand-heath communities throughout the south-western corner of WA. It is not very common in cultivation, but should adapt to most well drained soils, with dappled shade or partial sun. Propagate from seed or cuttings.

Comesperma ericinum DC.

(similar to the genus *Erica*)

Qld, NSW, Vic, Tas	Heath Milkwort
0.5-1.5 m x 1-2 m	Oct-Feb

A small **shrub**; **leaves** to 1.5 cm long, linear to

Comesperma confertum   T.L. Blake

oblong, strongly recurved margins and apex, glabrous, crowded; **flowers** small, borne in short racemes, profuse, pink to purple; **capsule** to 0.8 cm long, wedge-shaped.

One of the more widely cultivated species. It occurs naturally in a wide range of soil types, and has proved most adaptable in cultivation. It prefers relatively well drained soils, with dappled shade or partial sun. Withstands most frosts and extended wet periods. Pruning after flowering produces compact growth. Propagate from seed or cuttings.

Comesperma ericinum, × ·75

Comesperma retusum Labill.
(retuse)
Qld, NSW, Vic, Tas   Mountain Milkwort
1-2 m x 1-2 m   Nov-Jan
    Dwarf **shrub**; **branches** upright and glabrous; **leaves** to 1.5 cm long, linear to oblong, flat, blunt, glabrous, crowded; **flowers** small, borne in short racemes, profuse, pink to purple; **capsule** about 1 cm long, wedge-shaped.
    This species is very similiar to *C. ericinum*, and just as ornamental, but it is not well known in cultivation. It occurs from sea level to mountain areas, often in swampy heathlands, and should prove to be most adaptable. It will need relatively well drained soils that do not dry out readily. Grows in dappled shade, partial sun or full sun, provided there is some root protection. Propagate from seed or cuttings.

Comesperma scoparium Steetz
(broom-like)
Vic, SA, WA   Broom Milkwort
0.5-1.5 m x 0.5-1.5 m   Aug-Dec
    A small **shrub**; **branches** upright, appear leafless; **leaves** reduced to minute scales; **flowers** about 0.6 cm long, borne on short stalks, surrounded by small bracts, blue, wings about 0.6 cm long; **capsule** about 0.5 cm long.
    Not widely cultivated, this species usually occurs naturally on sandy soils in low-rainfall areas. It will grow best in well drained soils, with partial sun. Propagate from seed or cuttings which can be difficult to strike.

Comesperma virgatum Labill.
(twiggy)
WA
1-2 m x 1-2 m   Aug-Feb
    Small **shrub**; **branches** many, usually erect; **leaves** to 1.5 cm long, narrow, more or less obtuse, prominent midrib on undersurface, distant or crowded; **flowers** small, bluish to pink-purple, profuse, forming racemes from 2.5 cm to 10 cm long; **capsule** about 0.7 cm long, truncate or 3-toothed.
    Not commonly cultivated, this showy species is a dweller of gravelly soils and swampy regions over a large area of the South-West Botanical Province. Requires relatively well drained soils, but tolerates waterlogging for short periods. Best suited to partial or full sun. Withstands light frosts. Propagate from seed or cuttings.

Comesperma volubile Labill.
(twining)
Qld, NSW, Vic, Tas, SA, WA   Love Creeper
1-3 m x 1-3 m   Sept-Dec
    A twining **perennial**; **stems** furrowed; **leaves** about 0.5-1 cm long, linear to lanceolate, few only, scattered along branches; **flowers** to about 1 cm long, borne in many racemes along branches, blue, mauve or rarely white; **capsule** to 1.5 cm long, oblong.
    A most decorative light twiner, which commonly scrambles through small shrubs in nature. It grows in a wide range of soil types, and prefers dappled shade or

Comesperma volubile, × ·5

partial sun, with root protection. Withstands light to moderate frosts and dry periods. Not widely cultivated to date. It is difficult to establish as there seems to be a symbiotic relationship required for successful growth. Early settlers used the root as an alternative to sarsaparilla. Propagate from seed or cuttings which are usually difficult to strike.
    C. ciliatum Steetz is very similar, but leaves are smaller and hairy. Capsule is also smaller. It occurs in SA and WA.

COMMELINA L.
(after J. & G. Commelin, 17th-century Dutch botanists)
Commelinaceae
    Weak, spreading or straggly **herbs**; **stems** creeping or erect, fleshy, jointed; **leaves** sheathing at the base, spreading, thin-textured; **flowers** borne in a complex leafy sheath, showy, with 2-3 flimsy petals, usually blue, produced in succession, each lasting only a few hours before collapsing.
    Species about 230 with about 7 indigenous to Australia. All are weak perennial herbs, showy when in flower and worth considering for cultivation. Some

species make excellent groundcovers, forming a dense mat of interlacing stems and leaves. The stems root at the nodes, and under good conditions the clumps spread rapidly. They are easily propagated from cuttings or by division of the clump.

Commelina acuminata Ewart & McLennan
(long-pointed)
Qld, NT
0.3-0.5 m tall   Jan-June
Stems erect or straggling; **leaves** 4-8 cm x about 1 cm, linear, roughened, bright green; **spathes** very conspicuous, on long peduncles, heart-shaped at the base (6-10 cm wide) then drawn out into a long point (3-4 cm long); **flowers** held above the spathes on long peduncles, about 3 cm across, bright blue, with yellow stamens.

A very interesting species, showy when in flower. Forms small patches along the margins of creeks. An attractive rockery plant for a sheltered position in tropical areas, or pot plant for glasshouses in southern Australia. Very frost sensitive. Propagate from stem cuttings which strike easily.

Commelina cyanea R. Br.
(blue)
Qld, NSW, NT   Scurvy Weed
0.2-0.6 m tall   Jan-Aug
Stems weak, decumbent or erect, straggly, fleshy; **leaves** 3-7 cm long, ovate-lanceolate, dark green, sheathing at the base; **flowers** arising from a green spathe, about 2.5 cm across, bright blue, showy, each made up of 3 flimsy petals.

Found in moist, protected situations, sometimes actually growing in water. An underrated plant, useful as a groundcover or small plant for protected or open sunny situations. Very attractive when in flower. A useful basket plant. May be damaged by frost but usually regenerates from the base or below ground. Will tolerate wet conditions and could be grown as an aquatic. Useful for coastal conditions and will even tolerate some inundation by salt water. Propagate from stem cuttings which strike easily.

Commelina ensifolia R. Br.
(sword-shaped leaves)
Qld, NT
0.3-0.5 m tall   Jan-Aug
Stems erect or straggling, fleshy; **leaves** 5-10 cm x 0.4-0.8 cm, linear-lanceolate, dark green, sheathing at the base; **flowers** borne in a leafy sheath, about 1 cm across, bright blue, each made up of 2 rounded, flimsy petals.

Found in moist areas such as along the margins of swamps, usually in small, spreading patches. A useful groundcover for a sheltered position in tropical and subtropical areas. Very frost sensitive and unsuitable in southern Australia except in glasshouses. Propagate from stem cuttings which strike easily.

Commelina lanceolata R. Br.
(lanceolate leaves)
Qld, WA, NT
0.1-0.3 m tall   April-Aug
Stems creeping or semi-erect, fleshy; **leaves** 6-12 cm

Commelina cyanea   D.L. Jones

long, narrow-lanceolate, dark green, with a short sheath at the base; **flowers** borne in leafy sheaths, 1.5-2 cm across, bright blue, each made up of 3 rounded, flimsy petals.

A straggly species found along the margins of swamps and in shallow water, drains etc. Showy when flowering in mass. Will tolerate wet conditions, and could be grown as an aquatic or emergent plant in shallow water. Very frost sensitive and unsuitable in southern Australia. Propagate from stem cuttings which strike readily.

Commelina undulata R. Br.
(undulate margins)
Qld, NT
0.1-0.2 m tall   Dec-March
Stems weak, prostrate, straggly, fleshy; **leaves** 6-10 cm long, lanceolate, bright green, with strongly crisped or undulate margins, sheathing at the base; **flowers** borne in a green spathe, 2.5-3 cm across, bright blue, each made up of 2 rounded, flimsy petals and a bunch of yellow stamens.

A showy little herb well worthy of cultivation. Found naturally in rainforest. Grows well in a protected situation, tolerating morning sun or filtered sun. Very showy when in flower. Makes an attractive basket plant. Frost tender but will reshoot from dormant buds along the stems. Propagate from stem cuttings which strike easily.

Commelinaceae R. Br.

A family of monocotyledons consisting of 38 genera and about 500 species widely distributed throughout the world in tropical, subtropical and warm temperate regions. They are succulent annual or perennial herbs, often with a creeping growth habit, and many species produce showy, colourful flowers in succession over a long period. The family is represented in Australia by 6 genera and about 14 species, most of which occur in the tropics. Many of the Australian species are worthy of cultivation. Australian genera: *Aneilema*, *Commelina*, *Cyanotis*, *Floscopa*, *Murdannia* and *Pollia*.

COMMERSONIA Forster & Forster f.
(after P. Commerson, 18th-century French doctor and naturalist)
Sterculiaceae

Shrubs or small **trees**; **branches** bearing stellate hairs; **leaves** simple, toothed or lobed, hairy; **inflorescences** of axillary or terminal cymes; **flowers** small, the calyx large and conspicuous; **fruit** a rounded capsule, often bearing bristles.

Species about 9, eight of which are endemic in Australia. The eastern states species are tall shrubs or trees, whereas those from southern and western areas are small shrubs. All have excellent potential for cultivation and some species are becoming widely grown. They are easy to grow, favouring a sunny situation in well drained soil. Propagate from seed or cuttings.

Commersonia bartramia (L.) Merr.
(after J. & W. Bartram)

| Qld, NSW | Brown Kurrajong |
| 5-12 m x 3-5 m | Nov-Jan |

Tall **shrub** or small, spreading **tree**; young branches covered with white, woolly hairs; **leaves** 6-15 cm x 2-5 cm, ovate, entire or usually with irregularly toothed margins, dull green above, whitish beneath; **cymes** axillary, erect, along the branches; **flowers** 0.4-0.6 cm across, cream; **capsule** 2-2.5 cm across, rounded, bearing numerous soft-pointed bristles.

Widespread along stream banks and rainforest margins from the Clarence River to Cape York Peninsula. The leaves are palatable and eaten by cattle. An attractive species with long branches which spread almost horizontally. This gives a layered effect when the trees are in flower, with the leaves hanging and the flower cymes erect. Commonly cultivated in coastal subtropical districts as a small shade tree. Requires well drained soil and a sunny aspect. Decorative when in fruit. The Aborigines used the bark to make fibre for nets and fishing lines. Previously known as *C. echinata*. Propagate from seed or stem cuttings.

Commersonia bartramia D.L. Jones

Commersonia echinata Forster & Forster f. =
C. bartramia (L.) Merr.

Commersonia fraseri Gay
(after C. Fraser)

| Qld, NSW, Vic | Blackfellow's Hemp |
| 2-6 m x 1-2 m | Sept-Jan |

Tall slender **shrub**; **branches** bearing woolly hairs; **leaves** 7-15 cm x 2-5 cm, cordate or ovate-lanceolate, with coarsely toothed margins, dull green above, whitish beneath; **cymes** axillary, branched; **flowers** 0.8-1 cm across, white or cream; **capsule** 1.5-2 cm across, rounded, bearing numerous soft bristles.

Widespread along stream banks and moist areas near rainforest from south-eastern Qld to eastern Vic. The Aborigines used the bark to make fibre. A useful shrub that has been in cultivation for over 50 years but is not widely grown. The foliage is attractive and the species has a long flowering period. Prefers a semi-shady position but will grow in full sun if given regular water during dry periods. Responds to side dressing with slow-release fertilizers. Propagate from seed or stem cuttings.

Commelina ensifolia D.L. Jones

Commersonia gaudichaudii

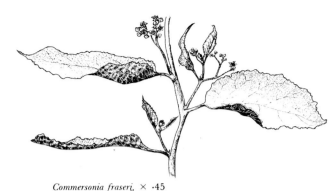

Commersonia fraseri × ·45

Commersonia gaudichaudii Gay
(after Charles Gaudichaud-Beaupré)
WA
0.2-0.6 m x 0.5-1 m   Sept-Dec
   Dwarf, compact **shrub**; young shoots bearing stellate hairs; **leaves** 2-2.5 cm long, asymmetrical, irregularly lobed, dark green above, whitish or brownish beneath; **cymes** few-flowered, axillary; **flowers** 0.6-0.8 cm across, white, with pointed sepals and blunt petals; **capsule** 0.6-0.8 cm across, globular, bearing a few small bristles.
   A hardy little shrub from sandplain country. Recently introduced into cultivation and has proved to be attractive and hardy. The growth habit is compact, sometimes almost prostrate, and flowering continues with the new growth. Prefers well drained, acid soil in a protected situation receiving morning sun or filtered sun. Propagate from stem cuttings.

Commersonia leichhardtii Benth.
(after Ludwig Leichhardt, explorer)
Qld
1-4 m x 1-2 m   Nov-Jan
   Medium to tall **shrub**; **branches** densely covered with woolly hairs; **leaves** 5-8 cm long, narrow to ovate-lanceolate, irregularly toothed, harsh-textured, covered with velvety hairs on both surfaces; **cymes** axillary, short, few-flowered; **flowers** about 1 cm across, reddish or purple.
   Found along stream banks in central Qld. Not known to be in cultivation but has all the attributes to make it a desirable shrub for tropical and subtropical areas. Foliage is attractive, and the flowers are small but colourful. Propagate from seed. Stem cuttings would probably also be successful.

Commersonia pulchella Turcz.
(beautiful)
WA
0.6-1.5 m x 0.5-1 m   Sept-March; also sporadic
   Dwarf to small, compact **shrub**; young shoots bearing stellate hairs; **leaves** 1.5-2.5 cm long, oblong or lanceolate, with irregularly lobed and toothed margins, dull green above, brownish beneath; **cymes** terminal

on branchlets; **flowers** about 1 cm across, white or pink, with blunt sepals and small, pointed petals.
   Found in heathland and sandplain country north of Perth. A compact, attractive shrub flowering continuously during the summer months. Prefers a well drained soil in a protected situation receiving morning sun or filtered sun, but also tolerates a sunny location. Will adapt to clay soils. Can sucker lightly, once established. Excellent container plant. Responds well to regular pruning. Propagate from stem cuttings.

Commersonia tatei F. Muell.
(after Prof. Ralph Tate)
SA
0.1-0.5 m x 1 m   Sept-Dec
   Dwarf prostrate to rounded **shrub**; **branches** covered with small stellate hairs; **leaves** 0.5-1 cm long, oblong or wedge-shaped, lobed, green above, white beneath; **flowers** about 1 cm across, white, borne in groups of 2-3 in axillary cymes, showy, long-lasting.
   A hardy little shrub from coastal scrubs. Recently introduced into cultivation and has proved to be attractive and well worth growing. Under good conditions the foliage is bright green and contrasts with the satiny white flowers. The growth habit is compact and the flowering continues with new growth. Prefers a well drained, acid soil in a protected situation receiving morning sun or filtered sun. Propagate from stem cuttings.

Compositae — see **Asteraceae**, Volume 2, page 244

Connaraceae R. Br.
   A family of dicotyledons containing about 350 species in 16 genera, widely distributed throughout the tropical regions of the world. Only 2 species belonging to the family occur in Australia, and they are restricted to the tropics where they are mainly found in rainforest. Australian genera: *Connarus* and *Rourea*.

CONNARUS L.
(derivation unknown)
Connaraceae
   Tall or shrubby **climbers**; **stems** twining; **leaves** simple or more usually pinnate; **flowers** small, borne in large, branched panicles; **fruit** a woody capsule, each containing a single large seed.
   Species about 100 widespread around the world with one endemic in Australia.

Connarus conchocarpus F. Muell.
(shell-fruited)
Qld   Scarlet Shell Vine
5-15 m   Jan-March
   Medium sized, woody **climber**; young stems covered with rusty hairs, roughened when mature; **leaves** pinnate, of 3-5 leaflets; **leaflets** 1.5-2.5 cm long, oblong, dull green; **panicles** large, arising in the upper leaf axils; **flowers** small, about 0.5 cm across, pale, fragrant; **capsule** 2-2.5 cm long, bright red.
   Confined to scrub along rainforest margins and in moist gullies. An unusual climber useful for tropical

areas. Not well known in cultivation but occasionally included in botanical collections because of its interesting relationship to legumes. Attractive when in flower, and the fruits are quite decorative. Propagate from stem cuttings which are slow to strike or from seed which must be scarified or treated with boiling water before sowing.

CONOSPERMUM Smith
(from the Greek *conos*, a cone; *sperma*, a seed; referring to the shape of the nut)
Proteaceae

Shrubs; **branches** few; **leaves** entire, alternate, sometimes crowded; **flowers** tubular, small, often woolly, in short dense spikes, either together in dense heads or solitary on axillary peduncles, or in panicles, or axillary on terminal peduncles, each flower within a broad, persistent bract, white, grey or bluish; **fruit** a small, cone-shaped nut, with hairs around base.

An endemic genus with approximately 50 species. The majority are confined to the South-West Botanical and South Eremaean Provinces of WA, but species are also distributed throughout temperate Qld, NSW, Vic, Tas and SA. They generally occur in well drained, sandy soils which can have gravel incorporated, and sometimes grow on the edges of swamps. Usually they grow amongst plants of approximately their own height, where they receive plenty of sunshine.

Conospermums have been eagerly sought by enthusiasts for garden cultivation, but not many species have been available, because of lack of propagating material. It is, however, most likely that this situation will change in the near future.

Most species have masses of well displayed flowers, and have excellent horticultural potential. The profuse flowers give the appearance of smoke, from which they

Commersonia tatei, × ·8

Conospermum amoenum   B.W. Crafter

have gained the common name of smokebush. The flowers are very useful for decorative purposes once cut, as they retain their features for a long period. It is hoped that suitable species can be cultivated for cut flowers, and their removal from natural stands prevented.

Cultivation

Their main requirement is for very well drained soils, with a preference for lighter types such as sand and gravels. Most *Conospermum* species are ideal for semi-arid areas. They do not appreciate hot, humid climatic conditions. One of the easiest ways to grow them is in containers such as large pots, with a very open medium. In general, they prefer partial or full sun. Most species are frost tolerant. Nearly all respond favourably to light pruning after flowering, which helps produce bushy growth. Light application of a slow-release fertilizer low in phosphorus can be beneficial in producing more vigorous growth if required.

Conospermums do not seem to suffer attacks from many pests. The main diseases are ones that tend to attack the root system, eg. cinnamon fungus, and can result in the death of plants. It is imperative that well drained conditions be provided if plants are to be grown in heavy soils. This can help to alleviate damage caused by the outbreak of such diseases.

65

Conospermum acerosum

Propagation has proved to be difficult from seed. One method of pre-sowing treatment is to heat the seed in a wire strainer over a flame until all the hairs have been removed. There is much research still to be done regarding the success of this and other methods of assisting germination. In most cases plants can be grown from cuttings, using firm cuttings taken from vigorous shoots.

Conospermum acerosum Lindl.
(needle-shaped)
WA
1-2 m x 1-2 m   Aug-Oct

A dwarf **shrub**; **branches** rigid; **leaves** to 5 cm long, terete, rigid, often pungent-pointed; **flowers** tubular, about 0.8 cm long, borne in axillary spikes shorter than leaves, or the upper spikes longer than leaves, not woolly, profuse, white; **bracts** broad, half as long as tube.

This species is generally found on sandplains north of Perth. It is little-known in cultivation, but requires very well drained soils, with partial or full sun. Tolerates light frosts and extended dry periods. Propagate from seed or cuttings.

Conospermum amoenum Meisner
(pleasant)
WA   Blue Smokebush
0.5-1 m x 0.5-1.5 m   Aug-Nov

A dwarf, spreading **shrub**; **branches** slender; **leaves** to 2 cm long, linear-terete, acute but not pungent, spreading, numerous but not crowded; **flowers** tubular, to 0.5 cm long, borne in axillary spikes along branches, crowded, blue; **bracts** broad, more than half as long as tube.

A species which has had limited cultivation. It occurs naturally in the south-west corner of WA, in sand-heath associations. Its requirements are for well drained, light to medium soils, with partial or full sun, and it will tolerate dappled shade in semi-arid areas. Pruning after flowering will produce dense growth. Suitable for gardens or containers. Propagate from seed or cuttings.

Conospermum bracteosum Meisner
(conspicuous bracts)
WA
0.5-1 m x 1-1.5 m   Sept-Nov

A dwarf **shrub**; **branches** covered in silky hairs, greyish; **leaves** to 2.5 cm long, lower leaves oval to orbicular, narrowed at base, 3-nerved, silky, upper leaves oval, bract-like, stem-clasping, overlapping, pointed or blunt, usually glabrous; **flowers** tubular, about 0.6 cm long, borne in spikes from upper axils, dense, silky hairs, white; **bracts** shorter than tube, hairy.

An ornamental species in foliage and flower. It grows in sandheath plant associations in the Stirling district, and appreciates very well drained soils, with partial or full sun. Withstands light frosts and limited dry periods. Little is known of other cultivation results, but it should be suitable for gardens or containers. Propagate from seed or cuttings.

Conospermum brownii

Conospermum brownii Meisner
(after Robert Brown, 18-19th-century English botanist)
WA   Blue-eyed Smokebush
0.5-1 m x 0.5-1.5 m   Sept-Nov

A dwarf **shrub**; **branches** erect; **leaves** to 7 cm long, oblong, pointed, rigid, 3-nerved, glabrous; **flowers** tubular, about 0.8 cm long, glabrous, borne in terminal panicles on leafless scapes to 30 cm long, buds blue, opening to white.

A low shrub that occurs in the wheatbelt area east of Perth. It grows in sandy or gravelly heathlands, and will require well drained soils, with partial or full sun. Tolerates light frost and extended dry periods. Suitable for gardens or containers. Propagate from seed or cuttings.

Conospermum burgessiorum L. Johnson & MacGillivray
(after Rev. C. Burgess and his son P. Burgess)
Qld, NSW
1.5-4 m x 1.5-3 m   Sept-Dec

Small to medium, upright to spreading **shrub**; **branches** many; **branchlets** initially hairy, becoming more or less glabrous; **leaves** 10-25 cm x 0.4-1 cm, linear to narrowly obovate, flat, slightly curved, ascending, acute apex; **flowers** tubular, to 1 cm long, cream to white, borne in clusters of spikes near ends of branchlets.

A recently named species that occurs on both sides of the Qld-NSW border, growing in granitic, sandy soils. Cultivation requirements should be for well drained soils, with dappled shade to full sun. Tolerant of light to medium frosts. Propagate from seed or cuttings.

Closely allied to *C. mitchellii*, which has leaves to 15 cm long, and flowers usually displayed above the foliage.

Conospermum caeruleum R. Br.
(dark blue)
WA
0.5-1.5 m x 1-2 m   Sept-Oct

A decumbent **shrub**; **rootstock** woody; **stems** decum-

66

bent or ascending; **leaves** to 15 cm long, oblong, nar-rowed each end, basal, with smaller ones near middle of stem, slightly veined; **flowers** tubular, to 0.8 cm long, only slightly hairy, borne on leafless, branched stems, in short oblong spike, deep blue; **bracts** broad, woolly, about as long as tube, white.

A species of the woodlands, common in sandy soils around Albany. It needs relatively well drained soils, with partial or full sun. Will tolerate dappled shade if in a warm location. Suitable for gardens or containers. Propagate from seed or cuttings.

Conospermum capitatum R. Br.
see under *C. petiolare* R. Br.

Conospermum crassinervium Meisner
(thick nerves)
WA   Summer Smokebush
0.5-1 m x 1-1.5 m   Oct-Jan
A dwarf **shrub**; **rootstock** woody; **leaves** to 30 cm or longer, radical or at ends of short branches, linear, acute, thick nerve-like margins, midrib promin-ent underneath, densely hairy; **flowers** tubular, about 0.5 cm long, woolly, borne in branched clusters at ends of stems to 60 cm long, white; **bracts** broad, about as long as flower, long silky hairs, white.

This species grows north of Perth in the open, sandy woodlands from Gin Gin to Eneabba. It is not common in cultivation, but should grow in very well drained soils with partial or full sun. Will probably tolerate dappled shade in warm climates. Suitable for gardens or containers. Excellent cut flower. Propagate from seed. There could be difficulty in obtaining suitable cutting material from plants.

Conospermum densiflorum Lindl.
(dense flowers)
WA
0.5-1 m x 0.5-1.5 m   Oct-Dec
A dwarf **shrub**; **stems** with long, spreading hairs; **leaves** to 5 cm long, thread-like, hairy, dense on lower part of branches; **flowers** tubular, about 1 cm long, slightly hairy, borne in a terminal spike or corymb, on leafless, erect stalks about 30 cm long, white.

An outstanding ornamental, not well known in culti-vation. In nature it occurs in sand, and therefore needs very well drained soils. It will grow best in partial or full sun. Suitable for gardens or containers. Propagate from seed or cuttings.

Conospermum distichum R. Br.
(arranged in 2 opposite rows)
WA
1-1.5 m x 1-2 m   Sept-Nov
A bushy small **shrub**; **branches** glabrous; **leaves** to 7.5 cm long, terete, incurved, crowded; **flowers** tubular, about 0.8 cm long, very woolly, borne in dense terminal clusters, white to blue.

A species found in deep sand of the Stirling Range and southern sandplains. It requires very well drained soils, with partial or full sun, and should withstand ex-tended dry periods.

It is not well known in cultivation, but has potential

Conospermum ellipticum, × ·8

for use in gardens or containers. Propagate from seed or cuttings.

C. floribundum Benth. is closely related to *C. distichum,* but differs in having shorter leaves which can often become silvery in appearance.

Conospermum dorrienii Domin
(after Captain A. A. Dorrien-Smith)
WA   Stirling Range Smokebush
0.5-1 m x 0.5-1 m   Oct-Nov
A dwarf, erect **shrub**; **leaves** 1-2 cm long, terete, in-curved, glabrous; **flowers** tubular, to 0.8 cm long, more or less hairless, borne in dense clusters on upper part of branchlets, blue.

This ornamental species is restricted to the sandy or rocky soils on the slopes of the Stirling Range. Its culti-vation requirements are for very well drained soils, with partial or full sun. Suitable for gardens or con-tainers. Propagate from seed or cuttings.

Conospermum ellipticum Smith
see under *C. taxifolium* Smith

Conospermum ephedroides Kippist ex Meisner
(similar to the genus *Ephedra*)
WA
0.3-1 m x 0.5-1.5 m   Aug-Oct
A dwarf **shrub**; **stems** erect, rush-like, slightly bran-ched, with silky hairs; **leaves** to 15 cm long, terete, rush-like, on lower part of stems only, reduced to scales on upper parts; **flowers** tubular, small, borne in short, sessile spikes scattered along upper part of stems, pale blue.

Usually found in sand or gravel, this interesting low species requires very well drained soils, with partial or full sun. For gardens or containers.

Propagate from seed. Because of the growth habit of this species, there could be difficulty in obtaining suit-able cutting material.

Conospermum ericifolium

Conospermum ericifolium, × ·8

Conospermum ericifolium Smith
see under *C. taxifolium* Smith

Conospermum flexuosum R. Br.
(flexuose)
WA   Tangled Smokebush
1-1.5 m x 1-2 m   Sept-Jan
A spreading, open, small **shrub**; **branches** flexuose, tangled; **leaves** 15-30 cm long, tapering to base, rigid, prominent margins; **flowers** tubular, small, glabrous, borne in small clusters at ends of branchlets, white to pale blue.

A species with interesting foliage. Although it sometimes occurs naturally in swampy soils, it is more common in the woodlands, and in cultivation needs well drained soils. It will grow well in dappled shade or partial sun, and also tolerates full sun. Propagate from seed or cuttings.

Conospermum floribundum Benth.
see under *C. distichum* R. Br.

Conospermum glumaceum Lindl.
(refers to the thin papery bracts enclosing the flowers)
WA   Hooded Smokebush
1-2 m x 1-2 m   Aug-Oct
A small **shrub**; **branches** glabrous; **leaves** to about 2-6 cm long, linear to linear-lanceolate, acute or pungent-pointed, with nerve-like margins; **flowers** tubular, very small, yellow-and-white, scented, concealed by straw coloured bracts 0.6-1.2 cm long, borne on simple or branched stalks of up to 35 cm long.

A most unusual species with prominent bracts. It is found in open jarrah forests or lateritic gravels. Not common in cultivation. Its requirements are for well drained soils, with dappled shade or partial sun and it will possibly tolerate full sun. For gardens or containers. Propagate from seed or cuttings.

Conospermum huegelii R. Br.
(after Karl van Hugel, 19th-century Austrian botanist)
WA   Slender Smokebush
0.3-1 m x 0.5-1.5 m   July-Oct
A dwarf **shrub**; **stems** glabrous; **leaves** 5-20 cm long, narrow, crowded on lower parts of stems, glabrous; **flowers** tubular, about 0.75 cm long, borne in terminal spike on a simple, erect, leafless stalk often above 30 cm long, blue.

This ornamental species inhabits the coastal plains north and south of Perth, and the Darling Scarp. It often occurs in moist but well drained situations. In cultivation, requirements are similar, with a preference for partial or full sun. Suitable for gardens or containers. Propagate from seed. Difficulty can be experienced in obtaining suitable cutting material from plants.

Conospermum incurvum Lindl.
(incurved)
WA   Plume Smokebush
1-2 m x 1-1.5 m   Sept-Oct
A small **shrub**; **branches** erect; **leaves** to 2.5 cm long, narrow, almost terete, spreading, incurved, glabrous, crowded; **flowers** tubular, about 0.6 cm long, lavender to white, borne in numerous spikes about 2.5 cm long, along a terminal stalk.

This species occurs in deep sand, both in woodlands and heath vegetation. It requires very well drained soils, and will grow in dappled shade, partial sun, and also tolerate full sun. Withstands moderate frosts and extended dry periods. For gardens or containers. Propagate from seed or cuttings.

Conospermum leianthum (Benth.) Diels
(smooth flowers)
WA
1-1.5 m x 1-1.5 m   July-Oct
A small **shrub**; **branches** erect; **leaves** to about

Conospermum incurvum, × ·55

Conospermum incurvum   B.W. Crafter

Conospermum mitchellii Meisner
(after Sir Thomas Mitchell, 19th-century explorer and botanist)
Vic, SA   Victorian Smokebush
1-3 m x 1-3 m   Oct-Dec
   Shrub; **branches** erect; **leaves** 5-15 cm x 0.3 cm, erect, stiff; **flowers** tubular, about 0.5 cm long, woolly, borne in dense, flat-topped clusters about 10-15 cm diameter beyond foliage, white to grey.
   This species occurs in both sandy woodland and heathland, and is endemic to the south-west of Vic and south-east of SA. In cultivation it requires well drained soils, although it appreciates moisture. It is extremely sensitive to poor drainage, and susceptible to attack by cinnamon fungus. It will grow in dappled shade or partial sun, as well as tolerating full sun. Withstands moderate frosts. Suitable for gardens. Propagate from seed or cuttings.
   Allied to *C. burgessiorum* which usually has flowerheads shorter than the foliage.

Conospermum nervosum Meisner
(prominent nerves)
WA
1-1.5 m x 1-1.5 m   June-Aug
   A small **shrub**; **branches** erect; **leaves** variable, upper ones usually under 3 cm long, elliptic, blunt or with a recurved point, 3 prominent nerves, lower ones much longer; **flowers** tubular, about 0.75 cm long, only slightly hairy, borne in small clusters in upper axils, pink or pale blue.
   An inhabitant of the northern sandplains. It requires very well drained soils, with partial or full sun. Withstands moderate frosts and extended dry periods. It is not common in cultivation, but should prove to be useful in low-rainfall gardens, or as a container plant. Propagate from seed or cuttings.

12 cm, slender, terete, crowded; **flowers** tubular, about 0.8 cm long, borne in panicles of small spikes, white.
   A species from the southern sandplains and heath vegetation. It requires very well drained soils, with partial or full sun. Not well known in cultivation, but has potential for gardens or containers. Propagate from seed or cuttings.

Conospermum longifolium Smith
(long leaves)
Qld, NSW
1-2 m x 1-2 m   Sept-Nov
   A small **shrub**; **branches** erect, densely hairy; **leaves** to 25 cm x 2 cm, flat, tapering to base; **flowers** tubular, about 1 cm long, usually hairy, borne in dense spikes on terminal or axillary stalks to 30 cm long, white.
   The typical form of species occurs mainly north of Port Jackson, and grows in heath, woodland, and sandstone vegetation communities. It is not widely cultivated, but will grow in well drained soils, with partial or full sun. Propagate from seed or cuttings.
   The ssp. *angustifolium* (Meisner) L. Johnson & MacGillivray occurs mainly south of Port Jackson, and has leaves about 0.4 cm wide. It hybridizes in nature with *C. taxifolium*.
   The ssp. *mediale* L. Johnson & MacGillivray has leaves to about 0.8 cm wide and occurs mainly in the Blue Mountains and south of Sydney.
   C. sphacelatum Hook. differs from *C. longifolium* in its narrow, hairy leaves, and the flower-spikes are usually less than 15 cm long.

Conospermum mitchellii, × ·5

69

Conospermum patens Schltdl.
(spreading)
Vic, SA　　　　　　　　　Slender Smokebush
0.5-1 m x 1 m　　　　　　　Aug-Nov

A dwarf **shrub**; **branches** erect; **leaves** to 2 cm long, narrow, spreading widely, hairy while young, otherwise glabrous, crowded; **flowers** tubular, about 0.75 cm long, usually borne in dense, small spikes above foliage, grey to blue.

This species usually occurs naturally in drier areas than those inhabited by *C. mitchellii*, but at some times is co-extensive. It is not widely cultivated. Requirements are for very well drained soils, with partial or full sun, and it will withstand moderate frost and extended dry periods. For gardens or containers. Propagate from seed or cuttings.

Conospermum petiolare R. Br.
(with petioles)
WA
0.5-1 m x 0.5-1.5 m　　　　　Sept-Dec

A dwarf **shrub**; **stems** woody; **leaves** 15-30 cm long, lanceolate, tapering to base, often hooked at tip, leathery; **flowers** to 1.3 cm long, tubular, with narrow lobes longer than tube, borne in dense, terminal flower-heads amongst leaves, white.

This species is evidently not in cultivation. It is found in the south-west corner, where it grows in sandy soils. It should do best in well drained soils, with partial or full sun. For gardens or containers. Propagate from seed. It may be difficult to obtain suitable cutting material.

A similar species is *C. capitatum* R. Br., with smaller heads of flower, and lobes about as long as the tube.

Conospermum polycephalum Meisner
(many heads)
WA
0.7-1.5 m x 1-1.5 m　　　　　Aug-Oct

A dwarf to small **shrub**; **branches** glabrous; **leaves** to about 15 cm long, on lower part of stems, terete, very narrow, glabrous; **flowers** tubular, about 0.75 cm long, only slightly hairy, more or less globular; **spikes** borne on branched stems often longer than 30 cm, blue.

In nature this species occurs in moist sandy soils in the Darling District. It is not common in cultivation, and will require well drained soils, in partial or full sun. For gardens or containers. Propagate from seed or cuttings.

Closely allied to *C. leianthum*, which has white, glabrous flowers.

Conospermum sericeum C. Gardner
(silky)
WA
0.3-1 m x 0.5-1.5 m　　　　　Nov-Jan

Dwarf **shrub**; **rootstock** woody; **basal leaves** silky; **stem-leaves** about 12 cm long, broad, erect, 1-nerved, glabrous, nerve-like margins, acute apex; **flowers** tubular, about 0.4 cm long, densely hairy lower lobes, borne on leafless, branched stem, pale blue, bracts narrow, not woolly.

An ornamental species from the Irwin District. It

Conospermum stoechadis　　　　　　　B.W. Crafter

requires very well drained, light to medium soil, with a preference for sand. Will grow in partial or full sun. Propagate from seed or cuttings.

It has affinities to *C. caeruleum*, but differs by the narrow, non-woolly bracts, and densely hairy lobes of the lower lip.

Conospermum sphacelatum Hook.
　　　　　see under *C. longifolium* Smith

Conospermum stoechadis Endl.
(refers to similarity to French Lavender)
WA　　　　　　Common Smokebush; Smoke Grass
1-2 m x 1-2 m　　　　　　　Sept-Nov

A small **shrub**; **branches** erect; **leaves** to about 15 cm long, terete, rigid, often with sharp tips, juvenile growth silky, otherwise glabrous; **flowers** tubular, about 0.5 cm long, densely woolly, borne on simple or branched woolly stalks in upper axils, white to grey.

A very widely spread species of the South-West Botanical Province, where it occurs amongst sand-heath and sandy woodland vegetation. In cultivation it needs well drained soils, with partial sun or preferably full sun. Withstands moderate frost and extended dry periods. Suitable for gardens or containers. Propagate from seed or cuttings.

Conospermum taxifolium Smith
(leaves similar to those of the genus *Taxus*)
Qld, NSW, Vic, Tas　　　　Variable Smokebush
0.5-2 m x 1-2 m　　　　　　Sept-Nov

A variable dwarf to small **shrub**; **branches** erect, slender; **leaves** 1-3 cm x 0.2-0.3 cm, lanceolate to linear, flat, crowded, can be slightly twisted, with silky hairs; **flowers** tubular, about 0.7 cm long, hairy, lobes shorter than tube, borne in dense clusters on single or branched stems, white to cream.

This species occurs in a wide range of soils. It prefers well drained soils, with partial sun but will also tolerate dappled shade or full sun. Not widely cultivated. Propagate from seed or cuttings.

C. taxifolium is known to hybridize readily with *C. ericifolium* and *C. longifolium*, in its natural habitat.

C. ellipticum Smith is similar to *C. taxifolium*, but has elliptical or broad-lanceolate foliage.

C. ericifolium Smith is also similar to *C. taxifolium*, but differs in having narrower leaves, about 0.1 cm wide.

Conospermum tenuifolium R. Br.
(slender leaves)
NSW

0.1-0.2 m x 1.5-2 m   Sept-Nov

Dwarf, spreading **shrub**; **branches** procumbent; **leaves** to 1.7 cm long, linear, almost terete, ascending, channelled above; **flowers** tubular, small, blue to lilac, borne on long peduncles, in clusters near ends of branchlets.

A species not widely grown in cultivation, but with potential because of its low, cascading growth habit. It occurs naturally in relatively well drained, sandy soils that are moist for most of the year. Suitable for situations with shade to partial sun, and will probably tolerate full sun. Frost tolerant. Should also do well as a container plant. Propagate from cuttings.

Conospermum teretifolium R. Br.
(terete leaves)
WA

1-1.5 m x 1-2 m   Nov-Dec; also sporadic

A small **shrub**; **branches** glabrous; **leaves** to about 30 cm long, terete, rigid, rush-like; **flowers** about 2 cm long, tubular, with long lobes, glabrous, borne in dense, terminal panicles, white to cream.

A species widespread, from the Stirling Range to Israelite Bay, along the coastal areas. It requires very well drained soils, with partial or full sun. Not common in cultivation, but suitable for gardens or containers. Propagate from seed or cuttings.

Conospermum triplinervium R. Br.
(3-nerved)
WA   Tree Smokebush

2-4 m x 2-3 m   Sept-Oct

A medium to tall **shrub**; **branches** erect, with

Conospermum teretifolium   B.W. Crafter

Conospermum triplinervium   A. Gray

minute hairs when young; **leaves** to about 10 cm long, narrow to broad, glabrous or with silky hairs; **flowers** tubular, about 0.6 cm long, densely woolly, borne in spikes to 10 cm long, on stalks to about 30 cm long, white to grey.

This is one of the most commonly cultivated western species. It usually occurs in sandplain country, and requires very well drained soils, with partial or full sun. Withstands moderate frosts and extended dry periods. For gardens or large containers. Ideally suited for coastal regions with sandy soils. Propagate from seed or cuttings.

CONOSTEPHIUM Benth.
(from the Greek *conos*, cone; *stephanos*, crown; alluding to the concealed corolla)
Epacridaceae

Dwarf to small **shrubs**; **leaves** small, alternate; **flowers** tubular, small, solitary, usually pendulous, tube partially or fully enclosed by calyx, conical near tip; **fruit** a drupe.

An endemic genus of 6-7 species, mainly represented in the South-West Botanical Province of WA. One species is found on Kangaroo Island, SA. They usually occur on sandy soils, amongst other shrubs. The genus is closely related to *Astroloma*, but readily distinguished by the corolla tube. Their low growth habit and decorative flowers, which are produced over long periods, give the various species horticultural potential for gardens and also containers, but little is known to date of their reaction to cultivation. As with many other epacrids, propagation can be a problem, but it seems that by using firm new growth as cutting material, results could be successful.

Conostephium drummondii (Stschegl.) C. Gardner
(after James Drummond, 1st Government Botanist of WA)
WA

0.5-1 m x 0.5-1 m   Sept-Dec

A dwarf **shrub**; **branches** erect; **leaves** to about 1 cm

Conostephium halmaturinum

Conostephium pendulum, × ·75

long, broad to narrow-lanceolate, flat or concave, ending in acute point, prominent nerves; **flowers** tubular, about 1 cm long, cylindrical to near tip, then tapers, red-purple, surrounded by numerous bracts, borne on small pendulous stalks.

This species occurs naturally in the South-West Botanical Province, growing in well drained soils, amongst other plants. It is evidently not well known in cultivation, but should require similar conditions to those of its natural habitat. Propagate from cuttings.

The species from Kangaroo Island, SA, *C. halmaturinum* J. Black, is closely allied, but has smaller flowers.

Conostephium halmaturinum J. Black
    see under *C. drummondii* (Stschegl.) C. Gardner

Conostephium minus Lindl.
(smaller)
WA   Pink-tipped Pearl Flower
0.3-0.5 m x 0.3-0.5 m   July-Sept
An erect, dwarf **shrub**; **leaves** to 2 cm long, linear, small pungent point, margins much revolute; **flowers** tubular, about 1 cm long, inflated in middle, white tipped pink, borne on short stalks, initially upright becoming pendulous.

Evidently not in cultivation, this species will require well drained soils. It prefers dappled shade or partial sun, but could adapt to full sun. An ornamental species with horticultural potential. Propagate from cuttings.

Conostephium pendulum Benth.
(pendulous)
WA   Pearl Flower
0.2-0.7 m x 0.5-1 m   May-Nov
A dwarf **shrub**; **branches** erect, glabrous; **leaves** 2-3 cm long, linear-oblong, acute or blunt, with a short, pungent point, convex or with recurved margins, glabrous; **flowers** tubular, 1.5-2 cm long, inflated in

middle, surrounded by numerous bracts, borne on pendulous stalks; **calyx** white; **corolla tip** purple.

Mainly confined to coastal regions north and south of Perth, this is the most common *Conostephium* species in these areas. In cultivation it will require very well drained soils, with dappled shade or partial sun. Will make an excellent container plant. Propagate from cuttings.

Conostephium preissii Sonder
(after Ludwig Preiss, 19th-century German botanist)
WA
1-2 m x 1-1.5 m   Dec-May
Small **shrub**; **branches** erect; **leaves** to about 2 cm long, oblong, blunt or with small pungent point, flat or margins slightly recurved, green above, glaucous below; **flowers** tubular, about 1.5 cm long, cylindrical except for tapering tip, red, surrounded by numerous bracts, borne on pendulous stalks to about 0.6 cm long; **calyx** white.

Widespread but not common over a large area of the south-west of WA, this ornamental species requires well drained soils, with dappled shade or partial sun. Ideal as a container plant. Propagate from cuttings.

Conostephium roei Benth.
(after J. S. Roe, 1st Surveyor-General of WA)
WA
0.5-1.5 m x 0.5-1 m   June-Aug
Dwarf to small **shrub**; **branches** erect; **leaves** to about 1 cm, ovate to wedge-shaped, blunt, with short, pungent points, convex, with recurved margins, short ones usually flat; **flowers** tubular, about 1 cm long, inflated in the middle, white, surrounded by numerous bracts, sessile or borne on short peduncles, pendulous.

This species is not well known, but certainly seems worthy of cultivation. Requirements would be for well drained soils, with dappled shade or partial sun. Propagate from cuttings.

CONOSTYLIS R. Br.
(from the Greek *conos*, cone; *stylos*, column; refers to the conical base of the style)
Haemodoraceae
Clump-forming perennial **herbs**, usually low; **clumps** can be solitary or are connected by stolons or rhizomes to other clumps; **stems** short and undivided or long and sometimes branched, or growing near or above ground level, producing roots or bundles of leaves from nodes; **leaves** flat or terete, linear, straight or falcate, tapering to a point, can be glabrous or have hairs on margins, or hairs on surface giving a whitish appearance, longitudinal nerves prominent; **scape** arises from stem-apex surrounded by sheathing leaves, simple or branched; **inflorescence** terminal head or cyme; **flowers** tubular, campanulate, 6 lobes, on stalks, outside covered with woolly indumentum, yellow, reddish yellow, brick red, purplish, white or cream; **stamens** 6; **fruit** a 3-celled capsule which opens and disperses the small dark coloured seeds.

The genus *Conostylis* is endemic to the South-West Botanical Province of WA. At present there are 29 named members of this most interesting group of

Conostylis aculeata ssp. *aculeata,* × ·45

plants which is undergoing botanical revision. They are very floriferous, have eye-catching foliage, and all have potential for landscaping and garden planting. They are ideal for planting near bases of tree trunks or nestled in next to large boulders, or in informal groups that will tend to give a natural appearance to any planted area.

Many species are in cultivation and are proving to be adaptable to a wide range of soil and climatic conditions. They generally prefer fairly well drained, light to medium soils. Some species will withstand extended wet periods and others are tolerant of frost. Conostylis prefer a sunny situation but many will also grow well in light shade. Light applications of a slow-release fertilizer can help to promote new growth if plants are not growing well. They do not seem to be affected by pests or diseases. Some blackening of foliage can occur with some species, as with *Anigozanthos* spp. This is usually caused by poor drainage or lack of air movement (see also *Anigozanthos*, Volume 2, page 198).

Hybrids between *C. aculeata* and *C. candicans*, and also *C. aculeata* and *C. pauciflora* occur south of Perth. Some of these hybrids have possibly been introduced into cultivation by nurseries.

Propagation

These plants are not readily available from nurseries, due mainly to the difficulty encountered in propagation. The two most commonly used methods are from seed and by division. Usually seed does not develop fully in cultivated plants and it is difficult to obtain viable seed from the wild. In some species, seed, when available, germinates well, eg. *C. aculeata, C. juncea* and *C. setosa*. There is no need for treatment of seed before sowing. The seed is sown on a well drained

medium, and covered by its own depth. It should be kept moist until germination, which is usually between 3 and 6 weeks.

The best time to divide plants seems to be in June or July. At this time they are beginning to form new roots underground, and also in some cases from aerial nodes. It is recommended that the newly potted plants be placed in a warm, protected area for 4-6 weeks (eg. polythene house or glasshouse). This will accelerate new root growth, and produce a vigorous plant that will be ready for planting out within 3-6 months.

Conostylis aculeata R. Br.
(furnished with prickles; refers to the leaves)
WA
0.3-0.4 m x 0.5-1 m   Aug-Feb
Clump-forming **perennial**; **stems** usually short; **leaves** 15-35 cm x up to 0.5 cm, with rigid and more or less erect spine 0.2-0.3 cm long on well defined margins, prominent longitudinal nerves; **scape** unbranched; **flowers** about 1 cm long, borne in loose raceme or cyme shorter than the leaves, yellow.

This species has proved to be very adaptable in cultivation. It will grow in a wide range of soil and climatic conditions, with a preference for well drained soils and partial or full sun. It seems to be tolerant of most frosts, and also withstands extended dry periods. It is one of the less showy species, but is still very desirable because of its hardiness and growth habit. Propagate from seed or by division.

It is extremely variable, and there are a further 5 subspecies, all of which are worthy of cultivation:
C. aculeata ssp. **bracteata** (Lindl.) J. Green

A clump-forming **perennial**; **leaves** 30-40 cm x about 5 cm, prominent fibrous margins which have only a few spines near the apex; **scape** is about the same length as leaves; **flowers** borne in a dense, globular head, often a small scape and head arising from upper bract on main scape.
C. aculeata ssp. **breviflora** S. Hopper

A clump-forming **perennial**; **stolons** to about 20 cm long; **leaves** 10-20 cm x up to 0.3 cm; **marginal spines** erect, glabrous, along length of leaf; **flowers** yellow, 0.6-0.9 cm long, borne in dense, globular heads. This ssp. is one of the parents in natural hybrids, with *C. aculeata* ssp. *aculeata* and *C. candicans*.
C. aculeata ssp. **bromelioides** J. Green

Very similar to ssp. *aculeata*, but usually has shorter leaves, and the leaf-margins are very prominent, with rigid, spreading or deflexed spines.
C. aculeata ssp. **preissii** (Endl.) J. Green

This differs from ssp. *aculeata* in having more-branched stems. **Leaves** are about 15-25 cm x up to 3-4 cm; **marginal spines** are less rigid than in other subspecies; **scape** is shorter than the leaves, and has 1-2 bracts; **inflorescence** is a loose head of flowers.
C. aculeata ssp. **rhipidion** J. Green

A hard-foliaged **perennial** of about 0.6 m x 0.6 m; **leaves** up to 20 cm x about 0.3 cm, very spiny, arranged with fan-like appearance on branching stems; **scape** about same length or longer than leaves; **flowers** borne in dense head, with 1 or 2 leafy brown bracts about 2 cm long below head.

Conostylis aculeata

Conostylis aculeata ssp. *rhipidion*, × ·45

Conostylis aculeata ssp. **robusta** J. Green =
C. robusta Diels

Conostylis androstemma F. Muell.
(refers to the crown-like arrangement of the anthers)
WA
0.1-0.3 m x 0.3-0.5 m   May-Sept
A small, tufted **perennial**; **stem** short, unbranched; **leaves** 10-30 cm x 0.1 cm, terete, glabrous, prominent longitudinal veins; **flowers** tubular, 3-4 cm long, hairy outside, glabrous inside, solitary, on very short stalks; **lobes** narrow, spreading, pale yellow, often translucent.

Evidently confined in nature to near-coastal areas north of Perth, where it grows in sand. It is not well known in cultivation at present. Appreciates very well drained soils, with partial or full sun. Ideal for gardens or containers.

The var. *argentea* differs in having flat, silvery-hairy leaves. It occurs south-east of Perth. It has been more commonly grown, but has not been widely tested to

Conostylis aff. *aurea*   W.R. Elliot

date. Cultivation requirements are similar to the above.

Propagate both varieties from seed or by division. This species and the varieties are most likely to be included in a different genus on finalization of the botanical revision.

Conostylis aurea Lindl.
(gold)
WA
0.3 m x 0.3-0.5 m   July-Nov
Small, tufted **perennial**; **stem** short, unbranched; **leaves** to 30 cm x 0.5 cm, striate, often sticky, small hairs on margins; **flowers** tubular, to 2 cm long, densely woolly outside, glabrous inside, gold to pale yellow, sometimes with purplish red, borne in clusters at end of scape.

This ornamental species occurs around Perth and north to the Murchison River, where it grows on laterite or sand. It has adapted well to cultivation, and does best in well drained soils, with partial or full sun. Withstands extended moist or dry periods, and light frosts. Suitable for gardens or containers. Propagate from seed or by division.

Conostylis bealiana, × ·5

Conostylis candicans   B.W. Crafter

Conostylis aff. aurea
WA
0.25 m x 0.3 m   April-Sept

Dwarf, tufting **perennial**; **stem** short; **leaves** to 25 cm x 0.4 cm, curved, faintly striate, covered in dense grey tomentum, longer hairs scattered along margins; **scape** to 30 cm long, usually woolly, with 2 bracts; **flowers** tubular, to 1 cm x 1.2 cm, pale yellow, sessile, dense long hairs outside, glabrous inside, spreading acuminate lobes, borne in dense clusters at end of scape.

Recently introduced to cultivation, this species is ornamental both in foliage and flower. It occurs naturally in lateritic sands near Eneabba. Requires well drained soils, with partial or full sun. Tolerates light to medium frosts. Responds well to applications of slow-release fertilizers. An excellent container plant. Propagate from seed or by division.

Conostylis bealiana F. Muell.
(after A. Beal, 19th-century fern grower)
WA
0.2 m x 0.3 m   May-Sept

Small, tufted **perennial**; **stem** shortly branched; **leaves** to 15 cm x 0.2 cm, longitudinally nerved, glabrous except for hairy margins; **flowers** tubular, to 3.5 cm long, hairy outside, sparsely hairy inside, yellow to orange, rarely greenish purple or dark red, solitary on short stalks amongst leaves.

Although flowers can be slightly hidden by the foliage, this is still an outstanding species. It occurs naturally in a wide range of soils, from sand to laterite gravel. In cultivation it grows best in well drained soils, with dappled shade or partial sun. Withstands limited wet or dry periods and tolerates light to medium frosts. Excellent for gardens, especially rockeries or con-

tainers. Responds well to moderate applications of fertilizer. Propagate from seed or by division. This species can be difficult to divide successfully. The recommended time is June to August when new roots are beginning to form from the base of the stems.

Conostylis breviscapa R. Br.
(short scape)
WA
0.2-0.3 m x 0.3-0.5 m   Aug-March

Dwarf, tufting **perennial**; **stems** very short, somewhat branched; **leaves** to 30 cm x 0.3 cm, flat, striate, glabrous except for woolly margins; **scape** 2-5 cm long; **flowers** about 1 cm long, densely hairy outside, loose hairs inside, divided to ovary, lobes spreading, about 0.5 cm long, pale yellow.

This conostylis is easily identified by its deeply divided corolla lobes. It occurs along the south coast, near Esperance and Hopetoun, in sandy soils. It is not well known in cultivation, but should do best in very well drained soils, with dappled shade or partial sun. Suitable for gardens or containers. Propagate from seed or by division.

Conostylis candicans Endl.
(shining white)
WA   Grey Cottonheads
0.3-0.5 m x 0.5 m   July-Feb

Tufting **perennial**; **stems** sometimes bearing off-shoots; **leaves** 15-50 cm x 0.3-1 cm, entirely covered in dense greyish white hairs, sometimes glabrous when aged; **scape** to 75 cm tall, with 1 or 2 leaf-like bracts; **flowers** tubular, about 1.2 cm long, yellow, densely hairy outside, sparsely hairy inside, borne in dense heads on branched rhachis.

This ornamental species is becoming increasingly well known in cultivation. It will grow in most well drained soils, with partial or full sun, and tolerates light to medium frosts and extended dry periods. Excellent for gardens or containers. Propagate from seed which germinates readily or by division.

This species has a wide distribution from Shark Bay, north of Carnarvon, to south of Perth. In 1699, Dampier collected specimens of this species from near Shark Bay and this is therefore one of the earliest records of an Australian plant collected from its natural habitat.

Natural hybrids of this species and *C. aculeata* have possibly been introduced into cultivation.

Conostylis caricina Lindl.
(rush-like)
WA
0.2-0.3 m x 0.3-0.5 m   Aug-Oct

Dwarf, tufting **perennial**; **stems** very short; **leaves** 10-30 cm x about 0.3 cm, striate, glabrous except for prominent fibrous margins, rarely hairy on leaf surface; **scape** 4-10 cm long; **flowers** about 1.5 cm long, densely hairy outside, cream and yellow, lobes narrow, producing a claw-like appearance, borne on short stalks in a loose cluster of 6-8 flowers in 2-3 rows.

This species is mainly concentrated in a small area of the Darling Ranges, north-east of Perth, where it grows in lateritic, sandy soils. It is not well known in cultivation, but should adapt to a wide range of well drained soils, with partial or full sun. Suitable for gardens and containers. Propagate from seed or by division.

Conostylis crassinervia J. Green
(thick nerves)
WA
0.1-0.15 m x 0.2-0.3 m   April-Dec
Dwarf, tufting **perennial**; **stem** short; **leaves** to 15 cm x 0.3 cm, prominent fibrous margins, glabrous or with a few soft hairs; **scape** about as long as leaves, or shorter, bearing a long, sheath-like bract near base, and two shorter bracts beneath flower-heads; **flowers** 1-1.2 cm long, densely hairy outside, creamy yellow or reddish, borne in a loose head.

Known only from a few areas in the Hill River district, north of Perth, this ornamental species adapts well to cultivation in well drained soils, with partial or full sun.

An undescribed variety from Mogumber to Dongarra differs from the type form in having the scapes longer than the foliage, and the perianth is ribbed. Flowers age to a reddish brown in colour. It is an excellent container plant. Can flower for some months. Propagate from seed or by division.

Closely allied to *C. setigera*, which has long hairs on leaf-margins.

Conostylis deplexa J. Green
(embracing or clasping)
WA
0.15-0.25 m x 0.1-0.2 m   Sept-Oct
Dwarf, tufting, perennial **herb**; **stems** short; **leaves** 15-20 cm x up to 0.3 cm, flat, longitudinally veined, glabrous except for enclasping bristles on margins; **scape** 2-5 cm long, densely hairy, with 2-3 cm long bract near middle, bearing 10-20 flowers in a terminal cluster to 3.5 cm across, with 1 or 2 brown bracts; **flowers** to 1.5 cm long, yellow, hairy interior and exterior.

A recently described species from the Fitzgerald River area, where it grows on sand or sandy loam over-lying laterite, amongst heath, shrubs and trees of various sizes. Some locations are wet for extended periods during the winter. Should prove adaptable in cultivation, growing in a wide range of relatively well drained soils. Suited to a situation that receives dappled shade to partial sun, and it may also tolerate full sun. Hardy to most frosts. Propagate from seed or by division.

C. deplexa is easily distinguished from other species, as it is the only one with enclasping leaf-bristles.

Conostylis dielsii W. Fitzg.
(after F. L. E. Diels, 19-20th-century German botanist)
WA
0.1-0.2 m x 0.2-0.3 m   Aug-Sept
Dwarf, tufting **perennial**; **stems** shortly rhizomatous; **leaves** to 10 cm x 0.1 cm, flat, covered with soft, white, woolly hairs; **scape** about same length as leaves; **flowers** about 1 cm long, creamy yellow, densely hairy on outside, glabrous or slightly hairy inside, lobes spreading, borne in a dense cyme or cluster.

In nature this outstanding species occurs south-east of Geraldton, where it grows in deep sandy soils, and withstands light to medium frosts as well as extended dry periods. It is not well known in cultivation, but should grow in most well drained soils, with partial or full sun. For gardens or containers. Propagate from seed or by division.

Conostylis festucacea Endl.
(similar to the genus *Festuca*)
WA
0.3-0.5 m x 0.5-1 m   Sept-Feb
Tufting **perennial**; **stems** often branching; **leaves** to about 40 cm x 0.1-0.2 cm, terete, glabrous; **scape** much shorter than leaves; **flowers** about 1.2 cm long, yellow, densely hairy outside, glabrous inside, lobes about 0.5 cm long, borne in loose cluster.

This species occurs in sandy soils north of Perth. It should grow in well drained soils, with partial or full sun, and tolerate light frosts and extended dry periods. Its spreading habit of growth will make it an ideal addition to the garden in association with rocks or logs, or suitable also as a container plant. Propagate from seed or by division. *C. filifolia* F. Muell. is synonymous.

Conostylis juncea, × ·5

Conostylis juncea Endl.
(rush-like)
WA
0.2-0.5 m x 0.3-0.6 m   June-Dec

Tufting **perennial**; **stems** unbranched; **leaves** to 50 cm x 0.4 cm, terete or flat, prominent longitudinal nerves, glabrous or sparsely hairy; **scape** 5-15 cm long, woolly; **flowers** 1.5-2 cm long, bright yellow to yellow-green, long rigid hairs outside, glabrous inside, lobes equal, about 1 cm long, borne in a few-flowered cluster.

A most ornamental species, from the coastal plains north and south of Perth. It grows best in well drained soils, with partial or full sun. Also tolerates dappled shade. Ideal for gardens or containers. Propagate from seed or by division.

It has been known as *C. involucrata*, and plants have also been available under the incorrect label of *C. setigera*.

Conostylis misera Endl.
(sickly; starved)
WA
0.1-0.15 m x 0.2-0.3 m   Nov-Jan

Dwarf, tufting **perennial**; **stems** short; **leaves** 5-10 cm x 0.2-0.5 cm, often falcate, not very prominently striate, margins thin, with small, distinct rigid hairs; green; **scape** very short, with 2-3 pointed brown bracts; **flowers** 1.5 cm long, cream, densely hairy on outside, lobes about 0.8 cm long, borne at base of leaves, often partly hidden.

This species has proved adaptable in cultivation. It grows best in well drained soils, with dappled shade or partial sun, and appreciates moisture. Although flowers tend to be hidden, it is a decorative species suitable for gardens or containers. In nature it occurs in the sandy soils of the Stirling district. Propagate from seed or by division.

Conostylis neocymosa S. Hopper
(new; cymes; refers to connection with another species)
WA
0.1-0.3 m x 0.2-0.4 m   June-Nov

Dwarf, tufting, perennial **herb**; **stems** if present are short, can have slender adventitious roots; **leaves** to 10-30 cm x up to 0.5 cm, green except for yellow-green at base, many short spines along margins; **scape** to 5-15 cm tall, once or twice-divided; **flowers** tubular, with spreading lobes, to 1.5 cm long, yellow, short tomentum on exterior, glabrous within, borne in a loose cyme of up to 10 flowers.

A recently described ornamental species from sand-plains north to west of Perth. Evidently not well known in cultivation, but certainly shows horticultural potential. Probably best suited to well drained, light to medium soils, with partial to full sun. Suited to cultivation in containers. Tolerant of light frost. In nature seed appears to germinate readily, so propagation from seed should be possible. Division of clumps should also be successful, as adventitious roots are often present.

Conostylis pauciflora S. Hopper
(few-flowered)
WA
0.2-0.3 m x 0.5 m   Aug-Jan

Clump-forming **perennial**; **stolons** to about 15 cm

long; **leaves** 10-25 cm x 0.1-0.2 cm, green, flat, sword-shaped; **marginal spines** about 0.1 cm long, flexible, hairy, along length of leaf; **scape** simple, as long or longer than leaves; **flowers** 0.8-1.5 cm long, yellow, borne in terminal head, densely woolly outside, slightly hairy within.

A newly named species that occurs near the coast south of Perth, where it grows on sandy slopes. It is not known whether it is in cultivation at present. Requirements will be similar to that of *C. aculeata*. Propagate from seed or by division.

Conostylis petrophiloides F. Muell. ex Benth.
(resembling the genus *Petrophile*)
WA
0.1-0.2 m x 0.3-0.5 m   July-Dec

Dwarf, tufting **perennial**; **stems** short, unbranched; **leaves** 15-25 cm x 0.1-0.4 cm, flat, prominently nerved, glabrous except for small marginal hairs; **scape** to about 20 cm long; **flowers** about 1.5 cm long, whitish to deep cream, loose woolly hairs outside, sparingly hairy inside, lobes about 0.7 cm long, equal, borne in a head 2-3 cm diameter.

An ornamental species that grows in a wide range of soils, occurring in coastal areas near Hopetoun and also inland. It should prove to be adaptable in cultivation, growing best in well drained soils, with partial

Conostylis prolifera, × ·5

or full sun. Suitable also for containers. Propagate from seed or by division.

Conostylis phathyrantha Diels
(sun flower)
WA

0.2-0.3 m x 0.1-0.2 m   Aug-Oct

Tufting, perennial **herb**; **stems** short; **leaves** to 25 cm x up to 0.5 cm, falcate, in 2 rows, faintly nerved, shiny black at base, glabrous except for faintly hairy margins; **scape** short; **flowers** tubular, 1-1.5 cm long, yellow, hairy exterior, in a loose cluster.

This species is not well known. It occurs in sandy soils, inland from Israelite Bay to Esperance. Has potential for cultivation. Should do well in light to medium soils with good drainage, in an aspect that has dappled shade to partial sun. Hardy to light frosts. Worthy of trial as a container plant. Propagate from seed or by division.

Conostylis prolifera Benth.
(bearing off-shoots)
WA

0.1-0.3 m x 0.3-1 m   Aug-Feb

Tufting **perennial**, often forming an extensive mat; **stems** branching, stoloniferous; **leaves** 1-10 cm x 0.1-0.2 cm, flat, limp, green or sometimes hairy, margins sometimes bearing minute erect bristles; **scape** undivided, as long as, or much longer than leaves, with 1 leaf-like bract; **flowers** to 1.2 cm long, cream to yellow, densely hairy outside, glabrous inside, lobes about 0.5 cm long, borne in a dense head or loose raceme.

This species is widely cultivated. It prefers moist soils that have good to excellent drainage. Situations with dappled shade or partial sun are preferred to full sun. Tolerates light to medium frosts. It is well suited to both garden and pot culture. Propagation is from seed or readily by division when stolons are producing new roots.

In 1960 this species was included under *C. stylidioides*, but it is once again separated. *C. stylidioides* differs in having short, rigid, grey, hairy leaves and golden yellow flowers.

Conostylis pusilla Endl.
(very small)
WA

0.05-0.1 m x 0.1-0.3 m   Aug-Dec

Dwarf, compact, tufting, perennial **herb**; **stems** very short; **leaves** to 0.7 cm x 0.1 cm, glabrous except for prominent white bristles on margins, erect to slightly spreading; **scape** about as long as leaves, with 1-2 leaf-like bracts; **flowers** about 0.5 cm long, cream to yellow, in small clusters, usually ageing to reddish, hairy, profuse.

Plants of this species are the smallest in the genus, and form tight, compact tufts. They are usually found in sandy soils, and occur over a wide range of the south-west. The species is sought after for cultivation, but is not readily available due to difficulties in propagation. Has proved adaptable, with a preference for well drained sands or loams. Likes dappled shade or partial sun. Tolerant of light to medium frosts, and withstands limited dry periods. Propagate from seed or by division. *C. psyllium* Endl. is synonymous.

Conostylis robusta flower-head   W.R. Elliot

This is considered a separate species from *C. setigera*, to which it is very closely related. *C. setigera* has longer leaves and a lax growth habit.

Conostylis robusta Diels
(robust)
WA

0.4-0.5 m x 0.5-1 m   Aug-Feb

Clump-forming **perennial**; **leaves** to 45 cm x 1 cm, usually green, sometimes slightly glaucous, margins with small erect spines; **flowers** tubular, golden yellow, in a dense terminal head of 5 cm diameter or larger, borne on stems longer than leaves.

This ornamental species, which occurs from south of Geraldton to north of Kalbarri, is not common in cultivation. It prefers well drained soils, with partial or full sun, and will possibly tolerate a shaded location in areas that have frequent high temperatures. Withstands light to medium frosts. The flower-heads have potential as cut-flower material as they are borne on stems to about 60 cm long. Propagate from seed or by division. Previously known as *C. aculeata* ssp. *robusta*.

Conostylis seorsiflora F. Muell.
(single-flowered)
WA

prostrate x 0.3-0.5 m   Sept-Dec

Mat-like **perennial**; **branches** stoloniferous; **leaves** to about 15 cm x 0.3 cm, flat, sometimes falcate, prominent nerves, can be hairy, but usually glabrous when mature; **scape** very short; **flowers** about 1.3 cm long, inflated, yellow, usually solitary, densely hairy outside, loose hairs inside, lobes about 0.7 cm long; **perianth** is green in the fruit.

Fairly well known in cultivation, this matting species makes an ornamental garden plant. It grows in a wide range of reasonably well drained soils, with dappled shade or partial sun, and will tolerate full sun if growing in moist soils. Suitable also for containers. In nature this species occurs along and near the southern coast of WA, where it grows in sandy soils that are moist for extended periods. Propagate by division.

Conostylis serrulata R. Br.
(finely serrate)
WA
0.2-0.5 m x 0.3-0.5 m   Aug-Feb

Tufting **perennial**; **stem** is shortly branched rhizome; **leaves** 20-50 cm x 0.2-0.7 cm, prominent longitudinal nerves, glabrous except for minutely toothed, pale margins, slightly glaucous; **scape** short, with 2 bracts about 2 cm long; **flowers** about 1.5 cm long, cream or dull yellow, dense mat of hairs outside, glabrous inside, lobes equal, divided to ovary, about 0.7 cm long, borne in compact cymes, near the base of leaves or up to 10 cm tall.

Although not widely planted at present, this species should prove to be most adaptable. It will grow best in well drained, light to medium soils, with dappled shade or partial sun, but will also tolerate permanently moist, heavier clay-loam soils. It occurs over a wide

Conostylis setosa   W.R. Elliot

area of the south-western coastal regions of WA. Propagate from seed or by division.

Conostylis setigera R. Br.
(many bristles)
WA
0.1-0.3 m x 0.15-0.5 m   Aug-Dec

Dwarf, tufting **perennial**; **stems** very short; **leaves** 7-30 cm x about 0.2 cm, prominent longitudinal nerves, glabrous except for long white or rarely black bristles on margins; **scape** shorter or larger than leaves, with 1-2 leaf-like bracts; **flowers** about 1 cm long, cream to yellow, often ageing to reddish, dense woolly hairs outside, slightly hairy inside, lobes about 0.5 cm long, spreading, borne in profuse clusters 2-3 cm across.

A variable species from the south-west corner, where it is probably the most common and widespread of the genus. It usually occurs in sandy soils. It is most ornamental, with its bristly foliage and well displayed flowers. Will grow in most well drained soils, with dappled shade, partial or full sun. Withstands light to medium frosts and extended dry periods. Ideal for gardens or containers. Propagate from seed or by division.

Conostylis setosa Lindl.
(abounding in bristles)
WA
0.2-0.3 m x 0.1-0.3 m   Sept-Dec

Tufting **perennial**; **stem** short; **leaves** 10-30 cm x about 0.3 cm, flat, often twisted, glabrous except for 2 or more ranks of long, spreading, white bristly hairs on margins; **scape** 20-30 cm long, with 1-2 leaf-like bracts; **flowers** about 2 cm long, purplish cream to creamy white, dense long silky hairs outside, slightly woolly inside, lobes about 1 cm long, slightly spreading, borne in dense heads about 5-6 cm diameter.

An ornamental species commonly found in forested areas of the Darling District around Perth. It has adapted well to cultivation, growing in most well drained soils. It prefers dappled shade, but will also grow in partial or full sun. Ideal for gardens or containers.

A desirable feature of this species is the variable flower colour. Propagate from seed or by division.

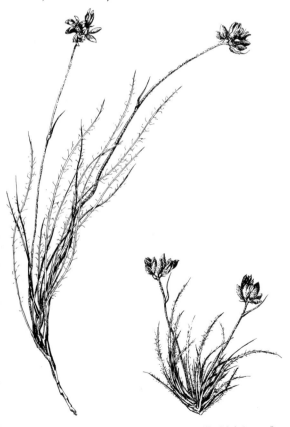

Conostylis setigera ssp. *setigera* (left), *C. pusilla* (right) × ·5

79

Conostylis stylidioides

Conostylis teretifolia

W.R. Elliot

Conostylis stylidioides F. Muell.
(similar to the genus *Stylidium*)
WA   Mat Cottonheads
0.1-0.3 m x 0.3-1 m   May-Feb

Tufting **perennial**; **stems** branching, stoloniferous; **leaves** 1-5 cm x 0.1 cm, flat to nearly terete, rigid, prominent longitudinal nerves, hairy especially when young, can become less hairy with age, margins usually with minute bristles; **scape** undivided, as long as or much longer than leaves, with 1 leaf-like bract; **flowers** to 1.3 cm long, golden yellow, densely hairy outside, glabrous and golden yellow inside, lobes about 0.5 cm long, slightly spreading, borne in profuse small clusters.

In nature this species frequents the *Acacia*-mallee scrub from Geraldton, north to the Shark Bay region. It prefers well drained soils, with dappled shade or partial sun. Tolerates light to medium frosts. A useful groundcovering species which spreads readily by rooting at branch nodes. It has not been widely cultivated, and the species most commonly grown under this name is now known as *C. prolifera*. Suitable for gardens or containers. Propagate from seed or by division.

The main distinguishing features between *C. stylidioides* and *C. prolifera* are that *C. prolifera* has limp, green, woolly, flat leaves that can be hairy while young, but become glabrous with age, and its flowers are pale yellow inside the floral tube.

Conostylis teretifolia J. Green
(terete leaves)
WA
0.1-0.15 m x 0.15 x 0.3 m   Aug-Oct

Dwarf, tufting **perennial**; **stems** short, or shortly rhizomatous; **leaves** 5-10 cm x 0.1 cm, terete, with long, spreading bristles; **scape** as long as or much longer than leaves, bearing small flower-heads; **flowers** about 1 cm long, yellow or rarely reddish, densely hairy outside and inside, lobes about 0.4 cm long, spreading.

This decorative species is from the Irwin and Avon Districts. It is not widely known in cultivation. Will grow in most well drained soils, with dappled shade or partial sun, and will possibly tolerate full sun. In some cases it has been slow to become established. Very suitable as a container plant. Propagate from seed or by division.

Conostylis vaginata Endl.
(sheathed)
WA   Sheath Conostylis
0.1-0.15 m x 0.2-0.5 m   Aug-Nov

Clumping **perennial**; **stems** much-branched, rhizomes rooting at intervals; **leaves** 3-10 cm x 0.1 cm, terete, grooved, pungent, glabrous; **scape** short, bearing small heads of rarely more than 5 flowers, surrounded at base by hairy, papery, sheathing bracts; **flowers** about 1-1.3 cm long, bright yellow, densely hairy outside, glabrous or slightly hairy inside, lobes equal, about 0.5 cm long.

An ornamental species, with aerial roots that act as props for the branches. It grows naturally in sandy soils, between Albany and Esperance. It has potential for wider horticultural use than at present, and will grow in well drained soils, with dappled shade or partial sun. Will tolerate full sun if in moist soils, and withstands light frosts. Very useful as a groundcover or container plant. Propagate by division as new aerial roots begin to form.

Conostylis villosa Benth.
(long soft hairs)
WA
0.15-0.25 m x 0.2-0.3 m   Oct-Dec
Dwarf, tufting **perennial**; **stems** short; **leaves** 12-25 cm x 0.1-0.2 cm, flat, covered in loose, long, silvery white hairs; **scape** 6-12 cm long, densely hairy when young, with 1-2 leaf-like bracts, bearing a many-flowered head; **flowers** 1.2-1.5 cm long, yellow, sometimes tinged or wholly purple-red, densely hairy inside and outside, lobes 0.5-0.7 cm long.

This species is closely allied to *C. setigera*, which differs in that its leaves have hairs only on the margins. It is evidently not in cultivation, but should grow in most well drained soils, with dappled shade or partial sun. With its densely hairy foliage and colourful flowers, it has potential as a garden or container plant. Propagate from seed or by division.

Conostylis wonganensis Hopper
(occurring in the Wongan Hills)
WA
0.2 m x 0.1 m   July-Sept
Dwarf, tufting, perennial **herb**; **rhizome** and **stems** short; **leaves** to 17 cm x 0.1 cm, terete, erect, grooved, glabrous except for hairy spines on margins; **scape** to 3 cm tall, hairy; **flowers** tubular, to about 0.8 cm long, creamy yellow, with spreading lobes, densely hairy, in loose heads of less than 10 flowers.

A recently described species from the Manmanning-Wongan Hills region, where it usually grows in mallee heath on yellow sand overlying clay or laterite. It has the most restricted range of any named *Conostylis* species, and is considered a rare endemic in the above region. Should adapt to cultivation in well drained, light to medium soils with a warm to hot situation, although it will tolerate dappled shade or partial sun. Propagate by division. As yet, seed has not been collected.

This species has affinities to *C. caricina* which has glabrous, flat leaves, *C. dielsii* which has slightly hairy, flat to terete leaves and *C. teretiuscula* which has slightly hairy, terete leaves. All species have creamy yellow to cream flowers.

Conostylis species (Eneabba)
WA
0.2-0.3 m x 0.3 m   April-Sept; also sporadic
Dwarf, tufting **perennial**; **stem** short; **leaves** to 30 cm x 0.3 cm, dark green, slightly curved, striate, lamina glabrous, small, slightly pungent, hairs on

Conostylis vaginata   T.L. Blake

margins; **scape** to 25 cm long, with one bract just above middle and 2 papery bracts at base of flowers; **flowers** tubular, to 1 cm long, pale yellow, densely hairy on outside, spreading cream lobes, borne on peduncles about as long as corolla, in loose clusters at end of scape.

An endemic, unnamed species from Eneabba, north of Perth, where it grows in lateritic sands. Needs well drained soils, with partial or full sun, and will tolerate light to medium frosts. Responds well to applications of slow-release fertilizers. An ornamental, long-flowering species, excellent for gardens or containers. Propagate from seed or by division.

CONOTHAMNUS Lindl.
(from the Greek *conos*, cone; *thamnos*, bush or shrub)
Myrtaceae
Small **shrubs**; **leaves** opposite, small, 1-3 nerved, rigid; **flowers** in terminal, small globular heads; **petals** 5, orbicular, spreading, or none, stamens prominent; **capsules** 3-celled, enclosed in enlarged calyx or nearly free, forming woody clusters.

An endemic genus of 3 species, confined to the South-West Botanical Province of WA. It is closely related to *Melaleuca*, and only differs by having 1 ovule in each cell of the ovary. The only species cultivated to any degree to date is *C. aureus*, which produces flowers for a long period. Propagation is from seed or cuttings.

81

Conothamnus aureus

Conothamnus aureus, × ·8

Conothamnus aureus (Turcz.) Domin
(gold)
WA
0.3-0.5 m x 0.5-1 m   Aug-Nov
Dwarf **shrub**; **branchlets** many, hairy, often ending in spines; **leaves** about 1 cm x 0.5 cm, ovate-lanceolate, opposite, rigid, blunt, 1-nerved, transversely veined, hairy; **flower-heads** globular, about 1 cm diameter, terminal, profuse, golden yellow; **petals** none.
This ornamental species occurs naturally, in deep sandy soils. For cultivation it requires very well drained soils, and will grow in partial or full sun. Withstands light to moderate frosts and extended dry periods. Responds well to pruning. Suitable for gardens or containers. Propagate from seed or cuttings.

Conothamnus divaricatus Benth. =
                              C. aureus (Turcz.) Domin

Conothamnus neglectus Diels
(neglected)
WA
0.3-0.6 m x 0.5-1 m   Aug-Oct
Dwarf **shrub**; **branchlets** many; **leaves** about 1 cm x 0.2-0.3 cm, oblong-lanceolate, opposite, initially silky-hairy, becoming glabrous; **flower-heads** globular, about 1-2 cm across, yellow, terminal, petals present.
An ornamental species extending from the south coast to the Stirling Range-Porongurup Range region. Uncommon in cultivation. Best suited to well drained, light to medium soils, with dappled shade to full sun. Certainly has potential as a container plant. Propagate from seed or cuttings.

Conothamnus trinervis Lindl.
(3-nerved)
WA
0.3-1 m x 1 m   Aug-Oct
Dwarf **shrub**; young growth silky-hairy; **branches** many, rigid; **leaves** 1-3 cm x about 0.5 cm, lanceolate to oblong-lanceolate, 3-nerved, rigid, narrowed at base, pungent apex, margins thickened; **flower-heads** globular to ovoid, to 3 cm across, white, terminal; **calyx tube** hairy.
An inhabitant of the sandplains north of Perth. It requires very well drained soils, with partial to full sun. Some plants are recorded as having white and purple flowers. Not common in cultivation. Propagate from seed or cuttings.

CONTAINER PLANTS — see Volume 1, page 101

Convolvulaceae Juss.
(Morning Glory Family)
A family of dicotyledons containing about 1800 species in 55 genera, widely distributed throughout most regions of the world. The climbing mode of growth is particularly well developed, but they may also be herbs, shrubs or occasionally trees. The flowers are characteristic, having 5 fused petals resulting in a trumpet shape. These are often colourful, and are borne in profusion over long periods. Many exotic species are grown as garden ornamentals, particularly in tropical regions. The family is well represented in Australia, with about 75 species in 12 genera, about 40 of these being in the genus *Ipomoea*. Members of the family are common in a variety of situations throughout Australia but are particularly well developed in rainforests and in dry inland regions. Many species have horticultural merit. Australian genera: *Argyreia, Bonamia, Calystegia, Convolvulus, Dichondra, Ipomoea, Lepistemon, Merremia, Operculina, Polymeria, Porana* and *Wilsonia*.

CONVOLVULUS L.
(from the Latin *convolvulo,* to twine around)
Convolvulaceae
Prostrate, erect or small climbing **shrubs**; **leaves** alternate, entire or toothed; **flowers** axillary, solitary or in cymes, funnel-shaped, lasting a few hours or a couple of days; **fruit** a rounded capsule.
A large genus of 250 species with 3 indigenous to Australia, 2 of which are endemic. The seeds are surrounded by a hard, woody coat which must be sliced to expose the seed before it will germinate.

Convolvulus erubescens Sims
(reddish flowers)
all states   Blushing Bindweed
0.1-0.3 m x 0.5 m   Sept-Feb
Small trailing or climbing **shrub**; **stems** hairy; **leaves** 2-6 cm x 0.5-1 cm, lanceolate, cordate to hastate at the base, entire or lobed, hairy; **flowers** 1.5-2.5 cm across, bright rosy pink to white, funnel-shaped, borne 1-3 in the axils of the leaves, on peduncles 1-3.5 cm long; **capsule** about 0.5 cm across, globular.

Widespread throughout Australia in a variety of habitats from the coast to well inland. Very showy when in flower. A useful small climber or trailer for a hot, sunny position in well drained soil. Plants are extremely variable, with some climbing vigorously, while others form an appressed mat of foliage. Deserves to be more widely cultivated, and can be a useful groundcover in areas of competition. Propagate from seed or cuttings which strike easily. Aborigines used extract, after boiling plants, for treatment of diarrhoea and stomach-ache.

Convolvulus parviflorus Vahl
(small flowers)
Qld, NT
0.1-1.5 m tall   Jan-March
Small to medium sized, slender **climber**; **leaves** 5-8 cm long, ovate, cordate at the base, entire, pointed, thin-textured, on long petioles; **flowers** 1-1.5 cm across, white or pink, funnel-shaped, in dense axillary cymes.

A widespread pantropical species extending to northern Australia. Not well known in cultivation but could be a useful light climber for tropical or coastal regions. Propagate from seed.

Convolvulus remotus R. Br.
(remote; isolated)
Qld, NSW, SA, WA, NT
prostrate x 0.3-0.6 m   Sept-Dec
Small, trailing, prostrate **shrub**; **stems** hairy, appressed to the ground; **leaves** 3-7 cm x 0.4-1 cm, linear-lanceolate to oblong, hastate at the base, blunt to pointed, bright green; **flowers** 1.5-2 cm across, pink, funnel-shaped, singly or in pairs in the axils of the leaves, on peduncles 1.5-5 cm long, the petals hairy; **capsule** about 0.5 cm across, globular.

Widespread in sandy soils of inland districts, usually in an open, sunny situation. The stems show no tendency to climb, and the species could be an excellent groundcover for inland districts with a warm climate, or drier tropical regions. Propagate from seed or cuttings.

COOPERNOOKIA Carolin
(after Coopernook, near Taree, NSW)
Goodeniaceae
Perennial **herbs** or dwarf **shrubs**; **stems** more or less woody, sometimes with rigid hairs; **leaves** simple, spirally arranged, with star-like hairs, resinous when young; **flowers** solitary, arranged in terminal racemes, sessile or on short stalks; **sepals** 5; **corolla** mauve-pink, blue-purple or white; **lobes** winged, with sword-shaped hairs; **stamens** free; **fruit** a capsule; **seed** ovoid, strophiolate.

An endemic genus of 6 species which, prior to revision in 1967, were included under *Dampiera* or *Goodenia*. The main distinguishing characteristics are the star-like hairs on leaves and stems, and the strophiolate seeds. It seems only *C. barbata* and *C. polygalacea* are in cultivation at present, although the others have potential. Propagate from cuttings of firm new growth. It is recommended that all flowers be removed from cutting material because *Botrytis cinerea* (grey mould) can spread on the spent flowers and damage growth-buds and leaves.

Coopernookia barbata (R. Br.) Carolin
(bearded)
Qld, NSW, Vic, Tas   Purple Goodenia
0.3-0.6 m x 0.2-0.5 m   Sept-Dec
Erect dwarf **shrub**; **stems** glandular, hairy, with scattered star-like hairs when young, becoming rough with age, rigid; **leaves** 1-3 cm x 0.1-0.5 cm, spirally arranged, linear to narrow-oblong, sessile, hairy on both surfaces, margins entire or slightly toothed, revolute; **flowers** to 1.5 cm long, blue to pink-purple, in loose terminal racemes.

This showy species usually inhabits dry sclerophyll forest and scrubland. It requires reasonably well drained, light to medium soils, and prefers partial sun. Will tolerate full sun if root system has some protection. Suitable for gardens or containers. As it can be short-lived it should be propagated regularly. Propagate from cuttings.

Previously known as *Goodenia barbata* R. Br.

Coopernookia chisholmii (Blakely) Carolin
(after E. C. Chisholm)
NSW
1-1.5 m x 1-1.5 m   Sept-Dec
Small **shrub**; **stems** weak, with dense covering of star-like hairs while young, or glabrous; **leaves** 4-9 cm x 2.5 cm, narrow-elliptic to elliptic, spirally arranged, almost sessile, upper and lower surface densely hairy, irregularly toothed margins, slightly recurved to flat, with blunt apex; **flowers** to 1.5 cm long, mauve-pink, borne in terminal leafy racemes.

An inhabitant of wet sclerophyll forests along the NSW coast from the Hunter Valley to Port Macquarie. Evidently not well known in cultivation. It grows as a showy shrub and requires well drained soils that do not dry out too readily. Suited to dappled shade and partial sun. Propagate from cuttings of firm new growth.

Differs from *C. barbata*, which is smaller and has shorter leaves with revolute margins.

Previously known as *Goodenia chisholmii* Blakely.

Coopernookia georgei Carolin
(after A. S. George, contemporary botanist, editor of Flora of Australia)
WA
1-2 m x 1-1.5 m   Sept-Nov
Small **shrub**; **stems** becoming glabrous; **leaves** 2-5 cm x 0.8-2.5 cm, elliptic-obovate, spirally arranged, more or less sessile, margins toothed and slightly recurved, glabrous when mature; **flowers** to 2 cm long, mauve with white throat, peduncles long, borne in terminal leafy racemes.

An ornamental species that is confined in nature to about 50-60 km south-west of Ravensthorpe, where it grows in sheltered gullies. Should do best in relatively well drained soils, with dappled shade or partial sun. May require pruning to promote bushy growth. Propagate from cuttings.

Readily distinguished by its broad-toothed leaves and large flowers.

Coopernookia polygalacea (De Vries) Carolin
(similar to the genus *Polygala*)
WA
0.3-1 m x 0.3-1 m   Sept-Nov; also sporadic

Dwarf to small **shrub**; young growth often greyish green; **stems** covered with white, woolly tomentum; **leaves** 2-4 cm x 0.2-0.5 cm, more or less linear, undersurface densely hairy, upper surface becoming glabrous, margins revolute; **flowers** about 1.5 cm long, white to purple, borne in terminal leafy racemes.

This decorative species has had limited cultivation, but reports show that it is fairly adaptable. A form with purplish pink flowers is the most commonly cultivated. It grows as a dense bush. Prefers relatively well drained, light to medium soils, with a fair amount of sunshine. Is prone to attack from grey mould if grown in shady locations. It occurs in the Eyre, Roe and Coolgardie Districts. Has potential as a container plant. Propagate from cuttings which strike readily. There has been limited success in germinating seed.

Easily distinguished from other species by its dense, narrow leaves.

Previously known as *Goodenia phylicoides* F. Muell. and *Dampiera polygalacea* De Vries.

Coopernookia scabridiuscula Carolin
(minutely scabrid)
Qld
0.5-1 m x 0.5-1 m   Sept-Jan

Dwarf **shrub**; **branches** slender; **stems** hairy when young; **leaves** 4-8 cm x 0.5-1 cm, narrow-elliptic to oblong-elliptic, nearly sessile, minutely scabrid-toothed, with pointed apex; **flowers** about 1.5 cm long, pink-purple, borne in terminal leafy racemes.

A recently described species that is restricted to the Mount Maroon region near the NSW border. Evidently it is not in cultivation. Should do best in well drained soils, with dappled shade or partial sun. Propagate from cuttings of firm new growth.

Differs from *C. barbata* which has smaller leaves, and *C. chisholmii* which has stellate hairs on the leaves.

Coopernookia strophiolata (F. Muell.) Carolin
(with strophioles)
SA, WA
0.5-1 m x 0.5-1 m   Sept-March

Dwarf to small **shrub**; **branches** many; **stems** initially angular and hairy, becoming round and glabrous; **leaves** 1-3.5 cm x 0.2-1.5 cm, obovate, spathulate or elliptic, spirally arranged, viscid when mature, undersurface more or less glabrous, varnished appearance, margins toothed, sometimes slightly recurved; **flowers** to 1.2 cm long, bluish white to white, borne in terminal leafy racemes.

An inhabitant of sandy heaths in western SA and southern WA. Often occurs in areas of low rainfall, so is probably best suited to a warm to hot location with plenty of sun, in well drained soils. Could need regular tip pruning to promote bushy growth. Propagate from cuttings of firm young growth. Some success gained in germinating seed.

Differs from *C. polygalacea*, which has narrower leaves with a densely hairy undersurface, and very narrow corolla lobes.

Previously known as *Goodenia strophiolata* F. Muell.

Coopernookia polygalacea   B.W. Crafter

COPROSMA Forster & Forster f.
(from the Greek *copros*, dung; *osme*, a smell; an allusion to the fetid smell of some species when bruised)
Rubiaceae

Prostrate to tall **shrubs**; **branchlets** often spiny; **leaves** simple, opposite, shiny or dull; **flowers** small, dull, male and female borne on separate plants; **fruit** a small succulent drupe.

Species about 90 with 7 in Australia, 6 of which are endemic. A further 3 species occur on Lord Howe Island. All are compact, twiggy shrubs found in cool, moist situations. The female plants are very decorative when in fruit. Propagate from seed, cuttings or in some species by division. Seed is best sown soon after collection. Stem cuttings seem to strike easily.

Coprosma hirtella Labill.
(minutely hairy)
NSW, Vic, Tas   Rough Coprosma; Coffee Berry
1-2.5 m x 0.5-1.5 m   Aug-Oct

Small to medium, slender **shrub**; **leaves** 1-7 cm x 1-1.5 cm, broadly elliptical to orbicular, roughened, dull green, with a long point; **flowers** about 0.6 cm long, greenish, in clusters; **drupes** about 1 cm across, orange, red or brownish, in clusters.

Widespread in open forest country, often in small scattered colonies. The showy fruit are edible and sweetish, but leave an unpleasant aftertaste. Easily grown in a protected position in the garden, such as where it receives morning sun or dappled sun. Must have good drainage and needs extra water during dry periods. Propagate from seed or stem cuttings which strike easily.

Coprosma lanceolaris F. Muell.
(shaped like a lance-head)
NSW (Lord Howe Island)
3-5 m x 1-3 m   Aug-Oct
Medium to tall **shrub**; **leaves** 5-7 cm x 2-3 cm, ovate-lanceolate, leathery, dark green, the midrib prominent, the margins undulate; **flowers** small, greenish, in axillary clusters; **drupe** about 1 cm long, oblong, orange.
Endemic on Lord Howe Island, where it grows in mossy forests. Not known to be in cultivation but could be a useful plant for a protected situation in coastal districts. Propagate from seed or cuttings.

Coprosma moorei F. Muell. ex Rodway
(after T. B. Moore)
Vic, Tas   Turquoise Coprosma
prostrate x 0.5 m   Oct-Nov
Small prostrate **shrub**; **stems** much-branched, rooting at the nodes; **leaves** 0.3-0.5 cm x 0.2 cm, ovate-lanceolate, thick, glabrous, shiny, acute; **flowers** about 0.6 cm across, solitary, greenish, terminating short branches; **drupe** 0.5-0.6 cm across, blue.
Restricted to subalpine habitats above 300 m alt., often growing in sphagnum moss. Little-known in cultivation but grown occasionally by alpine plant enthusiasts. Makes an attractive container plant. Useful as a rockery plant in cool positions. Must not be allowed to dry out during warm weather. Needs well drained soil. Propagate from seed, cuttings or by division.

Coprosma hirtella flowers and fruit, × ·65

Coprosma nitida Hook.f.
(shining)
Vic, Tas   Shining Coprosma
1-2.5 m x 0.5-1 m   Dec-March
Small to medium, rigid **shrub**; **branchlets** spine-tipped; **leaves** 1-2 cm x 0.3-0.5 cm, elliptic to lanceolate, shiny green, thick-textured, with recurved margins; **flowers** greenish, small, inconspicuous, male flowers bell-like, with conspicuous protruding stamens; **drupes** about 1 cm long, oval, bright red, shining.
Confined to highland areas above 900 m alt., and often locally common. May be a prostrate shrub in exposed situations. Little-known in cultivation but grown occasionally by alpine plant enthusiasts. The plants are very decorative when in fruit. The plants need well drained soils and plenty of water during dry periods. Propagate from seed or stem cuttings.

Coprosma nivalis W. Oliver
(growing in snow)
Vic   Snow Coprosma
prostrate x 0.5 m   Oct-Nov
Small, matted or prostrate **shrub**; **stems** branched, rooting only sparsely; **leaves** 0.5-1 cm x 0.1-0.2 cm, linear, oblong, shiny; **flowers** solitary, small, greenish; **drupe** 0.6-0.8 cm across, orange or red.
Restricted to subalpine conditions above 1000 m alt., usually in rocky situations. Grown occasionally by alpine plant enthusiasts; requiring conditions similar to *C. pumila* and *C. moorei*. Propagate from seed or cuttings.

Coprosma prisca W. Oliver
(derivation unknown)
NSW (Lord Howe Island)   Goatwood
1-5 m x 2-4 m   Aug-Oct
Small to tall **shrub**; **branches** slightly hairy; **leaves** 5-10 cm x 1.7-3.5 cm, narrow-oblong to ovate-lanceolate, glabrous, fleshy, shiny green above, paler beneath, the veins prominent, the margins revolute; **flowers** small, greenish; male flowers 8-12, in heads; female flowers in 2-flowered heads; **drupe** about 0.7 cm long, ovoid, dark brown.
This species is endemic on Lord Howe Island, where it grows in coastal situations. It is one of a group of 5 similar coprosmas that grow in coastal districts around New Zealand and Norfolk Island. It is similar in general appearance to the New Zealand species *C. repens*, which is widely cultivated in Australia and has become naturalized in some southern areas. *C. prisca* is not known to be in cultivation, but should prove to be a very useful plant for coastal areas in temperate and subtropical Australia. Propagate from seed or cuttings.

Coprosma pumila Hook.f.
(dwarf)
NSW, Vic, Tas   Creeping Coprosma
prostrate x 0.3-1 m   Oct-Nov
Small, matted or prostrate **shrub**; **stems** densely matted, rooting at the nodes; **leaves** 0.5-1 cm x 0.2-0.3 cm, elliptical, glabrous, shiny; **flowers** about 0.8 cm long, solitary, terminal on short branches, greenish; **drupe** 0.6-0.8 cm across, orange or red.
Restricted to subalpine habitats above 500 m alt.

Grown occasionally by alpine plant enthusiasts. Requires cool, moist conditions and must not be allowed to dry out during hot, dry spells. Best planted amongst rocks, to give a cool root-run. Makes an attractive container plant. Needs rich, well drained soil. The species is variable and embraces a couple of distinct forms. Propagate from seed, cuttings or by division.

Coprosma putida C. Moore & F. Muell.
(smelling evilly)
NSW (Lord Howe Island)   Stinkwood
4-7 m x 3-4 m   Oct
    Tall **shrub** or small **tree**; **wood** with a disagreeable odour; **leaves** 8-12 cm x 3-5 cm, ovate-elliptical, dark green above, pale green beneath, the margins entire, on petioles about 2 cm long; **flowers** pale cream to green, about 0.5 cm long, in axillary clusters; **drupe** 2-2.5 cm x 1-1.5 cm, ovoid, greenish, shiny, on stalks about 2 cm long.
    Endemic on Lord Howe Island, where it grows in forests from sea level to the summits of the mountains. Not known to be in cultivation but could be a useful plant for a protected situation in coastal districts. Propagate from seed or cuttings.

Coprosma quadrifida (Labill.) Robinson
(cleft halfway into 4 segments)
NSW, Vic, Tas   Prickly Currant Bush
2-4 m x 1-2 m   Sept-Nov
    Medium to tall, slender **shrub**; **branchlets** spiny; **leaves** 0.5-1.2 cm x 0.3-0.5 cm, lanceolate, thin-textured, dull, glabrous; **flowers** about 0.4 cm long, solitary, terminating short shoots, greenish; **drupe** 0.5-0.8 cm long, bright red, oval.
    A small twiggy species commonly found in cool, moist forests and gullies. Seedlings frequently germinate on the trunks of tree-ferns and grow as epiphytes. The fruits are edible but not very tasty. The species is not widely grown because of its prickly nature, but it makes a useful plant for a wet, shady or cool position, and blends well with ferns. A form with much larger leaves (to 5 cm x 1.3 cm) is known from the head of the Werribee River, Vic. Propagate from seed or stem cuttings.

Corananthera australiana C. White =
        Lenbrassia australiana (C. White) G. Gillett

CORCHORUS L.
(from the Greek *corchoros*; an obscure derivation alluding to supposed medicinal properties)
Tiliaceae
    Herbs or **shrubs**; **leaves** simple, alternate, with serrate margins; **flowers** small, yellow, borne solitary or in axillary cymes; **petals** 5; **capsule** long or globular, woody.
    A genus of about 100 species with about 25 species in Australia, the majority of which are endemic. The stems of many herbaceous species are fibrous and yield the jute of commerce. Propagate from seed.

Corchorus cunninghamii F. Muell.
(after A. Cunningham, 19th-century botanist)
Qld, NSW
0.5-1 m x 0.3 m   Nov-March
    Annual or perennial **herb**; **leaves** 5-10 cm x 1-2 cm, ovate-lanceolate, dark green, with coarsely serrate margins; **flowers** 1-1.5 cm across, yellow, 3-8 together in an axillary cyme; **capsule** 1-2 cm long, woody, 3-4 angled.
    Widespread in moist gullies and along the margins of rainforests, often in colonies. A weedy species, occasionally grown as an annual. Suited to a semi-shady position but will grow in sunny positions if well watered. Propagate from seed.

Corchorus walcottii F. Muell.
(after Pemberton Walcott, naturalist on Gregory's expedition, 1861)
WA, NT
0.3-1.5 m x 0.5-1.5 m   July-Oct; also sporadic
    Dwarf to small **shrub**; young shoots softly woolly; **leaves** 3-6 cm x 1-2 cm, broadly ovate to oblong, rather thick, densely and softly hairy, grey, the margins coarsely toothed; **flowers** 1-1.2 cm across, yellow to orange, the petals broad, the sepals woolly, in axillary clusters of 3-6; **capsule** 1-2 cm long, erect, tomentose.
    A showy shrub with small hibiscus-like flowers. It is found in open, rocky situations and is often sparse. Could be an attractive shrub for tropical and inland areas, and would undoubtedly become bushier in cultivation. Requires a well drained soil. Propagate from seed.

CORDIA L.
(after Euricius and Valerius Cordus, 16th-century German botanists)
Boraginaceae
    Shrubs or **trees**; shoots glabrous or hairy; **leaves** opposite, entire or toothed; **flowers** in cymes or panicles, usually terminal, showy, funnel-shaped; **fruit** a hard drupe.
    Species about 250 with 4 extending to tropical Australia. They are handsome tropical plants with excellent potential for cultivation. As a general rule, they are very sensitive to cold and are best suited to coastal districts of the tropics and subtropics. Those species in Australia are in need of revision and name changes are likely. Propagate from seed which germinates readily.

Cordia aspera Forster f.
(rough)
Qld
7-10 m x 3-5 m   Nov-March
    Small **tree**; young shoots bearing long, rusty hairs; **leaves** 7-20 cm x 4-8 cm, ovate, thin-textured, dark green, hairy, the margins irregularly toothed, the apex drawn out into a long point; **cymes** dense; **flowers** about 1 cm across, orange or yellowish, funnel-shaped, the lobes crisped; **fruit** 1-1.2 cm long, ovoid, whitish.
    A widespread species that extends to coastal districts of north-eastern Qld. It is usually dense and bushy with decorative leaves. An excellent shelter plant for

Cordyline manners-suttoniae   B. Gray

flowers 3.5-5 cm long, 3.5-4.5 cm wide, orange, funnel-shaped, with 5-7 broad lobes; **drupe** 2-3 cm x 1.5-2.5 cm, globular, contracted at the top, with a persistent calyx, hard.

Widespread throughout Asia, extending to a few coastal areas of tropical Australia, where it grows from the beachfronts to rainforest. An excellent species for cultivation in tropical or subtropical areas, especially in coastal districts. Requires well drained soils. Extremely sensitive to frost. Propagate from seed which germinates readily.

Cordia wallichii G. Don
(after Nathaniel Wallich, orchidologist)
Qld
4-7 m x 2-4 m   Aug-Jan

Medium **shrub** to small, spreading **tree**; **leaves** 6-15 cm x 6-15 cm, broadly ovate to orbicular, thin-textured, blunt, dark green; **cymes** fairly dense, in the upper leaf axils; **flowers** 1.5-2 cm long, cream, funnel-shaped; **fruit** about 2 cm long, ovoid.

A widespread species that extends to north-eastern Qld, where it is scattered and fairly uncommon in dry sclerophyll forest. It is extremely bushy and should be an excellent screening and shelter plant for tropical and subtropical regions. Its requirements are largely unknown but are probably for well drained soils in a sunny situation. Propagate from seed.

CORDYLINE Comm. ex Juss.
(referring to the club-like form of the stems of some species)
Agavaceae

Small to tall **herbaceous plants** forming spreading clumps; **stems** mostly unbranched, arising from underground rhizomes; **leaves** sheathing at the base, spreading or recurved, overlapping at the base; **inflorescence** a panicle; **flowers** small, tubular at the base, with spreading segments; **fruit** a fleshy berry.

Species about 15 with about 8 found in Australia including one unnamed species. All are popular subjects for cultivation, and many exotic species, forms and hybrids are grown, especially in tropical areas or in glasshouses. Some species are excellent for indoor decoration, tolerating darkness and neglect. The Australian species are deserving of much wider cultivation. All species are ideal plants for mingling with ferns and palms. They like well drained, organically rich soil. Propagation is by division or from seed which should be sown in warm, moist conditions. Stem cuttings also strike readily.

Cordyline cannifolia R. Br.
(leaves like those of the genus *Canna*)
Qld
1-3 m tall   Nov-Jan

Stems slender, forming compact clumps; **leaves** 30-45 cm x 5-6 cm, oblanceolate, dark green, drooping, the basal ones often shredded; **panicles** 20-35 cm long, much-branched; **flowers** about 0.6 cm long, white, sessile or on very short pedicels; **berries** about 0.6 cm across, globular, reddish.

A poorly known species from north-eastern Qld,

coastal districts, tolerating considerable exposure to salt-laden winds. Best suited to tropical regions, as it is very cold sensitive. Requires well drained soils in a sunny situation. Propagate from seed.

Cordia dichotoma Forster f.
(forked branches)
Qld, NT
5-20 m x 5-12 m   Jan-May

Small to medium, spreading **tree**; young shoots glabrous; **leaves** 5-8 cm x 2-4 cm, ovate to orbicular, blunt, with 3-5 conspicuous veins, the veins on the underside hairy, on glabrous petioles 3-4 cm long; **cymes** loose, sometimes branched and paniculate; **flowers** about 5 cm across, orange, funnel-shaped at the base, male or hermaphrodite on separate trees; **fruit** 1-1.5 cm long, yellow or pinkish, ovoid.

A widespread species that extends to northern Australia where it grows in near-coastal rainforests. An excellent tree for cultivation in tropical or subtropical areas, and especially useful for coastal conditions. Tolerant of wind and salt spray but very sensitive to frost. Requires well drained soils in a sunny situation. Propagate from seed which must be sown fairly soon after collection.

Cordia subcordata Lam.
(leaves nearly heart-shaped)
Qld, NT   Sea Trumpet
2-15 m x 3-10 m   Feb-May

Tall **shrub** or small **tree**; young shoots hairy; **leaves** 10-20 cm x 5-15 cm, broadly ovate, acute, cordate at the base, shiny green, scabrous, on petioles 3-5 cm long; **cymes** loose, terminal, bearing 6-20 flowers;

where it grows in rainforests. It is a useful species for a container or shady, moist situation in the garden. Best suited to tropical and subtropical regions. Propagate from seed, suckers or stem cuttings.

Cordyline fruticosa (L.) A. Chev. =
C. terminalis (L.) Kunth

Cordyline haageana Koch
(after J.N. Haage, 19th-century seed-grower)
Qld
0.3-1 m tall   Nov-Jan
A small species forming sparse, slender clumps; **leaves** 7-15 cm x 3-7 cm, lanceolate, spreading or slightly curved, dark green, abruptly contracted into a fairly long petiole; **panicle** 10-20 cm long, with a few short branches; **flowers** about 0.8 cm long, white, reddish or purple; **berries** 0.6-0.8 cm across, globular, red.
A small species found in rainforests and along stream banks between Rockhampton and Cape York Peninsula. Fairly rare and not commonly grown. Best suited to tropical or subtropical areas, but can be grown in a glasshouse in southern Australia. Prefers a shady position in well drained soil, and responds to regular watering during dry periods. An attractive species for mixing with ferns. Propagate from seed, suckers, stem cuttings or by division of the rhizomes. The species was previously known as C. murchisoniae F. Muell.

Cordyline manners-suttoniae F. Muell.
(after Domina Manners-Sutton)
Qld
1-5 m tall   Nov-Jan
Tall species forming large, spreading clumps; **leaves** 30-90 cm x 5-18 cm, broad-lanceolate, tapering at the base, with a pointed tip, light to dark green, erect or spreading, the **petiole** strongly inrolled; **panicle** 20-90 cm long, much-branched; **flowers** 1-1.5 cm long, white, in clusters, on pedicels 0.8-1 cm long; **berries** about 1 cm across, reddish, densely crowded.
A large, handsome species found in rainforests of central and north-eastern Qld. Grows readily and is a handsome, tall species, quite showy when in flower or fruit. Best suited to tropical or subtropical conditions, or as a glasshouse plant in southern Australia. Needs well drained soil in a semi-shady position protected from wind. Likes plenty of organic mulch and responds to fertilizer. Propagate from seed, suckers, stem cuttings or by division.

Cordyline murchisoniae F. Muell. =
C. haageana Koch

Cordyline rubra Huegel ex Kunth
(red)
Qld
2-4 m tall   Jan-May
A slender species forming sparse clumps of a few stems; **leaves** 20-40 cm x 2.5-6 cm, oblanceolate, reddish when young, contracted gradually into a **petiole** which is deeply channelled and reddish; **panicle** 10-40 cm long, narrow, distinctly pedunculate; **flowers** 6-8 cm long, lilac with yellow anthers; **berries** globular,

Cordyline rubra   D.L. Jones

deep scarlet, shiny, borne in dense clusters.
Widespread in rainforests and moist, open forests throughout eastern Qld from the north-east to the southern border. Little-known in cultivation but it can be easily grown in a shady position in subtropical or temperate areas. Should prove to be a useful tub plant. Prefers well drained, acid soils rich in organic material. The long, dense, drooping clusters of fruit are very showy. Propagate from seed, suckers, stem cuttings or by division of the rhizomes.

Cordyline stricta (Sims) Endl.
(stiff; upright)
Qld, NSW   Slender Palm Lily
2-5 m tall   Dec-Feb
Stems slender, erect, forming spreading clumps; **leaves** 20-60 cm x 1-2.5 cm, linear-lanceolate, drooping, dark green, the margins strongly toothed; **petiole** flat, widening gradually to the blade; **panicle** 30-70 cm long, branched in the upper two-thirds; **flowers** 0.8-1 cm long, tubular at the base, purple or violet, on very short pedicels; **berries** 0.3-0.5 cm across, globular, purple or blackish.
Widespread and common in rainforests and along stream banks, even extending into moist places of open forest. Very popular in cultivation, and widely grown as a garden plant in glasshouses or for use indoors. Very hardy and grows well in southern Australia. Slightly frost tender but damaged growths are replaced by suckers. Prefers a shady or semi-shady position in well drained soil. Quite drought resistant but responds to extra water during dry periods and regular applications of fertilizers. Makes an excellent indoor plant, tolerating dark situations. Propagate from seed, suckers, stem cuttings or by division of the rhizomes.

A widespread species found on many Pacific islands and extending to north-eastern Qld (Cook and North Kennedy Districts), where it grows in rainforests, along rainforest margins and in moist areas of open forest. A hardy and attractive species, especially decorative when in fruit. Distinguished by the short, broad-channelled petiole. Best suited to tropical or subtropical regions and prefers a semi-shady position in well drained soil. Useful as an indoor plant. There are many coloured leaf forms of this species grown in tropical and subtropical gardens. These are not of Australian origin but appear to have arisen as 'sports' or by deliberate hybridization in Hawaii. Propagate from seed, suckers, stem cuttings or by division.

Cordyline terminalis (L.) Kunth
var. **petiolaris** Domin

(terminal; with a prominent petiole)

Qld, NSW   Palm Lily

2-7 m x 1-3 m   July-Sept

Tall species forming large, spreading clumps; **trunks** woody, with fibrous bark; **leaves** 30-50 cm x 3-8 cm, broad-lanceolate, tapering at the base with a bluntly pointed tip, dark green, thin-textured and becoming lacerated, erect or spreading, on long (up to 25 cm), narrow, rolled **petioles**; **panicle** 30-60 cm long, much-branched, arching; **flowers** about 1 cm long, varying from white to purplish; **berries** 0.8-1 cm across, bright red, succulent.

A widespread and common species distributed in south-eastern Qld and north-eastern NSW, in rainforest, along rainforest fringes and moister areas of open forests. Clumps may get very tall, with individual stems tree-like. Leaves reduce in size up the stems. This species grows very easily and is a hardy and adaptable garden plant. It will succeed quite well in temperate regions but may be damaged by heavy frosts. A semi-shady position is most suitable and the plants respond to mulching and applications of complete fertilizers. It is an excellent plant for mingling with palms or ferns. Large clumps transplant easily. It also makes a very useful indoor plant, tolerant of dark and neglect. Forms with variegated leaves are known. One described as 'Baileyi' has striking cream variegations in the leaves, and was collected at Pimpama. The other is more subtle, with thin golden stripes. Propagate from seed, suckers, stem cuttings or by division. This variety is sufficiently distinct to be raised to species level.

COROKIA Cunn.

(Latinized version of a Maori name, korokio)

Escalloniaceae

Small to large, much-branched **shrubs**; **branches** wiry, spreading, clad with silvery hairs; **leaves** alternate, sparse, often in groups; **flowers** yellow or greenish, hairy, in axillary or terminal racemes or panicles; **fruit** a drupe.

Species about 6 with 2 in Australia, one of which is found in eastern Australia and the other on Lord Howe Island. The remaining 4 species are from New Zealand, and at least one is commonly grown in cultivation in south-eastern Australia. Propagate from seed which has a stony endocarp and is not easy to germinate. Stem cuttings are more successful.

Cordyline stricta   D.L. Jones

Cordyline aff. stricta

Qld, NSW

2-4 m tall   Dec-Feb

Stems slender, erect, forming spreading clumps; **leaves** 20-60 cm x 1.5-3 cm, linear-lanceolate, drooping in the upper part, the margins entire; **petiole** flat, widens abruptly to the blade; **panicle** 30-60 cm long, narrow, branched in the upper part, congested; **flowers** 0.8-1 cm long, tubular, white to purple; **berries** 0.3-0.5 cm across, globular, black.

An apparently undescribed species which is widely distributed in moist, shady situations in open forest and dry rainforest, often in rocky sites. It is quite a handsome plant and is useful in cultivation with similar remarks as for *C. stricta*. It can be propagated from seed, suckers, stem cuttings or by division.

Cordyline terminalis (L.) Kunth

(terminal)

Qld   Palm Lily

2-5 m x 1-2.5 m   Aug-Oct

Tall species forming large, spreading clumps; **leaves** 0.5-1 m x 5-10 cm, broad-lanceolate, tapering at the base with a bluntly pointed tip, dark green, thick-textured, erect or spreading; **petiole** short, broad and channelled; **panicle** 30-50 cm long, narrow, branched; **flowers** about 0.8 cm long, purplish; **berries** about 1.5 cm across, round, red.

Corokia carpodetoides

Correa alba

Corokia carpodetoides (F. Muell.) L. S. Smith
(resembling the genus *Carpodetus*)
NSW (Lord Howe Island)
2-4.5 m x 1-2.5 m   Oct

Tall spreading **shrub**; **branches** interweaving, angular, with appressed silky hairs; **leaves** 5-8 cm x 0.5-2.5 cm, ovate-lanceolate, leathery, dark green and glossy above, pale green beneath, the surface roughened, sparse on the branches, racemes or panicles in the upper axils; **flowers** about 0.8 cm across, pale yellow, with silky hairs; **drupe** about 1.2 cm long, globose, succulent.

An intricate shrub endemic on Lord Howe Island, where it is most common on the mountains. Little-known in cultivation but could have potential in warm temperate and subtropical regions. Requires a shady, moist position. May be a useful protective shrub for ferns. Propagate from seed or cuttings.

Corokia whiteana L. S. Smith
(after C. T. White)
NSW
2-4 m x 1-2 m   Nov-Jan

Tall **shrub** or small **tree**; **branches** bearing appressed silky hairs; **leaves** 2-8 cm x 0.5-3 cm, lanceolate, blunt, spirally arranged; **racemes** axillary or terminal; **flowers** 0.8-1 cm across, yellow, having silky hairs; **drupe** about 1 cm long, obovoid, red, juicy and succulent.

Found in highland rainforests in the ranges of north-eastern NSW and possibly also south-eastern Qld. Little-known in cultivation but grown by enthusiasts. Its cultivation requirements are largely unknown. Probably requires a protected position in well drained, acid soil. Propagate from stem cuttings.

CORREA Andrews
(after José Francisco Correa de Serra, 18-19th-century Portuguese botanist)
Rutaceae

Woody plants; from prostrate **shrubs** to small **trees**; **leaves** opposite, simple, bearing stellate hairs, aromatic when crushed; **flowers** conspicuous, usually pendulous, 4 petals (in most species fused to form a bell), solitary or in clusters from leaf axils; **fruit** split into 4 cocci.

Correas are endemic to Australia. There are 11 species (plus varieties), mainly restricted in their natural habitat to the eastern states and SA, although *C. reflexa* has been recorded about 500 km west of Eucla, WA. They grow in a variety of habitats, from exposed coastal situations to mountainous areas that receive snow. Natural hybridization is quite common, resulting in many forms that can be confusing.

This complex genus is currently undergoing botanical revision, which will result in reclassification of some species and varieties.

90

All species, forms and hybrids have considerable horticultural potential, and are becoming well known in cultivation, proving most adaptable for gardens. They are also excellent container plants. As an added bonus, the flowers of most species provide a copious supply of food for honey-eating birds.

In general, correas need well drained situations. Most species adapt to a wide variety of soils, provided the drainage is adequate. It is a versatile group of plants, as many species are happy growing in shaded locations, or where there is plenty of sun, eg. *C. alba, C. baeuerlenii* and *C. reflexa*. There is a preference for protection from hot sun on the root area. All species are capable of withstanding damage from frost. They respond well to light pruning. *C. reflexa* is one of several species which will accept hard pruning when well established, and this correa also has the ability to regenerate from its rootstock after bushfires.

Light application of a slow-release fertilizer may be beneficial. Correas can be subject to attacks of scale, with sooty mould often growing on the exudate. The scale is usually controlled with one or more sprays of white oil. Collar rot may be troublesome, usually occurring as a result of overwatering, and often if plants are placed too deep when planted.

C. pulchella and forms of *C. reflexa* which occur on the south-east coast of SA often grow on alkaline soils. Such plants can languish in acidic soils, and a light application of ground limestone or dolomite can help to promote growth.

Propagation

Seed of correas is not readily available, and is usually very difficult to germinate. Each seed has a hard, waxy coat and also contains germination inhibitors. Some success has been obtained by leaching seeds with a continuous stream of water for several weeks. Fortunately, plants are generally easily propagated from cuttings of firm young growth. This ensures trueness to type, whereas propagation from seed can result in the production of hybrids.

Rooted cuttings of *C.* 'Mannii', *C. pulchella* and some forms of *C. reflexa* can lose vigour shortly after removal from propagation unit or glasshouse. New growth can be encouraged by light application of ground limestone or dolomite. Rooted cuttings have also been noted to respond unfavourably if the medium is too acid.

When correas were introduced to Europe early in the 19th century, they created great interest, and artificial hybrids were produced. Similar work in Australia in recent years has resulted in a number of new hybrids being introduced to cultivation.

Correa aemula (Lindl.) F. Muell.
(rivalling or emulating)

Vic, SA	Hairy Correa
1-2 m x 1-3 m	Sept-Dec

A spreading, small **shrub**, dense to ground level; **leaves** 2-4 cm long, oval, pointed, green, very hairy; **flowers** tubular, 2-3 cm long, green turning blue-green to purple with age, sometimes splitting lengthwise, pendulous on short stalks, tips reflexed; **calyx** with 4 pointed lobes.

This species has a preference for moist, well drained light soils, with dappled shade to partial sun. It is or-namental, and useful also for light screening. An excellent bird-attracting plant. Propagate from cuttings.

Natural hybrids occur freely between *C. aemula* and *C. reflexa*, and have been introduced into cultivation. One from the Grampians, Vic, has soft pink flowers, and is proving to be an adaptable garden plant. Forms from the Barossa Ranges, SA, tolerate loamy soils.

Correa alba Andrews
(white)

NSW, Vic, Tas, SA	White Correa
0.5-2 m x 1-2 m	Nov-May

A dense, spreading dwarf to small **shrub**; **leaves** 2-4 cm long, oval, green, glabrous above, fine hairs below; **flowers** waxy, white (can have pink tonings), on short stems; **corolla tube** split into 4 petals.

This species differs from other correas, in that the corolla tube is split into 4 separate petals, giving a star-like appearance. A rare form with 5 petals is in cultivation. It is most adaptable, tolerating a range of conditions from moist to dry, well drained soils, in semi-shade to full sun. Withstands extended dry periods and salt spray. Useful as a garden plant, and also ideal for coastal planting or in the control of soil erosion. It can be induced to climb and cover fences. It is not regarded as a bird-attracting species. Forms with pink tonings occur in Vic, but the deepest pink flower forms are found on the Tas north coast. The foliage was used by early settlers as a substitute for tea. Propagate from cuttings. This species has proved to be an excellent grafting stock for various genera of the family Rutaceae.

The variety *C. alba* var. *pannosa* Paul G. Wilson, from south-eastern SA and south-western Vic, has smaller leaves which can attain a rusty appearance due to the dense hair covering. It can be prostrate to 1 m high and is floriferous and showy. Most Vic forms have white flowers, whereas pink-flowered forms occur west of Victor Harbour. Previously known as *C. alba* var. *rotundifolia* (Lindl.) Benth.

Natural hybrids between *C. alba* and *C. reflexa* occur on the south-west coast of Vic.

In England, as early as the late 18th century, *C. alba* was used as a rootstock for the grafting of other *Correa, Crowea* and *Eriostemon* species.

Correa backhousiana Hook.
(after James Backhouse, English Quaker missionary and botanical collector)

Vic, Tas	
1-2 m x 2-3 m	May-Nov

Dense, spreading small **shrub**; **branches** and stems covered with rusty hairs; **leaves** about 3 cm long, oval, green, leathery, upper surface smooth, underside hairy; **flowers** tubular, about 2.5 cm long, pendulous, solitary or in small clusters, cream to pale green or rusty, sometimes splitting.

This species will adapt to well drained, light to heavy soils, withstanding short wet periods. Tolerates semi-shade to partial sun. An ornamental garden plant or screen, with the flowers being prominent in overcast weather. Suitable for coastal exposure. Plants can grow up to 5 m tall in shaded, moist locations. Excellent bird-attraction. Propagate from cuttings.

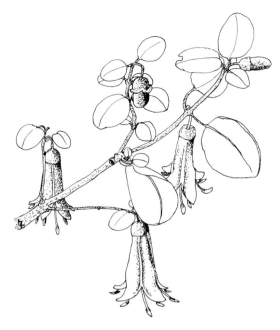

Correa backhousiana, × ·7

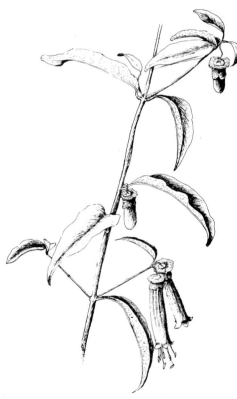

Correa baeuerlenii, × ·5

In garden hybrids, *C. backhousiana* is one of the most common parents. See also *C. alba* and *C. reflexa*.

Correa baeuerlenii F. Muell.
(after W. Baeuerlen, 19th-century botanical collector)
NSW   Chef's Cap Correa
1-2 m x 2-3 m   March-Aug
Dense, rounded small **shrub**; **stems** and **branchlets** brown; **leaves** 2-7 cm x 1-2 cm, green, smooth, with prominent glands on surface, sweetly aromatic; **flowers** tubular, greenish yellow, 2-3 cm long, pendulous, solitary on short stems; **calyx** prominent, inflated, with a flattened ridge.
In nature this species is restricted to the south-east coastal region of NSW. It is hardy, and adaptable to most well drained soils. Withstands short periods of dryness. It is best grown in dappled shade or partial sun. Widely cultivated as an ornamental plant, and useful also in low windbreaks and informal hedges, or for screen planting. Flowers and seed capsules are interesting and unusual, with the common name of Chef's Cap Correa relating to the tubular flowers and prominent flattened ridge on the calyx. Propagate from cuttings.

Correa calycina J. Black
(prominent calyx)
SA
1-3 m x 2 m   May-Nov
Small to medium **shrub**; **leaves** 2-4 cm x 1-2 cm, green, smooth to slightly hairy on the upper surface, undersurface slightly hairy; **flowers** tubular, to 3 cm long, pendulous, lime green; **calyx** deeply lobed.
A very adaptable correa. It prefers moist, well drained soils, with dappled shade or partial sun, but withstands short periods of dryness. A valuable garden or screen plant which will grow in a range of situations, even in shade beneath other plants. Flowers are inconspicuous and sometimes hidden by the leaves, but are excellent for bird-attraction. This species is restricted to the Fleurieu Peninsula in SA, where it is fairly rare. Propagate from cuttings.
An unnamed species with strong affinities to *C. calycina* has recently been introduced to cultivation.

Correa decumbens F. Muell.
(decumbent)
SA
0.2-1 m x 1-3 m   Nov-Feb
A dwarf, spreading **shrub**; **branches** prostrate; **leaves** to 3.5 cm x 1 cm, dark green and shiny above, grey to brown below; **flowers** narrow-tubular, to 2.5 cm long, erect, red with green tips; **calyx** can be deeply lobed.
An adaptable species which will grow best in moist, well drained, light to heavy soils. Will withstand extended wet periods. Tolerates dappled shade through to full sun. It is excellent for covering exposed areas such as embankments, or for use as a living mulch beneath other plants. The main representation of this species is in the Lofty Ranges east of Adelaide. Propagate from cuttings.
There is an interesting form from Kangaroo Island, SA where it grows in the mallee scrubs. It has a more

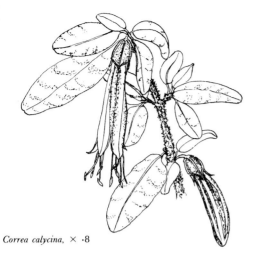

Correa calycina, × ·8

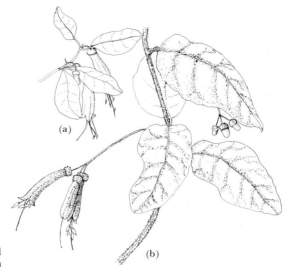

(a)

(b)

(a) Correa lawrenciana var. *lawrenciana,* × ·45
(b) Correa lawrenciana var. *cordifolia,* × ·45

upright habit of growth than the mainland forms, and the red corolla is around 2 cm long, with a style to 4 cm long.

Correa glabra Lindl.
(glabrous)
Qld, NSW, Vic, SA   Rock Correa
2-3 m x 1-3 m   May-Aug
   A dense, upright to spreading, medium sized **shrub**; **leaves** 1-4 cm x 0.5-2 cm, dark green, often with a wavy edge, usually glabrous; **flowers** tubular, pendulous, 1.5-3 cm long, pale green, green, golden or pink to red.
   This variable species is adaptable to most soil types and will tolerate dappled shade, partial or full sun. Withstands moderate coastal exposure. It is useful in the garden, and also suitable for roadside planting. Propagate from cuttings.
   C. schlechtendalii is closely related to this species.

Some botanists believe that it is conspecific and should only be recognized as a variety of *C. glabra.* However, current research has shown distinct differences, especially in regard to their natural habitats. *C. glabra* occurs only on rocky, hilly sites, while *C. schlechtendalii* is restricted to mallee sands.

Correa lawrenciana Hook.
(after R. W. Lawrence, an early Tas botanist)
NSW, Vic, Tas   Mountain Correa
2-8 m x 2-5 m   March-Nov
   A dense, small **shrub** to upright small **tree**; **branchlets** have dense grey hairs; **leaves** variable in size, 2-8 cm x 1-4 cm, leathery; **flowers** tubular, 1.5-3 cm x about 0.5 cm, pendulous, on stems from leaf axils, usually buff cream to green, but can be red; **calyx** cup-shaped, light rusty brown.
   This correa grows well in cultivation, preferring moist, well drained, medium to heavy soils. It will withstand extended wet periods, likes full shade to partial sun and is frost tolerant. An ornamental species, useful also as a screen. A Tas form from the east coast and the Asbestos Range has reddish stems and shows potential. Excellent bird-attraction. Propagate from cuttings.
   The Grampians, Vic form of *C. lawrenciana* will grow in full sun, provided it has a moist root system.
   Further varieties of *C. lawrenciana* are:
C. lawrenciana var. **rosea** Paul G. Wilson
   Restricted in nature to the highlands of north-eastern Vic, and south-eastern NSW. It is not common in cultivation, but is a desirable shrub with shiny leathery leaves. Some forms have dull red flowers, others a mixture of lime-yellow and red. It is possible that this variety may be included under *C. lawrenciana* var. *lawrenciana* in the future. There is strong evidence that flowers gain more red colouration with increase in alt.

Correa decumbens, × ·45

93

Correa pulchella

Correa pulchella × *aemula* at Wittunga Botanic Gardens, SA

available, varying in foliage, flower colour and shape. Some forms will tend to grow as dense climbers if planted against a rigid background. Prostrate forms are ideal for under-planting. An excellent form from Desert Camp, near Naracoorte, provides an outstanding display of bright orange flowers. At present it is threatened with extinction in nature, due to quarrying of limestone. The white-flowered form has an upright and fairly open growth habit.

Propagate from cuttings. Unthrifty rooted cuttings can respond well to light applications of ground limestone.

C. 'Harrisii', with pinkish red flowers, originated during the 19th century in England and has *C. pulchella* as one parent, with *C. reflexa* possibly being the other. It is also recorded as grown in USA, but at present is not thought to be cultivated in Australia under that name. There is strong evidence to suggest that *C.* 'Harrisii' and *C.* 'Mannii' are the same cultivar, as characteristics of growth, leaf and flower are extremely similar.

Correa reflexa (Labill.) Vent.
(sharply turned back)

Qld, NSW, Vic, Tas, SA, WA	Common Correa
0.3-3 m x 1-3 m	March-Sept

Prostrate to upright **shrub; leaves** to 5 cm x 3 cm, lanceolate to broadly ovate, often cordate, smooth to

C. lawrenciana var. **cordifolia** Paul G. Wilson

A shrub of about 2-4 m; leaves broad, thin; corolla usually red, rarely greenish yellow. This variety has proved to be most adaptable under cultivation.

C. lawrenciana var. **genoensis** Paul G. Wilson

A low shrub of 1-3 m high; calyx with prominent glandular dots, more or less glabrous, with lobes about half as long as the tube. This variety may also be included under var. *lawrenciana* in the future.

C. lawrenciana var. **glandulifera** Paul G. Wilson

A tall shrub to small tree, similar to var. *genoensis*, but calyx lobes small and tooth-like.

C. lawrenciana var. **macrocalyx** (Blakely) Paul G. Wilson

A shrub of 1-4 m, similar to var. *cordifolia* but with a very hairy greenish yellow corolla.

Gradation occurs between var. *cordifolia*, var. *genoensis* and var. *macrocalyx*, which can make it difficult to separate them as varieties.

Correa pulchella Mackay ex Sweet
(beautiful)
SA

prostrate-1.5 m x 1-3 m   April-Sept; also sporadic

Prostrate to small **shrub; leaves** 1-2 cm long, oval to elliptic, smooth, green; **flowers** tubular, 1.5-2.5 cm long, broadest at tip, horizontal or pendulous, orange to vermilion, pink, rarely white.

An adaptable species that is often abundant in calcareous soils and in very exposed situations. It grows in moist, well drained, light to heavy soils, and withstands extended periods of dryness. Tolerates semi-shade to full sun. Recommended for well drained alkaline soils. Popular in cultivation as an ornamental plant for gardens or containers. There are many forms

Correa reflexa 'Anglesea'

Correa reflexa a dwarf form from south-eastern coast of SA
B.W. Crafter

hairy above, with short hairs below; **flowers** tubular, 2-4 cm long, pendulous, tips sharply reflexed, hairy on outside, colours of cream, green, pink, red, or red with cream or green tips; **calyx** not usually pointed.

A most adaptable and widely cultivated ornamental species. It is suited to well drained, moist, light to heavy soils, but can withstand extended periods of dryness when established. Will grow in a range of situations, tolerating full shade or full sun, but has a preference for dappled shade or partial sun. There are many forms, most of which are worthy of cultivation. These are often available from nurseries, labelled with the area from which they originate, eg. *Correa reflexa* 'Grampians Vic'. The flower shape, particularly the width and length of bell, differs from area to area, as does flower colour. Forms from forested areas usually have long and narrow bells, while others from semi-desert or coastal conditions usually have a shorter, inflated corolla. It is common for sporadic flowering to occur throughout the year.

Propagate from cuttings. In some cases, rooted cuttings of the forms from south-eastern coastal regions lack vigour. A light application of ground limestone or dolomite usually helps rectify this situation.

The form from the forest areas, with green bells, is likely to be separated as a further variety or subspecies in the future.

The following varieties are also cultivated:

C. reflexa var. **cardinalis** (F. Muell. ex Hook.) A. B. Court

It occurs naturally in Gippsland, Vic, and grows 0.5-1.5 m x 1-2 m. The leaves are oblong, and the flowers are bright red with green tips. There is quite a variation in growth habit of this variety. Does best in

well drained, light to medium soils, with dappled shade or partial sun. Also tolerates full sun. This variety will be absorbed into var. *reflexa* as a result of the botanical revision.

C. reflexa var. **coriacea** Paul G. Wilson

This variety occurs naturally on the Eyre Peninsula in SA, and to about 500 km west of Eucla, WA. The leaves are more or less glabrous and the flowers are smaller than those of other varieties. Forms from the Eyre Peninsula have widely splayed corollas, whereas the form from near Eucla has more or less 4-sided corollas, and the foliage has an interesting fragrance when crushed.

C. reflexa var. **nummulariifolia** (Hook.f.) Paul G. Wilson

Usually a dwarf shrub of 0.2-0.5 m x 1-2 m, occurring naturally on the western coast of Vic, and to some of the Bass Strait islands. The leathery leaves are round to oval, can be glabrous or hairy above, but are densely hairy below. The hairs often give a rusty appearance. Flowers are greenish white, about 2 cm long, and very abundant. This variety is an ideal ground-cover, and is extremely well suited to coastal conditions. It is available sometimes as *C. reflexa* prostrate form from south-west Vic. Another form in cultivation grows to over 1 m tall. In the revision, the form from Kangaroo Island, SA, is most likely to become a variety of *C. glabra*. The other forms may be absorbed into var. *reflexa*.

A form of *C. reflexa* with large bells and hairy leaves is sometimes sold as *C.* 'Cane's Hybrid'. It forms erect branches and displays the flowers well.

A cultivar *C. reflexa* 'Fat Fred' has large flowers about 3 cm x 1.3 cm that are red with greenish yellow tips.

Natural hybrids with *C. reflexa* as one parent have been recorded with the following species: *C. aemula, C. alba, C. decumbens* and *C. pulchella*.

C. aemula x *C. reflexa* is the natural hybrid most commonly cultivated. It is most variable in form and flower colour.

A recent introduction is a hybrid of *C. reflexa* with the other parent possibly being *C. lawrenciana*. It occurs naturally in East Gippsland, Vic, and has old-gold flowers.

C. reflexa is also a very common parent of garden hybrids.

Correa schlechtendalii Behr
(after D. F. L. von Schlechtendal)
SA
0.5-2.5 m x 1-2 m   Nov-April

An upright, dwarf to medium **shrub; branches** covered with rusty hairs when young; **leaves** to 4.5 cm x 1.5 cm, broadly elliptic, blunt tip, aromatic, glabrous above, glabrous or hairy below; **flowers** tubular, to 2.5 cm long, solitary, borne on short stalks on lateral branchlets, pink to red with green lobes or whitish.

A very adaptable species that occurs naturally only in the mallee area of SA. It grows in most well drained soils, with dappled shade or partial sun, and will tolerate full sun. Withstands light to medium frosts and lengthy dry periods. Responds well to pruning if required. The plants have a long flowering period and are ideal for ornamental purposes, or can also be used as a low screen or windbreak. Propagate from cuttings.

The form from Vic that has been included in this species previously is applicable to *C. glabra*. It has been suggested by some botanists that this species is conspecific with *C. glabra* as there are many similarities, with minor differences. See *C. glabra*.

Previously known as *C. turnbullii* Ashby.

Correa turnbullii Ashby = *C. schlechtendalii* Behr

CORREA CULTIVARS and HYBRIDS

Correa 'Betts Red'
Cultivar
1.5-2 m x 2 m   March-Oct
Small, open **shrub**; **branchlets** scabrous; **leaves** to 2 cm x 1 cm, cordate, tapering to tip, scabrous above, undersurface rusty to light green, with dense covering of hairs; **flowers** tubular, to 3-5 cm long, deep pink, partially reflexed tips.

A recent spontaneous hybrid, not yet common in cultivation. It is thought to be of *C. reflexa* and *C.* 'Mannii' parentage. Should grow in most relatively well drained soils, with partial to full sun. Will probably tolerate dappled shade. Propagate from cuttings.

Differs from *C.* 'Mannii', which has glabrous leaves and the flowers have strongly reflexed tips.

Correa 'Carmine Bells' see *C.* 'Dusky Bells'

Correa 'Dusky Bells'
Cultivar
0.3-1 m x 2-4 m   March-Sept; also sporadic
Dwarf to small **shrub**; **branches** often reddish brown, many, spreading horizontally, with stellate hairs; **leaves** to 3.5 cm x 2 cm, broadest near base,

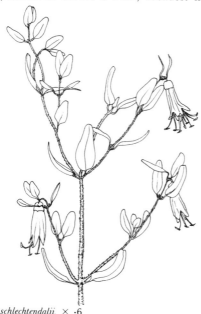

Correa schlechtendalii. × ·6

Correa reflexa prostrate form from near Nelson on the western Vic coast   W.R. Elliot

stellate hairs on both surfaces when young, becoming glabrous on upper surface; **flowers** tubular, pale carmine-pink, to 4 cm x 1 cm.

A cultivar of proven adaptability to a large range of soils. It is an excellent groundcover, with dense growth to ground level, and is long-lived. It prefers dappled shade or partial sun, rather than full sun. Excellent for shaded situations. Flowers can be partly hidden by foliage. Propagate from cuttings.

This correa is thought to have been in cultivation for over 50 years. It is known also by other names including *C.* 'Carmine Bells', *C.* 'species Pink', *C.* 'Pink Bells' and *C.* 'Rubra'; however, the Australian Cultivar Registration Authority has selected *C.* 'Dusky Bells' as the preferred name.

Correa 'Gwen'
Cultivar
1 m x 1-2 m   April-Sept
Dwarf **shrub**; **leaves** to 2.5 cm x 2 cm, oval, shiny dark green upper surface, light green below; **flowers** initially tubular, about 2 cm x 1.5 cm, white, suffused with pink near calyx, corolla splitting into 5 lobes which are strongly recurved, arranged in terminal clusters, profuse.

A recent man-made cultivar with parents *C. alba* and *C. reflexa*, that originated from Canberra. It should prove adaptable to a wide range of climatic conditions,

Comparative table of Correa cultivars

The recording of parents is not necessarily in genetic sequence of pollen donor and receiver, because many hybrids were unplanned.

Cultivar	Dimensions	Leaves	Flowers	Flowering time
C. 'Betts Red'	1.5-2 m x 2 m	To 2 cm x 1 cm, cordate, rough, green above, rusty to light green below.	Deep pink, 3-5 cm long, partially reflexed tips.	March-Oct
C. 'Dusky Bells' (probable parents C. pulchella and C. reflexa)	0.3-1 m x 2-4 m	To 3.5 cm x 2 cm, broadest near base, more or less glabrous on upper surface, pale to medium green.	Carmine-pink, to 4 cm x 1 cm.	March-Sept; also sporadic
C. 'Gwen' (C. alba and C. reflexa)	1 m x 1-2 m	To 2.5 cm x 2 cm, oval, shiny dark green above, light green below.	White, with pink near calyx, 2 cm x 1.5 cm, splitting into 5 lobes, strongly recurved.	April-Sept
C. 'Harrisii' — see C. 'Mannii', thought to be same cultivar.				
C. 'Ivory Bells' (C. alba and C. backhousiana)	1-2 m x 2-3 m	To 3 cm x 2 cm, elliptic to oval, dark green and glabrous above, pale greenish tan and hairy below.	Ivory to tan, about 2.5 cm long, recurved lobes.	June-Dec; also sporadic
C. 'Mannii' (probable parents C. pulchella and C. reflexa)	1-2.5 m x 2-3 m	About 3 cm x 2 cm, cordate, shiny dark green above, light green below, more or less glabrous on both surfaces.	Red exterior, pink interior, 4 cm x 1 cm, reflexed tips.	March-Sept; also sporadic
C. 'Pinky' (C. alba and C. reflexa)	1 m x 1-2 m	2-3 cm x 1.5-2 cm, oval.	Dusky pink, about 2 cm x 1.5 cm, splitting into 5 lobes to calyx.	April-Sept
C. 'Poorinda Fantasy' (C. backhousiana and C. reflexa)	1-1.5 m x 1-2 m	About 2.5 cm x 1 cm, shiny dark green above, short white hairs below.	Pink with yellow-green tips, about 3 cm x 1 cm, broadest near tip.	May-Nov; also sporadic
C. 'Poorinda Grace' (C. backhousiana and C. reflexa)	1-1.5 m x 1-2 m	To 3 cm x 1.5 cm, shiny dark green above, buff hairs below.	Pink with yellow-green tips, about 3 cm x 1 cm, broadest near tip.	May-Nov; also sporadic
C. 'Poorinda Mary' (C. backhousiana and C. reflexa)	1-1.5 m x 1-2 m	About 2.5 cm x 2 cm, ovate, shiny dark green above, hairy, light green below.	Pink with lime green tips, 3.5 cm x 1 cm, more or less cylindrical.	May-Nov; also sporadic
C. hybrid (C. backhousiana and C. reflexa)	1-2 m x 1.5-3 m	About 3 cm x 2.5 cm, undersurface with rusty hairs on margins and midrib.	Pale pink upper half, pale green lower half, to 3 cm x 1.5 cm, spreading lobes.	Mainly Feb-Oct
C. hybrid (probable parents C. backhousiana and C. reflexa)	0.5 m x 1-2 m	To 3 cm x 2 cm, cordate, shiny green above, pale green and hairy below.	Rose pink with pale cream-green tips, to 3 cm x 1 cm.	March-Sept
C. hybrid (C. backhousiana and C. 'Mannii')	1-2 m x 1.5-3 m	To 3 cm x 2 cm, shiny dark green above, pale green and hairy below.	Dull pale pink, to 3 cm x 1.5 cm.	Mainly Feb-Oct

Correa cultivars

Cultivar	Dimensions	Leaves	Flowers	Flowering time
C. hybrid (C. baeuerlenii and C. lawrenciana)	2-4 m x 1.5-3 m	7 cm x 3.5 cm, shiny dark green, broadest below middle.	Pale green, 3 cm x 1 cm.	April-July
C. hybrid (C. decumbens and C. sp.?)	1-1.5 m x 1-2 m	2-4 cm x 0.6-1 cm, oblanceolate, erect or spreading, concave, shiny, with few hairs.	2.5 cm x 1.7 cm, pinkish red with strongly reflexed greenish tips.	May-July; also sporadic

provided it has a relatively well drained location. Could prove to be a very useful landscaping plant. Propagate from cuttings.

Correa 'Harrisii' see *C.* 'Mannii' and *C. pulchella*

Correa 'Ivory Bells'
Cultivar

1-2 m x 2-3 m   June-Dec; also sporadic
Small **shrub**; **branches** many, spreading, densely hairy; **leaves** to 3 cm x 2 cm, elliptic to ovate, dull dark green and glabrous above, pale greenish tan and hairy below; **flowers** about 2.5 cm long, ivory to tan, recurved lobes.

This cultivar originated in San Francisco, California, USA, but to date is not cultivated in Australia. It is reputed to have *C. alba* and *C. backhousiana* as its parents. A vigorous cultivar that should have a wide application because of the adaptability of its parents. It will grow in most relatively well drained soils, with dappled shade or partial or full sun. Responds well to pruning. Propagate from cuttings.

Correa 'Mannii'
Cultivar

1-2.5 m x 2-3 m   March-Sept; also sporadic
Small to medium **shrub**; **branches** many, glabrous; **leaves** about 3 cm x 2 cm, cordate, tapering to tip, dark green above, light green below, nearly glabrous on both surfaces; **flowers** tubular, to 4 cm x 1 cm, red exterior, pale pink interior, reflexed tips.

A widely grown ornamental cultivar, thought to have originated in a Melbourne garden. Probable parents are *C. pulchella* and *C. reflexa*. Adapts to most situations with relatively well drained soils. Prefers dappled shade and partial sun, rather than full sun. Responds well to light pruning. Propagate from cuttings.

In some cases this hybrid has been confused with *C.* 'Dusky Bells', which is lower-growing, has light pink flowers, and the leaves are paler in colour.

This cultivar has very similar characteristics in growth, leaf and flower to *C.* 'Harrisii', and further investigation may result in the two being classed as identical cultivars. If that is the case, the name of *C.* 'Harrisii' will have precedence over *C.* 'Mannii' as it was introduced into cultivation during the 19th century.

Correa 'Marian's Marvel'
            see *C.* hybrid (*backhousiana* × *reflexa*)

Correa 'Pink Bells' see *C.* 'Dusky Bells'

Correa 'Pinky'
Cultivar

1 m x 1-2 m   April-Sept
Dwarf **shrub**; **branchlets** covered with rusty hairs; **leaves** 2-3 cm x 1.5-2 cm, oval, shiny dark green upper surface, hairy and greyish green below; **flowers** initially tubular, about 2 cm x 1.5 cm, splitting into 5 lobes to calyx, dusty pink, arranged in terminal clusters, profuse.

A recent man-made cultivar, with parents *C. alba* and *C. reflexa*, from Canberra. It is similar in many aspects to *C.* 'Gwen', except for flower colour. It has similar cultivation requirements. Propagate from cuttings.

Correa 'Poorinda Fantasy'
Cultivar

1-1.5 m x 1-2 m   May-Nov; also sporadic
Small **shrub**; **leaves** about 2.5 cm x 1 cm, broadest near base, shiny dark green above, short white hairs below; **flowers** tubular, about 3 cm x 1 cm, broadest near tips, pink with yellow-green tips.

A hardy cultivar that withstands a wide range of soil and climatic conditions. Frost tolerant, and withstands limited coastal exposure. Propagate from cuttings. Par-

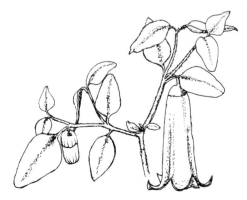

Correa 'Mannii', × ·8

ents are reported to be *C. backhousiana* and *C. reflexa*.

Differs from *C.* 'Poorinda Grace', which has larger leaves and buff hairs on undersurface of the leaves.

Correa 'Poorinda Grace'
Cultivar
1-1.5 m x 1-2 m   May-Nov; also sporadic
Small **shrub**; **leaves** to 3 cm x 1.5 cm, broadest near base, shiny dark green above, short buff hairs below; **flowers** tubular, to 3 cm x 1 cm, broadest at tips, pink with yellow-green tips.

This cultivar is of the same parentage as *C.* 'Poorinda Fantasy'. Cultivation requirements are also similar.

Correa 'Poorinda Mary'
Cultivar
1-1.5 m x 1-2 m   May-Nov; also sporadic
Small **shrub**; **leaves** about 2.5 cm x 2 cm, ovate, cordate at base, shiny green above, hairy and light green below; **flowers** tubular, about 3.5 cm x 1 cm, more or less cylindrical, pink with lime green tips.

A cultivar that should prove adaptable to many situations. Propagate from cuttings.

Is similar to *C.* 'Poorinda Fantasy', which has a more narrow floral tube.

Correa 'Rubra' see *C.* 'Dusky Bells'

Correa 'species Pink' see *C.* 'Dusky Bells'

Correa hybrid (*backhousiana* × *reflexa*)
Cultivar
1-2 m x 1.5-3 m   Feb-Oct; also sporadic
Small **shrub**; **branches** hairy; **leaves** to about 3 cm x 2.5 cm, broadest near base, dark green, upper surface slightly hairy or glabrous, undersurface densely hairy, with rusty hairs on margins and midrib; **flowers** tubular, to 3 cm x 1.5 cm, pale pink upper half, pale lime green lower half, 1-3 per axil, often with long peduncles, spreading lobes; **stamens** exserted.

This very adaptable garden cultivar is also known under the name of *C.* 'Marian's Marvel'. It will grow in most relatively well drained soils. It does well in a situation with shade to partial sun, and tolerates full sun. Frost tolerant. Withstands pruning, and can be useful as a low screening plant. Propagate from cuttings.

A similar cultivar is in cultivation, and thought to be *C. backhousiana* × *C.* 'Mannii'. The flowers are a pale dull pink. Little is known yet regarding its range of tolerance in gardens. It has been known as *C.* 'Beek's Beauty'. To date, neither of these names has been registered by the Australian Cultivar Registration Authority.

Correa hybrid (*baeuerlenii* × *lawrenciana*)
Cultivar
2-4 m x 1.5-3 m   April-July
Medium to tall **shrub**; **branchlets** brown and slightly hairy; **leaves** to 7 cm x 3.5 cm, broadest below middle, shiny dark green and slightly hairy upper surface, light green with stellate brown hairs below;

flowers tubular, about 3 cm x 1 cm, pale green; **calyx** usually slightly ribbed.

A vigorous man-made cultivar that withstands a wide range of climatic conditions. Needs relatively well drained soils. Useful as a screening plant. Propagate from cuttings.

Correa hybrid
Cultivar
0.5 m x 1-2 m   March-Sept.
Dwarf **shrub**; **branches** hairy, spreading horizontally; **leaves** 3 cm x 2 cm, cordate, dark green and glabrous above, pale with stellate hairs below; **flowers** tubular, to 3 cm x 1 cm, rose pink with pale cream-green tips.

A recent introduction that occurred as a hybrid in a garden near Sale, Vic, with the probable parents *C. backhousiana* and *C. reflexa*. The low, spreading growth habit will make it most useful as a landscaping plant. Not widely tested to date, but it is likely that it will prove hardy to most conditions, as are other hybrids with the same parents. Propagate from cuttings.

Correa hybrid
Cultivar
1-1.5 m x 1-2 m   May-July; also sporadic
Dwarf to small **shrub**; **branches** many, erect; **branchlets** with stellate tomentum; **leaves** 2-4 cm x 0.6-1 cm, oblanceolate, petiolate, erect or spreading, concave, shiny, with few stellate hairs above, light green and few stellate hairs below; **flowers** tubular, corolla to 2.5 cm x 1.7 cm, pinkish red with strongly reflexed greenish tips, usually initially erect becoming pendent; **calyx** with only minute teeth; **stamens** about 3 cm long.

This correa is a recent introduction to south-eastern Australia, where to date it is growing well. Should do well in dappled shade to partial sun, in a well drained situation. Frost tolerant. Responds favourably to regular pruning. The initially erect flowers suggest *C. decumbens* may be one of the parents. Propagate from cuttings.

Corsiaceae Becc.
A small family of monocotyledons containing about 25 species in 2 genera. They are small saprophytic herbs with unusual flowers, and are usually found in moist, shady positions on the floor of rainforests. A solitary species of *Corsia* is occasionally found in north-eastern Qld, but being a saprophyte, it cannot be cultivated.

CORYBAS Salisb.
(from the Greek *Korybas*, one of the dancing priests of Phrygia)
Orchidaceae
Small terrestrial **orchids**; **tubers** rounded, subterranean; **leaf** solitary, flat, rounded; **flower** solitary, sessile or stalked, delicate, conspicuously hooded; **fruit** a capsule.

Species about 50 with about 12 found in Australia, most of which are endemic. They are usually found in

Corybas abellianus

colonies in cool, sheltered places. The flowers are fairly short-lived.

They are a popular group with orchid enthusiasts, and are frequently cultivated in bushhouses or cool glasshouses, with varying degrees of success. They have a period of active growth over spring and then become dormant over summer. They are best grown in a pot of well drained mixture containing about one third of partially rotted leaf mould. They need plenty of water while growing and become dormant over summer, usually dying back to the tubers. Most species increase their tuber number by vegetative reproduction and will need repotting when they become too crowded. Repotting is carried out over the summer while the tubers are dormant. The tubers are placed 2-3 cm deep in the new potting mix. When in growth the orchids respond to regular applications of weak animal manures or liquid fertilizers. They are severely attacked by slugs and snails, and must be protected continually by baits.

Flowering is generally shy on plants in cultivation, and frequently the flower-buds abort before opening. This is caused by fluctuations in humidity, and can be overcome by inverting a small bottle over the bud.

Corybas abellianus Dockr.
(after Warren Abell, school-teacher)
Qld
Feb-May
Leaf 0.5-2 cm x 0.5-1.5 cm, cordate, dark bluish green, with conspicuous white veins; **flower** 0.5-1 cm long, nodding, reddish purple, with a conspicuous white patch on the labellum, hood-shaped and not opening widely.

Confined to highland rainforests of north-eastern Qld, and uncommon. The leaf is very attractively marked. Fairly difficult to grow, requiring an open soil mixture containing plenty of partially decomposed leaf mould. Requires an abundant supply of water when active growth is taking place, usually Dec-March, and needs cool conditions with protection from hot sun. Vegetative reproduction is fairly slow and the tubers are usually quite small.

Corybas aconitiflorus Salisb.
(flowers like those of the genus *Aconitum*)
Qld, NSW, Vic, Tas   Spurred Helmet Orchid
March-July
Leaf 1.5-3.5 cm x 1-2 cm, cordate, grey-green above, purplish beneath; **flowers** 2-3 cm long, distinctly hood-shaped and strongly arched, whitish with red tinges, or purple; **labellum** tubular, hidden by the dorsal sepal.

A widespread but rarely common species. Plants from north-eastern Qld frequently have much larger flowers than those from southern areas. It is fairly difficult to grow and is very sensitive to soil moisture fluctuations when in active growth, and must not be allowed to dry out. It seems very difficult to flower successfully in cultivation.

Corybas despectans D. Jones & R. Nash
(look down upon)
Vic, SA, WA
July-Aug
Leaf 1-2.5 cm x 1-3 cm, cordate or orbicular,

Corybas despectans   E.R. Rotherham

sometimes lobed; **flower** 0.7-1.2 cm long, reddish purple, sessile, strongly hooded; **labellum** tubular, expanded at the apex, with slightly denticulate margins.

Found in inland areas, usually under open forest or scrub. The smallest-flowered of the Australian species. It grows vigorously and reproduces freely, quickly building up in numbers. Responds to the use of liquid fertilizers and is fairly easy to flower. Seems to prefer a sandy mixture containing leaf mould.

Corybas diemenicus (Lindl.) Rupp & Nicholls
(from Van Diemens Land)
NSW, Vic, Tas, SA   Slaty Helmet Orchid
June-Sept
Leaf 2-3 cm x 1.5-2 cm, ovate, sometimes lobed; **flower** 0.1-0.15 cm long, hooded, stiff-textured, dark prune coloured and greyish; **labellum** incurved, purple outside, with a conspicuous white patch inside, the margins bearing short, blunt teeth.

A common species within its range, growing in large colonies in heathland and open forest. It grows vigorously and reproduces very freely, quickly filling up a pot with tubers. It needs plenty of water while growing and responds to the use of liquid fertilizers. It is fairly easy to flower but only a small proportion of plants produce buds. Successfully grown and flowered in a terrarium.

Corybas dilatatus (Rupp & Nicholls) Rupp & Nicholls
(widened into a blade)
NSW, Vic, Tas, SA   Veined Helmet Orchid
June-Sept
Leaf 1.5-2.5 cm x 1.5-2 cm, orbicular, green; **flower**

1.5-2.5 cm long, held erect on an ovary and pedicel 0.5-0.7 cm long, dark reddish purple; **labellum** opening widely, with a large, conspicuous white central patch, margined with short, fringed teeth.

A widespread, common species usually found in colonies in moist, sheltered places. The flowers are quite large for the genus and very showy. It is fairly easy to grow, although slow to reproduce and does not like being allowed to dry out while in active growth. It is, however, hard to flower in cultivation, with few buds being produced. It requires an open soil mixture containing plenty of partially decomposed leaf mould.

Corybas fimbriatus (R. Br.) Reichb.f.
(with a fringed margin)
Qld, NSW, Vic, Tas   Fringed Helmet Orchid
May-Aug

Leaf 1.5-2.5 cm x 1.5-2 cm, orbicular, bright sparkling green; **flower** 1.5-3 cm long, dark purplish red or crimson; **labellum** opening widely, purple-red inside and margined with numerous red teeth.

A widespread species which is common in coastal areas, often in large colonies. It grows very easily in cultivation and reproduces quite freely, soon building up to a small colony. The flowers abort very early while they are still buds, unless kept in humid conditions or covered with a bottle. It grows well in a sandy soil mixture containing plenty of partially decomposed organic material.

Corybas fordhamii (Rupp) Rupp
(after F. Fordham)
Qld, NSW, Vic, SA
July-Aug

Leaf 0.5-1.5 cm x 0.5-1 cm, cordate or ovate, green on both sides; **flower** 1.5-2.5 cm long, erect, on a

Corybas dilatatus in terrarium   W.R. Elliot

pedicel 1-1.5 cm long, reddish purple with conspicuous white striations; **labellum** tubular, with a small opening.

An unusual species found on tussocks in swampy habitats. It is a difficult species to maintain in cultivation, generally fading out after one season.

Corybas hispidus D. Jones
(bearing hispid hairs)
Qld, NSW, Vic
March-May

Leaf 1.5-4 cm x 1.5-3.5 cm, orbicular, green; **flower** 1.5-2.5 cm long, reddish purple; **labellum** opening widely, with a conspicuous white central patch, bearing numerous hispid hairs, the margins with numerous red teeth.

Found in inland areas, usually in large colonies. Showy when in flower. It grows very easily in cultivation and reproduces quite freely, soon building up to a potful. It is one of the easiest species to flower, the buds aborting only when subject to severe drying out. Prefers a well drained mixture containing plenty of organic material.

Corybas pruinosus (R. Cunn.) Reichb.f.
(waxy, powdery secretions)
NSW
April-July

Leaf 2-3 cm x 1.5-2.5 cm, orbicular, bright green; **flower** 1-1.8 cm long, greyish green with purplish markings; **labellum** opening very widely, margined with conspicuous white and red teeth.

Endemic in NSW, where it is confined to the central region. An easy species to grow and flower. It increases fairly slowly but responds to regular watering and the use of liquid fertilizers when growing. It is also quite hardy and tolerant of neglect. It seems to grow better in a mixture made up of well textured, heavy loam rather than sandy soil.

Corybas undulatus (R. Cunn.) Rupp & Nicholls
(undulate margins)
Qld, NSW
May-July

Leaf 1.5-2 cm x 1-1.5 cm, cordate, greyish green, purplish beneath; **flower** 1-1.5 cm tall, dark purplish red, with prominent veins; **labellum** opening widely, studded with numerous glands, the margins undulate and toothed.

Endemic in NSW, where it grows in small colonies under stunted scrub, often on soils of sandstone formation. It is a difficult species to maintain in cultivation, generally fading out after a couple of seasons.

Corybas unguiculatus (R. Br.) Reichb.f.
(having a claw)
NSW, Vic, Tas, SA   Small Helmet Orchid
May-Oct

Leaf 0.5-3.5 cm x 0.5-2 cm, ovate or cordate, sometimes lobed, greyish green, sometimes reddish beneath; **flower** 1-2 cm long, erect, on a pedicel 1-2 cm long, deep purple; **labellum** tubular, with a small opening.

Widespread in heathland, usually in scattered

colonies. It is fairly difficult to grow, tending to fade out after a couple of seasons. It usually grows in sandy soil, but in cultivation seems to prefer heavier soil mixes.

CORYMBORKIS Thouars
(from the Latin *corymbus*, a cluster of flowers; *orchis*, a Greek word for testicle but also applied to an orchid)
Orchidaceae

Small terrestrial **herbs**; **stems** erect, slender, arising from a subterranean rhizome; **leaves** thin-textured; **panicles** axillary; **flowers** congested, fragrant; **fruit** a capsule.

Species about 6 with a widespread one extending to Australia.

Corymborkis veratrifolia Thouars ex Blume
(leaves like those of the genus *Veratrum*)
Qld
0.5-1.5 m tall   Dec-March

Stems 0.5-1.5 m tall, slender (less than 0.5 cm), wiry; **leaves** 6-15, 10-30 cm x 5-10 cm, scattered along the stems, ovate, thin-textured, dark green, with irregular margins; **panicles** bearing 20-60 congested flowers; **flowers** 2-3 cm across, white, fragrant.

Confined to lowland rainforests of north-eastern Qld, usually near streams. The stems are palm-like and dissimilar to any other Australian orchid. The flowers give off a sweet cinnamon scent. Resents disturbance and is slow to re-establish. Easily grown in a pot of well drained soil mixture containing plenty of leaf mould. Requires humid, shady conditions. Can be grown as a garden plant in tropical areas but in southern Australia must be grown in a heated glasshouse with a minimum temperature of 10°C. Propagate from seed.

CORYNANTHERA J. Green
(from the Greek *coryne*, club; *anthera*, anther; referring to the club-like anthers)
Myrtaceae

An endemic, monotypic genus restricted to WA.

Corynanthera flava J. Green
(yellow)
WA
0.3-2 m x 0.3-1 m   Sept-Feb

Dwarf to small **shrub**; **branches** slender, erect; **leaves** to 0.5 cm long, narrow-elliptic, sessile, concave above, convex or more or less keeled below, overlapping on young branches, usually only on upper branches, margins usually hairy, prominent glands; **flowers** about 0.4 cm long, yellow, solitary, sessile in upper axils, forming a leafy spike to 7 cm long.

A recently named species from about 200 km north of Perth. It inhabits heaths and shrublands in loamy sands over laterite. Not well known in cultivation, but is very popular as a cut flower; therefore it needs to be farmed, as it is known only from a small area. Should do best in well drained, light to medium soils, with partial to full sun. Tolerates light frosts and extended dry periods. Propagate from cuttings and possibly seed.

Closely allied to *Micromyrtus* and *Thryptomene*, which do not produce the flowers in terminal leafy spikes.

Corynocarpaceae Engl.
A small family of dicotyledons containing only 5 species in the solitary genus *Corynocarpus*. They are trees or shrubs with entire leaves, and bear flowers in terminal panicles. Two species found in Australia.

CORYNOCARPUS Forster
(from the Greek *coryne*, a club; *carpos*, fruit; alluding to the shape of the fruit)
Corynocarpaceae

Trees; **leaves** alternate, simple, entire; **inflorescences** of panicles or racemes; **flowers** small, the parts in fives; **fruit** a large, fleshy drupe.

Species about 5 with 2 spp. indigenous to Australia.

Corynocarpus cribbianus (Bailey) L. S. Smith
(after J. G. Cribb)
Qld
10-20 m x 3-8 m   Oct-Jan

Small to medium sized **tree**; **bark** grey, with numerous longitudinal cracks; **leaves** 6-17 cm x 3-6 cm, oblong, dark green, thin-textured but firm, the apex acuminate; **racemes** 5-10 cm long, borne in the axils at the ends of the branches; **flowers** greenish; **drupe** 3-4 cm across, pinkish red, ovoid.

Found in rainforests of north-eastern Qld, both in coastal and highland areas. A handsome tree, very decorative in fruit. Not known to be in cultivation; however, it has many features to recommend it for tropical or subtropical areas. Propagate from seed.

Corynocarpus species
Qld, NSW
3-6 m x 1-3 m   Oct-Nov

Medium to tall **shrub**; **bark** smooth, brown, with corky patches; young shoots dark green; **leaves** 5-18 cm x 3-4 cm, oblong-elliptical, dark green and shiny above, pale beneath, blunt, thick and leathery; **panicles** 5-10 cm long, terminal, sparse; **flowers** about 0.3 cm across, pale green; **drupe** black.

An undescribed species restricted to dry rainforest on the slopes of Glenugie Peak in north-eastern NSW and a couple of peaks in south-eastern Qld. It has been brought into cultivation but at this stage little is known of its requirements. Plants apparently rarely set seed but can be propagated from cuttings.

CORYNOTHECA F. Muell. = *CAESIA* R. Br.

CORYPHA L.
(from the Greek *coryphe*, summit or top; probably a reference to the spectacular terminal inflorescence)
Arecaceae

Tall **palms**; **trunk** solitary, stout, spineless; **leaves** in a terminal crown, costapalmate, divided into numerous linear-lanceolate segments; **petiole** spiny on the margins; **inflorescence** terminating the trunk as a large, erect, spectacular panicle; **fruit** fleshy or fibrous.

Species about 8 with a single widespread species extending to northern Australia. The gigantic inflorescence terminates the trunk's growth, and after ripening fruit the plant dies.

Corypha elata Roxb.
(tall; lofty)
Qld, NT   Corypha Palm
10-15 m tall   Aug-Oct

Tall **palm**; **trunk** stout, up to 60 cm in diameter, woody, with spiral ridges; **leaves** 2-3 m across, dull green or glaucous, with 80-100 stiff-textured segments; **petiole** stout, 2-4 m long, with black margins and curved spines; **inflorescence** 4-8 m long; **flowers** small, pale; **fruit** about 2 cm across, fleshy, olive green.

A spectacular palm found in colonies along a few watercourses in open country of Cape York Peninsula, and NT. Plants grow 40-50 years before flowering, fruiting and dying. Not widely cultivated but found in a few tropical palm collections. Frost tender when young. Very slow-growing in southern Australia. Requires a warm sunny situation in well drained soil.

Propagate from seed which must be sown while fresh.

COSMELIA R. Br.
(from the Greek *cosmeo*, to arrange; adorn.)
Epacridaceae

An endemic, monotypic genus confined to swampy areas of the South-West Botanical Province of WA.

Cosmelia rubra R. Br.
(red)
WA   Spindle Heath
0.5-2 m x 0.3-1 m   July-March

An erect dwarf to small **shrub**; **branches** few, completely covered by the sheathing bases of the leaves, glabrous; **leaves** to about 2 cm long, broad, tapering to a pungent point, rigid, glabrous, shining; **flowers** tubular, about 2.5 cm long, tapering to tip, tube scarcely exceeds the calyx, borne on stalks, with bracts slightly pendulous, red.

A most ornamental species. Although it grows in swampy areas, in cultivation it requires relatively well drained soils, with dappled shade or partial sun. Will probably tolerate full sun if roots are protected and moist. Suitable for gardens or containers. There is little information on propagation. It is probably best to use cuttings of firm new growth, which has proved successful with most other epacrids. It is recorded that seed germination has occurred within 3-7 weeks.

Costaceae (K. Schum.) Nakai

A family of monocotyledons containing 4 genera and more than 200 species, distributed throughout the tropical regions of the world. They are closely related to the gingers and are often included in the family Zingiberaceae. The Australian representative consists of a solitary species in each of the genera *Costus* and *Tapeinochilos*.

COSTUS (K. Schum.) Nakai
(from an old Arabic name)
Costaceae

Herbs; **rhizome** creeping, subterranean; **stems** erect, leafy, fleshy; **spikes** dense, terminal; **flowers** tubular, with a conspicuous protruding labellum; **capsule** 3-celled.

Species about 250 with one endemic to north-eastern Australia.

Costus potierae F. Muell. =
                    C. speciosus (Koenig) Smith

Costus speciosus (Koenig) Smith
(showy or handsome)
Qld
1-3 m x 1-2.5 m   Nov-Jan

Stems about 3 m tall; **leaves** 10-18 cm x 3-6 cm, oblong or lanceolate, on short petioles, sheath extending above the petiole; **spike** 5-7.5 cm long, dense, globose; **bracts** scarlet, overlapping, shorter than the calyx; **calyx** 2.5 cm long, 3-toothed; **corolla** 1.6-2 cm long, yellow; **labellum** 2.5-3 cm long, convoluted, striate down the centre; **capsule** 1.2 cm long.

Confined to rainforests of north-eastern Qld. Fairly rare and little-known in cultivation. Suited to tropical gardens or glasshouse culture in southern Australia. Sensitive to cold, and damaged by temperatures less than 8°C. Needs well drained soil and protection from hot sun. Propagate from seed or by division of the rhizome. Seed may take a few months to germinate.

For many years this species has been known as *C. potierae* F. Muell.

COTULA L.
(from the Latin *cotula*; derived from the Greek *cotile*, a cup; alluding to the shape of the flower-head)
Asteraceae

Perennial or annual **herbs**; **stems** prostrate or decumbent; **leaves** alternate, entire, lobed or dissected; **flower-heads** hemispherical, small, solitary, borne on long, slender stalks; **fruit** an achene, flattened, sometimes winged, without pappus.

A large genus, distributed throughout the warmer and temperate regions of the world. Australia has a representation of 10 species, with 6 of these endemic. They occur from the coastline to alpine areas, and often frequent moist soils, even growing in inundated hollows, eg. *C. coronopifolia*. Although they are not strikingly ornamental, a few species have been cultivated as they can be useful mat or groundcovering plants.

Propagation is usually from cuttings and layered divisions but they also grow well from freshly collected seed.

Cotula alpina (Hook.f.) Hook.f.
(alpine)
NSW, Vic, Tas   Alpine Cotula
prostrate x 0.5 m   Oct-Feb

Dwarf, creeping perennial **herb**, stoloniferous, apparently glabrous; **leaves** 1-3.5 cm long, pinnatifid, lobes entire or toothed, in rosettes; **flower-heads** to 0.7 cm diameter, solitary, borne on peduncles shorter than leaves but which elongate to 3-5 cm when in fruit.

A useful matting plant that is relatively slow-growing. Will grow in a wide variety of moist soils, and tolerates shade to full sun. Hardy to frost. Propagate from cuttings or by division.

Cotula australis (Sieber ex Sprengel) Hook.f.
(southern)
all states   Common Cotula
prostrate x 1 m   Sept-Feb
Spreading annual or perennial **herb**; **stems** covered in long, soft hairs, rooting at nodes; **leaves** pinnate, 1-2 cm long, hairy; **flower-heads** about 0.5 cm diameter, borne on slender stalks longer than leaves, white and yellow-green.

A very adaptable species which can spread very quickly, even to an undesired extent. It is, however, useful as a groundcover in shaded areas where it is hard to develop other plants. Tolerates most frosts. Propagate from seed or cuttings.

Cotula coronopifolia L.
(leaves similar to those of the genus *Coronopus*)
all states   Water Buttons
0.1-0.3 m x 0.5-1 m   Sept-Feb
A creeping **perennial**; **branches** decumbent, succulent, glabrous; **leaves** 2-5 cm long, entire, toothed or lobed, glabrous, green; **flower-heads** flattish, about 1 cm diameter, borne on stalks longer than the leaves, yellow.

A common plant of edges of swamps, lakes and streams. It grows well in moist soils, with partial or full sun, often regenerating freely. Ideal for beside pools or in bog gardens, or for any area where water tends to lie after rains. Propagate from seed or cuttings.

Cotula coronopifolia, × ·7

Cotula cotuloides (Steetz) Druce
(similar to the genus *Cotula*)
WA   Smooth Cotula
0.1-0.25 m x 0.1-0.2 m   Sept-Nov
Dwarf, slender, annual **herb**; **branches** single or divided, erect; **leaves** 0.2-4 cm long, linear, filiform, entire, glabrous; **flower-heads** to 1 cm diameter, yellow, terminal.

This herb is not well known in cultivation, as to date seed has been available only through wildflower societies. It occurs over a wide range of southern WA, including the Archipelago of the Recherche. Should grow in most well drained soils, with partial to full sun. Suited to containers. Suited to coastal situations where it could be used as a colonizer of bare, sandy soils. Propagate from seed or cuttings.

C. gymnogyne F. Muell. ex Benth. is now considered conspecific with this species.

Cotula filicula (Hook.f.) Hook.f. ex Benth.
(fern-like)
NSW, Vic, Tas   Mountain Cotula
prostrate x 1 m   Nov-Feb
Spreading or tufted **perennial**; **stems** procumbent, hairy; **leaves** pinnate, 2-4 cm long, hairy; **flower-heads** button-like, to 1 cm diameter, borne on short stalks rarely exceeding the leaves, yellow-green.

Restricted in nature to the subalpine and alpine areas, this is an excellent groundcover for shaded areas. It will grow in most soils, but does not like to dry out. Will tolerate full sun and frost. Propagate from seed, cuttings or by division.

Cotula gymnogyne F. Muell. ex Benth. =
                              C. cotuloides (Steetz) Druce

Cotula longipes (Hook.f.) W. M. Curtis
(long peduncles)
Qld, NSW, Vic, Tas   Creeping Cotula
prostrate x 1 m   throughout year
Dwarf, spreading, herbaceous **perennial**; **stems** self layering, more or less glabrous; **leaves** to 10 cm long, pinnate; **pinnae** entire, toothed or deeply lobed, green; **flower-heads** to 0.5 cm diameter, solitary, on slender peduncles, yellowish.

This species occurs naturally on a wide range of soils, including those that are permanently wet. It is very adaptable and prefers an open situation with partial or full sun. Frost tolerant. Propagate from cuttings or by division.

The name of *C. reptans* (Benth.) Benth. var. *major* Benth. was misapplied to this species.

Cotula reptans (Benth.) Benth.
                    see *C. longipes* (Hook.f.) W. Curtis

CRASPEDIA Forster f.
(from the Greek *craspedon*, a border or hem; alluding to the woolly fringe of the leaves)
Asteraceae
Annual, biennial or perennial **herbs**; **leaves** radical and cauline, alternate, entire; **flower-heads** numerous,

Craspedia species C (see genus introduction)   T.L. Blake

clustered in dense, ovoid to globular compound heads, solitary and terminal on a long scape; **fruit** an achene, usually silky-hairy.

A genus also occurring in New Zealand. Australia has about 64 species, 40 of which are endemic. There is much confusion, even amongst botanists, regarding this genus, and it is in need of a complete revision.

The different species are distributed over a wide range of conditions and soils, being found on plains and also mountains, in heavy clay-loam soils or peaty, sandy soil types.

C. glauca, a most ornamental species, is one of the most commonly cultivated, but at this stage is still not grown to a great extent. A number of ornamental unnamed species occur on Mount Kosciusko. They certainly have potential for cultivation. The species tentatively described as *Craspedia* species C has silvery to silvery grey leaves and well displayed orange flower-heads.

Propagation is from seed or cuttings, or some clumps can be divided. The ripe seed is gathered from the spent flower-heads as they begin to disperse with the wind. Germination should occur within 1-3 weeks.

Craspedia chrysantha (Schltdl.) Benth.
(gold flowers)
Qld, NSW, Vic, SA   Golden Billy-buttons
0.25 m x 0.3 m   June-Dec

A tufting **herb**, annual or perennial; **leaves** 1-2 cm long, linear, silvery; **scape** to about 25 cm tall, simple; **flower-heads** globular, about 1 cm diameter, or ovoid, to 2.5 cm long, bright yellow.

This species usually occurs in heavier soils on the plains. It is not well known in cultivation, but should grow in most soils, with partial or full sun. Tolerates wet soils for extended periods. Propagate from seed, cuttings or by division.

Craspedia glauca (Labill.) Sprengel
(glaucous)
all states   Common Billy-buttons
0.3 m x 0.5-1 m   Oct-March

A tufted **perennial**; **leaves** 5-25 cm x 1-3 cm, oblong to lanceolate, tapering to base, radical, rough hairs on surface, green; **scape** to 50 cm tall, simple; **flower-heads** 1.5-3.5 cm diameter, pale yellow to orange.

Widespread from sea level to alpine areas, this very ornamental species requires relatively well drained soils, with partial or full sun. Withstands frosts.

Ideal for gardens or containers. The orange forms from subalpine regions are very ornamental. Propagate from seed or by division. Formerly known as *C. uniflora* Forster f.

Also, the following varieties are recognized:
C. glauca var. *alpina* (Backh. ex Hook.f.) Ewart

From Vic and Tas, has the foliage and stems covered in white woolly hairs.
C. glauca var. *glabrata* Hook.f. ex Hook.

From Tas, a slender, tufted plant with very narrow leaves which appear glabrous.
C. glauca var. *gracilis* Hook.f.

From Tas, has narrow leaves, whitish, with glandular hairs. The flower-heads are white, about 1 cm diameter.
C. glauca var. *macrocephala* (Hook.) Benth.

From Tas, a vigorous plant with green, hairy leaves; flowering stems are to 1 m high, with flower-heads to 3.5 cm diameter.

Craspedia globosa (Bauer ex Benth.) Benth.
(globular)
Qld, NSW, Vic, SA   Drumsticks
0.1-0.3 m x 0.3-0.5 m   Nov-Feb

A tufting **perennial**; **leaves** to 30 cm x up to 2 cm, ribbon-like, dense hairs giving a silvery appearance; **scape** to 1 m tall, simple; **flower-heads** globular, about 2.5 cm diameter, bright yellow.

This species is ornamental both in flower and foliage. It commonly occurs in low-lying heavy soils, and will grow in most soils, with partial or full sun. An excellent cut flower because of its long-lasting qualities. The flower-heads dry extremely well to enable further use. Propagate from seed or by division.

Craspedia leucantha F. Muell.
(white flowers)
NSW
0.1-0.7 m x 0.2-0.5 m   Dec-March

Loosely tufted perennial **herb**; **rootstock** branched, ascending; **leaves** 1-12 cm x 0.3-2.5 cm, oblanceolate to ovate, with narrowed petiole, acute or blunt, with rigid tip, green, more or less glabrous; **flowering stems** glabrous, with smaller leaves; **flower-heads** globular, 1-2.5 cm diameter, white.

Probably not known in cultivation, this species was originally named by Mueller about 1855, but has only recently been fully described. It is restricted to the alpine area around Mount Kosciusko, where it frequents the moist margins of springs and small creeks. In cultivation it should grow in moist, well drained soils, with partial or full sun. It will have application both for gardens and as a container plant. Propagate from seed or by division.

Craspedia pleiocephala F. Muell.
(many compound heads)
Qld, NSW, Vic, SA, WA   Soft Billy-buttons
0.5-1 m x 0.5-1 m   Aug-Jan
    A small annual, biennial or perennial **herb**; **leaves** to
5 cm long, those on stems shorter, slightly hairy, green-
grey; **flower-heads** globular, about 2 cm diameter,
borne on leafy stems, bright yellow.
    This ornamental species occurs on sandy loam or
heavier soils, and is not well known in cultivation. It
will grow well on any type of soil, with partial or full
sun. Suitable for gardens or containers. Propagate
from seed or cuttings.

Craspedia uniflora Forster f. =
                    C. glauca (Labill.) Sprengel

CRASSULA L.
(thick)
Crassulaceae
    Annual or perennial **herbs**; **stems** erect or spreading;
leaves entire, opposite, fleshy, joined at the base;
flowers closed or widely expanding, solitary or in clusters, sessile or with pedicels, with 3-20 segments.
    Species about 300 with about 10 in Australia, 7 of
which are endemic. The Australian species are mostly
small annuals or biennials with few features to recommend them for cultivation.

Crassula colorata (Nees) Ostenf.
(colourful)
NSW, Vic, SA, WA   Dense Crassula
1-10 cm tall   Oct-Feb
    Small annual **herb**; **stems** erect, succulent; **leaves**
0.2-0.3 cm long, fleshy, crowded, reddish or purplish;
flowers 0.3 cm across, with 5 petals, pinkish, in dense
clusters.
    A tiny, colourful species usually found in small
colonies on rock ledges etc. Occasionally grown as a
rockery plant. If planted in gravelly soil in a sunny
position, it can be induced to re-establish itself from
seed each year.

Crassula helmsii (T. Kirk) Cockayne
(after Richard Helms, naturalist)
Qld, NSW, Vic, Tas, SA, WA   Swamp Stonecrop
10-30 cm tall   Nov-April
    Aquatic or marsh plant; **stems** slender, much-
branched, creeping or floating; **leaves** 0.4-1.5 cm,
linear, succulent, opposite, green or reddish; **flowers**
0.3-0.4 cm across, white, solitary in the upper leaf axil.
    A succulent perennial found growing in or alongside
fresh water, usually rooting in mud. Occasionally
grown as an aquatic in ponds or dams, or as a bog
plant. Spreads fairly rapidly and may need to be
regularly checked. Propagate by division or from stem
cuttings.

Crassula sieberana (Schultes) Druce
(after Franz Wilhelm Sieber, 19th-century Austrian
botanist)
NSW, Vic, Tas, SA, NT
1-10 cm tall   Oct-Feb

Small annual or biennial **herb**; **stems** erect, succulent; **leaves** 0.2-0.4 cm long, fleshy, greenish; **flowers**
about 0.4 cm across, pink, with 4 petals, in dense
clusters, elongating and spike-like.
    A tiny annual, occasionally grown as a rockery plant.
Requirements and propagation as for *C. colorata*.

Crassulaceae DC.
    A family of dicotyledons consisting of 35 genera and
1500 species distributed throughout the warm, dry
regions of the world, with a prolific development in
South America. They are mainly small perennials with
thick, fleshy leaves, and many exotic species are grown
and esteemed as ornamentals. The family is very
poorly represented in Australia by a few lowly species
of the genus *Crassula*.

CRATEVA L.
(possibly from the Greek *cratos*, strength; hardness)
Capparaceae
    Small to medium **trees**; **leaves** trifoliolate, on long
petioles; **leaflets** entire, thin-textured, sessile; **racemes**
terminal, like a corymb; **flowers** large, showy, bisexual;
petals papery, on stalks; **stamens** numerous; **fruit** a
berry.
    A small genus of 6 species widespread through the
Pacific region with a solitary widespread species extending to northern Australia.

Crateva religiosa Forster f.
(of religious significance)
Qld, NT
3-5 m x 3-4 m   Sept-Nov
    Medium to tall, bushy **shrub**; **leaves** trifoliolate;
leaflets 5-27 cm x 3-10 cm, obovate to oblong, thin-
textured, light green, on stalks 2-9 cm long, the apex
shortly acuminate; **racemes** 3-14 cm long; **flowers** 5-
7 cm across, conspicuous; **petals** papery, white, fading
orange or brown, broadly ovate, stalked; **stamens** 10-
30, pink or purple, long, curved; **berry** 6-15 cm x 5-
9 cm, yellowish, obovoid.

Craspedia glauca in the Vic alps   T.L. Blake

Craspedia pleiocephala

T.L. Blake

Recently discovered in Australia, where it is confined to a few streams of Cape York Peninsula, growing in deep, gravelly alluvium along the stream banks and in the bed of the stream itself. It has religious significance in Asia, where it is planted around temples and religious sites. It is a showy, small tree which becomes deciduous for short periods during the dry season. Small suckers are produced from exposed roots. An attractive species for tropical areas although it is very slow-growing. Very cold tender. Requires a well drained soil in a sunny position, and responds to water in summer. Propagate from seed which should be sown while fresh or from root suckers which transplant very easily. Root cuttings would probably be successful.

CRATYSTYLIS S. Moore
(from the Greek *cratys*, strong; *stylos*, pillar; in reference to the thick, rigid style)
Asteraceae

Dwarf to small **shrubs**; **branches** hairy; **leaves** simple, alternate, sessile, dense covering of woolly hairs; **flower-head** solitary, terminal on short branches, involucre of several rows of **bracts**; **receptacle** very small; **fruit** an achene.

An endemic genus of 3 spp. with representation in western NSW, north-western Vic, SA and WA.

Only one species is known in cultivation at present.

Cratystylis subspinescens (F. Muell & Tate) S. Moore (somewhat spiny)
WA   Spiny Grey Bush
0.6-1.5 m x 0.5-1.5 m   Oct-Jan

Dwarf to small **shrub**; **branches** slightly hairy, spreading, often entangled, usually ending in spines; **leaves** to about 1 cm x 0.3 cm, more or less linear, slightly hairy giving a greyish appearance, scattered; **flower-head** to 2 cm x 0.5 cm, white to pale yellow, pendulous, comprising 3-5 flowers.

An inhabitant of inland regions, where it is often found on the edges of salt flats. The greyish appearance may make it popular in cultivation. It should be grown in very well drained, light to medium soils, and needs maximum sunlight. It is also worthy of trial in salt-affected areas. Has potential as a container plant. Propagate from seed or cuttings.

CREPIDOMANES (C. Presl) C. Presl
(from the Greek *crepis*, a boot; *manos*, soft)
Hymenophyllaceae

Small **ferns** forming mats; **fronds** erect, membranous, 1-3 times divided, shiny green; **sori** trumpet-shaped.

A small genus of about 12 species with 4 found in north-eastern Australia.

Crateva religiosa

D.L. Jones

Crepidomanes bipunctatum (Poiret) Copel.
(with 2 spots)
Qld
3-7 cm tall

Rootstock thin, wiry; **stems** winged; **fronds** longer than wide, twice-divided, dark green, shiny.

Restricted to far north-eastern Qld but there common on boulders in rainforest streams. Fairly easily grown in a pot of sphagnum moss and coarse mixture. Requires constant high humidity such as on the floor of a glasshouse, in a bottle or under a bell jar. See Filmy Ferns, Volume 4.

Crepidomanes kurzii (Beddome) Tag. & Iwatsuki
(derivation unknown)
Qld
1-2 cm tall

Small **fern** growing on rocks or trees; **rootstock** thin, hair-like; **stems** slender, winged; **fronds** erect, lobed, irregular in shape, dark green.

A widespread species that extends to highland rainforests of the Atherton Tableland. A tiny fern that is easily overlooked as it usually grows with mosses, liverworts etc. Easily grown in a pot of coarse mixture but requires constant high humidity and shade. Can be grown in a bottle, terrarium or aquarium. See Filmy Ferns, Volume 4.

Crepidomanes walleri (Watts) Tind.
(derivation unknown)
Qld
2-4 cm tall

Small **fern** growing on rocks or trees; **rootstock** slender, wiry, creeping; **stipes** shortly winged; **fronds** erect, pinnately lobed, irregular in shape, dark green.

Widely distributed throughout eastern Qld, on mossy rocks and trees in rainforest. A delicate fern that is easily grown in a pot of coarse mixture, but requires constant high humidity and shade. Can be grown in a bottle, terrarium or aquarium. See Filmy Ferns, Volume 4.

CRINUM L.
(from the Greek *crinon*, a lily)
Amaryllidaceae

Bulbous perennial **herbs**, with emergent or subterranean bulbs; **leaves** erect or limp, evergreen or deciduous; **flower-stems** arising from the axils of basal leaves or from underground bulbs; **flowers** in a terminal umbel, large, short-lived, white, pink or yellowish, often fragrant; **fruit** a rounded capsule, viviparous.

Species about 110 with about 12 extending to Australia. A very popular genus in cultivation, and a number of exotic species and hybrids are commonly grown

throughout Australia. Crinums are usually found naturally along stream banks or on flood plains, and the
seeds are often distributed by rivers and even coastal
currents. Hybridization occurs very frequently. The
Australian species have been poorly studied, mainly
because they grow in inaccessible areas and flower
during the wet season when transport is very difficult.
It is highly likely that a botanical study will result in
name changes. A number of obscure species are
believed to be merely forms of widespread species such
as *C. flaccidum*. Crinums are very easy to grow and
make excellent specimens for pots and tubs. They
thrive in a sunny situation in the garden, and seem to
flower best if neglected. Some species may need annual
flushings of water to remove toxic materials which
build up in the soil around the bulbs. In nature these
toxins are removed by floods. The seeds germinate
while the capsule is still on the plant, and take root
rapidly after they fall on to the soil.

Crinum angustifolium Bailey
(narrow-leaved)
Qld   Field Lily
0.3-1 m tall   Nov-Jan
    Leaves 30-80 cm x 2-3 cm, linear, erect and spreading
in a tussock, or flaccid, stiff, dark green, with roughened
margins; **flower-stalks** 30-50 cm long, arising in the
axils of the lower leaves; **flowers** 12-14 cm across,
white, with purple anthers, in clusters of 6-12; **capsule**
rounded.
    Found throughout Qld, along stream banks in
inland and coastal districts, usually on soils subject to
inundation. Occasionally grows in colonies but more
often as scattered plants. Grows best in warm conditions, and does not thrive in cold districts. A good
species for tropical areas or as a large glasshouse plant.
Grows well in shade but needs some sun to flower. Will
grow in a variety of soils and is tolerant of coastal conditions, but very frost sensitive. Propagate from seed
which may have to be collected while still in the capsule.

Crinum asiaticum   B. Gray

Crinum asiaticum L.
(Asian)
Qld, WA, NT
0.5-1 m tall   Dec-March
    Leaves 40-60 cm x 2-4 cm, erect and spreading in a
tussock, green or bluish green, tapering to a point;
flower-stalks 40-90 cm long, arising beside the leaves
from below ground; **flowers** 12-16 cm across, white or
pinkish, with dark red stamens and narrow petals, in
groups of 6-20; **capsule** pale green, rounded.
    Common along stream banks in sandy areas of nor-
thern Australia. A very attractive bulb for tropical or
warm inland areas, where it is often planted amongst
shrubs or in rockeries. Cold sensitive and difficult to
grow outside in southern Australia. Can be grown as a
glasshouse plant. Propagate from seed or by division of
the bulbs.

Crinum brachyandrum Herbert
(short anthers)
Qld
1-2 m tall   Jan
    Leaves 1-2 m x 4-6 cm, erect, spreading in a large
tussock, sword-like, dark green; **flower-stalks** 30-
60 cm long, arising from the axils of the lower leaves;
flowers 12-14 cm across, white, with slender segments,
in dense clusters of 10-20; **capsule** rounded.
    A little-known species found along the banks of a
few inland streams. Can be recognized by its stiff,
sword-like leaves. Hardy in tropical or warm inland
areas but not well suited to cold areas. A toxic leachate
from the plants builds up in the soil around them and
eventually kills them unless washed from the soil by
flooding. The soil around plants in cultivation,
especially those in containers, should be thoroughly
flushed at least twice a year to remove the toxin. Propagate from seed.

Crinum brevistylum Bailey
(short style)
Qld
0.1-0.5 m tall   Dec-Feb
    Leaves 60-90 cm x 3-6 cm, sword-shaped, erect and
spreading, pale green or grey, about 12-14 per bulb,
dying back each year during the dry; **flower-stalks** 60-
80 cm long, compressed, thick, arising from the axils of
the basal leaves as they emerge from the ground;
flowers 10-12 cm across, white, with broad segments,
in umbels of 10-20; **capsule** green.
    A little-known species apparently restricted to nor-
thern Cape York Peninsula and nearby islands. Introduced into cultivation in the early 1900s but not
very widespread. An attractive bulb suitable for
tropical or warm inland areas. Would probably not
grow outside in southern Australia, although could be
grown in a glasshouse. Propagate from seed.

Crinum brisbanicum Bailey
(from Brisbane)
Qld   Brisbane Lily
0.3-0.5 m tall   Nov-Jan
    Leaves 30-50 cm x 0.5-1.5 cm, dark green, linear,
erect or flaccid, about 5-6 per bulb; **flower-stalks** 30-
40 cm long, flattened, arising from the bulb before the

Crinum douglasii

emergence of leaves; **flowers** 8-10 cm across, white with green tips, segments narrow, in clusters of 8-12; **capsules** rounded, green.

Widespread in well drained, gravelly loams in hilly areas near Brisbane and also extending over much of south-eastern Qld, often on alluvial soils. An attractive species with fragrant flowers produced in succession. Not widely cultivated and it has been reported as being difficult to keep growing, although some people say it is easy. Cultivated plants are certainly shy of flowering. Best planted amongst small shrubs in a sunny situation. Propagate from seed.

Crinum douglasii Bailey
(after J. Douglas)
Qld
0.3-1 m tall   Dec-Feb
Leaves 60-80 cm x 10-14 cm, dark green, erect and spreading in a tussock, blunt; **flower-stalks** 40-60 cm long, flattened, darkly mottled, arising from the axils of basal leaves; **flowers** 12-15 cm across, white, with purple stamens, in umbels of 10-20; **capsule** green.

Apparently known only from Thursday Island. Very close to *C. asiaticum* but differing by its columnar flower-stems and broader leaves. Not known to be in cultivation, but an attractive species suitable for cultivation in tropical areas. Tolerant of coastal conditions. Would probably be unsuitable outside in southern Australia. Propagate from seed.

Crinum flaccidum Herbert
(limp)
Qld, NSW, Vic, SA, NT   Murray Lily; Darling Lily
0.5-1 m tall   Oct-Jan
Leaves 30-60 cm x 2-3 cm, weakly erect or spreading over the ground, with roughened margins; **flower-stalks** 30-70 cm long, flattened, arising from the bulb before the emergence of leaves; **flowers** 12-14 cm across, white or yellow, with broad petals, in groups of 6-14; **capsule** 1-3 cm wide, shiny green, rounded.

Found in inland areas along stream banks and plains, in areas subject to periodic flooding. Frequently grows in colonies. Leaves die back to the bulb over summer or during long dry spells. Yellow-flowered forms are known from SA (see *C. luteolum*). The flowers are strongly fragrant near dusk. Grows easily in a variety of soil types but will not tolerate bad drainage or shade. Not affected by frost, and grows in a warm protected position such as under eaves. A good rockery plant. The seeds may germinate in the capsule before being shed. Propagate from seed which germinates rapidly when kept moist.

Crinum luteolum Traub & L. S. Hannibal
(yellowish)
SA
0.5-1 m tall   Aug-Nov
Leaves 60-120 cm x 3-5 cm, somewhat flaccid, dark green, thick-textured; **flower-stalks** 0.5-1 m tall, flattened, arising from the bulb while the leaves are present; **flowers** 10-15 cm across, light primrose yellow, in heads of 10-12; **capsule** 1-3 cm wide, shiny green, rounded.

Crinum pedunculatum   B. Gray

A somewhat controversial species which may be doubtfully distinct from *C. flaccidum*. The plants are apparently evergreen, with larger leaves than *C. flaccidum*, and a different arrangement of the stamens. Yellow-flowered forms of *C. flaccidum* are very desirable subjects for cultivation, and this species would be no exception. Cultivation and propagation as for *C. flaccidum*.

Crinum pedunculatum R. Br.
(with a peduncle)
Qld, NSW   Swamp Lily; River Lily
1-3 m tall   Nov-March
Large, bulbous, perennial **herb**; **leaves** 0.6-2 m x 10-15 cm, erect and spreading in a large tussock, thickened, green or bluish, with blunt points; **flower-stalks** 30-80 cm long, flattened, arising from the axils of the lower leaves; **flowers** about 10 cm across, white, with slender segments, in dense clusters of 10-25, pleasantly perfumed; **capsule** 2-5 cm across, rounded, with a prominent beak.

A very large species found throughout coastal Qld and NSW as far south as Newcastle. Often grows in colonies along streams and tidal areas. The plants are evergreen. Commonly cultivated throughout its range. Very hardy and will grow successfully under established eucalypts. Grows in a variety of soils including clay, and will tolerate bad drainage. Not affected by frost. Will grow in full sun. A good species for coastal areas. Container plants usually flower well. Propagate from seed which germinates readily if kept moist. Sometimes seed germinates while still attached to the parent plant. Crushed leaves have been used in treatment of the very painful stings from the box jellyfish.

110

Crinum pestilentis Bailey
(ill effects of flower perfumes)
Qld
0.3-0.5 m tall   Dec-Feb
   Leaves 40-80 cm x 2-3 cm, linear, dark green, thick-textured, with roughened margins, dying back each year during the dry; **flower-stalks** 30-50 cm long, flattened, thick; **flowers** 15-20 cm across, white, very broad (2.5 cm), in umbels of 6-15; **capsule** about 3 cm across, globular, flattened.
   Found in colonies along the banks of a couple of inland streams. The flowers give off a very strong, sickening odour towards dusk and have been reported to cause vomiting. Not recommended for cultivation for this reason.

Crinum uniflorum F. Muell.
(single-flowered)
Qld, NT
20-50 cm tall   Nov-Feb
   Leaves 10-15 cm x 2-3 cm, green, weakly erect or spreading over the ground; **flower-stalks** 30-50 cm long, arising from the bulb before the emergence of leaves; **flowers** 10-12 cm across, white or pinkish, with red stamens and narrow petals, in groups at the end of the flower-stalk, a solitary flower open at any one time; **capsule** green, rounded.
   Common in open forest country subject to periodical flooding. Usually grows in large scattered colonies. The flowers are short-lived and are produced in succession. A useful bulb for tropical, subtropical or warm inland areas, tolerating heavy soils. Makes a nice rockery plant. Cold sensitive and difficult to grow outside in southern Australia. Can be grown as a glasshouse plant. Propagate from seed.

Crinum venosum R. Br.
(numerous veins)
Qld
0.3-0.5 m tall   Nov-Jan
   Leaves 40-60 cm x 1-2 cm, dark green; **flower-stems** 30-50 cm long, flattened, arising from the bulb before the emergence of leaves; **flowers** 7-10 cm across, white, with slender, pointed segments, in clusters of 10-20; **capsule** globular, green.
   A little-known species from Cape York Peninsula. Not known to be in cultivation but would be a useful bulb for tropical or warm inland areas. Suggested cultivation as for *C. angustifolium*, to which it is closely related. Propagate from seed.

CROTALARIA L.
(from the Greek *crotalon*, a rattle; a reference to the seeds rattling in the pods)
Fabaceae
   Herbs or **shrubs**; **leaves** simple, unifoliolate, trifoliolate or pinnate, with 1-7 leaflets; **leaflets** blunt, glabrous or hairy; **inflorescence** an axillary or terminal raceme; **flowers** pea-shaped, yellow, yellowish green or blue, crowded, showy; **pods** turgid, glabrous or hairy.
   A large genus of about 550 species with about 30 species indigenous to Australia. Approximately an-

other 15 species are naturalized as roadside or pasture weeds. The native species are mainly found in tropical and warm inland areas. Some are shrubby, but a few are herbaceous with a tendency to become weedy and colonize new areas. A number contain a poison which is very toxic to grazing animals, and a frequent cause of stock losses. As a group they are commonly known as rattlepods because the seeds come loose in the dry pods and rattle in the wind. Most species are readily propagated from seed after treatment with boiling water, or scarification, but a few species sucker and may possibly also be propagated from root cuttings.

Crotalaria acicularis Buch.-Ham. ex Benth.
(needle-shaped)
Qld
1-2 m x 1-2 m   Oct-March
   Spreading, small **shrub**; **stems** slender, flexuose, wiry, densely clothed with pale brown, silky hairs; **leaves** simple, 1-2.5 cm x 0.8-1.2 cm, linear-obovate, blunt, thin-textured, pale green above, glaucous beneath; **flowers** about 3 cm long, yellow, almost hidden by the silky calyx; **pods** 0.6-1 cm long.
   Widespread in near-coastal districts throughout Qld. A rather unimposing shrub with few features to recommend it for cultivation. Requires a sunny situation in well drained soil. Propagate from scarified seed.

Crotalaria calycina Schrank
(prominent calyx)
Qld, NT
0.1-0.5 m x 0.5-1 m   Oct-Jan
   Decumbent or prostrate annual **herb**; **branches** sparse, hairy; **leaves** simple, 5-15 cm x 2-5 cm, variable in shape from linear to lanceolate, glabrous and green above, grey and silky-hairy beneath; **racemes** 10-12 cm long, terminal; **flowers** about 3 cm long, pale yellow, dominated by the large, inflated, pendulous calyx which is thickly covered with long rusty hairs; **pods** 2-3 cm long, glabrous.
   Widespread in near-coastal districts of tropical Qld and NT. An unusual plant, interesting because of the large, curious calyx. Tends to be sparse and of little use as a groundcover, although it may become more bushy with regular watering and pruning. Requires well drained soil and a sunny aspect. Propagate from seed.

Crotalaria crassipes Hook. = *C. novae-hollandiae* DC.

Crotalaria crispata F. Muell. ex Benth.
(curled)
Qld, WA, NT
0.1-0.5 m x 0.5-1 m   May-Jan
   Dwarf, spreading **shrub**; **stems** prostrate or semi-erect, softly hairy; **leaves** simple, 1-2.5 cm x 0.5-1 cm, narrow-oblong to obovate, blunt, grey-green, hairy on both surfaces; **racemes** 5-10 cm long, few-flowered, terminal; **flowers** about 0.6 cm long, yellow, the calyx curling after flowering; **pods** about 1 cm long, containing a single shiny black seed.
   Widespread in tropical areas, usually in near-coastal districts and extending to islands of the Gulf of Carpentaria. Although lacking conspicuous flowers, it is

Crotalaria cunninghamii

an interesting shrub and a useful groundcover for tropical and subtropical regions. Requires a sunny situation in well drained soil. Propagate from scarified seed.

Crotalaria cunninghamii R. Br.
(after Allan Cunningham, 19th-century botanist)
Qld, NSW, WA, NT   Green Bird Flower
1-3 m x 0.5-2 m   May-Sept
    Small to medium **shrub**; **branches** softly hairy; **leaves** 3-8 cm x 1-4 cm, ovate, blunt, softly hairy on both surfaces; **racemes** 5-12 cm long, terminal, dense; **flowers** 2.5-3 cm long, yellowish green, streaked with fine black lines; **pods** 2-4 cm long, turgid, hairy.
    Widespread in inland areas of Australia, extending into desert regions. Popular in cultivation because of the large, unusually coloured flowers. Difficult to grow in cold, moist areas, and intolerant of heavy, wet soils. Prefers a hot position in well drained soil. Damaged by heavy frosts. Propagate from seed or stem cuttings. Boiled extracts of bark and also leaves were used by Aborigines to treat swellings on the body and as an eyewash respectively.

Crotalaria dissitiflora Benth.
(flowers scattered)
Qld, NSW, WA, NT   Grey Rattlepod
0.3-0.6 m x 0.5-1.5 m   Jan-April; also sporadic

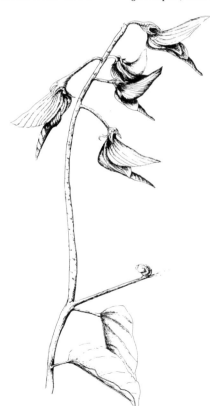

Crotalaria cunninghamii, × ·45

112

Dwarf, spreading **shrub** forming clumps by ascending congested suckers; **branches** covered with long silky, silvery hairs; **leaves** trifoliolate; **leaflets** 1.5-3 cm x 1-1.5 cm, elliptical to obovate, grey-green, blunt, glabrous above, covered with silky hairs beneath; **racemes** 10-15 cm long, loose, terminal; **flowers** about 1 cm long, roundish in outline, golden yellow; **standard** nearly circular; **pods** 1.5-2.5 cm long, bearing short hairs.
    Found on heavy clay soils along the flood plains of inland streams. An excellent plant for a rockery or low shrubbery in warm inland or tropical regions. Very hardy but responds to regular watering, which extends its flowering period. Showy when well grown. Intolerant of cool, moist conditions. Spreads by suckers, and if straggly, the shrubs can be rejuvenated by cutting back to ground level.
    The ssp. *rugosa* (Benth.) A. Lee differs from the typical form by its longer, loose, shiny hairs. It is found in inland regions of Qld, WA and NT, and is absent from NSW.
    Propagate both forms from seed, root cuttings or by transplanting suckers.

Crotalaria eremaea F. Muell.
(desert)
Qld, NSW, SA, WA, NT   Bluebush Pea
0.2-1 m x 0.5-1.5 m   Jan-May; also sporadic
    Dwarf, bushy **shrub**, spreading by horizontal sparse suckers; **stems** much-branched, hairy; **leaves** with 1-3 leaflets; **leaflets** 2-5 cm x 0.5-2 cm, elliptic, ovate or oblong, blunt, silvery grey-green or blue-grey, glabrous to densely hairy; **racemes** 10-15 cm long, loose; **flowers** 1-1.5 cm long, bright yellow; **standard** nearly circular; **pods** 2-3 cm long, inflated, silvery grey.
    A very handsome species found on sand-hills and along the flood plains of inland streams, but always in sandy soils. An excellent shrub for arid or warm inland areas. Extremely showy when well grown. Responds to regular watering. Intolerant of cool, moist conditions. Spreads by suckers, and if straggly, the shrubs can be rejuvenated by cutting back to ground level.
    The ssp. *strehlowii* (Pritzel) A. Lee differs from the typical form by its ovate rather than oblong leaflets which are nearly glabrous. It is co-extensive with the typical form but is absent from NSW and extends into WA.
    Propagate both forms from seed, root cuttings or by transplanting suckers.

Crotalaria incana L.
(hoary)
Qld, NSW, WA, NT   Woolly Rattlepod
0.5-1 m x 0.5-1 m   Dec-April
    Annual or biennial **herb**; **branches** covered with white or rusty-woolly hairs; **leaves** trifoliolate; **leaflets** 1-2.5 cm long, obovate, blunt, glabrous above, hairy beneath; **racemes** terminal or opposite the leaves, short; **flowers** about 1 cm long, yellow; **pods** 2-4 cm long, inflated, hairy.
    Found in dry coastal areas, extending inland along rivers. A very attractive species, especially when well grown. An annual under harsh conditions but longer-

Crotalaria dissitiflora

T.L. Blake

lived when regularly supplied with water. Suited to warm inland, or subtropical and tropical areas. Intolerant of cool, moist conditions. Propagate from seed.

Crotalaria juncea L.
(rush-like)

Qld, WA, NT	Sunn-hemp
0.5-3 m x 0.5-1 m	March-June

Annual, erect **herb**; **branches** striate, bearing a few silky hairs; **leaves** 2-8 cm x 1-3 cm, simple, oblong or linear, blunt, glabrous above, hairy beneath; **racemes** terminal, long; **flowers** large (about 2 cm long), yellow, sometimes with purplish markings; **pods** 2-5 cm long, densely hairy.

Widespread throughout the Pacific region and extending to northern Australia. Cultivated in Asia as an important source of fibre (sunn-hemp). Showy when in flower, and a hardy ornamental for inland areas. Best treated as an annual. Has become naturalized in NSW. Propagate from seed.

Crotalaria laburnifolia L.
(leaves like those of the genus *Laburnum*)

Qld, WA, NT	Bird Flower
1-3 m x 1-1.5 m	March-July

Medium **shrub**; **leaves** trifoliolate; **leaflets** 2.5-5 cm long, ovate, on long petioles, greyish green; **racemes** long, terminal; **flowers** 2.5-3.5 cm long, greenish yellow, with a long beak-like keel; **pods** 3-4 cm long, turgid.

Fairly restricted in its natural occurrence to sandy patches along a few large rivers of inland areas. Hardy in tropical, subtropical or inland regions and especially useful for coastal planting. Frost tender. Intolerant of cool, moist conditions. Requires well drained soil in a sunny aspect. The plant commonly grown and sold under this name in southern Australia is in fact a South African species, *C. agatiflora* Schweinf. Propagate from seed.

Crotalaria linifolia L.f.
(leaves like those of the genus *Linum*)

Qld, NSW, WA, NT	Rattlepod
0.2-0.6 m x 0.5 m	Jan-June

Annual or perennial **herb** or small **shrub**, with a thick taproot; **branches** clothed with long hairs; **leaves** 2-5 cm x 0.5-1 cm, oblong or obovate, sometimes linear, blunt; **racemes** terminal, loose; **flowers** about 1 cm long, bright yellow; **pods** about 1 cm long, glabrous.

Common in a variety of habitats from coastal to inland. Very common in Qld. Easily recognized by the very short pod. A useful ornamental for tropical areas. Showy when in flower. An annual under harsh conditions, but becoming longer-lived when supplied with water during dry periods. Easily becomes naturalized. Propagate from seed.

113

Crotalaria medicaginea Lam.
(resembling the medic genus *Medicago*)
Qld, NSW, WA, NT   Rattlepod
0.2-1 m x 0.5-1 m   March-July

Dwarf, perennial **shrub**, erect or spreading; **stems** hairy; **leaves** trifoliolate; **leaflets** 1-3 cm x 0.5-1 cm, oblong, linear or obovate, blunt, glabrous above, hairy beneath; **racemes** 2-8 cm long, terminal; **flowers** small (less than 1 cm), although some forms may have flowers 2.5 cm long, bright yellow; **pods** about 1.5 cm long, broad, turgid.

Occurs in large, often dense colonies and is very showy when in flower. A neat, compact species with attractive trifoliolate leaves. Hardy in tropical, subtropical or inland areas and especially useful for coastal planting. Intolerant of cool, moist conditions. Prefers a sunny, well drained situation. Extremely variable, and detailed studies may show that more than one species is involved. Appears to sucker and may possibly be propagated from root cuttings. Propagate from seed.

Crotalaria mitchellii Benth.
(after Sir Thomas Mitchell)
Qld, NSW   Rattlepod
0.3-1 m x 0.5-1 m   Aug-Nov; also sporadic

Dwarf **shrub**; **stems** glabrous or hairy; **leaves** 3-5 cm x 1-2 cm, simple, elliptical, glabrous or hairy; **racemes** terminal, up to 20 cm long; **flowers** about 1 cm long, deep yellow; **standard** circular, with plate-like thickenings at the base; **pods** 2-3 cm long, turgid, bearing soft hairs.

Widespread in coastal districts and western slopes between Cairns and Port Macquarie, often forming colonies in moist places, usually in sandy soils. An adaptable, somewhat weedy ornamental for tropical, subtropical or temperate areas. Prefers a sunny position but will grow in a variety of soils, even tolerating some waterlogging.

The ssp. *laevis* A. Lee differs from the typical form by its narrower, often smaller blunt leaves and a less crowded raceme.

Propagate both forms from seed.

Crotalaria novae-hollandiae DC.
(of New Holland)
Qld, WA, NT   Rattlepod
0.5-1.5 m x 0.5-1 m   April-July

Dwarf, erect **shrub**; **branches** glabrous or more usually covered with short woolly hairs; **leaves** unifoliolate; **leaflets** 5-8 cm x 1-3 cm, oblong, blunt, grey-green, covered in soft hairs; **racemes** terminal; **flowers** 1-1.5 cm long, yellow; **standard** nearly circular, with plate-like thickenings at base; **pods** 2.5-3.5 cm long, covered with woolly hairs, turning sharply downwards.

Widespread in warm inland areas, usually along stream banks and flood plains, and always in sandy soils, sometimes extending to near-coastal situations. A useful but weedy ornamental suited to warm inland or tropical areas such as around Darwin. Very drought resistant, but responds to extra water during dry periods. Cold sensitive.

The ssp. *lasiophylla* (Benth.) A. Lee differs from the typical form by its leaflets which are triangular with a cordate base and densely hairy on both surfaces. It is found mainly in inland areas of NT but also extends into Qld and WA.

C. crassipes Hook. is regarded as being merely a glabrous form of ssp. *novae-hollandiae*.

Both forms are propagated from seed.

Crotalaria pallida Aiton
(pale coloured)
Qld, NT
0.5-1.5 m x 0.5-1 m   Oct-April

Dwarf to small **shrub**; **branches** erect, silky-hairy when young; **leaves** trifoliolate, on petioles 3-9 cm long; **leaflets** 5-8 cm x 2-4 cm, oblong to obovate, blunt, grey-green, silky-hairy beneath; **racemes** 15-20 cm long, terminal; **flowers** 1-1.5 cm long, golden yellow, marked with prominent striations; **pods** 2.5-3.5 cm long, linear.

A small, weedy species widely distributed throughout coastal districts of eastern Qld and NT. Very easily grown in a sunny situation, and inclined to become naturalized, as it has in NSW. Tends to become straggly unless pruned regularly. Requires a sunny aspect, in well drained soil. Propagate from scarified seed.

Crotalaria quinquefolia L.
(5 leaflets)
Qld
0.5-1.5 m x 0.5-1 m   Aug-Jan; also sporadic

Erect annual or biennial **herb**; **stems** hollow, glabrous or clothed with silky hairs; **leaves** compound, of 3-7 leaflets; **leaflets** 3-8 cm x 1-3 cm, lanceolate to linear, blunt, glabrous or hairy; **racemes** 10-15 cm long, loose, terminal; **flowers** 2-3.5 cm long, yellow, the standard broad, the keel long and beak-like; **pods** 4-6 cm long, glabrous, on a peduncle about 1.2 cm long.

A widely distributed species that extends to coastal districts of north-eastern Qld, often in alluvial flood plains. An attractive but somewhat weedy species with large colourful flowers. Could be an excellent annual for cultivation and may be longer-lived if pruned and watered regularly. Requires well drained soils in a sunny situation. Best suited to tropical and subtropical areas. Propagate from scarified seed.

Crotalaria retusa L.
(blunt and notched)
Qld, WA, NT   Wedge-leaf Rattlepod
0.5-1.5 m x 0.5 m   Jan-May

Small annual or biennial **herb**; **branches** stiff, shortly hairy, ribbed; **leaves** 3-8 cm x 1-2.5 cm, simple, oblong or obovate, green above, pale beneath, rounded at the tip; **racemes** 15-30 cm long, sparse, terminal; **flowers** 1.5-2 cm long, bright yellow, with a broad standard; **pods** 2-3 cm long, black when ripe, spreading at right angles.

Commonly found in moist, sandy patches along the flood plains of larger rivers. Very attractive when in flower and worth growing in warm inland or tropical areas. An annual under harsh conditions but longer-lived when regularly supplied with water. Intolerant of cold. Has been successfully grown as a summer annual in southern Australia. Prefers a hot position. Propagate from seed.

Crotalaria smithiana A. Lee
(after L. Smith, 20th-century Qld botanist)
Qld, NSW, SA, NT   Rattlepod
0.3-0.5 m x 0.5-1 m   Aug-May; also sporadic

Dwarf **shrub** with ascending, procumbent or prostrate branches arising from a thick taproot; **stems** hairy, bluish grey; **leaves** unifoliolate; **leaflets** 1.5-5 cm x 1-3 cm, obovate to elliptical, densely hairy, especially when young; **racemes** terminal, up to 12 cm long, the flowers crowded; **flowers** 0.5-1 cm long, yellow, often with reddish markings; **standard** circular, with fine stripes; **pods** 1.5-2 cm long, glabrous, truncate.

Widespread on sandy and gravelly soils of inland regions, sometimes in colonies. A very drought resistant ornamental perennial suitable for areas with a hot, dry climate. Requires well drained soils in a sunny situation. Propagate from seed.

Crotalaria trifoliastrum Willd. —
Australian populations of this species are probably referable to *C. medicaginea* Lam.

Crotalaria verrucosa L.
(rough; warty)
Qld, NT
0.5-1.5 m x 0.5-1 m   Dec-May

Small annual or biennial **shrub**; **branches** spreading, prominently winged; **leaves** 5-15 cm x 3-6 cm, ovate to lanceolate, hairy, blunt; **stipules** prominent, lunar-shaped, spreading; **racemes** erect, 15-30 cm long, terminal; **flowers** 1.5-2 cm long, pale blue to violet, with a broad, recurved standard; **pods** 2.5-5 cm long, oblong, inflated, hairy.

Widespread in coastal and near-coastal districts, sometimes extending inland along riverine plains. A very attractive species with flowers of an unusual colour for the genus. An annual under harsh conditions but longer-lived when regularly supplied with water. Best suited to tropical and inland regions, in a sunny situation. Propagate from seed.

CROTON L.
(from the Greek *croton*, tick; from the appearance of the seeds of an exotic species)
Euphorbiaceae

Shrubs or small **trees**; **stems** glabrous or bearing stellate hairs; **leaves** alternate, entire or toothed, glabrous or hairy; **inflorescence** a raceme; **flowers** unisexual, lower flowers on **raceme** chiefly female, upper flowers chiefly male; **fruit** a 2-celled capsule.

Species about 750 with about 23 found in Australia, most of which are endemic and 6 of which are undescribed. The Australian species are mostly fairly drab shrubs and have few features to recommend them for cultivation. Only 2 of the Australian species appear to be cultivated to any extent. They are very easy to grow in well drained soil in a protected situation. They respond to fertilizers and to regular watering during dry periods. Some species are inclined to become leggy and should be pruned regularly to encourage a bushy habit. A few make good pot plants and could be useful for indoor decoration. This genus should not be con-

Croton arnhemicus   D.L. Jones

fused with the common name of 'croton' which is used for the colourful-leaved shrubs commonly grown in tropical and subtropical areas and which belong to the genus *Codiaeum*. Species of the genus *Croton* are propagated from seed or cuttings.

Croton acronychioides F. Muell.
(like the genus *Acronychia*)
Qld, NSW
6-10 m x 3-5 m   Oct-Dec

Tall **shrub** to small **tree**; **bark** thick, corky; young shoots bearing a few scaly hairs; **leaves** 5-10 cm x 2-3 cm, ovate to oblong, blunt, leathery, shiny dark green above, dull beneath, the margins entire or sinuate; **racemes** 3-6 cm long; **male flowers** in clusters, numerous; **female flowers** few, usually only in the cluster nearest the stem; **capsule** about 1 cm x 0.8 cm, scaly.

Widespread in coastal rainforest from north-eastern Qld to north-eastern NSW. A useful shrub or tree for subtropical and temperate regions. It is fast growing, with dark green, shiny foliage. Requires well drained soil and some protection when small. Young plants could make a useful specimen for indoor decoration. Propagate from seed or cuttings.

115

Croton armstrongii S. Moore
(after John Armstrong, 19th-century collector, NT)
NT
1-3 m x 1-2 m   May-Nov
Small to medium, slender **shrub**; **branches** erect, slender, hairy when young, in clusters or whorls; **leaves** 3-6 cm x 1-2 cm, ovate to oblong, dark green above, bearing stellate hairs beneath, the margins dentate, on short petioles, in clusters towards the ends of the branches; **racemes** 4-7 cm long, hairy, bearing separate male and female flowers; **flowers** about 0.4 cm across, greenish.

A little-known species from coastal and near-coastal areas and adjacent islands. Not known to be in cultivation but could be a very valuable, hardy plant for coastal districts in tropical regions. Propagate from seed or cuttings.

Croton arnhemicus Muell. Arg.
(from Arnhem Land)
NT
2-4 m x 1-1.5 m   July-Sept
Small to medium **shrub**; young shoots covered with whitish tomentum; **leaves** 6-15 cm x 4-8 cm, broadly cordate to nearly orbicular, dark green above, pale grey to white beneath, the veins prominent, the margins minutely and irregularly toothed, on **petioles** 4-6 cm long; **racemes** 8-20 cm long, axillary or terminal; **flowers** about 0.5 cm across, greenish, separate male and female mixed on the raceme; **capsule** about 0.5 cm across, rounded or slightly 3-lobed, covered with brownish hairs, mature Dec-Jan.

An attractive shrub that is widespread through open forest and sandstone escarpments of NT. It is rarely seen in cultivation but is a useful screening plant. The leaves are an interesting shape and have bronze tonings when young. It is a hardy species that may become partially deciduous during long dry periods. Suited to tropical or inland regions with a warm climate. Could make an attractive bonsai plant. Propagate from seed or cuttings.

Croton brachypus Airy-Shaw
(with a short foot or base)
Qld
2-4 m x 1-2.5 m   June-Aug
Small to medium **shrub**; **leaves** 5-17 cm x 2.5-6 cm, obovate, narrowly cuneate at the base, the apex blunt to shortly acuminate, shiny green above, paler beneath, the margins somewhat sinuate; **racemes** 3-5 cm long, terminal; **male flowers** white, distinctly pedicellate; **female flowers** nearly sessile; **capsule** 0.6-0.7 cm long.

Restricted to Cape York Peninsula where it is found in dry rainforest and heathland of the McIlwraith and Iron Range areas. It is virtually unknown in cultivation but could have some potential as a garden shrub for tropical and subtropical regions. Propagate from seed or cuttings.

Croton byrnesii Airy-Shaw
(after N. Byrnes, contemporary botanist, Qld)
NT
2-4 m x 1-2 m   Oct-Feb
Small to medium, slender **shrub**; **branches** terete,

glabrous; **leaves** 6-15 cm x 2-9 cm, ovate to elliptical, the base round, the apex slender, acuminate, thin-textured, bright green and glabrous, the margins toothed and/or crenate, the young ones very shiny; **racemes** 7-15 cm long, terminal; **male flowers** numerous, on slender pedicels 0.3-0.6 cm long; **female flowers** sparse, on short, thick pedicels; **capsule** about 0.8 cm across, pale green, with white stellate hairs.

An interesting species restricted to sandstone outcrops of the Top End. Recently introduced into cultivation and may have some potential as an ornamental in tropical areas. Its bushy habit may allow its use for screening. Requires well drained soil and a sunny situation. Propagate from seed or cuttings which strike fairly easily.

Croton densivestitus C. White & Francis
(densely hairy)
Qld
2-4 m x 1-2.5 m   Oct
Small to medium **shrub**; **branchlets** with a dense white covering of stellate hairs; **leaves** 9-14 cm x 3-7 cm, lanceolate to oblanceolate, acuminate, the margins entire or serrate, upper surface green and glabrous, the underside densely covered with stellate hairs; **racemes** 5-7.5 cm long, terminal, the **male flowers** in the upper part, the **females** beneath; **capsule** 0.6-0.8 cm across, bearing stellate hairs.

A striking species found in lowland rainforests of north-eastern Qld. Should have excellent potential as an indoor or glasshouse plant and is a useful garden shrub for tropical and warm subtropical areas. Requires a shady situation in organically rich soil. Very frost sensitive. Propagate from seed or cuttings.

Croton dockrillii Airy-Shaw
(after A. W. Dockrill, contemporary botanist, Qld)
Qld
2-3 m x 1-2 m   Dec-May
Medium **shrub**; young shoots bearing white stellate hairs; **leaves** 5-15 cm x 2-5.5 cm, elliptic, thin-textured, green and shiny above, dull beneath, the margins slightly sinuate, revolute; **racemes** 1-5 cm long, terminal, bearing separate male and female flowers; **flowers** white or green, with white stamens; **capsule** 0.6-0.7 cm across, bearing numerous stellate hairs.

Confined to dryish monsoonal rainforests of Cape York Peninsula. Little-known in cultivation. Requires well drained soil in a sunny situation. Very cold sensitive. Propagate from seed or cuttings.

Croton insularis Baillon
(inhabiting islands)
Qld, NSW   Queensland Cascarilla
2-6 m x 1-3 m   Nov-Dec
Medium to tall **shrub** or small **tree**; young branches silvery white from stellate hairs; **leaves** 3-8 cm x 1-3 cm, oval, green above, silvery beneath, clustered towards the ends of branchlets; **racemes** 6-10 cm long; **flowers** green, about 0.2 cm across; **capsule** about 0.6 cm across.

Widespread in or near rainforests between the Blue Mountains and the Atherton Tableland. The bark is

Croton verreauxii

D.L. Jones

long; **male flowers** numerous, in clusters; **female flower** solitary, among the males; **capsule** 0.6-0.8 cm across, densely hairy.

Widespread in rainforest from north-eastern Qld to north-eastern NSW. An interesting, fast growing tree with lax, drooping branches and a silver reverse to the leaves, which provides a contrast when the wind blows. Rarely encountered in cultivation. Requires well drained soil and protection from sun and wind when small. Propagate from seed or cuttings.

Croton stigmatosus F. Muell.
(conspicuous stigmas)
Qld, NSW — White Croton
6-12 m x 1-2 m — April-Dec

Tall **shrub** or small **tree**; **bark** brown, bearing numerous small pustules; **leaves** 8-20 cm x 3-6 cm, lanceolate to oblong, dark green above, silvery grey and densely hairy beneath, the apex long and drawn out to a fine point, on petioles 0.5-4 cm long; **racemes** 6-14 cm long, terminal; **flowers** about 0.5 cm across, creamy brown, hairy, male and females mixed on the raceme; **capsule** 0.6-0.8 cm across, round or oblong, brownish, mature Dec.

Widespread in rainforests and moist forests of south-eastern Qld and north-eastern NSW. Young plants have much larger leaves than adult plants, and these are densely silvery beneath. They make a particularly handsome pot subject. The species is easily grown in well drained soil, but requires protection when small. Propagate from seed or cuttings.

Croton triacros F. Muell.
(3 lobes)
Qld
6-8 m x 3-4 m — Oct-Dec

Tall **shrub** or small **tree**; young shoots glabrous or with a few scaly hairs; **leaves** 10-25 cm x 4-6 cm, ovate, leathery, smooth and shiny above, dull beneath, the margins slightly toothed; **racemes** 5-10 cm long, in clusters at the ends of the branches; **male** and **female flowers** on separate racemes; **capsule** about 0.8 cm across, 3-lobed, bearing stellate hairs.

Widespread in rainforests and open scrubs along the margins of streams in coastal districts and the tablelands of north-eastern Qld. An interesting, bushy, small tree with large shiny leaves. Rarely encountered in cultivation. Best suited to tropical and subtropical regions. Requires well drained soil in a semi-shaded situation. The long, straight stems were used by the Aborigines to make spears. Propagate from seed or cuttings.

Croton verreauxii Baillon
(after M. Verreaux)
Qld, NSW, Vic — Native Cascarilla
2-6 m x 2-4 m — Nov-Jan

Medium **shrub** to small slender **tree**; **leaves** 5-10 cm x 2-4 cm, ovate or lanceolate, dull green, with bluntly toothed margins; **racemes** 4-8 cm long, slender, the lower flowers chiefly female, the upper male; **flowers** 0.2-0.4 cm across, green, hairy; **capsule** globular.

Widespread in moist habitats including rainforest

---

fragrant when bruised. A straggly bush, not commonly cultivated. Requires semi-shade in a cool, protected position. The silvery underside to the leaves can be quite decorative. Propagate from seed or stem cuttings.

Croton maidenii R. Baker = *C. phebalioides* F. Muell.

Croton magneticus Airy-Shaw
(after Magnetic Island)
Qld
1-3 m x 2-4 m — June-Aug

Small to medium, straggling **shrub**; young shoots densely covered with tawny hairs; **leaves** 3-10 cm x 2-4 cm, oblong-elliptical, crowded towards the ends of the branches, bearing brownish hairs; **racemes** 6-8 cm long, densely hairy, bearing separate male and female flowers; **flowers** small, greenish.

Apparently restricted to Magnetic Island, but there common along the rocky seashore. Not known to be in cultivation; however, it could be an extremely useful plant for coastal districts in tropical and subtropical regions, tolerating considerable exposure to salt-laden winds. Propagate from seed or cuttings.

Croton phebalioides F. Muell. ex Muell. Arg.
(like the genus *Phebalium*)
Qld, NSW
10-15 m x 4-8 m — Aug-Nov

Small to medium **tree**; young shoots silvery white and bearing a scaly tomentum; **branches** slender, pendulous, silvery; **leaves** 8-15 cm x 2-4 cm, lanceolate to ovate-lanceolate, green above, silvery beneath, the margins entire or bearing small teeth; **racemes** 2-8 cm

117

and along stream banks. A rather drab tree, hardly worthy of cultivation, but fairly frequently met with in old Melbourne gardens and various botanical gardens. Requires semi-shade, good drainage and plenty of water during dry periods. Frost tender while small. Propagate from seed or stem cuttings.

CROWEA Smith
(after James Crowe, 18-19th-century English surgeon and plant enthusiast)
Rutaceae

Woody **perennials** or **shrubs**; **branchlets** angular; **leaves** alternate, simple, glandular, glabrous; **flowers** axillary or terminal, on peduncles; **petals** 5, spreading, persistent; **stamens** 10, free; **fruit** a capsule with 5 segments.

An endemic genus with 1 species from WA, and 2 from NSW and Vic. All species are ornamental, and have been cultivated for many years, with *C. exalata* and *C. saligna* proving to be adaptable to a wide range of soils and climatic conditions.

Croweas are not prone to any pests or diseases. The main pest is probably scale, with the resultant sooty mould. This can be controlled by spraying with white oil. Plants can be prone to root-fungus attack in poorly drained soils.

Plants are usually propagated from cuttings, using healthy, firm new growth, with or without the addition of a root-promoting hormone. Good results are experienced with all species. As early as 1790, plants of *C. saligna* were raised from seed in England, but this is not at all a common practice at present. *Correa alba* was used as a grafting stock for *Crowea* species in England during the early part of the 19th century.

Crowea angustifolia Smith
(narrow leaves)
WA
0.5-1 m x 0.5-1.5 m   Aug-Nov

A dwarf to small **shrub**; **branches** slender, glabrous; **leaves** to about 5 cm x 0.5 cm, linear to narrow-oblong, margins toothed, darker green on upper surface; **flowers** about 1.5 cm diameter, axillary, on short stalks, usually solitary, white to pink, buds usually pink.

This species is found mainly in moist, peaty soils of the south-western corner. In cultivation it seems adaptable, growing in most well drained soils but is susceptible to cinnamon fungus attack. It prefers dappled shade or partial sun. Responds well to light pruning, which will result in denser growth. Suitable for gardens or containers. Propagate from cuttings.

The var. *dentata* (Benth.) Paul G. Wilson can grow to 3 m high. It differs in having elliptic to obovate leaves to about 5 cm x 1 cm, and the flowers are usually white. It needs the same growing conditions as above.

Crowea exalata F. Muell.
(without wings)
NSW, Vic   Small Crowea
0.2-2 m x 0.5-1.5 m   Oct-June

Dwarf to small **shrub**; **branches** glandular; **leaves**

Crowea angustifolia var. *dentata*   W.R. Elliot

1.5-5 cm x 0.2-0.5 cm, narrow, flat, entire, glabrous, scent of aniseed when crushed; **flower** about 2 cm diameter, solitary, axillary or terminal, borne on short peduncle, profuse, pink, rarely white.

A very ornamental species, widely cultivated due to its extended flowering period and adaptability. It will grow in most relatively well drained soils, and does best in dappled shade or partial sun. Will also tolerate full sun if the root system is kept cool and moist. Frost tolerant. An ideal plant for gardens or containers. It has long-lasting qualities as a cut flower, and pruning will help promote dense growth. Propagate from cuttings.

There are many forms available commercially. The most common grows as a low, dense shrub, while another form sold as *C. exalata* 'form' or 'hybrid' is taller, usually more open and has larger flowers than the above. It originated from south-eastern NSW, and is sometimes incorrectly labelled as *C. exalata* × *saligna*. Prostrate or low, spreading forms are also available, eg.

C. exalata 'Austraflora Green Cape', which has thick, blunt-tipped, light green leaves and pale pink flowers. A low, compact form with deep pink flowers from the Victorian alps should prove to be popular in cultivation.

Crowea saligna Andrews
(willow-like)
NSW   Willow-leaved Crowea
1-2 m x 1-2 m   Dec-June; also sporadic
Small **shrub**; **branches** smooth, narrow wings from leaf-base; **leaves** 4-8 cm x 1-2 cm, elliptic to lanceolate, sessile, entire, flat, or margins recurved, midrib prominent below; **flower** to about 3.5 cm diameter, solitary, axillary, petals thick, waxy, pink, rarely white.

An outstanding ornamental species that is limited in nature to sandstone regions surrounding Sydney. It will grow in most relatively well drained soils, with dappled shade or partial sun. Tolerates full sun if provided with root protection. Frost tolerant. For gardens or containers, and like *C. exalata* is a good cut flower. Propagate from cuttings which can be slow to form roots. Best results are obtained if firm new growth is used in combination with root-promoting hormone.

Crowea 'Festival'
Cultivar
1-1.5 m x 1-1.5 m   Nov-March; also sporadic
Small dense **shrub**; **leaves** 0.3-0.4 cm x about 0.5 cm, narrow-elliptical to elliptical; **flowers** about 2 cm diameter, deep pink, axillary.

A recently registered cultivar. It is a chance seedling which is said to be a hybrid between *C. exalata* and *C. saligna*. Responds well to heavy pruning and is a good cut flower. Strikes readily from cuttings.

Crowea 'Poorinda Ecstasy'
Cultivar
1 m x 1 m   Dec-May
Shrub; **branches** smooth; **leaves** about 3-4 cm long, obovate; **flowers** about 2.5 cm diameter, axillary, pale pink.

Crowea exalata (2 forms), × ·6

A hardy ornamental hybrid, with parents stated to be *C. exalata* and *C. saligna*. It grows best in well drained soils, with dappled shade or partial sun. Withstands light pruning after flowering, if required, and is also good as a cut flower. Propagate from cuttings.

Crowea 'Poorinda Glory'
This cultivar has large flowers with narrow petals. It has strong affinities to *C. exalata*. Not well known in cultivation. It seems the name *C.* 'Poorinda Beauty' has been wrongly placed on this cultivar.

Cruciferae — see **Brassicaceae**

CRYPSINUS C. Presl
(from the Greek *cryptos*, hidden; *sinos*, a recess or rounded curve)
Polypodiaceae
Small **ferns** forming spreading patches; **rootstock** creeping, much-branched; **fronds** entire, erect or pendulous, with thickened margins, veins conspicuous; **sori** large, rounded.

Species about 40 with one extending to Australia.

Crypsinus simplicissimus (F. Muell.) S. Andrews
(very simple)
Qld
5-15 cm tall
Fronds erect, entire, on long stalks, lanceolate, narrowing to a tail, the margins shallowly toothed, dark green, with conspicuous venation; **sori** large, round, brown.

A small fern found in north-eastern Qld, on rainforest trees, usually at high elevations. Makes an attractive little basket fern. Resents soil and must be grown in a coarse epiphyte mixture. Can be grown on a fern slab. Hardy in a bushhouse or unheated glasshouse in southern Australia. Propagate by division of the rhizomes or from spore.

CRYPTANDRA Smith
(from the Greek *cryptos*, hidden; *andros*, a man; referring to the anthers being hidden by the hood-shaped petals)
Rhamnaceae
Small **shrubs**, usually hairy; **branches** twiggy, glabrous or hairy; **leaves** usually very small, alternate, in clusters at nodes; **flowers** small, white to pink, crowded, in small heads at ends or branches, surrounded by persistent brown bracts; **sepals** 5, persistent; **petals** 5, minute, hood-shaped, enclosing the anthers; **fruit** a capsule, enclosed in the persistent floral tube.

An endemic genus of approximately 40 species distributed from south-eastern Qld to NSW, Vic, Tas, southern SA and southern WA. Many species occur in very well drained soils of low-rainfall regions, eg. *C. propinqua* and *C. tomentosa*, where they usually grow with some protection from other plants.

Only a few species are cultivated to any extent, with the variable *C. amara* being the most common in gardens. Some are useful because they flower in the gloom

Cryptandra amara dwarf form, × 1   flower, × 2·8

of winter. The flowers are small, and while in most species they are white, some have the property of deepening to pink with age, eg. *C. tomentosa*. Most species require relatively well drained soils, whether they be light or heavy in texture. There is also a preference for semi-shade or partial sun locations, although they will grow in full sun, provided there is some protection for the root area. Some species, eg. *C. arbutiflora* and *C. tomentosa*, show potential as container plants. Light pruning after flowering will create dense plants. Propagation is commonly carried out from cuttings of firm young material such as lateral branchlets. Some species from WA have been propagated from seed.

Blackallia C. A. Gardner is a closely allied genus. Its species usually have their flowers arranged in racemes or bundles.

Cryptandra alpina Hook.f.
(alpine)
Tas
0.1-0.5 m x 0.2 m   Feb-June
Low **shrub**; **stems** prostrate or decumbent, wiry; **leaves** 0.1-0.3 cm long, terete, scattered or in groups, dark green; **flower** 0.2 cm long, hairy, white, solitary on the ends of branchlets.

Restricted to highland regions of central and western Tas, above 1000 m alt. Not common in cultivation, but grown by enthusiasts. Makes an attractive pot plant. Needs cool, moist conditions and is a useful small shrub for a protected rockery. Responds to regular water during dry periods. Propagate from stem cuttings.

120

Cryptandra amara Smith
(bitter)
Qld, NSW, Vic, Tas, SA   Bitter Cryptandra
0.3-0.6 m x 0.5 m   April-Oct
Small wiry **shrub**; **branchlets** spiny, stiff; **leaves** 0.2-0.6 cm x 0.1-0.2 cm, linear to obovate, flat, or with recurved margins, glabrous or slightly hairy; **flowers** 0.3-0.4 cm long, white or reddish, several together in short, leafy spikes or clusters.

Widespread in open country, especially in heathland. An attractive small shrub, particularly useful because it flowers during the autumn and winter months. Suitable for a small shrubbery or a larger rockery. Flowers are very long-lasting. May be slow-growing. Tolerates light or heavy soils and although it prefers semi-shade it will grow happily in full sun. A dwarf, compact form is sometimes sold as *C. amara* 'Nana'. It is an excellent container plant.

The species is variable and made up of 3 varieties. The typical form, found in NSW, Vic, Tas and SA, has flat leaves. *C. amara* var. *floribunda* Maiden & Betche has recurved margins to the leaves and is known from south-east Qld, NSW and Vic.

C. amara var. *longiflora* F. Muell. ex Maiden & Betche, with tubular flowers up to 0.5 cm long, is from NSW, Qld, Vic and Tas.

All forms can be propagated from stem cuttings which may be slow to strike.

Cryptandra arbutiflora Fenzl
(flowers similar to those of the genus *Arbutus*)
WA
0.6-1.5 m x 1-1.5 m   June-Nov
Small **shrub**; **branches** twiggy, slightly hairy; **branchlets** sometimes spiny; **leaves** to 0.7 cm long, narrow-linear, blunt or with recurved point, strongly revolute margins; **flowers** about 0.5 cm long, white, fragrant, more or less sessile, borne on smaller branches, profuse.

This species is from the Darling, Eyre and Irwin Districts, where it occurs on sandy soils. It is evidently not well known in cultivation, but has potential, with its profuse, fragrant flowers. Best suited to well drained, light to medium soils, with dappled shade or partial sun. Should withstand light to medium frosts and limited dry periods. For gardens or containers.

Propagate from cuttings.

The variety *tubulosa* (Fenzl) Benth. is recorded as being in cultivation, but no clear definition of this variety is made in WA.

Cryptandra buxifolia Fenzl
(leaves like those of the genus *Buxus*)
NSW
0.1-0.3 m x 0.2-0.4 m   July-Oct
Dwarf, compact **shrub**; **branchlets** slender, wiry, erect; **leaves** 1-2 cm x 1-1.2 cm, lanceolate, spreading, stiff, dark green and smooth above, velvety white beneath, the margins abruptly recurved near the apex; **flowers** about 0.6 cm long, white or pink, tubular or bell-shaped, bearing long velvety hairs, in terminal clusters of 2-6.

Found in dry situations of central and north-eastern NSW, often in mountainous areas. A very interesting species that should make an excellent rock-garden subject for temperate regions. Needs well drained soil in a sunny situation. Propagate from cuttings which seem to be slow and difficult to strike.

Cryptandra ericifolia Rudge = *C. ericoides* Smith

Cryptandra ericoides Smith
(like the genus *Erica*)
Qld, NSW
0.2-0.6 m x 0.5 m   March-July
Small wiry **shrub**; **branches** stiff, bearing appressed hairs; **leaves** 0.4-0.8 cm long, linear or terete, dark green, often in clusters, glabrous or with a few hairs; **flowers** 0.3-0.5 cm long, narrowly tubular, white, in small compact heads, surrounded by leafy bracts.

Widespread in heathlands and rocky areas from southern Qld to south of Sydney. An attractive small shrub, well suited to a rockery. Prefers a sunny aspect, in well drained soil. Grows very well amongst sandstone and is fairly commonly grown around Sydney. Useful because of its autumn-winter flowering. Propagate from stem cuttings.

Cryptandra gracilipes (Diels) C. Gardner
(slender peduncles)
WA
0.5-1 m x 0.5-1 m   July-Aug
Dwarf **shrub**; **branchlets** hairy; **leaves** 1-2 cm x 0.6-1.5 cm, obovate to oblong-obovate, truncate apex, more or less glabrous on upper surface, densely hairy below, densely hairy margins; **flower-heads** in loose cymes longer than leaves; **flowers** tubular, about 0.5 cm long, white.

A decorative species, with its hairy foliage and prominent flower-heads. It needs very well drained, light to medium soils, with dappled shade or partial sun. Suited to container cultivation. Propagate from cuttings.

Cryptandra grandiflora C. Gardner
(large flowers)
WA
0.5-1 m x 1-1.5 m   Aug-Oct
Small **shrub**; **branches** many, spreading; **branchlets** at first hairy, becoming glabrous; **leaves** 0.5-1.5 cm long, oblong to oblong-obovate, narrowest at base, often in clusters, with dense, silky hairs when young, slightly hairy or glabrous when mature; **flowers** about 0.7 cm long, white maturing to dull pink, borne in terminal clusters about 2 cm diameter.

The clustered flowers make this an ornamental species deserving of cultivation, although it is evidently not being grown much at present. It requires well drained, light to medium soils, as it grows naturally in sandy soils near Carnamah in the Irwin District. Is suited to dappled shade, or partial or full sun. For gardens or containers. Propagate from cuttings.

Cryptandra hispidula Reisseck & F. Muell.
                                                    ex Reisseck
(roughened with small bristles)
SA   Rough Cryptandra
0.2-0.5 m x 0.5-1 m   Aug-Dec; also sporadic

Dwarf **shrub**; **branchlets** roughened, bearing stellate hairs; **leaves** 0.3-0.6 cm x 0.1 cm, terete, thick, glabrous, dark green, the margins strongly revolute, in small clusters; **flowers** about 0.6 cm long, white, tubular, bearing silky hairs and stellate hairs, in terminal clusters of 2-8 flowers, each flower subtended by 4-5 brown, hairy bracts.

Confined to SA, where it grows in coastal and near-coastal areas, often in swampy conditions. A useful small shrub, tolerant of heavy soils. Interesting when in flower, with the brown bracts contrasting with the white, hairy flowers. Propagate from cuttings which may be very slow to strike.

Cryptandra lanosiflora F. Muell.
(woolly flowers)
Qld, NSW   Woolly Cryptandra
0.5-1 m x 0.5-0.8 m   Aug-Dec
Dwarf **shrub**; **branches** thin, wiry, flexuose or zig-zagged; **leaves** 0.3-0.6 cm x 0.1-0.2 cm, oblong, terete, dark green, blunt; **flowers** about 0.4 cm long, white, bell-shaped, hairy, in dense, rounded, terminal woolly clusters, each flower subtended by several shiny brown bracts.

Restricted to highland areas of south-eastern Qld and north-eastern NSW, where it grows in gorges and on rocky escarpments. Hardly spectacular but an interesting shrub. Requires well drained soil in a sunny or semi-shady position. Propagate from cuttings.

Cryptandra leucophracta Schltdl.
(having white protection)
NSW, Vic, SA, WA   White Cryptandra
0.3-0.6 m x 0.5 m   Aug-Nov
Dwarf, spreading **shrub**; young branches bearing rusty hairs; **leaves** 0.4-0.6 cm x 0.3-0.5 cm, obovate, flat, recurved at the tip, whitish on each surface from woolly hairs; **flowers** 0.4-0.5 cm long, white, silky-hairy, narrowly tubular, in dense clusters at the ends of branchlets, each cluster subtended by 2-3 brown, hairy bracts.

Confined to inland areas, on sand-hills and in mallee scrub. Not widely cultivated but useful in hot, dry areas. Requires a well drained position. May become straggly, and requires regular pruning to keep bushy. Propagate from stem cuttings which may be very slow to strike.

Cryptandra longistaminea F. Muell.
(long stamens)
Qld, NSW
0.2-1 m x 0.3-0.5 m   Sept-Nov
Dwarf **shrub**; **branchlets** bearing short hairs; **leaves** 0.2-0.6 cm long, ovate, glabrous above, silky-hairy beneath, the margins revolute; **flowers** about 0.4 cm long, numerous, crowded on the smaller branches, white, with numerous brown bracts; **petals** recurved; **stamens** erect, exserted.

Common in southern Qld and north-eastern NSW, usually on rock faces or on cliffs. A hardy, small shrub for subtropical or temperate areas. Tolerates full exposure to sun but needs very well drained soils. Frost hardy. Propagate from cuttings.

Cryptandra parvifolia Turcz.
(small leaves)
WA
0.5-1 m x 1-1.5 m   July-Aug; also sporadic
    Low, spreading, stiffly branched **shrub**; **branchlets** spiny, bearing stellate hairs; **leaves** 0.1-0.2 cm long, terete, the margins recurved, clustered on side-shoots, glabrous; **flowers** about 0.5 cm long, white, cylindrical, in small clusters of 2-6 at the branch-tips, the sepals densely hairy; **stamens** not exserted.
    A hardy, wiry shrub from desert areas, usually growing in sandy soils. It would be a useful ornamental for inland districts, in a well drained soil. Propagate from cuttings.

Cryptandra pimeleoides Hook. =
                    Stenanthemum pimeleoides Benth.

Cryptandra polyclada Diels
(many branches)
WA
0.3-1 m x 0.5-1 m   May-July; also sporadic
    Dwarf **shrub**; **branches** many; **bark** greyish; **leaves** 0.25 cm long, revolute, obtuse, in scattered clusters; **flowers** to 0.4 cm long, white to pink, in terminal clusters.
    A species with potential, as it has well displayed flowers. Occurs naturally in sandy clay-loam and loam soils. Requirements are for well drained, light to medium soils, with dappled or partial sun. Ideal for container cultivation. Propagate from cuttings.

Cryptandra propinqua Cunn. ex Fenzl
(related)
Qld, NSW, Vic, SA   Silky Cryptandra
0.2-0.4 m x 0.3 m   June-Oct
    Dwarf, wiry, much-branched **shrub**; **branchlets** rigid, spiny; **leaves** 0.2-0.4 cm long, dark green, linear, in small clusters; **flowers** 0.5-0.7 cm long, white, covered with silky hairs, in clusters of 3-8 at the ends of branchlets.
    Found in coastal and inland plains, and mallee areas. A very hardy little shrub, useful for a rockery or planting with other small-growing species. Usually remains compact and bushy, and is very showy when in flower. Flowers are very long-lasting. Prefers a sunny position and is quite drought resistant. Propagate from stem cuttings.

Cryptandra pumila (F. Muell.) F. Muell.
(little)
WA
0.1-0.3 m x 0.3-0.5 m   Sept-Nov
    Dwarf **shrub**; **stems** many; **leaves** about 0.5 cm x 0.1 cm, linear, hairy, blunt apex, crowded; **flowers** small, covered in dense white wool, sessile.
    Not well known in cultivation. This species grows as a dense dwarf shrub and will probably do best in well drained, light to medium soils, with dappled shade or partial sun. Has potential as a container plant. Propagate from cuttings.
    Is also known as *Spyridium pumilum*.

Cryptandra scortechinii F. Muell. =
                    Stenanthemum scortechinii F. Muell.

Cryptandra tomentosa red flower form   W.R. Elliot

Cryptandra spinescens Sieber. ex DC.
(ending in a spine)
Qld, NSW
0.2-1 m x 0.5-1 m   March-Oct
    Dwarf, rigid **shrub**; **branchlets** very spiny, crowded, short; **leaves** 0.4-0.6 cm long, linear or oblong, dark green; **flowers** 0.3-0.4 cm long, white, hairy, crowded along the branches but not grouped into heads.
    Scattered in open forest and rocky areas from southern Qld to south of Sydney. Little-known in cultivation. A hardy little shrub, suitable for a rockery or planting amongst other small shrubs. Requires a sunny, well drained position. Frost tolerant. Propagate from stem cuttings.

Cryptandra tomentosa Lindl.
(covered with short, soft hairs)
Vic, SA, WA   Prickly Cryptandra
0.2-0.6 m x 0.2-0.4 m   Aug-Oct
    Dwarf, wiry, much-branched **shrub**; **branchlets** wiry, pointed; **leaves** 0.2-0.6 cm long, terete, dark

green; **flowers** 0.2-0.3 cm· long, white, tubular or star-like, bearing both simple and stellate hairs, in dense clusters at the ends of branchlets.

Found in coastal heathlands and a few inland areas. A hardy, rigid little shrub, useful for a rockery or planting with other small-growing plants. Prefers a sunny or semi-shaded, well drained position.

A form from the Grampians, Vic, has striking deep red flowers. It is not common in cultivation and should prove to be a very desirable cultivar.

Propagate from stem cuttings which may be slow to strike.

Cryptandra tubulosa Fenzl =
C. arbutiflora var. *tubulosa* (Fenzl) Benth.

Cryptandra waterhousii (F. Muell.) F. Muell.
(after G. A. Waterhouse)
SA (Kangaroo Island)
0.5-1 m x 1-1.5 m   Aug-Oct
Dwarf, erect, sticky **shrub**; **leaves** 1-2.5 cm x 0.2-0.3 cm, linear, dark green, with loose, silky hairs, the margins revolute; **flowers** about 0.3 cm long, white, hairy, in small cymes of 1-3 flowers.

Confined to Kangaroo Island but there plentiful, usually in deep sandy soils. A useful shrub for coastal situations in temperate regions, tolerating some exposure to salt-laden winds. Requires well drained soil in a sunny or semi-shady situation. Propagate from cuttings.

CRYPTANTHEMIS Rupp
(from the Greek *cryptos*, hidden; *anthos*, a flower; alluding to the subterranean flowering)
Orchidaceae
A monotypic genus endemic in eastern Australia.

Cryptanthemis slateri Rupp
(after E. W. Slater, the original collector)
Qld, NSW   Underground Orchid
                                         Oct-Nov
Subterranean, saprophytic **orchid**; **rhizome** up to 15 cm long, stout, fleshy, branched; **bracts** subtending the inflorescence; **inflorescence** a capitulum; **flowers** 15-30, small, crowded, waxy white, sepals and petals free.

An extremely rare orchid that has been found only on about 4 occasions in open forest. It is completely subterranean, even in its flowering. Nothing is known about its behaviour in cultivation, but being a leafless saprophyte it probably cannot be grown by any known method.

CRYPTOCARYA R. Br.
(from the Greek *cryptos*, hidden; *caryon*, a nut; in reference to the fruit hidden by the style)
Lauraceae
Shrubs or **trees**; **leaves** simple, alternate, thin-textured or leathery, full or shiny, often hairy; **inflorescence** an axillary or terminal panicle; **flowers** bisexual, small, greenish or white; **fruit** rounded or oval, succulent or semi-succulent, enclosing a single large seed.

A very large genus of 250 species with 23 being described in eastern Australia, and in addition there are several unnamed species. They are handsome, usually dense shrubs or trees, often with attractive foliage, and are found in or near rainforests. The majority of species have some potential for cultivation but few if any have become well known. Some are too large for the average garden but are well suited to parks or for acreage planting. A number, unfortunately, have strongly smelling, malodorous flowers, which precludes their planting near houses. At least one and possibly two species have extremely toxic sap, and perhaps others should be handled carefully. Plants may need some protection when young, and as a group they prefer free-draining soils rich in organic matter. Most species have handsome foliage when young and make excellent indoor plants. The are often called walnuts from the wrinkled coat of the seed. Propagation is from seed which should be sown while fresh and may take 12 months to germinate. Some species can be propagated from cuttings of semi-mature wood.

Cryptocarya angulata C. White
(having corners)
Qld   Ivory Walnut
20-30 m x 10-15 m   Jan-Feb
Medium to tall **tree**; **bark** corrugated; young branches glabrous or with a few hairs; **leaves** 7-10 cm x 2-4 cm, elliptical, pale green beneath; **panicles** about as long as the leaves; **fruit** 2-2.5 cm long, ovoid, bright purple.

Restricted to highland ranges and tablelands of north-eastern Qld, in dense rainforest. An attractive, large tree suitable as a shade tree for tropical or subtropical areas, especially for large parks and gardens. Fruits are very decorative, ripening in Sept. Propagate from seed.

Cryptocarya bidwillii Meisner
(after J. C. Bidwill)
Qld, NSW
8-15 m x 5-10 m   Oct-Dec
Small to medium **tree**; young shoots clothed with short, rusty hairs; **leaves** 9-15 cm x 3-4 cm, lanceolate to elliptical, dull green above, paler beneath; **flowers** small, greenish, borne in short racemes in the axils of the upper leaves; **fruit** about 1.2 cm across, ovoid, black, mature Jan-Feb.

Found along rainforest margins and in moist forests from Proserpine to north-eastern NSW, often on gravelly ridges. Usually a small, straggling tree but sometimes assuming forestry proportions. Growth habit is interesting and the species grows easily in well drained soils. Appreciates water during dry periods, and the application of mulches and fertilizers. Young plants are excellent for indoor decoration. Propagate from seed or cuttings.

Cryptocarya bowiei Hook. —
see *C. laevigata* var. *bowiei* (Hook.) Kosterm.

Cryptocarya cinnamomifolia Benth.
(leaves like those of the genus *Cinnamomum*)
Qld — Cinnamon Laurel
10-20 m x 8-12 m — Oct-Dec

Medium **tree**; young shoots bearing short, rusty hairs; **leaves** 7-15 cm x 2-5 cm, ovate to lanceolate, thick and leathery, prominently 3-veined, dark green and glossy above, glaucous or ash grey beneath; **flowers** small, greenish, in panicles near the ends of branches; **fruit** about 1.2 cm across, round, brownish black, mature Sept-Nov.

Widespread in highland and lowland rainforest between Mount Spec and Mount Spurgeon. A handsome, dense tree with attractive leaves. Best suited to large gardens, parks, or as a street tree. Seedlings make suitable plants for indoor decoration. Requirements are for well drained soil rich in organic matter, and a situation that offers some protection when young. Propagate from seed which should be sown while fresh.

Cryptocarya corrugata C. White & Francis
(wrinkled; corrugated)
Qld — Oak Walnut; Corduroy
15-25 m x 10-12 m — June-Aug

Medium to tall **tree**; **bark** brown, longitudinally corrugated; young shoots clothed with dense, short brown hairs; **leaves** 5-8 cm x 2-3 cm, ovoid, glossy above, paler beneath, the apex drawn out and blunt; **flowers** small, greenish; **fruit** about 1 cm long, round, black, mature Oct-Nov.

Widespread in highland rainforests of north-eastern Qld between Eungella and the Atherton Tableland. The corrugated bark gives rise to the local name of washing-board tree. Not widely grown, and because of its size best suited to parks and large gardens. Requires well drained soil rich in organic matter. Propagate from seed which should be sown while fresh.

Cryptocarya cunninghamii Meisner
(after Alan Cunningham, prominent early botanical collector)
WA, NT — Coconut Laurel
6-15 m x 2-4 m — Nov-Jan

Small to medium, spreading **tree**; **leaves** 8-11 cm x 2-4 cm, lanceolate, leathery, dark green and glossy, dense, the main veins somewhat prominent; **cymes** and **panicles** terminal or axillary; **flowers** about 0.4 cm across, greenish, numerous, with an offensive odour; **fruit** about 1.5 cm across, bluish black, globular or somewhat flattened.

An attractive, dense tree with glossy foliage, widespread in moister areas of the Top End but not necessarily along stream banks. The trees may become deciduous during the dry season. It is little-known in cultivation, and its further use may be restricted following suspicious toxic reactions to its sap noted by botanists and foresters. These reactions are apparently not as severe as in *C. glabella* but should be considered.

Cryptocarya erythroxylon Maiden & Betche
(red wood)
Qld, NSW — Rose Maple; Pigeonberry Ash
20-50 m x 10-20 m — Dec-Feb

Cryptocarya bidwillii — D.L. Jones

Tall, spreading **tree**; young branches slightly hairy; **leaves** 8-13 cm x 3-5 cm, elliptical, dark green, dull or greyish beneath; **panicles** about as long as the leaves; **flowers** 0.1-0.2 cm across, greenish; **fruit** about 1 cm across, ovoid, red.

A very important timber tree. The wood is pink-brown and has a strong fruity odour when freshly cut. The fruit is avidly sought after by fruit pigeons. The species is restricted to north-eastern NSW and south-eastern Qld, in rainforest or moist gullies. A handsome, spreading tree suited to large gardens or parks in subtropical areas. Hardy in a protected position in southern Australia, but slow-growing. Frost tender while young. Requires regular water during dry periods. Propagate from seed.

Cryptocarya floydii Kosterm.
(after A. G. Floyd, contemporary NSW botanist)
Qld, NSW — Gorge Laurel
6-15 m x 1-2.5 m — Oct-Nov

Tall **shrub** or small, slender **tree**; **trunk** crooked; **bark** grey to black, rough; **branchlets** greenish black; **leaves** 3-6 cm x 0.8-2 cm, broad-lanceolate, leathery, tapering to a long point, dark green and shiny above, dull beneath, with a pale margin; **cymes** 1-2 cm long, axillary; **flowers** about 0.3 cm across, pale green, fragrant; **fruit** about 1.2 cm across, globular, black, pointed, shiny, with prominent longitudinal ribs.

An uncommon species usually found on skeletal soils in dry rainforest, or in rocky gorges of south-eastern Qld and north-eastern NSW. The plants are handsome, with a very dense crown, and have potential for garden cultivation or as a container plant. Tends to be slow-growing. Young plants may require protection from direct sun. Propagate from seed or possibly cuttings.

Cryptocarya foetida R. Baker
(fetid)
Qld, NSW — Stinking Cryptocarya
8-12 m x 3-5 m — Oct-Dec

Small **tree**; young shoots bearing short hairs; **leaves** 9-13 cm x 4-6 cm, ovate to elliptical, leathery, shiny green, the apex drawn out, blunt; **flowers** small, green-

124

Cryptocarya glaucescens   D.L. Jones

cases of foresters, botanists and timber workers suffering severe skin blistering and burning after contacting the timber or sap. The effects from the alkaloids (pleurospermine in particular) are so bad that they have been likened to mustard-gas burns. Of particular significance is the delayed reaction — it is usually 1-3 days before the first symptoms appear, and these get progressively worse and do not subside until 11-12 days after contact. Initial symptoms are patchy redness and itchy blisters on the hands, and later a widespread itchy rash, swelling of the face and scrotum, depression and a continual dull headache. The poison from the sap can enter the bloodstream direct through the unbroken skin of the hands and arms. It is advisable to remove trees of this species already planted in gardens. Care should be taken to avoid any contact with sap during the removal operation. The species has also been known as *C. pleurosperma* C. White & Francis.

ish, offensive, in small cymes; **fruit** about 1 cm across, globular, black.

An uncommon species that grows in littoral rainforests and sheltered areas behind coastal sand-dunes of northern NSW and southern Qld. The flowers have an offensive smell which precludes its use near dwellings; however, it has significance as a coastal plant for beach reclamation work. Will not tolerate full exposure to coastal winds, and must be established in a protected situation. Propagate from seed which must be sown fresh.

Cryptocarya foveolata C. White & Francis
(with hairy domatia or foveoles)
Qld, NSW
20-40 m x 10-20 m

Medium to tall, spreading **tree**; young branches covered with fine hairs; **leaves** 4-8 cm x 2-4 cm, elliptical, blunt, green on both surfaces; **flowers** small, green or white; **fruit** about 1 cm across, globular, black.

Restricted to coastal rainforests and ranges of south-eastern Qld and north-eastern NSW. Common at high elevations. Little-known in cultivation. Suited to large gardens or parks in temperate coastal or subtropical areas. Frost tender while young. Propagate from seed.

Cryptocarya glabella Domin
(bearing ribbed seeds)
Qld   Poison Walnut
8-15 m x 3-7 m   Dec-Jan

Medium **tree**; young shoots angular, the buds covered with silky hairs; **leaves** 9-14 cm x 1-4 cm, elliptical, bluntly acuminate, dark green, glossy above, dull beneath, with 3 veins prominent; **fruit** 3.5-4.5 cm x 2.5-3 cm, ovoid, globular, bright red; **seed** large, conspicuously ribbed.

Restricted to highland rainforests of north-eastern Qld. All parts are extremely poisonous, and the sap is very toxic, so cultivation cannot be recommended. Despite this severe drawback it is to be found in enthusiasts' collections. On no account should nurseries stock it for sale to the general public. There are many

Cryptocarya glaucescens R. Br.
(greyish)
Qld, NSW   Silver Sycamore; Brown Beech
10-30 m x 5-15 m   Nov-Jan

Small to tall, spreading **tree**; **leaves** 8-14 cm x 3-5 cm, lanceolate or elliptical, leathery, glaucous beneath; **panicles** terminal, or in axils; **flowers** about 0.2 cm across, numerous, hairy; **fruit** 1-1.5 cm across, black, globular, flattened.

Found in rainforest, frequently in drier situations. The timber is light coloured and useful for a variety of purposes. Makes an attractive tree, although probably too large for home gardens. Quite cold tolerant, and can be grown in areas with a moderate to severe winter. Fairly fast growing. Prefers a semi-shady position and regular water during dry periods. Propagate from seed.

Cryptocarya gregsonii Maiden
(after Jesse Gregson, botanical collector)
NSW (Lord Howe Island)   Black Plum
5-10 m x 3-6 m   Jan-March

Tall **shrub** to small **tree**; young shoots bearing rusty hairs; **leaves** 4-8 cm x 3-5 cm, ovate, elliptic, glabrous, glossy dark green above, paler beneath, fairly crowded; **panicles** terminal on branchlets and in the axils of upper leaves; **flowers** about 0.4 cm across, pale yellow, the calyx bearing rusty hairs; **fruit** 1-1.5 cm long, ovoid, green to black.

Apparently endemic on Lord Howe Island, where it is common in the forests above 300 m alt. Rare in cultivation, but a shapely, small tree well suited to coastal areas in subtropical and temperate regions. Requires well drained soil and protection from frosts and excessive sun when young. Propagate from seed or cuttings.

Cryptocarya hypospodia F. Muell.
(ash grey beneath)
Qld
20-30 m x 10-12 m   Dec-March

Medium to tall **tree**; young shoots fluted, bearing brown hairs; **leaves** 15-20 cm x 4-8 cm, ovate-lanceolate, glabrous and shiny above, pale beneath,

125

Cryptocarya laevigata

Cryptocarya laevigata var. *bowiei,* × ·5

hairy; **panicles** short; **flowers** small, greenish, with a strong offensive odour; **fruit** 1.5-2 cm long, reddish, ribbed.

Widespread in rainforests between Gympie and the Atherton Tableland, and cut as a timber tree. Too large for the average home garden but well suited as a specimen tree for parks. A drawback is the strongly smelling flowers. Needs well drained soil and a protected situation when small. Propagate from seed which should be sown when fresh.

Cryptocarya laevigata Blume
var. **bowiei** (Hook.) Kosterm.
(smooth; slippery)
Qld, NSW   Grey Sassafras
4-6 m x 2-3 m   Sept-Nov

Tall **shrub** to small **tree**; **leaves** 8-12 cm x 3-4 cm, lanceolate, glossy above, with a prominent vein on either side of midrib; **panicle** short; **flowers** 0.2-0.3 cm across, greenish; **fruit** about 2 cm across, yellow to bright red, rounded.

Widespread in coastal scrubs between Cairns and the Clarence River. Grown in glasshouses in England since the early 19th century. Not widely cultivated, but bushy to ground level so would be a useful shrub for coastal, subtropical or tropical districts. Also hardy in temperate Australia. Decorative when in fruit and always attractive, with shiny leaves. Frost tender. Requires regular watering during dry periods, and responds to fertilizers. Young plants in pots are excellent for indoor decoration. Occasionally, yellow-fruited plants are found and these are quite decorative. Propagate from seed. Stem cuttings are also successful.

Cryptocarya mackinnoniana F. Muell.
(after L. Mackinnon)
Qld   Koonjoongaroo
10-30 m x 10-15 m   Dec-Feb

Medium to tall **tree**; young branches densely clothed with brown hairs; **leaves** 10-20 cm x 4-6 cm, elliptical, blunt, with prominent veins; **panicles** shorter than the leaves, dense, terminal, profuse; **flowers** small, green-cream, somewhat strong-smelling; **fruit** 2.5-3 cm long, oval, purple with brown spots, mature May-June.

Found in coastal and highland rainforest. Trees become covered with masses of flowers in good years. Fruits also are very decorative. Suitable as a shade tree for tropical or subtropical areas, but requires plenty of room to spread. Flushes of new growth are bright red and are a very decorative feature, especially on young trees, which make excellent indoor plants. Needs well drained soil. Propagate from seed which must be sown fresh.

Cryptocarya meisnerana Frodin
(after K. F. Meisner, 19th-century Swiss botanist)
Qld, NSW
3-8 m x 2-3 m   Nov

Medium **shrub** to small, slender **tree**; **branches** dark reddish brown; **leaves** 5-8 cm x 2-3 cm, elliptical, thin-textured, dark green above, paler beneath; **panicles** axillary, small; **flowers** small; **fruit** 1-1.2 cm long, elliptical, black, smooth.

Widespread in cool, moist rainforests of south-eastern Qld and north-eastern NSW. Very fast growing and may be short-lived, as it colonizes roadsides and forest clearings. Prefers cool, moist conditions in shady or semi-shady situations in well drained soils. Hardy to most frosts. Propagate from seed.

Cryptocarya microneura Meisner
(very small veins)
Qld, NSW   Murrogun
10-25 m x 5-12 m   Sept-Jan

Medium sized, spreading **tree**; **leaves** 5-12 cm x 2-4 cm, lanceolate, dark green, thin-textured, with prominent veins; **panicles** terminal on branchlets and in the axils of upper leaves; **flowers** 0.2-0.4 cm across, white; **fruit** 1-1.5 cm long, black, globular.

Found in near-coastal rainforests from north-eastern Qld to south-eastern NSW. A drab, small bushy tree occasionally grown in temperate and subtropical areas, but with few features to recommend it. Frost tender while young. Slow-growing but responds to regular watering and fertilizers. Seedlings have a prominent white mid-vein in the leaves and make attractive indoor plants, tolerating very dark conditions. Propagate from seed.

Cryptocarya murrayi F. Muell.
(after J. J. Murray)
Qld
15-25 m x 8-12 m   Oct-Dec

Medium to tall **tree**; young shoots clothed with long brown hairs; **leaves** 15-25 cm x 4-8 cm, ovate-elliptical, glabrous and shiny above, hairy and slightly glaucous beneath, blunt or shortly pointed; **flowers** about 0.2 cm across, greenish, bearing long brown hairs; **fruit** about 1 cm long, ovoid, glossy, black, mature April-May.

Widespread in coastal and highland rainforests of north-eastern Qld. The flowers exude a heavy perfume which tends to be sickly and overpowering. Rarely seen in cultivation, but suitable as a shade tree or ornamental for parks and large gardens. Seedlings have large leaves, and could be used for indoor decoration. Needs well drained soil rich in organic matter, and some protection when young. Propagate from seed which must be sown fresh.

Cryptocarya triplinervis   D.L. Jones

Cryptocarya oblata Bailey
(oblate; orange-shaped)
Qld
20-30 m x 10-15 m   May-June
   Medium to tall **tree**; young shoots covered with
bright rusty red hairs; **leaves** 10-15 cm x 4-6 cm, ovate-
lanceolate, shiny, with long, drawn out points, membranous, the upper surface concave; **flowers** small,
greenish; **fruit** 3-5 cm across, globular, red, mature
Nov-Dec.
   Widespread in coastal rainforests of north-eastern
Qld, and cut as a timber tree. Rarely grown because of
its size, but suitable for parks and large gardens.
Seedlings grow rapidly to about 60 cm tall and make
attractive pot plants suitable for indoor decoration.
The trees need well drained soil rich in organic matter,
and need protection from direct sun when young. Best
suited to tropical or subtropical regions. Propagate
from seed which must be sown fresh and germinates
within 2 months.

Cryptocarya obovata R. Br.
(obovate leaves)
Qld, NSW   White Walnut; Pepperberry Tree
20-40 m x 10-15 m   Jan-April
   Medium to tall **tree**; young branches covered with
brown, velvety hairs; **leaves** 5-14 cm x 3-4 cm, ovoid,
leathery, glabrous and shiny above, glaucous beneath,
hairy; **panicles** about as long as the leaves; **flowers**
about 0.2 cm across, greenish, hairy, with a strong offensive smell; **fruit** about 1.2 cm across, globular,
black.
   Common in rainforest and along stream banks between Port Stephens and Gympie. A useful timber tree;
however, the wood is light coloured and not very
desirable. An excellent shelter tree for tropical areas or
coastal districts, but not suited to dry areas. A
drawback is the strong-smelling flowers. Frost tender
and slow-growing in southern areas. Requires a regular
supply of water and responds to fertilizers. Propagate
from seed.

Cryptocarya pleurosperma C. White & Francis =
                              C. glabella Domin

Cryptocarya rigida Meisner
(rigid leaves)
Qld, NSW   Brown Beech
10-30 m x 5-15 m   Dec-Feb
   Small to tall **tree**; young branches clothed with rusty
hairs; **leaves** 5-10 cm x 2-3 cm, lanceolate, acuminate,
green above, whitish beneath; **panicles** shorter than
the leaves, very broad, hairy; **flowers** about 0.2 cm
across, greenish, very hairy; **fruit** 1-1.5 cm long, oval,
black.
   Found in rainforest and moist areas of open forest
from near Newcastle to the Atherton Tableland. In the
southern part of its range it rarely gets over 10 m tall. A
useful small tree for ornamental purposes. Fairly slow-
growing and hardy. Requires regular watering and
responds to fertilizers. Propagate from seed.

Cryptocarya triplinervis R. Br.
(3-nerved)
Qld, NSW
5-10 m x 2-5 m   Oct-Jan
   Small to medium sized **tree**; **leaves** 5-10 cm x 2-
4 cm, ovate-elliptical, pointed, glabrous above, hairy
beneath; terminal **panicles** branched, axillary panicles
short and dense; **flowers** about 0.2 cm across, greenish;
fruit about 1.5 cm long, black, ovoid.
   A variable species that grows in rainforest and along
stream banks from north-eastern NSW to the tip of
Cape York Peninsula. It is also recorded from Lord
Howe Island. In cultivation it becomes a tall bush or
small bushy tree. It has attractive foliage and growth
habit, but is not commonly grown. Very slow-growing
and frost tender while young. Hardy as far south as
Melbourne if given some protection. Prefers semi-
shade. Propagate from seed.

Cryptocarya species
NSW   Dorrigo Laurel
6-20 m x 1-2 m   Oct-Nov
   Tall multi-stemmed **shrub** to small **tree**; **bark**
greyish brown; **branchlets** bearing soft rusty hairs;
young shoots silvery; **leaves** 3-8 cm x 2-3 cm, ovate-
lanceolate, dark green and shiny above, dull grey-green
beneath; **panicles** 2-3 cm long, in the upper axils;
flowers about 0.2 cm across, creamy yellow, hairy;
fruit 1.3-1.7 cm across, bluish black, fleshy, mature
Jan-May.
   Confined to the highland rainforests of the Dorrigo
region, on skeletal soils, often in exposed positions. It is
hardly known in cultivation but is a hardy plant that
could have use in temperate or subtropical gardens.
Tends to be slow-growing. An attractive container
plant that could be useful for indoor decoration. Propagate from seed or perhaps cuttings.

Cryptocarya species
NSW
3-6 m x 1-1.5 m   Oct-Nov
   Medium to tall **shrub**; **trunk** crooked; **bark** dark
brown; new shoots pinkish red; **leaves** 1.5-2.5 cm x 1.3-

Cryptostylis erecta   J. Fanning

1.6 cm, ovate to broadly elliptical, leathery, glossy dark green on both surfaces; **racemes** 0.5-1 cm long, axillary, bearing 1-3 flowers; **flowers** about 0.2 cm long, white, silky-hairy; **fruit** about 1 cm across, oval to globular, black, somewhat shiny.

A little-known species confined to limestone areas of north-eastern NSW, where it grows in colonies. It is a wiry, gnarled plant with a dense head of glossy, leathery foliage. It has excellent garden potential and could also make an ideal pot plant. Its gnarled growth habit lends it well to bonsai culture. It will grow in acid as well as limestone soils, and the plants will take full sun even when quite small. It is well suited to temperate regions. Propagate from seed or cuttings.

Cryptocarya species
NSW   Mountain Laurel
6-15 m x 1.5-3 m   Dec-Jan
Tall **shrub** to small **tree**; **bark** brown; **branches** thick, spreading horizontally; **branchlets** orange, hairy; new shoots pink; **leaves** 4-8 cm x 2-2.5 cm, ovate-lanceolate, drawn out to a blunt tip, dark green and shiny above, dull grey beneath; **panicles** up to 1 cm long, axillary; **flowers** grey-fawn; **fruit** about 1 cm long, globular, black.

An uncommon species restricted to cool, moist highland rainforests of north-eastern NSW, above 1000 m

alt. Not known to be in cultivation but could be a decorative plant for temperate or subtropical regions. Would probably require cool, moist conditions, and would most certainly adapt to a shady position. Propagate from seed or cuttings.

CRYPTOSTYLIS R. Br.
(from the Greek *cryptos*, hidden; *stylos*, a column; in reference to the nearly hidden column)
Orchidaceae

Small terrestrial **orchids**; **roots** fleshy, rhizome-like, subterranean; **leaves** in clumps, erect, entire, evergreen; **inflorescence** a tall **raceme**; **flowers** few to numerous, the labellum large and conspicuous, the rest of the segments small and thin; **fruit** a capsule.

A small genus of about 20 species with 5 species in Australia, 4 of which are endemic.

They are popular subjects for cultivation in bush-houses or unheated glasshouses. *C. hunterana* Nicholls is a leafless saprophyte and cannot be grown by any known method. The other species have a period of active growth during summer but retain their leaves throughout the year. They have an extensive root system and must be grown in fairly large pots. Watering should be carried out regularly while the plants are in active growth. Potting mixtures consist of bush loam, coarse sand and partially rotted leaf litter. The plants are severely attacked by slugs and snails, and must be protected continually by baits. The flower-stems are tall and need staking, because if they break off they can damage the rhizomes, leading to rotting. Propagation can be by offsets produced along the rhizomes.

Cryptostylis erecta R. Br.
(upright)
Qld, NSW, Vic   Bonnet Orchid
                                                   Sept-Feb
Leaves 1-5 per clump, 7-14 cm x 1-2 cm, ovate-lanceolate, stiff, erect, dark green above, reddish purple beneath; **flower-stems** 20-45 cm tall; **flowers** 2-12, green with purplish and reddish stripes and markings; **labellum** large, erect, inflated.

Widespread from south-eastern Qld to eastern Vic, in open forest or heathland. A handsome species with very attractive flowers. Fairly difficult to grow, tending to die out easily. Needs regular watering throughout the year, especially in summer when in active growth. Responds slowly to the use of liquid fertilizers.

Cryptostylis leptochila F. Muell.
(narrow lip)
Qld, NSW, Vic   Small Tongue Orchid
                                                   Dec-April
Leaves 4-8 cm x 2-3 cm, ovate, on stout petioles, dark green above, purplish beneath; **flower-stem** 15-35 cm tall; **flowers** 6-12, purplish or deep red; **labellum** thin, curved upwards, with dark, rounded calli.

Found in open forest country, usually in small colonies. Resents disturbance and is slow to re-establish, but once growing is fairly easy to maintain, although slow to increase. Prefers a mixture containing a loam rich in organic matter. Responds to applications of old animal manure or liquid fertilizer.

Cryptostylis ovata R. Br.
(ovate)
WA   Slipper Orchid
                                              Oct-Feb
Leaves 8-15 cm x 3-6.5 cm, ovate-lanceolate to broadly ovate, dark green above, purplish beneath; **flower-stem** 20-45 cm tall; **flowers** 2-10, reddish, with pale patches; **labellum** large, horizontal or oblique, the margins folded.

Widespread in forests, often found in swampy conditions. Fairly difficult to grow, tending to die out easily, although often plants grow vigorously for 1-2 years. Resents disturbance and is slow to re-establish. Seems to prefer a sandy soil mixture and probably responds to an annual dressing of leaf litter or sawdust.

Cryptostylis subulata (Labill.) Reichb.f.
(awl-shaped)
Qld, NSW, Vic, Tas   Large Tongue Orchid
                                              Oct-May
Leaves 5-15 cm x 1-2 cm, ovate-lanceolate, stiffly erect, yellowish green; **flower-stems** 20-60 cm tall; **flowers** 2-20, green with brownish or reddish tinges; **labellum** large, horizontal or oblique, the margins folded, a hump on the upper surface.

A common species usually found in moist swampy situations amongst rushes and sedges. It is popular in cultivation as it grows and flowers very easily. It is not demanding as to the soil mixture, provided it is well drained. Enhanced growth can be obtained over the summer by standing the base of the pot in water. Responds to the use of liquid fertilizers. Could probably be established in the garden as a bog plant, provided the plants can be protected from slugs and snails.

CTENOPTERIS Blume
(from the Greek *ctenos*, a comb; *pteris*, a fern)
Grammitidaceae

Small **ferns** growing on rocks or trees; **fronds** erect or weeping, leathery, once-divided, broad or narrow; **sori** rounded.

A large genus of about 200 species with 7 extending to eastern Australia. All are very difficult, if not impossible, to maintain in cultivation. The usual response is slow death of the plant, each succeeding frond getting smaller. There is no record of the species having been successfully raised from spore.

Ctenopteris contigua (Forster f.) Holttum
(adhering together)
Qld
10-30 cm long
Fronds semi-weeping, narrow, pale green, leathery, lobed; **segments** linear, toothed at the ends, where the sori are borne.

Found in the rainforests of Cape York Peninsula. Impossible to maintain in cultivation. Has been tried without success, in coarse epiphyte mixtures, tree-fern fibre and sphagnum moss. May last up to 2 years before fading away.

Ctenopteris fuscopilosa (F. Muell. & Baker)
                                              S. Andrews
(brownish hairs)
Qld
10-30 cm long
Rootstock fairly stout, green; **fronds** semi-erect or pendent, dull green, deeply lobed and fishbone-like; **lobes** thick, rounded; **sori** circular, black.

Scattered on rocks and trees in lowland rainforest, sometimes in small colonies. Resents disturbance, and difficult to establish in cultivation. Requires a coarse mixture. Partial success has been obtained by growing in a bottle or sealed plastic bag.

Ctenopteris gordonii (Watts) S. Andrews
(derivation unknown)
Qld
5-10 cm long
Small **fern**; **fronds** semi-erect, pinnately lobed, broadest towards the middle; **lobes** decurrent; **sori** rounded, up to 8 per lobe.

A small species found on rocks and trees in rainforest. Very difficult to grow, and tends to linger and fade away slowly. Best results have been obtained in a sealed plastic bag of moss.

Ctenopteris heterophylla (Labill.) Tind.
(variable fronds)
Vic, Tas   Gypsy Fern
10-30 cm long
Fronds weeping, leathery, broad, dull, lobed, winged, broadest towards the middle; **sori** rounded or oblong.

Found on trees or rocks in wet gullies. Has proved very difficult to keep growing, and has been tried with a range of treatments in various conditions. Plants have survived for fairly long periods in a large bottle.

Ctenopteris maidenii (Watts) S. Andrews
(after J. H. Maiden)
Qld
10-25 cm long
Small **fern** growing in slowly spreading patches; **fronds** semi-erect, pinnately lobed, broadest at the middle, narrowed to a tail at the apex; **lobes** linear, irregular at the apex; **sori** circular, crater-like.

Found in highland rainforest, growing on trees or rocks. Resents disturbance and is very difficult to re-establish. Requires cool, moist conditions, in a coarse mixture.

CUCUMIS L.
(an old Latin name)
Cucurbitaceae

Annual or perennial **climbers** or **trailers**; **stems** fleshy, hollow, hairy; **leaves** small to large, often lobed; male and female **flowers** borne separately, the males in axillary clusters or solitary, the females solitary in leaf axils; **fruit** succulent, variously shaped.

Species about 25 with one indigenous to Australia. A number of exotic species are widely cultivated as important horticultural crops, eg. *Cucumis melo* (Melon) and *C. sativa* (Cucumber). A couple of species have

129

Cucumis melo

become naturalized and are now common weeds, the most familiar of which is the Paddy Melon (*C. myriocarpus*). Propagate from seed which germinates readily in warm conditions.

Cucumis melo L. ssp. **agrestis** (Naudin) Grab.
(a melon; pertaining to fields or cultivated land)
Qld, NSW, NT — Native Cucumber; Ulcardo Melon
prostrate — Nov-March
Slender annual **trailer** or **climber**; **stems** angular, coarsely hairy, brittle; **leaves** 0.3-0.8 cm x 0.2-0.4 cm, cordate, 3-7 lobed, light green, the margins finely toothed; **male flowers** in clusters, on peduncles 0.5-2 cm long; **female flower** larger than the male flowers, solitary, yellow; **fruit** 2-5 cm long, globular or oval, glabrous or hairy.
Widespread along stream banks and flood plains in inland areas. Grows rapidly after rains, but the vines wither and die during the dry season, leaving the area strewn with fruit. The fruit was an important item of food for the Aborigines and early white settlers. The rind is bitter but the flesh is palatable, and quite refreshing and sweet when ripe. Occasionally cultivated by enthusiasts. Best suited to warm inland areas, but can be grown as an annual in southern Australia during the warm months. Very frost tender. Prefers rich soils and regular watering during dry periods. The species is also widely distributed in Asia and Africa. Formerly it was confused with *C. trigonus* Roxb., a species not native to Australia. Propagate from seed which germinates easily under warm, moist conditions.

Cucumis trigonus Roxb. —
see under *C. melo* ssp. *agrestis*

Cucurbitaceae Juss.
An important family of dicotyledons containing about 700 species in 100 genera, mainly centred in tropical zones. They are chiefly soft-wooded climbing plants with well developed tendrils and unisexual flowers. They may be annuals or perennials, and the fruits of many species are of major importance as a source of food for man. The family is very poorly represented in Australia with 9 genera and about 15 species, few of which have any horticultural merit. Main Australian genera: *Cucumis, Diplocylos, Luffa, Melothria, Sicyos* and *Trichosanthes*.

Cudrania cochinchinensis Lour. =
Maclura cochinchinensis (Lour.) Corner

CULCITA C. Presl
(a pillow; in reference to its use by campers)
Dicksoniaceae
Coarse **ferns** spreading or growing in clumps; **fronds** tall, lacy, soft, very hairy, broadly triangular, 3-4 times divided; **sori** rounded.
A small genus of about 9 species with 2 extending to Australia. *C. villosa* C. Chr. is unknown in cultivation. It is a rare species confined to Mount Spurgeon in north-eastern Qld. Propagate by division of the rhizomes or from spore.

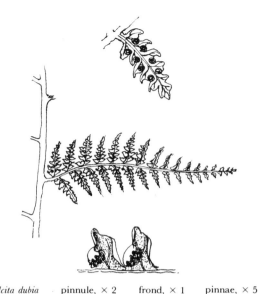

Culcita dubia — pinnule, × 2 — frond, × 1 — pinnae, × 5

Culcita dubia (R. Br.) Maxon
(doubtful)
Qld, NSW, Vic, Tas — Rainbow Fern
0.5-1.5 m tall
Medium to tall **fern** forming large, spreading patches; **fronds** erect or arching, yellowish, hairy, triangular, thrice-divided; **segments** triangular, lobed.
A very common fern of eastern Australia, forming extensive patches in open forest, frequently in exposed situations. Popular in cultivation as it is hardy and will grow in exposed sites. Grows especially well in clay soil. Drought resistant but develops a better appearance with regular watering. Must be given room to spread. Can be kept in check by regular pruning and removal of the creeping rhizomes. Will grow in sunny or shady positions. Propagate by division of the rhizomes or from spore.

CULTIVARS

A cultivar is a horticultural variety of a plant or crop, which differs from the usual strain and other cultivars by some significant feature. It is usual for horticultural cultivars to be perpetuated by vegetative means; however, many agricultural cultivars such as crop plants come true from seed and thus the strain can be maintained on a large scale. A few horticultural cultivars also come true from seed.

Horticultural cultivars are registered by various bodies such as Nurserymen's Associations and Orchid Societies, but the major governing body in Australia concerned with native plants is the Australian Cultivar Registration Authority. This Authority was formed in 1963 to register cultivars resulting from flora endemic to Australia. It was initially established at the National Herbarium in Melbourne, but in 1973 was relocated at the National Botanic Gardens, Canberra (formerly Canberra Botanic Gardens). The Authority consists of representatives from botanic gardens and herbaria throughout Australia, as well as concerned groups such as the Society for Growing Australian Plants and the Federation of Australian Nurserymen's Associations. The Authority is recognized by the International Commission for the Nomenclature of Cultivated Plants. The Authority meets annually and publishes details of new cultivars that have been registered. To date, the list contains more than 100 cultivars. As well as new cultivars there is a considerable backlog of applications to be processed. Other aspects include the tracing and checking of details of old cultivars and the checking of cultivar names being used in the nursery trade.

When a cultivar is cited in literature, the abbreviation cv. should be placed before the cultivar name, or the cultivar name enclosed within single quotation marks. Capital letters must be used to start all words of a cultivar name. Thus, a cultivar can be written *Acacia pravissima* 'Golden Carpet' or *Acacia pravissima* cv. Golden Carpet. Where the cultivar is known to be a form of a species, the cultivar is shown as such, as in the example above. Where the cultivar is of hybrid origin or its identity is uncertain, it is listed as a cultivar of the genus, eg. *Callistemon* 'Mauve Mist' or *Callitris* 'Golden Zero'.

The naming of horticultural cultivars is governed by the International Code of Nomenclature for Cultivated Plants. Generally, popular names are chosen for cultivars and botanical terms are not used, to avoid confusion with infraspecific taxa such as subspecies and varieties. Some cultivars, however, do retain a botanical flavour, eg. *Christella subpubescens* 'Keffordii', and these arose at a time when the concept of cultivars was unknown. Last century and earlier this century it was common practice to name unusual forms, which we now recognize as cultivars, as botanical varieties. According to the International Code, such an epithet, correctly published according to the Botanical Code, is to be retained as a cultivar name unless it duplicates an existing cultivar name for that species. As the earlier names were correctly published, they still retain the botanical spelling. Some botanical terms have become adapted popularly by horticulturists and thus purple-leaved forms are known as 'Purpurea', golden-leaved forms as 'Aurea', variegated-leaved forms as 'Variegata' and double-flowered forms as 'Flore-Pleno'. Such terms are to be avoided where possible in the naming of new cultivars.

The registration of cultivars of Australian plants is gathering momentum in Australia. There is a tendency to register any variation without first investigating thoroughly its worth to Australian horticulture. This is caused in part by a desire to be first with a novelty, and also by the potential of monetary gains with the commercial release of new cultivars accompanied by an advertizing campaign. This is a very short-sighted approach. A form or hybrid is only of significance and worth registering as a cultivar if it is superior or significantly different in some important feature to those strains already being grown.

Horticultural cultivars can arise in a number of ways.

1) From variation within a species. Widely distributed species tend to vary as the plants adapt to the particular circumstances of their environment. These changes may not be significant in a botanical sense but may be important in horticultural features such as flower size, colour and floriferousness, growth habit and foliage features. Such variations can be perpetuated by vegetative propagation and are often distributed as a form of the species, eg. *Grevillea alpina* 'Mt. Slide' form and *G. alpina* 'Pomonal' form. Some species vary little, while others such as *Grevillea alpina*, *G. glabella*, *G. lavandulacea* and *G. rosmarinifolia* produce a multitude of forms. Such forms can be registered as cultivars, eg. *Grevillea glabella* 'Lara Dwarf' and 'Limelight'.

In some cases the changes in a species have resulted from geographical isolation, and the forms may eventually be considered as subspecies. Such forms frequently come true from seed. An example is *Eucalyptus caesia* 'Silver Princess'. This is a striking cultivar which differs from the usual form of *E. caesia* by a strong, weeping habit and very large flowers. It occurs naturally in some isolated granite outcrops of WA, and seed collected from the wild comes true to type. Eucalypts, however, are naturally outcrossing, and seed from plants grown in gardens, plantations or windbreaks may not come true because of hybridization, especially if the normal form is growing nearby.

Apart from obvious morphological differences, species can vary by less obvious but equally vital features such as vigour and resistance to pathogens.

2) As a mutation or 'sport'. Such an occurrence usually shows up as a single branch or section of a plant which is different from the rest of the plant because of some genetic change. Common examples are a change of foliage shape, colour and variegation. By propagating vegetatively from the mutation, the 'sport' can be perpetuated. *Callitris* 'Golden Zero' is an example of a mutation, as well as various variegated 'sports', eg. *Agonis flexuosa* 'Variegata', *Tristania conferta* 'Perth Gold' and 'Variegata'.

3) From seedling variation. Many native plants are well established in the nursery trade, and are raised from seed in large quantities each year. Variations which would probably die out in nature are frequently put aside by nurserymen and grown on to test their worth. Such variations can then be perpetuated by vegetative propagation techniques. Forms of *Agonis flexuosa* and *Baeckea virgata* with distinct foliage and/or growth habit have arisen in this way. Some variegated forms have also originated by the same means.

4) From hybridization. Hybrids occur naturally where some species grow together in the bush, in collections (eg. gardens) where species from different areas are planted together, and as a result of deliberate hybridization by man. Many cultivars that have been registered are of hybrid origin, most being spontaneous garden hybrids. Deliberate hybridization occurred with Australian natives introduced to England in the early 1800s but there has been little interest in this activity in Australia. Even at present, with the tremendous interest in the cultivation of Australian native plants, there is very little emphasis on a deliberate,

Cunoniaceae

planned hybridization programme based on scientific genetic principles. This is despite the obvious commercial potential that natives have in the nursery, landscape and cut flower industries in Australia and overseas.

5) From artificial manipulation such as by radiation or chemical treatment. Apart from a few orchids which have been treated with colchicine to improve flower size and texture or to overcome fertility barriers, such techniques are mainly applied to crop plants.

Cunoniaceae R. Br.

A family of dicotyledons comprising 250 species in 26 genera, mostly confined to the southern hemisphere. They are mainly shrubs or trees found in moist forests or rainforests, and have leathery leaves which are usually compound and borne in whorls. They always have paired stipules, and these may be enlarged and conspicuous. The family in Australia is very diverse, with 12 genera, but none of these is large (they total about 24 species). Most are handsome shrubs or trees, and are well worthy of cultivation. Australian genera: *Anodopetalum, Aphanopetalum, Caldcluvia, Callicoma, Ceratopetalum, Geissois, Gillbeea, Pseudoweinmannia, Pullea, Schizomeria, Spiraeanthemum* and *Vesselowskya*.

CUPANIOPSIS Radlk.

(from the Latin *Cupania*, a tropical genus not found in Australia; and the Greek *opsis*, resemblance to)
Sapindaceae

Small, slender or spreading **trees**; **leaves** pinnate, alternate; **leaflets** leathery, entire or toothed; **inflorescence** a raceme or small panicle; **flowers** inconspicuous, usually greenish; **fruit** a lobed, leathery capsule; **seeds** black, bearing a fleshy aril.

Species about 60 with 7 endemic in eastern Australia. All are rainforest plants, frequent in coastal areas. Previously included in the genus *Cupania*. The various species are found in widely divergent conditions, from exposed sea-fronts to cool, protected positions in moist rainforests. All have horticultural merit but only one species is commonly encountered in cultivation. The flowers are generally small and insignificant but the foliage is interesting and the fruits sufficiently large and numerous to be decorative. The plants are easily grown in a park or garden situation, favouring some protection when young. They require soils of unimpeded drainage, and may be slow-growing, especially in the initial stages. Some species make decorative pot plants and could be useful for indoor decoration. Propagate from seed which must be sown fresh. Frequently, many capsules are parthenocarpic and contain no seeds.

Cupaniopsis anacardioides (A. Rich.) Radlk.
(like the genus *Anacardium*)
Qld, NSW, NT   Tuckeroo; Cashew-leaf Cupania
8-15 m x 6-15 m   March-July

Small to medium, spreading **tree**; **leaflets** 1-6 pairs, 7-10 cm x 3-5 cm, ovate to oblong, leathery, dark green, dull; **panicles** terminal or axillary, numerous; **flowers** 0.4-0.6 cm across, greenish white, fragrant;

capsule 1-2 cm across, 3-6 lobed, bright orange, leathery, mature Nov-Dec.

Found in coastal scrubs of the east coast between Townsville and Port Hacking, south of Sydney, often extending into the dunes where it is tolerant of salt-laden winds. Also found in coastal districts of NT. Commonly cultivated in subtropical areas, especially along the coast. May be slow-growing, but responds to regular watering and fertilizer. Very hardy and will grow successfully in southern Australia. Makes an excellent shade or shelter tree and can be trained as a street tree. Very decorative when in fruit, but not all trees fruit well, and there is scope for selection and vegetative propagation. Grafting would probably be successful. Seed should be sown while fresh and takes 4-6 weeks to germinate. Seed collection is often difficult as in some years the majority of seeds are destroyed by caterpillars.

Cupaniopsis flagelliformis Bailey
(thin, supple shoots; in this species referring to the panicle branches)
Qld, NSW
2-6 m x 1-2 m   Oct-Nov

Tall, slender **shrub** or small **tree**; young shoots bright red, bearing grey, woolly hairs; **leaves** 30-45 cm long; **leaflets** 8-12 pairs, 8-15 cm x 2.5-3.5 cm, cuneate, the margins coarsely toothed, the veins prominent on the lower surface, on very long petioles; **panicles** 30-60 cm long, terminal, each of a few supple branches; **flowers** about 0.8 cm across, pink-mauve, in small clusters; **fruit** 2-2.7 cm across, obovate, rusty brown, with a velvety tomentum, mature Oct-Jan.

This species has a curious disjunct distribution, being known from the border range area of north-eastern NSW and south-eastern Qld, and there are also a few collections from north-eastern Qld. It is a very slender species, and in the southern part of its distribution it grows in dry rainforests and along rainforest margins. Plants make a very interesting

Cupaniopsis anacardioides. × ·45

Cupaniopsis flagelliformis　　　　　　　　　　D.L. Jones

green, the apex rounded; **panicles** 7-10 cm long, axillary; **flowers** about 0.3 cm long, greenish white, hairy; **capsule** about 1 cm across, brownish, densely covered with rusty hairs; **seeds** dark brown, almost covered by a large, fleshy orange aril, mature Sept and Jan.

Distributed in dry rainforests between Gloucester in NSW and Gympie in Qld. Similar in general appearance to *C. anacardioides*. Requires well drained soil and some protection from hot sun when young. Has proved to be fairly tolerant of dryness. Propagate from seed which must be sown fresh.

Cupaniopsis serrata (F. Muell.) Radlk.
(toothed)
Qld, NSW   Rusty Tuckeroo
5-12 m x 2-5 m   Oct-Jan
Tall **shrub** or small, slender **tree**; young branches clothed with rusty hairs; **leaflets** 3-8 pairs, 6-10 cm x 2-4 cm, ovate, leathery, with sharply and coarsely toothed margins, dark green, glossy and smooth above, paler beneath where often hairy, on very short petioles; **panicles** terminal, sparse, hairy; **flowers** 0.6-0.8 cm across, white, on short stalks; **capsules** 1.5-2.5 cm across, pale to dark brown, ovoid, wrinkled, covered with dense, fine hairs, 2-3 lobed; **seeds** brown, glossy, half covered by an orange aril, mature Nov-Dec.

Confined to dry rainforests in near-coastal regions between Ingham and the Richmond River. Has very attractive foliage and decorative fruits. Well worth growing in temperate or subtropical regions. Frost tender when young. Requires a protected, semi-shady position and water during dry periods. Young plants in pots should be excellent for indoor decoration. The var. *tomentella* (F. Muell.) Domin has the underside of the leaflets shortly hairy, very small teeth on the leaflets and large fruit. It also has, typically, flushes of rusty brown new growth. It is more common than the typical form in NSW and also occurs in the southern part of the species range in Qld. Propagate from seed. This species is easily confused with *C. flagelliformis* which has 8-12 pairs of long-stalked leaflets.

Cupaniopsis shirleyana (Bailey) Radlk.
(after J. Shirley)
Qld   Kooraloo
5-10 m x 3-5 m   Aug-Oct
Tall **shrub** or small **tree**; young branches covered with rusty hairs; **leaves** 5-10 cm long; **leaflets** 3-8 pairs, 3-5 cm long, wedge-shaped, hairy beneath, with coarsely toothed margins; **racemes** 5-10 cm long, slender; **flowers** small, greenish; **capsule** 1-2 cm long, brown, hairy, 3-lobed.

Confined to coastal rainforests between Brisbane and Bundaberg. Little-known in cultivation. A useful shrub or small tree for subtropical areas. Frost tender. Requires a protected, semi-shady position. Propagate from seed.

Cupaniopsis wadsworthii (F. Muell.) Radlk.
(after R. Wadsworth)
Qld
5-12 m x 3-5 m   May-June
Small, slender **tree**; **leaflets** 1-2 pairs, 3-9 cm x 2-

acquisition to the garden and are distinctive, with their stiff, coarsely toothed compound leaves. They also make very decorative container plants but are unsuitable for indoor use. Require well drained soils and respond to mulches, fertilizers and watering during dry periods. Best suited to subtropical regions. Propagate from seed which must be sown fresh.

The species is easily confused with *C. serrata* which has 3-8 pairs of shortly stalked leaflets.

Cupaniopsis foveolata (F. Muell.) Radlk.
(with small pits)
Qld, NSW   Toothed Tuckeroo
6-12 m x 2-4 m   Feb-March
Small **tree**; **bark** smooth, grey to brown; **leaflets** 5-6 pairs, 7-12 cm x 2-4 cm, lanceolate, with coarse, bluntly toothed margins, glossy dark green above, paler beneath, the apex rounded or notched; **panicles** axillary, up to 10 cm long; **flowers** about 0.2 cm long, greenish, the pedicels covered with rusty hairs; **capsule** 1.5-2 cm across, red to brown, leathery, 3-lobed, hanging in large bunches; **seeds** blackish brown, almost covered by a fleshy orange aril, mature Nov-Dec.

Widespread between Rockingham Bay in Qld and the Macleay River in NSW, usually along stream banks and rainforest margins. An attractive species, impressive when in fruit, and with decorative foliage. Forms a dense crown and would make a useful shelter tree. Easily grown in well drained soils, but prefers those soils rich in organic matter, and responds to the use of organic fertilizers and manures. Young plants in pots should be excellent for indoor decoration. Propagate from seed which must be sown fresh.

Cupaniopsis parvifolia (Bailey) L. Johnson
(small leaves)
Qld, NSW   Small-leaved Tuckeroo
10-20 m x 5-8 m   April; Aug-Sept
Small to medium **tree**; young shoots hairy; **leaflets** 3-4 pairs, 3-8 cm x 2-4 cm, elliptical to obovate, dull

133

3 cm, triangular, truncate at the apex, deep green and shining; **panicles** axillary, narrow, shorter than the leaves; **flowers** about 0.4 cm across; **capsule** 2.5 cm across, 2-3 lobed, brown, covered with velvety hairs.

Found in coastal rainforests between north of the NSW border and Rockhampton. Not known to be in cultivation, but would make a handsome small tree for subtropical or coastal temperate areas. Propagate from seed.

Cupressaceae Bartling

A family of gymnosperms containing about 130 species in 19 genera, widely distributed in most regions of the world. They are shrubs or trees, with the leaves scale-like and decussate, or in whorls. The flowers are small and unisexual, and the plants may be monoecious or dioecious. The seeds are borne in a cone which usually has woody, overlapping scales. The family is not large in Australia (3 genera, 20 species) but is well represented by the genus *Callitris* which forms extensive stands in the drier regions. The other Australian genera are *Actinostrobus* and *Diselma*.

CUPULANTHUS Hutch.

(from the Latin *cupula*, small tub; and the Greek *anthos*, flower; refers to the prominent cup formed by the bracteoles below the calyx)
Fabaceae

A monotypic, endemic genus confined to WA.

Cupulanthus bracteolosus (F. Muell.) Hutch.

(prominent bracteoles)
WA

prostrate x 1-3 m   July-Aug

Dwarf **shrub**; **branches** spreading; **leaves** to 10 cm long, alternate, linear to lanceolate, blunt apex, becoming glabrous, margins revolute, underside silky-hairy when very young; **flowers** pea-shaped, to 1.5 cm long, red, solitary or in pairs, on slender pedicels; **wings** shorter than keel; **keel** twice as long as calyx; **calyx** with 5 equal lobes, covered in silky hairs; **bracteoles** 2, orbicular.

Evidently not well known in cultivation, this species from the southern coastal areas of WA will need very well drained soils. It appreciates partial or full sun, and will probably tolerate dappled shade. Propagate from seed or cuttings.

Previously known as *Brachysema bracteolosum* F. Muell.

CURCULIGO Gaertner

(from the supposed resemblance of a seed to a weevil)
Hypoxidaceae

Stout **herbs** with a lily-like or palm-like appearance; **rhizomes** thick, fibrous, with numerous coarse roots; **leaves** erect, radical, sheathing at the base, plicate; **inflorescence** a compressed spike, hairy; **flowers** lily-like, of 6 spreading perianth segments, hairy, the base united into a tube; **fruit** a succulent capsule.

A small genus of 10 species with 2 found in Australia. They are interesting plants confined to moist, shady sites in tropical and subtropical regions. They are very easy to grow and are fairly commonly

Curcuma australasica   B. Gray

grown. Propagate from seed which is best sown while fresh and by division of the rhizomes. There is some disagreement amongst botanists as to whether the species should be placed in this genus or *Molineria*.

Curculigo capitulata (Lour.) Kuntze

(like a head)
Qld   Weevil Lily
0.3-1 m tall   Feb-May; also sporadic

Rhizome short-creeping, thick, with numerous coarse roots; **leaves** 0.3-1 m x 10-14 cm, radical, palm-like, erect, oblong-lanceolate, dark green, pleated longitudinally, sheathing at the base; **scapes** 5-15 cm long, woolly, curved, ending in a dense, clustered spike; **flowers** about 2 cm across, starry, yellow, woolly outside; **capsule** about 0.6 cm across, globular, succulent.

A widespread species that extends to tropical rainforests of north-eastern Qld. It grows in shady, protected places, often near streams, and in organically rich soils. It needs similar conditions if it is to be grown to perfection; however, it is very hardy and adaptable, especially once established. Grows well out of doors in southern Australia. It responds to fertilizers, mulches and water during dry periods. An excellent plant for mingling with ferns, and also successful indoors. Propagate from seed or by division of the rhizomes.

Curculigo ensifolia R. Br.
(sword-shaped leaves)
Qld, NSW
0.2-0.5 m tall   Dec-March
   Rhizome short-creeping, erect, stout, with numerous coarse roots; **leaves** 20-50 cm x 1-2 cm, erect, grass-like, dark green, with prominent veins; **spikes** short, erect, few-flowered, bearing numerous papery bracts; **flowers** about 1.5 cm across, yellow, hairy, with a long, slender perianth tube; **capsule** about 0.5 cm long, oblong, enclosed in a bract.
   Widespread in moist areas and rainforests almost throughout Qld, and in NSW confined to the north coast, generally growing on rocky slopes where litter has accumulated. It often grows in scattered colonies. An interesting plant, useful in the garden because of its tufted growth habit. Requires well drained soils and a protected situation. Responds to mulches, fertilizers and water during dry periods. Large plants have a carrot-like taproot which was apparently roasted and eaten by the Aborigines. Propagate from seed or by division of the rhizomes.

CURCUMA L.
(from an Arabic name for one species)
Zingiberaceae
   Perennial **herbs** with a subterranean fleshy rhizome; **leaves** erect, sheathing at the base, then spreading into a broad, thin-textured lamina; **spikes** crowded, bearing numerous overlapping, sac-like bracts; **flowers** tubular, yellowish; **fruit** a 3-celled capsule.
   Species about 5 with one endemic in Australia.

Curcuma australasica Hook.f.
(Australian)
Qld, NT   Cape York Lily; Native Tumeric
1-1.5 m x 0.5-1 m   Dec-March
   Rhizome fleshy, with a cluster of fleshy white roots; **leaves** 30-60 cm long x up to 12 cm across, erect, pleated, thin-textured, narrowing to a long, fleshy petiole, sheathing the flower-stem; **flower-stem** fleshy; **spike** 15-30 cm long, dense, with large conspicuous bracts, the lower bracts greenish, the upper ones bright rosy pink; **flowers** yellow, in clusters of 3-5 in the base of each bract.
   A spectacular plant found along rainforest margins and in shady spots, often amongst rocks. The flower-spikes appear before the leaves, and the plants die back to the perennial rhizome during the hot, dry season. Grows very well in tropical or subtropical areas, preferring a rich, well drained soil in a sunny position. Spectacular when in flower. Needs a heated glasshouse for successful growth in southern Australia. Plants of the exotic Indian species *C. inodora* are often mistakenly grown and sold as the Australian species. *C. inodora* is much more vigorous, with leaves over 1 m long that have prominent reddish bands and tonings. Propagate by division of the rhizome or from seed.

Cuscutaceae Dumort.
   A cosmopolitan family of dicotyledons consisting of the solitary genus *Cuscuta* which consists of about 170 species. They are leafless and rootless parasites with slender, twining stems, and obtain their nourishment through haustorial attachment to their host. Flowers are borne in small, dense clusters. The genus has no ornamental value and in fact some species may be problem weeds in gardens.

CUSHION PLANTS
   Cushion plants derive their popular name from the unique moss-like growth which develops into slowly spreading, dense mats or cushions.
   In Australia, the subalpine regions of Tas provide ideal conditions for the growth and proliferation of such plants. It is here that *Abrotanella forsteroides*, *Chionohebe ciliolata* var. *fiordensis*, *C. densifolia*, *Donatia novae-zelandiae*, *Dracophyllum minimum*, *Phyllachne colensoi* and *Pterygopappus lawrencei* are commonly found. Although these plants have similar growth habits and form co-extensive colonies, only two are related. *Abrotanella forsteroides* and *Pterygopappus lawrencei* are members of the Asteraceae family.
   The NSW and Vic subalpine area also has plants which are sometimes grouped collectively under the name of cushion plants. The species are *Colobanthus nivicola*, *C. pulvinatus*, *Ewartia nubigena*, *Juncus antarcticus*

The cushion plant *Donatia novae-zelandiae* with *Celmisia saxifraga* in Tas alps   T.L. Blake

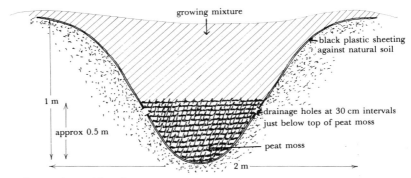

Preparation of area for growing cushion plants

and *Oreomyrrhis pulvinifica. Scleranthus biflorus* also occurs in this region, as well as being recorded from Qld and Tas, and is now widely grown as a rockery and container plant in southern Australia.

Some of the species mentioned also occur in other regions such as New Zealand and the south of South America, eg. *Donatia novae-zelandiae, Phyllachne colensoi* and *Chionohebe ciliolata.* Their natural habitat is the mountain summits above 1300 m, where snow lies for many months of the year. The roots find anchorage in acid, peaty soil deposits that are continually moist. The moss-like growth of the different species provides an undulating surface that can cover large areas.

These intriguing plants have horticultural potential, not only for their ornamental foliage, but also in some cases for the floral display, which can be most decorative, eg. *Donatia novae-zelandiae.*

There has been some success in establishing these slow-growing plants in cultivation by simulating the conditions of their natural habitat, which provide constant moisture in the area of the root system.

One such method is to prepare a cone-shaped hole to a depth of about 1 m, with a diameter of about 2 m. (These dimensions could be varied, depending on the area required and/or available). The hole is then lined with heavy-duty black plastic sheeting, and drainage holes are provided at about 0.5 m deep. Peat moss is placed in the hole to the depth just above the drainage holes, and the hole is filled with a mixture of equal parts of good quality loam, sand and peat moss (see illustration). Regular watering during hot weather may be required to provide a constant moisture level for the root systems.

Bog gardens or other situations which are constantly moist can also provide suitable conditions for cultivation of cushion plants.

These plants are best suited to temperate regions. In areas of high temperatures, it is necessary to provide some shade over the plants.

The cultivation of these intriguing cushion plants in containers, with a constant supply of moisture being provided by capillary action from water trays, is well worth trying.

For detailed information, including propagation, see genus and species descriptions.

CUTTINGS, Propagation from —
see Volume 1, page 217

CUTTSIA F. Muell.
(after J. Cutts, who retained the collections of explorer Ludwig Leichhardt)
Escalloniaceae
A monotypic genus which is endemic in Australia.

Cuttsia viburnea F. Muell.
(like the genus *Viburnum*)
Qld, NSW   Native Elderberry
3-10 m x 1-3 m   Oct-Feb
Medium **shrub** to small bushy **tree; leaves** 8-15 cm x 2-5 cm, lanceolate, alternate, thin-textured, dark green, shiny, the margins toothed, the apex drawn out; **panicles** large, conspicuous, terminal; **flowers** about 0.6 cm across, snowy white; **capsule** about 0.3 cm across, ovoid.

Found in cool, moist forests of south-eastern Qld and north-eastern NSW, often at high elevations. An attractive plant for cultivation, requiring well drained soils and a cool position protected from long exposure to hot sun. Needs plenty of water during dry periods. Showy when in flower. May become straggly but responds to regular pruning. Frost hardy and easily grown in southern Australia. Propagate from seed or cuttings.

CYANOSTEGIA Turcz.
(from the Greek *cyanos*, dark blue; *stege* or *stegos*, shelter; alluding to the calyx covering the corolla)
Dicrastylidaceae
Small to medium **shrubs; branches** many, glabrous, resinous; **leaves** decussate, undivided, glabrous, resinous; **inflorescence** loose terminal panicle; **calyx** blue to purple, very prominent, disc-like, campanulate, papery, toothed or 5-lobed; **corolla** blue to purple, small, campanulate, 5-lobed, 2 upper lobes longer than 3 lower; **stamens** 4; **fruit** a dry drupe.

An endemic genus comprising 5 species which grow over a wide range of WA, with a limited occurrence in southern NT. They are very prominent in the drier areas, amongst gravelly soils. From late spring to early summer they provide a massed display of bluish purple flowers. The coloured calyces remain on the plant for an extended period after the spent corollas fall.

The genus has excellent horticultural potential, especially for areas with warm to hot climatic conditions. At this stage they have been grown only on a limited scale, but species like *C. angustifolia* and *C. lanceolata* should find wider acceptance as plants become available commercially.

Plants can be readily propagated from freshly collected seed or from cuttings of firm young growth. They respond well to light applications of slow-release fertilizer, once established. Light pruning after flowering is recommended.

Cyanostegia angustifolia Turcz.
(narrow leaves)
WA
1-2 m x 1-2 m   Sept-Dec

Small **shrub**; **branches** upright; **branchlets** glabrous, resinous; **leaves** 2.5 cm x 0.4 cm, flat or concave, decussate, folded lengthwise, margins usually toothed, glabrous, resinous; **flowers** about 1.5 cm diameter, borne in a loose pyramidal panicle; **calyx** blue to purplish blue, hairy outside; **anthers** gold and just protruding from **corolla** which is deep blue to dark purple; **fruit** globular, small, warty, densely hairy.

It has been recorded that this ornamental species

Cuttsia viburnea, × ·4

does not flower as densely under cultivation as in its natural habitat. This could be due to excess growth caused by overwatering and overfertilizing, and/or climatic and soil conditions. It will grow best in well drained soils, with full sun, but should also tolerate partial sun. Withstands frost and extended dry periods. Ideal for gardens, and useful also as a low screening or windbreak shrub. Propagate from seed or cuttings.

This species can be distinguished from other *Cyanostegia* spp. by its leaves, which tend to fold lengthwise.

Cyanostegia corifolia Munir
(leathery leaves)
WA
1-2 m x 1-2 m   Aug-Dec

Small **shrub**; **branches** glabrous, resinous; **leaves** 2-4 cm x 0.5-1.5 cm, narrow-elliptic, oblong or wedge-shaped, dentate, resinous, leathery; **flowers** about 2 cm diameter, borne in a loose pyramidal panicle; **calyx** bright blue outside, lower half hairy; **corolla** deep blue; **fruit** warty, wrinkled, not winged or lobed.

This recently named species from south-western WA is most ornamental. Its requirements in cultivation are similar to those of *C. angustifolia*, *C. lanceolata* and *C. microphylla*. Propagate from seed or cuttings. It has previously been confused with both *C. cyanocalyx* and *C. lanceolata*, but can be distinguished from *C. cyanocalyx* which has a glabrous calyx, and *C. lanceolata* which has leaves which are not as broad or as leathery.

Cyanostegia cyanocalyx (F. Muell.) C. Gardner
(blue calyx)
WA
1-2 m x 1-2 m   July-Oct; also sporadic

Small **shrub**; **branches** upright to spreading; **leaves** 2-6.5 cm long, wedge-shaped to oblong, decussate, toothed near apex, glabrous, very resinous; **flowers** about 2.5 cm diameter, borne in a long, narrow panicle; **calyx** blue or purplish blue, only slightly lobed margin, glabrous; **corolla** deep blue; **fruit** distinctly 4-winged.

This species has the largest flowers of the genus. It occurs in sandy loams on the north-west coast near Roebuck Bay, and inland to the east. It is a most ornamental species, but evidently difficulties have been encountered in trying to establish it in cultivation. It needs very well drained soils, preferably with full sun, although it may tolerate partial sun. Best suited to warm to hot locations. Propagate from seed or cuttings.

C. cyanocalyx is easily distinguished by its large flowers and leaves, glabrous calyx and 4-winged fruit.

Cyanostegia lanceolata Turcz.
(lanceolate leaves)
WA   Tinsel Flower
1-2.5 m x 1.5-3 m   Aug-Dec; also sporadic

Small to medium **shrub**; **branches** many, upright; **branchlets** glabrous, resinous; **leaves** 2-6 cm x about 1 cm, decussate, flat or slightly convex, margins with rounded teeth, glabrous, resinous; **flowers** about 1.5 cm diameter, borne in a loose pyramidal panicle;

Cyanostegia microphylla

Cyanostegia lanceolata

A. Cudmore

calyx blue or purplish blue, hairy on outside, distinctly 5-lobed on the margin; **corolla** deep blue; **anthers** gold, just protruding from corolla; **fruits** warty and wrinkled, densely hairy.

An ornamental species that forms a dense shrub, and is suited to well drained soils, especially gravelly types. It appreciates full sun but will tolerate partial sun. This species has the widest distribution of the genus, growing in the Avon, Coolgardie, Irwin and Roe Districts, where it withstands light to medium frosts and extended dry periods. Recommended for semi-arid regions and for warm to hot situations in cooler climates. Propagate from seed or cuttings.

This species has affinities to *C. corifolia* which has broader leaves and dentate margins.

Cyanostegia microphylla S. Moore
(small leaves)
WA
1-1.5 m x 1-2 m   Aug-Nov

Small **shrub**; **branches** upright; **branchlets** glabrous, resinous; **leaves** 0.5-1.6 cm x 0.2-0.4 cm, decussate, flat, regularly toothed margins, glabrous, very resinous; **flowers** about 1 cm diameter, deep blue to bluish violet, borne in a loose pyramidal panicle up to 22 cm long; **calyx** blue to purplish blue; **corolla** blue, anthers gold; **fruit** warty, to 0.3 cm long, nearly as broad.

This is perhaps the species with the flowers nearest to blue. It is most decorative, with the blooms well displayed above the foliage. Not well known in cultivation. Requirements are for well drained soils, with full sun, or it may tolerate partial sun. It occurs in the Avon and Coolgardie Districts, so should do well in a warm to hot location. Propagate from seed or cuttings.

C. microphylla is easily distinguished from *C. angustifolia* and *C. lanceolata*, as both have larger leaves.

CYANOTIS D. Don
(from the Greek *cyaneos*, blue; *otis*, an ear; in reference to the shape and colour of the flowers)
Commelinaceae

Annual or perennial **herbs**; **stems** creeping or ascending; **leaves** simple, entire, alternate, sheathing at the base; **inflorescence** a spike or cluster enclosed in a leafy spathe or loose leaf-sheaths; **flowers** regular, in 2 whorls, the inner ones united at the base and with 3 spreading lobes; **fruit** a 3-valved capsule.

A pantropical genus of about 50 species with a solitary widespread species extending to Australia. It can be propagated from seed or cuttings.

Cyanotis axillaris (L.) D. Don
(axillary flowers)
Qld, WA, NT   Blue Ears
prostrate x 0.5-1.5 m   Dec-Feb

Annual **herb**; **stems** glabrous, slender, creeping or

climbing; **leaves** 5-10 cm x 0.5-1.5 cm, linear to linear-lanceolate, dark green; **flowers** about 0.6 cm across, 2-3 together in an axillary cluster, the outer segments pale, the inner ones deep blue; **stamen filaments** with a tuft of hairs.

A slender annual that grows rapidly with the onset of the wet season. It is mainly found in soakage areas where the soil stays moist for several months. Its flowers are attractive, and the species could have potential in moist situations such as around the margins of ponds, bog gardens etc. Propagate from seed which is fine and is best sown on the surface of mud or from cuttings which root easily.

CYATHEA Smith
(from the Greek *cyathos*, a cup; in allusion to the sori)
Cyatheaceae   Tree-ferns

Tree-ferns with a thick or slender, woody trunk topped by a spreading crown of fronds; **fronds** finely divided, widest towards the middle, dull or shiny green; **sori** rounded, usually conspicuous, in some species covered by indusia.

A group of ferns of about 800 species with about 15 species extending to the east coast of Australia. This includes a well developed endemic group on Lord Howe Island. The genus *Cyathea* has been split into 5 genera, of which *Alsophila* and *Sphaeropteris* occur in Australia. As the work has not been completed for the Australian species, nor is it widely accepted, all species have been retained in *Cyathea*.

A very popular group, widely grown in Australia and overseas, doing best in well drained soils in shady or semi-shady situations. Some tropical species are frost tender and may be difficult to grow in areas with severe winters. Propagate from spore or in a few cases offsets. For further details on cultivation see Ferns.

Cyathea australis (R. Br.) Domin
(southern)
Qld, NSW, Vic, Tas   Rough or Hard Tree-fern
5-10 m tall
Trunk large, to 10 m tall x 1 m diameter, fibrous at

Cyathea brevipinna   D.L. Jones

the base, covered with roughened, rasp-like frond-bases in the upper half; **scales** shiny brown, conspicuous on the upper part of the trunk; **fronds** up to 5 m x 1 m, finely divided, dark green or bleached in exposed conditions; **sori** without indusia.

Very adaptable, tolerating a range of conditions from wet soils to dry situations. Attractive and widely grown in southern Australia. Will withstand more sun than most species. Immature fronds may be burnt by frost, but are quickly replaced by new fronds in the spring. Propagate from spore.

Cyathea baileyana (Domin) Domin
(after F. M. Bailey)
Qld   Wig Tree-fern
2-5 m tall
Trunk slender, to 5 m tall x 10 cm diameter, fibrous at the base, woody above and covered with frond-bases; **scales** black; **fronds** up to 3 m x 50 cm, bipinnate, dark glossy green, a curious wig-like growth at the base of each frond; **sori** without indusia.

Rare in nature, being confined to rainforests of north-eastern Qld at high elevations, and only recently introduced into cultivation. Prefers well drained soils in shady situations. Slow-growing and very frost tender, requiring a cool glasshouse for protection in areas with severe winters. Propagate from offsets. This species is extremely difficult to raise from spore.

Cyathea brevipinna Baker
(short pinnae)
NSW (Lord Howe Island)   Tree-fern
2-4 m tall
Trunk woody, to 4 m tall, slender, covered with frond-bases in the upper half; **scales** papery, dark brown, shiny; **fronds** up to 2.5 m x 0.4 m, finely divided, with crowded segments.

A little-known species endemic to Lord Howe Island. Requires cool, moist conditions and protection from hot sun and drying wind. Propagate from spore.

Cyathea celebica Blume
(from the Celebes)
Qld   Tree-fern
2-6 m tall
Trunk fairly stout, up to 6 m tall x 20 cm diameter, fibrous at the base, woody above and covered with prickly frond-bases; **scales** white to pale brown; **fronds** up to 4 m x 1 m, bipinnate, dull green above, covered with cobwebby hairs beneath; **sori** covered by cup-shaped indusia.

In Australia this tree-fern is restricted to highland areas of north-eastern Qld. It is a striking species with a conspicuous mass of white scales on the top of the trunk. Plants adapt exceedingly well to cultivation but are not commonly grown. They are very frost hardy and can be successfully grown in southern Australia. A semi-shady position in moist, well-drained, organically rich soil is suitable. Propagate from spore.

Cyathea cooperi (Hook. ex F. Muell.) Domin
(after Daniel Cooper)
Qld, NSW   Scaly Tree-fern
5-12 m tall

Cyathea cunninghamii

Trunk slender, woody, up to 12 m tall x 20 cm diameter, patterned with fallen leaf-scars, fibrous at the base; **scales** of 2 types, long and white or short and reddish brown; **fronds** up to 4 m x 1 m, bipinnate, dull to shiny green above, paler beneath; **sori** covered by circular indusia.

A common tree-fern found in moist forests from north-eastern Qld to south of Sydney. It is a very handsome species and is extremely popular in cultivation, thousands being raised and sold in the nursery industry each year. It thrives in tropical and subtropical regions but can also be grown as far south as Melbourne, where plants may suffer frond damage in heavy frosts. It is one of the fastest growing of the tree-ferns and will withstand considerable exposure to sun, especially if the roots are kept cool and moist. It is an excellent species for the home garden, and it mingles well with other plants. It can also be used to provide protection for other, more delicate ferns. Propagate from spore.

Cyathea cunninghamii Hook.f.
(after Allan Cunningham)
Qld, NSW, Vic, Tas   Slender Tree-fern
10-20 m tall

Trunk tall and slender, to over 20 m tall x 12 cm diameter, fibrous at the base, woody above and covered with frond-bases; **scales** brown, papery; **fronds** up to 3 m x 50 cm, finely divided; **sori** covered by indusia.

Very stringent in its cultural requirements of com-

Cyathea howeana   D.L. Jones

Cyathea celebica frond   D.L. Jones

plete protection from hot sun and wind, with plenty of water throughout the summer months. Young sporelings are very susceptible to damping off and are difficult to establish. Propagate from spore.

Cyathea felina (Roxb.) Morton
(cat-like)
Qld   Tree-fern
2-8 m tall

Trunk slender, woody, up to 8 m tall x 15 cm diameter, patterned with fallen leaf-scars; **scales** pale, thin; **fronds** up to 3.5 m x 0.7 m, bipinnate, bright green; **sori** without indusia.

In Australia this is a rare species confined to a couple of near-coastal areas of central Cape York Peninsula. Plants have proved to be somewhat difficult to maintain in cultivation and would appear to be best suited to lowland tropical regions. They prefer a shady situation, and are very sensitive to drying of the root system. Propagate from spore.

Cyathea howeana Domin
(from Lord Howe Island)
NSW (Lord Howe Island)   Tree-fern
3-10 m tall

Trunk slender, woody, patterned with fallen leaf-scars; **scales** small, pale; **fronds** 1-2.5 m x 0.7 m, fine and lacy, pale green; **sori** covered by quadrant-shaped indusia.

A little-known but very attractive species endemic to Lord Howe Island, where it often grows in exposed situations. It is similar in general appearance to *C. robertsiana* and requires warm, humid conditions. Propagate from spore.

Cyathea leichhardtiana (F. Muell) Copel.
(after L. Leichhardt, early explorer)
Qld, NSW, Vic   Prickly Tree-fern
3-7 m tall

Trunk slender, woody, up to 7 m tall x 12 cm thick, covered with persistent leaf-bases which are very prickly; **scales** pale to white; **fronds** up to 3 m x 0.7 m, bipinnate, dark shiny green; **sori** without indusia.

Cyathea macarthurii showing the distinctive frond scales T.L. Blake

A widely distributed tree-fern extending from central Qld to eastern Vic. It is often locally common and may grow in colonies. Its popularity in cultivation is reduced somewhat by its prickly nature. It prefers shady conditions, in well drained, humus-rich soil. Young plants are very slow-growing, and may be damaged by heavy frosts. Propagate from spore.

Cyathea macarthurii (F. Muell.) Baker
(after M. Macarthur)
NSW (Lord Howe Island)   Tree-fern
2-4 m tall
  Trunk woody, to 4 m tall, covered with frond-bases in the upper half; **scales** papery, long, narrow, twisted, brown; **fronds** up to 3 m x 0.5 m, finely divided, dark green above, silvery or glaucous beneath.
  A little-known species endemic to Lord Howe Island. Requires cool, moist conditions and protection from hot sun and drying winds. Very slow-growing. Propagate from spore.

Cyathea marcescens Wakef.
(withering and persistent; in reference to the skirt of dead fronds)
Vic   Skirted Tree-fern
5-10 m tall
  Trunk thick, up to 10 m tall x 40 cm diameter, fibrous at the base, covered above with roughened, rasp-like frond-bases and often persistent dead fronds in a skirt; **scales** shiny brown; **fronds** up to 4 m x 80 cm, finely divided; **sori** with small indusia.
  A rare species that may be a natural hybrid between *C. australis* and *C. cunninghamii.* Spores appear to be sterile. Relatively untried in cultivation because of its rarity, but requires conditions similar to those for other species. It can grow quite vigorously.

Cyathea rebeccae (F. Muell.) Domin
(after a lady friend of Baron von Mueller)
Qld   Tree-fern
3-7 m tall
  Trunk slender, up to 7 m tall x 10 cm diameter, fibrous at the base, woody above and covered with frond-bases; **scales** dark brown, shiny; **fronds** up to 3 m x 50 cm, bipinnate, dark glossy green; **sori** without indusia.
  Common in many areas of north-eastern Qld, but still not well known in cultivation. Established plants are most attractive. Needs a well drained soil and a semi-shaded position. Slow-growing and very frost tender, requiring a cool glasshouse for protection in areas with severe winters. Propagate from spore or offsets.

Cyathea robertsiana (F. Muell.) Domin
(after W. G. Roberts)
Qld   Tree-fern
3-7 m tall
  Trunk very thin, fibrous and buttressed at the base, up to 7 m tall x about 6 cm diameter, woody above and fleshy green near the apex; **scales** brown, few; **fronds** up to 2 m x 70 cm, fine and lacy, pale green; **sori** with indusia.
  Fast growing and with all attributes that make it a most desirable plant. Very adaptable to a range of conditions. Frost tender, and requires a protected position in areas with severe winter. Under good conditions it is fast growing, producing a succession of handsome lacy fronds. Propagate from spore.

Cyathea robusta (C. Moore) Holttum
(robust; strong)
NSW (Lord Howe Island)   Tree-fern
2-4 m tall
  Trunk stout, woody, to 4 m tall, fibrous at the base; **scales** pale brown or whitish, in masses at the top of the trunk; **fronds** 1.5-2.5 m x 0.5 m, dark green, fine and lacy, spreading; **sori** without indusia.
  A very attractive species endemic to Lord Howe Island. It is not well known in cultivation, but grows easily and requires warm, humid conditions. It will stand considerable exposure to sun, provided it is kept well watered. Propagate from spore.

Cyathea woollsiana (F. Muell.) Domin
(after Rev. W. W. Woolls)
Qld   Tree-fern
3-7 m tall
  Trunk slender, up to 7 m tall x 18 cm diameter, fibrous at the base, woody above and covered with roughened frond-bases; **scales** brown, conspicuous near the apex; **fronds** up to 2 m x 60 cm, finely divided, with crowded segments; **sori** covered by small indusia.
  Very attractive and adaptable to cultivation, requiring shaded conditions, well drained soils and an abundance of water in warm weather. Slightly frost tender, especially on young fronds. Rare in nature, being confined to rainforests of north-eastern areas. Propagate from spore.

141

Cyatheaceae

Cyatheaceae Reichenb.
(Tree-fern Family)

A family of ferns which contains about 650 species in 2 genera. The main genus is *Cyathea*, with 600 species in tropical and subtropical regions of the world. This is sometimes further subdivided into a number of other genera; however, the split is controversial and not yet widely recognized. They are tree-ferns which develop woody trunks and have excellent horticultural potential. About 14 species are found in Australia, and all have been tried in cultivation.

CYATHOCHAETA Nees
(from the Greek *cyathos*, a cup; *chaite*, long hair; in reference to the long bristles below the inflorescence)
Cyperaceae

Perennial rush-like plants with a spreading rhizome; **leaves** simple, erect, linear, flat to concave, sheathed at the base with scarious bracts; **inflorescence** a narrow panicle; **spikelets** narrow; **fruit** a nut.

An endemic genus of 5 species, 4 of which are found in WA and the remaining one is from the eastern states. Propagation is by division of the rhizomes.

Cyathochaeta avenacea Benth.
(like the genus *Avena*)
WA
0.3-0.6 m x 0.3-0.5 m   Oct-Jan

Rhizomatous, perennial **herb**; **leaves** 20-35 cm x 0.1-0.2 cm, narrow, erect, dark green, with inrolled margins, with narrow brown, entire sheaths; **culms** 30-60 cm tall, rigid, terete, the floral bracts long and narrow; **panicle** 10-25 cm long, slender, sparsely branched; **spikelets** 2-2.5 cm long, in groups of 3-5.

Endemic in the south-west, where it grows in colonies in moist soils. Could have potential as a container plant or around dams, bog gardens etc. Propagate by division.

Cyathochaeta clandestina Benth.
(hidden)
WA
1.5-2 m x 0.3-1 m   Oct-Jan

Tall, rhizomatous, perennial **herb**; **leaves** 20-35 cm x 0.2 cm, erect, rigid, crowded, with spreading brown **sheaths** 5-7.5 cm long, the margin lacerated; **culms** 1.5-2 m tall, rigid, with long, loose, sheathing bracts on the upper part; **panicle** 10-20 cm long, narrow, brown; **spikelets** about 5 cm long, in pairs.

Endemic in the south-west, where it forms tall thickets in moist, peaty soil. An interesting rush that could be tried as a container plant or around dams etc in temperate regions. Propagate by division.

Cyathochaeta diandra (R. Br.) Nees
(2 anthers)
Qld, NSW, Vic   Sheath Rush
0.3-0.6 m x 0.3-0.5 m   Oct-Jan

Rhizomatous, perennial, rush-like **herb**; **leaves** 20-30 cm x 0.1-0.2 cm, erect, rigid, flat or concave, pale green with prominent brown, papery sheaths; **culms** 30-60 cm tall, erect, terete, furrowed and striate; **panicle** 10-15 cm long, very narrow, shiny brown to reddish brown.

Cyathea rebeccae   B. Gray

A perennial rush-like plant that grows in spreading colonies in peaty soils of coastal heathland. It is an attractive species for planting around dams, ponds etc, and can even be induced to grow in a moist position in the garden, provided it is watered regularly. Propagate by division of the rhizomes.

CYATHODES Labill.
(from the Greek *cyathodes*, cup-like; allusion to the cup-shaped, toothed disc)
Epacridaceae

Prostrate to tall **shrubs**; **branches** many; **leaves** alternate, often overlapping, sometimes in clusters, striate, often white undersurface; **flowers** bisexual or unisexual, tubular, small, axillary, solitary or rarely 2-3 together; **bracts** small, upper ones largest and usually surrounding the base of calyx; **sepals** 5; **petals** 5, joined to form corolla tube which is longer than calyx; **lobes** spreading or recurved; **fruit** a fleshy drupe.

A genus of about 15 species. Australia has 8 representatives, 7 of which are endemic. *C. juniperina* also extends to New Zealand. Other species occur in the Hawaiian Islands. The genus has developed to its fullest extent in Tas, with all species recorded there. *C. juniperina* is the only Australian species to occur outside that state, with an isolated occurrence on Wilson's

142

Promontory in Vic. Generally the species are subalpine or alpine, with a few species growing on or near the coast and foothills.

They have not been widely cultivated to date, but have definite potential. Although the flowers are small, the foliage is decorative, often having brightly coloured new growth. Most species have ornamental fleshy fruits which can be white, blue, purple, pink or red. Some plants may not have bisexual flowers or separate male and female flowers on the same plant, in which case the formation of the fruits will be restricted.

Cyathodes are generally slow-growing, and do best in moist, well drained soils, with dappled shade or partial sun. They are probably best suited to cool climates with high rainfall, and are useful for general garden cultivation or as container plants. All species are frost tolerant. One of the main reasons for their lack of cultivation to date is the unavailability of suitable propagating material, or difficulty experienced in propagation. There has been limited success with seed which may take 1-3 years to germinate. Propagation of cuttings of firm young growth has been successful with many species, although in most cases roots are slow to form. A possible alternative method of propagation is aerial layering.

Cyathodes abietina (Labill.) R. Br.
(similar to the genus *Abies*)
Tas
1-3 m x 1-3 m   Dec-Feb
Small to medium **shrub**; **branches** many, rigid; **leaves** 1-1.5 cm x 0.2 cm, linear-lanceolate, flat, alternate, crowded, erect or spreading, pungent apex, lower surface striate; **flowers** tubular, about 0.6 cm long, white, solitary, in upper axils, profuse; **lobes** spreading; **fruit** about 1.2 cm diameter, pink to red.

Although not very ornamental in flower, this species is transformed from Feb-May by the brightly coloured fleshy fruits. To date there has been great difficulty in propagating and growing plants. It is certainly worthy of cultivation. In nature it grows in exposed coastal areas of the west, south-west and south. The best conditions for growth are in moist but very well drained soils, with a cool situation in dappled shade. Propagate from seed or cuttings. This species has been reported as being self layering; therefore successful propagation could be achieved by aerial layering.

Cyathodes acerosa R. Br. =
C. *juniperina* var. *oxycedrus* (Labill.) Allan

Cyathodes dealbata R. Br.
(whitened)
Tas
prostrate x 0.5-1 m   Dec-Feb
Prostrate, mat-like **shrub**; **branches** many, slender, wiry; **leaves** 0.4 cm x 0.1-0.2 cm, narrow-oblong, alternate, blunt apex, green above, white with a prominent dark midrib below; **flowers** about 0.5 cm long, white, nearly sessile, solitary, in leaf axils near ends of branchlets; **lobes** reflexed; **fruit** about 0.5 cm diameter, bright red.

A very attractive matting species that shows potential for cultivation. Scattered fruits decorate the surface

of the plant from Feb-May. Only limited success has been achieved with cultivation to date. It requires moist, well drained, light acid soils, with dappled shade or partial sun. It is grown as a greenhouse plant in England. Propagate from cuttings or aerial layers.

Cyathodes divaricata (Hook.f.) Hook.f.
(spreading)
Tas
0.3-1 m x 0.3-1 m   Dec-March
Dwarf **shrub**; **branches** erect, many; **leaves** about 1 cm x 0.1 cm, narrow-linear, alternate, spreading or reflexed, apex with long, pungent point, recurved margins, glaucous and grooved below; **flowers** about 0.5 cm long, male and female usually on same plant, white to cream, solitary, in upper leaf axils, borne on short, reflexed stalks; **fruit** about 0.8 cm diameter, bright pinkish red.

This species occurs on the foothills of Mount Wellington, and on the east coast. It requires moist, well drained soils, with dappled shade or partial sun. Evidently not in cultivation, but suitable for gardens and containers; also useful for sheltered coastal planting. Propagate from cuttings or aerial layers.

Cyathodes glauca Labill.
(glaucous)
Tas   Cheeseberry
1-10 m x 1-4 m   Dec-Feb
Small **shrub** to small **tree**; **branches** many; **branchlets** glabrous or slightly hairy; **leaves** 1.5-3 cm x 0.3-0.5 cm, narrow-elliptical to oblong, often clustered in false whorls, slightly convex, apex acute or pungent-pointed, striate and glaucous below, often reddish new growth; **flowers** tubular, to 1 cm long, white, solitary, in axils of uppermost leaves; **lobes** spreading or reflexed; **fruit** about 0.8 cm, red, pink or purple, rarely white.

This is one of the best-known species, and occurs on hills and mountains over a large area of Tas. It is usually a shrub, but grows to a small tree on the west coast. During late summer and autumn, the plants are very decorative with the fleshy fruits. Although not widely grown, there are plants in cultivation which are now quite a few years old. It does best on moist, well drained, light to medium soils, with dappled shade or partial sun. Suitable for both gardens and containers. Propagate from cuttings.

Cyathodes juniperina (Forster) Druce
                          var. **oxycedrus** (Labill.) Allan
(similar to the genus *Juniperus*; sharp; cedar-like)
Vic, Tas   Crimson Berry
1-10 m x 1-4 m   Aug-Jan; also sporadic
Small **shrub** to small **tree**; **branches** spreading, many; **leaves** 1-2 cm x 0.15 cm, narrow-linear to lanceolate, crowded, spreading or reflexed, tapering to a pungent point, margins recurved, striate and glaucous below; **flowers** tubular, about 0.5 cm long, white, fragrant, solitary, in upper axils, profuse; **lobes** spreading; **fruit** about 1 cm diameter, pale to deep pink, rarely white.

This cyathodes has the widest natural distribution of

Cyathodes juniperina var. *oxycedrus*

D.L. Jones

the Australian species, usually growing in coastal situations, but can be at altitudes of up to around 1000 m. It does best in moist, well drained, light to medium soils, with dappled shade or partial sun. Will withstand some coastal exposure, but is best grown in the protection of salt spray resistant plants. Suitable for gardens or containers. Propagate from cuttings.

Previously known as *C. acerosa* R. Br.

Cyathodes nitida Jarman
(shining)
Tas
0.2-0.5 m x 0.5-1 m   Sept-Oct

Dwarf **shrub**; **branches** spreading and ascending; **leaves** to about 1 cm x 0.2 cm, lanceolate, tapering to slender point, can be pungent, dark green and shiny above, glaucous and striate below, margin recurved; **flowers** tubular, about 0.5 cm long, white, solitary, in axils near branchlet tips; **lobes** spreading; **fruit** about 0.7 cm diameter, purple-black, flattened.

This rare species occurs at over 1100 m alt. on the central plateau, growing in shallow stony soils. It is evidently not in cultivation. Should grow well in well drained soils, with dappled shade or partial sun. Propagate from cuttings.

Closely allied to *C. petiolaris* which has blunt leaves and red fruits.

Cyathodes parvifolia R. Br.
(small leaves)
Tas   Pink Mountain Berry
0.5-1.5 m x 0.5-1.5 m   Dec-Feb

Dwarf to small **shrub**; **branches** erect, rigid; **leaves** about 0.5 cm x 0.15 cm, alternate, linear-lanceolate, often convex, spreading, tapering to a pungent point, glaucous or striate below; **flowers** dioecious, tubular, very small, white, solitary, on short, usually reflexed stalks, mainly in upper axils, profuse; **lobes** spreading; **calyx** and sepals about as long as tube; **fruit** on female plant only, to 0.8 cm diameter, pink to red, rarely white.

Similar in many ways to *C. juniperina*, which is a larger species and not dioecious. It is common on rocky hillsides throughout a large part of the state. A most ornamental species in fruit, from late summer to winter. It has had limited cultivation, but with successful results. Grows best in moist, well drained, light to medium soils, with dappled shade or partial sun. It could probably tolerate full sun in cool climates, provided it has a cool and moist root-run. Suitable for gardens and containers. Propagate from cuttings.

Cyathodes petiolaris (DC.) Druce
(with petioles)
Tas
0.2-0.5 m x 0.5-1 m   Dec-Feb

Cyathodes straminea   D.L. Jones

Dwarf **shrub**; **branches** spreading and ascending; **leaves** to about 0.7 cm x 0.25 cm, oval, more or less erect and overlapping, apex usually blunt, rarely pointed, dark green and shiny above, glaucous and striate below; **flowers** tubular, about 0.5 cm long, white to pinkish, sessile, can be solitary but usually 2-3, in axils near the branchlet tips; **lobes** spreading; **fruit** about 0.5 cm diameter, red.

An abundant species of the mountains at altitudes of over 1100 m. It is thought not to be in cultivation at present, but its growth habit and fruit make it an ideal plant for gardens and containers. Should do best in moist, well drained, light to medium soils, with dappled shade or partial sun. Propagate from cuttings or aerial layers.

Cyathodes straminea R. Br.
(straw coloured)
Tas
0.2-1.5 m x 0.5-1.5 m   Dec-Feb

Dwarf to small **shrub**; **branches** in false whorls; **leaves** 0.5-1.5 cm x 0.3 0.4 cm, elliptical, apex rounded, sometimes with a blunt point, clustered in false whorls, usually flat, green above, glaucous and striate below; **flowers** tubular, to about 1 cm long, white, solitary or in axillary clusters at ends of branchlets; **lobes** spreading, narrow, acute; **fruit** about 0.8 cm diameter, rounded and flattened, usually red.

Although this species is very similar to *C. glauca*, it can be readily identified by the delicate lobes of the flower-tube. It usually occurs above 1100 m, where it is very widespread. It requires moist, well drained, light to medium soils, with dappled shade or partial sun. Propagate from cuttings.

Cycads

Cycadaceae Pers. (emend L. Johnson)

A family of Gymnosperms containing about 20 species in the solitary genus *Cycas*. They are sporadically distributed in Madagascar, eastern Asia and Malesiana and are well represented in Australia by 11 species. One of these, *C. media*, is particularly widespread in northern areas and frequently grows in extensive stands. All are woody plants with a crown of leathery, pinnate fronds. The male flowers are borne in woody cones and the female flowers in a terminal, cone-like structure which collapses at maturity to shed the large seeds.

CYCADS

Cycads are primitive palm-like plants with woody trunks which may be above-ground or subterranean. The trunk is usually unbranched (but branched specimens are sometimes found) and terminates in a crown of pinnate leaves or fronds. Each leaf consists of a stiff stalk, to which are attached many leaflets that in most species are narrow, stiff, flat or folded, with the basal ones frequently being reduced to teeth or spines. The growing apex is protected by scales. Flowers and fruits are borne in cones, with the male and female flowers being in separate cones on different parts of the plant or on different plants.

Australian cycads belong to the genera *Cycas* (11 species) in the family Cycadaceae, *Bowenia* (2 species), *Lepidozamia* (2 species) and *Macrozamia* (16 species), in the related family Zamiaceae, are sometimes also called cycads but are probably best called zamiads. For further information, see Zamiad entry. Cycads are widely distributed throughout northern and central Australia, with no species reaching southern areas. They grow mainly as scattered individuals but occasionally large, sparse colonies are encountered. Most species grow in open habitats, in poor, shallow or rocky soils but a couple of species are found in richer soils near rainforests.

The cycads are very primitive plants that reached their peak of development some 200 million years ago. Their present status is a fraction of the numbers which covered the earth during that peak, and now they are scattered mainly through the tropical regions. They have features in common with both the ferns and the palms, and are sometimes referred to as fern-palms.

Most species are extremely slow-growing, producing only a few new leaves each season. Large specimens are of great age and indeed may be the oldest living plants on earth.

The Aborigines ate the large seeds of various cycads but not without considerable preparation, because without treatment they are quite poisonous. The preparation usually consisted of baking followed by leaching in running water for 5-20 days. Preparation also included a fermentation-type process. The trunks of cycads contain up to 40% starch, and at one time they were harvested commercially as a source of this material. Cycad seeds may cause poisoning of livestock if eaten, and the leaves may cause rickets or staggers.
Cultivation

Cycads are popular in cultivation, being mainly grown for their graceful, arching fronds and unusual growth habit. They can be grown as specimens or plan-

145

Cycas armstrongii

D.L. Jones

ted amongst low shrubs to provide a contrast in foliage. They make excellent container or tub plants as they survive periods of neglect without major detriment.

Most species are very hardy and will stand long periods of drought without ill effects. They can be successfully established under large trees, surviving the competition well, although perhaps they may be slower-growing than normal. They will grow in virtually any position including shade, but grow best where they receive some sun. Most species are quite tolerant of heavy frosts, but young fronds that have not hardened off sufficiently may be damaged or killed.

As a group they are slow-growing plants and respond to light applications of slow-release fertilizers such as blood-and-bone. Soil type is unimportant, provided drainage is good, although they do not generally grow well on calcareous soils. They do not suffer from attacks by many pests, although young fronds and developing cones may be damaged by caterpillars and locusts.

Propagation

The seeds are large and woody. They are sown whole at a shallow depth in a sandy, well drained mixture, with the top part of the seed visible above the soil. The pots should be placed in a warm, protected position and kept moist but not overwet. Germination is very slow, taking 10-30 months. The seedlings should be potted up as soon as the first leaf develops. At this stage the young plants have usually developed a strong taproot, and transplant quite readily. They should be kept moist until they become established, but not overwatered at any stage.

Transplanting

Large specimens can be successfully transplanted, but require some of their intricate root system for best results. Transplanted specimens may not develop new fronds for 1-2 years after moving.

CYCAS L.

(from the Greek *coicos*, a name for a kind of palm)

Cycadaceae

Palm-like plants; **trunk** slender to thick, woody; **leaves** borne at the trunk apex, in a spreading crown, pinnate; **pinnae** numerous, spreading longest in the middle of the frond, the basal ones often spine-like; **male flowers** forming catkins or cones; **female flowers** on petiolate spikes in loose cones; **fruit** large, woody, borne on marginal notches on the female spikes or megasporophylls.

Species about 20 with 10 or 11 being found in Australia, although some species are under revision. They belong to a very primitive group of plants having links with palms and ferns. As a genus, they are popular in cultivation and are generally easy to grow, although slow. Their distinctive silhouette makes them a very useful horticultural subject. Propagation is from seed which is best sown while fresh. Germination usually takes 12-18 months and the plants are very slow to develop. Advanced specimens usually transplant fairly readily but may take 1-2 years to settle down after disturbance.

Cycas angulata R. Br.
(angular)
NT   Zamia Palm
1-5 m tall

Trunk 1-5 m tall, 35-60 cm thick, massive, black, often branched; **leaves** 1-2 m long, green; **leaflets** 100-150 pairs, flat, acute, linear, dull to somewhat shiny; **male cones** about 20 cm long, brown; **female spikes** very large, hairy; **fruit** 2-4, about 4 cm long, globular, brown.

This large cycad is distributed through the islands adjacent to Darwin and the Top End. It is a very distinctive species, recognized by its massive black trunk and green leaves. Not well known in cultivation but probably has requirements similar to those of *C. armstrongii*. Propagate from seed.

Cycas armstrongii Miq.
(after John Armstrong, 19th-century collector for Kew Gardens, resident in NT)
WA, NT   Zamia Palm
1-4 m tall

Trunk 1-4 m tall, 10-15 cm thick, slender, dark grey; **leaves** 0.5-1.2 m long, leathery; **leaflets** 10-13 cm x 0.5-1 cm, entire, linear, straight, pale green, shiny, flat, with yellowish brown tips, decurrent at the base, the lower ones spine-like; **female spikes** flattened, glabrous, the blade triangular, with 2 basal lobes, the margins and apex with sharp spines; **male cones** 12-20 cm long, oval, rusty brown; **fruit** 4, 2-4 cm long, globular, yellowish green, shiny, covered with bluish powder.

A widespread and locally common species that grows in stunted open forest, usually in sandy soil. It often grows in scattered colonies in areas where forest fires occur almost annually. Frequently, a large plant is surrounded by younger ones including numerous seedlings 1-2 years old. The leaves are poisonous to stock and the seeds are poisonous without preparation. The Aborigines soaked the seeds in water to remove the poison prior to cooking. The plants are very well suited to tropical regions and inland areas with a warm, dry climate but can be tricky in other regions. In areas with cold winters the plants may defoliate completely, and regrow with the resumption of warm weather. While dormant they are very sensitive to overwatering and rotting. A sunny situation in well drained soil is most suitable. Seedlings are very slow-growing but advanced specimens transplant readily. Propagate from seed which takes 6-18 months to germinate. The species was previously confused with *C. media*, from which it can be distinguished by the more slender trunk and the smaller leaves with flat, softer, pale green leaflets. It was also previously known as *C. lanepoolei* C. Gardner.

Cycas basaltica C. Gardner
(growing on soils of basalt origin)
WA   Zamia Palm
1-3 m tall

Cycas cairnsiana showing male cone   B. Gray

Cycas cairnsiana with female cone   W.R. Elliot

Trunk 1-3 m tall, 30-60 cm thick, rough, dark grey, swollen at the base; **leaves** 0.5-1.5 m long, covered with hoary hairs, especially when young; **leaflets** linear, rigid, sharply pointed, hairy beneath, with recurved margins, decurrent, a few lower ones spine-like; **male cones** narrow, conical; **female spikes** petiolate, densely covered with rusty hairs, the blade narrow and entire; **fruit** 4, about 2 cm long, globular, slightly flattened.

Apparently confined to open forest of the Kimberleys, on basalt soils. Close to *C. media* but with very different female spikes, and leaflets with recurved margins. Fruit has been recorded in Aug. Not well known in cultivation. Requirements as for *C. armstrongii*. Propagate from seed.

147

Cycas cairnsiana F. Muell.
(after Sir W. Cairns)
Qld   Zamia Palm
1-5 m tall

Trunk 1-5 m tall; **leaves** 1-2 m long, glaucous, the petiole slightly hairy; **leaflets** 6-15 cm long x 0.4 cm, linear, with recurved margins, the lower ones reduced to short teeth; **male cones** about 20 cm long, narrow, rusty brown, the scales about 2.5 cm x 0.8 cm; **female spikes** petiolate, the blade lanceolate, entire or weakly toothed, rusty brown; **fruit** 2, about 3 cm long, shiny brown.

A relatively uncommon species confined to the ranges of north-eastern Qld, where it grows in shallow, stony soils under open forest. It is close to *C. basaltica* but produces 2 fruits instead of 4 on each megasporophyll. Plants become very attractive, with a crown of bluish grey fronds, and the flushes of new rusty fronds can be quite spectacular. Hardy and will grow successfully in mild areas of southern Australia. Requires conditions similar to those of *C. media*. Propagate from seed. Large specimens transplant readily, although they may take 6-18 months to produce new leaves.

Cycas calcicola Maconochie
(growing on limestone)
NT   Zamia Palm
1-3 m tall

Trunk 1-3 m tall, 17-30 cm across; **leaves** 0.6-1.2 m long, flat or arched in a crown, dark green; **leaflets** 8-12 cm x 0.2-0.3 cm, linear, 150-300 per leaf, straight or slightly curved, glabrous or hairy above, with prominent revolute margins; **male cones** 17-26 cm x 5-6 cm, narrow, ovoid, bearing grey hairs; **female spikes** loose, the blade to 15 cm long, entire or with 7-8 papery marginal teeth, glabrous above, rusty-hairy beneath; **fruit** 2-6, 3-3.5 cm long, brownish.

A recently described species from isolated limestone outcrops of the Top End. It can be distinguished by the leaflets, which are usually pubescent above and with prominently revolute margins. At this stage it is not well known in cultivation, but has similar requirements to those of *C. armstrongii*. Propagate from seed.

Cycas furfuracea W. Fitzg.
(covered with loose, woolly scales)
WA   Zamia Palm
1-2 m tall

Trunk 1-2 m tall, stout; **leaves** 0.6-0.9 m long, stiff; **leaflets** 10-15 cm x 0.4-0.6 cm, entire or bifid, linear, rigid, straight, ending in dark tips, strongly keeled, glabrous above, conspicuously furfuraceous beneath; **male cones** 30-35 cm x 6-8 cm, narrow-ovoid, covered with woolly brown hairs; **female spikes** 15-20 cm long, the blade ovate, with prominent spiny teeth and a blunt apex, covered with woolly brown hairs; **fruit** 2-3 cm across, yellow.

A little-known species confined to sandy soil in rocky areas of the King Leopold Range, Kimberley region. It can be recognized by the prominent, loose woolly scales which clothe the undersides of the leaflets. Its cultivation requirements are probably as for *C. armstrongii*.

Cycas calcicola   D.L. Jones

Cycas kennedyana F. Muell.
(after Sir A. Kennedy)
Qld   Zamia Palm
1-4 m tall

Trunk 1-4 m tall, 20-35 cm thick; **leaves** 1-2 m long, green above, glaucous beneath, flexuose; **leaflets** up to 100 pairs, flat, acute, linear, decurrent at the base, shiny, glabrous, the lower ones spine-like; **male cones** 20-30 cm x 10-15 cm, oval, covered with velvety hairs; **female spikes** very large, covered with velvety hairs; **fruit** 4, about 4 cm long, globular, brownish.

Widely distributed in the ranges of coastal and nearby inland central Qld between Rockhampton and Bowen. It grows in scattered colonies, in shallow stony soil, under open forest. A distinctive species recognized by the large, spreading crown of flexuose leaves with glabrous leaflets. Not well known in cultivation but probably has similar requirements to those of *C. media*. Propagate from seed.

Cycas aff. kennedyana
Qld   Zamia Palm
1-3 m tall

Trunk 1-3 m tall, 10-15 cm thick, with a small neat crown; **leaves** 1-1.5 m long, markedly glaucous; **leaflets** flat, acute, linear, decurrent at the base, dull and glaucous, the lower ones spine-like; **male cones** 10-15 cm x 5-10 cm, linear, rusty brown; **female spikes** hairy; **fruit** 4, 2-4 cm long, globular, yellow-brown.

An apparently undescribed species from the Many Peaks Range south of Gladstone, where it grows in open forest. It is distinguished from *C. kennedyana* by the leaflets being glaucous on both surfaces, and the smaller, narrower cones. Suited to cultivation in tropical, subtropical and warm temperate regions at least as far south as Sydney. Requirements as for *C. media*. Propagate from seed.

Cycas lane-poolei C. Gardner = *C. armstrongii* Miq.

Cycas media R. Br.
(medium)
Qld   Zamia Palm
1-3 m tall
    Trunk 1-3 m tall, thick, usually unbranched; **leaves** 0.5-1.5 m long, leathery; **leaflets** 10-13 cm x 1-1.5 cm, linear, straight or falcate, blunt or pointed, stiff, the lower ones spine-like; **male cones** 30-45 cm long, slender, woolly; **female spikes** about 40 cm long, the blade broad-lanceolate, with a sharp apex and toothed margins; **fruit** 4-6, 2-4 cm long, oval, orange.
    Widespread and very common in tropical regions, extending south as far as Sarina. It usually grows in ex-tensive colonies in open forest or rocky areas, but is oc-casionally found in rainforest. Very old specimens with trunks up to 8 m tall are occasionally encountered, as also are branched specimens. The trunks are very fire resistant and although the fronds are destroyed, they rapidly produce a new crown. At this stage of growth the fronds are a delicate green and the plants look quite spectacular. New fronds normally uncoil and are produced in flushes. The seeds are very poisonous but were regarded as an important food source by the Aborigines, who developed a specialized technique for their preparation. The trunks of this species were at one time harvested commercially for starch.
    This zamia palm is popular in cultivation, being very hardy and adaptable to most conditions except bad drainage and deep shade. Quite frost hardy and will grow successfully in most areas of southern Australia. Very slow-growing and drought resistant. Large speci-mens can be transplanted successfully but may take 12-18 months to re-establish. Propagate from seed.

Cycas normanbyana F. Muell.
(after the Marquis of Normanby)
Qld   Curly Pine Palm
1-3 m tall
    Trunk 1-3 m tall, thick, woody; **leaves** 1-2 m long, shiny green, prominently arched or curved; **leaflets** numerous, opposite, up to 1.5 cm x 0.6 cm, the basal leaves becoming spine-like; **male cones** 20-30 cm long, tawny yellow; **female spikes** large; **fruit** 2, 2.5-3.5 cm long, orange, rounded.
    An ill-defined and little-known species confined to the ranges of central Qld. It is apparently distinguished by the crown of prominently arching leaves, and the sporophylls never bearing more than 2 rounded fruit. Not known to be in cultivation. Requirements are probably as for *C. media*. Propagate from seed.

Cycas pruinosa Maconochie
(bearing white powder)
WA   Zamia Palm
1-2 m tall
    Trunk up to 2 m tall; **leaves** 0.8-1 m long, straight, grey-green; **leaflets** up to 120 pairs, opposite, linear, entire, glabrous, with revolute margins; **male cones** 40-50 cm long, narrow, deltoid, grey, hairy; **female spikes** to 30 cm long, rusty brown, the blades strongly toothed; **fruit** 4, 3-4 cm long, globular, brown, covered in white powder.
    A recently named species restricted to the Kim-berley region, where it grows on rocky slopes on the ranges around the Ord River. Quantities of seeds of cycads have been collected from the Kimberleys, and this species is probably in cultivation. Requirements as for *C. armstrongii*. Propagate from seed.

CYCLOSORUS Link
(from the Greek *cyclos*, circle; *soros*, a heap; usually in-terpreted as the fern sorus)
Thelypteridaceae
    Medium to large **ferns** forming clumps or spreading patches; **fronds** erect or semi-weeping, narrow, yellow-ish green, hairy, once-divided, bearing papery scales on

Cycas media   B. Gray

Cyclosorus interruptus

the rhachises and veins; **sori** rounded, with a kidney-shaped cover.

A small genus of about 3 species with one widespread species extending to Australia. The genus has recently been revised and split into a number of distinct genera (see *Amphineuron*, *Christella*, *Pneumatopteris* and *Sphaerostephanos*).

Cyclosorus interruptus (Willd.) H. Ito
(with the symmetry broken)
Qld, NSW, WA, NT
0.3-1 m tall

Medium sized **fern** forming extensive spreading patches; **rootstock** creeping, much-branched; **fronds** erect, leathery, dark green; **segments** lobed, lanceolate to linear, largest at the base of the frond, papery scales on **rhachis** and veins; **sori** in a nearly continuous marginal band.

A coarse species found in swamps, usually in large, dense colonies. Not widely grown because it is not a particularly attractive fern. It is, however, a very hardy fern that is easy to grow and useful for wet clay positions, dam banks or soaks exposed to sun. Frost hardy and can be successfully grown in southern Australia. Propagate by division of the rhizomes or from spore.

CYMBIDIUM Sw.
(from the Greek *cymbe*, a boat; *idium*, a diminutive suffix; there is a small, hollow, boat-shaped recess in the lip of some species)
Orchidaceae

Epiphytic **orchids**; **pseudobulbs** large and crowded, or slender; **leaves** long-linear, thick or thin-textured, terminating the pseudobulbs; **racemes** arising from the pseudobulbs, erect or pendulous; **flowers** numerous, small to large, showy; **fruit** a capsule.

Species about 40 with 3 endemic in Australia. They grow as epiphytes, rarely lithophytes, and may form huge clumps. All are popular subjects for cultivation, although a couple are fairly difficult to grow. All native species are free-flowering, with attractive, small fragrant flowers and are popular subjects for hybridization, especially as parents for the commercial pursuit of miniature-flowered cymbidiums. They grow in mixtures of coarse material such as charcoal, pinebark, sand, leaf mould, rice hulls etc. They are well suited to pot culture, but can also be grown in baskets or in stumps. They generally flower best if exposed to some sun during the day. Propagation is by division or from seed which requires special sterile conditions to germinate.

Cymbidium canaliculatum R. Br.
(stalks with longitudinal grooves)
Qld, NSW, WA, NT   Channel Leaf Cymbidium
Aug-Dec

Pseudobulbs 5-15 cm long, short and stout, in dense clusters; **leaves** 2-6, 20-60 cm long, thick, rigid, keeled, triangular in cross-section, greyish green; **racemes** 20-40 cm long, pendulous, numerous; **flowers** 2.5-3.5 cm across, variable in colour, green, brownish, red-brown or dark red, with or without blotches, fragrant; **capsule**

Cymbidium canaliculatum   B. Gray

2-3 cm long, ovoid, swollen.

A very hardy orchid usually found on eucalypts in drier inland areas, tolerating considerable exposure to the sun. Rhizomes roam through the decaying heartwood of the host tree and produce new plants where there is a hole in the trunk or branch. The plants are very floriferous, producing numerous racemes each season. Aborigines chewed the bulbs as a cure for dysentery and for the starch they contain. Pods and seeds are also edible.

The species is very popular in cultivation, especially in areas with hot climates. Established plants resent disturbance and are very difficult to re-establish. Small plants are much easier to handle, and establish well in cultivation. The species is always very tricky to grow, requiring warm conditions and abundant water while in active growth (usually Oct-March),and no water while dormant. Drainage must be excellent. The mixture must be coarse, and is made up of large pieces of pinebark, decaying eucalypt heartwood and charcoal. The species grows fairly well in warm inland areas but in cool, moist areas needs the protection of a heated glasshouse. Once established, the plants should not be divided, and should be repotted with care.

The species is variable and a number of forms are found in cultivation. The most popular form has deep red flowers which may be so dark that they appear black, and is known as 'Sparkesii'. A rare form has clear apple green flowers with a white labellum. Good forms bear exceptionally long racemes. Propagate from seed.

The species has been crossed with *C. madidum* (see *C.* 'Little Black Sambo'), and many other species and hybrids (see Cymbidium Hybrids).

150

Cymbidium madidum Lindl.
(damp; moist)
Qld, NSW   Aug-Feb

Pseudobulbs 6-25 cm x 2-6 cm, stout, large, in large clusters; **leaves** 3-9, 20-90 cm x 2-3 cm, narrow, linear, dark green; **racemes** 20-60 cm long, pendulous; **flowers** 2-3 cm across, green or brown, fragrant; **capsule** 3-4.5 cm long, swollen, turgid.

A widespread species usually favouring moist areas in open forest or rainforest on trees or rocks, sometimes forming huge clumps. Aborigines chewed the bulbs as a cure for dysentery and for the starch they contain. Popular in cultivation because it is very easy to grow and flower. The flowers are long-lived and fragrant. Established plants transplant readily and can be divided or repotted with ease. They need a coarse epiphyte mixture and can be successfully grown in a bushhouse as far south as Melbourne, or in an unheated glasshouse. The species is variable and some excellent forms are grown. Some have particularly long racemes with clear apple green flowers.

Propagate from seed or by division. The old bulbs can be separated and induced to shoot by planting in pots of fresh sphagnum moss in a bottom heat unit.

The species has been crossed with both *C. canaliculatum* and *C. suave* (see *C.* 'Kuranda' and *C.* 'Little Black Sambo'). It has also been crossed with many other exotic species and hybrids, in the search for compact, small-flowered plants (see Cymbidium Hybrids).

Cymbidium suave R. Br.
(sweet)
Qld, NSW
                                            Aug-Jan

Stems 10-35 cm long, slender, covered with fibre; **leaves** 15-45 cm x 1-1.5 cm, dark green, linear, keeled; **racemes** 10-25 cm long, pendulous; **flowers** 2-3 cm across, olive green, fragrant; **labellum** dark red; **capsule** 2-3 cm long, swollen.

A widespread and common species usually growing

Cymbidium suave, × ·2

in large tussocks in hollowed branches of eucalypts. The pseudobulbs are covered with fibrous bark, are long-lived and become elongated when old. Rhizomes roam through the decaying heartwood of the tree and produce new plants where there is a hole in the trunk or branch.

The species is popular in cultivation and is primarily grown for its masses of deliciously fragrant flowers. Established plants resent disturbance and are very difficult to re-establish in cultivation unless whole limb sections are removed from the tree. Seedlings are frequently found on logs or stumps in the years after bushfires, and these are relatively easy to transplant and establish in cultivation. They need a coarse epiphyte mixture containing some charcoal and decaying heartwood from eucalypts. They can be successfully grown in a bushhouse as far south as Melbourne, or in an unheated glasshouse. Once established, they should not be divided, and should be repotted with care. Propagate from seed.

C. suave has been hybridized with exotic species and hybrid cymbidiums in an attempt to produce miniature-flowered forms. It has also been crossed with *C. madidum* (see *C.* 'Kuranda').

CYMBIDIUM HYBRIDS

Cymbidiums are amongst the most popular orchids cultivated in Australia. They are easy subjects to grow and flower, and are successful over a very large area of the continent. Most of those grown are of hybrid origin, and their ancestry can be traced to exotic species from India, Thailand and other Asian countries. The hybrids are generally much more flamboyant and colourful than the original species, although often they lack attributes such as fragrance. The 3 Australian species have relatively small flowers but what they lack in size they make up for in floriferousness and perfume. There is considerable interest at present in the production of compact, miniature-flowered hybrids, and the Australian species have participated in this trend. Up until Nov 1979 there had been 12 primary crosses (these are crosses between species) and 20 others involving Australian species, the progeny of which have been registered. These are shown in the accompanying tables.

List of primary hybrids involving Australian cymbidium species

Registered Name	Parents
Alcor	*C. simulans* × *C. canaliculatum*
Cricket	*C. devonianum* × *C. madidum*
Ensi Canal	*C. ensifolium* × *C. canaliculatum*
Francis Hunte	*C. madidum* × *C. finlaysonianum*
Iris Bannochie	*C. finlaysonianum* × *C. canaliculatum*
Kuranda	*C. madidum* × *C. suave*
Little Black Sambo	*C. canaliculatum* × *C. madidum*
Pee Wee	*C. madidum* × *C. pumilum*
Pied Piper	*C. devonianum* × *C. canaliculatum*
Scallywag	*C. pumilum* × *C. suave*
That's It	*C. virescens* × *C. madidum*
Kevin Ragg	*C. eburneum* × *C. suave*

Cymbidium

List of hybrids involving Australian cymbidium species other than primary crosses

Registered Name	Parents
Abundance	Lyoth × C. canaliculatum
Bob Norton	C. madidum × Miretta
Canterbury	Princess Mary × C. canaliculatum
Darkie	Jean Brummit × C. madidum
Evonne	Esmeralda × C. suave
Fifi	C. madidum × Argonaut
Impish	C. madidum × Egret
Lamorack	Charm × C. canaliculatum
Madelene Madsen	C. madidum × Lucy
Mitzi	Angela × C. madidum
Nonna	C. madidum × Alexanderi
Odyssey	Eagle × C. canaliculatum
Pat Ann	Apollo × C. madidum
Red Orange	C. madidum × Anna
Scamp	Flirtation × C. madidum
Sunshine Falls	King Arthur × C. madidum
Sweet Lime	Esmeralda × C. madidum
Torette	C. madidum × Dorchester
Yellow Scamp	C. madidum × Balkis
Little Nugget	C. madidum × Greenwood

Cymbopogon ambiguus   flowering stem, × ·3   flowers, × ·4

Cymbidium 'Kuranda'

A man-made hybrid between *C. madidum* and *C. suave*. Popular in cultivation because it is easy to grow and flower. The pseudobulbs are large like those of *C. madidum*, with long, pendulous flower-spikes bearing fragrant yellowish green flowers. Will grow successfully in an unheated glasshouse as far south as Melbourne. Culture as for *C. madidum*. Propagate by division or meristem culture.

Cymbidium 'Little Black Sambo'

A man-made hybrid between *C. madidum* and *C. canaliculatum* 'Sparkesii'. Easier than *C. canaliculatum* to grow and flower. The pseudobulbs are large, with pendulous flower-spikes bearing dark red flowers. In southern areas it grows best in a heated glasshouse. Propagate by division or meristem culture.

CYMBONOTUS Cass.

(from the Greek *cymbe*, a boat; *notos*, black; alluding to the convex back of the achenes)
Asteraceae

A small genus of 2 species.

Cymbonotus lawsonianus Gaudich.

(after William Lawson, 19th-century NSW explorer)
Qld, NSW, Vic, Tas, SA, WA   Bear's Ear
prostrate x 0.1-0.3 m   Sept-Nov

Dwarf, perennial **herb**; **leaves** to 10 cm long, ovate to lanceolate, margins wavy and sometimes toothed, arranged in a rosette; **flower-heads** yellow, about 1 cm diameter, borne on stalk shorter than leaves.

A dainty species that will grow in a wide range of soils and climatic conditions. Likes soils rich in humus. Suitable for a rockery. It will, however, not survive in permanently wet soils. Frost hardy. Propagate from seed or by division. Early settlers used an ointment made from the leaves mixed with lard as a treatment for wounds.

CYMBOPOGON Sprengel

(from the Greek *cymbe*, a boat; *pogon*, a beard; in reference to the boat-shaped, bearded spathes)
Poaceae

Densely tufted, perennial **grasses**; **leaves** flat, usually basal, often aromatic; **stems** slender or stout, erect; **panicles** loose or dense, often brittle; **spikelets** flattened, sometimes bearing awns.

Species about 60 with about 8 extending to the drier parts of Australia. They are tussock grasses, often with interesting fragrances in the leaves. It is recorded that Aborigines used *C. obtectus* and *C. refractus* for medicinal purposes.

Cymbopogon ambiguus A. Camus

(changeable; doubtful)
Qld, NSW, SA, WA, NT Scent Grass; Lemon Grass
0.3-1 m tall   July-Nov

Tufted, perennial **grass** forming erect, leafy tussocks; **leaves** 20-45 cm long, folded, bluish green, sometimes reddish; **stems** slender, erect or arching; **panicle** up to 35 cm long, very narrow; **spikelets** blue-green, very hairy; **awns** 1-2 cm long.

Widespread on shallow, rocky soils. The leaves give off a strong lemon or citron odour when crushed. A very attractive grass deserving of wide cultivation. Useful as a rockery plant or for mingling amongst low shrubbery. Needs a sunny position in well drained soil.

Drought resistant but develops a better appearance if watered regularly. Propagate from seed or by division.

Cymbopogon bombycinus (R. Br.) Domin
(inflorescence resembles masses of silk)
Qld, WA, NT   Silk Grass
0.3-1 m tall   Oct-March

Tufted, perennial **grass** forming erect, rigid tussocks; **leaves** in a basal rosette, narrow, flat, rigid; **stems** erect, rigid, slender; **panicle** 8-15 cm long; **spikelets** concealed by dense, long silky hairs with a woolly appearance.

Widespread in northern areas, usually along stream banks and alluvial flats, in sandy or stony soil. A tough grass with attractive flower-heads. Hardy in cultivation in a sunny situation. Propagate by division of the rhizome or from seed.

Cymbopogon obtectus S. T. Blake
(protected)
all states except Tas   Silky Heads
0.5-1 m tall   Oct-Jan

Tall, leafy, perennial **grass** forming erect tussocks; **leaves** numerous, in a basal rosette, flat, smooth, hairy; **panicle** short, compact, dense; **spikelets** reddish, densely covered with white, silky hairs.

A common grass of drier inland areas. A useful, slender, tall grass which develops a neat tussock and is ornamental when in flower. The leaves are aromatic when crushed. Very hardy but responds to regular watering. Suited to warm districts but can be successfully grown in southern areas. Propagate by division of the rhizome or from seed.

Cymbopogon procerus (R. Br.) Domin
(tall; slender)
WA, NT
1-2 m tall   Sept-April

Tufted, perennial **grass** forming erect, slender tussocks; **leaves** 20-50 cm x 0.5-1.5 cm, in a basal rosette, tough, grey-green; **panicle** 40-90 cm long, with many spreading branches, hairy; **spikelets** glaucous, very hairy.

A robust grass, widespread in open forest, often on lateritic soils or in rocky situations. The leaves have a citron odour when crushed. It is a very hardy grass, useful for inland areas with a hot climate. Propagate by division of the rhizome or from seed.

Cymbopogon refractus (R. Br.) A. Camus
(bent back sharply from the base)
Qld, NSW, Vic, NT   Barbed Wire Grass
0.5-1.5 m tall   Oct-Jan

Tufted, perennial **grass** forming erect, slender tussocks; **leaves** in a basal rosette, narrow, rough; **stems** slender, erect, stiff; **panicle** very narrow, with several branches which become reflexed; **spikelets** blue-green to red-brown, covered with waxy secretions.

Found in dry, open forests. Also known as turpentine grass and ginger grass. The leaves give off a ginger-like odour when crushed. A coarse grass, not widely cultivated but nevertheless an interesting species with its tall, fragile heads. Very hardy but responds to extra water during dry periods. Propagate by division of the rhizome or from seed.

Cymodoceaceae Norman Taylor
A family of monocotyledons which are marine grasses that grow submerged in shallow sea-water. The family consists of 5 genera and 16 species mostly found in tropical areas. They grow in beds and are important spawning and shelter areas for marine fish.

CYNANCHUM L.
(from the Greek *cynos*, dog; *agcho*, to strangle; literally would strangle a dog; an apparent reference to poisonous properties)
Asclepiadaceae

Slender or shrubby **climbers**, **stems** exuding milky sap if damaged; **leaves** opposite, entire; **inflorescences** of cymes or umbels; **flowers** small, deeply divided into 5 segments, with a corona around the anthers; **fruit** a large follicle.

A large genus of about 150 species with 9-10 endemic in Australia. Propagation is from seed which should be sown within a month of collection or from cuttings.

Cynanchum carnosum (R. Br.) Domin =
Ischnostemma carnosum (R. Br.) Merr. & Rolfe

Cynanchum elegans (Benth.) Domin
(graceful)
Qld, NSW
2-6 m tall   Aug-Oct

Slender **climber**; **leaves** 2-3 cm x 1.5-3 cm, ovate to broad-ovate, truncate or crowded at the base, light to dark green; **umbels** axillary; **flowers** about 1 cm across, white; **follicle** about 5 cm long.

A slender species occasionally found in botanical collections. The flowers are attractive and the pods are decorative. Requires a cool root-run and is best planted amongst shrubbery. Propagate from seed.

Cynanchum erubescens R. Br.
(pale red)
Qld
1-3 m tall   Nov-Feb

Small, slender **climber**; **stems** twining; young shoots slightly hairy; **leaves** 2.5-7.5 cm x 2-3 cm, ovate, prominently cordate at the base, distinctly petiolate; **cymes** open and lax, pedunculate; **flowers** about 0.4 cm across, pinkish white, shortly hairy outside; **corona** folded; **follicle** 3-6 cm long, prominently angled, winged.

A little-known climber from north-eastern Qld, with interesting fruit. It has potential for planting amongst shrubs in tropical and subtropical regions, and would add interest to a garden. Requires well drained soil. Propagate from seed or cuttings.

Cynanchum floribundum R. Br.
(numerous flowers)
Qld, NSW, SA, WA, NT   Native Pear
0.5-1 m tall   June-Sept

Shrubby small **climber**; young shoots hairy; **leaves** 2.5-5 cm x 2-4 cm, ovate to cordate, rounded at the base, drawn out into a long point, dark green and glabrous when mature, hairy when young; **cymes** dense;

Cynanchum pedunculatum

flowers about 0.6 cm long, white; **follicles** 3-5 cm long, spindle-shaped to ovoid, acuminate; **seeds** with bristles 1-2.5 cm long.

A widespread climber of inland regions, usually on sand-dunes. The unripe green milky pods were collected and avidly eaten by Aborigines. They are moist and astringent. The plant is not really showy but is a useful, hardy climber for arid areas. Propagate from seed. Cuttings would probably strike readily.

Cynanchum pedunculatum R. Br.
(pedunculate)
Qld, WA, NT
2-4 m tall   Aug-Nov
Slender **climber**; young shoots twining, glabrous; **leaves** 3-6 cm x 1-2 cm, ovate, deeply cordate at the base, on fairly long petioles, the apex acuminate; **cymes** leaf-opposed, on peduncles 3-8 cm long; **flowers** about 0.5 cm across, mauve-purple, the corona divided into acute lobes; **follicles** about 6 cm long, sharply angled.

A slender climber confined to the tropical districts and adjacent islands, often in rocky situations. An interesting free-flowering species with excellent potential for tropical areas, where it could be suitable for planting amongst shrubs. Propagate from seed.

Cynanchum puberulum F. Muell. ex Benth.
(with a few downy hairs)
NT
1-3 m tall   Nov-Feb
Small, slender **climber**; **stems** twining, softly hairy; **leaves** 3-6 cm x 1-3 cm, ovate, cordate at the base, softly hairy all over, drawn out at the apex; **cymes** dense, on peduncles 4-6 cm long; **flowers** about 0.6 cm long, white, the corona lobes with long points; **follicles** about 10 cm long, fusiform.

A little-known tropical climber found in rocky situations. It could have application for gardens in tropical areas. Propagate from seed.

CYNODON Pers.
(from the Greek *cynos*, dog; *odontos*, a tooth; a translation of the French name for the plant)
Poaceae

Stoloniferous, spreading **grasses**; **stems** rooting at nodes; **inflorescence** erect, of digitate spikes radiating from the end of the peduncle; **spikelets** one-flowered, sessile, in 2 rows on either side of the spikes.

A small genus of about 10 species with 2 found naturally in Australia. About 3 other species have become naturalized as roadside or pasture weeds.

Cynodon dactylon (L.) Pers.
(finger-like)
all states   Couch Grass
prostrate   Feb-May
Stems prostrate, rooting at nodes and forming a mat; **leaves** 1-2.5 cm long, linear, green or often glaucous; **inflorescence** held erect, of 3-5 digitate spikes, often purplish; **spikelets** sessile.

A cosmopolitan grass that is widely planted for lawns, as it forms an excellent dense mat if mown,

watered and fertilized regularly. A number of selections are available commercially but it is uncertain whether these originated in Australia or overseas. Some are obviously from overseas, eg. Bermuda Grass. Couch also makes a good, productive pasture grass and is a useful sand-binder for coastal districts. It requires well drained soils and seems to favour sandy loams. Growth is checked in winter, especially by frost. Propagate from seed or by transplanting pieces (sprigging).

CYNOGLOSSUM L.
(from the Greek *cynoglossum*, name of a plant in Dioscarides; from *cynos*, dog; *glossa*, tongue)
Boraginaceae

Annual, biennial or perennial **herbs**; **branches** erect or spreading; **leaves** entire, lower ones with stalks, upper ones sessile, clothed with stiff hairs, sometimes reduced to scattered tubercles; **flowers** in 1-sided simple or forked racemes, blue, purplish or white to yellow; **calyx** deeply divided into 5 segments; **corolla** with tube as long as calyx, throat closed with scales opposite the lobes; **lobes** spreading, rounded; **anthers** enclosed in tube; **fruit** dry, dividing into 4 fruitlets, flat or convex, the upper surface more or less covered with short, hooked spines.

A genus of about 60 species, occurring in temperate and subtropical regions of the world. Australia has about 4 species, of which 3 are thought to be endemic. They occur over a wide range, with one species growing mainly in forests, and the others in more open and drier country. They are not at all well known in cultivation, and are mainly grown by enthusiasts. *C. suaveolens* is renowned for its strong fragrance. Propagate from seed, cuttings or by division.

Cynoglossum australe R. Br.
(southern)
NSW, Vic, Tas, SA, WA, NT
                                     Australian Hound's Tongue
0.3-1 m x 0.5-2 m   Oct-Feb
Perennial **herb**; **stems** spreading or erect, rough; **leaves** 6-12 cm long, lanceolate or oblong, wavy margins; **flowers** small, light blue, only slightly fragrant, borne in leafless cymes, with stalks about 0.5 cm long; **fruitlets** flattened, winged, about 0.6 cm long, hooked spines all over.

This species usually occurs on dry or well drained soils, and will appreciate similar conditions in cultivation. Will grow well in dappled shade, or partial or full sun. The small flowers are slightly fragrant, which adds to its use in gardens. Regular light pruning is recommended. Propagate from seed or cuttings.

The var. *drummondii* (Benth.) Brand differs with its larger flowers, which can be light blue, white or pink, and it flowers from Sept-Nov.

Cynoglossum latifolium R. Br.
(broad leaves)
Qld, NSW, Vic, Tas   Forest Hound's Tongue
0.5-1.5 m x 0.5-2 m   sporadic
Dwarf to small **perennial**; **stems** weak, spreading, rough with prickles or hairs; **leaves** upper ones 5-8 cm long, oval, acute, 5-veined, lower ones become smaller,

Cynanchum floribundum

T.L. Blake

hairy, tubercles above, hairy, on 5 prominent veins below; **flowers** small, bluish, solitary, on slender, recurved stalks about 1.5-2.5 cm long, or in leafy racemes; **corolla** shorter than calyx; **fruitlets** about 0.3 cm long, outer surface with hooked spines, stalk not recurved.

Usually found in moist forests, this species will probably not have wide application in horticulture, but is useful in moist, shaded areas, as a groundcover and soil-binder. It prefers reasonably well drained soils. Tolerates light to medium frosts, and responds well to light pruning. Propagate from seed or cuttings.

Cynoglossum suaveolens R. Br.
(sweet-scented)

Qld, NSW, Vic, Tas, SA — Sweet Hound's Tongue
0.1-1 m x 0.5-1.5 m — most of the year

Perennial spreading by underground stems; **stems** erect to spreading, rough with short hairs; **leaves** 4-12 cm long, on basal rosettes and along stems, upper ones becoming smaller and sessile, lanceolate-elliptic, surfaces rough with stiff hairs; **flowers** about 0.6 cm diameter, white to pale yellow, strongly fragrant, borne in leafy cymes; **fruitlets** ovoid, wingless, convex face, densely covered in hooked spines.

The flowers of this species are not highly ornamental, but the fragrance they emit compensates fully for their lack of size. Plants will grow best in well drained soils, and will tolerate dappled shade, or par-

tial or full sun. Excellent as groundcover for embankments, especially in crumbly clay soils. Suitable also as a container plant which can be moved to a selected location when in flower. Responds well to regular light pruning. Propagate from seed or cuttings.

CYNOMETRA L.
(from the Greek *cynos*, dog; *metra*, core; heart; the shape of a dog's womb; an obscure reference)
Caesalpiniaceae

Trees or **shrubs**; **leaves** pinnate, with 1-3 pairs of leaflets; **flowers** small, pea-shaped, reddish, in axillary clusters or racemes; **pods** thick, fleshy, turgid.

Species about 60 with a widespread one extending to northern Australia.

Cynometra iripa Kostel.
(derivation obscure)

Qld — Wrinkle Pod Mangrove
3-5 m x 1-3 m — Sept-Dec

Medium to tall, straggly **shrub**; young branches brown, roughened; **leaflets** 4, 3-7 cm long, obovate, blunt or emarginate, thick, unequal-sided; **flowers** about 0.5 cm long, red, in short axillary clusters; **pods** 1-2 cm x 1-2 cm, turgid, wrinkled, each containing a single seed.

Found along the edges of mangroves in tropical areas. The wood is light brown and tough, and stains

155

water purple. Not widely cultivated but important for coastal stabilization and reclamation work. Attractive and worth growing in saline areas. Propagate from seed which needs scarification for satisfactory germination.

Cyperaceae Juss. (Sedge Family)

A large family of monocotyledons containing about 4000 species in 90 genera, widely distributed in moist areas around the world. They are mainly annual or perennial herbs, and usually grow in wet or marshy areas. They have tufted, grass-like leaves and solid culms which bear a compound inflorescence made up of small spikelets. The family is well represented in Australia with about 47 genera and 650 species. These are widely distributed over the continent wherever suitable moist soils occur. Few species have any horticultural appeal. Main Australian genera: *Bulbostylis, Carex, Caustis, Chorizandra, Cladium, Crosslandia, Cyperus, Fimbristylis, Gahnia, Lepidosperma, Schoenus* and *Scirpus*.

CYPERUS L.

(from the Greek *cypeiron*, a sedge)
Cyperaceae

Annual or perennial **herbs**; **roots** fibrous, occasionally bearing tubers; **leaves** long, grass-like, basal; **culms** erect, 3-cornered; **umbel** compound, loose or dense, subtended by spreading bracts; **spikelets** oppositely arranged, flattened; **fruit** a nut.

A large widespread genus of about 550 species, about 80 of which are found in Australia, 30 being endemic. A number of others are naturalized as weeds. They are commonly known as sedges. Few are cultivated, the majority being small annuals or weedy type plants. A cosmopolitan species found in Australia is a serious weed of pasture and horticulture. It is *C. rotundus* L. and is commonly known as nut grass because of the small tubers found on the roots. It is highly invasive and extremely difficult to eradicate once established. A similar species is the ground-nut *C. esculentus* L. which has edible root tubers. This species is not native but is sometimes naturalized. It is also weedy but much less troublesome than nut grass. The root tubers of *C. bulbosus* were roasted and eaten by the Aborigines. Some species are quite ornamental, and useful for wet positions. Propagation is by division or from seed which should be sown under glass on moist soil.

Cyperus alopecuroides Rottb.

(like the genus *Alopecurus*)
Qld, WA, NT   Sedge
0.8-1.3 m tall

Perennial **herb** forming stout tussocks; **leaves** 40-60 cm x 0.3-0.4 cm, numerous and dense; **culms** 0.8-1.3 m tall, erect, stout, 3-angled, bright green; **inflorescence** spreading, large and much-divided; **bracts** numerous, 30-60 cm long, spreading stiffly; **spikelets** about 0.4 cm long, pale brown, densely crowded in short, cylindrical spikes.

A large sedge found in seasonally inundated, heavy soils beside roadsides, drains, streams etc. Very impressive when in flower, and could be a useful garden or pot plant. Propagate from seed or by division.

Cyperus alterniflorus R. Br.

(alternate flowers)
Qld, NSW, SA, WA   Sedge
0.5-1 m tall   Oct-March

Coarse, tufted, perennial **herb**; **leaves** 30-140 cm x 0.4-1 cm, leathery, broad, erect, dark green, with roughened edges; **culms** stout, robust, acutely 3-angled, 0.5-1 m tall; **inflorescence** spreading, rigid, much-branched; **bracts** about 6, 30-60 cm long, spreading; **spikelets** 1-2 cm long, rich brown, shiny, in short globular clusters.

Widespread along creek banks and soaks, often in inland districts, usually in heavy soils. A very attractive, robust sedge, well suited to cultivation in tropical and subtropical regions and inland districts. Requires a moist situation, and will grow with the roots immersed in water. Could be a useful emergent for ponds or dams. Propagate by division.

Cyperus bulbosus Vahl

(bearing bulbs or tubers)
Qld, NSW, SA, WA, NT   Nalgoo
0.1-0.4 m tall   Aug-Nov

Rhizomes slender, producing shiny brown to black tubers at the tip; **leaves** 10-20 cm x 0.2-0.3 cm, flat, erect, shiny green; **culms** 20-40 cm tall, erect; **umbel** spicate, of a few loose clusters of spikelets; **bracts** 10-15 cm long, slender, spreading; **spikelets** 1-3 cm long, compressed, oblong.

A widely distributed species which extends to Australia where it is most common in inland tropical regions, growing in colonies around the margins of claypans and salt lakes. The tubers were collected and eaten by the Aborigines. May become invasive. Propagate by division or tubers.

Cyperus clarus S. T. Blake

Qld, NSW   Sedge
30-50 cm tall

Perennial **herb** forming loose tussocks; **leaves** 20-80 cm x 0.8-1 cm, bright green, broad and flat, acuminate; **culms** 30-50 cm tall, erect, rigid, triangular, roughened; **inflorescence** fairly compact, with only about 6 branches; **bracts** 3-5, up to 15 cm long, spreading; **spikelets** 0.8-1.5 cm long, yellow to brownish, in globular clusters.

A handsome sedge found in moist depressions in inland regions, sometimes in grassland. Tussocks are reminiscent of a small *Gahnia* sp., and this plant could have landscaping potential. Propagate from seed or by division.

Cyperus concinnus R. Br.

(neat; trim)
Qld, NSW, Vic   Trim Sedge
20-70 cm tall

Perennial **herb** forming erect tussocks; **leaves** 10-50 cm x 0.1-0.2 cm, slender, rigid, the midrib prominent; **culms** 15-60 cm tall, erect, rigid, striate; **inflorescence** loose and open; **bracts** 1-3, up to 40 cm long, slender; **spikelets** 0.4-1 cm long, dark brown, dense.

An attractive sedge that is distributed mainly in inland and adjacent regions, usually in heavy clay soils subject to periodic inundation. Grows readily in a moist but sunny garden situation, or as a pot plant. Propagate by division or from seed.

Cyperus dactylotes Benth.
(finger-like)
Qld, NSW, SA, NT   Sedge
0.3-1 m tall

Perennial **herb** forming dense tussocks; **leaves** 15-40 cm x 0.1-0.2 cm, narrow, flat, dark green; **culms** 0.3-1 m tall, erect, smooth, striate, the corners rounded; **inflorescence** somewhat loose and spreading; **bracts** 4-7, some up to 60 cm long, spreading, with roughened margins; **spikelets** 1-2 cm long, rich yellow-brown.

An inland sedge found growing on heavy clay soils which become very wet following rain. Tussocks are very dense, and as they spread slowly are easily contained. An attractive sedge for a moist position in a rockery or bog garden. Propagate by division or from seed.

Cyperus diffusus Vahl = C. laxus Lam.

Cyperus digitatus Roxb.
(finger-like)
Qld, NT
0.3-1 m tall   July-Sept

Leaves 30-60 cm x 1 cm, arching, dark green, in a basal rosette; **culms** stout, 3-angled, almost winged; **umbel** compound, subtended by spreading bracts, 2 longer than the others; **spikelets** in elongated spikes, pale coloured.

An attractive sedge common in some rainforests, and often growing in colonies along stream banks. A useful plant for a shady position in tropical or subtropical areas. Requires plenty of moisture. Frost tender and in southern areas does best as a glasshouse plant. Propagate from seed.

Cyperus disjunctus C. B. Clarke
(disconnected; separated)
Qld, NSW   Sedge
30-60 cm tall   July-Oct

Rhizome shortly creeping, bearing clusters of culms at intervals; **leaves** 6-10 cm x 0.2 cm, flat or the margins revolute, leathery, the apex toothed; **culms** 30-60 cm tall, erect, stiff; **panicle** congested, subtended by 2-3 bracts 6-8 cm long; **spikelets** 0.5-0.6 cm long, linear.

This sedge of northern NSW and south-eastern Qld grows in shady spots, often in rainforest and usually in spreading patches or colonies. It could be an excellent groundcover for a shady position and should mix well with ferns. Plants are very cold hardy and well suited to temperate regions. Propagate by division or from seed.

Cyperus enervis R. Br.
(veinless or apparently so)
Qld, NSW   Sedge
15-30 cm tall

Perennial **herb** forming loose, slender tussocks; **leaves** 10-20 cm x 0.1-0.2 cm, slender, filiform, dark green, entire or scabrous; **culms** 15-30 cm tall, erect;

inflorescence loose and open; **bracts** 4-6, very slender, about 6 cm long; **spikelets** numerous, about 1 cm long.

A slender sedge, one form of which is found in rainforest, often growing in rocky soils and forming loose colonies. The typical form grows in coastal woodland habitats. It has some potential for planting in a shady garden position amongst ferns etc. Propagate by division or from seed.

Cyperus exaltatus Retz.
(tall)
all mainland states   Tall Flat Sedge
0.3-1 m tall   Jan-April

Leaves 20-60 cm long, grass-like, bright shiny green, basal; **culms** stout, erect; **umbel** compound, subtended by large, spreading bracts; **spikelets** very numerous, congested, rich shiny brown.

A handsome sedge found along river-banks, and margins of lagoons and swamps. Can be grown as a bog plant or as an emergent plant in a pond, in a submerged pot of soil (see Aquatics, Volume 2, page 217). Frost hardy. Flowers over a long period. Responds to fertilizers. Propagate from seed or by division.

Cyperus gracilis R. Br.
(slender)
Qld, NSW   Slender Sedge
0.1-0.3 m tall   Nov-Jan

Stems tufted; **leaves** 10-15 cm long, thin, dark green, with brown sheaths at the base; **culms** slender, tufted; **umbel** clustered, dense; **bracts** 10-12 cm long; **spikelets** brown or green.

Widespread on dry, often rocky sites. A very hardy species useful for a sunny rockery, and should do well in a container. Propagate by division.

Cyperus gymnocaulos Steudel
(naked stems)
all mainland states   Spiny Flat Sedge
0.5-1.3 m tall   Nov-March

Rhizome creeping; **leaves** reduced to small, membranous, sheathing bracts; **culms** slender, cylindrical, dark green, prominently striate; **umbel** densely clustered, forming a globose head; **bracts** 2.5-4 cm long, rigid, with a pungent point; **spikelets** dark brown.

Widespread in seasonally moist, boggy inland areas, usually in sunny situations. Makes a very handsome plant for contrast, but can only be grown in moist soil. Useful to grow in shallow water around the margins of ponds, and provides excellent shelter for water-life. Can also be grown in a pond if planted in a pot of soil with the top of the pot about 4 cm below the water surface (see Aquatics, Volume 2, page 217).

Cyperus haspan L.
(Haspan is a Sinhalese name for the plant)
Qld, NSW, WA, NT   Sedge
0.15-0.6 m tall

Stems tufted or short-creeping; **leaves** reduced to sheathing scales; **culms** 15-60 cm tall, flat or acutely 3-angled; **umbel** simple or compound, of a few loose clusters of spikelets; **bracts** 5-7.5 cm long, spreading, numerous, bright green; **spikelets** small, shiny brown.

A widespread sedge which extends to tropical parts

of Australia and down the east coast to central NSW. It makes a handsome pot plant and could also be useful as a bog plant, or around dams and ponds. Propagate from seed or by division.

Cyperus ixiocarpus F. Muell.
(sticky fruit)
Qld, WA, NT   Sedge
0.5-1 m tall   Aug-Oct
   Rigid perennial forming **tussocks**; **leaves** 20-30 cm long, rigid, channelled, with scabrous margins; **culms** 50-70 cm tall, rigid; **umbel** compound, of spreading clusters of spikelets; **bracts** 10-14, up to 10 cm long, spreading; **spikelets** 1-2 cm long, flattened, sticky, dark red-brown.
   An interesting sedge with sticky spikelets and long, spreading bracts. It is found in tropical regions along stream banks, and is worthy of cultivation as a container plant. Propagate by division.

Cyperus laxus Lam.
(loose; open)
Qld, NT   Sedge
0.3-0.8 m tall
   Perennial **herb** forming tussocks; **rhizomes** short, stout; **leaves** 15-80 cm x 0.5-2 cm, flat, flaccid, with 3 prominent nerves, channelled at the base, the margins roughened; **culms** 30-80 cm tall, erect, 3-angled; **inflorescence** up to 30 cm across, diffuse, lax; **bracts** 4-10, up to 50 cm long, spreading or reflexed; **spikelets** up to 1 cm long, solitary or clustered.
   An interesting sedge from the Top End of NT, and north-eastern Qld, and also many overseas countries. It is very uncommon in cultivation but could have potential as a semi-aquatic for tropical regions. Makes an attractive pot plant. Propagate by division or from seed. It was previously known as *C. diffusus* Vahl.

Cyperus leucocephalus Retz. see *C. pulchellus* R. Br.

Cyperus lhotskyanus Boeck.
(after Johann Lhotsky, 19th-century Austrian physician and naturalist)
NSW, Vic, SA   Flat Sedge
0.4-0.6 m tall   Oct-Feb
   Rhizomatous, perennial **herb**; **leaves** 50-80 cm long, slender, bright green, sometimes with reddish tonings; **culms** slender, about the same length as the leaves; **umbels** loose, not clustered; **bracts** 8-10 cm long; **spikelets** reddish brown.
   Widespread in swampy situations. Useful in moist areas, and will tolerate full sun and heavy frosts. Propagate by division. Previously known as *C. rutilans* (C. B. Clarke) Maiden & Betche.

Cyperus lucidus R. Br.
(shining)
Qld, NSW, Vic, Tas   Leafy Flat Sedge
0.6-1.6 m tall   Sept-Nov
   Leaves 20-80 cm x 1-1.5 cm, dark green, shiny; **culms** stout, 3-angled, erect; **umbel** compound, subtended by large bracts 5-15 cm long; **spikelets** loosely arranged, pale.
   Common along watercourses and around swamps.

Can be grown as a bog plant or in a wet position in the garden. Can also be grown as an emergent aquatic (see Aquatics, Volume 2, page 217). Responds to fertilizers. Propagate from seed or by division.

Cyperus platystylis R. Br.
(broad styles)
Qld, NSW, NT   Sedge
15-35 cm tall
   Perennial **herb** forming slowly spreading clumps; **leaves** 10-25 cm x 0.4-0.6 cm, broad, flattened, the midrib keeled; **culms** 15-35 cm tall, stout; **inflorescence** compact and dense; **bracts** 2-4, with 2 long and spreading; **spikelets** 0.8-1.2 cm long, brown, flat.
   Found in near-coastal and adjacent areas, usually in heavy soils subject to inundation after rain. Makes an attractive, neat tussock for a moist position in a garden, or a good-looking pot plant. Propagate from seed or by division.

Cyperus pulchellus R. Br.
(beautiful)
Qld, WA, NT   Sedge
10-40 cm tall
   Perennial **herb**; **rhizomes** short; **leaves** slender, stiff and rigid, smooth or somewhat roughened, sparse; **culms** 10-40 cm tall, erect, smooth, 3-angled; **inflorescence** a dense, globular, white or cinnamon coloured head up to 1 cm across; **bracts** up to 10 cm long, spreading or reflexed; **spikelets** up to 0.5 cm long, numerous, strongly compressed.
   Widespread in tropical Africa and Asia, and in Australia restricted to northern tropical regions. It is a rather interesting and attractive sedge that seems to flower almost continuously when grown as a potted plant in a glasshouse. Should also be suited to wet soil conditions in the tropics. Propagate by division or from seed. Australian plants were previously thought to be *C. leucocephalus* Retz., but this is a distinct Asian species.

Cyperus rutilans (C. B. Clarke) Maiden & Betche = *C. lhotskyanus* Boeck.

Cyperus sanguinolentus Vahl
(spotted with red)
Qld, NSW, Vic, SA   Dark Sedge
0.1-0.5 m tall   Oct-Feb
   Stems tufted; **leaves** variable, sometimes very short and almost filiform (less than 10 cm tall), up to 50 cm tall, usually dark green; **culms** slender, about the same length as the leaves; **umbels** loose; **bracts** 8-10 cm long; **spikelets** reddish brown.
   Widespread in moist sites along stream banks, often in shady locations. It is a somewhat weedy species that may be an annual under adverse conditions. Easily grown in a moist situation such as a bog garden, and readily naturalizes itself from seed.

Cyperus subpinnatus Kukenthal
(somewhat pinnate)
Qld, NSW, NT   Sedge
10-30 cm tall
   Perennial **herb** forming slowly spreading clumps;

leaves 20-30 cm x 0.2-0.3 cm, rigid, erect, dark green, the margins toothed; **culms** 10-30 cm tall, rigid, triangular, roughened, deep green; **inflorescence** a dense, almost globular head, deep green to black; **spikelets** about 1.5 cm long, reddish, clustered.

A stiff sedge found in drainage channels, along stream banks and in low-lying areas, often in the water. Clumps are a rich green, and the plant could make a very useful emergent for pond and dam margins. Propagate by division or from seed.

Cyperus tetraphyllus R. Br.
(4 leaves; probably in reference to the inflorescence-bracts)
Qld, NSW   Sedge
15-40 cm tall
     Perennial **herb** forming tussocks, sometimes spreading by rhizomes; **leaves** 10-20 cm x 0.3-0.5 cm, dark green, entire, flat, blunt; **culms** 30-45 cm tall, erect, rigid, 3-angled; **inflorescence** loose and open; **bracts** 4, prominent, 4-8 cm long, narrow and tapering; **spikelets** in clusters of 3-6, black, about 1 cm long.

A slender sedge found in rainforest, often growing in rocky soils or along stream banks, forming loose colonies. It has potential for planting in a shady position amongst ferns. Propagate by division or from seed.

Cyperus vaginatus R. Br.
(sheathed)
all mainland states   Flat Sedge
0.3-1 m tall   Sept-Dec
     Rhizome creeping; **stems** erect, terete, striate, leafless, brown; **heads** globular, about 1 cm across, subtended by 5-6 bracts, 2-8 cm long, rigid; **spikelets** rich chestnut brown.

Found along streams in damp places. A handsome sedge, useful in a moist position. Closely related to *C. gymnocaulos*, but generally found in wetter situations. Propagate readily by division.

CYPHANTHERA Miers
(from the Greek *cyphos*, bent or stooping; *anthera*, anthers)
Solanaceae
     Small to tall **shrubs**, more or less glabrous to densely tomentose; **leaves** usually entire, often thick; **flowers** bell-shaped to tubular, white to yellow with purple streaks, arranged in axillary clusters of 1-3, or in terminal panicles or racemes; **calyx** 5-toothed or 5-lobed; **stamens** 4; **anthers** 1-celled; **fruit** a capsule.

This genus was recently reinstated. Previously, its species were included in *Anthocercis*. *Cyphanthera* comprises 9 species. The main differences between the genera are that *Cyphanthera* has unilocular anthers and a chromosome number of 30, whereas *Anthocercis* has bilocular anthers with a chromosome number of 36.

A few species have horticultural potential, due to their floriferous display and their foliage characteristics. The main requirements are for relatively well drained soils, with adequate sun. Not much work has been carried out on propagation from seed, but cuttings of young firm stem growth generally produce roots.

Cyphanthera albicans (Cunn.) Meirs
(whitish)
Qld, NSW, Vic   Hoary Ray-flower
2-3 m x 2-3 m   March-Nov
     Medium **shrub**; **branches** many, covered with whitish stellate hairs; **leaves** to 2 cm long, ovate to oblong, entire, margins can be recurved, densely covered in white stellate hairs; **flowers** tubular, about 1 cm long, white to creamy white with purple streaks inside, with narrow, spreading lobes to 2 cm across, borne in clusters of 2-3 in upper axils; **capsule** globular.

When plants are growing well, the foliage of this species has an outstanding silvery grey appearance. Has proved adaptable in cultivation, doing best in well drained soils. Tolerates heavy clay-loam, provided drainage is adequate. Best suited to a location with partial or full sun. Frost and drought tolerant. Responds well to pruning. Prone to attack by grey mould in sheltered and shaded locations. Propagate from cuttings.

Previously known as *Anthocercis albicans* A. Cunn.

There are a further two subspecies:

C. albicans ssp. **notabilis** Haegi
     Has larger leaves and corolla than the other subspecies. The flowers can reach 2.2 cm in length. Is restricted in nature to the Warrumbungle Ranges, NSW.

C. albicans ssp. **tomentosa** (Benth.) Haegi
     Has yellow flowers and a dense tomentum on branches and leaves. It occurs in western NSW.

C. scabrella (Benth.) Miers from NSW is closely allied to *C. albicans*. It has white flowers that are borne on longer stalks, and the leaves are not as hairy.

Cyphanthera anthocercidea (F. Muell.) Haegi
(similar to the genus *Anthocercis*)
Vic   Large-leaf Ray-flower
2-4 m x 2-4 m   June-Nov
     Medium **shrub**; **branches** many, minutely hairy; **leaves** 3-10 cm x 1-2 cm, flat, lanceolate, prominent central nerve, usually glabrous, deep green, strong odour when crushed; **flowers** tubular, bell-shaped, 1-1.5 cm x about same diameter, white with purple streaks inside, spreading triangular lobes, borne in panicles, profuse, strongly scented.

This ornamental species is not well known in cultivation. Flowers can literally cover the outer foliage. It requires well drained soils, having a preference for sandy types. Withstands some extended dry periods, but does better on soils that are moist for the greater part of the year. Grows well in dappled shade, or partial or full sun. Frost tolerant. Pruning after flowering can be quite hard, and this helps to promote bushy growth. Propagate from cuttings.

Previously known as *Anthocercis frondosa* (Miers) J. Black. Was originally recorded as present in NSW, but plants there are now considered to be hybrids between *Cyphanthera albicans* and *Duboisia myoporoides*. Records of its occurrence in SA are now also doubted.

Cyphanthera odgersii (F. Muell.) Haegi
(after W. H. Odgers, 19th-century Victorian public servant)
WA   Woolly Cyphanthera
1-2.5 m x 1-2 m   Sept-Dec
Small to medium **shrub**; **branches** densely covered with whitish hairs; **leaves** to 2 cm x 1 cm, orbicular-ovate to linear-oblong, dense whitish tomentum; **flowers** tubular, about 1.5 cm x 1 cm, white with violet streaks, sessile in upper axils.

A species with ornamental foliage, from the Helms and Coolgardie Districts. It will need very well drained soils, with a hot location receiving plenty of sun. Frost and drought tolerant. Has potential as a container plant for temperate regions. Propagate from cuttings.

The ssp. *occidentalis* Haegi has larger leaves and longer hairs on the branches. It occurs to the west of ssp. *odgersii*, in the Avon and Austin Districts.

Previously known as *Anthocercis odgersii* F. Muell.

Cyphanthera scabrella (Benth.) Miers
           see under *C. albicans* (Cunn.) Miers

Cyphanthera tasmanica Miers
(from Tasmania)
Tas
3-5 m x 2-3 m   Oct-Dec
An erect, medium to tall **shrub**; **branchlets** hairy; **leaves** to 8 cm x 1-2.5 cm, elliptical, upper surface usually convex, margins slightly recurved, hairy, light green; **flowers** tubular, to 1 cm x 2 cm, white with purple streaks, borne in clusters of 2-3 per upper axil, profuse; **capsule** globular.

Evidently not in cultivation, this Tas endemic is a most ornamental species. It should do best in moist, well drained soils, with a location that provides dappled shade or partial sun. It may even tolerate full sun. Propagate from cuttings.

Previously known as *Anthocercis tasmanica* (Miers) Hook.f.

CYRTANDRA Forster & Forster f.
(from the Greek *cyrtos*, arched; *andros*, male; in reference to the arched stamens)
Gesneriaceae
Small to large **shrubs**, often hairy; **leaves** opposite or alternate, entire; **flowers** white or yellow, solitary or in bunches, tubular, with protruding stamens; **fruit** fleshy, indehiscent.

A large genus of about 350 species, one of which is endemic in Australia.

Cyrtandra baileyi F. Muell.
(after F. M. Bailey)
Qld
1-3 m x 0.5-1 m   ? April-May
Small to medium **shrub**; **branches** hairy when young; **leaves** 15-20 cm x 4-8 cm, ovate to lanceolate, alternate, dull green, with shortly denticulate margins; **flowers** about 1.5 cm long, white, borne singly opposite a leaf, tubular, with conspicuous stamens; **fruit** about 1.5 cm long, ellipsoid.

Confined to highland rainforest above 1200 m alt. in north-eastern Qld (Mount Bellenden Ker), and rarely collected. A very attractive shrub worthy of cultivation, and probably suited to bushhouse or glasshouse culture. Requires cool, moist conditions and protection from frost. Propagate from seed or stem cuttings.

CYSTOPTERIS Bernh.
(from the Greek *cystis*, a bladder; *pteris*, a fern)
Athyriaceae
Small **ferns** forming erect tufts; **fronds** erect, flimsy, thin-textured, once-divided; **segments** lobed; **sori** minute.

Species about 18 with a widespread one extending to Australia.

Cystopteris fragilis (L.) Bernh.
(brittle fern)
NSW, Vic, Tas   Brittle Bladder Fern
5-30 cm tall
Fronds delicate, thin-textured, brittle, pale green, once-divided; **segments** irregularly lobed and toothed; **sori** black.

A tiny fern found in wet rocky crevices at high altitudes. Very difficult to grow, and resents disturbance. Needs cool, moist, humid conditions. Best grown in a bottle, aquarium or terrarium. Propagate from spore.

Drosera stolonifera — close-up of foliage showing the beads of liquid which have led to the descriptive name 'sundews'.   W.R. Elliot

D

DACRYDIUM Solander ex Lamb. emend de Laub.
The Australian species referable to this genus was recently placed in the new genus *Lagarostrobus* C.J. Quinn.

Dacrydium franklinii Hook.f. =
Lagarostrobus franklinii (Hook.f.) C. J. Quinn

DALBERGIA L.f.
(after the brothers Nils Dalberg, 18-19th-century Swedish botanist, and Carl Gustav Dalberg, 18th-century Swedish officer)
Fabaceae
Trees, shrubs, subclimbers or **lianes; leaves** pinnate, the leaflets alternate; **flowers** pea-shaped, small, numerous, in axillary or terminal cymes or panicles; **fruit** a thin, indehiscent pod.
A large genus of about 300 species widely distributed in the world's tropics, with a strong development in South Africa. In Australia there are 3 species, neither endemic. Propagation is from seed which is hard and needs treatment.

Dalbergia candenatensis (Dennst.) Prain
(an overseas name)
Qld
3-5 m tall   Nov-Dec
Medium **climber**; **stems** twining, with prominent lenticels; **leaves** 8-15 cm long, pinnate; **leaflets** 3-7, 1.5-4 cm x 1-2.5 cm, obovate, glabrous above, glaucous beneath, the apex rounded or notched; **racemes** 0.5-5 cm long, in the upper axils; **flowers** 0.6-1 cm long, white, pea-shaped; **pods** 2-3.5 cm x 1-1.5 cm, oblong, falcate.
A climber of many countries, occurring in Australia in near-coastal vegetation of north-eastern Qld, including mangroves. It has limited horticultural appeal, although it could be a very useful species for tropical and subtropical regions, especially in coastal districts. Will tolerate fairly heavy soils that do not drain freely. Propagate from scarified seed.

Dalbergia densa Benth.
(dense)
Qld
4-10 m x 2-4 m   Nov-Dec
Tall **shrub** or small **tree; stems** sometimes climbing; **leaves** 5-10 cm long, pinnate; **leaflets** 3-15, 2-8 cm x 0.8-4 cm, oblong to elliptical, bright green and glabrous above, sparsely hairy beneath; **racemes** 4-6 cm long, crowded in panicles or clusters, densely rusty-hairy; **flowers** about 0.4 cm long, cream to pink, pea-shaped, sweetly scented; **pods** 3-8 cm x 1-1.2 cm, oblong, thin-textured, blunt, golden brown.
An unusual plant found in rainforests of the Cook District of north-eastern Qld and also New Guinea. Plants in the open are shrubby, while in shaded places the stems may climb, the short lateral shoots bending in response to contact. Plants are not showy and have limited horticultural application, although they may be of botanical interest. They can be successfully grown in the tropics and perhaps the subtropics, in an open, sunny situation, in well drained soil. Propagate from scarified seed. The typical form has 3-11 leaflets up to 8 cm x 4 cm, and occurs in lowland rainforest, grassland and woodland. The var. *australis* has 9-15 leaflets less than 4 cm x 1.7 cm and is found in stream bank vegetation.

DAMASONIUM Miller ex Schreber
(from the Greek *damasis*; a reference to supposed antidotal properties)
Alismataceae
Aquatic herbs; leaves erect, radical, on long petioles; **inflorescences** erect, whorled; **flowers** bisexual; **perianths** of 6 segments, the 3 inner ones flimsy and shedding early; **stamens** 6; **carpels** joined by their bases into a fruit, each carpel bearing 2 seeds.
A small genus of 5 species with one endemic in Australia. Propagation is from seed which must be sown under glass.

163

Damasonium minus

Damasonium minus (R. Br.)Buchenau
(small; less)
Qld, NSW, Vic, Tas, SA, WA   Star Fruit
0.3-0.6 m tall   Oct-July

Annual or perennial emergent **aquatic**; **leaves** radical, erect, in many whorls; **petiole** 40-60 cm long, fleshy; **leaf-blades** 6-10 cm x 3-6 cm, lanceolate to ovate, cordate at the base, with 3-5 main veins, the margins slightly undulate; **inflorescence** to 60 cm tall; **flowers** about 0.6 cm across, white or pale pink, very flimsy; **fruit** 1-1.2 cm across, star-like, composed of flattened carpels, each containing 2 seeds.

A widely distributed, endemic aquatic found in shallow expanses of fresh water to about 15 cm deep. Plants are extremely variable in size, the small stunted ones generally being annuals, the perennials more robust. The species is an attractive aquatic and makes a useful addition to ponds or dams. It can be planted direct into the basal mud or grown in a submerged pot (see Aquatics). Tropical, subtropical and temperate regions are suited to its culture. Propagate from seed which must be sown on moist peat under glass or by transplants.

DAMPIERA R. Br.

(after William Dampier, 17-18th-century English navigator and explorer)
Goodeniaceae

Herbaceous, perennial **herbs** to small erect **shrubs**, often suckering; **leaves** usually cauline but can be basal, entire, toothed or sinuate, usually alternate, glabrous to dense tomentum, usually of stellate hairs; **flowers** irregular, in racemes or heads, or solitary to clustered in axils; **corollas** deeply split on upper sides, various shades of purple and blue, can be pink or white, rarely yellow; **lobes** 5, upper ones with auricles; **anthers** coherent around style; **ovaries** 1-celled, but 2-celled in one species.

An endemic genus of about 66 species. The greatest representation is in the South-West and Eremaean Botanical Provinces of WA. Species also occur in the arid regions of WA and NT, with limited extensions into the temperate regions of the eastern states, and a few species are found in tropical regions.

The genus has recently undergone botanical revision, but at the time of compiling this entry it had not been published. It is proposed to add 22 species, and about 16 of the presently recognized species have been absorbed into other species.

The natural habitat of dampieras ranges from sea level to mountainous regions, where they grow on all kinds of soils from sands through to heavy clay-loams. Some species frequent semi-shaded situations, while others are in full sun. They often grow in scattered colonies, due to their suckering habit, and are very showy when in flower.

Dampieras have great potential for cultivation. Many species are highly floriferous, eg. *D. diversifolia* and *D. linearis*. Flower colour is commonly blue, with varying degrees of purple and mauve, giving a wide range of shades. Blue-flowering plants are eagerly sought for cultivation, and the hardiness of most *Dampiera* species has also increased their popularity. Many have a suckering habit of growth and, while not vigorous enough to cause trouble, they add a pleasing natural effect to the landscape by appearing in unexpected places.

The low, spreading dampieras are excellent as groundcovering mat plants, eg. *D. diversifolia*. They are also useful for small embankments, and add charm as they spill over rocks or logs. Clumping species such as *D. stricta* are ideal for planting near boulders or on edges of paths.

Some have proved to be excellent container plants, and others have potential for use in hanging baskets, eg. *D. stenophylla* which grows as a small, open clump with arching branches.

The suckering species can be rejuvenated by removing most of the growth above ground level, provided they have become well established. Vigorous new growth is usually the result of this action.

Quite a few species tolerate long periods of dryness, eg. *D. rosmarinifolia*, while others withstand heavy frosts, eg. *D. lanceolata*.

Caterpillars may cause limited damage to leaves, but generally pests do not present any major problem. Some of the species from well drained, sandy soils, eg. *D. wellsiana*, can be subject to attacks by root-rotting fungi.

Propagation

Dampieras are commonly grown from cuttings or by division of creeping underground stems, as seed is rarely available. With propagation from seed, it seems that species that have basal leaves and produce their flowers in showy heads are more readily germinated,

Damasonium minus, × ·1   flower, × 2

eg. *D. eriocephala, D. plumosa* and *D. wellsiana*. There is certainly a need for more experimentation with propagation from seed.

The majority of species can be propagated from cuttings of firm new growth. Some prove difficult, eg. some forms of *D. linearis, D. spicigera* and *D. teres*, and more favourable results are obtained by using suckering new growth as cutting material. Suitable material for cuttings is often lacking on species that have a single rootstock, eg. *D. eriocephala*; however, there has been some experimentation using leaf cuttings from those species which have large basal or stem-leaves. It seems that the leaf cuttings used have not had basal buds attached, and roots have formed, followed by successful development of plants. Further trials on a wider basis are needed with this method, to ensure that it can be regarded as a viable alternative form of propagation with *Dampiera* species. Experimentation with leaf-bud cuttings is also worthy of trial (see Volume 1, page 218).

Dampiera adpressa Cunn.
(closely flattened)
Qld, NSW   Purple Beauty-bush
0.2-0.5 m x 0.3-0.5 m   Sept-Oct

Dwarf, perennial **herb**; woody underground **rootstock**; **stems** erect, covered with white hairs when young; **leaves** to 3 cm x 2 cm, ovate to lanceolate, entire or with small teeth, leathery, thick, hairy becoming glabrous; **flowers** about 2.5 cm wide, bluish mauve, axillary or terminal, profuse.

Evidently not well known in cultivation, this outstanding clumping species should adapt to a wide range of conditions. It occurs naturally in moisture retentive sandy clay that can also dry out for limited periods. Suitable for dappled shade to full sun. Tolerant of light to medium frosts.

Propagate from cuttings.

Dampiera alata Lindley
(winged)
WA   Winged-stem Dampiera
0.5-1 m x 0.5-1.5 m   July-Nov

Dwarf, perennial **herb**; **stems** erect to arching, many, glabrous, angled, winged; **wings** to 1 cm wide; **leaves** 1-6 cm x up to 1 cm, oblanceolate to obovate, sometimes toothed, glabrous; **flowers** to 2 cm wide, purple-blue with yellow centre, dense grey tomentum on outside, 1-3 per peduncle; **corolla lobes** broadly winged.

An ornamental species that occurs over a wide range in south-western WA in the Avon, Darling and Irwin Districts. It has adapted well to many soil types but does need good drainage. Grows in dappled shade to full sun, and is tolerant both of frost and extended dry periods.

Plants grow as loose clumps, and the flowers are well displayed. There can be dead stems, but regular hard pruning will overcome this problem by promoting new growth.

Propagate from cuttings of firm new growth. It is wise to use cuttings of about 12-15 cm long, as this allows for 3-4 leaf nodes above the cutting-medium level. On some occasions, although the cuttings root,

there is no development from the leaf-buds. New sucker growth can eventuate, but if this does not happen the rooted cutting will not produce new top growth as all the buds have died.

D. epiphylloidea Vriese is now included in this species.

Dampiera altissima F. Muell.
(highest)
WA   Tall Dampiera
1-2.5 m x 1-2 m   June-Dec

Small to medium **shrub**, usually erect; **branches** covered in white hairs; **leaves** many, to 3 cm x 1 cm, obovate, scabrous above, dense white tomentum below, distinctly toothed; **flowers** dark blue, about 1.5 cm wide, borne in loose cymes from upper axils, profuse; **corolla** glabrous inside, dense white tomentum outside.

The tallest of the genus, and a most decorative species. It inhabits low-rainfall regions, both inland and near the coast, in the Irwin and northern Darling Districts, usually occurring on calcareous soils. Requires well drained soils, with plenty of sun, although it may tolerate semi-shaded locations in warm climates. Tolerant of light to medium frosts and extended dry periods. Propagate from cuttings.

Dampiera brownii F. Muell. = *D. purpurea* R. Br.

Dampiera candicans F. Muell.
(shining white)
WA, NT
0.5-1 m x 0.5-1 m   April-Oct

Dwarf **shrub**; **lower branchlets** with dense white hairs; **upper branchlets** glabrous; **leaves** to 2.5 cm long, ovate to rhomboid, scattered; **flowers** about 0.8 cm long, deep violet to blue, exterior densely hairy, borne in loose spikes on short branchlets.

A relatively unknown species, with similarities to *D. spicigera*. It occurs in mountain ranges of central-eastern WA and to slightly east of the border in NT. Worthy of experimentation in cultivation. Is probably best suited to arid or semi-arid regions, but could do well as a container plant in other areas. Propagate from cuttings.

Dampiera carinata Benth.
(keeled)
WA   Summer Dampiera
0.2-0.5 m x 0.1-0.3 m   Sept-April

Dwarf, perennial **herb**; **stems** erect, sparse, glabrous except for silky, woolly hairs at base, often leafless; **basal leaves** 3-6 cm x about 0.6 cm, linear, thick, in rosette; **stem-leaves** to 0.5 cm long, linear-terete, glabrous; **flowers** to 1 cm wide, purple-blue, solitary, terminal, silvery hairs on exterior.

This decorative species occurs in sandheath of the Darling and Irwin Districts. It is not well known in cultivation. Requires very well drained, light to medium soils, with semi-shade to full sun. Tolerates light frosts. Excellent potential as a container plant. Propagation could prove difficult, but young stem cuttings may root successfully.

D. mooreana E. Pritzel is now included in this species.

Dampiera carinata, × ·55

Dampiera cauloptera DC. = *D. coronata* Lindley

Dampiera cinerea Ewart & Davies
(ashy)
WA, NT
0.5-1 m x 0.5-1 m   July-Aug; also sporadic
  Dwarf **shrub**; **branches** terete, densely hairy; **leaves** 3.5-6 cm long, oblong-lanceolate, tapering to a petiole, thick, concave, entire, densely hairy giving an ashy grey appearance; **flowers** small, blue or pinkish mauve, sessile, solitary, borne in spikes 10-15 cm long which can be branched.
  Apparently not in cultivation, this species is from arid regions. It has ornamental foliage, and is worthy of trial. May be best suited to arid or semi-arid regions; however, it could do well as a container plant in other areas. It is thought to need extremely well drained, light to medium soils, with plenty of sunshine. Propagate from cuttings.

Dampiera conospermoides W. Fitzg.
(resembling the genus *Conospermum*)
WA, NT
0.3-1 m x 0.5-1 m   Sept-Oct; also sporadic
  Dwarf spreading **shrub**; **branches** and **stems** terete, slender, densely covered in white hairs; **leaves** 2-6 cm long, oblong to broad-lanceolate, petiolate, apex blunt, margin entire or with few teeth, initially very hairy on both surfaces, becoming glabrous above; **flowers** small, blue, almost sessile, solitary or 2-3 together, dense covering of whitish hairs on petal exterior.
  Little is known of this species, which occurs in the Gardner District in WA, and in the Darwin and Gulf District of NT. It is often found on quartzite. Evidently not in cultivation, but should be best suited to tropical and subtropical regions. Will need well drained soils, with partial or full sun. Propagate from cuttings.

Dampiera coronata Lindley
(crowned)
WA
0.2-0.6 m x 0.5-1.5 m   July-Oct
  Dwarf, perennial **herb**; **stems** simple or branched, slightly winged, providing 3-4 angles; **leaves** to 3 cm x 2 cm, cuneate-oblong, toothed, basal leaves petiolate and larger than upper leaves, glabrous; **flowers** about

2 cm wide, pale to deep blue, borne on long terminal or axillary peduncles; **corolla lobes** pointed.
  Uncommon in cultivation, this desirable species occurs in the Irwin and Darling Districts, in soils that are moist for long periods. It grows best in relatively well drained, light to medium soils, with plenty of sunshine. Will tolerate semi-shade and light to medium frosts. Can sucker and form colonies. Propagate from cuttings.
  D. cauloptera DC. is now included in this species.

Dampiera cuneata R. Br. = *D. linearis* R. Br.

Dampiera curvula Krause = *D. tenuicaulis* E. Pritzel

Dampiera dentata Rajput
(toothed)
WA, NT
0.3-0.5 m x 0.3-0.5 m   Aug-Oct
  Dwarf, perennial **herb**; **stems** ascending; **leaves** 5-16 cm x 0.3-1.5 cm, basal, sessile or petiolate, oblong-lanceolate or spathulate, glabrous, deeply toothed margins; **flowers** about 0.6 cm across, blue, loosely arranged in heads that elongate into spikes to 40 cm long.
  This showy species has recently been named. It occurs over a wide range in the arid regions of WA, and is found just east of the WA-NT border. Will need very well drained soils, with plenty of sunshine. Should tolerate most frosts and certainly withstands extended dry periods. Propagate from cuttings and possibly seed.
  D. dentata is distinguished from other species that have basal leaves, eg. *D. eriocephala*, *D. plumosa* and *D. wellsiana*, by its glabrous, toothed leaves, and glabrous flower-stems.

Dampiera coronata, × ·6

Dampiera diversifolia (2 colour forms)

T.L. Blake

Dampiera discolor Krause
(usually 2 distinct colours)
Qld
1.5-2 m x 1-1.5 m   Aug-Oct

Small **shrub**; new growth hairy; **branches** erect; **leaves** 6-12 cm x 2-3 cm, lanceolate to elliptic, dark green, shiny and glabrous above, with soft pale grey hairs below, margins entire; **flowers** about 1 cm long, bright purple-blue, in long, slender spikes forming large open panicles.

A shrubby and most showy species from the Burnett, Leichhardt and Mitchell Districts, where it is common on stony ridges. Becoming popular in cultivation, and should grow successfully in most well drained soils, in a situation that receives partial or full sunshine. Hardy to light frosts. May need pruning to promote bushy growth. Propagate from cuttings.

Dampiera diversifolia Vriese
(variable leaves)
WA
prostrate x 1-2 m   Aug-Dec

Dwarf, perennial **herb**, sometimes suckering; **stems** spreading, glabrous, angular, grooved, usually not self layering; **basal leaves** to 5 cm long, narrow, lanceolate or cuneate-lanceolate, somewhat concave, usually slightly toothed, glabrous; **stem-leaves** about 1 cm x 0.3 cm, oblanceolate, usually grouped; **flowers** to

1.5 cm across, purple-blue or rarely pale blue, with strong scent during hot weather, glabrous, borne on short peduncles from axils of bundled leaves along stem.

A widely cultivated species that has proved hardy through a range of soils and climatic conditions. Does best in moist, well drained locations, and grows rapidly in clay-loams. Tolerates full sun but prefers dappled shade or partial sun. Withstands light to medium frosts. Excellent for containers, including hanging baskets. One form with small flowers does not usually sucker. Propagate from cuttings or suckers.

D. glabriflora F. Muell. is now included in this species.

Dampiera epiphylloidea Vriese = *D. alata* Lindley

Dampiera eriantha Krause
(woolly flower)
WA
0.3-0.6 m x 0.3-0.5 m   Sept

Dwarf, erect **shrub**; **stems** striate; **leaves** to 5 cm long, many, linear-lanceolate to linear-terete, margins revolute, initially hairy, becoming glabrous, not crowded; **flowers** to 1.5 cm wide, blue to purple, can be in loose irregular cymes, outsides covered with silky grey hairs.

A most decorative species from the northern region

167

Dampiera eriocephala

of the Eucla District, where it is often found in calcareous soils. Evidently not cultivated, its requirements are for well drained, light to medium soils, with a fair amount of sunshine. It is probably best suited to hot, low-rainfall regions, but may do well in temperate regions, in a protected, north-facing location. Tolerant of light to medium frosts and extended dry periods. Usually does not sucker and form colonies. Propagate from cuttings.

Dampiera eriocephala Vriese
(woolly heads)
WA   Woolly-headed Dampiera
0.2-0.3 m x 0.2-0.5 m   Sept-Dec
   Dwarf, clumping, perennial **herb**; **central rootstock**; **leaves** to 15 cm x 3 cm, basal, spathulate, dark green and glabrous above, densely grey-white, woolly below; **scapes** much-branched, woolly, striate, leafless; **flowers** to 1.5 cm wide, blue, densely hairy outside, borne in dense terminal heads.
   An outstanding ornamental species, with large heads of brilliant blue flowers. It has had limited cultivation, due to unavailability of propagation material. Occurs naturally in sandy soils of low-rainfall areas, over a wide range of south-western WA. Needs very well drained, light to medium soils, with partial or full sun. Will sucker after fire or if roots are damaged. Grows extremely well as a container plant. Frost tolerant, and withstands extended dry periods when well established. Propagate from seed or possibly by division. Experimentation using leaf or leaf-bud cuttings is worthwhile.

Dampiera fasciculata R. Br.
(in bundles or clusters)
WA   Bundled-leaf Dampiera
0.3-1 m x 0.5-3 m   July-Jan
   Dwarf **shrub**; **stems** trigonous, glabrous, many, lightly suckering; **leaves** to 5 cm x 1.5 cm, obovate-oblong, or cuneate-oblong, thick, sessile, in bundles on upper part of stems, glabrous on both sides; **flowers** about 1.5 cm wide, pale to deep blue, almost sessile, in axillary clusters, profuse.
   A most variable species in leaf and flower. It has adapted well to cultivation. In nature, it is found over a large area of the south-western region, occurring in the Darling and Eyre Districts. It grows on clay, with or without gravel, and also on sandy gravels. In cultivation, it will grow in most soils, provided there is reasonable drainage, and preference is for partial or full sun. In an open situation, plants can spread up to 3 m wide. Tolerant of most frosts and extended dry periods.
   Propagate from cuttings. Some forms seem to be more difficult than others. Young growths from underground stems usually strike readily.

Dampiera ferruginea R. Br.
(rust coloured)
Qld   Velvet Beauty-bush
0.5-1 m x 0.5-1 m   June-Aug
   Dwarf **shrub**; young shoots rusty; **branches** erect or ascending, terete, sulcate, densely hairy; **leaves** to 5 cm x 2 cm, obovate to rhomboidal, entire or toothed,

Dampiera eriocephala   D.L. Jones

becoming glabrous; **flowers** to 2 cm wide, blue with yellow throat, outside densely tomentose, more or less sessile in the leafy bracts of the lower axils, or peduncles of several flowers in upper axils.
   An ornamental species with attractive new growth. It occurs amongst granite boulders in the Cook, North and South Kennedy, Port Curtis and Wide Bay Pastoral Divisions. Not commonly cultivated, but worthy of further trial. Should do well in moisture retentive, well drained soils, with dappled shade to partial sun. May tolerate full sun. Can lightly sucker. Propagate from cuttings.

Dampiera glabrescens Benth.
(becoming glabrous or almost glabrous)
WA
0.3-1 m x 0.3-1 m   Aug-Sept; also sporadic
   Dwarf, perennial **herb**; **stems** erect, many; **lower leaves** to 4 cm x 0.7 cm, lanceolate to oblong, with finely appressed hairs, numerous along length of stem; **upper leaves** smaller; **flowers** about 1.5 cm across, pale to medium blue, exterior covered with white silky hairs; **bracts** 2, small, subtending flower; **peduncles** 1-3 flowered, in upper axils.
   A decorative species from the northern Darling District, where it occurs in sandheath plant communities. Requires well drained, light to medium soils, with partial or full sun. May tolerate dappled shade in warm to hot climates. Propagate from cuttings and probably root suckers.

Dampiera glabriflora F. Muell. = *D. diversifolia* Vriese

168

Dampiera hederacea R. Br.
(similar to the genus *Hedera*)
WA   Ivy-leaved Dampiera; Karri Dampiera
0.3-1 m x 0.5-1.5 m   June-Jan

Dwarf, scrambling perennial **herb**; usually 1 main **stem**, can sucker lightly; **branches** spreading, trailing; **leaves** to 5 cm x 3 cm, ovate to ovate-oblong, petiolate, often with prominent lobes; **upper leaves** much smaller; all **leaves** becoming glabrous on upper surface, densely woolly below; **flowers** about 1 cm across, pale to rich blue, densely hairy on outside, borne on long hairy peduncles, profuse.

A dense, scrambling species that is commonly found as an underplant of the karri and jarrah forests in the south-west, in the Darling District. It also occurs in coastal situations. Has adapted well to cultivation in semi-shaded to partial sun locations. Tolerates full sun, provided the soil is moist. Does best in relatively well drained soils. Useful as a dense groundcover in shady places. Successfully used in hanging baskets. Responds well to hard pruning, and tolerates light frosts. Propagates readily from stem cuttings.

The name *D. hederacea* was wrongly placed on a form of *D. linearis* by the nursery trade many years ago. It seems this error has now been rectified.

Dampiera helmsii Krause = *D. tenuicaulis* E. Pritzel

Dampiera humilis E. Pritzel = *D. wellsiana* F. Muell.

Dampiera incana R. Br.
(grey)
WA
0.2-0.4 m x 0.3-1 m   Aug-Sept

Dwarf **shrub**; **branches** longitudinally grooved, densely covered with whitish grey hairs; **leaves** 1.5-3.5 cm x 0.5-1 cm, obovate or oblong-cuneate, usually entire, dense white hairs on both surfaces; **flowers** about 1.5 cm across, pale blue to deep blue, yellow throat, densely hairy exterior, 1-2 per peduncle.

The first collection of this species was made by William Dampier. Evidently not well known in cultivation, but a species with decorative foliage. It occurs naturally in the Austin, Carnarvon, Fortescue, Irwin and Keartland Districts. It appreciates well drained, light to medium soils, with plenty of sunshine. Propagate from seed or stem cuttings which may prove difficult due to dense hairs.

The var. *fuscescens* Benth. has a dark tomentum, not white as in the species. It can flower as early as March. A white-flowered form is in cultivation.

Dampiera juncea Benth. = *D. oligophylla* Benth.

Dampiera lanceolata Cunn.
(lanceolate)
Qld, NSW, Vic, SA   Grooved Dampiera
0.3-0.6 m x 0.5-2 m   Sept-Nov; also sporadic

Dwarf, perennial **herb**; young shoots hairy; **stems** erect or procumbent, slightly grooved, nearly glabrous; **leaves** 1-5 cm x 0.5-2.5 cm, lanceolate to obovate-oblong, thick, sessile, not often in clusters, margins flat or slightly recurved, entire or slightly toothed, glabrous above, densely hairy below; **flowers** about 1.5 cm

Dampiera lanceolata, × ·6

across, deep blue with yellow throat, exterior covered in long dark hairs, borne on hairy peduncles in groups of 2-5, in axils along stems; **bracts** and **bracteoles** to 0.3 cm long.

A hardy and decorative species that usually inhabits sandy soils in relatively low-rainfall regions. Best suited to well drained, light to medium soils, with plenty of sunshine, but will tolerate partial sun. Frost and drought tolerant, and withstands limited coastal exposure. It can sucker to form colonies and will scramble through other plants if given the opportunity. An excellent container plant. Propagate from stem cuttings.

Dampiera lavandulacea Lindley
(resembling lavender)
WA   Lavender Dampiera
0.2-0.5 m x 0.5-3 m   Aug-Dec

Dwarf, perennial **herb**; young shoots hairy; **stems** erect or spreading, stiff, many, grooved, covered in white hairs; **leaves** 1-2.5 cm x 0.2-1 cm, oblong to linear-lanceolate, sessile, usually clustered, usually entire, margins recurved, becoming glabrous above, densely hairy below; **flowers** about 1.5 cm across, blue, purple-blue, lilac or pink, with light yellow throat, exterior covered with short hairs, clusters of 2-5 in the upper axils, profuse; **bracts** and **bracteoles** to 0.7 cm long.

This attractive clumping species can form large colonies. It occurs in a wide range of soil types in the Austin, Avon, Coolgardie, Darling, Irwin and Roe Districts. Well known in cultivation. Prefers well drained, light to medium soils, with partial or full sun, but also tolerates dappled shade. Frost and drought tolerant. Suitable for slightly alkaline soils. Can also be grown as a container plant. Propagate from cuttings.

Similar to *D. lanceolata* which usually has leaves with flat, toothed margins. *D. preissii* Vriese and *D. rupicola* S. Moore are now included in this species.

169

Dampiera leptoclada

Dampiera lavandulacea, × 1

Dampiera lindleyi   W.R. Elliot

Dampiera leptoclada Benth.
(slender branches)
WA
0.3-0.6 m x 0.3-1 m   Aug-Nov

Dwarf, erect or spreading **herb**, can sucker; **branches** often ascending or erect, can be long and slender, angled, especially near base, glabrous; **leaves** to 5 cm x 1 cm, oblong to linear, sessile, many, 1-2 teeth on each margin, glabrous; **flowers** to 2 cm across, blue, exterior covered in greyish stellate hairs; **peduncles** glabrous, 1-3 in upper axils.

In WA this species has a wide distribution in the south-western corner, in the Darling District. It is not well known in cultivation, but should grow in relatively well drained soils, with partial or full sun. It is also recorded as inhabiting swampy regions, so moist soils may also be to its liking. Propagate from stem cuttings.

D. subspicata Benth. is now included with this species.

Dampiera lindleyi Vriese
(after John Lindley, 19th-century English botanist)
WA
0.3-0.6 m x 0.3-0.6 m   June-Oct

Dwarf, perennial **herb**; **stems** winged, erect, not flexuose, glabrous, many arising from the base; **leaves** to about 1.5 cm long, linear, entire, often absent; **flowers** about 1 cm across, blue-violet, with whitish throat, profuse; **peduncles** long, with 1-3 flowers, glabrous.

An outstanding species when in full flower. It occurs over a wide range in the Ashburton, Austin, Darling and Irwin Districts, where it is often found growing in gravelly sands. Not common in cultivation, but suited to well drained, light to medium soils, with partial to full sun. Tolerant of most frosts. May sucker but usually develops as a single clump. Has potential as a container plant. Propagate from stem cuttings (see *D. alata*).

Dampiera linearis R. Br.
(linear)
WA   Common Dampiera
0.2-0.6 m x 0.5-2 m   May-Nov

Dwarf, perennial **herb**, spreads by suckering; young shoots hairy; **stems** initially more or less terete, usually angular when mature, becoming glabrous; **leaves** very variable, 1-4 cm x 0.1-1 cm, linear to cuneate, sometimes irregularly toothed, initially hairy, usually glabrous when mature; **flowers** 1.5-2 cm across, pale blue to purple with white or lemon yellow throat, exterior densely covered with grey hairs, 1-3 flowers per hairy peduncle.

A most variable and decorative species that is very popular in cultivation. It occurs in the Avon, Darling, Eyre, Irwin and Roe Districts. Has adapted very well to a wide range of soils and climatic conditions. In general, it does best in well drained soils, with partial or full sun. Does not like full shade or extended waterlogging. Tolerant of most frosts, although some of the hairy-foliaged forms can suffer damage in heavy frost. Withstands extended dry periods. Some of the spreading forms are excellent for growing over logs or low rocks, and on small embankments. Ideally suited as a plant for containers, including hanging baskets. Some selected forms flower profusely, while other forms have been selected for their greyish appearance which is due to a dense covering of white to greyish hairs. Propagate from stem cuttings or by root divisions. Cuttings from young suckering stems form roots much more quickly than those from older growth.

For many years plants have been sold as *D. cuneata*, which is now included in this species. At one time plants were also sold as *D. hederacea*. This has now been rectified. Since about 1966, a plant known as *D.* species nova 'WA' has been grown and it is now recognized as a form of *D. linearis*. It suckers vigorously, has leaves of 1-2 cm x 0.3-0.6 cm and the flowers are a bright deep blue, about 1.3 cm across. The same form has also been sold under the incorrect name of *D. humilis*.

Dampiera linschotenii F. Muell. =
                    Linschotenia discolor Vriese

Dampiera loranthifolia F. Muell.
(leaves similar to those of the genus *Loranthus*)
WA
0.3-0.6 m x 0.3-1 m   Sept-Jan

Dwarf, perennial **herb**; **stems** erect, angular, glabrous; **leaves** to 4 cm x 1.5 cm, obovate-lanceolate to cuneate, sessile, often toothed, glabrous; **flowers** 1-1.5 cm across, mauve-blue to deep blue, densely hairy exterior; **peduncles** 1-3 flowered, in upper axils.

An interesting species from the hills of the Barren Ranges and other southern regions in the Eyre District. It is not well known in cultivation, but should grow in well drained soils, with dappled or partial sun. Will probably tolerate full sun. Propagate from stem cuttings.

Closely allied to *D. fasciculata*. A form of *D. linearis* has been incorrectly available as *D. loranthifolia* in the eastern states.

Dampiera luteiflora F. Muell.
(golden yellow flowers)
WA   Yellow Dampiera
0.3-0.6 m x 0.3-0.6 m   Aug-Dec
Dwarf, suckering perennial **herb**; **stems** terete,

Dampiera linearis form, × ·55

Dampiera linearis form, × ·55

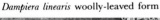

Dampiera linearis woolly-leaved form   B. Crafter

Dampiera linearis forms, × ·55

Dampiera marifolia

densely hairy; **leaves** 2-5 cm x about 1 cm, elliptical, entire, densely hairy on both surfaces; **flowers** about 1 cm across, yellow, borne in a terminal panicle.

This is the only *Dampiera* species that has all-yellow flowers. It is most decorative, with densely hairy stems and leaves. Known naturally from the Austin, Coolgardie, northern Darling, Eyre and Roe Districts, where it grows in the gravelly, sandy soils of the wheatbelt region. It has had limited cultivation, with some success. Needs well drained, light to medium soils, with plenty of sunshine, but will probably tolerate dappled shade in warmer climates. Suited to container cultivation. Propagate from stem cuttings or suckers.

Dampiera marifolia Benth.
(leaves similar to those of *Teucrium marum*)
Vic, SA   Velvet Dampiera
0.2-0.5 m x 0.3-1 m   Aug-Dec

Dwarf **shrub**; young shoots hairy; **branches** velvety, slender, terete; **leaves** to 2 cm x 0.4-1 cm, ovate to elliptic, oblong, sessile, erect to spreading, not bundled, sometimes appearing glabrous above at maturity, undersurface densely hairy, margins thickened; **flower** to 1.5 cm across, initially deep mauve-blue, fading to pale mauve, with pale lime-yellow throat, solitary, forming short, leafy spikes.

A most decorative species. It usually occurs in sandy soils, and will grow in well drained, light to medium soils, with partial or full sun. Tolerant of most frosts and extended dry periods. An excellent container plant. Propagate from stem cuttings.

Dampiera mooreana E. Pritzel = *D. carinata* Benth.

Dampiera oligophylla Benth.
(sparse leaves)
WA   Sparse-leaved Dampiera
0.2-0.4 m x 0.2-1 m   Sept-Nov

Dwarf **shrub**; young shoots hairy; **stems** numerous, terete, sulcate, erect, slender but rigid, can be branched; **leaves** to 2 cm but usually under 1 cm long, ovate to oblong, few, scattered along stems, initially hairy, becoming glabrous; **flowers** to 2 cm across, blue to deep violet, borne on short peduncles which can be 1-2 in upper axils.

A decorative dampiera that occurs over a wide range of south-western WA, both in coastal and inland situations. Probably best grown in well drained, light to medium soils, with partial or full sun. Tolerates most frosts. Has high potential as a garden or container plant. Propagate from stem cuttings.
D. stowardii S. Moore is now included in this species. *D. juncea* Benth., with its rush-like stems, will become a subspecies in the revision.

Dampiera parvifolia R. Br.
(small leaves)
WA   Many-bracted Dampiera
0.3-0.5 m x 0.5-1 m   Oct-Dec

Dwarf, perennial **herb**; young shoots woolly; **basal leaves** to 5 cm x 2 cm, obovate to cuneate, sessile, sometimes toothed, thick, glabrous when mature; **stem-leaves** to about 1.5 cm, narrow, entire; **flowers**

Dampiera oligophylla ssp. *oligophylla*   W.R. Elliot

about 1.5 cm across, dark purple-blue, exterior has white, silky hairs, terminal and usually solitary; **bracts** prominent, many, often reddish.

This dampiera is distributed in the Eyre and eastern Roe Districts of south-western WA. It occurs mainly in coastal situations. A most attractive species that has not been cultivated to any extent. Should grow successfully in relatively well drained, light to medium soils, with partial or full sun. Recommended for coastal gardens and also for container cultivation. Should tolerate most frosts. Propagate from stem cuttings, suckers or by root division.

Dampiera plumosa S. Moore
(feathery; plumed)
WA
0.1-0.2 m x 0.3-0.5 m   Oct-Nov

Dwarf, perennial **herb**; **leaves** 3-8 cm x 0.5-1 cm, basal, oblanceolate to oblong, glabrous above, densely white, tomentose below, more or less entire; **flowers** about 0.6 cm across, dark blue, borne in clustered heads; **calyx lobes** long, densely hairy.

This is one of a group of 4 species that display their flowers in terminal heads. A very decorative dampiera from the Austin, Avon and Coolgardie Districts. In cultivation, it prefers very well drained soils, with optimum sunshine. Has potential as a container plant. Propagate from seed or possibly by division. Worth experimenting with leaf and leaf-bud cuttings.

Dampiera preissii Vriese = *D. lavandulacea* Lindley

Dampiera prostrata Vriese
        see under *SCAEVOLA* L.

Dampiera purpurea R. Br.
(purple)
Qld, NSW, Vic
1-1.5 m x 0.5-2 m   Sept-Dec; also sporadic

Small to tall, suckering, perennial **herb**; **stems** erect, sparsely branched, densely hairy, can be slightly

Dampiera purpurea broad-leaf form, × ·7

angular; **leaves** 1-6 cm x 0.5-2.5 cm, lanceolate to or-bicular, distantly toothed or entire, slightly hairy above, densely hairy below; **flowers** to 2.5 cm across, pale blue to bluish purple, light yellow throat, exterior has dense dark grey hairs, axillary or terminal, either solitary or in loose, short racemes.

A most variable species in foliage and flower. Many forms are in cultivation, probably the most spectacular being one with oval leaves and bright bluish purple flowers. Hardy under a wide range of conditions, it usually grows best in medium to heavy soils, where it suckers freely. Will grow in dappled shade to full sun.

Dampiera purpurea broad-leaved form　　　　W.R. Elliot

Frost tolerant. Propagate from cuttings (young suckering growth strikes readily) or by division.
D. brownii F. Muell. is considered conspecific.

Dampiera rosmarinifolia Schltdl.
(leaves similar to those of the genus *Rosmarinus*)
Vic, SA　　　　　　　　　　　　　Wild Rosemary
0.2-0.5 m x 0.5-2 m　　　　　　　　　Aug-Dec

Dwarf, suckering, perennial **herb**; **stems** terete, striate, densely hairy, many, erect or can be arching; **leaves** to 2 cm x 0.5 cm, oblong to linear-oblong, often in 3s, glabrous above, dense white tomentose under-surface, margins revolute; **flowers** 1.5-2 cm across, usually blue but can be mauve, purple, pink or white, with yellow throat, exterior with white or grey hairs, often in clusters near ends of branchlets.

A widely cultivated and most ornamental species. It has adapted to a wide range of soils and conditions, and grows well in light or heavy soils, with plenty of sunshine. Frost and drought tolerant. An excellent container plant. Growth can be slow initially, but plants often form a large colony. Readily kept under control by removal of suckering roots. A range of flower colours adds interest to this species. Propagation from stem cuttings is usually slow but the firm young suckering growth strikes readily.

Dampiera plumosa　　　　　　　　　T.L. Blake

173

Dampiera roycei Rajput
(after R. D. Royce, 20th-century botanist, WA)
WA
0.3-0.5 m x 0.2-0.5 m   Sept-Nov

Dwarf **shrub**; **stems** terete, ribbed, densely covered with brownish grey hairs; **leaves** 0.7-3.5 cm x 0.3-1.3 cm, sessile along stems, lower ones spathulate, upper ones oblong-obovate to oblong-elliptic, initially hairy on both surfaces, becoming glabrous above, margins entire, with small projections; **flowers** about 1.5 cm across, blue, exterior densely covered with grey hairs, borne on hairy peduncles.

A recently named species that occurs in scattered locations over a wide range of semi-arid and arid WA. Evidently not in cultivation, it will require very well drained soils, with partial or full sun. Tolerant of extended dry periods and most frosts. Propagate from cuttings.

This species is allied to *D. linearis*, which differs in its glabrous stems and its toothed or slightly lobed leaves.

Dampiera rupicola S. Moore =
                                        D. lavandulacea Lindley

Dampiera sacculata F. Muell.
(little pouch)
WA   Pouched Dampiera
0.2-0.4 m x 0.5-1 m   June-Oct

Dwarf, suckering, perennial **herb**; **stems** angular, not winged, wiry, glabrous, can be flexuose; **leaves** to 4 cm x 0.3 cm, linear, thick, obtuse, glabrous, scattered along stems; **flowers** about 1.5 cm across, deep purple-blue to dark blue, yellow-and-white throat, borne on peduncles of variable length, exterior with dense, dark hairs.

An ornamental species that is widespread in the south-western corner of WA, occurring in the Austin, Avon, Darling, Eyre and Roe Districts, where it commonly inhabits low-lying heathlands or swampy areas. It is not commonly cultivated, but should adapt to a wide range of soils and climates. Light to medium soils, with partial to full sun should be suitable. Suckers readily and can form large colonies. Propagate from stem cuttings or by division.

Dampiera sericantha F. Muell.
(silken flower)
WA
0.2-0.3 m x 0.5-2 m   Sept-Dec

Dwarf, perennial **herb**; young shoots hairy; **stems** slender, triangular, glabrous; **leaves** to 1.5 cm x 1 cm, oblong-cuneate, sessile, glabrous when mature; **flowers** to 1 cm across, blue, exterior silky white; **peduncles** hairy, 1-3 flowered.

An ornamental species from the south-western area of WA. Not commonly cultivated. It should grow in well drained soils, with partial or full sun. Suited to container cultivation. Propagate from young stem cuttings.

There has been confusion between *D. linearis* and *D. sericantha*. Plants sold as *D. sericantha* in eastern Australia are undoubtedly a form of *D. linearis*, which is currently undergoing botanical revision. *D. linearis* has terete stems, and the flowers have grey hairs on the petal exteriors.

Dampiera spicigera Benth.
(bearing flower-spikes)
WA   Spiked Dampiera
0.3-0.5 m x 0.5-3 m   June-Dec; also sporadic

Dwarf **shrub**; young shoots hairy; **branchlets** hairy; **leaves** 1-4 cm x 0.5-1 cm, oblanceolate to obovate, flat or concave, often toothed on upper margins, becoming glabrous above, usually hairy below but can be glabrous on both surfaces; **flowers** about 2 cm across, pale blue to blue-violet, yellow throat, exterior with grey hairs, borne in a leafless terminal spike 4-5 cm or more in length, lowest flowers open first.

An attractive, small shrubby species that can sucker over a wide area. It is long-flowering, as the flowers open progressively along the stems. It occurs in the sandplains of the Austin, Avon, Darling and Irwin Districts, and is best suited to light or medium soils with good drainage, and optimum sunshine. Tolerant of most frosts. Pruning after flowering promotes bushy growth. Suitable for cultivation as a container plant. Propagate from stem cuttings which can be slow to form roots. Experimentation with leaf and leaf-bud cuttings is worth trying.

Dampiera stenophylla Krause
(narrow leaves)
WA
0.2-0.3 m x 0.3-1 m   Sept-Feb; also sporadic

Dwarf **shrub**; young shoots densely hairy; **stems** terete, striate, often arching; **leaves** to 1.5 cm x 0.3 cm, oblong to narrow-elliptical, entire or distantly toothed, margins recurved to revolute, glabrous above, densely hairy below, scattered or in bundles at nodes; **flowers** about 1.2 cm across, pale to deep blue, hairy exterior, borne on short peduncles at nodes along branchlets.

A graceful species from low-rainfall areas of the Keartland and Coolgardie Districts. It is not commonly grown in cultivation, but seems very adaptable. Does well in light to medium soils, with dappled shade to full sun. Tolerates most frosts. Old stems can die back. Pruning after flowering stimulates new stems. Excellent for hanging baskets and pots.

Propagates readily from stem cuttings of firm new growth.

Dampiera stenostachya E. Pritzel
(narrow flower-spikes)
WA
0.3-0.6 m x 0.5-1 m   July-Oct; also sporadic

Dwarf **shrub**; **branches** usually erect; **branchlets** flexuose, initially densely covered in white hairs, becoming glabrous; **leaves** to 2 cm x up to 0.8 cm, oblong, obtuse, densely hairy or becoming glabrous above, dense white hairs below, many; **flowers** about 1.25 cm across, blue to violet-blue, borne in leafless terminal spikes.

Occurs in the Austin, Avon, Coolgardie, northern Darling and Roe Districts. Prefers very well drained soils, with plenty of sun, but will probably tolerate partial sun. It has cultivation potential because of its long flower-spikes, but evidently is not widely grown. Tolerant of frosts and extended dry periods. Propagate from stem cuttings.

Has similarities to *D. spicigera*, which has a more hairy calyx and generally longer leaves.

174

Dampiera trigona var. *latealata*   B. Crafter

Dampiera stowardii S. Moore = *D. oligophylla* Benth.

Dampiera stricta R. Br.
(erect; upright)
Qld, NSW, Vic, Tas   Blue Dampiera
0.3-0.8 m x 0.3-2 m   May-Jan
    Dwarf, suckering perennial **herb**; **stems** angular,
usually erect but can be arching, more or less glabrous;
leaves 2-6.5 cm x 1-2.5 cm, linear to elliptical, entire or
with a few coarse teeth, becoming glabrous; **flowers** to
2.5 cm across, sky blue to deep mauve-blue, rarely
white, with pale yellow to white throat, exterior with
rusty hairs; **peduncles** to 7 cm long, 1-3 flowered, in
upper axils.
    A most variable and adaptable species from heath-
lands and forests of the eastern states. It grows well in a
wide range of soils, with dappled shade to full sun. As
it suckers lightly, it can add an attractive natural
feeling to the garden. Frost tolerant. Hard pruning
after flowering is useful for promoting new growth.
    A form from Benedore River in East Gippsland, Vic,
is more vigorous than other forms. It has large flowers
of light blue.
    Propagate from young suckers or by division. Stem
cuttings can be used but are often slow to form roots.

Dampiera subspicata Benth. = *D. leptoclada* Benth.

Dampiera tenuicaulis E. Pritzel
(slender stems)
WA
0.1-0.3 m x 0.2-1 m   Sept-Oct
    Dwarf, perennial **herb**; **stems** slender, more or less
without leaves, striate, hairs in grooves; **leaves** very
small, scattered at a distance along stems; **flowers**
about 1 cm across, blue to bright blue, exterior with
dark grey hairs, borne in loose panicles.
    An inhabitant of the sand heathlands in the Avon,
Coolgardie and Roe Districts. This decorative species
is highly worthy of cultivation. Requirements are for
well drained, light to medium soils, with partial or full
sun. Tolerant of most frosts and extended dry periods.
Propagate from stem cuttings.
    Some plants sold under this name are in fact a form
of *D. linearis*.
    D. curvula Krause and *D. helmsii* Krause are now in-
cluded in this species.

Dampiera teres Lindley
(terete)
WA   Terete-leaved Dampiera
0.3-0.5 m x 0.5-1 m   Aug-Jan; also sporadic
    Dwarf **shrub**, can sucker; **branches** many; **leaves** 1-
4 cm x 0.1-0.3 cm, linear, fleshy, usually terete when
young, can become flat with maturity, clustered along
branchlets, with minute hairs on both surfaces, often
appear glabrous; **flowers** about 2 cm across, mauve-
blue or rarely pink, with lime-yellow-and-white throat,
dense grey hairs on exterior, borne on short peduncles
forming a spike of up to 15 flowers and to 30 cm long,
lower flowers open first.
    A most ornamental species from the northern
Darling and Irwin Districts. It has been in cultivation
for many years but is not widely known. Best suited to
well drained soils, with plenty of sunshine. Tolerates
frost and extended dry periods. Pruning after flowering
promotes bushy growth. An excellent container plant.
A pinkish-flowered form is most appealing. Propagate
from firm young stem cuttings which can be slow to
form roots or by transplanting suckers. Experimentation with leaf, leaf-bud and root cuttings is worth
trying.

Dampiera stricta 'Benedore River', × ·6

175

Dampiera tomentosa

Dampiera trigona forms, × ·55

Dampiera tomentosa Krause
(cottony)
WA   Felted Dampiera
0.2-0.5 m x 0.3-1 m   Sept-Oct

Dwarf **shrub**; **branchlets** covered in greyish white hairs; **leaves** 2-3.5 cm x 0.8-1.2 cm, oblong to ovate-oblong, densely hairy on both sides, entire; **flowers** blue, about 1 cm across, axillary or in simple slender spikes, grey hairs on exterior.

An interesting species with decorative foliage. It occurs in the dry regions of the Avon, Coolgardie, Darling and Roe Districts. May be difficult to keep growing in the more temperate zones because of the hairy leaves and branchlets. Best suited to well drained soils, with full sun. Worthy of trial as a container plant. Propagate from stem cuttings.

Dampiera trigona Vriese
(3-angled)
WA   Angled-stem Dampiera
0.3-0.7 m x 0.5-1 m   July-Feb

Dwarf, tufting, perennial **herb**; **stems** more or less 3-angled, glabrous, initially upright, often becoming arched; **leaves** 1.5-4 cm x 0.2-0.5 cm, lanceolate to linear-lanceolate, more or less sessile, margins entire or with odd small teeth, glabrous, scattered along stems, **flowers** to 2 cm across, blue to blue-violet, with pale yellow-and-white throat, exterior only slightly hairy; **peduncles** long, glabrous, 1 or usually 2-flowered; **ovary** 2-celled.

A widely cultivated ornamental species that has proved hardy in many situations and conditions. It occurs in the Darling District, where it is found mainly in near-coastal areas. Grows in most relatively well drained soils, with dappled shade to full sun. An excellent container plant. Pruning after flowering helps

to produce new stems. Usually develops as a single clump. Responds well to light applications of fertilizer. Hardy to most frosts. Propagate from stem cuttings which strike readily.

The var. *latealata* E. Pritzel has widely winged stems and can have prominently toothed leaves. It will be raised to species level in the revision.

Dampiera triloba Lindley
(3-lobed)
WA
0.3-0.5 m x 0.3-0.8 m   Aug-Dec

Dwarf, tufting, perennial **herb**; young shoots with rusty hairs; **stems** many, more or less angular, initially covered in rusty hairs but becoming glabrous with maturity; **leaves** 2-5 cm x 1-2 cm, ovate to orbicular, usually 3-lobed at apex but can be entire, sparsely hairy to glabrous, thick; **basal leaves** usually petiolate; **stem-leaves** more or less sessile; **flowers** about 1 cm across, blue, exterior with rusty hairs; **peduncles** long, covered in rusty hairs, 1-2 flowered.

This species occurs in the Darling District, and is thought to not be in cultivation. The combination of the rusty stems and blue flowers makes it worthy of attention. Requirements are likely to be the same as for many other *Dampiera* species, ie. well drained soil, with partial or full sun. Propagation should be tried from firm young shoots.

Dampiera wellsiana F. Muell.
(after Julia S. Wells, WA botanical collector)
WA   Wells' Dampiera
0.1-0.3 m x 0.5-3 m   Aug-Nov

Dwarf, suckering, perennial **herb**; **leaves** 7-20 cm x 1.5-2.5 cm, basal, oblanceolate to spathulate, fleshy, glabrous on both sides, can have crenate margins; **flowers** small, light blue, clustered in dense terminal heads, surrounded by prominent bracts.

176

An outstanding ornamental that occurs in the Austin, Avon, Darling, Eyre, Irwin and Roe Districts, where it grows on sand or coarse sandy loam, amongst other shrubs. It has proved difficult to maintain in garden cultivation, and seems to require very well drained, light to medium soils, with partial or full sun. Suckers readily if roots are damaged. Has excellent potential as a container plant. Propagate from underground stems or suckers. Experimentation with leaf and leaf-bud cuttings is worthy of trial. *D. humilis* E. Pritzel is now included in this species.

Dampiera species nova 'WA'

see under *D. linearis* R. Br.

Dampiera species

Qld, NSW — Blue Beauty-bush
0.3-0.6 m x 0.3-1 m — Aug-Oct

Dwarf, perennial **herb**; **stems** soft, angled, more or less erect; **leaves** to 4 cm x 1 cm, oblanceolate, coarsely toothed; **flowers** about 1.5 cm across, blue to purple-blue with yellow throat, greyish hairs on exterior; **peduncles** long, 1-3 flowered.

This as yet undescribed species is from forests in the coastal regions of south-eastern Qld and north-eastern NSW. It occurs in well drained, sandy to sandy clay soils that can be moist for extended periods. Should grow in light to medium soils, with dappled shade to full sun. Not well known in cultivation, as there have been difficulties encountered in establishing plants. Propagate from stem cuttings.

This species is sometimes known as *D.* 'sylvestris'.

Dampiera species (Katanning)

WA
0.1-0.5 m x 1-2 m — July-Dec

Sprawling dwarf, perennial **herb**; **branchlets** long, spreading, densely covered with long, plumose hairs; young shoots with deep cream hairs; **leaves** to 1.5 cm x 1 cm, elliptic to cuneate, scattered, becoming glabrous above, hairy below, sometimes few teeth on young

Dampiera species (Katanning), × ·6

leaves; **flowers** to 1.5 cm wide, light mauve-blue with white-and-lime-yellow throat, exterior densely covered with creamish hairs, 1-2 flowers per hairy peduncle.

This apparently unnamed species occurs in lateritic gravel with sand. It appears to be adapting well to cultivation, although to date experience is limited. Prefers well drained soils, with dappled shade or partial sun. Should tolerate full sun in temperate regions. Can tend to become open, but responds well to pruning. Tolerant of light to medium frosts. Potential for cultivation in containers, including hanging baskets. Propagate from cuttings.

DANSIEA N. B. Byrnes
(after S. J. Dansie, forester and botanical collector)
Combretaceae

A monotypic genus endemic to north-eastern Qld.

Dansiea elliptica N. B. Byrnes
(elliptical)
Qld
15-30 m x 5-15 m — Jan-March

Medium to tall **tree**; **bark** flaky; **leaves** 3-8.5 cm x 1-3.5 cm, elliptical, spirally arranged, densely hairy when young, the margins entire, 2 marginal glands present; **flowers** about 2 cm long, solitary, on axillary peduncles, greyish white; **calyx** tubular at the base, ending in 5 calyx lobes; **petals** 5, about 1 cm long, hairy, broadly elliptical, deciduous; **fruit** about 1 cm long, with 2 broad papery wings attached, each 1.8-2.8 cm x 2.3-3 cm.

An uncommon to rare tree restricted to rainforests near Mount Bartle Frere in north-eastern Qld. It has only recently been described and so far fruiting material has been seldom collected. The species is unknown in cultivation, but could be an attractive tree for parks and acreage planting in the tropics. Flowers are interesting and attractive. Trees may have a short deciduous period. Cultivation requirements are unknown, but the trees would require well drained soil and some protection when young. Seeds probably have a short viability period.

DANTHONIA DC.
(after Etienne Danthoine, French botanist of the late 18th century)
Poaceae

Perennial **grasses** forming small to large, discrete clumps or tussocks; **leaves** flat or inrolled, glabrous or hairy, lax or firm and bristle-like; **culms** straight or geniculate, smooth, striate or hairy; **inflorescence** a panicle or raceme; **spikelets** solitary or in clusters, often becoming straw coloured with age; **lemma** conspicuously awned.

A genus of about 33 species with 32 species found in Australia, at least one of which is undescribed. *D. decumbens* (L.)DC. is naturalized in Vic and Tas. As a group, they are commonly known as wallaby grasses, and are a common feature of open forest, woodland and grassland, almost throughout Australia. They are a particularly important component of native pastures, and are still of extreme significance in areas where pastures have not been upgraded by practices such as liming, fertilizing and the introduction of exotic species. Wallaby grasses generally cannot compete with improved pasture, and slowly die out. They are eagerly eaten by introduced stock as well as native animals, and species such as *D. duttoniana* and *D. richardsonii* are held in high esteem for their palatable, nutritious feed. Many species have been evaluated in pasture plot and pot trials in Australia and overseas, and there have been attempts to establish them as pasture plants in New Zealand and some states of the USA. Most species of *Danthonia* are very hardy grasses, and will grow and produce feed on infertile soils and under dry conditions.

Being tussock grasses with appealing characteristics when in flower, the group possesses some useful ornamental features. With a growing interest in using indigenous species for regeneration schemes on areas that have been abused, the use of native grasses such as *Danthonia* should become an important aspect of these schemes. They are hardy once established in a suitable situation, and seem to thrive on neglect. They mix well with other small plants and have a definite affinity with rocks. Some of the smaller species make interesting container plants. Most species require a sunny situation, in well drained soil. After a few years their tussocks tend to become full of dead leaves and culms, and present a tired appearance. They can be rejuvenated by lifting and dividing, or by burning the tussock. A heavy watering soon after the tussock is burnt will ensure fresh green shoots within a week.

Propagate by division of the clumps or from transplants. Danthonias generally resent disturbance and may take 6 months to settle in their new home. A much quicker method is to pot the divisions or transplants, and plant them in the garden when new growth is satisfactory. Propagation from seed is somewhat tricky. Fresh seed will not germinate; it should be stored for at least one year before sowing. The seed is best sown on the surface of a pot and covered lightly.

Danthonia alpicola Vick.
(mountain-dweller)
NSW, Vic   Crag Wallaby Grass
10-40 cm tall   Dec-April

Small, tufted, perennial **grass**; **leaves** 5-10 cm x 0.2-0.4 cm, thick, stiff and rigid, flat or inrolled, the apices somewhat pointed; **culms** up to 40 cm tall, smooth, geniculate at the nodes; **panicle** 3-5 cm x 2-3 cm, dense, straw coloured; **spikelets** crowded, hairy, with long exserted awns.

A dwarf species confined to subalpine areas, where it forms large communities in rocky situations. It is somewhat prickly but nevertheless makes an interesting grass for amongst rocks. It is best in an open situation, and grows well in a soak or seepage area. Flowering is somewhat reduced at low altitudes. The plants also make an interesting pot specimen and should be grown with the base of the container in a saucer of water. Propagate from seed or by division of the clumps.

Danthonia bipartita F. Muell. =
Monachather paradoxa Steudel

Danthonia caespitosa Gaudich.
(growing in tufts)
NSW, Vic, Tas, SA, WA   Common Wallaby Grass;
                                        White Top
20-80 cm tall   Oct-Feb

Perennial **grass** forming dense tussocks; **leaves** 20-30 cm x 0.1-0.4 cm, narrow, soft, hairy, sometimes flexuose; **culms** erect, smooth, slender to robust; **panicle** 8-15 cm long, compact, soft and bristly; **spikelets** light green, becoming white.

This abundant grass is one of the most important components of natural pastures and grasslands. It is extremely useful in the drier regions because of its palatable, nutritious feed, and can withstand heavy grazing. Sporadic leaf growth is produced throughout the summer, with the bulk in spring and autumn. Flower-heads, which are soft and bristly, bleach white and are quite decorative, especially when placed amongst dark rocks. Propagate from seed or by division of the clumps.

Danthonia carphoides F. Muell. ex Benth.
(like the genus *Carpha*)
NSW, Vic, Tas, SA   Short Wallaby Grass
10-30 cm tall   Aug-Dec

Small perennial **grass** forming short tufts; **leaves** 0.3-2 cm x 0.1-0.2 cm, usually inrolled, hairy, scabrous; **culms** smooth, slender, erect; **panicle** 2-5 cm long, short, compact, dense; **spikelets** green or purplish, becoming straw coloured with age, hairy.

Widespread in grasslands and open forest, this species is often locally abundant in sparse swards. It is quite attractive when in flower, but unfortunately the spikelets lose colour quickly. It is useful amongst rocks or small shrubs and can make an interesting container plant. It is extremely hardy. Propagate from seed or by division.

Danthonia clelandii Vick.
(after J. B. Cleland)
SA   Wallaby Grass
50-75 cm tall   Oct-Feb

Perennial **grass** forming erect tussocks; **leaves** 15-30 cm x 0.1-0.2 cm, flat or inrolled, green and smooth on the upper surface, scabrous beneath; **culms** stout, erect; **panicle** 8-13 cm long, loose, spreading, many-flowered, sparse; **spikelets** green or purplish, with prominent red, hanging anthers.

An attractive tussock grass restricted to the Lofty Ranges of SA, where it grows in open forest. It is highly ornamental when in flower and useful for mingling amongst low shrubs or in a rockery. Prefers poor soils. Propagate from seed or by division.

Danthonia duttoniana Cashm.
(derivation unknown)
NSW, Vic, SA   Brown-back Wallaby Grass
30-80 cm tall   Oct-Feb

Perennial **grass** forming robust tussocks; **leaves** 10-20 cm x 0.3-0.4 cm, flat, glabrous, green or glaucous, with blunt apices; **culms** stout, smooth, shining; **panicle** 4-12 cm long, the lower branches widely spaced from the others and prominently reflexed;

spikelets numerous, crowded towards the end of the panicle branches, pale green tinged with purple, the anthers orange.

This species is usually found in lowland grasslands or open forest country. It is a very leafy wallaby grass that is highly valued as a pasture species because of its palatable and nutritious feed. It has been widely tested in pot trials, and introduced with success into some areas. It has ornamental properties because of its leafy bright green tussocks. It grows well in clay soils. Propagate by division or from seed.

Danthonia eriantha Lindley
(bearing woolly flowers)
NSW, Vic, SA   Wallaby Grass
20-70 cm tall   Aug-Dec

Perennial **grass** forming erect, dense tussocks; **leaves** 10-25 cm x 0.1-0.2 cm, strongly inrolled, hairy, somewhat wiry; **culms** slender, smooth; **panicle** 2-7 cm long, loosely compact, hairy; **spikelets** green tinged with purple, the anthers pale yellow.

A widespread and often common grass that forms a very significant component of native pastures of southeastern Australia. It has some ornamental appeal but is not as attractive as other species. The plants are very hardy and grow easily, tolerating quite dry conditions. Propagate from seed or by division.

Danthonia frigida Vick. =
     Chionochloa frigida (Vick.) Conert

Danthonia geniculata J. Black
(bent like a knee)
Vic, Tas, SA   Kneed Wallaby Grass
15-45 cm tall   Aug-Nov

Perennial **grass** forming very slender, erect tussocks; **leaves** 10-15 cm x 0.1 cm, filiform, inrolled, hairy, somewhat roughened; **culms** very slender, striate, geniculate; **panicle** 1-2 cm long, much contracted, with 3-15 spikelets; **spikelets** pale, crowded.

An interesting grass usually found in coastal and near-coastal districts, in open forest situations. It is a slender, rather graceful species that makes an interesting addition to a rockery. It requires exposure to some sun, and soil of free drainage. Propagate from seed or by division.

Danthonia gracilis Hook.f.
(slender)
Tas   Wallaby Grass
20-45 cm tall   Nov-Feb

Perennial **grass** forming erect tussocks; **leaves** 10-15 cm x 0.2-0.3 cm, flat or inrolled, green, conspicuously hairy, somewhat striate; **panicle** 4-7 cm long, compact, hairy; **spikelets** green or purple, becoming straw coloured, hairy, the anthers yellow.

A graceful grass, easily distinguished by the softly hairy leaves. It is endemic in Tas, where it appears to be confined to peaty soils in highland regions of the west. It appears to be a useful ornamental grass for gardens in temperate regions. Propagate from seed or by division.

Danthonia induta Vick.
(clothed with hairs)
Qld, NSW, Vic   Wallaby Grass
0.5-1.2 m tall   Nov-Feb

Robust perennial **grass** forming erect tussocks; **leaves** 15-30 cm x 0.2-0.4 cm, flat to loosely inrolled, with a somewhat rigid point; **culms** robust, thick, smooth; **panicle** 9-18 cm long, loose and spreading, bearing numerous spikelets; **spikelets** green to straw coloured, the anthers orange.

A vigorous grass often found in hilly to highland situations in open forest, and often in moist soils. It is quite attractive when in flower, and deserves cultivation. Looks good when grown amongst rocks. Needs a sunny situation. Propagate from seed or by division.

Danthonia linkii Kunth
(after J. H. F. Link, 18-19th-century German botanist)
Qld, NSW, Vic, SA   Wallaby Grass
40-70 cm tall   Nov-March

Tufted, leafy perennial **grass**; **leaves** numerous, 30-70 cm x 0.1-0.25 cm, flat or loosely rolled, glabrous or sparsely hairy; **culms** erect, smooth, slender; **panicle** 4-12 cm long, sometimes not far from the leaves, sparse to somewhat dense; **spikelets** greenish, becoming straw coloured.

This is an important pasture grass and is especially valuable on black soils, where it forms lush, leafy tussocks in good seasons. It was originally cultivated in

Danthonia longifolia leaves and flower-head, × 1
flower-spike, × ·25

the Berlin Gardens in the 1820s and was exhibited at the Franco-British Exhibition of 1914. Although it is a more than useful pasture grass, its ornamental qualities are limited. It grows readily in heavy garden soils and is suitable for subtropical or temperate regions.

Danthonia longifolia R. Br.
(long leaves)
Qld, NSW, Vic, Tas — Long-leaved Wallaby Grass
20-75 cm tall — April-Oct

Tufted perennial **grass**; **leaves** 10-35 cm x 0.1-0.2 cm, slender, lax or weeping, becoming inrolled and curled with age, glabrous; **culms** up to 75 cm tall, glabrous; **panicle** 5-15 cm long, dense, grey-brown; **spikelets** crowded, dense white hairs visible when open.

A wiry grass found in sandy or clay soils in rocky situations of open forest country, often in mountainous or tableland areas. In very dry conditions the leaves become inrolled, giving the tussocks a sparse, wispy appearance. The plant is a hardy tussock grass, useful for mingling in a shrubbery or rockery. It requires sunny conditions, in well drained soil. Prefers poor, infertile soils. Propagate from seed or by division of the clumps.

Danthonia nivicola Vick.
(growing in snow)
NSW, Vic, Tas — Snow Wallaby Grass
10-30 cm tall — Dec-April

Small, tufted, perennial **grass**; **leaves** 2-10 cm x 0.1-0.2 cm, tightly inrolled, smooth, green or glaucous, glabrous or sparsely hairy; **culms** up to 30 cm tall, slender, stiff and erect, usually conspicuously reddish; **panicle** 1.5-4.5 cm x 0.4-1 cm, fairly dense, becoming more open with age, purplish to straw coloured; **spikelets** hairy, the awns shortly exserted.

A common grass of subalpine tracts, where it grows in large expanses, often in moist situations. Some plants have a distinctly glaucous appearance and are desirable for cultivation. The reddish flowering stems are another attractive feature, especially when viewed in mass. This grass makes a decorative container plant and also looks appealing when planted amongst rocks in the garden. It requires a sunny aspect and appears to adapt to most free-draining soils. Flowering may be reduced at low altitudes. Propagate from seed or by division of the clumps. Parts of the division will frequently brown off and die, but some part usually remains green and survives.

Danthonia nudiflora P. Morris
(naked flowers)
NSW, Vic, Tas — Alpine Wallaby Grass
10-40 cm tall — Dec-April

Small, tufted, perennial **grass**; **leaves** 2-15 cm x 0.2-0.3 cm, smooth, glabrous, inrolled, somewhat stiff and spiny; **culms** up to 40 cm tall, erect; **panicle** 3-7 cm x 0.5-1.5 cm, linear, dense, straw coloured; **spikelets** crowded, very hairy, the awns shortly exserted.

A common grass of high mountains and subalpine areas, usually growing in colonies. The leaves are somewhat prickly, but the grass makes an attractive specimen for a small container. It can also be grown in a garden in cool temperate regions, and is best planted amongst rocks. It likes moist to wet soil conditions. Propagate from seed or by division.

Danthonia pallida R. Br.
(pale)
NSW, Vic, WA — Red-anther Wallaby Grass;
Silvertop Wallaby Grass
0.5-1.2 m tall — Aug-Nov

Perennial **grass** forming erect tussocks; **leaves** 20-70 cm x 0.3-0.6 cm, flat or more usually inrolled, somewhat rigid, green or glaucous, rarely purplish; **culms** very stout, erect, robust; **panicle** 20-35 cm long, loose, spreading, many-flowered; **spikelets** purplish when young, ageing straw coloured with prominent red, hanging anthers.

A widespread grass usually found on poor, acid soils in open forest. Clumps in flower are quite striking and this species has excellent ornamental qualities. It is very hardy but prefers poor soil to rich garden loam. It looks good in a rockery, and needs a sunny situation. Old clumps can be rejuvenated by burning. Propagate by division or from seed.

Danthonia paradoxa R. Br. =
Plinthanthesis paradoxa (R. Br.) S. T. Blake

Danthonia pauciflora R. Br.
(few-flowered)
Tas
5-15 cm tall — Dec-April

Perennial **grass** forming small spreading tufts; **leaves** slender, glabrous, strongly inrolled, fairly stiff; **culms** smooth, slender, wiry; **panicle** 1-2 cm long, crowded, sometimes partly enclosed by a sheath; **spikelets** purplish, crowded.

This Tas endemic grows in peaty soils in subalpine conditions. After the snow thaws in spring-early summer, the rhizomes produce new leaves and inflorescences. It is an interesting, small grass for a container. It likes wet, peaty soil and during summer the container should be held with its base in water. In cold districts it could probably be grown as a rockery plant. Propagate from seed or by division.

Danthonia penicillata (Labill.) P. Beauv.
(brush-like)
NSW, Vic, Tas — Slender Wallaby Grass
30-75 cm tall — Oct-Feb

Perennial **grass** forming open, somewhat straggly tussocks; **leaves** 10-15 cm x 0.1-0.2 cm, flat and narrow or inrolled, sparsely hairy; **culms** very slender, striate, shortly pubescent; **panicle** 5-15 cm long, slender, sparse, hairy; **spikelets** widely spaced, often purplish.

A distinctive grass found in shady, moist forests, usually in highland situations. It has an attractive habit and is a useful grass for shady situations. It blends well with ferns, and looks good beside water. It likes moist soil and can be propagated from seed or by division.

Danthonia pilosa R. Br.
(hairy)
Qld, NSW, Vic, Tas, SA, WA
Velvet Wallaby Grass
20-50 cm tall — Aug-Dec

Perennial **grass** forming slender tussocks; **leaves** 6-12 cm x 0.2-0.3 cm, flat or loosely inrolled, hairy; **culms** erect, striate; **panicle** 4-5 cm long, crowded, hairy; **spikelets** crowded, often overlapping, straw coloured with purple awns.

A widespread grass found in open forests and grasslands. It is quite palatable and is a useful pasture species. It has limited ornamental features but can look attractive amongst rocks or other small plants. It appreciates a sunny situation, in well drained soil. Propagate from seed or by division.

Danthonia procera Vick.
(tall and slender)
NSW, Vic, Tas   Tall Wallaby Grass
0.5-1.2 m tall   Oct-Feb

Robust perennial **grass** forming tussocks; **leaves** 20-30 cm x 0.2-0.3 cm, narrow, inrolled, pointed and somewhat bristle-like, glabrous; **culms** stout, smooth, straw coloured with purple dots; **panicle** 8-15 cm long, open and loosely spreading, the branches roughened; **spikelets** straw coloured, loosely arranged in small clusters.

A handsome grass which forms coarse, robust tussocks in a variety of situations from open forest to subalpine woodland. It is quite decorative when in flower, and worth growing amongst rocks or small shrubs. It is quite hardy once established. Propagate from seed or by division.

Danthonia purpurascens Vick.
(purplish)
Qld, NSW, Vic, Tas, SA   Wallaby Grass
0.3-1 m tall   Oct-Jan

Perennial **grass** forming erect tussocks; **leaves** 8-20 cm x 0.1-0.3 cm, flat or loosely inrolled, glabrous to hairy, green; **culms** slender, smooth, tall; **panicle** 8-15 cm long, narrow and loose, hairy; **spikelets** tinged with purple, becoming straw coloured with age, the anthers deep orange.

Widely distributed in open forest and grasslands. Ornamental when in flower, and a useful tussock grass for mingling with small shrubs. Requires a sunny situation. Propagate from seed or by division.

Danthonia racemosa R. Br.
(in racemes)
Qld, NSW, Vic, Tas, SA   Wallaby Grass
20-60 cm tall   Oct-Feb

Perennial **grass** forming slender tussocks; **leaves** 5-15 cm x 0.1 cm, narrow, strongly inrolled, glabrous to hairy, somewhat rough; **culms** slender, glabrous, smooth to striate; **panicle** 5-15 cm, slender, condensed; **spikelets** pale green to straw coloured, usually in clusters of 2-5.

A very widespread and common grass of open forest country. It is extremely variable and embraces a wide range of forms. It is a useful pasture grass but has limited ornamental appeal. Requires a sunny situation, in well drained soil. Propagate from seed or by division.

Danthonia richardsonii Cashm.
(after M. Richardson)
Qld, NSW, Vic, SA   Wallaby Grass
50-100 cm tall   Nov-Dec

Robust perennial **grass** forming erect tussocks; **leaves** 1-3.5 cm x 0.1-0.25 cm, green, striate, acuminate, flat or slightly inrolled, smooth; **culms** stout, erect, glabrous; **panicle** 4-10 cm long, dense, much-branched; **spikelets** pale green, occasionally tinged with purple, becoming straw coloured with age.

This is a robust native grass that forms large, leafy tussocks. It is readily eaten by stock and is quite nutritious, but is not an abundant species. It has been well tested as a pasture grass in Vic, and has some ornamental value. After fire, the tussock regrowth is lush and green. Propagate by division or from seed.

Danthonia rodwayi C. E. Hubb. =
  Plinthanthesis rodwayi (C. E. Hubb.) S. T. Blake

Danthonia setacea R. Br.
(bristly)
NSW, Vic, Tas, SA, WA   Bristly Wallaby Grass
15-60 cm tall   Oct-Feb

Erect perennial **grass** forming tussocks; **leaves** 5-20 cm x 0.1-0.2 cm, flat or mostly inrolled and appearing filiform, glabrous or more usually hairy, scabrous; **culms** geniculate at the base, smooth or hairy; **panicle** 3-9 cm long, linear to lanceolate, slender, dense, hairy; **spikelets** silvery, the awns slender and often purplish.

An abundant grass which is an important component of natural grasslands. It has some ornamental appeal when in flower. The tussocks are extremely hardy and drought resistant, and tolerate a variety of soils and situations. Propagate by division or from seed.

DAPHNANDRA Benth.
(from the Greek *daphne*, the bay laurel tree; *andros*, a man; the anthers resemble those of the bay laurel)
Atherospermataceae (alternative *Monimiaceae*)

Trees; **bark** usually bearing corky pustules, often aromatic; **branchlets** prominently flattened at the nodes, the larger branchlets breaking away in a ball and socket joint; **leaves** opposite, simple, entire, with coarsely toothed margins, often aromatic; **inflorescence** a panicle; **flowers** few; **sepals** and **petals** alike, in 2 whorls; **staminodes** alternating with stamens; **fruit** an enlarged perianth, splitting to release hairy fruitlets (carpels).

A small, unique genus of about 8 species, restricted to Australia and New Guinea. Three species in southeastern Qld are undescribed. A couple of species are important timber trees. They are rarely encountered in cultivation; however, they can be grown readily. Propagate from seed which has a limited life and is best sown when fresh.

Daphnandra micrantha (Tul.) Benth.
(small flowers)
Qld, NSW   Socketwood
20-30 m x 5-15 m   Sept-Oct

Daphnandra repandula

Daphnandra micrantha leaves and fruit, × ·45

Medium to large **tree** with a dense, spreading crown; **bark** grey, with numerous corky pustules; young shoots bearing woolly hairs; **leaves** 8-10 cm x 0.5-2 cm, broadly lanceolate to elliptic, glossy green above, paler beneath, drawn out into a point, the margins toothed, especially in the upper half; **panicles** 4-6 cm long, arising in the leaf axils or from the older wood; **flowers** 7-8 cm across, white, of 10-15 perianth segments in 2 whorls; **stamens** 5; **staminodes** 6, hairy; **fruit** an expanded perianth enclosing ovoid fruitlets 1.5-2.5 cm long, covered with long brown hairs.

A very common tree that is widely distributed in rainforests from south-eastern Qld to the Illawarra District of south-central NSW. It is widely cut as a timber tree, the yellow-grey wood being soft and used for cabinet work, joinery and veneers. The leaves and bark have a pleasant, aromatic smell. This species makes a handsome specimen tree for parks but is generally too large for the average home garden. It will grow in subtropical and temperate regions, in well drained soils rich in organic matter. Young plants need protection from direct sun until about 1 m tall. The species is extremely variable and some botanists believe that it could be split into 4 species based on the leaf midribs and inner bark colour. Propagation is from seed which must be sown fresh.

Daphnandra repandula F. Muell.
(somewhat bowed)
Qld   Northern Yellow Sassafras
12-18 m x 3-8 m   April-June
Medium **tree** with a spreading crown; **bark** grey, with small pustules in nearly longitudinal rows; young shoots bearing fine grey hairs; **leaves** 7-15 cm x 1-2.5 cm, lanceolate or ovoid, dark shiny green above, paler beneath, the margins toothed, the apex drawn out into a fine point; **panicles** 2-5 cm long, axillary in the upper leaves, the lower branches of the panicle bearing 3 flowers, the upper bearing 1; **flowers** 0.4-6 cm across, cream to white, of 13 perianth lobes, in 2

whorls, the inner of 9 segments with wavy margins; **stamens** 7; **staminodes** 7; **fruit** 2.5-3 cm long, narrow, clavate, enclosing 4-6 carpels which are 0.6 cm long and hairy.

Restricted to rainforests of north-eastern Qld but there distributed from the coast to the tablelands. Too large for the average home garden, except perhaps in temperate regions. This species is ideally suited to parks, large gardens and perhaps as a street tree in tropical and subtropical regions. It is somewhat slow-growing when young, and needs protection from direct sun. Soils must be acid, well drained and, for optimum growth, rich in organic matter. Propagate from seed which must be sown fresh.

Daphnandra tenuipes Perkins
(slender flower-stalks)
Qld, NSW   Socket Sassafras
10-20 m x 3-8 m   March-June
Small to medium **tree**, single-stemmed or with many slender trunks; **bark** smooth, cream to grey, with corky pustules; **leaves** 5-11 cm x 1-2.5 cm, elliptical, dark green above, paler and shiny beneath, toothed, drawn out into a long point; **panicles** 2-4 cm long, axillary; **flowers** about 0.5 cm across, dark red; **fruit** about 0.8 cm across, globular, orange-yellow, enclosing a number of hairy carpels, mature Nov-Jan.

A common tree extending from south-eastern Qld to north-eastern NSW, growing in rainforests and moist valleys beside streams. The branches are often covered with epiphytic mosses, lichens, ferns and orchids. The timber is deep yellow and useful for joinery, cabinet work and carving. This species is a handy tree for large gardens and parks. It could be grown in home gardens in temperate regions, where the climate would reduce the ultimate size. The canopy is quite dense and shady, but the trees may be slow-growing, especially in the young stages. Requirements are for some protection when young, and an acid, well drained soil. Applications of mulch and light dressings of fertilizer are beneficial. Propagate from seed which must be sown fresh.

DARLINGIA F. Muell.
(after Sir Charles Darling, 19th-century Governor of Victoria)
Proteaceae
Trees; **leaves** entire or variously lobed, large and spreading, glabrous or hairy; **inflorescence** an axillary spike or terminal panicle, generally hairy; **flowers** bisexual, profuse, waxy, often scented; **perianth** tubular, with spreading lobes; **anther connective** extending as a gland-like appendage; **style** long, with a terminal stigma; **fruit** a follicle; **seeds** thin and flat, winged.

A small genus of 2 species, both endemic in north-eastern Qld. They have many useful attributes for cultivation, but are not commonly seen. In a suitable situation they are easy and rewarding subjects. Propagation is from seed which is best sown fresh or from semi-hardwood cuttings which are generally slow to strike.

Darlingia darlingiana, × ·25

Darlingia darlingiana (F. Muell.) L. Johnson
(like the genus *Darlingia*)

Qld	Brown Silky Oak
15-20 m x 3-12 m	Sept-Nov

Medium **tree** with a bushy canopy; young shoots bright green, glabrous; **leaves** 15-45 cm x 6-12 cm, variable in shape, oblong, oblanceolate or deeply lobed with 3-7 long, lanceolate, acute, spreading or falcate lobes, dark green and somewhat shiny, glabrous, with numerous prominent veins; **spikes** 10-25 cm long, bearing small rusty hairs, in the upper axils or forming a terminal panicle, densely covered with flowers; **flowers** 2-3.5 cm long, white, sessile, the basal part tubular, then with recurved lobes and a long style; **follicles** 3-5 cm x 2-2.5 cm, brown, woody; **seeds** 2-4, 3-5 cm x 2 cm, flat, winged, pale brown, mature Jan.

Confined to rainforests of north-eastern Qld and there often locally common at intermediate and high altitudes. It is occasionally cut as a timber tree, with the tough wood being attractively patterned with rays, and useful for cabinet making. This species is a very handsome tree, well worth growing for its decorative leaves and attractive flowers, which are generally similar to *Buckinghamia*. It grows readily in subtropical and tropical regions and can even be grown in a warm, protected situation in southern Australia. Will grow in a sunny or semi-shady situation, in acid, well drained soil. It makes a useful specimen tree and should fit in well with park or street planting requirements. Young plants may need some protection from direct sun until about 1 m tall. Responds to water during dry periods, regular mulching and light dressings of nitrogenous fertilizers. Propagate from seed which should only be lightly covered. Seed germinates readily when fresh but loses its viability over 1-2 years. Stem cuttings are slow and difficult to strike.

Darlingia ferruginea J. F. Bailey
(rusty hairs)

Qld	Rose Silky Oak
15-20 m x 3-10 m	June-July

Medium **tree** with a bushy canopy; young shoots densely covered with rusty brown hairs; **leaves** 20-60 cm x 8-15 cm, variable in shape but generally deeply lobed with 3-7 spreading lobes, dark green above, with a dense covering of dark rusty brown hairs beneath, with prominent, coarse veins; **spikes** 15-40 cm long, rusty-hairy, in the upper axils; **flowers** 2.5-4 cm long, cream, hairy, strongly fragrant, sessile; **follicle** 4-6 cm x 2-3 cm, leathery, hairy; **seeds** 2-4, 3-6 cm x 2 cm, flat, winged, pale brown, mature Nov-Dec.

Confined to tablelands of north-eastern Qld in dense rainforests. It is a highly decorative tree with large, lobed, rusty leaves and masses of waxy, cream, fragrant flowers in long spikes. Rarely encountered in cultivation but deserves to be much more widely grown. Suited to tropical and subtropical regions. The plants are readily damaged by frost and direct exposure to hot sun while small. A semi-shady situation in deep, acid, well drained, humus-rich soil seems most satisfactory. The plants respond to water during dry periods, and regular mulching to cut down on water loss and weed competition. Propagate from seed which should be lightly covered. Seed germinates readily when fresh but loses its viability over 1-2 years.

Darlingia spectatissima F. Muell. =
Darlingia darlingiana (F. Muell.) L. A. S. Johnson

DARWINIA Rudge
(after Dr. Erasmus Darwin, 1731-1802, grandfather of famous naturalist Charles Darwin)
Myrtaceae

Dwarf to medium **shrubs**; **leaves** usually opposite, decussate, often keeled, trigonous, prominent oil-glands, aromatic; **flowers** small, arranged in terminal heads or pairs, erect or pendulous; **bracts** small and leaf-like or large and decorative; **bracteoles** usually in pairs at base of calyx, can enclose individual flowers; **calyx** smooth or ribbed, with 5 glabrous lobes; **stamens** 10, alternating with 10 staminodes; **style** often exserted beyond floral parts, brush of hairs just below stigma; **ovary** single-celled; **fruit** a nut.

An endemic genus of about 60 species, the majority of which are confined to WA. At present it is undergoing botanical revision and there will be further species named. Their natural habitat is generally sandy coastal heathland in temperate areas. The Stirling Range, in south-western WA, is renowned for the diversity of *Darwinia* species. There they occur in moist, peaty sands, on peaks, slopes and in gullies. In eastern Australia one species, *D. porteri*, occurs only on the Atherton Tableland, in north-eastern Qld, while 10 species occur on the eastern side of the Great Dividing Range. *D. homoranthoides* occurs on the Eyre Peninsula in SA, and *D. micropetala* inhabits low-lying sandy or clay soils from western Vic to the Eyre Peninsula. Both these species tolerate slightly alkaline soils.

In nature, hybridization has occurred in WA and NSW, resulting in quite a deal of variation, or hybrid swarms.

Several species in the genus are well known for their long-lasting, ornamental, bell-like flower-heads, and many are cultivated. In 1820 *D. fascicularis* was recorded as being grown in England, *D. diosmoides* was introduced in 1827 and *D. squarrosa* in 1860.

Another pleasing characteristic is that many have foliage with distinctive aromatic properties. The aroma is readily released by simply brushing past the foliage of species such as *D. diosmoides*.

Because of their varying growth habits they have many applications. Prostrate species are useful as low groundcovers which can act as living mulches beneath and around other plants. Of these prostrate species only 4, *D. glaucophylla*, *D. grandiflora* (both from NSW), *D. homoranthoides* (SA) and *D. repens* (WA), actually self layer. The use of these for soil erosion control in small areas is successful. Grown where they cascade down embankments or over boulders, they can provide a pleasant visual aspect. Taller species are suited as undershrubs beneath trees. Some are useful as low screening plants, eg. *D. citriodora* and *D. diosmoides*.

Their main requirements for successful cultivation are well drained, neutral to acid soils which can be from sandy to clay-loam. They generally appreciate moist soil for most of the year, but will tolerate short periods of dryness once established. A mulch of coarse sand or similar material is beneficial.

Usually plants do best in a situation that does not receive full sun all day; however, great success has been experienced by the University of California, Santa Cruz, USA, where over 25 species and varieties are grown in an open situation, watered by drip irrigation.

More trials are needed to determine their longevity in cultivation. Some, such as *D. collina*, *D. hypericifolia* and *D. squarrosa*, are prone to collapse after a few years, whereas *D. citriodora* and *D. diosmoides* are very hardy.

Darwinias are extremely well suited to container cultivation, and this method of growing the more difficult species is highly recommended (see Cultivation of Container Grown Plants, Volume 1, page 101).

All species respond well to light applications of a slow-release fertilizer if required. Yellowing of leaves can occur with most species. Application of chelated iron will usually remedy iron deficiency quickly, and nitrogen deficiency will be rectified by applying a suitable fertilizer.

Pruning is recommended if bushy plants are required. Some will withstand heavy pruning, eg. *D. citriodora*. Regular tip pruning is usually all that is needed with most species. Pruning at flowering time means the long-lasting flower-heads can be used for floral decoration.

Most darwinias provide a good flow of nectar during flowering. Honey-eating birds are frequent visitors to plants with bell-type flowers or upturned clusters with reddish flowers.

The pest most likely to attack darwinias is scale, with the resultant sooty mould growing on the exudate. Control by spraying with white oil on a cool day.

Propagation

This can be from seed, cuttings or by division of layered branchlets in some species. Seed is usually difficult to obtain and results are extremely spasmodic. Most species strike readily from cuttings, using material that has vigour and is just becoming firm. It is best to prepare cuttings immediately after collection because stripping of the cambium layer can occur when the lower foliage is removed. With some species, stripping occurs on material that is stored in a moist container such as a plastic bag for only a few hours. Results using cuttings of *D. lejostyla*, *D. meeboldii* and *D. oxylepis*, from which lower leaves have not been removed, have been encouraging. This method is worthy of further trial.

Darwinia acerosa W. Fitzg.
(needle-shaped)
WA   Fine-leaved Darwinia
0.1-0.4 m x 0.5-1.5 m   Sept-Nov

Dwarf **shrub**; **branches** crowded, spreading; **branchlets** whitish; **leaves** 0.2-1 cm, linear, trigonous, tapering to a fine point, often hooked, green to glaucous; **flower-heads** about 1.5 cm across, hemispherical, pendent, terminal on short branchlets, inner bracts green with reddish margins, outer bracts green with pale margins.

This spreading species is restricted to the Mogumber area, north of Perth, where it grows in rocky soils on granite outcrops. Not common in cultivation. Best suited to very well drained, light to medium soils, with dappled shade or partial sun. Should do well in an open situation in temperate regions. Suited to container cultivation. Propagate from cuttings.

Darwinia biflora (Cheel) B. Briggs
(2-flowered)
NSW
0.1-0.8 m x 1-2 m   Oct-Nov

Dwarf **shrub**; **branches** erect to spreading; **leaves** 0.6-1 cm x about 0.1 cm, trigonous, decussate, crowded; **flower-heads** to 1.5 cm long, green and purplish, upturned, usually in pairs but can be 1-4; **style** initially green, becoming red.

Confined to the Central Coast of NSW, from the Hawkesbury River south to Port Jackson. It grows in dry sclerophyll forest on sandstone. Very uncommon in cultivation. Requires very well drained soils, with dappled shade or partial sun. Propagate from cuttings.

Darwinia camptostylis, × ·6   flower and bracteole, × ·3

Darwinia carnea   C.A. Gardner

Darwinia camptostylis B. Briggs
(flexible style)
NSW, Vic
0.1-0.3 m x 0.5-1.5 m   Oct-Dec
   Dwarf **shrub**; **branches** usually spreading but can
be erect; **branchlets** often pendulous, whitish; **leaves**
0.6-1.5 cm x about 0.1 cm, decussate, crowded; **flower-
heads** composed of 2-4 flowers, in pairs, about 0.6 cm
long, green to reddish; **style** strongly recurved.
   Not as common in cultivation now as it was some
years ago. It occurs in sandy soils of the eastern NSW
and East Gippsland coasts. Best suited to well drained,
light to medium soils, with dappled or partial sun.
Does well amongst other plants. Will cascade when
grown in containers. Propagate from cuttings.

Darwinia carnea C. Gardner
(flesh coloured)
WA   Mogumber Bell
0.3-4 m x 1-2 m   Oct-Dec
   Dwarf to medium **shrub**; **branches** either spreading
or erect; **leaves** to 1.5 cm x 0.4 cm, linear-lanceolate,
opposite, decussate, trigonous, keeled, acute apex;
flower-heads bell-shaped, to 3 cm long, green to pink,
terminal, pendent.
   A most attractive species, with large pendent bells.
Very rare in its natural habitat of near Mogumber and
Narrogin, where it grows in stony, lateritic soils. There
are both procumbent and upright forms grown, but
they have proved difficult to maintain in cultivation.
Requires very well drained, light to medium soils, with
dappled shade to partial sun. Tolerates semi-shade and
is hardy to light and medium frosts. Suited to container
cultivation. Responds well to regular tip pruning.
Propagate from cuttings.

Darwinia citriodora (Endl.) Benth.
(lemon-scented)
WA   Lemon-scented Myrtle
0.5-2 m x 1-2.5 m   May-Dec; also sporadic
   Dwarf to small **shrub**; **branches** many; **leaves** 1-

2 cm x up to 0.9 cm, ovate-lanceolate, can be cordate,
mostly opposite, glaucous, margins usually recurved,
aromatic; **flower-heads** to 3 cm across, erect to pendulous, outer bracts orange-red and green; **flowers**
ageing to orange-red, 4-6 per head; **style** prominent, to
2 cm long, yellow to orange-red.
   The most common darwinia in cultivation. It has
proved reliable under a wide range of conditions. Its
natural habitat is the jarrah forest of the Darling scarp,
where it grows in moist soils, amongst granite outcrops. Does best in relatively well drained soils, with
dappled shade or partial sun. Tolerant of full sun if
root area protected. Withstands limited coastal exposure and is hardy to frost.
   There are at least three forms in cultivation. A form
with large leaves and flowers is a most desirable shrub,
but does not grow as densely as the smaller leaf form,
which is a useful, low windbreak. A prostrate form has
become popular. All respond very well to regular pruning. Plants are sometimes subject to leaf-drop caused
by fungal attack or waterlogging. This usually occurs
if growing in a shady, moist location. The flowers are
attractive to nectar-eating birds. Propagate from seed
or cuttings.

Darwinia collina C. Gardner
(inhabiting hills)
WA   Yellow Mountain Bell
0.5-1.5 m x 0.6-1.5 m   March-April; Aug-Nov
   Small bushy **shrub**; **branches** many; **leaves** to 1 cm x
0.5 cm, elliptic, crowded, minutely toothed margins,
yellowish green; **flower-heads** bell-shaped, to 3 cm
long, with large greenish yellow bracts, terminal on
branchlets, pendent.
   A most decorative species that occurs above 1000 m
alt. on the four eastern summits of the Stirling Range.
The natural habitat is moist, peaty sands overlying
shale. To date it has not proved to be long-lived in
cultivation, as it is subject to attack from root
pathogens. It requires very well drained, acidic to
neutral, light to medium soils. Best situated in a semi-
shaded to partial sun location. Some plants have grown
well with a southerly aspect; however, others have
done well in a relatively open situation, amongst other
shrubs. Recommended as a container plant. Responds
well to light pruning. Propagate from cuttings which
strike readily.

Darwinia diminuta B. Briggs
(diminished)
NSW
1-1.5 m x 0.6-1.5 m   Sept-Nov
   Small erect to spreading **shrub**; **branches** erect, few;
branchlets many, lateral, short; **leaves** to 1.2 cm x
0.1 cm, decussate, trigonous, crowded; **flower-heads** to
1 cm long, white with red tips, arranged in pairs, composed of 2-4 upturned flowers; **style** to 0.9 cm long,
nearly twice as long as calyx tubes.
   This species is evidently not in cultivation. It occurs
in heath and dry sclerophyll forest on the Central
Coast, where it grows in dappled shade to partial sun.
Should be best suited to well drained soils, with some
protection from the sun. Propagate from cuttings.

185

Darwinia diosmoides

Darwinia diosmoides (DC.) Benth.
(similar to the genus *Diosma*)
WA
1-3 m x 2-3 m   Aug-Jan
  Small to medium, bushy **shrub**; **branches** many; **leaves** to 0.6 cm x 0.1 cm, trigonous, crowded, dark green, highly aromatic; **flower-heads** 1-2 cm across, white to pink, terminal, composed of many upturned flowers.

  Although not a very ornamental species in flower, its pleasant aromatic foliage more than compensates. The fragrance is released by light brushing or crushing of the leaves, and the species may be planted near paths, where this characteristic can be used to advantage. It occurs in south-western WA, in rocky soils through to sand-dunes. Withstands coastal exposure and is suitable as a low windbreak. Adaptable to most soil types, provided drainage is adequate. Tolerates semi-shade to full sun. Propagate from seed or cuttings.

Darwinia fascicularis Rudge
(in bundles or clusters)
Qld, NSW
0.3-2 m x 0.5-1.5 m   Aug-Dec
  Dwarf to small **shrub**; **branches** slender; **leaves** 0.8-1.6 cm x up to 0.2 cm, more or less terete, grey-green to green, aromatic, crowded in clusters at ends of branchlets; **flower-heads** to 2 cm across, initially white, becoming red with maturity, terminal, erect, composed of 4-20 small flowers.

  An adaptable species from shallow soils in heathland and dry sclerophyll forests. Grows in a wide range of fairly well drained soils, with dappled shade to partial sun. Will tolerate full sun if the root region has some protection. The flower-heads are decorative as the colour changes from white to red. Plants can become leggy, but they respond well to pruning to promote dense growth. Frost tolerant. Does well in containers.

  The ssp. *oligantha* Briggs is also in cultivation. It grows as a low shrub up to 0.6 m high, with clusters of 4-6 flowers. The spreading branches can layer themselves, which does not occur in the ssp. *fascicularis*. It

Darwinia collina   T.L. Blake

prefers a situation with moist soil, and dappled shade or partial sun.

  Propagate both ssp. from cuttings which strike readily.

  In its natural habitat, hybridization of *D. fascicularis* ssp. *fascicularis* has occurred with the following species: *D. diminuta*, *D. glaucophylla*, *D. grandiflora* and *D. procera*.

Darwinia forrestii F. Muell.
(after Sir John Forrest, 19-20th-century WA explorer and parliamentarian)
WA
1-1.5 m x 1-2 m   Sept-Nov
  Small **shrub**; **leaves** 1.5-2 cm long, linear, more or less terete, blunt apex; **flowers** about 0.5 cm diameter, white ageing to red, pendulous, solitary in upper axils.

  An outstanding ornamental species with pendulous flowers. It occurs in areas of granite in the Warren Subdistrict of the south-western corner. Requires well drained soils, and likes partial or full sun, although it will tolerate shade. Appreciates summer watering. Propagate from cuttings.

Darwinia diosmoides, × ·6

186

In a current botanical revision this species will be included in the genus *Chamelaucium* as its style and anther characteristics belong to that genus.

Darwinia glaucophylla B. Briggs
(glaucous leaves)
NSW
prostrate-0.15 m x 1-2.5 m   Nov-Dec
Dwarf, spreading, prostrate **shrub**; **branches** many, self layering; **branchlets** often pinkish, short, erect; **leaves** to 1.7 cm x about 0.2 cm, falcate, decussate, glaucous, often with purplish tonings, aromatic; **flower-heads** about 2 cm long, insignificant, greenish cream, attaining pinkish tonings with maturity, erect, usually composed of 2-4 flowers; **style** over twice as long as floral tube.

A most decorative species with glaucous leaves, which during the cool months can gain purplish tonings. It can form a dense mat. The species occurs in the Gosford region where plants grow on sandstone in heath and light woodland. Has proved reliable in a wide range of conditions. Does best in well drained soils, with partial or full sun. Tolerates semi-shade but does not form a very dense cover. Excellent container plant. Withstands most frosts. Propagate from stem cuttings or by division of layered branches.

This species has been sold under the wrong name of *Homoranthus flavescens* 'Prostrate'.

Darwinia grandiflora Benth.
(large flowers)
NSW
0.1-0.5 m x 1.5-2.5 m   May-Dec
Dwarf, spreading **shrub**; **branches** usually prostrate and self layering, with the **branchlets** ascending; **leaves** to 1.8 cm x 0.2 cm, linear-lanceolate, trigonous, decussate, ending in fine mucro, aromatic; **flower-heads** to about 1 cm across, white ageing to dark red, erect, composed of 4-6 flowers; **calyx tube** to 1.2 cm long; **style** to 2 cm long.

This adaptable prostrate species occurs naturally in the sandy soils of the Central Coast. Grows well in most well drained soils, with dappled shade or partial sun. Excellent as a groundcover beneath other shrubs. Suited to embankments and scrambling around rock outcrops. It seems to tolerate extended dry periods once it is established. Grows well in various types of containers including hanging baskets. Propagate from cuttings or by separation of layered branches.

This species has been confused by the nursery trade with *D. taxifolia* ssp. *macrolaena*. The latter has larger flowers and does not self layer.

Darwinia helichrysoides (Meisner) Benth.
(similar to the genus *Helichrysum*)
WA
0.3 m x 0.3 m   Aug-Nov
Dwarf **shrub**; **branches** usually erect; **leaves** to 1 cm long, narrow, trigonous, scattered, crowded; **flower-heads** bell-shaped, narrow, to 2.5 cm long, grey and pinkish red to scarlet, pendent.

An inhabitant of the Irwin District, it grows in gravelly sands, usually in an open situation. Not well known in cultivation at present. Should do well in an open, warm situation, with very well drained, light to medium soils. Propagate from cuttings. There has been limited success in germinating seed.

Darwinia homoranthoides (F. Muell.) J. Black
(similar to the genus *Homoranthus*)
SA
0.05-0.1 m x 2-3 m   Nov-Feb
Dwarf, spreading **shrub**; **branches** prostrate, self layering; **branchlets** short, glandular, often with pinkish tonings; **leaves** 0.6-0.8 cm x about 0.1 cm, narrow, trigonous, greyish green, spreading to recurved when mature, glandular, ending in fine mucro, aromatic; **flowers** tubular, to 3 cm long (including style), white to cream with green, initially upright, solitary, axillary; **style** about 3 times longer than calyx tube.

This self layering, prostrate species occurs on sandy or lateritic soils of the Eyre Peninsula near Port Lincoln. It is often found in situations that receive full sun. In cultivation, it seems adaptable to a wide range of soils, withstanding extended dry periods and, to some extent, wet periods. Does well in semi-shade to full sun. An excellent container or rockery plant. Frost tolerant. Propagate from cuttings or separation of layered branches.

Homoranthus flavescens was sold for several years with

Darwinia hypericifolia   T.L. Blake

the wrong name of *D. homoranthoides. H. flavescens* is a dwarf shrub, and does not have prostrate branches that are self layering.

Darwinia hypericifolia (Turcz.) Domin
(leaves similar to those of the genus *Hypericum*)
WA
0.5-1 m x 0.5-1 m   Oct-Nov; also sporadic

Dwarf **shrub**; **branches** erect, few; **branchlets** short, often reddish; **leaves** to 2 cm x 0.3 cm, oblong, scattered, glabrous, margins recurved; **flower-heads** bell-shaped, to 3 cm long, narrow, pink to scarlet, pendent, terminal.

A particularly ornamental species from the valleys or peaks of the central Stirling Range. It is not as well known as many of the other species with bell-shaped flower-heads. At this stage it is not common in cultivation, but was grown in England in the mid-19th century. Has not proved to be hardy as yet. Requires very well drained, light to medium soils which are usually moist. Prefers dappled shade or partial sun, but will tolerate full sun, provided some protection is available for the root region. An excellent container plant. Responds well to pruning. Propagate from cuttings.

Darwinia lejostyla (Turcz.) Domin
(smooth style)
WA
0.2-1 m x 0.6-1 m   Aug-Dec

Dwarf, compact **shrub**; **branches** many; **branchlets** short, often with pendent tips; **leaves** to 1 cm x 0.1 cm, lanceolate, scattered, crowded, glabrous, aromatic; **flower-heads** bell-shaped, 3-4 cm x up to 2.5 cm, reddish pink and white, terminal, solitary, pendent.

This very ornamental species from the Stirling Range and Middle Mount Barren is the most commonly cultivated of the 'bell' type. It adapts to a wide range of soils from sands to clay-loams, provided there is good drainage. Grows well in an open situation, in full sun to semi-shade. Plants are sometimes prone to attack by grey mould if grown in still, shady, moist areas. Light pruning helps to promote bushy growth. An excellent container plant. Propagate from cuttings which strike readily. It is best to prepare cuttings immediately after collection, as the bark has a tendency to tear if cutting material is kept for long periods.

Until recently the species was better known as *D. leiostyla*.

Darwinia leptantha B. Briggs
(slender flowers)
NSW
0.2-0.8 m x 0.2-0.6 m   Sept-Nov

Dwarf, slender **shrub**; **branches** usually erect; **leaves** to about 1 cm long, grey-green, very narrow, decussate, trigonous; **flower-heads** small, composed of 2-8 flowers, initially white, ageing to red.

A little-known species from swampy heathlands of coastal districts. It occurs from Laurieton near Port Macquarie and extends south to the Clyde River near Batemans Bay. Best suited to moist, light to medium soils, with dappled shade to partial sun. Withstands periods of inundation. Frost tolerant. Propagate from cuttings.

Darwinia luehmannii F. Muell. & Tate
(after J. G. Luehmann, early Victorian Government Botanist)
WA
0.5 m x 0.5 m   July-March

Dwarf, compact **shrub**; **branches** many; **branchlets** whitish; **leaves** to 0.8 cm long, linear, grey-green, trigonous, crowded; **flower-heads** about 1 cm across, green and cream, composed of 4-6 upturned flowers; **style** very prominent.

A species with interesting foliage. It usually occurs on deep sand or lateritic soils north of Ongerup and Ravensthorpe, and also in the Widgiemooltha region. Has adapted reasonably well to cultivation, doing best in a warm, sunny location. Grows in well drained, light to heavy soils. Suited to container cultivation. Responds well to light pruning. Frost and drought tolerant. Propagate from cuttings.

Darwinia macrostegia (Turcz.) Benth.
(large floral bracts)
WA   Mondurup Bell
0.5-2 m x 0.5-1 m   May-June; Sept-Dec

Dwarf to small **shrub**; **branches** many; **branchlets** often with reddish tonings; **leaves** to 2 cm long, oblong to elliptical, scattered, minutely toothed margins, glandular undersurface; **flower-heads** bell-shaped, to 6 cm long, terminal, pendent, bracts red with white blotches or rarely all red.

Perhaps the most spectacular representative of the genus. It occurs on well drained, stony soils in the western regions of the Stirling Range and in the Porongurup Range. Has proved troublesome to establish in cultivation in the ground, and often does not flower well. It has, however, been grown with considerable success as a container plant, producing flower-heads for extended periods. Needs very well drained, light to medium soils, with dappled shade to partial sun. Some success has been gained in full sun, with protection for the root area. Plants respond well to regular light pruning. Frost tolerant. Excellent cut flower. Propagate from cuttings which can be slow to

Darwinia luehmannii, × ·6

Darwinia macrostegia   W.R. Elliot

An outstanding ornamental species from the northern region of the Stirling Range, where it occurs on the moist, peaty soil slopes. Has not proved to be reliable in cultivation. Requires very well drained, light to medium soils, with dappled shade to partial sun. May adapt to an open situation, provided there is protection for the root area. Is suited to container cultivation. Responds well to pruning. Frost tolerant. Propagate from cuttings which strike readily.

For a limited period *D. oxylepis* was sold as a form of *D. meeboldii*. It differs in the longer, almost terete to trigonous leaves and the nearly all-red bracts.

Darwinia micropetala (F. Muell.) Benth.
(small petals)
Vic, SA
0.3-0.8 m x 0.3-1 m   May-Oct
Dwarf twiggy **shrub**; **branches** many; **branchlets** many, silvery to whitish; **leaves** to 0.3 cm x 0.1 cm, linear, decussate, trigonous, grey-green; **flower-heads** small, white and pale pink, comprising 2-4 flowers, erect, borne in terminal compound heads.

A hardy and adaptable species from poorly drained, sandy or clay soils. Occurs from western Vic, where it is extremely rare, and across to the Eyre Peninsula. Grows well in most soil types, including those that are alkaline. Tolerates semi-shade to full sun. Withstands extended wet periods. Responds well to pruning. Although it is not a spectacular species, it makes an excellent container plant. Propagate from cuttings which strike readily.

Darwinia nieldiana F. Muell.
(after J. E. Nield, 19th-century Vic academic)
WA   Fringed Bell
0.5-1 m x 0.5-1 m   Aug-Dec
Dwarf **shrub**; **branches** many; **leaves** to 1 cm x 0.1 cm, linear, tapering to a fine point, margins strongly recurved, crowded, especially at ends of branchlets; **flower-heads** 2-5 cm across, initially green, ageing to crimson, pendent, terminal; **flowers** small, up to 60 in each head, outer bracts light green, margins fringed.

strike. Best results are gained from vigorous but firm young growth. Very limited success in germinating seed.

A *D. macrostegia* × *D. meeboldii* hybrid which originated in Wittunga Botanic Gardens, SA, has had limited cultivation.

Darwinia masonii C. Gardner
(after Denis Mason, 20th-century botanical collector)
WA
1.5-2 m x 1-1.5 m   Aug-Nov
Small upright **shrub**; **branches** usually erect; **leaves** about 1 cm x 0.1 cm, lanceolate, almost trigonous, thick, crowded; **flower-heads** somewhat bell-shaped, to 2.5 cm x 1.5 cm, purplish red, pendent, outer bracts with serrated margins.

There is some doubt about whether this species is in cultivation, although in the 1850s it was grown from seed in England. It occurs in the Austin and Irwin Districts, where it grows on deep, sandy soils. Should do best in a warm to hot situation, in light to medium soils that are very well drained. Propagate from seed or cuttings.

Has some similarities to *D. lejostyla*, which has larger bell-shaped flowers and flattish leaves.

Darwinia meeboldii C. Gardner
(after Alfred Meebold, 19-20th-century German botanical collector)
WA   Cranbrook Bell
1.5-3 m x 1-1.5 m   Sept-Nov
Small to medium, upright **shrub**; **branches** erect; **branchlets** small; **leaves** to 1 cm x 0.2 cm, linear to oblong, erect, crowded, glabrous, margins recurved; **flower-heads** bell-shaped, to 3 cm long, green to white with red tips, or rarely red with white tips, pendent.

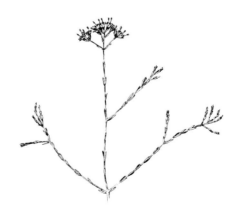

Darwinia micropetala. × ·6

189

Darwinia oederoides

Darwinia nieldiana   W.R. Elliot

Darwinia oldfieldii   B. Crafter

A most ornamental species which at present includes a wide range of forms. It occurs from Perth north to Dongara, where it is found on gravelly and sandy soils. Current revision will undoubtedly lead to changes within this complex species. One commonly grown form has reddish flower-heads with a spicy scent, and this is almost certain to be known as a new species. Another recently introduced form shows excellent potential, with its vigour and large flower-heads. Cultivation requirements are for very well drained soils, preferably light to medium types, in a warm to hot location, with dappled shade to full sun. If grown in full sun, some protection should be provided for the root system. Excellent for container cultivation. Plants can be subject to attack by grey mould during still, humid weather, so are best situated where there is adequate air movement. Propagate from seed which can take more than 3 months to germinate or from cuttings of firm young growth which strike readily.

Darwinia oederoides (Turcz.) Benth.
(similar to the genus *Oedera*)
WA
0.2-0.5 m x 0.5-1 m   Oct-Dec

Dwarf, spreading **shrub**; **branches** many, diffuse; **leaves** 0.4-0.6 cm x about 0.1 cm, linear, scattered, trigonous to semi-terete, spreading, crowded; **flower-heads** bell-shaped, 2-3 cm long, yellow or green to red, terminal, pendent, inner bracts fringed.

This is a widespread species of sandy soils, north of Perth and extending to the Albany area in the south. Not common in cultivation. Requires well drained soils, with dappled shade to partial sun. Tolerates full sun if it has a protected root system. Withstands limited coastal exposure. Does well as a container plant. Propagate from cuttings.

Has been confused with *D. nieldiana*, which has leaves with strongly recurved margins and small flowers with fringed petals and calyx lobes.

Darwinia oldfieldii Benth.
(after Augustus Oldfield, 19th-century English botanist and zoologist)
WA
0.5-1 m x 1-2 m   Aug-Nov

Dwarf twiggy **shrub**; **branches** many; **branchlets** cream-green, short; **leaves** 0.2-0.6 cm x up to 0.2 cm, oblong, alternate, grey-green, crowded, sweetly aromatic, margins ciliate, usually recurved; **flower-heads** to 2.5 cm across, initially cream-green, ageing to deep purplish red, comprising 10-12 upright flowers.

A species with ornamental foliage and the bonus of colourful flower-heads. It occurs in the Murchison River region, on sandy soils or in limestone outcrops. At least two foliage forms are in cultivation. The one with larger leaves and compact growth usually flowers better in temperate regions. The species seems to adapt to a wide range of soils, provided there is adequate drainage. Tolerates semi-shade and dappled shade, but plants generally grow better in partial or full sun. Excellent container plant. Frost and drought tolerant. Responds well to pruning. Propagate from cuttings which strike readily. Seed usually begins to germinate within 8 weeks.

D. purpurea has some affinities, but the flower-heads are smaller and the leaves do not have fringed margins.

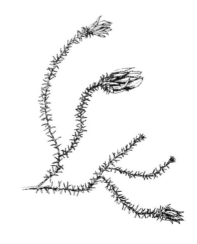

Darwinia oederoides, × ·6

190

Darwinia oxylepis

T.L. Blake

Darwinia oxylepis (Turcz.) N. Marchant & Keighery
(sharp scale)
WA
1-1.5 m x 1 m   Oct-Nov
Small upright **shrub**; **branches** erect; **branchlets** short; **leaves** about 1 cm x 0.1 cm, linear, trigonous to almost terete, scattered, glabrous, initially erect, often spreading to recurved when mature; **flower-heads** bell-shaped, 3 cm x 2-3 cm, nearly all-red with some white, terminal, pendent, on short recurved branchlets.

A most decorative species that grows north-west of the Mondurup Peak in the Stirling Range. It occurs on rough, rocky soils, in partial shade or full sun. In cultivation in the ground, it has proved difficult to maintain alive for long periods, yet it does well as a container plant. It needs fairly constant soil moisture, but must have excellent drainage. Tolerates dappled shade to full sun, when it needs root protection. Plants can become leggy, but respond well to regular tip pruning, which promotes bushy growth. Propagate from cuttings which strike readily.

Has been sold as a form of *D. meeboldii*.

Darwinia pauciflora Benth.
(few-flowered)
WA
0.3-1.5 m x 0.5-1.5 m   Aug-Oct
Dwarf to small, bushy **shrub**; **branches** short, many; **leaves** 0.2-0.5 cm x about 0.2 cm, obovate or oblong, appressed, crowded on smaller branches, margins usually serrulate; **flower-heads** narrow, to 1.5 cm long, white with pink, terminal heads composed of 3-6 stalked flowers, lateral heads 1-flowered.

Evidently uncommon in cultivation, this small species inhabits sandy, calcareous soils north of Perth, between the Moore and Murchison Rivers. Should do best in well drained, light to medium soils, with dappled shade to partial sun. May tolerate full sun. Suitable for coastal areas, provided it receives some protection from excessive salt spray. Propagate from cuttings.

Has affinities to *D. vestita* which has larger flower-heads, and *D. polycephala* which has more or less sessile flowers.

Darwinia peduncularis B. Briggs
(with peduncles)
NSW
0.6-1.5 m x 1-2 m   Sept-Nov
Dwarf to small, spreading **shrub**; **branches** many; **leaves** 0.7-1.2 cm x up to 0.1 cm, linear, trigonous, decussate, spreading to slightly reflexed; **flower-heads** to about 2 cm long, white with reddish pink, erect, terminal, usually comprising 1-2 flowers but can have 4.

Inhabits the Central Coast from the Hawkesbury River to Hornsby, where it grows in sclerophyll forest on shallow soil over sandstone. Evidently not in cultivation, but should adapt to well drained, light to medium soils, with dappled shade or partial sun. Propagate from cuttings.

Darwinia pimelioides Cayzer & F. Wakefield
(similar to the genus *Pimelea*)
WA
0.25-0.6 m x 0.5-1 m   May-June; Sept-Oct
Dwarf, spreading **shrub**; **branches** slender; **leaves** to 0.8 cm x 0.3 cm, oval to oblong, opposite, margins recurved, glabrous; **flower-heads** bell-shaped, about 2.5 cm x 1.5 cm, pendent, terminal, outer bracts purplish red, inner bracts yellowish green.

An especially ornamental darwinia that is rarely encountered in cultivation. It occurs in granitic soils of the Darling Ranges. Should do best in well drained, light to medium soils, with dappled shade to partial sun. Has potential as a container plant. Propagate from cuttings.

Has been confused with *D. speciosa* which is a sprawling shrub of about 0.15 m in height, with reddish brown bracts.

Darwinia pinifolia (Lindley) Benth.
(leaves similar to those of the genus *Pinus*)
WA
0.5-1.5 m x 0.5-1.5 m   Sept
Dwarf to small **shrub**; **branches** usually erect, slender; **leaves** to 1.5 cm x 0.1 cm, linear, trigonous to semi-terete, alternate; **flower-heads** about 1.5 cm long, terminal, many flowers.

Little is known about this species, as it is now thought to be extinct. Originally collected in the Swan River region. It has closest affinity to *D. fascicularis*, from the eastern coast. Recent collections have brought forth a similar species in the Wittenoom Hills, north-east of Esperance. Propagate from cuttings if available.

191

Darwinia porteri, × ·6

Darwinia polycephala C. Gardner
(many heads)
WA
0.3-1 m x 0.5-1.5 m   Sept; also sporadic

Dwarf, diffuse **shrub**; **branches** slender; **leaves** 0.2-0.3 cm x 0.1 cm, obovate, sessile, appressed, crowded on smaller branches, margins ciliate; **flower-heads** 1.5 cm x about 1 cm, purple to crimson, terminal, erect, many.

Occurs naturally in the Coolgardie and Roe Districts, where it grows as an undershrub. Not well known in cultivation, but should be suited to well drained soils, with dappled shade to partial sun. May tolerate full sun in temperate climates. Withstands light to medium frosts. Propagate from cuttings.

Similar to *D. pauciflora* which has white with pink flowers.

Darwinia porteri C. White
(after Mr Charles Porter, 20th-century botanical collector)
Qld
0.5-1 m x 0.5-1 m   Sept-May

Dwarf **shrub**; **branches** many, twiggy; **leaves** to 1 cm x about 0.1 cm, linear to falcate, greyish green, opposite, crowded at ends of branchlets, glabrous;

Darwinia procera, × ·6

flower-heads 2.5 cm x about 1 cm, bright purplish pink and pale salmon pink, pendent, composed of 2 flowers with very long pink styles.

A decorative species from north-eastern Qld where it is found in sandstone hill country. Has had only limited cultivation to date, but shows potential. Needs well drained, light to medium soils, with a warm aspect, but certainly tolerates dappled shade or partial sun. Withstands light frosts. Excellent container plant. Pruning after flowering produces bushy growth. Propagates readily from cuttings.

Is very similar to *Homoranthus darwinioides*; the small difference in the size of the calyx lobes separates them. The flowers of *H. darwinioides* do not usually have the vibrant pink colouration of *D. porteri*.

Darwinia procera B. Briggs
(very tall)
NSW
1-3 m x 0.6-2 m   Oct-Dec

Small to medium, erect **shrub**; **branches** usually upright; **leaves** 1-2.5 cm x about 0.1 cm, glaucous, linear to falcate, decussate, erect, glabrous; **flower-heads** to 2.5 cm long, reddish purple, terminal, erect, composed of 2-8 (usually 4) flowers; **style** more than twice the length of calyx tube.

A species with ornamental foliage. Native to the sheltered sandstone slopes of sclerophyll forest on the Central Coast between Gosford and Hornsby. Not widely cultivated but seems hardy in most well drained, light to medium soils, with dappled shade or partial sun. Tolerant of frost. Responds well to pruning. Suitable as a container plant. Propagate from cuttings which strike readily.

Darwinia purpurea (Endl.) Benth.
(purple)
WA   Rose Darwinia
0.3-0.7 m x 0.6-1 m   Aug-Dec

Dwarf, erect to spreading **shrub**; **branches** many; **leaves** 0.2-0.5 cm long, oblong, blunt apex, faintly toothed, upper surface flat, undersurface convex, glaucous, glandular, crowded; **flower-heads** 1.5-2 cm across, crimson to dark purplish red, erect to semi-erect, terminal, inner bracts crimson, slightly longer than flowers, outer bracts leaf-like.

A showy species from between Mullewa and Merredin, where it occurs on sandy soils near granite outcrops, or in heathland. Cultivation requirements are for a warm to hot location, with very well drained, light to medium soils. Suited to dappled shade to full sun. Appreciates moist soils for most of the year. It is not common in cultivation but certainly has potential, particularly as a container plant. Propagate from cuttings which strike readily.

This species has some affinities to *D. acerosa*, which differs in its pendent, greenish yellow flower-heads.

Darwinia repens A. S. George
(creeping)
WA
prostrate x 1-1.5 m   Aug-Sept

Dwarf, spreading, glabrous **shrub**; **branches** often rooting at nodes; **branchlets** short, erect; **leaves** to

Darwinia purpurea

T.L. Blake

1 cm x 0.1 cm, linear, more or less terete, opposite, crowded on branchlets; **flowers** to about 1.5 cm long, reddish purple, erect, axillary; **calyx tube** about 0.5 cm long; **style** 2-3 times as long as calyx tube.

A recently described species from the sandy soils east of Mingenew in the Irwin District. Evidently not in cultivation, but should grow in very well drained, light to medium soils, with dappled shade to partial sun. May tolerate full sun. Withstands light to medium frosts. Propagate from cuttings.

Darwinia rhadinophylla F. Muell.
(slender leaves)
WA
0.1-0.3 m x 1-2 m   Sept-Dec; also sporadic

Dwarf, spreading **shrub**; **branches** slender, many; **leaves** to 1.5 cm x 0.1 cm, linear, terete to trigonous, grey-green, spreading to erect, crowded, glabrous, soft, sweetly aromatic; **flower-heads** to 2 cm x 2 cm, bright purplish red, erect, terminal, composed of up to 20 flowers; **style** about twice as long as calyx tube.

A widely cultivated species that usually is ground hugging. Occurs over a wide range north of Perth in the Darling and Irwin Districts. It is usually found in moist areas near permanent water. Requires well drained, light to medium soils, with partial to full sun. Tolerates extended dry periods and most frosts, although it will suffer damage from heavy frosts. Sometimes prone to fungal attack during still, humid

weather. An excellent container plant. Propagates readily from cuttings and successful germination of seed is recorded.

Darwinia sanguinea (Meisner) Benth.
(blood red)
WA
0.1-0.3 m x 0.5-1.5 m   Aug-Dec

Dwarf, spreading **shrub**; **branches** slender; **leaves** to 0.7 cm x about 0.1 cm, linear-oblong to lanceolate, opposite, recurved, ciliolate margins; **flower-heads** 1-2.5 cm across, dark red-purple, terminal, erect, comprising many partial heads of 1-4 flowers each.

This decorative prostrate species inhabits heathlands between the Moore and Murchison Rivers, where it grows in sandy and lateritic soils. Not well known in cultivation at present, but certainly has potential. Requirements are for very well drained, light to medium soils, with partial to full sun. Does well in a container, with flowering stems well displayed and cascading over the rim. Tolerant of light to medium frosts. Propagate from cuttings which strike readily.

Darwinia speciosa (Meisner) Benth.
(showy or handsome)
WA
0.1-0.3 m x 0.3-1 m   July-Oct

Dwarf, spreading **shrub**; **branches** often prostrate, but can be erect; **leaves** to 0.7 cm x 0.4 cm, oblong, op-

193

posite, concave, glaucous, entire; **flower-heads** bell-shaped, about 3 cm long, deep reddish brown, terminal, more or less pendent.

A very showy species from the Irwin District, where it occurs in the sandheath between the Hill River and Mingenew. Has had limited cultivation, with varying success. Needs very well drained, light to medium soils, with dappled shade to partial sun. It should tolerate full sun if the root area has some protection. Does well in containers, which must receive regular water. Susceptible to root fungal diseases. Tolerant of frost. Propagate from cuttings.

Similar to *D. pimelioides* which grows much taller and has purplish red and yellow-green flower-heads.

Darwinia squarrosa (Turcz.) Domin
(rough; scurfy)
WA   Pink Mountain Bell
0.5-1 m x 0.5-1 m   Sept-Dec

Dwarf to small **shrub**; **branches** more or less erect; **leaves** to 0.8 cm x 0.4 cm, oblong to elliptical, keeled, crowded, glabrous except for fringed margins, sweetly aromatic; **flower-heads** bell-shaped, 2 cm x about 1 cm, pink, terminal, pendent, bracts fringed.

From the eastern peaks of the Stirling Range, this outstanding ornamental species inhabits the moist, peaty sands of the rocky upper slopes. In cultivation, results have been mixed. It requires very well drained soils, with dappled shade to partial sun. May tolerate full sun, provided the root region has some protection. Withstands medium frosts and may even be hardy to heavy frost. Dislikes alkaline conditions. An excellent container plant. Responds favourably to light pruning. Originally introduced to cultivation in England during 1864. Propagate from cuttings.

Natural hybrids of this species with *D. lejostyla* are recorded from Mount Success.

Darwinia taxifolia Cunn.
(leaves similar to those of the genus *Taxus*)
Qld, NSW
0.2-1.5 m x 1-2 m   Oct-Dec; also sporadic

Dwarf to small **shrub**; **branches** spreading to erect; **branchlets** short; **leaves** to 1.2 cm x about 0.1 cm, linear to falcate, trigonous, keeled, decussate, grey-green, crowded; **flower-heads** to 1.5 cm long, initially white, ageing to red, erect, axillary, in opposite pairs, composed of 2-4 flowers.

A variable and adaptable darwinia from the Central Tablelands, where it grows above 950 m alt., on shallow soils overlying sandstone. Does well in relatively well drained, light to heavy soils. Tolerates semi-shade to partial sun. Frost hardy. Plants can become open but they respond well to pruning.

The ssp. *macrolaena* B. Briggs has larger flower-heads, with very prominent pink to red bracts. It has a prostrate growth habit, and is very popular as a groundcover. Adapts to a wide range of conditions. Decorative as a container plant, with its cascading branches. This ssp. has been confused with *D. grandiflora* which has smaller flower-heads and is often self layering, unlike ssp. *macrolaena*.

Propagate both varieties from cuttings which strike readily.

Darwinia taxifolia ssp. *macrolaena*   W.R. Elliot

Darwinia thomasii (F. Muell.) Benth.
(after Dr. David J. Thomas, 19th-century Vic medical practitioner)
Qld
1.5-2 m x 1-1.5 m   April-Aug; also sporadic

Small slender **shrub**; **branches** usually erect; **leaves** 0.6-1.5 cm x 0.2-0.5 cm, obovate-falcate, opposite, grey-green, glabrous, terminating in short, recurved point or acute angle; **flower-heads** about 1.5 cm long, pink, pendent, in upper axils.

Not well known, this decorative species occurs in sandstone escarpments of central inland Qld. Has proved hard to maintain in cultivation. Best results have been obtained in containers. Needs very well drained soils, with dappled shade to partial sun, in a warm location.

Propagate from cuttings which are reported as often slow to produce roots. Probably best to use firm young growth.

Darwinia thymoides (Lindley) Benth.
(similar to the genus *Thymus*)
WA
0.3-0.4 m x 0.5-1 m   July-Nov

Dwarf, spreading **shrub**; **branches** spreading, can be cascading; **leaves** to 1.2 cm x 0.3 cm, linear-lanceolate,

Darwinia thomasii, × ·6

Darwinia vestita   T.L. Blake

Darwinia vestita, × ·6

mostly opposite, usually spreading, margins revolute, glabrous; **flower-heads** to 2 cm long, white, terminal, erect, composed of 4-8 flowers.

A little-known species from moist, swampy regions of the south-west corner. Best suited to moist, well drained, acidic soils, in dappled shade to partial sun. May prove to be a useful species as it can develop a dense ground hugging cover of foliage. Potential for container cultivation. Propagate from cuttings. Seed has been germinated successfully.

Darwinia verticordina Benth.
(similar to the genus *Verticordia*)
WA
0.5-2 m x 1-2 m   Sept-Dec

Dwarf to small, dense **shrub; branches** many; **leaves** to 0.8 cm x about 0.1 cm, linear, semi-terete or trigonous, opposite, glandular; **flower-heads** flat-topped leafy corymb, white to yellowish, profuse, inner ring of white hairs at base of calyx; **style** very prominent.

An inhabitant of open heathlands such as along the granite coastal region east of Esperance. Evidently not in cultivation. This species produces a massed display of flat-topped flower-heads and merits trial. Should do well in well drained, light to medium soils, with dappled shade to partial sun. May tolerate full sun in temperate areas. Propagate from cuttings.

Darwinia vestita (Endl.) Benth.
(clothed)
WA   Pom-pom Darwinia
0.3-1 m x 0.2-0.6 m   July-Oct

Dwarf **shrub; branches** usually erect; **leaves** to 0.4 cm x 0.2 cm, heath-like, oblong to elliptic, scattered, crowded, appressed, green, aromatic; **flower-heads** globular, to 3 cm across, white ageing pink to red, terminal, composed of many flowers.

A widespread species of the coast and slightly inland areas of the south-west corner. It has adapted well to cultivation, although it is not widely grown. Suited to relatively well drained, light to heavy soils, with dappled shade to full sun. Withstands extended periods of

dryness. Responds well to pruning. An excellent container plant, especially the more compact dwarf form from north of Denmark. Propagate from cuttings which strike readily.

Has some affinities to *D. pauciflora* which has much smaller flower-heads.

Darwinia virescens Benth.
(light green)
WA   Murchison Darwinia
0.1-0.3 m x 0.5-1 m   Sept-Jan

Dwarf **shrub; branches** spreading; **branchlets** erect, except for flowering ones, which are prostrate; **leaves** 0.5-1 cm x about 0.1 cm, linear, terete to trigonous, light green, alternate, crowded; **flower-heads** 2.5-4 cm across, green to reddish, upturned, terminal, comprising many flowers.

Nearly always occurs on deep, well drained sand near the lower Murchison River. It is a striking species that has potential for cultivation. Not well known to date. Must have excellent drainage and plenty of sunshine. Suited to container cultivation. Propagate from cuttings which strike readily. There has been limited success from seed.

Has affinities to *D. sanguinea* which has smaller red flower-heads.

Darwinia virescens   B. Crafter

195

Darwinia wittwerorum

Darwinia species (Hopetoun), × ·6

Darwinia wittwerorum N. Marchant & Keighery
(after Magda Wittwer, and Superintendent of Kings
Park and Botanic Gardens, WA, Ernest Wittwer)
WA
0.3-0.8 m x 0.2-0.5 m   Sept-Dec
    Dwarf **shrub**; **branches** more or less erect; **leaves**
0.5-1 cm x less than 0.1 cm, linear, trigonous, scattered; **flower-heads** ovoid, about 2 cm long, pink upper
part and cream lower part, terminal, solitary, inner
bracts not inflexed.
    A recently described species from the central Stirling
Range area. At present it is not well known in cultivation. Its requirements are well drained, light to
medium soils, with dappled shade to partial sun.
Should make an attractive container plant. Propagate
from cuttings.
    Closely allied to *D. lejostyla*, which has flat leaves and
pink to red spreading bracts that form a bell-shaped
flower-head.

Darwinia species (Hopetoun)
WA
0.2-0.3 m x 0.2-0.4 m   March-May
    Dwarf pyramidal **shrub**; **branches** many; **leaves** to
0.5 cm x 0.2 cm, oblong, decussate, keeled, crowded,
slightly glandular, margins denticulate; **flower-heads**

to 1.3 cm x 1 cm, pale cream-green, terminal, pendent,
profuse; **style** about twice as long as calyx tube.
    This apparently unnamed species occurs in the
sandy flats east of Ravensthorpe to Hopetoun, and in
white quartzite sand at the foot of East Mount Barren.
Apparently not in cultivation, but its natural habitat
suggests it should grow in well drained, light to
medium soils, with dappled shade to partial sun. It
may tolerate full sun. Propagate from cuttings.

Darwinia species (Whicker Range)
WA
1-1.5 m x 1-1.5 m   March-April
    Dwarf to small **shrub**; **branches** many, compact;
leaves to 2 cm x about 0.1 cm, more or less terete, often
falcate, decussate, ending in a fine soft point; **flowers** to
1.5 cm x 0.5 cm, pink, erect, solitary, in upper leaf
axils, bracts and bracteoles absent; **calyx tube** lightly
ridged.
    An apparently unnamed species that has single,
bractless flowers. It is from the Whicker Range east of
Busselton, and occurs in well drained but moist, sandy
to gravelly soils. Not thought to be in cultivation at
present. Is probably best suited to well drained, light to
medium soils, with dappled shade to partial sun.
Propagate from cuttings.

Darwinia species
WA
0.5-1 m x 0.5-1 m   Sept-Dec
    Dwarf **shrub**; **branches** many; **branchlets** short;
leaves 0.2-0.8 cm x about 0.1 cm, linear, sometimes
falcate, 3-angled, pale grey-green, spreading to erect,
crowded, glabrous, pointed apex; **flower-heads** about
1 cm long, greenish, with some reddish bracts, terminal, styles exserted.
    An apparently undescribed species from the south-
western corner of WA. It has recently been introduced
into cultivation. It is not very showy in flower, but
develops into a dense, low mass of light greyish green
foliage. Should prove to be a useful garden plant.
Tolerates an open, sunny to semi-shaded situation, and
requires fairly good drainage. Hardy to light frosts. Has
potential as a container plant. Propagate from cuttings
which strike readily.

Darwinia species (Whicker Range), × ·6

Darwinia species, × ·6   flowers, × 3·5

Dasypogon bromeliifolius flower-head

T.L. Blake

Darwinia hybrid
WA
0.3-0.6 m x 0.5-1 m   Aug-Oct

Dwarf, semi-prostrate **shrub**; **branches** many; **branchlets** ribbed below leaves; **leaves** 0.6-1 cm x about 0.1 cm, linear, 3-angled, greyish green, crowded, glabrous, flat above, keeled below, broadest near base, apex acute; **flower-heads** to 3.5 cm x 2.5 cm, initially lime green, gaining dull reddish tonings, terminal, semi-pendent; **bracts** tapering to narrow apex which is often reflexed, margins fringed.

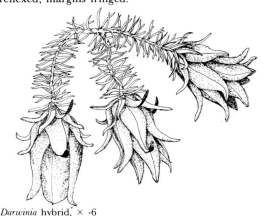

Darwinia hybrid, × ·6

This decorative darwinia is a recent introduction to cultivation. It has adapted well to freely drained soils, and will grow in a situation that receives plenty of sunshine, or in dappled shade. Tolerates light to medium frosts. Is an excellent container plant. The origin of this darwinia is doubtful. There is evidence to suggest it is a hybrid, with *D. carnea* as one of the parents. Further research is required to ascertain the actual parents. Propagate from cuttings.

DASYPOGON R. Br.
(from the Greek *dasys*, thick with hair or rough with hair; *pogon*, a beard; refers to the perianth which has bristles)
Xanthorrhoeaceae

Perennial **herbs**; **trunk** simple or slightly branched; **leaves** grass-like, crowded at apex of trunk and/or bases of flowering stems; **flowers** small, in globular heads on long terminal peduncles; **perianth** persistent, with bristles; **stamens** 6; **fruit** a 1-seeded globular capsule.

An endemic genus of 2 species restricted to southwestern WA, where plants are long-lived in their natural habitat. The genus is known mainly because of the decorative flower-heads that are eagerly sought for floral decoration.

Propagation is from seed which does not need any

197

pre-sowing treatment. The seed is shed when ripe. Plants need to be kept under observation so that seed can be collected before dispersal.

Dasypogon bromeliifolius R. Br.
(leaves similar to those of the genus *Bromelia*)
WA   Pineapple-leaved Dasypogon
0.5-1 m x 0.5-1 m   Aug-Dec
Perennial **herb**; **trunk** short and slender; **leaves** to 30 cm x 0.5 cm, tapering to a fine point, rigid, margins slightly toothed, sheath-like at base; **flower-heads** globular, about 2.5 cm diameter, cream, terminal on hairy peduncles to 40 cm long; **perianth** has deciduous, long bristles.
This decorative species has had only very limited cultivation. It occurs in the sandy soils of the coastal plain of the south-west corner and in the jarrah forests. Requires very well drained, light to medium soils, with partial or full sun. Will possibly tolerate dappled shade. Has potential as a container plant. Available as a cut flower. Propagate from seed.

Dasypogon hookeri Drumm. ex F. Muell.
(after Sir Joseph Hooker, 19th-century British botanist)
WA   Pineapple Bush
1-4 m x 0.6-1.5 m   Oct-Jan
Small to medium, perennial **herb**; **trunk** tall, to 8 cm diameter; **leaves** to 1 m long, many at tip of trunk, with overlapping, sheathing bases, margins rough, tapering to a fine point; **flower-heads** globular, about 3 cm diameter, cream, terminal on long, rough, hairy peduncles; **perianth** has persistent, short bristles.
Evidently not in cultivation, this species is restricted to the most western region of south-western WA, where it occurs in sandy soils of the jarrah forests. Cultivation notes are as for *D. bromeliifolius*.

DATURA L.
(from Dhatura; a native name for *D. fastuosa*)
Solanaceae
Annual or perennial **shrubs**; **leaves** alternate, simple or lobed, petiolate, the margins entire, toothed or prickly; **flowers** tubular or funnel-shaped, solitary in branch junctions; **fruit** a spiny capsule.
A genus of about 10 species, principally American in origin. One species is found in Australia and another 5 species are widely distributed as naturalized weeds. They grow readily from seed.

Datura leichhardtii F. Muell. ex Benth.
(after L. Leichhardt, early explorer)
Qld, NSW, SA, WA, NT   Native Thorn Apple
0.5-1 m x 0.3-0.5 m   Feb-April; also sporadic
Small, fleshy annual or biennial **shrub**; **branches** succulent; **leaves** 5-8 cm x 2-4.5 cm, ovate, the margins lobed or toothed, sinuate, dark green; **flowers** 5-7 cm long, trumpet-shaped, white, sometimes with reddish tinges; **capsule** 2-2.5 cm across, globular, deflexed, the surface covered with numerous spines.
A weedy species widely distributed in inland regions, usually along stream banks and channels. It may have been introduced very early. The species is identical

with the central American *D. pruinosa*. The flowers are attractive, but the plant has limited horticultural appeal. Leaves generally have a strong, obnoxious smell. It can be propagated from seed and probably from cuttings.

DAVALLIA J. Smith
(after Edmund Davall, 18th-century Swiss botanist)
Davalliaceae
Epiphytic or terrestrial **ferns**; **rootstock** fleshy to woody, creeping, the young parts covered with scales; **fronds** borne in 2 rows, jointed to the rhizome, deeply dissected, the fertile fronds often more finely dissected; **sori** cup-shaped, on small marginal lobes, bearing an indusium.
This genus consists of about 40 species widely distributed through Asia, with 3 found in eastern Australia. All are popular subjects for cultivation in glasshouses or ferneries. They make attractive basket plants and can be readily grown into large specimens. Propagation is by division of the rhizomes or from spore.

Davallia denticulata (Burm.f.) Mett.
(small teeth)
Qld
0.5-1 m tall
Epiphytic or terrestrial **fern** forming spreading patches; **rootstock** creeping, about 1 cm thick, woody, the young parts covered with light brown scales; **fronds** 0.5-1 m tall, 3-4 times divided, triangular, dark shiny green, fine and lacy, the margins finely toothed; **sori** cup-shaped, marginal.
This fern, which is widely distributed through Asia, extends to north-eastern Australia, where it forms colonies in sandy soil or amongst rocks, rarely on trees. It is a very hardy species, the plants dying back to perennial rhizomes during the dry season. This fern is very easily grown as a pot or basket specimen or as a

Davallia pyxidata, × ·5

198

Davidsonia species, Currumbin Valley, Qld

D.L. Jones

garden plant in the tropics, where it requires a some-what sheltered situation. In a container it needs an acid, coarse, well drained mixture. In temperate regions the plants require the winter protection of a heated glasshouse. Propagation is by division or from spore.

Davallia pyxidata Cav.
(provided with a lid; a reference to the cup-shaped indusium)

Qld, NSW, Vic   Hare's Foot Fern

0.3-1 m tall

Epiphytic **fern** forming spreading clumps; **rootstock** creeping or spreading through the air, about 1 cm thick, woody, the young parts covered with brown, spreading, papery scales; **fronds** 0.3-1 m long, deltoid, sheathing, glossy dark green, 3-4 times divided, the fertile fronds with finer segments; **sori** marginal, as long as wide.

A common fern that grows on rocky outcrops or on trees, often amongst the roots of other large epiphytes. It is found in a variety of situations from moist, shaded rainforest to open forest and sunny rock crevices. On large clumps, the rhizomes spread through the air for up to 1 m, and give them a distinctive silhouette. This fern is very popular as a pot or basket subject, and grows readily in ferneries and sheltered verandahs etc. It requires a coarse, open epiphyte mixture. Propagation is by division of the clumps or from spore which is slow to raise.

Davallia solida (Forst.f.) Sw.
(solid; not hollow; free from cavities)

Qld

0.3-0.9 m tall

Epiphytic **fern** forming spreading clumps; **rootstock** creeping or spreading through the air, 1-1.4 cm thick, woody, the young parts covered with fine, grey, appressed scales; **fronds** 30-90 cm tall, triangular, purplish when young, dark green when mature, 3-4 times divided, the fertile fronds with much finer segments; **sori** marginal, much longer than wide.

A widespread species that in Australia is restricted

to Cape York Peninsula, where it grows as an epiphyte in moist, sheltered rainforests. The plants usually grow in the roots of other epiphytes such as elkhorns. This fern is a choice collector's item, and grows readily in tropical or subtropical regions. In temperate areas it requires a heated glasshouse with a minimum temperature of 10°C. A coarse, open epiphyte mixture is best, and the plants make an excellent basket subject. Propagation is by division of the clumps or from spore.

Davalliaceae Reichb.

A widespread family of ferns mainly developed in tropical and subtropical regions, with a few cold hardy species in the temperate zones. They have creeping rhizomes and fine, lacy fronds. They grow mostly as epiphytes or lithophytes, in spreading colonies, a few terrestrial species growing in organically rich soil. The family includes some very attractive ferns, and many exotic species are widely grown as basket or pot plants in Australia. The Australian representative comprises 3 genera and 6 species dispersed in tropical and temperate regions of the east coast. Most native species are popular in cultivation. Australian genera: *Davallia*, *Humata* and *Rumohra*.

DAVIDSONIA F. Muell.
(after J. E. Davidson, a pioneer sugar-grower at Rockingham Bay where the species was first discovered)

Davidsoniaceae

Trees; **stipules** present, prominent; **leaves** pinnate, alternate, large, spreading, winged throughout, glabrous or hairy, the margins irregularly toothed; **inflorescence** a pendulous raceme; **flowers** small; **calyx** 4-5 lobed; **petals** absent; **stamens** 8-10; **fruit** a drupe-like berry; **seeds** 2, flattened, fibrous.

A small endemic genus of 2 species, one of which is undescribed. They are very distinctive small trees with many desirable features for cultivation. Propagation is from seed which must be sown fresh.

Davidsonia pruriens F. Muell.
(itching; referring to the prickly hairs)

Qld, NSW   Ooray; Davidson's Plum

6-10 m x 1-3.5 m   Aug-Oct

Tall **shrub** to small slender **tree**, often unbranched; **bark** brown, corky, scaly; young shoots densely hairy; **stipules** large, prominent, toothed; **leaves** 30-80 cm long, pinnate, alternate, stiffly spreading, the rhachis winged and toothed throughout; **leaflets** 5-15, 7-25 cm x 3-9 cm, ovate-lanceolate, opposite, densely hairy, dark green, leathery, the margins sharply and irregularly toothed; **panicles** 30-70 cm long, pendulous, narrow, borne in the upper axils; **flowers** about 0.6 cm across, brown or reddish brown; **berry** 3-5 cm long, ovoid, purple to blue-black, hairy, resembling a plum, the pulp crimson and juicy but very acid; **seeds** 2, flat with lacerated margins, mature Feb-March.

A species with a disjunct distribution, found in north-eastern Qld and in north-eastern NSW, south to the Brunswick River. It is a very distinctive tree, usually of slender habit and with large, conspicuously

199

pinnate and prickly leaves. It is best known, however, for its large purple fruit which resemble plums and were highly prized by the early settlers for their edible qualities. This fruit has a soft, juicy purple flesh which is very acid, but when it is stewed with sugar it makes an enjoyable jam or dessert. This species is becoming popular as a garden plant and as a tub specimen for indoor or verandah decoration. It can be grown in tropical and subtropical regions, and will adapt also to temperate regions; there is a large plant which flowers annually in Melbourne's Royal Botanic Gardens. The species is very slow-growing when young, and demands protection from hot sun and strong winds. It is best planted amongst shrubs and other trees where its slender habit does not intrude on others. It needs acid soils of unimpeded drainage, and responds to regular watering, mulches and regular light dressings of a complete fertilizer. Solitary plants are known to bear fruit; however, to ensure success in the garden it is wise to plant 2 or 3 together. The timber of this plant is dark brown, tough and durable, and valuable for tool handles. The hairs on stems and leaves can cause an irritation if contacted. Propagation is from seed which must be sown soon after its extraction from the fruit.

The var. *jerseyana* F. Muell., restricted to NSW, is regarded as being the southern race of the species. It differs by being smaller in all its parts, is less hairy and bears glabrous fruit. Studies may show that it is distinct from the northern race, which is larger-growing and much hairier in its parts, including the fruit.

Davidsonia species
Qld, NSW
6-8 m x 2-3 m   Nov-Dec

Tall **shrub** or small bushy **tree** with a suckering habit; **bark** grey-brown, somewhat corky; young shoots glabrous, pink; **stipules** large, prominent, toothed; **leaves** 20-30 cm long, pinnate, alternate, spreading, glabrous, the rhachis winged and toothed throughout; **leaflets** 5-9, 7-20 cm x 3-6 cm, ovate-lanceolate, sometimes falcate, opposite, leathery, glabrous, dark green and glossy, the margins sharply and irregularly toothed; **panicles** 8-15 cm long, pendulous, in the upper axils; **flowers** about 0.6 cm across, reddish pink; **berry** 3-4 cm x 4-5.5 cm, oblate, purplish black, covered by short, loose brown hairs, the pulp crimson, juicy and acid; **seeds** 2, flat, with lacerated margins, mature Feb-June.

An apparently undescribed species confined to rainforests and wet sclerophyll forests between Mullumbimby in north-eastern NSW and the Currumbin Valley in Qld. It can be readily distinguished from *D. pruriens* by its bushy, spreading habit. Its fruit is edible (but sour) and can be used for jams, pies and jellies after stewing with sugar. This species is relatively unknown in cultivation because of its localization, and to date has only been grown by a few enthusiasts. Its requirements are similar to those of *D. pruriens*. Plants will grow well in a sunny position in well drained, organically rich soil. They are naturally very bushy, and flushes of new growth may be quite decorative and colourful. Young specimens could make an attractive pot subject. Propagation is from suckers which trans-

Daviesia alternifolia   T.L. Blake

plant readily. Root cuttings would be worth a try. Apparently, a large percentage of the seeds produced are infertile.

Davidsoniaceae Bange

A small family of dicotyledons consisting of the solitary genus *Davidsonia*. This comprises 2 species, one as yet undescribed. Both are endemic to eastern Australia. They are slender trees with alternate, pinnate leaves, prominent stipules and bisexual, apetalous flowers. The fruit is large and plum-like, with deep purplish red flesh and juice, and 2 flattened seeds. They are uncommon in cultivation. Some authorities include this genus in the family Cunoniaceae.

DAVIESIA Smith
(after Hugh Davies, 18-19th-century Welsh botanist)
Fabaceae

Dwarf to tall **shrubs**; **branchlets** can end in thorns; **leaves** alternate, simple, entire, rigid, reduced to thorns or completely absent; **stipules** usually absent but can be minute; **flowers** pea-shaped, small, solitary or a few in short axillary clusters, most commonly yellow and brown, but can be in various shades of red to orange, pink and purple; **calyx lobes** 5, short, upper 2 often connate; **standard** broad, notched, about as long as the wings and curved keel; **stamens** free; **fruit** a pod, more or less flattened, triangular, containing 2 seeds.

An endemic genus of approximately 75 species, the majority occurring in the South-Western Botanical Province of WA. They are also widely distributed in other parts of WA, in SA, Vic and Tas, but in NSW and Qld they are usually restricted to the eastern region. There are 1-2 species in northern Australia.

The triangular pods readily distinguish *Daviesia* from any other genus of pea-flowered plants.

Species of *Daviesia* generally reach their optimum development in sandy heathland situations, in company with other plants. They can also be found in open places, or as undershrubs in forests. Flowering can be profuse, with some plants literally covered with colourful small flowers, and this can compensate for their sometimes short flowering period. A most decorative aspect of some species is their pods, which change colour as they mature.

The foliage of species such as *D. epiphylla*, *D. incrassata*, *D. pectinata* and *D. quadrilatera* is ornamental and intriguing. The pungent, prickly, dried stems of *D. pectinata* were sometimes used by early settlers as a substitute for fencing wire. Many of the species with prickly foliage have potential for use as plants to direct human traffic.

Very few *Daviesia* species have been cultivated to any extent, probably due to lack of propagating material and interest.

General requirements are for well drained, light to medium soils, with plenty of sunshine.

Some plants, such as *D. brevifolia* which is often difficult to establish in gardens, respond well to hard pruning after flowering, resulting in the production of vigorous new shoots.

Pests and diseases do not seem to present any great problems with this genus. Root-rotting fungi can attack some species, but if drainage is unimpeded this problem should be reduced.

Propagation

The easiest method is to use seeds which ripen quickly during hot weather and are expelled from the triangular pods at maturity. The seeds have a hard outer casing, and therefore need treatment prior to sowing. See Volume 1, pages 196-8, for information on seed collection, and page 205, Treatment and Techniques to Germinate Difficult Species. Some species have been grown from cuttings, but it is often difficult to obtain suitable material. Vigorous young growth which has just started to become firm seems the most suitable.

Daviesia acicularis Smith
(needle-like)
Qld, NSW
0.5-1 m x 1-2 m   Sept-Nov

Dwarf **shrub**; **branches** many, terete or angular, often arching; **leaves** 1.5-3.5 cm x about 0.2 cm, linear, rigid, glabrous or hairy, margins revolute, apex

Daviesia arborea   D.L. Jones

pungent; **flowers** pea-shaped, small, brown and yellow, solitary or in clusters along ends of branchlets; **pods** 1 cm long, triangular.

A very adaptable species. It is most suitable for growing in shady, dry situations, but also does well in a sunny position. It needs well drained soils, as it is sensitive to root diseases. In nature plants inhabit stony soils, often with a canopy of tall trees, or cleared areas. Tolerates most frosts. Propagate from seed or cuttings.

Daviesia alata Smith
(winged)
NSW
0.5-1.5 m x 0.5-1.5 m   Sept-Nov

Dwarf to small **shrub**; **branches** flat and winged or trigonous, often arching from a woody base; **leaves** reduced to minute scales; **flowers** pea-shaped, small, red and orange, in small clusters along the branches; **pods** to 1 cm long.

An inhabitant of the Blue Mountains and the banks of the Hawkesbury River, where it grows on sandstone, indicating a preference for well drained situations. It is tolerant of dappled shade to full sun, and should withstand most frosts. This leafless species can have long, weak branches, but light pruning while the plants are young will promote bushy growth. Pruning of mature plants after flowering can also promote growth if this is desired. Propagate from seed and possibly cuttings.

Daviesia alternifolia Endl.
(alternate leaves)
WA
0.3-0.8 m x 0.5-1.5 m   Oct-Dec

Dwarf, spreading **shrub**; **branches** decumbent to ascending, slightly angular, often pubescent when young; **leaves** 3-5 cm long, narrow-linear to linear-oblong, or oblong-cuneate, alternate or sometimes opposite, in 3s, usually ending in a short rigid point, thick, leathery; **flowers** pea-shaped, about 0.8 cm across, borne in 3s on long peduncles; **standard** and **wings** flame coloured; **keel** purple; **calyx** pubescent; **bracts** enlarge after flowering; **pods** triangular, about 1.2 cm long.

This species occurs in sandy soils on and near the south coast in the Albany region. It has had only limited cultivation to date, but shows potential with its well displayed flowers. Best suited to well drained, light to medium soils, with partial or full sun. Propagate from seed.

Daviesia aphylla F. Muell. ex Benth. =
D. benthamii Meisner

Daviesia arborea F. Muell. & Scortech.
(tree-like)
Qld, NSW   Golden Pea
8-12 m x 3-5 m   Sept-Jan

Small **tree**; **trunk** erect; **bark** dark, deeply furrowed; **branches** strongly weeping, willowy; **branchlets** slender; **leaves** to 10 cm x 1 cm, linear-lanceolate, willow-like, prominent midrib, pendulous, glabrous, bright green; **flowers** pea-shaped, small, light golden yellow, borne in axillary racemes about 2-6 cm long, forming sprays up to 30 cm long, fragrant; **standard** reflexed; **pods** to 1 cm long, triangular, inflated.

This is probably the tallest *Daviesia* species, and is native to south-eastern Qld and north-eastern NSW. It is indeed a decorative species, with its pendulous growth and masses of racemes carrying bright yellow flowers. Superficially it resembles a wattle, and is often mistaken for one. An inhabitant of open forests and the edges of rainforests, it usually grows in poor, stony soil. Not well known in cultivation, but should grow in well drained, light to medium soils that are moist for extended periods. Suited to dappled shade or full sun. Will tolerate light to medium frosts. Excellent for planting on clay or stony banks. A very attractive species with pleasantly perfumed flowers. The fragrance is especially noticeable on warm mornings. Deserving of wider cultivation, but limited by difficulty of propagation. Cuttings are difficult to strike and most seed is destroyed by caterpillars before it ever ripens.

Daviesia arenaria Crisp
(growing in sandy places)
NSW, Vic, SA
0.5-2 m x 1-2.5 m   Aug-Oct

Dwarf to small **shrub**; **branchlets** many, spreading, rigid, spiny; **leaves** to 1 cm x 0.8 cm, usually broad-ovate and cordate, sometimes elliptic and narrowed at base, rarely obovate, sessile, ascending, concave, hairy or glabrous, apex rigid, prickly; **flowers** pea-shaped, small, about 0.5 cm across, 1-2 per upper axil; **standard** orange-pink with deep maroon back; **wings** and **keel** maroon; **pods** about 0.7 cm x 0.4 cm.

A recently named species from mallee regions. Requires very well drained, light to medium soils. Prefers a warm to hot situation, although not necessarily in full sun. Responds well to fairly hard pruning when established. Its prickly nature is useful for directing foot traffic, and plants provide a refuge for small native birds. Frost and drought tolerant. Propagate from seed or cuttings.

This daviesia has been confused with *D. ulicifolia* Andrews, which is usually a glabrous shrub with convex, but never cordate leaves, and often with up to 4 flowers in an umbel.

Daviesia arthropoda F. Muell.
(jointed peduncles)
WA, NT
1-2 m x 1-2 m   July-Oct; also sporadic

Small, prickly **shrub**; **branches** and **branchlets** angular; **leaves** to 3.5 cm x 0.5 cm, narrow-elliptic to linear, broadest near the pungent apex, usually crowded; **flowers** pea-shaped, about 0.8 cm across, yellow and brown, 1-3 flowers per axil near ends of branches; **pods** triangular, to 1.5 cm long.

A little-known species that inhabits sand-dunes in the Central Australian and Giles Districts. Evidently only grown by enthusiasts. Will need extremely well drained soils, with a warm to hot position. Propagate from seed.

This species has strong affinities to *D. ulicifolia*, which has smaller leaves.

Daviesia asperula Crisp
(slightly rough)
SA
1-2 m x 1-2 m   Sept-Oct

Small, compact or spreading **shrub**; **branchlets** slightly ribbed; **phyllodes** 0.5-2.5 cm x up to 0.5 cm, subulate to narrow-obovate, crowded, arranged spirally, spreading, sometimes reflexed, slightly rough, ending in a pungent point; **flowers** pea-shaped, about 0.8 cm across, 2-3 in axillary racemes; **standard** orange or yellow with deep red centre; **wings** and **keel** red; **pods** triangular, 1-1.5 cm x up to 1 cm.

Endemic to SA, where it grows on poor, sandy or lateritic soils in mallee or open forest communities on Eyre Peninsula, Kangaroo Island and Fleurieu Peninsula. It is a variable species and is separated into 2 subspecies. The typical form has narrow phyllodes which are broadest near the base, while the ssp. *obliqua* Crisp has falcate to obovate phyllodes which are broadest at or above their centres. Little-known in cultivation, but plants should grow in most soils that have adequate drainage. A position with partial or full sun will be suitable and dappled shade should be tolerated. Plants will withstand most frosts and extended dry periods. Propagate from seed or possibly cuttings.

Daviesia benthamii Meisner
(after George Bentham, 19th-century English botanist)
NSW, Vic, SA, WA
0.5-2 m x 1-1.5 m   Aug-Oct

Small **shrub**; **branches** usually leafless, terete, rigid, pungent; **leaves** 0.5-3 cm long, subulate, spreading, pungent; **flowers** pea-shaped, small, red to white, solitary or in racemes to 2.5 cm long; **pods** about 0.8 cm long.

Uncommon in cultivation, this broom-like shrub needs very well drained, light to medium soils, with partial to full sun. It is probably best suited to semi-arid regions. Frost and drought tolerant. Recorded as being slow-growing. Propagate from seed.

The ssp. *humilis* Crisp, from NSW, Vic and SA, grows to about 0.6 m tall and has leaves to 3 cm long, while the typical form is a taller shrub, usually with shorter leaves.

D. genistifolia is similar to this species, but has angular branches.

D. aphylla F. Muell. ex Benth. and *D. nudula* J. Black are considered conspecific with *D. benthamii*.

Daviesia brevifolia, × ·5

Daviesia brevifolia

T.L. Blake

Daviesia brevifolia Lindley
(short leaves)
NSW, Vic, SA   Leafless Bitter-pea
0.6-1.5 m x 0.5-1 m   Aug-Nov

Dwarf to small **shrub**; **rootstock** woody; **branches** broom-like, often flexuose, can be glaucous, rigid; **leaves** about 0.5 cm long, thorn-like, continuous with branchlets, few, scattered; **flowers** pea-shaped, small, apricot to maroon-red, borne in small clusters along branchlets, profuse; **pods** to 1.2 cm long, inflated at base, often attain reddish hues.

The combination of the delicately coloured flowers, the greyish stems and the colourful immature seed pods makes this a most decorative species. It has, however, proved difficult to maintain in cultivation. Its natural habitats are well drained gravels or sandy soils, and it often grows as an undershrub to tall trees. Should do best in very well drained soils, with dappled shade to full sun. In temperate regions it should be given a warm location. Propagate from seed or cuttings.

Daviesia buxifolia Benth.
(leaves similar to those of the genus *Buxus*)
Qld, NSW, Vic   Box-leaf Bitter-pea
1.5-2.5 m x 1-2 m   Sept-Dec

Small to medium **shrub**; new growth often bronze; **branches** erect, ribbed, often reddish, glabrous; **branchlets** can be pendulous; **leaves** 1.5-3.5 cm x 0.8-1.3 cm, ovate to oblong, sessile, concave, leathery, glabrous, shiny, blunt apex; **flowers** pea-shaped, small, yellow and red, in clusters or racemes to 3 cm long; **pods** about 1 cm long, triangular.

An ornamental shrub, both in foliage and flower. It is not well known in cultivation, but is best suited to well drained soils. Seems to adapt to light or heavy types. Prefers dappled shade or partial sun. Responds well to pruning. Tolerant of at least light to medium frosts. Propagate from seed or cuttings.

Daviesia cardiophylla F. Muell.
(heart-shaped leaves)
WA
0.6-1 m x 1-1.5 m   June-Aug

Dwarf, spreading **shrub**; **branches** many, glabrous, more or less terete or only slightly angular; **leaves** about 1 cm long, broadly cordate, sessile, ending in pungent point, prominent midrib, thick; **flowers** pea-shaped, about 0.8 cm long, red-purple, axillary, 1-3 on slender pedicels.

This prickly-foliaged species occurs over a wide range of the south-west corner, where it is usually found on sandy loam. Has had limited cultivation, but should adapt to most well drained, light to medium soils. It will probably tolerate clay-loams, provided drainage is excellent. Prefers a sunny situation. Tolerant of most frosts and extended dry periods. Propagate from seed.

203

Daviesia chordophylla Meisner
(long, slender, terete leaves)
WA
1-2 m x 1-2 m   Sept-Dec
   Small **shrub**; **branches** long, slender, furrowed, glabrous; **leaves** 7.5-30 cm long, slender, terete, striate, glabrous; **flowers** pea-shaped, small, yellow and orange, borne in loose racemes to 6 cm long; **pods** about 0.8 cm long.
   An interesting species, with distinctive, long, slender leaves. It occurs in the Darling District. Evidently not well known in cultivation. Should do best in well drained, light to medium soils, with partial or full sun. Could prove to be a decorative container plant. Propagate from seed or cuttings.

Daviesia colletioides Meisner
(similar to the genus *Colletia*)
WA
0.6-1.5 m x 1-1.5 m   July-Aug
   Dwarf to small **shrub**; **branches** many, terete, glabrous; **leaves** to 4 cm long, terete, narrow-falcate, erect or ascending, glabrous, pungent; **flowers** pea-shaped, about 0.8 cm long, red and yellow, borne in axillary clusters on slender pedicels; **calyx** about 0.4 cm long; **pods** about 2 cm long, triangular.
   A dweller of gravelly soils in the south-west corner of WA. It shows potential for cultivation, and is probably best suited to well drained, light to medium soils, with plenty of sunshine. Tolerant of frosts and extended dry periods. Propagate from seed.
   Has affinities to some forms of *D. incrassata*. *D. incrassata* often has flattened leaves, and the calyx is shorter than that of *D. colletioides*.

Daviesia cordata Smith
(heart-shaped)
WA   Bookleaf
1-2 m x 1-2 m   Sept-Nov
   Small **shrub**; **branches** many, angled, glabrous, usually erect; **leaves** 5-10 cm x 2-3 cm, cordate, stem-clasping, reticulate, pointed apex, glabrous; **flowers** pea-shaped, about 1 cm across, bright orange-yellow and purple-red, borne in umbels on axillary peduncles; **bracts** 3-6 cm across, orbicular, enlarge during and after flowering to enclose the pods.
   A very ornamental species that has had limited cultivation. It occurs naturally in sand or sandy gravels, as an undershrub to trees. Requires very well drained, light to medium soils (more trials needed in heavier soil types). Suited to dappled shade and partial or full sun. Tolerates medium frosts and responds well to pruning. The species has high potential as a container plant. The foliage with the enlarged bracts is used for floral arrangements. Propagate from seed or cuttings.

Daviesia corymbosa Smith
(in the shape of a corymb)
NSW
1-3 m x 1-2 m   Sept-Nov
   Small to medium **shrub**; **branches** usually ascending, slightly angular, glabrous; **leaves** 4-12 cm long, variable, linear-lanceolate to narrow-ovate, narrowed at base, strong reticulation, glabrous, pointed apex;

flowers pea-shaped, to about 1 cm across, red and orange, borne in corymbose racemes which are usually shorter than the leaves; **pods** about 1 cm across, triangular.
   A showy species of dry sclerophyll forest and sandstone areas. It likes good drainage, with partial to full sun and will probably tolerate dappled shade. Useful as a light screening plant. Has the ability to reshoot after fire. Propagate from seed or cuttings.

Daviesia crenulata Turcz.
(small convex teeth)
WA
0.6-1.3 m x 0.6-1.5 m   Sept-Oct
   Dwarf to small **shrub**; **branches** rigid, pubescent; **leaves** to 2 cm x 2 cm, orbicular-cordate, opposite or alternate, rigid, margins undulate and slightly crenulate, ending in a fine pungent point, both surfaces strongly reticulate, glabrous and shiny; **flowers** pea-shaped, small, orange, borne in umbels of 4-6 or on peduncles longer than the leaves; **pods** about 1.2 cm across, triangular.
   This ornamental species is found over a wide area of south-western WA, often both near the coast and inland. It has decorative foliage, and the flowers are well displayed beyond the leaves. Evidently not in cultivation to any great extent, but should grow well in a wide range of well drained soils, in dappled shade to partial sun. Should also tolerate full sun. Pruning will promote bushy growth. Propagate from seed or cuttings.

Daviesia daphnoides Meisner
(similar to the genus *Daphne*)
WA
0.6-1.2 m x 0.6-1 m   June-Sept
   Dwarf to small **shrub**; **branches** rigid, prominently angled, glaucous; **leaves** 2-8 cm x 0.5-1.5 cm, oblong-lanceolate, petiolate, tapering to a pungent point, flat, thick, rigid, glaucous, glabrous; **flowers** pea-shaped, small, reddish brown, borne in small axillary clusters or very short racemes.
   This species occurs over a wide range of south-

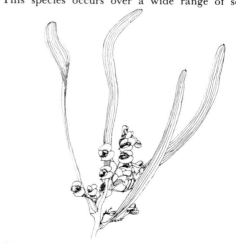

Daviesia corymbosa, × ·6

Daviesia epiphylla   B. Crafter

western WA. It is not well known in cultivation, but it will have appeal because of its glaucous foliage. Requirements are for a warm to hot location with plenty of sun. The soil must be well drained. Propagate from seed.

Daviesia debilior Crisp
(weak; refers to its growth habit)
WA
0.3-0.6 m x 1.5 m   May-July
Dwarf **shrub**; **branches** procumbent; **branchlets** many, weak, ascending, angular, prominently ribbed; **juvenile leaves** 2-5 cm x up to 0.8 cm, narrow-spathulate, flat, thick, thickened margins; **intermediate leaves** longer and narrower than juveniles, usually at base of mature plants; **adult leaves** to 12 cm x up to 0.2 cm, angular or compressed, many-ribbed, reduced to minute scales on upper parts of branchlets; **flowers** pea-shaped, about 0.5 x 0.6 cm, 2-4 flowers per axil; **standard** yellow with purple-black or red; **wings** orange-pink; **keel** dark purple-red; **pods** 1.4-1.7 cm x about 1 cm, triangular, compressed, maturing Sept.

A recently described species from the Darling and Irwin Districts, where it usually occurs on shallow sand overlying lateritic gravel and clay, in heath vegetation. In cultivation, it should adapt to a wide range of soil types, with a preference for good drainage. Will need plenty of sunshine, but may tolerate dappled shade.

The ssp. *sinuans* Crisp differs in not having any leaves; also, the weak branchlets become sinuous and sometimes glaucescent. It occurs in the Avon and Darling Districts.

Propagate both forms from seed and possibly cuttings.

Daviesia discolor Pedley
(usually 2 distinct colours)
Qld
0.8-1.5 m x 1-1.5 m   Aug-Sept
Dwarf to small **shrub**; **branchlets** angular, glabrous; **leaves** 7-15 cm x 0.4-0.8 cm, linear, flat or with margins slightly incurved, acute, slightly discolorous,

reticulate nerves; **flowers** pea-shaped, small, yellow and red, borne in short racemes, 1-2 per axil near ends of branchlets.

This recently named species was originally included in *D. mimosoides* which has broader leaves of one colour. It occurs in the Leichhardt District. Evidently not well known in cultivation. Should grow well in soils that have good drainage, in a situation that receives dappled shade to partial sun. The bicoloured foliage may have appeal to enthusiasts. Propagate from seed or cuttings.

Daviesia divaricata Benth.
(forked or spreading)
WA
1-2 m x 1-2 m   June-Oct
Small **shrub**; **branches** terete, leafless, glabrous; **branchlets** many, forked, spiny; **flowers** pea-shaped, small, yellow and brown to orange and red, on slender pedicels in short lateral racemes.

Although not well known, this daviesia is worthy of trial, with its good display of flowers on the spiny branchlets. It is also useful for directing foot traffic. Occurs naturally on well drained sands and sandy gravels in the Avon, Darling and Irwin Districts. The provision of a well drained situation, with partial or full sun, seems necessary for optimum growth. Tolerant of most frosts. Propagate from seed and probably cuttings.

Daviesia epiphylla Meisner
(flowers borne on the leaves)
WA   Staghorn Bush
0.6-1.5 m x 0.5-1.5 m   April-Aug
Dwarf to small **shrub**; **branches** leafless, flattened, glabrous; **branchlets** flattened, thick, glaucous; **lobes** triangular, pungent tips; **flowers** pea-shaped, to 2 cm long, coral pink to scarlet, borne in small clusters along the branchlets; **pods** to 3 cm long.

A species with highly ornamental foliage and the bonus of brilliantly coloured flowers. It occurs naturally on sandy, gravelly soils in the Northern Sandplains, between the Hill River and Eneabba. Prefers very well drained, light to medium soils, with a warm to hot situation. Frost and drought tolerant. Judicious pruning could provide more bushy growth if required. Has potential as a container plant. Propagate from seed.

Daviesia euphorbioides Benth.
(similar to the genus *Euphorbia*)
WA
0.6-1.5 m x 1-1.5 m   Sept-Oct
Dwarf to small, leafless **shrub**; **branches** erect and trailing, terete, thick and pithy, glabrous, green to glaucous; **branchlets** small, usually erect, about 0.8 cm diameter; **leaves** absent, replaced by small scales; **flowers** pea-shaped, small, red and yellow, borne in short racemes or clusters.

This is now a very rare species in its natural habitat of sand and gravel in the Avon District. Only a few plants remain and these are threatened by further clearing. It has had only limited cultivation, but should be suited to most well drained, light to medium soils. A situation with partial or full sun is recommended.

Grows very well in a container. Hardy to most frosts and extended dry periods. Propagate from seed or from cuttings which strike readily.

Daviesia flava Pedley
(yellow)
Qld
1-2 m x 1-1.5 m   June-Feb
Small **shrub**; **branchlets** angular, glabrous; **leaves** 5-12 cm x 0.3-1.5 cm, linear to narrowly oblong, flat, obtuse, mucronulate, glabrous; **flowers** pea-shaped, about 0.6 cm long, yellow, borne in racemes to 5 cm long, of 1-3 per axil near ends of branchlets; **pods** about 1 cm wide, triangular.

A decorative species from the Cook and North Kennedy Districts. It should prove hardy in cultivation for tropical and subtropical areas. Probably grows best in well drained, light to medium soils, with dappled shade to partial sun. Propagate from seed or cuttings.

Daviesia flexuosa Benth.
(zigzagged)
WA
0.6-1.5 m x 0.6-1.5 m   July-Oct
Dwarf to small **shrub**; **branches** flexuose, rigid, angular, glabrous; **leaves** to 3 cm long, linear, erect to spreading, lower ones flattened vertically, upper ones may be nearly terete, all tapering to a pungent point; **flowers** pea-shaped, small, yellow and red, borne in axillary clusters; **pods** about 2 cm long, triangular.

Most commonly found on white sand in the southern region of WA. Has had only limited cultivation. Best suited to very well drained soils, with plenty of sun. Should be a useful plant for directing foot traffic, due to its dense and prickly growth habit. Frost tolerant and withstands extended dry periods. Propagate from seed and possibly cuttings.

Daviesia genistifolia Cunn. ex Benth.
(leaves similar to those of the genus *Genista*)
Qld, NSW, Vic, SA   Broom Bitter-pea
1-2.5 m x 1-2 m   Sept-Nov
Small to medium **shrub**; **branches** slender, slightly grooved, glabrous; **leaves** 1-3 cm long, terete, pungent, spreading, glabrous; **flowers** pea-shaped, about 0.7 cm across, reddish orange and yellow, in axillary clusters of 3-5; **pods** about 0.8 cm long.

A dweller of sandy soils, this pungent-leaved species needs a very well drained situation. Tolerates dappled shade to full sun. Withstands frosts and extended dry periods. An excellent refuge plant for birds. Propagate from seed.

Daviesia hakeoides Meisner
(similar to the genus *Hakea*)
WA
0.3-1 m x 0.5-1 m   June-Oct
Dwarf **shrub**; **branches** erect, rigid, terete to slightly flattened, grooved, glabrous; **leaves** more or less terete, rigid and pungent, lower leaves to 20 cm long, almost erect, upper leaves to 5 cm long, varying from slender to stout, often spreading, glabrous; **flowers** pea-shaped, small, orange and red, borne in very short, sessile and axillary racemes; **bracts** imbricate, longer

than pedicels; **pods** about 1 cm long, triangular.

Not well known in cultivation. It occurs in low-rainfall regions of the Austin, Coolgardie and Darling Districts, where it is usually found on sandy soils. To grow it successfully, excellent drainage must be provided, in a situation that receives maximum sunshine. Frost and drought tolerant. Propagate from seed and possibly cuttings.

The var. *major* Benth. is distinguished by its larger bracts and flowers. The var. *subnuda* Benth. has only a few leaves, which are much smaller and prominently spreading to recurved.

Daviesia horrida Preiss ex Meisner
(very thorny)
WA
1-2 m x 1-2 m   Oct-Nov
Small **shrub**; **branches** rigid, often glaucous, glabrous; **branchlets** usually leafless, spreading, spiny, glaucous, glabrous; **leaves** 3-15 cm long, linear to linear-lanceolate, flat, rigid, glaucous, blunt or pointed, usually on main branches; **flowers** pea-shaped, small, red and orange, borne in loose racemes on leafless, spiny branchlets.

An interesting, usually erect, shrubby species. The phyllode-like leaves are on the main branches, none on the branchlets. Requires relatively well drained, light to medium soils, with partial to full sun. Tolerant of most frosts. Should withstand judicious pruning. Has potential as a container plant. Propagate from seed.

Daviesia incrassata Smith
(thickened)
WA
0.5-1 m x 1-2 m   Sept-Nov
Dwarf **shrub**; **branches** many, terete or slightly flattened, glabrous; **branchlets** leafless, short, terete or sometimes flattened, rigid, ending in a pungent point, green; **flowers** pea-shaped, about 0.8 cm across, red-

Daviesia genistifolia, × ·6

Daviesia latifolia   T.L. Blake

dish orange with reddish keel, borne in small clusters of 3-8, massed near ends of branches; **pods** to 1.4 cm long, slightly inflated.

A highly ornamental species, seeming to be in full flower for long periods, the flowers opening progressively, which is not the case with most daviesias. It occurs naturally on sand or sandy gravel, in open or shaded situations. Should adapt to a wide range of well drained soils, with partial or full sun. Tolerates dappled shade and withstands most frosts. Pruning could be useful to promote bushy growth. Propagate from seed which is difficult to collect. Experimentation with cuttings is required.

In a pending botanical revision, *D. reversifolia* will become a subspecies of *D. incrassata*.

Daviesia juncea Smith
(rush-like)
WA
0.3-1 m x 0.5-1 m   June-Aug

Dwarf **shrub**; **branches** long, erect, leafless, slightly branched, more or less terete, glabrous; **leaves** usually absent, reduced to minute scales; **flowers** pea-shaped, small, yellow and purple, borne in small axillary clusters; **bracts** imbricate, covering the short, flowering rhachis; **pods** about 2 cm long, triangular, very acute.

This rush-like species can develop into a dense plant. It occurs on sandy soil, over a wide area in the south-west. Should grow in very well drained, light to medium soils, with plenty of sunshine. Frost and drought tolerant. Propagate from seed. Experimentation is needed in propagation of cuttings of the leafless stems.

Daviesia lancifolia Turcz.
(lance-shaped leaves)
WA
0.2-0.5 m x 0.4-1 m   Aug-Dec

Dwarf, spreading **shrub**; **branches** decumbent, with soft hairs, or glabrous; **leaves** to 1.2 cm long, lanceolate to obovate, orbicular or oblong, crowded, often ending in a pungent point, glabrous; **flowers** pea-shaped, small, yellow, borne in small umbels; **bracts** do not enlarge after flowering; **pods** about 1.2 cm x 1.2 cm.

This is not a well known species. It occurs near the southern coast, eg. West Mount Barren, where it grows in very well drained soils. For cultivation, similar conditions are required, in a situation with partial or full sun. Propagate from seed or cuttings.

Previously known as *D. mollis* var. *minor* Benth. Has strong affinities to *D. mollis* which has larger obovate to orbicular leaves and bigger flowers.

Daviesia latifolia R. Br.
(broad leaves)
Qld, NSW, Vic, Tas   Hop Bitter-pea
1-3 m x 1-2 m   Sept-Jan

Small to medium **shrub**; **branches** more or less terete; **leaves** 5-10 cm x 2-3 cm, ovate-elliptic to ovate-lanceolate, reticulate, glabrous; **flowers** pea-shaped, about 0.6 cm across, yellow and brown, borne in racemes 3-6 cm long, fragrant; **pods** about 1 cm long.

A decorative shrubby species that grows in a wide range of relatively well drained soils. Prefers dappled shade and partial sun, but also tolerant of full sun. Frost hardy. In nature, it often forms dense thickets, and is likely to be useful as an informal hedge or windbreak. The flowers have an attractive perfume. The bitter-tasting leaves are recorded as having medicinal properties, and have been used as a substitute for hops. Propagate from seed or cuttings.

Daviesia longifolia Benth.
(long leaves)
WA
0.6-1.5 m x 0.6-3 m   Sept-Dec

Dwarf to small **shrub**; **branches** long, slender, angular or grooved, becoming terete, leafless; **branchlets** short, prominent midrib and marginal nerves; **flowers** pea-shaped, about 1 cm across, variable in colour from dull yellow to light orange-red, in pairs with long peduncles, near ends of branches; **pods** about 1 cm long, slightly inflated, often reddish before maturity, then black.

A dense, spreading species from the southern regions of the state, occurring in gravelly sands, amongst other shrubs. It is most decorative prior to the pods maturing, and back-lighting enhances the reddish colouration. Not widely grown, but should grow in well drained, light to heavy soils. Hardy to medium frosts. Useful as a low windbreak. Withstands pruning. Propagate from seed or cuttings.

Daviesia mesophylla Ewart
(middle leaf)
WA
0.8-1.5 m x 0.8-1 m   Sept-Oct

Dwarf to small **shrub**; **branches** grooved, glabrous, not pungent; **leaves** 0.6-1.5 cm x about 0.1 cm, linear to lanceolate, erect, rigid, thick, concave above, convex below, apex with sharp point; **flowers** pea-shaped, to 1 cm long, red and yellow, borne in leafy racemes.

From western and south-western WA. This species is not well known and has had only very limited cultivation. It should do best in well drained, light to medium soils, with partial or full sun. Suited to growing in containers. Tolerant of most frosts. Propagate from seed or cuttings.

Daviesia mimosoides R. Br.
(similar to the genus *Mimosa*)
Qld, NSW, Vic, SA   Blunt-leaf Bitter-pea
1-2.5 m x 1-2 m   Sept-Dec

Small to medium **shrub**; **branches** slightly angular, glabrous; **leaves** 3-15 cm x 1-2 cm, narrow-lanceolate to elliptical, blunt or acute, faintly reticulate, glabrous, can be glaucous; **flowers** pea-shaped, about 0.7 cm across, yellow and red, in axillary racemes to 3 cm long, 1 to several per axil, profuse, scented; **pods** to 1 cm long.

A hardy and adaptable species that is very showy when in full bloom. Grows naturally on poor, stony soils in open forest. Should do well in most relatively well drained soils, with dappled shade and partial or full sun. Frost tolerant, so suitable for cold locations. Plants can sucker. It was grown at Kew Gardens in 1809.

The var. *laxiflora* (J. H. Willis) J.H. Willis grows to about 5 m tall, and has broader (to 3 cm), often glaucous leaves. The flowers are all-yellow, and are displayed in loose racemes to 10 cm long. This ornamental variety is restricted to mountainous areas of Vic, and prefers cool, moist conditions.

Propagate both forms from seed or cuttings.

Daviesia mollis Turcz.
(softly hairy)
WA
0.2-0.5 m x 0.4-1 m   Sept-Dec

Dwarf **shrub**; **branches** decumbent, many, soft, spreading hairs; **leaves** to 2 cm long, obovate to orbicular, crowded, hairy, apex often ending in pungent point, prominent midrib, thick, leathery; **flowers** pea-shaped, about 0.8 cm long, yellow-purple, borne in small umbels; **bracts** do not enlarge after flowering; **pods** about 1.2 cm x 1.2 cm.

This species is virtually unknown in cultivation, but it is likely to become popular due to its useful, low and spreading growth habit. Should do best in very well drained, light to medium soils, with partial or full sun. Propagate from seed or cuttings.

D. mollis var. *minor* AC. is now *D. lancifolia* Turcz. It has narrower, less hairy leaves, and smaller flowers.

Daviesia mimosoides, × ·6

Daviesia nudiflora Meisner
(naked or bare flowers)
WA
0.5-1 m x 1-2 m   June-Oct

Dwarf **shrub**; **branches** many, grooved, glabrous; **leaves** to 2.5 cm x 0.8 cm, oblanceolate to ovate, margins recurved, tapering to a fine, pungent point, dull green; **flowers** pea-shaped, about 1 cm across, pale yellow with dull reddish keel, profuse near ends of branchlets; **peduncles** about 1.5 cm long, 1 per axil; **pods** about 1.5 cm long, inflated.

A species with ornamental foliage. It occurs on gravelly sands over a wide range of south-western WA, and should do well in well drained, light to heavy soils. Prefers full sun, but will possibly tolerate some shade in areas of high temperatures. Tolerant of light to medium frosts. May need pruning to promote bushy growth. Propagate from seed and possibly cuttings.

Daviesia obovata Turcz.
(obovate)
WA
0.5-1 m x 0.5-1 m   March-April; also sporadic

Dwarf **shrub**; **branches** terete to flattened, but not angular; **leaves** 5-12 cm long, elliptic to oblong, rounded apex, blunt to notched, petiolate, thick, leathery, veins obscure; **flowers** pea-shaped, about 1.2 cm long, yellow, in few-flowered, short racemes; **calyx** about 0.6 cm long, with narrow, acuminate teeth; **pods** to 3 cm long, triangular.

This is a showy species from the jarrah forests in the south-west. It is little-known in cultivation, but certainly has potential. Its flowers are relatively large for this genus, and are well displayed. Plants should succeed in a range of well drained soils, with a situation that receives partial or full sun. May tolerate dappled shade. Propagate from seed or cuttings.

Daviesia obtusifolia F. Muell.
(blunt leaves)
WA
0.6-1 m x 0.6-1 m   Sept-Oct

Dwarf **shrub**; **branches** usually erect, angular, glabrous; **leaves** 3-8 cm x up to 1.5 cm, narrow to broad-oblong, greyish green, petiolate, rounded or notched apex, leathery, more or less veinless; **flowers** pea-shaped, about 0.7 cm across, yellow and reddish brown, in short axillary racemes comprising a few flowers; **pods** triangular, to 3 cm long, very leathery.

This species occurs in coastal and near-coastal situations, as well as inland. Its distribution includes the Darling, Eyre, Roe and Coolgardie Districts. It has had only limited cultivation. Best suited to relatively well drained soils, with an open, sunny situation, but it will probably tolerate dappled shade. Propagate from seed.

The var. *parvifolia* E. Pritzel has smaller leaves.

Daviesia oppositifolia Endl.
(opposite leaves)
WA   Rattle Pea
0.5-1 m x 1-2 m   Sept-Jan

Dwarf **shrub**; **branches** thick, 3 or 4-angled; **leaves** 3-10 cm long, opposite or scattered, oblong-elliptical,

Daviesia pachyphylla   W.R. Elliot

usually blunt apex, thick, leathery; **flowers** pea-shaped, about 1 cm across, yellow and maroon-red, borne in terminal umbels; **bracts** 2-3, about 2.5 cm across, orbicular, enlarge during and after flowering to enclose the pods; **pods** about 1.2 cm long.

Like *D. cordata*, this species has ornamental bracts that enclose the pods. It occurs on gravelly sands from Albany to Ravensthorpe, where it grows in woodland and heathland communities. Probably best suited to well drained, light to medium soils, with partial to full sun. Has potential as a container plant. Propagate from seed and possibly cuttings.

Daviesia pachyphylla F. Muell.
(thick leaves)
WA   Ouch Bush
0.6-1.5 m x 1-2 m   July-Sept

Dwarf to small **shrub**; **branches** often arching, not dense, glaucous; **leaves** 1-2 cm long, very thick, terete, tapering to a hard, pungent point, glaucous, crowded along branches; **flowers** pea-shaped, about 0.7 cm across, yellow, orange and dull reddish brown, borne in loose racemes in axils along upper parts of branches; **pods** about 1.5 cm long, inflated.

A most ornamental species, with succulent-like foliage and a good display of flowers. Certainly has a relevant vernacular name. It occurs naturally on a variety of soil types between Ongerup and Ravensthorpe. Best suited to well drained soils, with full sun, but will also tolerate partial sun. Withstands light to medium frosts. Pruning while plants are young may help to promote early bushy growth. Has potential as a container plant. Well suited to semi-arid regions. Propagate from seed or cuttings which can be difficult to strike.

Daviesia pectinata Lindley
(comb-like)
Vic, SA, WA   Prickly Bitter-pea; Thorny Bitter-pea
0.5-2 m x 1-2 m   June-Nov

Dwarf to small **shrub**; **branches** rigid, glabrous; **leaves** 1-4 cm long, lanceolate, decurrent base flattened vertically, spreading, pungent-pointed, green to

grey-green; **flowers** pea-shaped, about 0.5 cm across, red and yellow to orange, borne in axillary clusters or short racemes; **pods** to 1.3 cm long.

A species that is renowned for its prickly foliage, which was used by early settlers in the construction of fences, thus preventing stock from wandering. It is not common in cultivation. Best suited to warm to hot locations, with well drained, light to medium soils, but will also tolerate dryish heavy soils. Withstands frost and extended dry periods. Worth experimentation as a container plant. Propagate from seed.

Daviesia pedunculata Benth. ex Lindley
(with peduncles)
WA
0.5-1 m x 1-2 m   Aug-Dec

Dwarf **shrub**; **branches** terete, short, slender, hairy while young; **leaves** 1-2 cm x about 0.5 cm, oblong to linear, narrowed both ends, pungent tip, convex, glabrous; **flowers** pea-shaped, small, yellow and brown, borne in terminal umbels to 2 cm across; **peduncles** much longer than leaves.

This decorative species occurs over a wide range of south-western WA, where it is usually found in sandy, well drained soils, growing amongst other plants. Evidently not in cultivation. Should be suited to light to medium soils, provided there is unimpeded drainage. Will tolerate dappled shade to partial sun. Withstands extended dry periods and light to medium frosts. Propagate from seed or cuttings.

Daviesia polyphylla Benth.
(many leaves)
WA
0.6-1.5 m x 0.6-1.5 m   April-Nov

Dwarf to small **shrub**; **branches** many, glabrous, angular or grooved when young; **leaves** to 3 cm long, linear to lanceolate, usually falcate, narrowed to base, spreading to somewhat erect, vertically flattened, thick, rigid, ending in a pungent point, glabrous; **flowers** pea-shaped, small, red and yellow, borne in short axillary clusters; **pods** triangular, about 1.3 cm long, inflated.

A species with prickly foliage. It is hardly known in cultivation, although it is recorded as having been introduced to England in 1822. It occurs in coastal situations, as well as in the jarrah forests in the south-west corner of WA. Should do best in well drained, light to medium soils, with dappled shade to full sun. Propagate from seed.

Daviesia preissii Meisner
(after Ludwig Preiss, 19th-century German botanist)
WA
0.5-1.5 m x 1-2 m   July-Dec; also sporadic

Dwarf to small, prickly **shrub**; **branches** many, rigid, glabrous; **leaves** 1-3 cm x 0.2-0.3 cm, linear-falcate, usually striate, vertically flattened near base, pungent point, glabrous, scattered along branches; **flowers** pea-shaped, about 0.7 cm across, yellow and reddish brown or yellow and pinkish red, axillary, 2-3 together, near ends of branchlets, profuse.

A showy species that usually occurs in well drained, sandy soils. It is not well known in cultivation, but should be suited to most well drained soils, with dappled shade to full sun. Tolerant of most frosts and extended dry periods. Its prickly foliage makes it useful as a protective plant. Propagate from seed and possibly cuttings.

Similar to *D. incrassata* which has more or less leafless branches.

Daviesia purpurascens Crisp
(refers to the purplish glaucescence of the foliage)
WA

0.5-1.5 m x 0.5-1.5 m   Sept-Jan

Dwarf to small **shrub**; **branches** many, glabrous; **branchlets** many, rigid, flexuose, grey-green to purplish, glaucescent; **leaves** to 0.5 cm x about 1 cm, terete, tapered, rigid, pungent tip, grey-green to purplish, glaucescent, spreading to ascending; **flowers** pea-shaped, about 0.5 cm across, yellow and maroon, borne in axillary racemes of 2-7 flowers; **pods** to 0.6 cm x 0.4 cm, not triangular, approximately ovoid.

This is a rare and endangered species. It has a limited occurrence in southern WA, where it grows in deep sands. The ornamental foliage and flowers make it highly worthy of cultivation. Should be suited to very well drained, light to medium soils, in a warm to hot location. Tolerant of frost and extended dry periods. Would possibly make an excellent container plant. It seems that the purplish colouration is the result of exposure to plenty of sun. Propagate from seed and possibly cuttings.

Has similarities to *D. benthamii* which never has glaucous foliage.

Daviesia quadrilatera Benth. ex Lindley
(4-sided)
WA

0.6-1.5 m x 1-2 m   June-Oct

Dwarf to small **shrub**; **branches** more or less terete, rigid, glaucous, glabrous; **leaves** 0.8-2.5 cm x 0.3-1 cm, oblong-rhomboidal, erect, more or less vertical against the branches, attached by narrow base, 2 pungent points on ends of outer edge, glaucous, glabrous; **flowers** pea-shaped, about 0.6 cm across, yellow to orange and red, borne in axillary umbels comprising 3-6 flowers; **pods** to 1.5 cm long, triangular.

The interesting glaucous leaves and branches alone make this a decorative species, but when combined with the many colourful flowers it is eye-catching. It occurs north of Perth in sandy heathlands. In cultivation, it is likely to do best in very well drained soils, with partial to full sun. Frost and drought tolerant. Propagate from seed and cuttings.

Daviesia reversifolia F. Muell.
see under *D. incrassata* Smith

Daviesia spiralis Crisp
(refers to the spirally twisted leaves)
WA

1-2 m x 1-2 m   Sept-Jan

Small, rounded **shrub**; **branchlets** ribbed, slightly roughened; **leaves** to 10 cm x 0.3 cm, linear, spirally twisted, alternate, ascending, surface roughened, apex narrow and hooked; **flowers** pea-shaped, to 1 cm

across, in short axillary racemes; **standard** yellow with red; **wings** reddish; **keel** reddish to maroon; **pods** 1-1.3 cm x up to 0.5 cm, triangular, compressed, maturing Dec-Feb.

This recently described species has interesting spiralled foliage. It is an uncommon endemic in the Wongan Hills of the Avon District. Performance in cultivation is as yet unknown, but it should prove adaptable as it occurs on laterite-derived clay and gravel. In its natural habitat it grows in mallee shrubland, where it does best in open situations. Should grow in dappled shade to full sun. Frost and drought tolerant. Has potential as a container plant. Propagate from seed and possibly cuttings.

Daviesia squarrosa Smith
(overlapping and recurved)
Qld, NSW

0.8-2.5 m x 0.8-2.5 m   Aug-Dec

Dwarf to medium **shrub**; **branches** often developing from near ground level, with long hairs; **branchlets** many, lateral, short, often arching; **leaves** to 0.8 cm x 0.5 cm, sessile, cordate or ovate-lanceolate, spreading or reflexed, tapering into a pungent point, crowded in a spiral arrangement; **flowers** pea-shaped, about 0.5 cm wide, golden yellow and purple-brown, usually solitary, on slender stalks in axils; **pods** to 1.25 cm long, triangular, initially tinged with red, maturing to pale brown.

An inhabitant of open eucalyptus forest, in sandy soils over sandstone, on rocky slopes or dry, stony hills. Occurs east of the Great Dividing Range, from north of Brisbane to near Batemans Bay in NSW. Evidently not well known in cultivation. Requires very well drained, light to medium soils. Suited to a situation with dappled shade or partial sun. Hardy to most frosts. Propagate from seed or cuttings.

Daviesia striata Turcz.
(finely grooved)
WA

0.6-1.5 m x 1-1.5 m   July; Dec-Jan

Dwarf to small **shrub**; **branches** thick, rigid, slightly angular, or terete and striate, glabrous; **leaves** about 1.2 cm x 0.6 cm, falcate, ovate to almost rhomboidal, vertically flattened, attached to the stem by broad base, pungent apex, usually striate; **flowers** pea-shaped, small, yellow, in short axillary racemes.

Very uncommon in cultivation, this species occurs on clay soils of the Irwin, Avon, Roe and Eyre Districts. It should prove to be adaptable to a wide range of soils, provided there is adequate drainage. Probably best suited to an open position, with plenty of sunshine. Propagate from seed.

Daviesia stricta Crisp
(erect; upright)
SA

1-1.5 m x 0.6-1 m   Aug-Sept

Small upright **shrub**; **branchlets** usually stiffly erect, 3-angled, smooth; **phyllodes** 1-10 cm x up to 1.5 cm, narrow to linear-elliptic, arranged spirally, erect or ascending, flat, apex blunt or acute, ending in a small point; **flowers** pea-shaped, about 0.7 cm across, on slender pedicels, in axillary clusters; **standard** orange with purplish centre; **wings** and **keel** purplish; **pods** triangular, to 1.5 cm x 0.7 cm, with narrow beak.

A recently described species, endemic to the Flinders Ranges, SA, where it occurs on ridge tops and precipitous mountain slopes, growing in stony soil derived from quartzite. It is found amongst other shrubs and *Triodia*, as an understorey to mallee eucalypts. This is a showy species that requires very well drained soils and a warm to hot position. It tolerates extended dry periods and most frosts. Worthy of cultivation in containers for temperate zones. Propagate from seed or possibly cuttings.

D. wyattiana is closely related to *D. stricta*, but differs in its longer pedicels and beakless pods.

Daviesia teretifolia R. Br. ex Benth.
(terete leaves)
WA

0.8-1.5 m x 1-1.5 m   Oct-March

Dwarf to small **shrub**; **branches** many, usually erect, glabrous; **leaves** 2-4.5 cm x about 0.2 cm, linear, terete, pungent apex, erect, usually glabrous; **flowers** pea-shaped, about 0.9 cm across, yellow with red, borne in loose axillary racemes of several flowers; **pods** to 2.5 cm long, leathery.

A plant of the sandheaths in the Eyre, Roe and Coolgardie Districts. It is not well known in cultivation. Suited to well drained, light to medium soils, with plenty of sunshine. Will probably tolerate dappled shade in hot regions. Frost tolerant. Propagate from seed.

Has some affinities to *D. genistifolia* which has smaller flowers and pods.

Daviesia trigonophylla Meisner
(leaves are triangular in cross-section)
WA

0.5-2 m x 1-2 m   Aug-Dec

Dwarf to small **shrub**; **branches** many, winged, glabrous or slightly hairy; **leaves** to about 1.5 cm long, continuously decurrent along branches, spreading, recurved and ending in a pungent point, upper edge broad, usually concave; **flowers** pea-shaped, about 0.8 cm across, orange and maroon, borne in axillary clusters near ends of branches; **pods** to 1.5 cm long, triangular, inflated.

This is a species with ornamental foliage, but it is not well known in cultivation. It occurs on sandy, stony soils in the Stirling Range. Should do best in well drained, light to medium soils, with partial or full sun. Withstands most frosts. Propagate from seed and possibly cuttings.

Similar to *D. pectinata*, the leaves of which do not have concave or channelled upper edges.

Daviesia ulicifolia Andrews
(leaves similar to those of the genus *Ulex*)
all states except NT   Gorse Bitter-pea
1-2 m x 1-2 m   July-Oct; also sporadic

Small, prickly **shrub**; **branches** and **branchlets** many, angular, usually ending in spines, glabrous or hairy; **leaves** to 2.5 cm long, cordate to narrow-lanceolate, prominent midrib, pungent tip, often

Daviesia ulicifolia, × ·6

crowded, glabrous or hairy; **flowers** pea-shaped, about 0.8 cm across, orange-yellow or yellow and brown, 1 to a few flowers per axil, near ends of branches; **pods** to 1.5 cm long, triangular, inflated.

This species occurs in a wide range of habitats and climatic conditions, and has proved hardy in cultivation, although it is not very popular. Best suited to relatively well drained, light to medium soils, but tolerates heavy soil types. Appreciates dappled shade or partial sun. Can be pruned very hard to promote dense growth. An excellent restrictive plant due to its prickly nature. Propagate from seed or cuttings.

Previously known as *D. ulicina* Smith.

Daviesia ulicina Smith = *D. ulicifolia* Andrews

Daviesia umbellulata Smith
(small umbel)
Qld, NSW

1-2 m x 1-2 m   Aug-Oct

Small, stiff **shrub**; **branches** slender, emanate from a short woody base, hairy; **leaves** 0.8-2 cm long, cordate to narrow-lanceolate, spreading or reflexed, glabrous or without hairs on margins, prominent midrib, pungent point; **flowers** pea-shaped, about 0.7 cm across, yellow and red, in axillary racemes, profuse; **pods** to about 1 cm long, triangular, inflated.

A most floriferous species, in which the branches can be weighed down by the profusion of flowers. Plants are found growing in open heathland or as undershrubs in light forest. Should do well in soils that have good drainage. Will adapt to dappled shade and partial or full sun. Tolerates light frosts. Propagate from seed or cuttings. This species has been available under the incorrect name of *D. umbellata*.

Daviesia uniflora D. Herbert
(1-flowered)
WA

0.6-1 m x 0.5-1 m   July-Oct

Dwarf **shrub**; **branches** rigid, with soft long hairs; **leaves** about 1 cm long, obovate to orbicular, crowded, obtuse, with pungent tip, hairy when young; **flowers** pea-shaped, about 0.7 cm across, yellow, solitary on slender stalks; **pods** about 1.2 cm long.

Not a well known species. It occurs on sandplains in the Coolgardie District. Has had very limited cultivation, and is suited to well drained soils, with plenty of

Daviesia virgata

sunshine. Adapts to container cultivation. Frost tolerant and should withstand extended dry periods. Propagate from seed or cuttings.

Has affinities to *D. mollis* which differs in its flowers being borne in umbels.

Daviesia virgata Cunn. ex Hook.
(twiggy)
NSW, Vic, SA   Narrow-leaf Bitter-pea
1-2 m x 1-2 m   Sept-Jan

Small, open **shrub**; **branches** slender, usually erect, often without leaves on upper parts; **leaves** to 12 cm long, oblanceolate to linear, thick, rigid, erect, usually blunt apex, glabrous, dark green, never glaucous; **flowers** pea-shaped, about 0.6 cm across, yellow and red, in axillary racemes to 2 cm long; **pods** to 1 cm long, triangular, inflated.

This species is particularly useful for growing in a wide range of relatively well drained soils. It does well in dappled shade to full sun, and even adapts to a fair amount of shade. Frost tolerant. Responds well to pruning if dense growth is required. Propagate from seed or cuttings.

Daviesia wyattiana Bailey
(after Dr. William Wyatt, promoter of botany and horticulture, SA)
Qld, NSW
1.5-2.5 m x 1-2 m   July-Nov

Small to medium **shrub**; **branches** slender, angular, often arching, glabrous; **leaves** 15-20 cm x up to 0.6 cm, linear, prominent midrib, glabrous; **flowers** pea-shaped, to 0.6 cm long, yellow with apricot to orange and purple, borne in axillary umbels of up to 5 flowers; **pods** about 1 cm long, triangular, golden brown.

A showy species that inhabits open eucalyptus forest, rocky ridges, mountain slopes and granite outcrops. Usually occurs in sandy or gravelly soils. Not well known in cultivation, but certainly has potential. Best suited to well drained soils, with dappled shade to partial sun. May tolerate full sun. Propagate from seed or cuttings.

Daviesia virgata, × ·8

Dawsonia superba, × ·6

DAWSONIA R. Br.
(after Dawson Turner, 18-19th-century English botanist)
Dawsoniaceae

Small to robust **mosses** forming clumps; **stems** erect, resembling miniature fir or pine trees; **leaves** simple, spirally arranged up the stem; **capsule** terminal, ovate, flattened.

A small genus of mosses found in Australia, New Zealand and New Guinea. It contains the largest of the mosses, some of which are handsome subjects occasionally grown by enthusiasts.

Dawsonia polytrichoides R. Br.
(resembling the genus *Polytrichum*)
Qld, NSW, Vic, Tas   Moss
5-20 cm tall

Erect **moss** forming spreading colonies; **stems** 5-20 cm tall, thin, erect, often branched; **leaves** 1-2 cm long, linear-lanceolate, acuminate, dark green, scattered spirally up the stems; **capsule** terminal, ovate in outline, flattened.

An attractive endemic moss found growing on exposed soil and clay, in partially shaded situations in forests etc. Plants form small, spreading patches of dark green foliage like miniature fir trees. Can be established in a suitable garden position or in a pot of friable clay soil. Clumps are deep-rooted, but can be transplanted or divided.

Dawsonia superba Grev.
(magnificent)
Qld, NSW, Vic, Tas   Giant Moss
10-50 cm tall

Robust **moss** forming spreading colonies; **stems** 10-50 cm tall, thin, erect; **leaves** 1-2 cm long, lanceolate, spreading, green with a glaucous bloom, scattered spirally up the stem; **capsule** terminal, ovate in outline, flattened.

Decaspermum fruticosum   D.L. Jones

This is one of the largest mosses known, and it is widespread in Australia on heavy clay outcrops in shady forests. (It also occurs in New Zealand.) These clays are friable, and the species grows in small, spreading colonies. Each stem resembles a miniature pine seedling. In dry times, the leaves close against the stems. This is quite an attractive moss and one which can be established in a suitable shady garden position or in a pot of friable clay soil. Clumps can be transplanted and even divided, but they are often deeply rooted in the soil.

Dawsoniaceae Broth.

A monogeneric family of mosses found in Australia, New Zealand and New Guinea. Some of the larger species have horticultural merit.

DECASCHISTIA Wight & Arn.

(from the Greek *decos*, ten; *schistos*, divided; in reference to the divisions of the involucre)

Malvaceae

Annual or perennial **shrubs**; most parts bearing stellate hairs; **leaves** alternate, simple or lobed, petiolate, with conspicuous stipules; **flowers** large and showy, hibiscus-like, axillary or in terminal racemes; **calyx** of 5 parts, supported by a persistent involucre of bracts; **petals** 5, flimsy, colourful; **stamens** borne on a central column; **fruit** a capsule.

A small genus of about 15 species mainly found in India and Malaysia, with a solitary species endemic in Australia. Propagation is from seed which retains its viability for 2-3 years or from cuttings of soft growth.

Decaschistia byrnesii Fryx.

(after N. B. Byrnes, contemporary Qld botanist)

NT

0.5-1.5 m x 0.5-1 m   Jan-April

Slender dwarf **shrub** with 1 to many stems; **stems** terete, densely covered with soft, stellate hairs; **leaves** 10-17 cm x 3-12 cm, simple or more often deeply 2-3 lobed, green on the upper surface, greyish white beneath, the lobes about 3 cm across; **stipules** about 1 cm long, the margins deeply lacerated; **flowers** about 15 cm across, bright yellow with a maroon centre (or about 10 cm across and lavender), opening widely, arising singly in the upper leaf axils or in a terminal, leafy raceme; **capsule** 0.6-0.8 cm long, oblate, covered with brown stellate hairs.

This attractive hibiscus-relative is endemic to NT where it is distributed around the Top End and Arnhem Land, growing on well drained, rocky slopes. Growth commences with the onset of the wet season, and after flowering the species matures seed and becomes dormant, or dies back to a perennial root system. Some plants may be annuals. Although somewhat straggly it is worth growing for its superb, colourful flowers. Each lasts a mere few hours, but they are produced in succession over many weeks. Cultural requirements of the species are a sunny situation, in freely draining soil. Climatically it is best suited to the tropics, but could also be expected to grow in warm subtropical areas. It can be propagated from seed which germinates readily if it is fresh or from cuttings of the soft growth, taken before flowering commences. Two forms are known. The typical subspecies has yellow flowers, and the upper leaves on the stems are distinctly 3-lobed. The ssp. *lavendulacea* Fryx. differs noticeably by its smaller flowers which are of an attractive lavender colouration, and its upper leaves on the stems, which are simple.

DECASPERMUM J. R. & G. Forster

(from the Greek *deca*, ten; *sperma*, a seed; referring to the fruit frequently having 10 seeds)

Myrtaceae

Shrubs or **trees**; young shoots hairy; **leaves** opposite, simple, entire, glabrous, bearing minute oil dots; **inflorescence** an axillary cyme or raceme; **flowers** small, numerous; **calyx** tubular, with 4-5 lobes; **petals** 4-5; **stamens** numerous, conspicuous; **fruit** a berry.

A small genus of about 30 species with a solitary one widespread in eastern Australia and also found in New Caledonia. It is a handsome plant deserving of wider cultivation. Propagation is from seed or cuttings.

Decaspermum humile

(low)

Qld, NSW   Silky Myrtle

6-12 m x 1-4 m   March-Aug

Tall **shrub** to small spreading **tree**; **bark** dark brown to black, hard, furrowed; young shoots silvery pink, bearing numerous silky hairs; **leaves** 2-5 cm x 1-2 cm, simple, ovate-lanceolate, entire, tapering to a long, fine point, shiny dark green on both surfaces; **racemes** 3-8 cm long, axillary, profuse; **flowers** about 0.6 cm across, white, pale mauve or pinkish, with numerous fluffy stamens, perfumed, on slender stalks 0.6-1.3 cm long; **berry** 0.3-0.6 cm across, globular, black, hard, mature April-Nov.

Widely distributed in drier rainforests from north-eastern Qld to near Gosford in NSW, often on shallow, skeletal soils. It is a handsome plant with excellent horticultural potential, and deserves to be widely cultivated. Its features include the dark, fissured bark, small shiny leaves, a spreading crown of foliage, silky new growth and masses of fluffy flowers followed by small black, currant-like berries. It grows easily in any acid, well drained soil, but young plants may be slow-growing and require some protection from full sun. Suited to temperate, subtropical or tropical regions. Young plants make very attractive pot specimens and could be suitable for indoor decoration. The species has an interesting flowering habit, with groups of flowers opening sequentially along the racemes, all on the same day. The flowers open early in the morning with the stamens widely spread, imparting a fluffy appearance. At this stage they exude a strong, sweet spicy perfume. By mid-morning the scent has generally gone and the flowers have lost their fluffy appearance, and by evening the flowers have completely withered. The next batch of flowers opens together a few days later. Propagation is from seed which should be sown fresh (the whole ripe berry can be sown) or from semi-hardwood cuttings which are slow to strike. This species has often been recorded under the name of *D. paniculatum* Kurz.

Decaspermum paniculatum Kurz.

see *D. fruticosum* Forster

DECAZESIA F. Muell.
(after Louis C. E. Amanieu, Duke of Decazes, 19th-century French politician and patron of botany)
Asteraceae

An endemic, monotypic genus, confined to the north-west coast of WA.

Decazesia hecatocephala F. Muell.
(a hundred heads)
WA

0.2-0.3 m x 0.1-0.2 m　　　　　Aug-Oct; also sporadic

Dwarf annual **herb**; **stems** many, erect from rootstock, few-branched, slightly hairy; **leaves** to 4 cm x 1 cm, oblong to lanceolate, nearly flat, woolly; **flowerheads** 1-1.5 cm across, terminal, white, scented.

In his original description Mueller describes this species as 'beautiful'. It occurs naturally in near-coastal areas of mulga scrub in the Austin and Ashburton Districts. Little is known of its reaction to cultivation, but it certainly has potential and is worth trying in containers, using 2 or more plants per container. Requirements are as for other annual species (see Volume 2, page 202 for further information). Propagate from seed or cuttings.

DECIDUOUS PLANTS

Deciduous plants are a feature of the flora in countries of the northern hemisphere, but in Australia the forests are composed principally of evergreen species. However, in tropical regions of northern Australia, there is a strong tendency in drier vegetation-types, for the trees to lose their leaves during the dry season. Such plants could be said to be facultatively deciduous rather than obligatorily but the dormant cycle persists even in cultivated plants. The period of deciduousness

in the relevant Australian species is generally short compared with deciduous species from overseas. Some may lose their leaves for 1-2 weeks; most for 4-6 weeks, while comparatively few, eg. *Melia azederach* var. *australasica*, may be bare for 3-4 months.

In the open forests of the Top End of NT, the Kimberleys and, to a lesser extent, Cape York Peninsula the majority of the eucalypts are deciduous, shedding their leaves during Sept-Oct and regreening with the first storms of the wet, eg. *Eucalyptus alba*, *E. clavigera* and *E. confertiflora*. Other trees in open forests are also deciduous, eg. *Brachychiton* spp., *Bombax ceiba* var. *leiocarpum* and the Baobab (*Adansonia gregorii*). The vegetation-type known as monsoonal rainforest or vine thickets is also dominated by deciduous species. Here the rainforest floor becomes thickly carpeted with leaves (of both trees and vines) during the dry season.

Deciduous plants are less noticeable in subtropical areas, and are a novelty in temperate Australia. In the autumn, subtropical rainforests are marked with gold splashes which are the shedding leaves of Red Cedar (*Toona australis*) and White Cedar (*Melia azederach* var. *australasica*.). In Aug-Sept the Deciduous Fig (*Ficus virens*) sheds its leaves but is bare for a period of only 2-3 weeks. Some trees may shed their leaves just before flowering. This means that in good seasons the flowers are well displayed; however, in many years a partial leaf-shedding and flowering makes the trees look unusual, eg. *Brachychiton acerifolium* and *Flindersia australis*. A notable deciduous plant in cold areas of Tas is the Deciduous Beech (*Nothofagus gunnii*). This is a common species and in the autumn the foliage turns yellow and bronze hues before falling.

Two aspects of deciduous plants are of importance for horticulture. The first of these concerns plants which may be growing beneath the trees. In tropical areas deciduous trees generally shed their leaves very rapidly, and this means a sudden increase in light-intensity and burning ultra violet rays on the plants sheltered by the canopy. Shade-loving plants such as ferns and gingers will generally suffer from these changed conditions, and even hardy species may be damaged. The second feature of deciduous Australian plants is the colourful flushes of new growth which follow the dormant period. Deciduous fig has light green to bronze growth enclosed in delicate stipules which are shed in clouds. Red cedar is often a colourful reddish bronze, while white cedar is always a delicate bright green. Many other rainforest species have bright red to purplish flushes.

DEERINGIA R. Br.
(after Dr. George Charles Deering)
Amaranthaceae

Slender to woody **climbers**; **leaves** simple, entire, alternate, glabrous or hairy; **inflorescence** a terminal panicle or axillary raceme; **flowers** unisexual or bisexual, small; **perianth** of 4-5 segments; **stamens** 4-5; **fruit** a berry.

A small genus of 12 species of climbers distributed in Madagascar, Indonesia and Malaysia, with 2 in eastern Australia. They have limited horticultural appeal and are rarely encountered in cultivation.

Propagate from seed which has a limited viability period and must be sown fresh or from stem cuttings.

Deeringia amaranthoides (Lamk.) Merr.
(like the genus *Amaranthus*)
Qld, NSW
3-4 m tall   Feb-April
Medium **climber** growing 3-4 m tall; young shoots slender, glabrous; **leaves** 5-9 cm x 2-4 cm, ovate to ovate-lanceolate, entire, bright green, thin-textured, drawn out into a long, acuminate point; **racemes** 5-25 cm long, pendulous, axillary or in terminal clusters; **flowers** about 0.4 cm across, greenish white, bisexual; **berry** 0.5-0.6 cm across, globular, red, fleshy, with 3 conspicuous furrows.
Widely distributed in rainforests and moist areas in sclerophyll forests from the Atherton Tableland to the south coast of NSW. Its most noticeable feature is the large clusters of colourful fruit which hang for fairly long periods. In dry seasons the plants may shed their leaves completely. The species has limited horticultural appeal and can be grown readily in gardens. It is not overly vigorous and can be safely trained over a bush or tree. Young plants should be placed in a sheltered area so that the stems can climb into a sunny situation and display their fruit. Propagate from cuttings which strike readily or seed which should be sown fresh. The species was previously known as *D. celosioides* R. Br.

Deeringia arborescens (R. Br.) Druce
(tree-like)
Qld, NSW
5-8 m tall   Oct-Jan
Tall, vigorous **climber**; young stems hairy; **leaves** 5-15 cm x 1.5-4.5 cm, ovate, thin-textured, drawn out into a long, acuminate point, often with a few rusty hairs on the underside; **racemes** 10-20 cm long, axillary or more usually clustered in terminal, leafy panicles; **flowers** unisexual; **male flowers** about 0.2 cm across, white; **female flowers** greenish, with a prominent ovary; **berry** about 0.3 cm across, globular, red.
A widely distributed woody climber extending from Cape York Peninsula to north-eastern NSW, usually growing in rainforests. It is rarely cultivated except by enthusiasts, and has limited horticultural appeal, being too vigorous for the average home garden. Propagate from seed which should be sown fresh. Cuttings would probably strike readily. The species was previously known as *D. altissima* F. Muell.

DELARBREA Vieill.
(after M. Delarbre, 18-19th-century French naturalist)
Araliaceae
Slender **shrubs** or small **trees**; **leaves** alternate, pinnate, crowded near the ends of branches; **leaflets** leathery, entire or crenulate; **inflorescence** a panicle; **flowers** bisexual; **calyx** bell-shaped, 5-lobed; **petals** 5, falling early; **stamens** 5; **fruit** ovoid, with a thin, fleshy layer.
A small genus of 6 species, all confined to the Australasian region. A solitary species is endemic in

north-eastern Qld. It is rarely encountered in cultivation but has considerable potential. Propagate from seed which must be sown fresh.

Delarbrea michieana F. Muell.
(after Archibald Michie)
Qld
4-8 m x 1-1.5 m   Sept-Nov
Slender, tall **shrub** or small **tree**; **petiole** bases expanding like a stipule; **leaves** pinnate, 0.6-1.2 m long; **leaflets** 20-30, 10-18 cm x 3-5 cm, lanceolate, stiff and papery, glossy green above, pale beneath, the tips acuminate; **panicles** 0.6-1.2 m long, narrow, sparsely branched; **flowers** about 0.4 cm across, yellowish; **sepals** divided into lobes; **fruit** 1.2-2 cm long, ovoid, dark shiny blue.
This attractive plant is confined to rainforests of north-eastern Qld, usually in near-coastal situations. The leaves are attractively patterned, and the large clusters of shiny dark blue fruit are most decorative. The species would appear to have excellent potential as an indoor plant for use in the home or office. It also makes an attractive garden subject for tropical and subtropical regions. It requires a shady, protected situation (especially when young) and deep, well drained, acid soils rich in humus. Propagate from seed which must be sown fresh.

DENDROBIUM Sw.
(from the Greek *dendron*, a tree; *bios*, life; the majority of species are epiphytes on trees)
Orchidaceae
Epiphytic, lithophytic or rarely terrestrial **orchids** forming clumps; **rhizome** creeping, usually branched; **pseudobulbs** reed-like or swollen, conical or angular, crowded or well-spaced, the reed-like pseudobulbs with leaves scattered, those with swollen pseudobulbs with leaves in a terminal group; **leaves** entire, small to large; **inflorescence** a raceme arising from a node; **flowers** numerous, small to large, colourful; **labellum** lobed, attached to the apex of the column-foot; **fruit** a capsule.
A very large genus of about 1400 species widely distributed through Asia and Polynesia to New Guinea and Australia. In Australia there are about 56 species, most of which are found in the eastern states, particularly Qld. Two species extend as far south as Vic, one of these reaching eastern Tas. Most species grow as epiphytes on the bark and branches of trees, but some are lithophytic on rock faces and boulders. A few terrestrial species are also known in the genus, and one of these occurs in Australia. Dendrobiums are generally to be found in sunny situations where the humidity is high at least for part of the year. All species have a distinctive cycle of vegetative growth and flowering.
Dendrobiums are very popular with orchid enthusiasts, and many exotic species and hybrids are grown in Australia. All of the native species have been tried in cultivation and most adapt well. Many species can be grown as garden plants in the tropics, either attached to trees or in pots on verandahs etc. Very hardy

Dendrobium adae

species, eg. *D. speciosum*, can be grown as garden plants (on trees, in pots or rockeries) as far south as Melbourne. The species from south-eastern Qld and NSW are cold hardy and can be grown in a bushhouse or unheated glasshouse in temperate areas. Species from highland areas of north-eastern Qld (such as the Atherton Tableland) will also grow in temperate areas under similar conditions but species from the tropical lowlands are cold sensitive and require a heated glasshouse to over-winter. Cold hardy species do not generally grow and flower well in the tropics.

Dendrobiums as a group prefer well lit situations with an abundance of air movement and humidity. Many species, especially the pendulous types, can be grown on slabs of material such as tree-fern fibre, cork or weathered hardwood. Other species are grown in pots with a coarse epiphyte mixture, and require repotting into new mixture every 2-4 years. All species respond favourably to regular, light applications of liquid fertilizers throughout the growing season. Water should be applied in abundance from Aug to March but the plants should be kept much drier from April to July to harden off the growths and encourage flowering. Dendrobium beetle is a very damaging pest in tropical and subtropical regions. There is considerable interest in the hybridization of native dendrobiums to produce new cultivars (see end of species section).

Propagation

Dendrobiums can be propagated from seed or by vegetative techniques such as division, from aerial growths or by tissue culture. Division of the clumps is best carried out in spring or early summer, and no clump should be split into fewer than 4 bulb sections. Some species, eg. *D. kingianum* and *D. discolor*, produce aerial growths at the top of the pseudobulbs, and when sufficiently large these can be removed and started as separate plants. Propagation of dendrobiums by tissue culture has met with considerable difficulty, but a few species and hybrids have been successfully propagated in this way. Seed must be sown on sterilized media and held in aseptic conditions until the plants are large enough to survive on their own. A useful technique for the enthusiast to try is the sprinkling of seeds on a pot of sphagnum moss to which has been added some root-tips of the orchid. The root-tips provide the fungus necessary for germination. This technique is successful on a limited scale (see Volume 2, page 214).

Dendrobium adae Bailey
(after Mrs J. W. R. Stuart)
Qld

July-Oct; also sporadic

Epiphytic **orchid** forming small to medium clumps; **rhizome** short, much-branched; **pseudobulbs** 15-60 cm x 0.4-0.7 cm, linear, cylindrical, becoming furrowed, yellowish; **leaves** 6-8 cm x 2-3.5 cm, 2-4 at the apex of the pseudobulb, ovate, dark green, keeled beneath; **racemes** 2-8 cm long, bearing 1-6 flowers on pedicels about 1.5 cm long; **flowers** 2-2.5 cm across, white, greenish, pale yellow or rarely apricot, waxy, opening widely, fragrant; **labellum** about 1 cm long, white with reddish markings, shortly and densely tomentose.

This slender orchid grows on trees of highland areas of north-eastern Qld, above 700 m alt. It is often locally common in humid areas of open forest and in rainforest. It is popular with orchid enthusiasts because of its ease of culture and neat, fragrant flowers. The form with apricot flowers (superficially resembling those of *D. fleckeri*) is a very attractive horticultural form. *D. adae* grows readily on slabs of tree-fern or in small terracotta pots of coarse material. It requires cool, humid conditions, and succeeds well in temperate regions. The plants can also be successfully attached to suitable garden trees, in a humid, shady position. Growth is generally slow but steady. The flowers of this orchid have a very pleasant fragrance in warm, sunny conditions. Propagate by division of the rhizomes into 4-5 pseudobulb sections or from seed.

Dendrobium aemulum R. Br.
(similar)
Qld, NSW   Ironbark Orchid; White Feather Orchid
July-Oct

Epiphytic **orchid** growing in small to medium clumps; **rhizome** woody, short; **pseudobulbs** 7-30 cm x 0.3-1 cm, extremely variable, linear to oblong, erect or curved upwards, clumped or radiating, dark brownish green to yellowish green, furrowed, hard and horny, narrowed at the base; **leaves** 2-4 at the apex of the pseudobulb, 2-5 cm x 1-3 cm, ovate to oblong, thick-textured, pale to dark green, dull or somewhat shiny; **racemes** 5-10 cm long, bearing 2-12 flowers on pedicels 1.2-2 cm long; **flowers** 1.5-2.5 cm across, white, cream or pinkish (all colour forms turn a deep pink before withering), opening widely, the segments slender and pointed, delicately fragrant; **labellum** 0.5-0.8 cm long, strongly curved, white with purplish markings, the callus bright yellow and crested.

This widespread and often locally common orchid extends from the Atherton Tableland in north-eastern Qld to south-eastern NSW. It grows in a variety of habitats and on a variety of host trees, and is quite variable in pseudobulb shape, leaf size and number, and flower dimensions and colour.

Dendrobium aemulum   J. Fanning

216

Dendrobium agrostophyllum

J. Fanning

In all, about 5 fairly distinct forms can be recognized:

(a) Brush Box form — common on the rough basal bark of large brush box trees (*Tristania conferta*) of north-eastern NSW and south-eastern Qld. The pseudobulbs, which grow up to 30 cm x 0.5 cm, are slender, produced in abundance and radiate like the spokes of a wheel. Each has 2-4 thick, dark green, shiny leaves and cream, white or pinkish flowers.

(b) Casuarina form — grows on casuarina trees in some open forests of the Atherton Tableland. The clumps are fairly small, and the pseudobulbs stout and erect, generally 5-8 cm x 1-1.3 cm, dark brown and prominently ribbed. Each has 2-3 thick, leathery, dark green leaves, and the flowers are large and crystalline white.

(c) Rainforest form — has straight, fairly stout pseudobulbs up to 20 cm x 0.8 cm, with 2-4 dark green, shiny leaves and cream or white flowers. It extends from north-eastern Qld to south-eastern NSW.

(d) Ironbark form — grows on ironbark trees (*Eucalyptus paniculata*) in open forests between south-eastern Qld and south-eastern NSW. It develops into fairly large clumps, and the pseudobulbs are crowded and dumpy, up to 7 cm x 1 cm, purplish brown, with 2-4 leathery, yellowish green leaves, and cream or white flowers.

(e) Callitris form — restricted to stringybark cypress pines (*Callitris macleayanus*) of the Atherton Tableland.

The clumps of this form are small, with a few very slender pseudobulbs 10-20 cm x 0.3 cm, usually 2 dark green leaves, and short racemes of 2-4 flowers which are only about 1 cm across and greenish cream.

The last form is slow-growing and resents disturbance. All of the other forms, however, are very amenable to cultivation and popular because of their showy displays of feathery, fragrant flowers. They can be grown in bushhouses or cool glasshouses, or attached to the trunks of suitable garden trees. The plants can be tied on to slabs of tree-fern fibre, weathered hardwood or cork, or can be grown in small pots of coarse mixture. They should be watered daily during summer weather, but kept much drier during winter, when watering every 7-10 days is sufficient. Plants respond to liquid fertilizers during summer. Propagate by division of the clumps into 4-5 bulb sections or from seed.

Dendrobium agrostophyllum F. Muell.
(grass-like leaves)
Qld   Buttercup Orchid
                                  July-Nov; also sporadic
   Slender epiphytic **orchid** forming erect linear clumps; **pseudobulbs** 15-60 cm x 0.5-1 cm, narrowly fusiform, erect, furrowed, dark brown, 2-6 cm apart on the rhizome, usually in lines rather than clumps; **leaves** 3-10 cm x 0.5-1.2 cm, narrowly ovate, dark green to yellowish, thin-textured, spreading, up to 20 per stem,

Dendrobium antennatum

the apex notched; **racemes** 2-5 cm long, axillary, bearing 2-10 flowers on pedicels 1-2 cm long; **flowers** 1.5-2.5 cm across, canary or buttercup yellow, widely expanding, sweetly perfumed, somewhat crowded on the raceme, usually partially drooping; **labellum** 7-9 cm long, prominently 3-lobed, the mid-lobe expanding suddenly.

A locally common orchid that is restricted to north-eastern Qld between the Burdekin River and the Annan River near Cooktown. It is most common at high elevations above 1000 m, in rainforest, but also extends to moist areas of open forest where it grows on casuarinas and small flaky-barked trees. Occasionally it grows on boulders, usually in situations where litter accumulates. This orchid flourishes in cultivation in temperate and cool subtropical regions, but is very difficult to grow in the tropics, where it will succeed only in highland areas. In suitable areas it can be grown as a garden plant attached to trees, but is most commonly grown in bushhouses or unheated glasshouses. The plants can be grown in small pots of coarse mixture or on slabs of tree-fern, cork or weathered hardwood. If grown in a pot, the rhizome should be on the surface and the plant may need staking until root growth is sufficiently established. Clumps can be induced to branch by severing into 3-4 bulb sections. This orchid is easy and rewarding to grow, requiring cool, humid conditions with adequate air movement. Propagate from seed or by division of the clumps.

Dendrobium baileyi   B. Gray

Dendrobium antennatum Lindley
(with antennae; in reference to the erect petals)
Qld   Antelope Orchid
March-Dec; also sporadic
Epiphytic **orchid** forming tall, slender clumps; **pseudobulbs** 15-60 cm x 1-2.5 cm, slightly fusiform, usually cylindrical in the upper half, erect; **leaves** 4-14 cm x 0.5-4 cm, usually about 12 scattered along the upper two-thirds of the pseudobulb, ovate, rigid, thick-textured, dark or yellowish green, the apex unequally notched; **racemes** 20-35 cm long, erect, usually arising from the apical nodes, bearing up to 15 flowers on pedicels about 3 cm long; **flowers** 2-6 cm across, opening widely, the sepals white or pale green, curly, the petals stiffly erect, white or pale green; **labellum** 2-2.5 cm long, 3-lobed, the disc with 5 keels, white with purple veins.

This widespread and common New Guinea orchid is rather rare in Australia, where it is confined to ranges of the northern part of Cape York Peninsula. It grows in well lit situations, usually high in sparse rainforest trees or in wattles along stream banks. This orchid is a very easy and rewarding species to grow. It is essentially tropical in its requirements, and large plants are rarely out of flower. A weathered hardwood slab is suitable for its culture, but the plants really thrive in pots of coarse material. It likes warm, humid conditions with plenty of light, and in temperate areas needs the protection of a heated glasshouse to overwinter. Plants respond vigorously to regular applications of liquid fertilizer over summer. Propagate from seed, by division of the clumps into 5-6 bulb sections or from aerial growths.

Dendrobium baileyi F. Muell.
(after F. M. Bailey, 19-20th-century Qld botanist)
Qld
sporadic; mainly Jan-Feb
Epiphytic **orchid** forming small to medium clumps; **pseudobulbs** 20-120 cm x 0.2-0.3 cm, linear throughout, slender and willowy, wiry, yellowish, especially when old; **leaves** 4-9 cm x 0.4-0.8 cm, linear to linear-lanceolate, thin-textured, pale green, spreading, falcate, somewhat grass-like; **racemes** 0.6-0.8 cm long, axillary, bearing 1-2 flowers on pedicels 0.7-0.8 cm long; **flowers** 2-3.5 cm across, expanding widely for only a few hours, yellowish green outside, densely spotted and blotched with red or purple on the inside, the segments long and spidery, becoming readily tangled as the flowers age; **labellum** 0.8-1 cm long, prominently 3-lobed, the mid-lobe acuminate.

Trees overhanging watercourses are the favoured habitat of this orchid which is uncommon in cultivation. It is found mainly in lowland tropical areas from the Burdekin River to central Cape York Peninsula, always in sheltered, humid situations. This orchid resents disturbance and is somewhat slow to re-establish in cultivation. It is very cold sensitive and dies quickly if the temperature drops below 8°C. The plants have very few reserves, and must be kept constantly moist in warm, humid conditions if cultivation is to be successful. A semi-shaded situation is desirable, and growth is continuous throughout the year. Cultivation is relatively easy in lowland tropical areas but the species has proved to be difficult in temperate regions, even in a heated glasshouse. Propagate from seed.

218

Dendrobium bairdianum Bailey
(after J. C. Baird, original collector)
Qld

Oct-Jan

Epiphytic **orchid** growing in small clumps; **pseudobulbs** 3-25 cm x 0.4-1 cm, linear or slightly swollen, erect, dark purplish brown, when young covered with pale, papery bracts, prominently furrowed; **leaves** 3-10 cm x 1-2 cm, 2-5 at the apex of the pseudobulb, narrow-ovate, dark green, unequally notched at the apex, keeled; **racemes** 2-8 cm long, bearing 2-7 flowers on slender pedicels 1-2 cm long; **flowers** 2-2.5 cm across, opening widely, pale green to yellowish green, becoming browner with age; **labellum** deep purple, about 1.5 cm long, curved, with a prominent expanded mid-lobe.

A localized orchid restricted to highland tropical regions of north-eastern Qld, where it grows on hardwood trees in open forest situations. It is a very distinctive orchid, with attractive flowers of an unusual colour combination. Unfortunately it has proved to be extremely difficult to grow, the majority of plants lingering and dying slowly. Plants have been tried in a variety of situations and conditions, with some measure of success being gained by potting them in a coarse mixture in very small terracotta pots and placing them in a humid but airy situation in an unheated glasshouse. Propagation from seed has proved to be difficult.

Dendrobium bigibbum   D.L. Jones

Dendrobium beckleri F. Muell.
(after Dr. H. Beckler, 19th-century botanist)
Qld, NSW   Pencil Orchid

Aug-Nov

Epiphytic **orchid** forming sparse, erect or semi-erect clumps up to 1 m long; **rhizome** thick, sparsely branched; **stems** about 0.3 cm thick, yellowish, tough; **leaves** 2-16 cm x 0.2-1.2 cm, terete, thick and fleshy, erect, deeply grooved, dark green, often with black markings, decreasing successively in size; **racemes** slender, bearing 1-4 flowers; **flowers** 2.5-3.5 cm across, white, pale green or mauve, with dark stripes on the petals and sepals, and purplish markings on the labellum; **labellum** 2-3 cm x 0.3-0.5 cm, white with purple markings, the margins crisped, the apex long and pointed.

A common orchid widespread from north-eastern Qld to central-eastern NSW, usually in lowland situations and often in *Casuarina* swamps and along stream banks. The plants are erect when small, but the longer stems generally hang outward from the clump. This species is a very easy orchid to grow, and succeeds well with scant protection in temperate regions. It can be grown in a terracotta pot of coarse mixture or mounted on a slab of tree-fern, weathered hardwood or cork. Plants can also be readily established on suitable rough-barked garden trees. Exposure to a fair amount of light is necessary for good flowering. Plants are quite tolerant of dryness once established and respond to applications of liquid fertilizers. Propagate by division or from seed.

Dendrobium bifalce Lindley
(2 beaks)
Qld

April-July

Epiphytic **orchid** forming medium sized clumps; **pseudobulbs** 10-40 cm x 1-2.5 cm, fusiform, furrowed, erect, yellowish or with purplish tonings; **leaves** 5-15 cm x 2-5 cm, 2-4 at the apex of the pseudobulbs, ovate, leathery; **racemes** 10-25 cm long, erect, the 8-10 flowers crowded near the end, on pedicels 1-2 cm long; **flowers** 2-3 cm across, pale yellowish green with numerous dark brown dots and blotches; **labellum** 1.5-2 cm long, deflexed, with a prominently notched mid-lobe.

A widespread orchid that in Australia is restricted to monsoonal rainforests of Cape York Peninsula, where it grows high on trees or in open, bright-light situations. It is comparatively rare in Australia, and much more common in New Guinea. It is an easy species to grow, but is encountered only in enthusiasts' collections. It is essentially a tropical orchid, as it is very sensitive to cold and requires hot, humid conditions. In temperate regions, it needs a heated glasshouse with a minimum winter temperature of 12-15°C. It needs a well lit situation, especially for flowering. Plants can be attached to hardwood slabs or grown in pots of a coarse, well drained mixture. Watering should be copious during the summer, but the plants are best kept fairly dry during the winter. Propagate by division of the clumps into 4-6 bulb sections or from seed.

219

Dendrobium bigibbum

Dendrobium bigibbum Lindley
(2-humped; referring to the spur)
Qld   Cooktown Orchid
March-July; also sporadic

Epiphytic **orchid** forming small to medium, slender clumps; **pseudobulbs** 15-120 cm x 1-1.5 cm, slender, terete, erect or semi-erect, green, pink or purplish, when young covered with papery bracts; **leaves** 5-15 cm x 1-3.5 cm, 3-5 scattered in the upper third of the pseudobulb, narrow-ovate, acuminate, dark green, often with purplish margins; **racemes** 10-40 cm long, arching, bearing 2-20 flowers on pedicels 2-3 cm long; **flowers** 3-5 cm across, widely opening, the segments sometimes recurved, the petals broad and rounded, white, lilac, mauve or magenta; **labellum** 1.5-2.5 cm long, deflexed, with a narrow mid-lobe and a prominent spur.

This colourful orchid is common and widespread on Cape York Peninsula and the Torres Strait islands, growing on stunted scrubby trees in well lit, often exposed situations. It is a very hardy orchid and the plants frequently shed all their leaves during the dry winter season, although those in more protected sites may retain them. This orchid is undoubtedly Australia's showiest and best-known species. It is Queensland's floral emblem and is very popular in cultivation. This popularity has led to its decimation in its native state. The orchid is easily grown, but is essentially tropical in its requirements. In coastal tropical regions it can be grown on trees, on stakes in the garden or in bushhouses, but in temperate regions it needs a heated glasshouse with a minimum temperature of 12-15°C. Strong light is essential for good flowering, and the buds are sensitive to cold snaps while developing, frequently yellowing and falling prematurely after such events. Plants can be attached to hardwood slabs or grown in small terracotta pots of a coarse, well drained mixture. Water should be applied copiously during the summer months when growth is strong, but the plants are best kept fairly dry while dormant over winter. Liquid fertilizers encourage strong growth if applied regularly over summer. Propagate by division of the clumps into 4-6 bulb sections, from aerial growths or from seed.

D. bigibbum hybridizes naturally with *D. discolor* to produce some attractive progeny (see *D.* × *vinicolor* and *D.* × *superbiens*). It is also widely used by orchid-breeders in hybridization programmes, and has given rise to many colourful hardcane hybrids which are popular in tropical regions.

D. bigibbum is an extremely variable orchid, the identity of which has been subject to much confusion and controversy. The names *D. phalaenopsis* Fitzg. and *D. schroederanum* (Hort. ex Masters) H. G. Reichb. have been applied to it at various times, but these are now known to be non-Australian species.

Botanically, there are 2 varieties of *D. bigibbum*. The typical variety is found in the upper part of Cape York Peninsula, and the Torres Strait islands. Its petals curl after opening.

The other variety (var. *superbum* Hort. ex H. G. Reichb.) consists of 2 subvarieties.

The subvar. *superbum* has slender pseudobulbs and fairly large flowers, the petals of which do not curl back. It is confined to the southern part of Cape York Peninsula.

The subvar. *compactum* (C. White) Dockr. may be immediately recognized by its short, stout, swollen pseudobulbs. It grows in colonies on rocks in the south-eastern part of Cape York Peninsula, often in harsh, exposed situations. It is very popular in cultivation and has been nearly exterminated by over-collecting. It grows readily and can be propagated from aerial growths. A white-flowered form has been reported. The cv. 'Blue Moon' is a very fine form with pale lilac or bluish flowers.

A range of other varieties has been named, but these are now recognized as natural hybrids or minor colour variations which are better classed as cultivars. The 2 natural hybrids are:

var. *georgei* White — this orchid is actually a natural backcross of *D. bigibbum* to *D.* × *superbiens*, but the flowers are uniformly pink-mauve, and lack any white margin to the segments. It appears to be of rare sporadic occurrence on Dayman Island in the Torres Strait.

var. *venosum* Bailey — this orchid is another natural backcross of *D. bigibbum* to *D.* × *superbiens*. The general growth habit is similar to *D.* × *superbiens*, but the flowers are smaller, and pink with white margins on the recurved segments. It is of rare sporadic occurrence in low scrubs near the top of Cape York Peninsula. Cultural requirements are as for *D.* × *superbiens*.

Minor colour variations which are best regarded as cultivars are as follows:
'Albomarginatum' — with white margins on the sepals and petals
'Album' — with white flowers
'Candidum' — a white-flowered form that originated on Torres Strait islands
'Hololeucum' — with white flowers
'Macranthum' — with large mauve-pink flowers lacking any white patch on the labellum
'W. Parton' — sepals mauve, petals pale lilac streaked with broad, darker bands.

Dendrobium canaliculatum R. Br.
(longitudinal grooves)
Qld, WA, NT   Antelope Orchid; Tea-tree Orchid
Aug-Nov

Epiphytic **orchid** forming small to medium, congested clumps; **pseudobulbs** 3-15 cm x 1.5-3 cm, fusiform, fleshy, very swollen about the middle, crowded, when young covered with white, papery bracts; **leaves** 5-25 cm x 0.5-1.2 cm, 2-6 at the apex of the pseudobulb, linear, fleshy, dark green, deeply grooved along the upper side; **racemes** 10-40 cm long, arching, bearing numerous flowers crowded towards the end, on pedicels 1-2.5 cm long; **flowers** 2-2.5 cm across, with twisted tepals which are white or green at the base and yellow, pink or brownish in the upper half; **labellum** 1-1.5 cm long, 3-lobed, white with rich purple markings, the keel crested.

A widely distributed and locally common orchid found almost exclusively on paperbarks in coastal and inland swamps. It is most abundant in Qld, where it extends from Cape York Peninsula to Mackay. The plants are found in open, bright-light but humid situations. The flowers are long-lasting and fragrant.

This orchid is rather difficult to cultivate, and poses many problems for beginners and enthusiasts in temperate regions. It is very cold sensitive and quickly rots if frosted or watered too frequently over winter. The plants are best attached to a material that dries out rapidly after watering, such as a piece of weathered hardwood or cork with no moss backing. They can also be grown on paperbark, but this usually rots fairly quickly under glasshouse conditions, and harbours pests. Plants should be hung in a well lit, airy position and watered daily during the summer months but kept almost dry over winter. A heated glasshouse with a minimum temperature of 10°C is necessary in temperate climates. Well grown specimen plants are extremely showy. In tropical and warm subtropical areas the plants can be hung in a bushhouse or attached to suitable garden trees. Propagate from seed. Division of the clumps is best avoided.

This orchid is quite variable in growth habit and flower colour. The var. *canaliculatum* from Qld has yellow-tipped tepals and mauve markings on the labellum. The var. *foelschei* (F. Muell.) Rupp & Hunt has longer, much less swollen pseudobulbs, slender, terete leaves and smaller flowers. It is found in NT and north-western WA. The var. *nigrescens* Nicholls, from Cape York Peninsula and the Torres Strait islands, has brown to chocolate-tipped tepals. The var. *pallidum* Dockr. has straight segments which are white with yellow tips, and an entirely white, broad labellum.

Dendrobium cancroides Hunt
(like a crab; in reference to the flower)
Qld

Dec-May; also sporadic

Epiphytic **orchid** forming small clumps; **pseudobulbs** 20-80 cm x 0.4-0.6 cm, semi-erect or pendulous, thin near the base, somewhat flattened in the upper two-thirds; **leaves** 5-10 cm x 1-3.5 cm, ovate to obovate, dark green, thin-textured, spreading, scattered in the upper half of the stem; **racemes** about 0.3 cm long, bearing 2 flowers which face each other, on pedicels about 0.4 cm long; **flowers** 1.5-2 cm across, not expanding widely, reddish brown with a yellowish centre, rough and tuberculate, lasting only a few hours; **labellum** about 1 cm long, the mid-lobe densely covered with cream cilia.

This orchid favours trees overhanging watercourses in lowland tropical areas of north-eastern Qld between the Johnstone River and central Cape York Peninsula. It always grows in shady, humid situations and is not commonly encountered. In cultivation it has proved to be a difficult and tricky species. The plants resent disturbance, and in unfavourable conditions they die very quickly. They require warm, moist, humid conditions with air movement. In lowland tropical areas they can be successfully grown in bushhouses if hung beneath the benches. They can be grown in small pots containing a coarse medium or on slabs of tree-fern, cork or weathered hardwood. Once suitable conditions are found, the plants should not be moved. Propagate from seed.

Dendrobium carrii Rupp & C. White
(after T. Carr, original collector)
Qld

Aug-Oct

Epiphytic **orchid** growing in small, slender strands; **rhizomes** about 0.4 cm thick, sparsely branched; **pseudobulbs** 1.5-3.5 cm x 0.8-1.5 cm, somewhat attenuated or constricted, yellow-green, furrowed with age, when young covered with bracts; **leaves** 5-10 cm x 0.7-1 cm, linear to oblong, 2 at the apex of the pseudobulb, dark green, sometimes twisted; **racemes** 5-8 cm long, erect, bearing 5-10 flowers on pedicels 0.5-1 cm long; **flowers** 0.8-1 cm across, dingy white or cream, with pointed segments, not opening widely; **labellum** 12-15 cm long, curved, white with a yellow or orange mid-lobe, the lateral lobes prominent.

This little-known orchid is confined to a few highland rainforests of north-eastern Qld, usually growing in small, slender clumps on the outer branches of rough-barked trees. It resents disturbance and is somewhat slow to re-establish. Because of its creeping growth habit it is best grown on slabs of hard material such as weathered hardwood or cork. The plants are generally very slow-growing and require cool, humid conditions. They can be successfully grown in a cool glasshouse in southern Australia, but prefer to be kept dry during the winter and hung near the glass for flowering. The flowers are never showy and the species is mainly for the ardent enthusiast. Backcutting the rhizome into sections of 4 pseudobulbs will induce new growth from old bulbs. Propagate by division of the rhizomes into 4-5 bulb sections or from seed.

Dendrobium carronii P. S. Lavarack & Cribb
(after William Carron, botanist on the Kennedy expedition)
Qld

July-Sept

Epiphytic **orchid** forming small, compact clumps; **rhizome** short; **pseudobulbs** 1.5-5 cm x 1-3 cm, globose, swollen, clustered, purplish; **leaves** 6-12 cm x 0.3-0.5 cm, 2-4 at the apex of the pseudobulb, fleshy, almost terete, grooved, dark green; **racemes** 8-20 cm long, bearing 2-12 flowers on pedicels about 2 cm long; **flowers** 2-2.5 cm across, dark brown to purple, and white; **sepals** white; **petals** with a solitary twist, white at the base, brown to purple at the apex.

A recently described species from New Guinea and north-eastern Cape York Peninsula. It grows on paperbarks in open forest, and on *Xanthostemon* and *Tristania* species in swampy areas. It is a colourful, very attractive orchid, best suited to growing in tropical zones. Plants have cultural requirements similar to those for *D. canaliculatum*, and like that species they are very cold sensitive and difficult to grow in temperate regions. They quickly rot if frosted or watered too frequently over winter. In temperate regions they should be kept dry over winter and hung near the glass in a heated glasshouse with a minimum temperature of 10°C. In warmer regions they can be grown in a bushhouse or unheated glasshouse, but prefer a well lit situation. Plants can be grown in a small pot of very coarse material or attached to a piece of paperbark, weathered hardwood or cork. They should be hung in

Dendrobium cucumerinum

Dendrobium cucumerinum   J. Fanning

an airy situation and watered daily during the summer months, but kept almost dry over winter. Propagate from seed or by division. The species is very closely related to *D. canaliculatum*, but has short, triangular sepals and a very different labellum.

Dendrobium cucumerinum Macleay ex Lindley
(resembling a cucumber)
Qld, NSW   Cucumber Orchid
                                          Nov-Feb
    Epiphytic **orchid** growing in spreading patches; **rhizomes** about 0.4 cm thick, woody, branched; **leaves** 1.5-3.5 cm x 0.7-1.2 cm, ovoid, blunt, gherkin-like, with the surface raised in numerous bumps and keels, thick and fleshy, alternate, widely spaced, dark green; **racemes** 2-5 cm long, bearing 2-10 flowers; **flowers** about 1.2 cm across, cream to greenish white, with purplish stripes towards the centre; **segments** slender, blunt, often not spreading widely; **labellum** 1-1.4 cm long, curved, with distinctly crisped margins, with 3 dark, undulate, longitudinal keels.
    A rather uncommon orchid extending from south-eastern Qld to central NSW. It is most often found on River Oaks (*Casuarina cunninghamiana*), usually on the upper side of the main branches. These trees grow exclusively along stream banks, and the orchid seems to prefer valleys which are well inland. This orchid resents disturbance, and may be somewhat difficult to

re-establish and tricky to grow. It is best tied on to a hard slab of material such as weathered hardwood, cork or branches of *Casuarina* or *Pittosporum*. Tree-fern seems unacceptable, the plants lingering and dying slowly. Plants should be hung in a well lit situation where humidity is fairly high. They should be watered every second day during summer, but kept much drier over winter until new growth commences and the weather warms up. In temperate areas they require the protection of an unheated glasshouse, or a dry situation in a bushhouse. Propagate by division or from seed. Small divisions of less than 10 leaves may be difficult to establish.

Dendrobium × delicatum (Bailey) Bailey
(delicious)
Qld, NSW
                                          Aug-Oct
    Lithophytic **orchid** forming large congested clumps; **pseudobulbs** 10-50 cm x 1-4 cm, broadest at the base, evenly tapered upwards, cylindrical or slightly flattened, covered with overlapping grey-brown bracts which are obvious when young; **leaves** 5-18 cm x 1-3.5 cm, 2-6 at the apex of the pseudobulb, narrow-ovate to obovate, thick and leathery, dark green, occasional aerial growths produced from apical buds; **racemes** 10-40 cm long, many-flowered, crowded, on pedicels 1.5-2 cm long; **flowers** 1.5-2.5 cm across, white, cream, pink or mauve, with pink or mauve blotches and suffusions, narrowly to widely open, waxy, fragrant; **labellum** 1-1.5 cm long, curved, white with pink or purple markings.
    This orchid is a sporadic and uncommon natural hybrid between *D. speciosum* and *D. kingianum*. It is likely to occur where the two species intermingle, and has been found in scattered localities between the Blackall Ranges in Qld and the Hunter River of central NSW. Because both parents are variable, the progeny vary greatly from area to area. Many forms have been collected and grown by orchid enthusiasts, and the superior types have been propagated vegetatively and sold or traded as clones, sometimes with a cultivar name. Some clones are particularly colourful and free-flowering, lending themselves well to specimen culture. This hybrid has been produced artificially many times by orchid-breeders and has been greatly improved by the careful selection of the parents used in the cross. It is interesting to note that this cross was the first recorded artificial hybrid using Australian native orchids. It was carried out in England in the late 1800s and was named *D.* 'Specio-kingianum'.
    D. × delicatum is a hardy and rewarding orchid to grow. It is vigorous and ideal for pot or tub culture, but can also be grown on slabs, in baskets or attached to garden trees and rocks. It requires abundant water during hot weather but should be kept drier throughout winter, prior to flowering. It will tolerate considerable sunshine, and flowers best in exposed situations. The potting mixture should contain some coarse pieces of broken brick or terracotta to keep the medium open and well drained. Like many temperate epiphytes, this hybrid is difficult to flower in tropical regions. Propagate by division of the clumps, from

aerial growths potted in sphagnum moss or from seed of hybridized parents.

Some selected cultivars of this orchid are:

'Apple Blossom' — this form is strong-growing and produces white flowers

'Coloratum' — this form has slender, dark pseudobulbs and delicate pink flowers which have fairly slender sepals. It is popular and rather common in cultivation

'Kestevenii' — this form was originally named as a distinct species. It has very thick pseudobulbs and large, thick, dark green leaves (up to 18 cm x 6 cm). The flowers are white or cream, slightly suffused with pink. It was originally collected in 1930 at Alum Mountain near Bulahdelah, NSW. This form is a very vigorous grower but is somewhat disappointing in flower-profusion and colour.

Dendrobium dicuphum F. Muell.

(in reference to the lobed spur at the base of the labellum)

WA, NT

March-Aug

Epiphytic **orchid** forming small to medium, slender clumps; **rhizomes** short, much-branched; **pseudobulbs** 5-50 cm x 1-2.5 cm, slender, terete, erect, green, when young covered with papery bracts; **leaves** 5-20 cm x 0.5-3 cm, 2-10 scattered in the upper half of the pseudobulb, narrow-ovate, acuminate, dark green, often recurved; **racemes** 15-50 cm long, arching, bearing 2-20 flowers on pedicels 2-4 cm long; **flowers** 2.5-5 cm across, white, rarely pale pink, widely opening, the segments sometimes recurved, the petals broader than the sepals; **labellum** 1.3-2 cm long, prominently 3-lobed, white with maroon markings on the lateral lobes.

Widely distributed from the Kimberley region of north-western WA to NT and some islands of the Gulf of Carpentaria. It commonly grows on paperbarks in semi-swampy woodland, around the margins of swamps or along streams. It always grows in situations of bright light, and the plants survive long dry periods, frequently shedding all leaves and the bulbs shrivelling severely in the process. In cultivation, the species succeeds best in dry tropical regions such as Darwin, where it can be grown as a bushhouse plant or on suitable garden trees. In temperate regions it needs a heated glasshouse with a minimum temperature of 12-15°C. Strong light is essential for flowering, and the buds are sensitive to cold snaps while developing. Plants can be tied to hardwood slabs or grown in small terracotta pots of coarse, well drained mixture. Water should be applied in abundance while the plants are in growth over summer, and then sparingly over winter. Propagate from aerial growths or seed.

Dendrobium discolor Lindley

(different colours)

Qld

Golden Orchid

April-Dec; also sporadic

Epiphytic, lithophytic or terrestrial **orchid** forming large, spreading clumps; **pseudobulbs** 30-70 cm x 1-6 cm, erect, swollen in the basal quarter, then tapering to each end, yellowish, brown or purple, when long the top part usually drooping; **leaves** 5-16 cm x 2-5 cm, ovate or elliptical, leathery, light or dark green, blunt, the tip often notched, scattered in the upper two-thirds of the stem; **racemes** 20-60 cm long, arching, borne in the upper nodes, bearing 10-40 flowers on pedicels 2-3 cm long; **flowers** 3-8 cm across, opening widely, yellow, yellowish brown, golden brown, bronze or dark brown, often with other colour suffusions, the perianth segments twisted in varying degrees; **labellum** 1.5-2.5 cm long, 3-lobed with 3 deep keels, often purplish.

This orchid is common and often widespread in north-eastern Qld, from near Rockhampton, extending north to adjacent coastal islands, through the Torres Strait islands to New Guinea. It grows in exposed situations, usually on large rocks but also on trees, and on some coastal islands may even grow in sand. It is a very tough, hardy orchid that forms untidy, sprawling clumps which spread by aerial growths produced from near the apex of the pseudobulbs. It is believed that some clones colonize areas by this vegetative technique and rarely produce fertile seed. The flowers are exceptionally long-lived and it is unusual to find large clumps without some flowers. Natural hybrids occur with other species such as *D. nindii, D. semifuscum* and *D. bigibbum.*

Golden orchid is a very hardy species and easy to grow. In the tropics and coastal areas of the subtropics it can be grown as a garden plant amongst rocks or attached to trees, walls etc. It likes plenty of sunshine and responds to feeding with liquid fertilizers or decayed animal manure. In temperate regions the species has proved to be cold sensitive, requiring a glasshouse with a minimum temperature of 10°C. It requires plenty of light and for best results should be hung near the glass. Plants can be mounted on a hardwood slab or potted in a mixture of coarse, chunky material.

This orchid is extremely variable, allowing scope for selection and cultivation, with some interesting colour forms and variations known.

The var. *discolor* is particularly variable and amenable to selection. A leading grower, Ken MacPherson of Proserpine, has devoted over 25 years to selection and breeding of this orchid. He selects strong-growing plants which have heavy-textured, dark bronze flowers. Five of his clones are:

'Beautiful' — sturdy plant; bronze flowers with a violet labellum

'Blue Eyes' — fawn coloured flowers with violet labellum; produces aerials very freely

'Brunette' — heavy-textured racemes of red-bronze flowers; rarely produces aerial growths

'Magnificent' — large racemes and heavy-textured flowers; produces aerial growths freely

'Salmon Gold' — sturdy plant with fawn to brown flowers; makes an excellent specimen plant; produces aerial growths freely.

Horticulturally, the var. *discolor* forma *broomfieldii* (Fitzg.) Dockr. is one of the most attractive colour forms. The plants are similar in most respects to var. *discolor* but the flowers are bright golden to canary yellow with a varnished lustre, and the labellum has a prominent white disc. This form grows readily under similar conditions to those for var. *discolor*, but is much slower. In some plants the flowers have a greenish tinge.

More than 20 variations of forma *broomfieldii* are known based on growth vigour, and flower size and colour. Three of them are:

'Golden' — a fine form with shiny yellow flowers, from Hamilton Island

'Green Gloucester' — flowers are greenish-tinged, originating from a selfing of 'Robusta'

'Robusta' — originated on Whitsunday Island. It is a strong-growing form with yellow flowers.

Two other botanical varieties have been named but little is known about them. The flowers of the var. *fimbrilabium* (H. G. Reichenb.) Dockr. have a yellow labellum with fringed lateral lobes, while in var. *fuscum* (Fitzg.) Dockr. the dark reddish brown flowers are only about 4 cm across and the sepals are flat, not undulate.

Propagate all forms by division of the clumps into 5-6 bulb sections, from aerial growths and in some cases from seed.

Dendrobium falcorostrum Fitzg.
(shaped like a falcon's beak)
Qld, NSW
Beech Orchid
Aug-Oct

Epiphytic **orchid** forming large clumps; **rhizome** thick, short, much-branched; **pseudobulbs** 10-50 cm x 1-1.5 cm, cylindrical or fusiform, erect, crowded, narrow at the base, yellowish green, ribbed; **leaves** 6-15 cm x 1.5-3 cm, 2-5 at the apex of the pseudobulb, ovate to obovate, dark green, leathery; **racemes** 8-16 cm long, bearing 4-20 flowers on pedicels 1.5-2.5 cm long; **flowers** 2.5-4 cm across, thick-textured, waxy, opening widely, crystalline white, strongly fragrant; **labellum** about 3.5 cm long, white with yellow and purple markings, curved, the mid-lobe ending in a prominent point.

This orchid is restricted in its natural state to highland beech forests of south-eastern Qld and north-eastern NSW, but is generally locally common and grows in large, congested colonies on the Antarctic Beech and adjacent trees. It is one of Australia's most beautiful orchids and is prized by enthusiasts. It grows readily, especially in temperate regions, and lends itself well to specimen culture. Large, well grown plants become smothered with flowers and provide an impressive and aromatic display. The plants are best suited to cultivation in pots but can also be attached to slabs or suitable garden trees. They need a coarse mixture and respond vigorously to weekly applications of liquid fertilizers during summer. They are hardy enough to grow in bushhouses in temperate regions, but need cool conditions in the tropics if they are to be flowered successfully. The species is somewhat variable, and two distinct growth forms occur. One has long, slender pseudobulbs 30-50 cm in length and the other has squat pseudobulbs, usually 10-17 cm long, which are prominently grooved. Both forms grow equally well. *D. falcorostrum*, because of its outstanding horticultural features, has been used extensively in breeding programmes with native orchids. Propagate by division of the clumps or from seed.

Dendrobium fleckeri Rupp & C. White
(after Dr. Hugo Flecker, original collector and noted north Qld orchidologist)
Qld
Aug-Jan; also sporadic

Epiphytic **orchid** forming small, slender clumps; **rhizome** short, much-branched; **pseudobulbs** 10-40 cm x 0.3-0.5 cm, linear, cylindrical, becoming furrowed, yellowish green, when young covered by brown, papery bracts; **leaves** 3-8 cm x 1-2.5 cm, 2-3 crowded at the apex of the pseudobulb, ovate, dark green, thin-textured, the apex notched; **racemes** 1-3.5 cm long, bearing 1-3 flowers on pedicels 1-1.5 cm long; **flowers** 2.5-3 cm across, cream to yellowish green but more usually a rich apricot hue, thick and fleshy, opening widely; **labellum** about 1 cm long, curved, white marked with purplish red, the margins of the mid-lobe densely hairy.

An unusual, slender orchid confined to mist and cloud zones of highland mountains and tablelands of north-eastern Qld, but often locally abundant and forming colonies on trees and boulders. The flowers are attractive individually, but are not produced in abundance. They exude a sweet musky fragrance during warm weather. The pseudobulbs produce aerial growths from the apical nodes, and colonies are built up by this technique of vegetative propagation. This orchid is fairly easy to cultivate, requiring cool, moist, humid conditions. It succeeds well in a bushhouse in temperate regions. The plants can be grown on slabs of tree-fern or weathered hardwood, but do best in small terracotta pots filled with a coarse mixture. Growth is generally slow. Plants in lowland tropical regions have proved difficult to flower. Propagate by division of the clumps, from aerial growths or from seed.

Dendrobium × foederatum St. Cloud
(evil-smelling)
Qld
Oct-Jan

Epiphytic **orchid** growing in pendulous clumps up to 30 cm long; **rhizome** thick, shortly creeping; **stems** about 0.4 cm across, horny, branched, pendulous; **leaves** 2.5-12 cm x 0.3-0.5 cm, terete or slightly flattened, dark green, straight or slightly curved; **racemes** 3-6 cm long, bearing 3-8 flowers; **flowers** 1.5-2 cm across, white or cream with purplish markings on the labellum, opening widely; **labellum** 0.9-1.2 cm x 0.4 cm, curved, prominently 3-lobed, the margins of the mid-lobe crenulate, keel of 3 ridges.

A rare orchid which is a natural hybrid between *D. rigidum* and *D. teretifolium* var. *fasciculatum*. It is known only from coastal swamps near the mouth of the Barron River in north-eastern Qld, and much of this habitat has been cleared since its discovery. This hybrid has been raised by artificial crossing of the natural parents. The plants grow readily but are somewhat cold sensitive and require a heated glasshouse over winter in temperate regions. They can be grown in a pot of coarse mixture or mounted on a slab of cork or weathered hardwood. Propagate by division or from seed. Because it is a hybrid, seed must be collected from hybridized flowers.

224

Dendrobium falcorostrum

J. Fanning

Dendrobium fusiforme (Bailey) Bailey =
D. *ruppianum* A. Hawkes

Dendrobium gracilicaule F. Muell.
(slender stems)
Qld, NSW

July-Sept

Epiphytic **orchid** growing in small to large, congested clumps; **rhizome** thick, short, much-branched; **pseudobulbs** 20-90 cm x 0.3-1 cm, swollen at the base, then constricted, then linear-cylindrical throughout, erect or deflexed, yellowish green, becoming prominently ribbed, when young covered with brown, papery bracts; **leaves** 5-13 cm x 2-4 cm, 3-6 at the apex of the pseudobulb, ovate to lanceolate, dark green, thin-textured, the tip often unequally bifid; **racemes** 5-12 cm long, bearing 5-30 flowers on pedicels about 1 cm long; **flowers** 1-1.5 cm across, thick and waxy, often drooping, dull yellow inside, irregularly blotched outside with red-brown, rarely all-yellow with no blotches, cup-shaped, not opening widely, fragrant; **labellum** 0.6-0.8 cm long, curved, with prominent lateral lobes.

A widespread and common orchid that extends from the Atherton Tableland in north-eastern Qld to the Hawkesbury River in central NSW. It grows in rainforests and open forests, usually on trees but occasionally on rocks and boulders. It is one of the hardiest of the native epiphytic orchids to cultivate, being very tolerant of adverse conditions and neglect. It can be attached to suitable garden trees, or grown in pots or on slabs in bushhouses or unheated glasshouses. It prefers to be underpotted in a coarse mixture, and will grow on a variety of slab materials. The plants need regular watering during the summer but can be kept quite dry in winter. Liquid fertilizers are appreciated over summer. Propagate by division of the clumps into 3-5 bulb sections or from seed.

A form from the Atherton Tableland is fairly vigorous, with thicker pseudobulbs than plants from southern areas. It tends to be difficult to flower in temperate regions. The var. *howeanum* Maiden, from Lord Howe Island, has stouter pseudobulbs, up to 6 large leaves and larger creamy yellow flowers which lack any blotches. These are generally borne on a shorter raceme and have a different perfume to that of the mainland variety. It grows and flowers readily in cultivation.

Dendrobium × gracillimum (Rupp) Rupp
(very slender; referring to the pseudobulbs)
Qld, NSW

Aug-Oct

Epiphytic or lithophytic **orchid** forming medium to large clumps; **rhizome** thick, short, much-branched; **pseudobulbs** 20-60 cm x 1-1.5 cm, erect, cylindrical to slightly swollen, erect or spreading, crowded, covered with grey-brown bracts when young; **leaves** 10-20 cm x 2-6 cm, 3-5 at the apex of the pseudobulb, oblong to obovate, dark green and leathery; **racemes** 10-30 cm

225

long, bearing numerous crowded flowers on pedicels about 1.5 cm long; **flowers** 1.5-2.3 cm across, waxy, not opening widely, often lax, greenish, cream, white or yellow, very fragrant; **labellum** 0.6-1.2 cm long, curved, with purplish markings, the lateral lobes prominent.

This orchid occurs in nature as a sporadic natural hybrid between *Dendrobium speciosum* and *D. gracilicaule*. It is widespread from south-eastern Qld to central NSW, growing on rocks or trees, but is not generally common. *D. speciosum* is quite variable from area to area and thus *D.* × *gracillimum* varies accordingly, particularly in the robustness of the pseudobulbs and the flower size and colour. The same cross has been produced many times artificially by hybridists using selected forms of *D. speciosum* and *D. gracilicaule* var. *howeanum* as the parents. Some of the progeny are robust and striking when in flower, particularly when the deep yellow form of *D. speciosum* has been used. *D.* × *gracillimum* grows readily and lends itself well to specimen culture. Large plants may carry in excess of 100 flower-spikes, and provide an impressive display. The orchid is best suited to cultivation in pots but can also be attached to slabs or suitable garden trees. It needs a coarse fibrous mixture and responds vigorously to weekly applications of liquid fertilizers during summer. It can be grown in bushhouses or cool glasshouses in temperate or subtropical regions. Propagate by division of the clumps or from seed of hybridized flowers.

Dendrobium × **grimesii** C. White & Summerhayes
(after B. D. Grimes, original collector)
Qld

April-Aug

Small epiphytic **orchid** forming semi-pendulous clumps up to 30 cm long; **rhizome** about 0.3 cm thick, creeping; **stems** yellowish, tough; **leaves** 5-15 cm x 0.5-1 cm, more or less terete or oval in cross-section, thick and fleshy, curved, grooved, the apex blunt; **racemes** 3-7 cm long, bearing 6-12 flowers; **flowers** 2-3 cm across, white, cream or pinkish with a few small purple marks near the centre; **labellum** 1-2 cm x 0.3-0.4 cm, curved, pointed, white with numerous fine mauve striations.

A rare orchid which is a natural hybrid between *D. linguiforme* var. *nugentii* Bailey and *D. teretifolium* var. *fasciculatum* Rupp. It is known only from a small area near the southern end of the Atherton Tableland. Man-made hybrids using the same species as parents (and various varieties within the species) are identical with or very similar to naturally occurring plants. These plants grow readily but are somewhat cold sensitive and require a glasshouse over winter in temperate regions. They can be grown in a heated glasshouse or mounted on a slab of cork, weathered hardwood or tree-fern. Propagate by division or from seed. Because it is a hybrid, seed must be collected from hybridized flowers.

Dendrobium insigne (Blume) H. G. Reichb. ex Miq.
(remarkable)
Qld (Saibai Island)
20-40 cm tall   sporadic all year
Epiphytic **orchid**; **pseudobulbs** 20-40 cm x 0.4-

0.8 cm, linear, stiff, wiry, brittle; **leaves** 2-7 cm x 0.5-3 cm, oblong, dark green and fleshy, unequally notched at the apex, scattered up the length of the pseudobulb when young, mature pseudobulbs shedding the upper leaves and remaining leafy in the lower half; **inflorescence** of paired flowers borne in the upper nodes; **flowers** 2.5-3.5 cm across, yellow with irregular orange-red markings, fragrant, the labellum white; **labellum** 1.5-1.8 cm long, with small, pointed lateral lobes.

This orchid is widespread and common in lowland areas of New Guinea and also extends to Saibai Island in the Torres Strait, which is part of Australia. It is an interesting species, readily recognized by the fleshy leaves grouped near the centre of the pseudobulbs. Flowers are produced at intervals throughout the year and last about 2 days. The species can be grown readily but is very cold sensitive, requiring tropical conditions. In temperate Australia a heated glasshouse with a minimum temperature of 15°C is essential. Plants require strong light to flower satisfactorily. A pot of coarse mixture is suitable for its culture. Propagate from seed.

Dendrobium johannis H. G. Reichb.
(after John Gould Veitch)
Qld

March-May

Epiphytic **orchid** forming slender clumps; **rhizome** short, branched; **pseudobulbs** 20-70 cm x 1.5-2.5 cm, fusiform, hard, becoming yellowish and furrowed; **leaves** 10-20 cm x 1-1.5 cm, 4-7 in the distal quarter, linear, dark green, thick-textured, channelled; **racemes** 30-50 cm long, bearing 6-20 flowers on pedicels 2.5-3 cm long; **flowers** 2.5-4 cm across, chocolate brown, the petals and sepals prominently twisted; **labellum** 1.5-2 cm long, brown with a yellow mid-lobe which is narrower than the lateral lobes.

This orchid is distributed on the islands of Torres Strait and the east coast of Cape York Peninsula. It grows in open woodland in well lit, humid situations. Plants are very free-flowering and the flowers release a strong fragrance during the mornings of warm days. Cultivation requirements are as for *D. semifuscum*. *D. johannis* and *D. semifuscum* have been confused botanically but can be separated readily on flower colour and flowering time.

Dendrobium johnsoniae F. Muell. — this spectacular New Guinea species has been recorded from Australia, but confirmatory specimens are lacking.

Dendrobium kestevenii Rupp =
D. × *delicatum* (Bailey) Bailey

Dendrobium kingianum Bidw. ex Lindley
(after Captain King, R.N., friend of J. C. Bidwill who described the species)
Qld, NSW   Pink Rock Orchid
Aug-Nov
Lithophytic **orchid** forming large congested clumps; **rhizome** short, thick, much-branched; **pseudobulbs** 8-35 cm x 1-2 cm, broadest at the base, then tapering upwards, cylindrical, covered with overlapping brown

Dendrobium lichenastrum   J. Fanning

D. kingianum is one of Australia's most variable orchids. In each area where it occurs naturally, it seems to vary in some feature of growth or flower form and colour. Hundreds of these forms have been collected and grown by enthusiasts, and the superior types have been propagated vegetatively and sold or traded as clones, sometimes with a cultivar name which is often related to the collector or area of origin. Frequently, selected clones with desirable flower features such as shape, texture, size or colour have been used as parents in hybridization programmes. Variations in pseudobulb shape and aerial growth production also occur in this orchid. There is a general trend for the northern populations to have longer pseudobulbs and to be more prolific in the production of aerial growths than those in the southern part of the range.

In all, about 5 varieties of this orchid have been named but only one of these (var. *pulcherrimum* Rupp) is considered to have any botanical significance. The others were based on flower colour which is an insignificant feature in a species of such variation, although they probably still have merit as cultivars.

Their principal features are:

var. *album* Bailey — crystalline white flowers with lemon marks on the labellum. The pure-white forms are prized by enthusiasts and are very showy when grown well. At least 2 white forms are known, one with short, compact pseudobulbs and the other much lankier

var. *aldersoniae* Bailey — white flowers with pale spots on sepals, and a purple labellum

var. *pallidum* Bailey — slender, weak pseudobulbs and pale lilac-stained flowers

var. *pulcherrimum* Rupp — short pseudobulbs (4-9 cm long), short, few-flowered inflorescences, and large flowers about 2.5 cm across. It is botanically distinct, and extends south of the Macleay River in north-eastern NSW to near Newcastle

var. *silcockii* Bailey — pure-white flowers and a striking purple labellum. This is a very showy horticultural form and extremely free-flowering.

Dendrobium lichenastrum (F. Muell.) Kranzl. emend Dockr.
(spreading and lichen-like)
Qld

sporadic all year

Small epiphytic **orchid** growing in spreading patches; **rhizomes** about 0.2 cm thick, woody, branched, clothed with scarious bracts; **leaves** 0.4-4 cm x 0.3-0.6 cm, extremely variable in shape from almost circular and flat to terete, always thick and fleshy, prostrate or erect, bright shiny green; **flowers** 0.4-0.8 cm across, solitary on peduncles 0.5-1.5 cm long, white, cream or pink with red stripes and a prominent orange labellum; **segments** blunt, not widely spreading; **labellum** 0.3-0.4 cm x 0.2 cm, oblong.

A common orchid confined to highland rainforests of north-eastern Qld. Plants form small patches on rocks or trees, often on the slender outer branches. It is a very easy and rewarding orchid to grow, tolerating disturbance and re-establishing quickly. It can be grown in a shallow pot or saucer of coarse epiphyte

bracts obvious when young; **leaves** 3-12 cm x 1-2 cm, 2-6 at the apex of the pseudobulb, linear-ovate to obovate, thin-textured, dark green and somewhat shiny above, paler beneath, aerial growths produced from apical buds; **racemes** 7-20 cm long, bearing 2-15 flowers on pedicels 1-1.5 cm long; **flowers** 1.5-2.5 cm across, variable in colour from white with a lemon or purple labellum through various shades of pink to red, mauve and purple, thick-textured, opening widely; **labellum** 1-1.5 cm long, spotted or blotched with red or mauve, curved, the lateral lobes prominent and pointed.

This orchid, although it is of fairly restricted range between the Carnarvon Gorge of central Qld and the Hunter River of central NSW, is locally common and often super-abundant. It colonizes rocks, boulders, cliff faces and crevices (rarely trees), and frequently forms extensive colonies to the exclusion of other plants. Clumps of the orchid are usually very congested, the new bulbs growing over the decaying growths of previous years. While the orchid grows on rocks, its roots wander far and wide over adjacent rocks, soil, litter and bark.

The pink rock orchid is very popular with orchid enthusiasts and non-specialists alike because of its hardiness, ease of culture and profusion of colourful flowers. It is best suited to temperate and subtropical regions because it flowers very poorly in the tropics. It can be grown in bushhouses or unheated glasshouses, or in the garden, attached to suitable trees, or on large rocks in a semi-shaded position. It prefers pot culture in a coarse mixture liberally fortified with leaf litter but can also be grown on a slab or in a wire or slat basket. Airy but humid conditions produce best growth and flowering. The plants respond vigorously to regular liquid feeding during the warm summer months. They can also be watered abundantly during this season but should be hardened off over winter to avoid rotting.

Propagation is by division of the clumps, from aerial growths (termed keikeis) which can be removed when sufficiently developed and potted in sphagnum moss, or from seed.

Dendrobium linguiforme

mixture, but succeeds best on a slab of weathered hardwood, cork or tree-fern. The plants should be watered frequently during summer and about once a week during winter. They prefer humid, shaded conditions and should be hung where they receive some air movement. In suitable humid areas they can be successfully established on garden trees which have permanent bark. Plants respond to liquid fertilizers during the warm growing-months. The species is hardy in temperate Australia, needing only the protection of a humid bushhouse. Propagate by division or from seed. The species is extremely variable in leaf shape and flower colour, offering considerable scope for selection. The typical variety has appressed, rounded or obovate leaves, while var. *prenticei* (F. Muell.) Dockr. has a thicker rhizome and longer but variable-shaped leaves.

Dendrobium linguiforme Sw.
(tongue-shaped)
Qld, NSW

Tongue Orchid
June-Sept

Epiphytic or lithophytic **orchid** growing in spreading patches; **rhizome** about 0.4 cm thick, woody, branched; **leaves** 2-4 cm x 0.8-1.5 cm, ovate, oblong or obovate, thick and fleshy, alternate, closely spaced, dark green, blunt, with a few prominent longitudinal furrows; **racemes** 5-15 cm long, bearing 6-20 flowers; **flowers** about 1 cm across, white or cream; **segments** slender, pointed, not spreading widely; **labellum** about 0.6 cm long, curved, white with purple margins, the surface with 3 longitudinal keels.

A common orchid widely distributed from north-eastern Qld to south-eastern NSW on trees or rocks in fairly open situations. In good conditions the plants spread rapidly into large patches and may cover trunks, large branches and rock faces. The leaves are very fleshy, and act as a water storage organ. The flowers are delicately fragrant during the warm part of the day. This orchid is very popular with orchid enthusiasts and is an easy and extremely rewarding species to grow. A specimen plant in full bloom is a delightful sight. The species can be successfully established on suitable trees or large rocks in tropical, subtropical and warm parts of temperate regions. It can also be established on slabs of material such as weathered hardwood, tree-fern, terracotta, cork or branches of *Casuarina*, *Pittosporum* or oak. In areas with a cold climate, the species benefits from the winter protection of a bushhouse or unheated glasshouse. Plants should be regularly watered throughout the year, and they respond to the application of liquid fertilizers during spring and summer. Propagate from seed or by division.

The var. *nugentii* Bailey differs by its larger, broader leaves and the smaller, blunter flowers which quickly age to cream after opening white. It is common in north-eastern Qld and grows readily. The var. *huntianum* Rupp is a doubtfully distinct variety from south-eastern Qld, with narrow, elongated leaves.

Dendrobium lobbii T. & B.
(after Thomas Lobb, orchid collector and horticulturist)
Qld, NT

March-May; also sporadic
Terrestrial **orchid** forming sparse, erect clumps;

Dendrobium luteocilium   B. Gray

pseudobulbs 25-100 cm x 0.1-0.3 cm, linear, slender, wiry, swollen at the base; **leaves** 2-6 cm x 0.4-0.6 cm, cordate at the base then tapering throughout, dark green, thin-textured, stiff, the apex unequally notched; **racemes** short, bearing 1-2 flowers on pedicels about 0.6 cm long; **flowers** about 1 cm across, cream or greenish yellow, sometimes with orange veins, the labellum white; **labellum** 1-1.3 cm long, straight, 3-lobed, the surface of the mid-lobe warty.

This orchid is unusual in that it is the only terrestrial member of the genus in Australia, where it is confined to north-eastern Qld in disjunct localities near Townsville, Cardwell and on Cape York Peninsula. It is also widely distributed throughout the Pacific region to Malaysia. In Australia it grows in swampy soaks amongst rushes, sedges, grasses etc, usually in sandy soil. This species has very little horticultural potential but is of interest to the enthusiast. It can be readily grown in a pot of sandy soil fortified with peat moss. The base of the pot should be stood permanently in a saucer of water in a warm, humid situation such as a glasshouse. The species is very cold sensitive, and in cold areas needs the protection of a heated glasshouse with a minimum temperature of 15°C. The flowers are short-lived and of a rather dingy colouration. Propagate from seed, by division of the clumps or from aerial growths.

Dendrobium luteocilium Rupp
(yellow cilia)
Qld

sporadic
Epiphytic **orchid** forming medium to large, coarse clumps; **pseudobulbs** 60-200 cm x 1-1.5 cm, erect to pendent, yellowish, tough, somewhat flattened; **leaves** 4-12 cm x 3-4.5 cm, ovate to ovate-lanceolate, yellowish green, tough and leathery, the apex prominently notched, scattered along the stems; **racemes** 0.6-1 cm long, bearing 1-2 flowers which usually face each other, on pedicels about 1 cm long; **flowers** 1.5-2 cm across, not expanding widely, dull yellow or greenish, with a honey fragrance, the

segments incurved, lasting only a few hours; **labellum** 0.7-0.9 cm long, prominently 3-lobed, yellowish with brown markings, a large central patch of yellow cilia on the mid-lobe.

A coarse orchid that may develop into enormous clumps on trees or rocks, usually in sunny situations. It grows in rainforests, most commonly in lowland areas, from the Tully River, throughout Cape York Peninsula to the Torres Strait islands and into New Guinea. In very sunny situations the plants are bleached yellowish green and bear thick, leathery leaves. Flowering occurs in flushes, and although the flowers last but a few hours, they are attractive and have a strong honey scent. *D. luteocilium* is best suited to cultivation in the tropics. It is very sensitive to cold, and because of its size is best suited to pot culture. In temperate regions it is a difficult species to maintain, requiring a heated glasshouse with a minimum temperature of 15°C and plenty of light. Propagate from seed.

Dendrobium malbrownii Dockr.
(after Malcolm Brown, original collector)
Qld
Dec-April; also sporadic
Epiphytic **orchid** forming grass-like clumps; **pseudobulbs** 12-30 cm x 0.1 cm, linear-terete throughout, wiry, semi-erect or arching; **leaves** 3-6 cm x 0.2-0.4 cm, linear, dark green, thin-textured, 10-20 scattered along the upper three-quarters of the stem, the apex deeply notched; **flowers** about 0.8 cm across, widely expanding, borne solitary on the nodes, shiny cream with a purple-and-yellow labellum; **labellum** about 0.5 cm long, the lateral lobes barely distinct.

A species of restricted distribution, confined to the McIlwraith Range of central Cape York Peninsula. It grows on trees in rainforest and is locally common. The species has little horticultural merit, but is of interest because of its grass-like clumps and the small solitary flowers which last 2-4 days. It grows readily, requiring warm, humid conditions. It can be successful in a small pot of coarse mixture, or attached to a slab of cork, weathered hardwood or tree-fern. In temperate regions, a heated glasshouse with a minimum temperature of 10°C is needed to over-winter this species. Growth is continuous throughout the year and the species is not very tolerant of long dry spells. Propagate from seed or by division of the clumps.

Dendrobium mirbelianum Gaudich.
(after Charles Francois Brisseau de Mirbel, 18-19th-century French botanist)
Qld
Aug-Nov; also sporadic
Epiphytic **orchid** forming small clumps; **pseudobulbs** 10-100 cm x 1-3 cm, nearly linear throughout, tapered to the apex, dark blackish brown; **leaves** 8-15 cm x 2-4 cm, 4-10 scattered in the upper half, ovate, dark greenish brown with purplish suffusions, concave; **racemes** 12-30 cm long, bearing 4-12 flowers, on pedicels 1-2 cm long; **flowers** 3-5 cm across, opening widely or barely opening at all, pale to dark brown with a yellowish labellum, the petals erect but not twisted; **labellum** 2-3 cm long, 3-lobed, with conspicuous red veins, the central part with 5 parallel keels.

This orchid, which in nature occurs in coastal swamps and mangroves, is rarely encountered in cultivation. It usually grows in sunny, very humid conditions and extends from the Johnstone River northwards to New Guinea. The flowers of some plants open widely, but in others they are self pollinating and rarely open at all. This orchid is essentially tropical in its cultural requirements, needing hot, humid conditions with exposure to plenty of sunlight and regular watering between Oct and April. It can be grown in a small pot of a coarse medium, or upon a slab of cork or weathered hardwood. In southern Australia it is difficult to grow and needs a well lit situation in a heated glasshouse with a minimum temperature of 15°C. Plants should be kept dry in winter to encourage flowering. Propagate from seed. The species was previously known as *D. wilkianum* Rupp.

Dendrobium monophyllum F. Muell.
(a single leaf)
Qld, NSW
Lily-of-the-Valley Orchid
Aug-Dec; also sporadic
Epiphytic or lithophytic **orchid** growing in small to large, spreading patches; **rhizome** about 0.4 cm thick, woody, branched, covered with brown overlapping bracts; **pseudobulbs** 2-12 cm x 2-3 cm, conical, erect, about 4 cm apart, furrowed with age, when young covered with brown overlapping, papery bracts; **leaves** 5-12 cm x 2-3 cm, 1-2 at the apex of the pseudobulb, oblong, dark green, erect; **racemes** 7-20 cm long, bearing 5-20 flowers on pedicels 0.5-1 cm long; **flowers** 0.8 cm across, bell-shaped to cup-shaped, dull to bright yellow, waxy, drooping, fragrant; **labellum** 0.5-0.7 cm long, erect in the base then curved.

A distinctive orchid, usually found growing in well lit, humid situations. It extends from north-eastern Qld to north-eastern NSW but the plants from the northern part of this range are usually small and not found below 600 m alt. The flowers are delicately fragrant during the warm part of the day. This orchid is very popular with orchid enthusiasts but it can be somewhat tricky to grow. The plants suffer a setback after disturbance such as division or repotting, and may take quite a period to re-establish. They are generally slow-growing and can be attached to hard slabs or grown in a pot of coarse material. Weathered hardwood is particularly successful for this species, which can also be established on garden trees in humid situations. Cultivated plants can be hung in a glasshouse or bushhouse and prefer humid, airy conditions. They should be kept fairly dry during winter and, preferably, allowed to dry out thoroughly between waterings. Propagate by division of the rhizomes into 5-6 bulb sections or from seed.

Dendrobium moorei F. Muell.
(after C. Moore, original collector)
NSW (Lord Howe Island)
Aug-Oct
Epiphytic or lithophytic **orchid** forming small clumps; **rhizome** short, much-branched; **pseudobulbs** 7-25 cm x 0.5-0.8 cm, erect, slightly narrowed at the base then linear throughout, prominently furrowed, when young covered with white bracts; **leaves** 5-12 cm

x 1-3 cm, 2-5 at the top of each pseudobulb, oblong, dark green, thin-textured but leathery; **racemes** 2-8 cm long, bearing 2-10 flowers on pedicels 0.5-1.2 cm long; **flowers** about 1.5 cm across, white, tending to droop; **labellum** about 1 cm long, narrow, with 2 short lateral lobes.

Endemic on Lord Howe Island, where it is common on trees and rocks in moist rainforest. Plants form aerial growths fairly freely and sometimes a clump will consist mainly of an untidy mass of aerial growths and their roots. This species grows easily in cultivation and likes cool, moist conditions with plenty of air movement. It succeeds best in temperate and subtropical regions, and can be grown in a bushhouse as far south as Melbourne. Most success is gained in a small pot of fairly coarse mixture containing some broken down leaf mould. Plants sometimes produce excessive vegetative and aerial shoots at the expense of flowering. Propagate from seed, from aerial growths or by division of the clumps into 5-7 bulb sections.

Dendrobium mortii F. Muell.
(after T. S. Mort, original collector)
Qld, NSW

Aug-Nov; Feb-June
Epiphytic **orchid** growing in semi-pendulous, sparse clumps up to 60 cm long; **rhizome** slender, sparsely branched; **stems** slender, tough, yellowish; **leaves** 2-15 cm x 0.2-0.4 cm, terete, straight or curved, dark green, shallowly grooved, sharply pointed; **racemes** slender, bearing 1-4 flowers; **flowers** 2-2.5 cm across, pale green, yellowish or brownish, with a conspicuous white labellum, the sepals and petals marked with a few reddish stripes; **labellum** 1.5-2 cm x 0.4 cm, the mid-lobe oval, with crisped margins.

A common orchid found in moist, open forest and dry rainforest from north-eastern Qld to north-eastern NSW. It grows mainly on trees with a sparse canopy. Young plants tend to grow erect but become more pendulous as they increase in size. Clumps often become tangled, and the leaves decrease successively in size. Plants also grow on rocks, and the species may extend a fair distance inland. It grows easily in cultivation, and can be established in a small terracotta pot of coarse mixture or on a slab of tree-fern, weathered hardwood or cork. It prefers cool, moist conditions with adequate air movement. In temperate areas it can be grown in a bushhouse, but seems to respond to the extra winter warmth of an unheated glasshouse. Regular applications of liquid fertilizer during the warm summer months are beneficial. Plants become untidy with age, and long shoots can be trimmed back. Flowering proceeds in flushes throughout the year. Propagate from seed or by division.

Dendrobium nindii W. Hill
(after P. H. Nind, original collector)
Qld

Blue Orchid
July-Oct
Epiphytic **orchid** forming tall, slender clumps; **pseudobulbs** 30-250 cm x 2-4 cm, erect, slightly swollen in the middle, dark brownish to blackish, ribbed, cylindrical in the upper half; **leaves** 5-15 cm x 3-8 cm, ovate, leathery, dark green, often concave, the tip

notched, scattered in the upper half; **racemes** 30-50 cm long, erect or arching, borne in the upper nodes, bearing 10-20 flowers on pedicels about 4.5 cm long; **flowers** 5-7.5 cm across, opening widely, white, usually with suffusions and stripes of violet and mauve, the petals erect and twisted; **labellum** 3-4 cm long, 3-lobed, with 3 parallel, deep purple ridges.

This handsome orchid is confined to north-eastern Qld, north of the Johnstone River. It is most commonly found on trees, in hot, humid coastal swamps but on Cape York Peninsula it extends some way inland and into other communities such as along stream banks. It grows on a variety of trees, including palms, but always in a situation where its roots dry out very rapidly following rain. The top of the plant is invariably in sunlight. The flowers of this orchid are amongst the most beautiful of Australian orchids and the species is eagerly sought after by collectors. Unfortunately it is not an easy species to grow, and most collected plants linger before dying, or die outright. It is best suited to culture in the tropics, requiring warm to hot, humid conditions with adequate air movement, and it must be grown on a hard slab or in a pot of very coarse material that dries out rapidly after watering. Chunks of charcoal, pinebark, cork, sandstone or scoria 2-4 cm across are suitable. Moss should not be placed around the base of the plant and the plants must not be over-potted. They should be repotted at the first sign of the compost breaking down. They need exposure to very strong light, and even in the tropics are best grown in full sunlight. They are very difficult to maintain in temperate regions, needing to be hung near the glass of a heated glasshouse, held at a minimum temperature of 15°C and kept bone dry over the winter months. Propagate from seed. Successful division of the clumps is very difficult. It was previously known as *D. tofftii* Bailey.

Dendrobium ophioglossum H. G. Reichb. =
D. smilliae F. Muell.

Dendrobium pugioniforme Cunn.
(dagger-shaped)
Qld, NSW

Dagger Orchid
Sept-Nov
Epiphytic **orchid** growing in pendulous clumps up to 2 m long; **rhizome** slender, creeping, much-branched; **stems** about 0.2 cm thick, woody, dark green to yellowish; **leaves** 1-7 cm x 0.5-2 cm, ovate, thick and fleshy, flat or sometimes triangular in section, dark green, with a long, rigid, sharp point; **flowers** 2-2.5 cm across, pale green with a white labellum; **labellum** 1.3-1.8 cm x 0.5 cm, distinctly 3-lobed, white with a few purple markings, the margins crisped.

This orchid, which is very common within its range, is widely distributed from south-eastern Qld to south-eastern NSW. It grows in fairly open situations, often forming tangled masses, and may be found on trees or rocks. The flowers are inconspicuous amongst the leaves, but the plants are popular with orchid enthusiasts because of their bright green mats of foliage. The species is readily adaptable to cultivation and is especially suited to temperate climates. It can be induced to grow on rocks or trees in moist, semi-shaded garden situations. It can also be grown as a basket subject in a coarse epiphyte mixture, or mounted on slabs of hardwood, cork or tree-fern fibre. Once established, the plants are tolerant of long periods of dryness, although they prefer regular watering and moist, humid conditions. The leaves remain greener if the plants are grown in a shady situation. Propagate from seed or by division of the clumps. Aerial stems can be induced to form roots by aerial layering, and the separation and potting of the clumps can be done when sufficiently established. Natural hybrids occur with *D. tenuissimum*.

Dendrobium purpureum Roxb. — this New Guinea orchid has been recorded from Australia but confirmatory specimens are lacking.

Dendrobium racemosum (Nicholls) Clemesha & Dockr.
(flowers in racemes)
Qld

Sept-Oct; also sporadic

Epiphytic **orchid** forming sparse, erect clumps up to 1 m long; **rhizome** thick, sparsely branched; **stems** about 0.4 cm thick, yellow-brown, tough; **leaves** 3-20 cm x 0.4-1.2 cm, terete, thick and fleshy, dark green, deeply grooved, decreasing successively in size; **racemes** 5-8 cm long, bearing 8-15 flowers; **flowers** 2.5-3 cm across, white or cream with mauve markings on the column, the segments strongly recurving; **labellum** 1.8-2 cm x 0.5 cm, strongly 3-lobed, curled, the apex drawn out into a long point.

An uncommon orchid of limited distribution, confined to the Atherton Tableland and adjacent areas of north-eastern Qld. It usually grows in bright-light situations along rainforest margins. It can be grown readily and is best mounted on a slab of weathered hardwood, tree-fern or cork, but can also be successfully grown in a small terracotta pot of a coarse mixture. In tropical and subtropical regions, a bushhouse is suitable, or plants can be established on garden trees, but in temperate regions they need the winter protection of an unheated glasshouse. The species was previously known as *D. beckleri* var. *racemosum* Nicholls. Propagate from seed or by division.

Dendrobium rigidum R. Br.
(rigid; in reference to the leaves)
Qld

sporadic all year

Epiphytic **orchid** growing in small to medium, semi-erect to pendulous clumps 10-40 cm long; **rhizomes** about 0.2 cm thick, tough; **stems** horny, yellowish, sparsely branched, tough, erect on small plants, pendulous on large plants; **leaves** 2-8 cm x 1-1.5 cm, ovate to obovate, thick and fleshy, dark to greyish green, often with reddish tinges, somewhat flattened, the apex blunt or pointed; **racemes** 2-5 cm long, bearing 2-7 flowers crowded near the apex; **flowers** 1-1.5 cm across, cream with reddish striations on the petals and sepals, not opening widely; **labellum** 0.6-1 cm x 0.3-0.5 cm, yellow with red markings, curved, the margins entire.

An attractive compact orchid which in Australia is restricted to Cape York Peninsula, but also extends to New Guinea. It is often locally abundant in coastal and estuarine mangroves but reaches inland, growing on trees in humid situations such as around the margins of creeks and swamps. It resents shade and is usually found in open, bright-light positions. This orchid has proved to be easy to grow and flower, but is very cold sensitive and needs warm to hot, humid conditions for success. Tropical and warm subtropical regions are most suitable for its culture, and in temperate areas it requires a heated glasshouse with a minimum temperature of 10-15°C. It is best tied on to a hard slab of material such as weathered hardwood or cork, but can also be grown successfully in a small pot of coarse material. Plants should be hung in a well lit situation, and will respond to liquid fertilizers over summer. Propagation by division is difficult; from seed is the most successful method.

Dendrobium ruppianum A. Hawkes
(after the Rev. H. M. R. Rupp, noted 20th-century Australian orchidologist)
Qld
Oak Orchid
July-Nov

Robust epiphytic **orchid** growing in large clumps; **rhizome** thick, woody; **pseudobulbs** 15-50 cm x 2-4 cm, erect, crowded, fusiform, dark brownish green, longitudinally ribbed; **leaves** 5-15 cm x 2.5-6 cm, 2-7 crowded at the apex of the pseudobulbs, ovate, thin but leathery, dark green; **racemes** 15-40 cm long, bearing numerous closely packed flowers; **flowers** 1.5-2.5 cm across, white or cream, becoming yellowish with age, strongly fragrant in the mornings, with pointed segments; **labellum** 0.8-1 cm long, curved, white with numerous purple stripes, the lateral lobes prominent.

This orchid of north-eastern Qld is widespread and common in a variety of habitats from coastal to montane, and also extends to New Guinea. Although it is common in rainforest, it usually perches in the well lit outer situations of the canopy. It is particularly frequent in open scrub bordering rainforest, and seems to prefer the rough bark of casuarinas, although it does grow on other trees as well as on rocks and boulders in sheltered situations. This orchid is extremely floriferous, and large plants in flower are an unforgettable spectacle. The species is very amenable to cultivation, and can be grown in a pot or on a slab of weathered hardwood or cork. If grown in a pot, it prefers to be underpotted in a coarse mixture of bark, leaf mould, charcoal and sand. It can be fairly easily established on suitable rough-barked garden trees. Plants from highland areas are particularly suited to culture in temperate regions, whereas those originating in lowland districts are best grown in subtropical regions. The plants can be grown in unheated glasshouses or bushhouses, and prefer humid conditions with plenty of air movement. By careful culture, plants can be built up into large specimens which are impressive when in flower. Propagate by division of the rhizomes into 4-5 bulb clumps or from seed.

This species is quite variable in growth habit and flower features, depending on the area of origin. Slen-

Dendrobium × ruppiosum

der-stemmed plants with white, fairly well spaced flowers are frequently the form encountered in lowland situations, whereas those of the highlands are robust, with crowded flower-spikes. The forma *magnificum* (Dockr.) Dockr. is a very robust form, with flowers up to 5 cm across. These open white but age to yellow. The var. *blackburnii* Nicholls is a little-known form with short, stout pseudobulbs and pale yellow flowers on pendulous racemes. It was originally collected on Font Hill Station, north-eastern Qld, but does not appear to have been found since.

Dendrobium × ruppiosum Clemesha
(a combination of the parental names)
Qld

Oct-Dec

Epiphytic or lithophytic **orchid** forming congested clumps; **rhizomes** very short, thick, much-branched; **pseudobulbs** 10-35 cm x 2-5 cm, broadest near the middle, swollen at the base, when young covered by white, sheathing bracts; **leaves** 10-15 cm x 3-4 cm, 2-4 at the apex of the pseudobulb, oblong, thick and leathery; **racemes** 30-40 cm long, bearing 20-40 flowers on pedicels 2-2.5 cm long; **flowers** 3-4 cm across, white to cream, with purple markings on the labellum.

This orchid is a natural hybrid between *D. speciosum* var. *curvicaule* and *D. ruppianum*, and is found sporadically where the parents grow in proximity to each other. Its characteristics are a combination of the parents'. It grows readily in a pot of coarse mixture, and can be grown from warm temperate to cool tropical areas. Propagate by division of the clumps or from hybridized seed.

Dendrobium schneiderae Bailey
(after Mrs H. Schneider, original collector)
Qld, NSW

Jan-April; also sporadic

Epiphytic **orchid** growing in small, crowded clumps; **rhizomes** about 0.2 cm thick, much-branched, clothed with scarious bracts; **pseudobulbs** 1-2.5 cm x 0.8-1.6 cm, ovate to conical, thickest at the middle, erect, densely crowded, furrowed with age, when young covered with brown, papery bracts; **leaves** 2.5-7 cm x 0.5-0.8 cm, 2 at the apex of the pseudobulb, narrow-oblong, dark green; **racemes** 6-17 cm long, semi-pendulous, bearing 5-25 flowers on pedicels 0.4-0.6 cm long; **flowers** 0.6-0.8 cm across, greenish yellow, often shiny, sometimes with pink or reddish tinges on the sepals, somewhat flattened; **labellum** 0.5-0.6 cm long, erect, curved, with long lateral lobes and a vestigial mid-lobe.

A dainty orchid that grows on the branches of emergent rainforest trees, and seems to be particularly fond of hoop pine (*Araucaria cunninghamii*). It is always found in humid situations exposed to plenty of light, and extends from the Eungella Range near Mackay to north-eastern NSW. The flowers are not strongly fragrant but may be very long-lasting. This dainty and rewarding orchid is easily grown in a small pot or on a slab of weathered hardwood, cork or tree-fern. It is well suited to glasshouse or bushhouse culture, and can be grown with moderate protection in temperate

Dendrobium smilliae

D.L. Jones

regions. It likes a humid, well lit situation and should be watered frequently during warm summer weather, and kept relatively dry in the winter. Propagate by division of the rhizomes into 5-6 bulb sections or from seed. The var. *major* Rupp is a robust form from the Eungella Range near Mackay. It is larger in all its parts, and has robust, multi-flowered racemes.

Dendrobium semifuscum (H. G. Reichb.) Lavarack & Cribb
(somewhat or partly brown)
Qld, NT

July-Nov

Epiphytic **orchid** forming slender clumps; **rhizome** short, branched; **pseudobulbs** 10-60 cm x 1-2 cm, fusiform, hard, becoming yellowish; **leaves** 8-15 cm x 1-1.5 cm, 3-6 in the distal quarter, linear, thick-textured, channelled; **racemes** 20-50 cm long, bearing 6-20 flowers on pedicels 2-3 cm long; **flowers** 2-3.5 cm across, yellow to yellowish brown, with reddish to purple stripes at the base of the lateral petals, the sepals and petals prominently twisted; **labellum** 1-2 cm long, the mid-lobe as wide as or wider than the lateral lobes.

A widely distributed orchid found in New Guinea, the Top End of NT, and Cape York Peninsula to just south of Cooktown. It grows on a variety of trees, always in well lit situations, but is most common in paperbarked *Melaleuca* woodland. It is a colourful, free-flowering orchid, well suited to cultivation in tropical areas. Plants grow rapidly during the wet season, and then harden off and flower during the dry. Plants are cold sensitive, and quickly rot if frosted or watered too frequently over winter. They can be grown in small terracotta pots containing a coarse mixture, but are best attached to slabs of weathered hardwood or cork with no moss backing. They should be hung in a well lit, airy position and watered daily during the summer months, but kept almost dry over winter. A heated glasshouse with a minimum temperature of 10°C is necessary in warmer climates. In warm tropical areas, the plants can be attached to suitable garden trees. Propagate from seed or by division of the clumps into 5-bulb sections. This orchid has been confused with *D. johannis*.

Dendrobium smilliae F. Muell.
(after Mrs E. J. Smillie)
Qld
Bottlebrush Orchid
Aug-Nov; also sporadic

Epiphytic **orchid** forming large, robust clumps; **pseudobulbs** 20-100 cm x 1-3 cm, distinctly fusiform, ribbed, somewhat spongy, when young clothed with white scarious sheaths; **leaves** 5-20 cm x 2-4 cm, oblong to lanceolate, thin-textured, often twisted, scattered in the upper half of the stems; **racemes** 4-15 cm long, erect, containing numerous tightly packed flowers in a bottlebrush-like cluster, on pedicels about 1 cm long; **flowers** 1-1.5 cm across, not expanding widely, the labellum held uppermost, the segments thick-textured, usually pink at the base and paler towards the end, the labellum with a prominent green apex, rarely all white; **labellum** 1-1.6 cm long, semicylindrical, slightly 3-lobed.

A widespread and common orchid of north-eastern Qld, extending from the Burdekin River to Cape York Peninsula, the Torres Strait islands and New Guinea. New growth is produced during the wet season, and these new shoots retain their leaves for about 12 months before defoliating and flowering. *D. smilliae* is a very robust orchid that grows to a large size when it finds suitable conditions. These include rotting hollows of eucalypts, or in clumps of other epiphytes such as elkhorns and basket ferns. It can be grown easily but is essentially tropical in its requirements. Collected plants, especially large specimens, may take several years to adjust to cultivation, mainly because of the loss of much of the root system. In temperate areas the plants need a large, airy glasshouse with a minimum winter temperature of 10-15°C. They should be hung close to the glass and kept completely dry over winter. Large plants look attractive in slatted wooden baskets and they can also be grown in pots of very coarse mixture, or on slabs of cork or weathered hardwood. Air movement is essential. Propagate from seed. Division is difficult. *D. ophioglossum* H. G. Reichb. is a synonym of this species.

Dendrobium speciosum Smith
(showy)
Qld, NSW, Vic
King Orchid; Rock Orchid
July-Oct

Epiphytic or lithophytic **orchid** forming medium to large, spreading clumps; **rhizome** thick, fibrous, much-branched; **pseudobulbs** 10-100 cm x 1-6 cm, variable in shape from linear to swollen at the base, then tapered, straight or curved, yellowish, green or purplish, when young covered with white bracts; **leaves** 4-

Dendrobium speciosum   T.L. Blake

25 cm x 2-8 cm, 2-5 at the apex of the pseudobulb, ovate to oblong or obovate, dark green, thick and leathery; **racemes** 10-60 cm long, bearing numerous flowers on pedicels 1-2 cm long; **flowers** 2.5-5 cm across, extremely variable, thick and fleshy, white, cream or yellow, strongly fragrant; **labellum** 1.5-3 cm long, prominently 3-lobed, white, cream or yellow with purple or red spots and striations.

A widely distributed orchid that extends from north-eastern Qld near Cooktown to eastern Victoria, growing on rocks or trees in rainforest, moist, open forest and even drier areas such as sandstone escarpments and gorges. It extends from coastal districts to some distance inland, and to high mountain peaks. Although mostly found in bright-light situations, plants will also grow in quite shady conditions. In some areas the species is extremely abundant, and is a familiar component of the vegetation. Plants can grow huge, with individual clumps reaching 2-3 m across. A specimen clump in flower is a magnificent sight. In open forest areas, plants of this species are often scarred and defoliated by bushfires but still survive and reproduce. Natural hybrids occur with other species such as *D. kingianum*, *D. gracilicaule* and *D. ruppianum*.

D. speciosum is a very hardy orchid and an easy

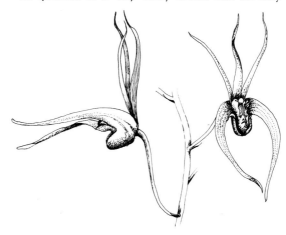

Dendrobium speciosum × *tetragonum*, × ·6

233

species to grow. Because of this and its massed display of flowers, it has achieved tremendous popularity with orchid enthusiasts and the general gardening public. It adapts well to a semi-protected garden situation and is a familiar sight in warm temperate and subtropical zones, on trees, rocks or stumps. In time plants develop into specimen sized clumps, and while in flower are attractive, with a strong but pleasant fragrance. The rock orchid also makes an excellent tub specimen for verandah decoration, and is grown in bushhouses or glasshouses by enthusiasts. The species is very adaptable to potting mixtures, almost any coarse material being suitable. A balanced mixture produces stronger growth and promotes better flowering. Strong light is essential for good flowering, and in southern Australia many growers move their plants into the sunshine during late autumn to harden off the new growth and promote flower-initiation. Most plants are grown in pots but they can be attached to hardwood slabs and baskets. Water should be applied in abundance over summer while growth is in progress. Liquid and solid fertilizers encourage growth. Blood-and-bone can be sprinkled over large clumps. Dendrobium beetle is a major pest of this orchid in subtropical zones, damaging the new growths, leaves and flowers.

Being a very widespread orchid, *D. speciosum* is extremely variable and consists of no less than 5 varieties in its range.

The typical form occurs on rocks, and is found in eastern NSW and Vic. It has stout, tapering pseudobulbs and racemes of fairly large white to cream flowers.

var. *capricornicum* Clemesha is a compact variety from central Qld, where it grows on rocks. Pseudobulbs are up to 15 cm tall, with hard, rigid leaves, and the flowers are white with purple markings on the labellum. Plants must be grown in the sun if they are to flower successfully.

var. *curvicaule* Bailey is common in north-eastern Qld between Mackay and Thornton's Peak, and extends from the coast to the highlands. It has curved, somewhat flattened pseudobulbs which narrow at the base, large leathery leaves, and racemes which bear numerous white to cream flowers.

var. *grandiflorum* Bailey has smaller pseudobulbs than most of the other varieties, and large pale to deep yellow flowers which become deeper yellow with age. This form is distributed in the ranges between Brisbane and Nambour. Some very fine plants of this variety are in cultivation, bearing large, deep butter yellow flowers.

var. *hillii* Bailey is a very common form in south-eastern Qld and north-eastern NSW. It always grows on trees, and forms large congested clumps with pseudobulbs up to 1 m long and nearly straight throughout. Flowers are fairly small, white to cream and crowded in the racemes.

var. *pedunculatum* Clemesha is an interesting form found on rocks in north-eastern Qld. It has short, stout, tapered pseudobulbs up to 16 cm long, stiff, leathery leaves, and racemes which are carried on a long, rigid peduncle. The flowers are white to cream, and frequently the leaves, the capsules and the bracts covering the new growths have purplish tonings.

Although this variety grows readily in southern Australia, it is very shy of flowering in these zones.

All forms can be propagated by division of the clumps or from seed.

Dendrobium striolatum H. G. Reichb.
(faintly striped or grooved)
NSW, Vic, Tas   Streaked Rock Orchid
                                              Sept-Nov
Lithophytic **orchid** growing in spreading, pendulous clumps up to 50 cm long; **rhizomes** about 0.2 cm thick, tough, horny; **stems** yellowish, sparsely branched, tough; **leaves** 4-12 cm x 0.2-0.3 cm, terete, thick and fleshy, curved or straight, slightly furrowed, pendulous; **racemes** slender, bearing 1-2 flowers; **flowers** 1.5-2 cm across, cream, yellow or greenish yellow with prominent brown stripes on petals and sepals, fragrant; **labellum** 0.7-1 cm x 0.3-0.5 cm, white with greenish markings, the margins deeply crisped.

A common orchid found on rocks near streams and along gorges, from north-central NSW to north-eastern Tas. It usually grows in fairly exposed situations, and the roots penetrate crevices and low areas where litter collects. In Tas it grows on granite boulders near the sea. Plants may grow as scattered individuals or in huge clumps covering boulders and whole cliff faces. The species is readily adaptable to cultivation and is especially suited to temperate climates. It can be induced to grow on rocks or even on trees in suitable garden situations, but must be protected from the depredations of slugs and snails. It also makes an attractive basket subject, with the leaves hanging through the wire or wooden slats. It can also be grown in a pot or saucer of coarse mixture, or on slabs of hardwood, cork or tree-fern. Once established, the plants are very tolerant of long periods of dryness. Applications of liquid fertilizer during the warm months are beneficial. Plants from Tas tend to be stronger growers and often have larger, less striate flowers. Propagate by division of the clumps or from seed. Small divisions establish readily in fresh sphagnum moss.

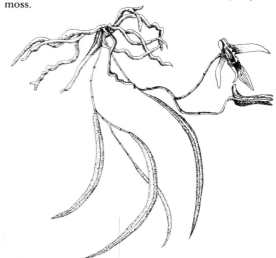

Dendrobium striolatum, × ·6

Dendrobium stuartii Bailey
(after J. W. R. Stuart, original collector)
Qld   Dec-March

Epiphytic **orchid** forming small, slender, pendulous clumps; **pseudobulbs** 15-60 cm x 0.2-0.5 cm, pendulous, somewhat zigzagged, bearing white, scarious, sheathing bracts when young; **leaves** 2.5-8 cm x 1.5-2.5 cm, ovate, falcate, thin-textured, pale green, somewhat acuminate apex; **racemes** short, slender, borne from the upper nodes, carrying 1-3 flowers on very slender pedicels 1.5-2 cm long; **flowers** about 2 cm across, opening widely, lime green ageing to yellow, with a prominent white, hairy labellum.

This fairly rare orchid is confined to rainforests of the ranges near the centre of Cape York Peninsula, where it grows on trees. New growth is produced after the wet season and these new shoots shed their leaves in the dry, prior to flowering. In cultivation, the plants tend to retain their leaves much longer. This orchid has proved easy to grow but is very cold sensitive, and in cold climates requires a heated glasshouse with a minimum winter temperature of 10°C. It responds to humid but airy conditions, and can be established on a slab of tree-fern, cork or weathered hardwood, or in a very small terracotta pot of coarse material. Plants should be kept dry in winter to encourage flowering. Propagate from seed.

Dendrobium × suffusum L. Cady
(tinged)
NSW   Aug-Nov

Epiphytic **orchid** forming small clumps; **rhizome** thick, short, branched; **pseudobulbs** 15-25 cm x 0.5-1 cm, cylindrical, erect, crowded, when young covered with greyish bracts; **leaves** 7-10 cm x 2-4 cm, 3-5 at the apex of the pseudobulb, ovate to obovate, dark green; **racemes** 6-12 cm long, bearing 3-10 flowers, on pedicels about 1 cm long; **flowers** about 1.5 cm across, cream to pale yellow suffused with mauve or pink on the labellum and outside, cup-shaped, fragrant; **labellum** about 1 cm long, cream with mauve markings, the lateral lobes prominent.

This rare orchid is a sporadic natural hybrid between *Dendrobium kingianum* and *D. gracilicaule*. It has been collected only about 5 times in nature, and always in north-eastern NSW, where the parents have been growing close together. The same cross has been produced many times artificially by hybridists, and the progeny are similar to those collected naturally. Variations in artificial hybrid progeny have been obtained by choosing different forms of *Dendrobium kingianum* as one parent. Thus, for example, those with short pseudobulbs produce compact progeny and those with deep pink-purple flowers produce colourful progeny. A further range of robust hybrids with colourful flowers has been obtained by using *D. gracilicaule* var. *howeanum* as one parent.

This orchid, whether naturally or artificially raised, grows easily in cultivation. It is well suited to pot culture in a coarse mixture but can also be grown on a hardwood or tree-fern slab. It likes cool, humid conditions and responds to liquid fertilizers during summer. Propagate by division of the clumps or from seed of hybridized flowers.

Dendrobium × superbiens H. G. Reichenb.
(magnificent)
Qld   Curly-Pink Orchid
                                Feb-June; also sporadic

Epiphytic or lithophytic **orchid** forming small to medium clumps; **pseudobulbs** 30-80 cm x 1-2.5 cm, erect, slightly swollen in the middle, when young covered with white, papery bracts; **leaves** 5-16 cm x 3-6 cm, 8-12 scattered in the upper half of the pseudobulb, ovate-lanceolate, thin-textured but leathery, dark green; **racemes** 25-40 cm long, arching, bearing 8-25 flowers on pedicels 2-3 cm long; **flowers** 4-5 cm across, in shades of pink or reddish mauve, the petals and sepals twisted and often with a white margin; **labellum** 2-3 cm long, curved, with prominent lateral lobes.

This orchid is a natural hybrid between *D. discolor* and *D. bigibbum*. In nature, it is restricted to stunted forests of the northern parts of Cape York Peninsula and adjacent Torres Strait islands. It grows on trees or rocks, usually in open, exposed situations. The plants are somewhat variable, depending on the form of the parents involved in the cross. Backcrosses also occur (see *D. × vinicolor* and *D. bigibbum* var. *georgei* and var. *venosum*). This hybrid has also been raised artificially, and some interesting variations have been produced by using superior colour variations in the parents. *D. × superbiens* is one of the most beautiful of Australia's orchids and is a very popular subject for cultivation, especially in tropical regions. The flowers are long-lasting and perform very well as cut flowers. Growth under ideal conditions is vigorous, and the plants are excellent for specimen culture. Such large plants, if well grown, are rarely out of flower. This orchid is well suited to tropical regions, and can be grown as a garden plant attached to a stake or suitable tree, in a sunny situation. In temperate regions it requires a sunny, heated glasshouse with a minimum temperature of 15°C. It is best grown in a pot of very coarse mixture such as large chunks of pinebark, terracotta and charcoal. Watering over summer should be daily, but the plants are best kept quite dry over winter. Liquid fertilizers promote vigorous growth if applied regularly over spring and summer. Dendrobium beetle may be troublesome in tropical regions, chewing the buds and new growths. Propagate by division of the clumps into 3-5 bulb sections, from occasional aerial growths or from seed of hybridized flowers.

Dendrobium tenuissimum Rupp
(very slender)
Qld, NSW

                                          Sept-Nov
Epiphytic **orchid** growing in slender, pendulous clumps up to 70 cm long; **rhizomes** sparsely branched, tough; **stems** slender, yellowish; **leaves** 1.5-10 cm x 0.2-0.4 cm, terete, thick, dark green, grooved, pendulous; **racemes** slender, bearing 1-3 flowers; **flowers** 1.5-2.2 cm across, brownish, dark green or pale green with a prominent white labellum, the sepals spreading or with distinctly curled tips; **labellum** 1.2-1.8 cm x 0.5-0.6 cm, white with a few purple markings, the lateral lobes distinct, the mid-lobe rounded, with crisped margins.

A slender orchid ranging from south-eastern Qld to

235

Dendrobium teretifolium

the Barrington Tops of central NSW. Although it is found occasionally in coastal districts, it reaches its best development in highland rainforests that are subject to frequent fogs and cloud cover. In cultivation, it grows readily in damp, humid conditions with plenty of air movement. It is readily grown in a bushhouse, even in temperate regions. Plants are best established on slabs of tree-fern or weathered hardwood, and seem to appreciate the presence of moss around the base of the plant, especially when being re-established. Propagate by division or from seed. This species occasionally hybridizes with *D. pugioniforme* when the two grow together.

Dendrobium teretifolium R. Br.
(terete leaves)
Qld, NSW   Bridal Shower; Pencil Orchid;
                            Rat's Tail Orchid
                            July-Oct

Epiphytic **orchid** forming small to medium sized, pendulous clumps up to 3 m long; **rhizomes** much-branched; **stems** about 0.4 cm thick, yellowish, tough; **leaves** 10-60 cm x 0.3-1 cm, terete, dark green, leathery, pendulous, decreasing successively in size; **racemes** 4-10 cm long, bearing 3-15 flowers on pedicels about 1 cm long; **flowers** 3-6 cm across, white, cream or pale yellow with red or purplish striations, spreading, spidery petals and sepals; **labellum** 2-3 cm long, lateral lobes fairly prominent, the mid-lobe drawn out into a long acuminate point and with irregularly crisped margins.

In Australia this orchid is widely distributed from northern Cape York Peninsula to the Bega region of south-eastern NSW. It grows most commonly on trees but is occasionally found on rocks, and is frequent in the peaty clumps of large epiphytic ferns. It grows in a variety of habitats including rainforest and trees fringing streams and swamps. Small plants are most common, but occasionally one sees quite large specimens, and when in flower these are a magnificent sight.

D. teretifolium is easy to grow, but large specimens usually suffer a setback on transplanting, and diminish in size until a strong new root system is established. Small plants establish very quickly. Because of their pendulous growth habit these orchids are best attached to a slab of weathered hardwood, cork or tree-fern, but they can also be grown in a hanging basket or a pot of coarse material. They succeed in a bushhouse or unheated glasshouse and can be successfully attached to garden trees. Plants respond strongly to regular applications of liquid fertilizers during the summer months.

D. teretifolium consists of 3 varieties, one of which consists of 2 forms.

The typical form occurs from central Qld to south-eastern NSW and is a fairly vigorous grower with numerous racemes bearing 4-15 large white or cream flowers. It avoids the ranges and tablelands and frequently grows on species of *Casuarina*. Specimen plants of this variety are very attractive when in flower.

var. *fairfaxii* (F. Muell. & Fitzg.) Bailey consists of 2 forms. Forma *fairfaxii* has very slender dark green leaves and short racemes which carry 1-4

Dendrobium teretifolium   J. Fanning

flowers. The flowers are quite large, cream to pale yellow with attractive purple stripes. Forma *aureum* (Bailey) Clemesha usually forms slender clumps and the flowers are greenish to golden yellow, striped and with prominently recurved segments. Both forms are found in NSW and south-eastern Qld.

var. *fasciculatum* Rupp is a very distinct variety from north-eastern Qld. Plants often have large, robust leaves and bear crowded racemes of white or cream flowers.

All varieties and forms can be propagated from seed. Dividing plants of this species can be tricky, and small divisions are generally unsuccessful. Aerial layering of the aerial stems is sometimes successful.

Dendrobium tetragonum Cunn.
(4-angled; referring to the pseudobulbs)
Qld, NSW   Tree Spider Orchid
                            May-Oct; also sporadic

Epiphytic **orchid** forming small to medium clumps; **pseudobulbs** 6-45 cm x 0.5-1.5 cm, semi-pendulous to pendulous, very slender and wiry at the base, then thickened and markedly 4-angled, usually brownish; **leaves** 3-10 cm x 1.5-2.5 cm, 2-5 near the apex of the pseudobulb, oblong to ovate-lanceolate, dark green, often partially twisted; **racemes** 1-3 cm long, bearing 2-5 flowers on pedicels 1.5-2.5 cm long, the racemes arising from amongst the leaves or nodes lower down on the pseudobulb; **flowers** 4-28 cm across, widely spreading, the segments thin and spidery, very fragrant, extremely variable in size and colour, green to greenish yellow or yellow with irregular reddish purple blotches; **labellum** 1-1.5 cm long, curved, with prominent lateral lobes.

A widespread and often locally common orchid found in a variety of forms from Cape York Peninsula in northern Qld to the Illawarra area in south-eastern NSW. It grows in moist, shady situations on the trunks of trees in rainforests, and along stream banks of open forest, from the coast to adjacent ranges and tablelands. The species is very popular in cultivation because of its graceful growth habit and showy,

236

Dendrobium tetragonum var. *giganteum*

J. Fanning

teristics are found in the subtropics. Plants of this form from Cape York Peninsula have large deep green, heavily blotched flowers.

The var. *hayesianum* Gilbert has greenish yellow to yellow flowers lacking any darker colouration. It occurs sporadically throughout the range of the species. Plants of this form from very high altitudes on the Atherton Tableland have clear yellow flowers with spidery segments.

Dendrobium tofftii Bailey = *D. nindii* W. Hill

Dendrobium toressae (Bailey) Dockr.
(after Toressa Meston)
Qld

                       sporadic all year

Small epiphytic **orchid** growing in dense, spreading patches; **rhizomes** about 0.05 cm thick, much-branched, densely clothed with papery bracts; **leaves** 0.4-0.8 cm x 0.2-0.4 cm, ovate, thick and fleshy, acute, deeply channelled above, keeled beneath, dark green with a glittering appearance; **flowers** about 0.6 cm across, cream, the labellum sometimes yellowish with red spots, sessile and cradled within the leaf channel; **labellum** about 0.6 cm x 0.2 cm, curved, grooved throughout.

A common orchid confined to north-eastern Qld, where it grows on trees and rocks in coastal and highland rainforests. Plants may grow in very shady situations but will also tolerate considerable exposure to sun. Clumps are usually neat and compact, and the leaves have a glittering or sparkling appearance which is especially noticeable when moist. Plants are never showy when in flower but are popular with orchid enthusiasts because of their neat growth habit. The species is easy to grow, although it may be a bit difficult to re-establish after disturbance, small patches

fragrant flowers produced in flushes throughout the year. The variability in flower size and colour allows scope for selection of clones for cultivation and hybridization. *D. tetragonum* has been used fairly extensively in the hybridization of temperate zone dendrobiums and has resulted in some excellent free-flowering, colourful hybrids with typical spidery flowers. The hybrids are generally very easy to grow but the various forms of *D. tetragonum* are more difficult. The plants generally resent disturbance and are very slow to re-establish after such an occurrence. Large plants in particular suffer a considerable setback and usually lose many pseudobulbs, the new growths then being weak. For successful growth the plants like moist, humid conditions with ample air movement. A shady situation is preferable, although flowering is improved in well lit positions. Plants should be mounted on slabs of tree-fern, cork or weathered hardwood, with a pad of sphagnum moss to encourage root growth. They can be grown in bushhouses or cool glasshouses, with the plants from the tropics appreciating the extra protection of a glasshouse in cold areas. They can also be established on suitable trees, in shady but humid situations. Propagate from seed. The typical variety of this orchid is widely distributed throughout the range of the species. It is characterized by fairly small flowers with sepals about 2-5 cm long. They are usually yellowish green with red or brown markings, but are occasionally all-green or cream.

The var. *giganteum* Gilbert has, as the name suggests, large flowers with sepals 5-9 cm long. The flowers are green and are usually heavily blotched with red, although pale green forms are known. This form is most common on the coast and in the ranges of the tropics, but occasional plants with similar charac-

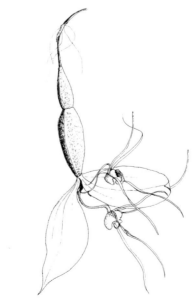

Dendrobium tetragonum, × ·5

237

tending to die out. It can be grown in a bushhouse as far south as Melbourne, and prefers cool, humid conditions, although it may need considerable light to flower well. The plants are best established on slabs of hardwood, cork or tree-fern fibre. They respond to applications of liquid fertilizers during the warm months. Propagate from seed or by division.

Dendrobium tozerensis P. S. Lavarack
(from Tozers Gap, Iron Range)
Qld

Sept-Feb; also sporadic

Epiphytic **orchid** forming medium to large, grasslike clumps; **pseudobulbs** 20-60 cm x 0.1-0.2 cm, linear, wiry, erect to semi-erect; **leaves** 3-8 cm x 0.4-0.8 cm, linear-lanceolate, thin-textured, dark green, the apex unequally notched; **flowers** about 2.5 cm across, white, expanding widely, the segments pointed and spidery, borne in pairs on a short peduncle arising from the nodes; **labellum** 0.8-1 cm long, 3-lobed, the mid-lobe sparsely hairy.

This orchid was recently discovered on Cape York Peninsula, where it is of restricted distribution although locally common. It grows on trees or rocks in open situations amongst rainforest. In cultivation, it grows readily in warm, humid conditions. It is very cold sensitive, and in temperate regions requires the protection of a heated glasshouse with a minimum temperature of 10°C. It can be grown in a small pot of coarse material, or attached to a slab of cork or weathered hardwood. Growth is almost continuous throughout the year and the plants should be kept moist. Flowering occurs sporadically, with all of the flowers on a plant opening simultaneously and lasting but a few hours. Propagate from seed or by division.

Dendrobium undulatum R. Br. =
D. discolor Lindley

Dendrobium × vinicolor St. Cloud
(wine coloured)
Qld

Dec-March

Epiphytic or lithophytic **orchid** forming small to medium, slender or compact clumps; **pseudobulbs** 20-100 cm x 1-2 cm, slender, terete, when young covered with white, papery bracts; **leaves** 5-12 cm x 1-2.5 cm, 5-17 scattered in the upper half of the pseudobulb, ovate, acuminate, dark green; **racemes** 20-50 cm long, arching, bearing 10-25 flowers on pedicels about 2.5 cm long; **flowers** 2.5-4.5 cm across, pink to wine red, the sepals and petals flat or twisted; **labellum** 1.5-2.5 cm long, deflexed, with prominent lateral lobes.

This species is a rare natural hybrid between *D. bigibbum* and *D. discolor*, of sporadic occurrence in two localities on Cape York Peninsula. In both localities the typical form of *D. discolor* was involved in the cross, but in each a different form of *D. bigibbum* was the other parent. Man-made hybrids using the same parents are very similar to naturally occurring plants. In cultivation, this orchid grows readily, requiring similar conditions to those for *D. × superbiens*. Propagate by division of the clumps into 4-6 bulb sections, from aerial growths or hybridized seed.

Dendrobium wasselli S. T. Blake
(after Fred Wassell, original collector)
Qld

May-Oct

Epiphytic **orchid** growing in spreading patches; **rhizomes** about 0.5 cm thick, woody, branched; **leaves** 5-10 cm x 0.7-1.2 cm, cylindrical, pale green to yellowish, fleshy, stiffly erect, with 5 longitudinal furrows; **racemes** 10-20 cm long, erect, bearing numerous crowded flowers; **flowers** about 2 cm across, white; **segments** very slender, linear, often recurved; **petals** often crossing; **labellum** curved, yellow with purple markings near the base, the margins undulate.

An uncommon orchid confined to central Cape York Peninsula where it forms spreading patches on trees on rainforest margins and stream banks. A frequent host is the rough-barked hoop pine (*Araucaria cunninghamii*). This orchid is a very easy and rewarding species to grow, flowering regularly and always attracting attention. It can be grown in a pot of coarse epiphyte mixture, but does best on a hard slab of weathered hardwood or cork. The plants should be watered every 2-3 days during the warm weather but once every 7-10 days is sufficient during winter. Tropical or subtropical regions are most suitable for its culture, and in temperate areas it requires a heated glasshouse with a minimum temperature of 8-10°C. The plants should be hung in a well lit spot over winter to promote flowering. They respond to liquid fertilizers at intervals over summer. Propagate by division of the rhizomes or from seed.

Dendrobium wilkianum Rupp =
D. mirbelianum Gaudich.

Dendrobium hybrid (*pugioniforme* × *tenuissimum*)
Qld, NSW

Sept-Dec

Epiphytic **orchid** growing in sparse, pendulous clumps 20-40 cm long; **rhizome** slender, creeping; **stems** about 0.2 cm thick, woody, dark green to yellowish; **leaves** 1-4 cm x 0.3-0.8 cm, terete to somewhat flattened, or 3-angled, pointed, dark green, smooth and not ribbed; **flowers** 1.5-2 cm across, pale green with a white labellum, sometimes with brownish suffusions in the tepals; **labellum** 1.2-1.8 cm x 0.5 cm, white with a few purplish markings, the lateral lobes distinct, the mid-lobe rounded with crisped margins.

This natural hybrid is occasionally found where *D. tenuissimum* and *D. pugioniforme* grow together, but its occurrence is sporadic and rare. Its major features are intermediate between the parents, although there is some slight variation in plants, especially in leaf shape. This hybrid is of novelty interest and mainly to be found in enthusiasts' collections. It grows easily, requiring similar treatment and conditions to those of its parents. Propagate by division or from seed of hybridized flowers.

DENDROBIUM CULTIVARS and HYBRIDS

Australian species of the genus *Dendrobium* have been hybridized by man more than any other genus of native plant. The hybridization programme has been carried out by professional nurserymen and enthusiasts alike. Many of the initial hybrids were made without much thought, but more recently hybridists have adopted a planned approach and have used superior clones as the parents. To be worth keeping, a hybrid must be superior to its parents in ease of culture and flower features such as size, colour and shape. Many registered hybrids do not meet these criteria. *Dendrobium* hybrids can be divided into those which tolerate cold and are hardy in southern Australia, and the tropical group which succeeds best in tropical regions. Plants of the latter group can be grown in temperate areas in a heated glasshouse, but cold hardy orchids do not succeed well in the tropics. There has been some effort to cross cold hardy orchids with the tropical types but the results have been generally disappointing, with the possible exception of *Dendrobium* 'Our Native'. The culture of orchid hybrids is similar to that for the species.

To date, about 88 hybrids of Australian dendrobiums have been registered as cultivars.

Dendrobium 'Alan Printer'
Cultivar

Aug-Oct

A man-made, second generation hybrid between *D.* 'Andrew Persson' and *D. kingianum*, which was registered in 1969. The progeny tend to look very much like *D. kingianum* in growth habit, and bear strongly erect racemes of pale pink to cream flowers, each about 2 cm across. This hybrid is not widely grown but is well suited to culture in a bushhouse or cool glasshouse in subtropical and temperate regions.

Dendrobium 'Albertine'
Cultivar

Nov-March

A man-made hybrid which is essentially for tropical conditions. It is the result of a cross between *D. bigibbum* and *D. antennatum*, and produces very colourful flowers in shades of mauve, with a greenish white labellum. They are carried on erect or arching, sturdy racemes and a large plant is showy. In temperate regions this hybrid needs a heated glasshouse with a minimum temperature of 10-15°C. It is best grown in a fairly small pot of coarse mixture.

Dendrobium 'Allyn Sands'
Cultivar

Nov-Jan

A man-made hybrid between *D.* 'Lowana Nioka' and *D. semifuscum*, registered in 1977. It has prominently twisted petals and the flowers are of brown and mauve tonings, sometimes with cream tips. It is a free-flowering, colourful hybrid which in the tropics can be grown as a garden plant. To flower in temperate regions, plants need a heated glasshouse with a minimum temperature of 10°C and plenty of light.

Dendrobium 'Allyn Star'
Cultivar

Aug-Nov

A second generation, man-made hybrid between *D.*

× *gracillimum* and *D. tetragonum*, registered in 1978. The plants are vigorous, with long, slender pseudobulbs which tend to be slightly angular. The racemes are erect and bear yellowish flowers which have spreading, somewhat pointed segments, and may be up to 2.5 cm across. The plants grow readily in pots of coarse mixture. A bushhouse or cool glasshouse is suitable in temperate and subtropical regions.

Dendrobium 'Andrew Persson'
Cultivar

Aug-Oct

A man-made hybrid, registered in 1960. It is a cross between *D. speciosum* and *D. falcorostrum*, and the progeny are strong-growing plants with stout pseudobulbs and thick leaves like those of *D. speciosum*. They are very easy to grow, tolerating pot culture very well in subtropical and temperate regions, but unfortunately many have proved to be shy flowerers. The flowers are about the size of *D. speciosum*, and vary from cream to white. Some clones have a pleasant to strong fragrance.

Dendrobium 'Andrew Upton'
Cultivar

Aug-Oct

An unusual second generation, man-made hybrid between *D.* 'Judy Leroy' and *D.* 'Star of Gold', registered in 1982. Plants have a compact growth habit and bear arching racemes of fairly large, attractive flowers.

Dendrobium 'Aussie Ira'
Cultivar

Aug-Oct; also sporadic

A second generation, man-made hybrid that is a backcross of *D.* 'Ellen' with *D. tetragonum*. It has a similar growth habit, flower shape and colour to those of *D.* 'Ellen', but the flowers tend to have more pointed perianth parts and may also have some blotching. Cultivation is as for *D.* 'Ellen'.

Dendrobium 'Australia'
Cultivar

Nov-April

A colourful tropical, man-made hybrid which is the result of a cross between *D. nindii* and *D. bigibbum*. The flowers are in shades of mauve and bluish purple, and the petals tend to have a twist. The plants like warm, humid conditions, and in temperate regions need a heated glasshouse with a minimum temperature of 10-15°C. In these conditions the plants are best hung near the glass.

Dendrobium 'Bardo Rose'
Cultivar

Aug-Oct

A man-made hybrid between *D. falcorostrum* and *D. kingianum*, registered in 1961. It is extremely popular in cultivation because it is very easy to grow and flower. It is best suited to temperate climates and has similar cultural requirements to those of its parents. Vegetative growth is most like a vigorous *D. kingianum*, but the flowers are 2.5-3 cm across and vary from white through pale pink to deep pink, and are highly perfumed. Some variation in flower size, shape and colour has been achieved by using different forms of *D. kingianum* in the hybridization process.

Dendrobium 'Blue Gloucester'
Cultivar

Nov-March

A man-made hybrid which produces very attractive bluish flowers with purple suffusions. The floral segments, with the exception of the labellum, are strongly twisted and the flowers are long-lasting and useful for cutting. It is the result of hybridizing *D.* × *superbiens* with *D. nindii*, and the plants are quite strong-growing, but best suited to tropical regions. The cross was registered in 1966. In temperate regions the plants need a heated glasshouse with a minimum temperature of 10-15°C.

Dendrobium 'Blue Tiki'
Cultivar

Nov-March

A second generation, man-made hybrid between *D. semifuscum* and *D.* 'Judy Leroy', registered in 1970. It is a colourful hybrid for the tropics, having bluish-suffused flowers which have twisted petals. The growth habit is compact and is suited to pot culture. The plants need plenty of light, and warm, humid conditions for optimum growth. In temperate regions a heated glasshouse is essential.

Dendrobium 'Bluey'
Cultivar

Nov-March

A second generation, man-made hybrid between the natural hybrid *D.* × *superbiens* and *D.* 'Peter Petersen'. It is a hardcane orchid, best suited to tropical conditions where, in suitable areas, it can be grown as a garden plant. To flower in temperate regions, plants need a heated glasshouse with a minimum temperature of 10-15°C and plenty of light. The floral segments are strongly twisted and are of blue to mauve tones.

Dendrobium 'Blushing Rose'
Cultivar

Aug-Oct

A second generation, man-made hybrid between *D.* × *suffusum* and *D.* 'Bardo Rose'. It is a hardy hybrid suited to temperate regions, with a similar compact growth habit to that of its parents, and with attractive pale pink flowers which may be blotched on the backs of the petals and sepals. Plants are free-flowering, and the flowers are attractively perfumed during warm weather.

Dendrobium 'Blushing Star'
Cultivar

Aug-Oct

A man-made hybrid which is the result of a cross between *D.* × *suffusum* and *D. tetragonum*. It was registered in 1978, and has a growth habit similar to *D.* × *suffusum* but forming a more open clump with angular pseudobulbs. The flowers have spidery segments and are an attractive pink colouration. This cultivar grows well in a bushhouse or cool glasshouse in temperate regions. Plants often flower twice a year.

Dendrobium 'Bush Chic'
Cultivar

Nov-Jan

A man-made hybrid between *D.* × *superbiens* and *D. semifuscum*. The plants are fairly strong growers, with long racemes of large flowers. The sepals and petals are prominently twisted, and the flowers are brownish with mauve tinges. Plants succeed best in tropical areas but they can also be grown in a heated glasshouse in temperate zones. Flowers are long-lasting and could be useful for cutting.

Dendrobium 'Chieno'
Cultivar

Oct-Feb

An unusual man-made hybrid which was registered in 1964. It is the result of a cross between *D. discolor* and *D. dicuphum*, and seems to be rarely grown. It is essentially tropical in its requirements, and the flowers are reported to be mainly brownish, with some twisting in the petals. The plants need plenty of light, and warm, humid conditions. In temperate regions a heated glasshouse is essential.

Dendrobium 'Colin'
Cultivar

Aug-Oct

A man-made hybrid which was registered in 1979. It is the result of a cross between *D.* 'Ellen' and *D. ruppianum*. The plants are vigorous, with slightly angled, ribbed pseudobulbs and dark green leathery leaves. The flowers are pink to cream, and starry, and are produced on a fairly long raceme. This hybrid grows and flowers well in subtropical and temperate regions.

Dendrobium 'Craig Hewitt'
Cultivar

Aug-Oct

A man-made hybrid of *D.* 'Emmy' and *D. speciosum*. It is not a particularly good hybrid as the plants tend to grow vigorously but flower poorly. It can be grown in a cool glasshouse or bushhouse in temperate regions.

Dendrobium 'Dang Toi'
Cultivar

Nov-Feb; also sporadic

This is a second generation, man-made hybrid which is really a backcross. It involves as parents *D.* × *superbiens* and *D. discolor*. It was produced in Singapore Botanic Gardens and registered in 1941. The same cross apparently occurs sporadically in nature. It is essentially a hybrid for the tropics, where it can be grown as a garden plant or on verandahs etc. In cooler regions it needs the protection of a heated glasshouse. It is best grown in a pot.

Dendrobium 'Delicate Falcon'
Cultivar

Aug-Oct

A man-made hybrid between *D. falcorostrum* and *D.* × *delicatum*, registered in 1979. It is a vigorous-growing plant which produces racemes of fragrant white flowers, and it is well suited to cultivation in temperate regions.

Dendrobium 'Ella Victoria Leaney'   D. Cannon

Dendrobium 'Double Two'
Cultivar

Aug-Oct

A second generation, man-made hybrid between *D.* × *gracillimum* and *D. adae*, registered in 1979. The plants are very compact, with tall, slender canes. They bear erect to arching racemes of 10-15 flowers about 1.5 cm across, which vary from pale to deep yellow. The flowers tend to be flatter, and the segments to spread more widely than in *D.* × *gracillimum*. The plants grow steadily but are slower than many other hybrids. They are best suited to subtropical or temperate regions.

Dendrobium 'Dusky Girl'
Cultivar

Nov-Jan

A man-made hybrid which is a backcross between *D.* 'Judy Leroy' and *D. canaliculatum*. It is similar in many respects to *D.* 'Judy Leroy' but has smaller, less colourful flowers. It is best suited to the tropics, and has similar requirements to *D.* 'Judy Leroy'.

Dendrobium 'Ella Victoria Leaney'
Cultivar

Aug-Oct

A man-made hybrid between *D. kingianum* and *D.* *ruppianum*, first produced many years ago but registered in 1980. This is an excellent hybrid that is very showy when in flower, and it is very easy to grow. The pseudobulbs tend to be fusiform and ribbed, like those of *D. ruppianum*, and the dark green, somewhat elongated leaves also resemble that species. Arching racemes carry drooping flowers in various shades of pink, and a large plant in flower is very attractive. Some superior clones of this hybrid have been produced using colourful, large-flowered forms of *D. kingianum* and the large-flowered *D. ruppianum* var. *magnificum*. Plants of *D.* 'Ella Victoria Leaney' are easy to grow in a pot of coarse mixture. In subtropical and warm temperate regions they grow well in a bushhouse, but in cooler areas do better in an unheated glasshouse.

Dendrobium 'Ellen'
Cultivar

July-Oct; also sporadic

A man-made hybrid between *D. kingianum* and *D. tetragonum*, registered in 1928. The plants flower very freely and are generally easy to grow, which makes them popular with amateurs and enthusiasts alike. With such variable parents, it is hardly surprising that the progeny also vary tremendously in flower size, colour and shape. *D. kingianum* mainly introduces colour, while *D. tetragonum* has a profound influence on shape and size. Progeny of crosses involving small-flowered *D. tetragonum* produce masses of small, starry flowers which are usually pale pink. Those from the large-flowered *D. tetragonum* var. *giganteum* produce much larger flowers with narrow segments, but in less profusion than those from the smaller forms. *D. tetragonum* var. *hayesianum* has been used to produce paler colours. Most large-flowered *D.* 'Ellen' are pale pink, but a few whites have also been recorded. One desirable aspect of this hybrid is the sporadic flowering induced by *D. tetragonum*. The hybrids have a peak flowering in spring, but tend to produce additional flushes of flowers sporadically throughout summer and autumn. Vegetative growth is basically similar to *D. kingianum* but with a tendency to longer, thinner pseudobulbs. *D.* 'Ellen' is best suited to temperate regions.

Dendrobium 'Ellen'   D. Cannon

241

Dendrobium 'Elsie Cox'
Cultivar

Aug-Oct

An unusual man-made hybrid which was registered in 1961. It arose from a cross between *D.* × *delicatum* and *D.* × *superbiens*, and is a second generation hybrid. It seems to be grown rarely and is reported to have growth and flower characteristics like *D.* × *superbiens*. The plants should tolerate some cold, but to over-winter in temperate regions, would probably require some heat.

Dendrobium 'Emma'
Cultivar

Aug-Oct

A man-made hybrid between *D.* 'Gillian Leaney' and *D. tetragonum*, registered in 1980. It is a fairly compact-growing plant with angled pseudobulbs, and racemes of pink flowers which tend to have pointed, somewhat spidery segments. It succeeds well in temperate and subtropical regions, growing in a bushhouse or unheated glasshouse.

Dendrobium 'Emmy'
Cultivar

Aug-Oct

A man-made hybrid which was registered in 1966. It is the result of a cross between *D. kingianum* and *D. aemulum*, and the progeny are most appealing. They are compact plants which bear spidery pink flowers about 2 cm across, in erect racemes. In some clones, the flowers do not open widely and tend to droop, but this is not always the case. The plants are quite hardy and easy to grow. They can be attached to garden trees or mounted on slabs, but are more usually grown in pots of a coarse mixture. They are well suited to cultivation in temperate or subtropical regions.

Dendrobium 'Esme Poulton'
Cultivar

Oct-Jan

An unusual hybrid that is the result of a man-made cross between *D. kingianum* and *D. bigibbum*. It was registered in 1978. The flowers are pinkish mauve, shaped like those of *D. bigibbum* but much smaller. The vegetative growth resembles *D. bigibbum*. Plants succeed well in the subtropics but tend to be cold sensitive, and in temperate regions need a heated glasshouse.

Dendrobium 'Eureka'
Cultivar

Aug-Oct

A man-made hybrid which was registered in 1980. It is the result of a cross between *D. fleckeri* and *D. ruppianum*, and is an excellent hybrid, producing racemes of yellow flowers with attractive, fringed lips. Plant growth tends to be slow, but the hybrid succeeds well in subtropical and temperate zones. It is best suited to pot culture. Large-flowered forms have been produced when *D. ruppianum* var. *magnificum* is used as one parent. These are very free-flowering, with erect racemes carrying up to 12 flowers which are large, open and flat.

Dendrobium 'Francis Henton'
Cultivar

Nov-Jan

A man-made hybrid which was registered in 1958. It arose as a result of a cross between *D.* 'Dang Toi' and *D. nindii*. It flowers freely, producing racemes of colourful mauve-and-brown flowers which have prominently twisted petals. The flowers are quite long-lasting, and could be useful as cut flowers. Plants are very cold sensitive, and grow best in lowland tropical regions, where they will even succeed as garden plants. To flower in areas with a cold climate, they need the protection of a heated glasshouse, and must receive plenty of light.

Dendrobium 'Gillian Leaney'
Cultivar

Aug-Oct

A man-made hybrid between *D. kingianum* and *D.* × *delicatum*, registered in 1964. The growth habit tends towards *D. kingianum*, and some plants are quite weak. The racemes are erect and bear pink flowers about 2.5 cm across. These may be lightly spotted and, apart from a couple of good clones, are rarely superior to *D.* × *delicatum*. They grow steadily and are best suited to temperate or subtropical regions. Some clones produce lightly perfumed flowers.

Dendrobium 'Glenys King'
Cultivar

Oct-Jan

A man-made hybrid which is a backcross of *D.* 'Lowana Nioka' and *D. canaliculatum*. It was registered in 1966. Plants tend to be more compact than *D.* 'Lowana Nioka', and have long sprays of cream-and-green flowers which have slightly twisted petals. This is a very free-flowering hybrid which is most successful in lowland tropical areas. It can be grown in a pot or on a slab, and needs plenty of light in order to flower. In temperate regions, it needs the protection of a heated glasshouse with a minimum temperature of 10°C.

Dendrobium 'Gloucester Gem'
Cultivar

Nov-Feb

A man-made hybrid between *D. bigibbum* and *D.* 'Gloucester Sands', registered in 1966. It is a free-flowering hybrid, well suited to garden or bushhouse culture in the tropics. The flowers are brownish with mauve tonings, and are carried on long racemes. They last well as cut flowers. To flower in temperate areas this cultivar requires a heated glasshouse and plenty of light.

Dendrobium 'Gloucester Sands'
Cultivar

Feb-April; also sporadic

A man-made hybrid between *D. canaliculatum* and *D. discolor*, registered in 1963. The same cross has been found occurring naturally in north-eastern Qld, but is sporadic and rare. *D.* 'Gloucester Sands' is a very free-flowering hybrid, well suited to tropical conditions, where it can be grown as a garden plant on suitable trees, or as a pot plant on verandahs etc. It likes exposure to plenty of light, and is somewhat difficult to

Dendrobium 'Hastings' D. Cannon

grow and flower successfully in temperate regions. It is cold sensitive, and during winter in temperate climates it must be kept dry, in a heated glasshouse with a minimum temperature of 10°C. It is best grown in a pot of coarse material. This hybrid exhibits characteristics of both parents, its pseudobulbs tending to be stubby, although they are elongated in some plants. The flowers, which are about 2.5 cm across, are various shades of brown, usually with a shiny appearance. They are quite long-lived and last well in water. There is some variation in flower colour, depending on the forms of the parents used in the cross.

Dendrobium 'Gold Stripe'
Cultivar

Nov-March
An unusual cultivar which was registered in 1973. It is the result of a cross between *D. bifalce* and *D. discolor*. Plants tend to be slow-growing, and produce long racemes with the flowers clustered towards the end. Flowers are a greenish brown, frequently with prominent yellow stripes on the sepals. Plants need warm, humid conditions to grow and flower.

Dendrobium 'Golden Fleck'
Cultivar

Aug-Oct; Feb-March
A man-made hybrid which was registered in 1980. It is the result of a cross between *D. fleckeri* and *D. gracilicaule* var. *howeanum*. It is a slow-growing hybrid with slender pseudobulbs and short racemes of golden yellow flowers which have attractive fringed lips. Plants of this hybrid grow well in a cool glasshouse or bushhouse in temperate regions. They flower prolifically, regularly twice a year.

Dendrobium 'Goose Bumps'
Cultivar

Aug-Oct
A novel, man-made hybrid which was achieved by crossing *D. cucumerinum* with *D. teretifolium*. It was registered in 1977. The leaves are fleshy and have

prominent bumps on their surfaces, hence the unusual name. The flowers are cream-green with purplish striations, and are displayed on short, semi-erect racemes. Plants have a fairly compact growth habit, and can be grown in pots or mounted on slabs. They will succeed in a bushhouse in temperate regions.

Dendrobium 'Gracious Cascades'
Cultivar

Aug-Oct
A man-made hybrid which was registered in 1982 and is the result of a cross between *D. speciosum* and *C. × gracillimum*. It is a strong-growing hybrid which is extremely free-flowering, carrying long racemes of cream, perfumed flowers. It is well suited to cultivation in subtropical and temperate regions.

Dendrobium 'Gracious Falcon'
Cultivar

Aug-Oct
A man-made hybrid which is the result of a cross between *D. falcorostrum* and *D. × gracillimum*, registered in 1979. It is a strong-growing plant which succeeds well in temperate zones in the protection of a bushhouse. It produces long racemes of fairly large, scented, cream to yellow flowers. One drawback is that plants tend to be a bit shy of flowering when young, but mature plants are reliable each year. The cross has been remade using a superior form of *D. × gracillimum*.

Dendrobium 'Graham Hewitt'
Cultivar

Aug-Oct
A man-made hybrid of *D.* 'Ellen' and *D. speciosum*. It is a strong-growing hybrid with rather disappointing flowers. These are small, somewhat pointed and of very pale colouration. Plants grow well in temperate regions.

Dendrobium 'Gresford'
Cultivar

Nov-Feb
A man-made hybrid between *D. dicuphum* and *D. nindii*, registered in 1977. The plants have erect pseudobulbs 10-25 cm long, with leaves scattered in the upper half. Long, erect racemes are borne from the upper nodes, and bear attractive flowers which are about 4 cm across. These vary from white to pale bluish, and have a prominent dark labellum. The petals and sepals are usually twisted. This orchid is best suited to tropical conditions, and in other areas must be grown in a glasshouse (heated, with a minimum temperature of 10°C in cold areas). It grows well in a small pot of very coarse material and likes to be kept dry over winter.

Dendrobium 'Hastings'
Cultivar

Aug-Dec
A man-made hybrid between *D. kingianum* and *D. fleckeri*, registered in 1970. It is a fairly vigorous grower, and the habit of the pseudobulbs is strongly biased towards *D. kingianum*, although they are usually at-

Dendrobium 'Hilda Poxon'

tenuated in the upper part. The flowers are pink to buff in colour, and have a hairy labellum like *D. fleckeri*. The plants grow readily in temperate regions, and flower regularly. They are best grown in a pot of coarse mixture, and are suited to cool glasshouse or bush-house culture.

Dendrobium 'Hilda Poxon'
Cultivar

sporadic all year

A man-made hybrid between *D. speciosum* and *D. tetragonum*, which has proved to be one of the best native orchid crosses so far made. It was registered in 1977. It is a fairly vigorous grower, its pseudobulbs tending to be squarish, as they are in *D. tetragonum*, but quite stout as in *D. speciosum*. Its best attribute is its ability to flower sporadically throughout the year, certainly with 2 or 3 flushes. The racemes carry 8-12 flowers which are each 3-5 cm across, varying from a clear greenish yellow to yellow heavily blotched with red-brown. They are quite attractive. The plants are best suited to pot culture in subtropical and temperate regions. They can be grown in a protected bushhouse or cool glasshouse.

Dendrobium 'Jamie Upton'
Cultivar

Aug-Oct

A recently registered man-made hybrid which is the result of crossing *D. tetragonum* with *D.* 'Golden Fleck'. It is a rather slow-growing hybrid with very slender, somewhat angular pseudobulbs. Fairly small, starry, yellow flowers are produced from amongst the leaves, on short racemes. This is a very attractive cultivar that is cold hardy and grows well in temperate zones.

Dendrobium 'Jane Leaney'
Cultivar

Aug-Oct

A man-made hybrid between *D. speciosum* and *D. × delicatum*, registered in 1964. It is rather uncommon in cultivation and the progeny have little to distinguish them from those of many other hybrids. They are strong-growing plants with long, stout pseudobulbs like *D. × delicatum*, and erect racemes of pale pink flowers, each about 2 cm across. The plants grow readily in pots of coarse mixture, and are best suited to subtropical and temperate regions.

Dendrobium 'John Upton'
Cultivar

Aug-Oct

A second generation, man-made hybrid which is the result of a cross between *D.* 'Kim' and *D.* 'Ellen'. It is a compact-growing hybrid with starry, pink flowers which have pointed segments. It grows best in a pot and will grow in temperate or subtropical regions.

Dendrobium 'Judy Leroy'
Cultivar

Oct-March

A man-made hybrid between *D. canaliculatum* and *D. dicuphum*, registered in 1961. The same cross has been found to occur naturally in NT, but it is sporadic and

Dendrobium 'Hilda Poxon'   D. Cannon

rare. This hybrid is a very colourful and free-flowering plant, best suited to tropical conditions. It likes warmth, high light-intensity and plenty of water during its growth period of Feb-May. It can be grown on a slab of weathered hardwood, cork or paperbark, or in a small pot of coarse mixture. In temperate regions it needs a heated glasshouse with a minimum temperature of 10°C. Growth habit is similar to that of the parents, and the flowers, which are up to 3 cm across, are cream with mauve tinges, and borne on arching racemes.

Dendrobium 'Julie Skillicorn'
Cultivar

Aug-Oct

A novelty man-made hybrid which is the result of a cross between *D. tenuissimum* and *D. striolatum*. Plants have terete leaves typical of the parents, and a more compact growth habit than that of *D. tenuissimum*. Flowers are small and greenish yellow with some reddish striation on the sepals. Plants grow easily in a bushhouse in temperate and subtropical regions. The cross was registered in 1982.

Dendrobium 'Kim'
Cultivar

Aug-Oct

A man-made hybrid which arose from the crossing of *D.* 'Peter' with *D. kingianum*. It is a compact-growing hybrid with racemes of pale pink flowers which have a prominent labellum. Plants grow well in a bushhouse or cool glasshouse in temperate or subtropical regions.

Dendrobium 'Kim Heinze'
Cultivar

Aug-Oct

A man-made hybrid which was registered in 1982. It is a second generation hybrid between *D.* 'Blushing Star' and *D.* 'Hilda Poxon', and bears large starry flowers which are yellowish green with dark maroon spots and blotches. It is reportedly an excellent free-flowering hybrid with a compact growth habit.

244

Dendrobium 'King Falcon'
Cultivar

Aug-Oct
A man-made cultivar which was registered in 1978. It is a cross between *D.* 'Kingrose' and *D. falcorostrum*. It is a strong-growing hybrid with racemes of quite large pink flowers which have a prominent lip and are strongly fragrant. A well-grown specimen is very showy. Plants can be grown in temperate and subtropical regions in a bushhouse or cool glasshouse.

Dendrobium 'Kingrose'
Cultivar

Aug-Oct
A man-made hybrid which was registered in 1966. It is a backcross of *D.* 'Bardo Rose' and *D. kingianum*, and has produced some very colourful progeny. The growth habit is similar to *D. kingianum*, as are the flowering habit and the flowers. The latter, however, may be larger and are usually of a much deeper colour. Some clones are excellent, being very free-flowering and having very dark purplish red flowers. They make excellent subjects for specimen culture. *D.* 'Kingrose' is very easy to grow, requiring conditions identical to those for *D. kingianum*.

Dendrobium 'Kit Murdoch'
Cultivar

Aug-Oct
A man-made hybrid which was registered in 1980. It is the result of a cross between *D. kingianum* 'Silcockii' and *D.* 'Susan'. Plants are very cold hardy and succeed well in a bushhouse or cool glasshouse in temperate regions. The flowers are white.

Dendrobium 'Kungara'
Cultivar

Aug-Oct
This man-made hybrid was registered in 1982 and is the result of a cross between *D. tetragonum* and *D. ruppianum*. Plants are somewhat slow-growing, with angular or ribbed pseudobulbs, and bear attractive starry flowers which are pinkish with red blotches. The hybrid is suited to cultivation in temperate and subtropical regions.

Dendrobium 'Ku-Ring-Gai'
Cultivar

Aug-Oct; also sporadic
A second generation, man-made hybrid between *D.* 'Specio-kingianum' (or *D.* × *delicatum*) and *D. tetragonum*, registered in 1972. The plants are much weaker-growing than *D.* 'Specio-kingianum' and their flowers have narrow, pointed segments which impart a somewhat spidery appearance. They vary in colour from greenish pink to yellowish, and are usually spotted or blotched. This hybrid is uncommon in cultivation and tends to be slower-growing than many others. It is best suited to pot culture in subtropical and temperate regions.

Dendrobium 'Leonie Jane'
Cultivar   Aug-Nov
A man-made hybrid between *D. adae* and *D.*

Dendrobium 'King Falcon'   D. Cannon

falcorostrum, which was registered in 1974. It is little-known in cultivation but some clones produce nicely shaped, strongly fragrant white flowers, each about 2 cm across. Plant growth tends to be weaker than in many other hybrids and flowering may be poor. The plants grow well in temperate and subtropical regions, in a bushhouse or unheated glasshouse.

Dendrobium 'Lowana Nioka'
Cultivar

Nov-March
A man-made hybrid between *D. canaliculatum* and *D. antennatum*, registered in 1962. The growth habit is compact, with thickened leaves, and the racemes are erect and bear long-lasting, colourful flowers which are useful for cutting. These are 1.5-2 cm across and are cream or greenish with mauve markings on the labellum. The plants grow well in the tropics, but in southern areas need a heated glasshouse with a minimum temperature of 10°C.

Dendrobium 'Lowana Tiki'
Cultivar

Nov-March
This second generation, man-made hybrid was registered in 1966. It is the result of crossing *D.* 'Judy Leroy' with *D. antennatum*, and is an excellent free-flowering hybrid for the tropics. The growth habit is

245

compact and the racemes are erect, bearing cream or pink-mauve, greenish-tinged flowers with a white labellum. The petals are erect and somewhat twisted as in *D. antennatum*. In temperate regions this hybrid must be grown in a heated glasshouse with a minimum temperature of 10-15°C.

Dendrobium 'Merritt Island'
Cultivar
Nov-Feb

A man-made hybrid between *D. bigibbum* and *D. dicuphum*, registered in 1969. Both parents are basically very similar in growth habit, and it is hard to tell the hybrid from either parent. The flowers are smaller than *D. bigibbum*, but are of similar shape and are light mauve in colour. Plants succeed best in lowland tropical areas, and in places with a cold winter, need a heated glasshouse.

Dendrobium 'Michael Jupp'
Cultivar
Aug-Oct

A novelty hybrid between *D. beckleri* and *D. linguiforme*, which was registered in 1980. Plants have a scrambly growth habit with fleshy, dark green, furrowed leaves. Greenish cream flowers with purple striations are borne on short racemes which arise from the base of the leaves. Plants can be grown in pots but are best mounted on slabs. They are very cold hardy and will succeed in a bushhouse in southern Australia.

Dendrobium 'Mini Pearl'
Cultivar
Oct-Feb

A man-made hybrid which was registered in 1974. It is the result of a cross between *D. canaliculatum* and *D. bigibbum*. The growth habit and flower shape of the hybrid are very similar to those of *D.* 'Judy Leroy', but the flowers tend to be larger and more colourful (lilac to mauve). Plants are very free-flowering and are excellent for tropical regions, where they can be grown in bushhouses or even as garden plants. To flower in temperate areas they need a heated glasshouse with a minimum temperature of 10°C and plenty of light.

Dendrobium 'Our Native'
Cultivar
Aug-Dec

A man-made hybrid between *D. semifuscum* and *D. speciosum*, registered in 1978. The growth habit is similar to that of *D. johannis*, with narrow leaves scattered up the stem. The racemes are erect and bear green to yellowish flowers with a strong yellow labellum. The flowers are 5-7 cm across and are fairly long-lasting. This attractive hybrid is versatile in its culture and can be grown in tropical, sub-tropical and warm temperate regions. It may require some heat to over-winter in colder areas.

Dendrobium 'Peach Glow'
Cultivar
Aug-Oct; Jan-April

A second generation, man-made hybrid which is a cross between *D.* 'Hastings' and *D.* 'Golden Fleck'. It

was registered in 1981. Plants are fairly slow-growing, with slender pseudobulbs. The flowers, which are carried on short racemes, are pink with orange tonings, and have an attractive fringed lip. Plants can be grown in a bushhouse or cool glasshouse, and will succeed in temperate or subtropical areas.

Dendrobium 'Peach Star'
Cultivar
Aug-Oct

A man-made hybrid between *D. tetragonum* and *D.* 'Hastings', registered in 1982. It produces attractive starry flowers which are yellow, flushed with red tonings and have prominent dark blotches on the labellum. This is an excellent free-flowering hybrid of easy culture.

Dendrobium 'Peewee'
Cultivar
Dec-Jan

A man-made hybrid which was registered in 1979 and is the result of a cross between *D. bigibbum* and *D. tetragonum*. It has unusual angular pseudobulbs, and bears racemes of reddish flowers which have fairly broad segments with pointed tips. Plants are cold sensitive and are subject to bud-drop. They require warm tropical conditions, and in cold districts, must be grown in a heated glasshouse.

Dendrobium 'Penny Ann'
Cultivar
Aug-Oct

A second generation, man-made hybrid between *D. kingianum* and *D.* × *gracillimum*, registered in 1974. The flowers are about 2 cm across and are in shades of pink, with red suffusions on the backs of the petals and sepals. The plants are fairly free-flowering and easy to grow, requiring conditions similar to those for *D. kingianum*.

Dendrobium 'Peter'
Cultivar
Aug-Oct

This hybrid is very popular because it is easy to grow and is free-flowering, producing attractive, colourful flowers. It is the result of a cross between *D. fleckeri* and *D. falcorostrum*, and it was registered in 1972. The flowers are large, with a waxy texture like *D. falcorostrum*, but have suffusions of yellow or orange, and a prominent, hairy labellum. The plants are easy to grow in a bushhouse or unheated glasshouse. They are best grown in pots of coarse mixture, and are suitable for temperate or subtropical regions.

Dendrobium 'Peter Petersen'
Cultivar
July-Dec; also sporadic

This hybrid has flowers 5-8 cm across, opening widely, brownish with bluish suffusions, the petals erect and twisted. It is a man-made hybrid between *D. nindii* and *D. discolor*. Natural hybrids involving the same parents have been found where the two species grow together, but they are very rare. The hybrid is a handsome orchid, having some of the colour of *D. nindii*, and is much easier to grow, as is *D. discolor*. It is essentially tropical in its requirements and, in suitable areas, can be grown outside as a pot plant or attached to garden trees etc. To flower in temperate regions, it requires a heated glasshouse with a minimum temperature of 15°C and plenty of light.

Dendrobium 'Phil Deane'
Cultivar
Aug-Oct

A man-made hybrid between *D.* 'Ella Victoria Leaney' and *D. tetragonum* var. *giganteum*, registered in 1982. It is reported to bear large spidery flowers which are a deep mahogany red with an attractive white labellum blotched with red.

Dendrobium 'Rachelle Simpson'
Cultivar
Oct-Feb

This man-made hybrid also occurs in nature, although very sporadically. It is the result of a cross between *D. canaliculatum* and *D. semifuscum*, and was registered in 1967. The growth habit is compact, like *D. canaliculatum*, but with slightly larger pseudobulbs. The racemes are erect, up to 45 cm long, and bear brown flowers with curled petals and sepals. The plants are very free-flowering and the flowers last quite well when cut. This hybrid grows well in tropical conditions but is very cold sensitive, and to promote flowering, needs to be kept dry over winter. In temperate areas the plants should be hung near the glass in a heated glasshouse with a minimum temperature of 10°C.

Dendrobium 'Red Wings'
Cultivar
Nov-Jan

A man-made hybrid between *D. semifuscum* and *D. antennatum*, registered in 1954. It is essentially a hybrid for the tropics, and in temperate regions needs a heated glasshouse with a minimum temperature of 10°C. The pseudobulbs are erect and somewhat drawn out at the apex, as in *D. antennatum*. The flowers are reddish brown with a greenish labellum. The perianth segments are twisted, as in both parents, and the flowers last well when cut.

Dendrobium 'Rosemary Jupp'
Cultivar
July-Oct

A man-made hybrid between *D. striolatum* and *D. teretifolium*, registered in 1975. It is primarily an interest-cross, having little commercial potential but providing a good conversation piece. The leaves are terete, as in both parents, and the growth is more compact than that of *D. teretifolium*. The flowers are borne in short racemes, and are quite colourful, being yellowish to buff with some red markings. Some variation in flower size and colour could be expected if different forms of *D. teretifolium* are used as parents. The plants grow readily in pots or on slabs, and are suited to temperate and subtropical regions.

Dendrobium 'Saskia'
Cultivar
Nov-March

A man-made hybrid which is really a backcross between *D.* × *superbiens* and *D. bigibbum*. This hybrid was registered in 1951. The same cross is believed to occur sporadically in nature. The plants resemble *D.* × *superbiens* in growth habit, but have shorter pseudobulbs. The flowers are in tones of mauve-purple and have broader, less twisted segments than *D.* × *superbiens*. This hybrid is essentially tropical in its requirements and should be cultivated in the same way as *D. bigibbum* and *D.* × *superbiens*. It is best grown in a small pot, and needs plenty of light to flower.

Dendrobium 'Shan Leaney'
Cultivar
Aug-Oct

A second generation, man-made hybrid which is really a backcross between *D.* × *gracillimum* and *D. gracilicaule*. It was registered in 1974. The plants have slender, tall pseudobulbs and bear relatively short, arching racemes which each carry 6-12 flowers. These are about 1.3 cm across and are greenish yellow, usually with some blotching on the reverse of the sepals and petals. This hybrid is not widely grown, and shows little improvement over *D.* × *gracillimum*. The plants grow readily in a bushhouse or cool glasshouse in subtropical or temperate regions.

Dendrobium 'Specio-kingianum'
Cultivar
Aug-Oct

This is the first man-made hybrid involving Australian native orchids. It was registered in 1892. The cross was made in England, where the parents were growing in a glasshouse owned by Sir Trevor Lawrence, the registrant. The same cross occurs naturally where the two parents grow together, and it is known botanically as *D.* × *delicatum* (q.v.). There have been many other man-made crosses of these species, especially in recent years, but using selected clones as parents. These have resulted in some interesting colour forms (mostly purple-spotted pinks, but a few yellowish) but the variation is fairly limited. This hybrid is a very vigorous plant that grows readily and flowers profusely in subtropical and temperate regions.

Dendrobium 'Star Delight'
Cultivar
Aug-Oct; also sporadic

A man-made hybrid which was registered in 1980. It has as its parents *D.* 'Star of Gold' and *D. ruppianum*. Plants are slow-growing, with fairly large, furrowed pseudobulbs and long racemes of pale yellow to cream flowers which have somewhat pointed segments. Plants are suited to cultivation in subtropical and temperate zones.

Dendrobium 'Star Imp'
Cultivar
Aug-Oct

A man-made hybrid between *D. tetragonum* and *D. adae*, registered in 1977. Plants are slow-growing and have thin, somewhat angular pseudobulbs. Small racemes each carry 2-3 flowers which are cream with starry, pointed segments. Plants are best grown in a pot of coarse mixture, and will succeed in a bushhouse or unheated glasshouse in temperate or subtropical regions.

Dendrobium 'Star of Dawn'
Cultivar

Aug-Oct

A man-made hybrid which is a cross between *D.* 'Star of Gold' and *D. kingianum.* It was registered in 1980. It is a fairly slow grower, well suited to temperate and subtropical zones. The pseudobulbs tend to be somewhat flattened or angular. The pink to reddish flowers have pointed segments, and tend to droop on the racemes.

Dendrobium 'Star of Gold'
Cultivar

Aug-Oct; also sporadic

A man-made hybrid between *D. falcorostrum* and *D. tetragonum,* which was registered in 1974. It has proved to be an excellent cross and is very popular with orchid enthusiasts. The growth habit is neat and quite manageable in a pot. The flowers are large, with spidery segments, and may be clear yellow or have some blotching. They are very strongly perfumed. Relatively few flowers are produced on the raceme, but the plants may flower more than once a year. *D.* 'Star of Gold' grows readily in a pot of coarse mixture, and thrives in temperate and subtropical regions.

Dendrobium 'Star of Riverdene'
Cultivar

Aug-Oct; also sporadic

A second generation, man-made hybrid between *D. speciosum* and *D.* 'Star of Gold', registered in 1981. The flowers open widely and have slender, spidery segments in clear yellow, or with some blotching. The plants grow strongly and may flower more than once a year. They can be successfully grown in pots of coarse mixture in a bushhouse or cool glasshouse in temperate or subtropical regions.

Dendrobium 'Star One'
Cultivar

Aug-Oct; also sporadic

A second generation, man-made hybrid between *D.* 'Specio-kingianum' (or *D.* × *delicatum*) and *D.* 'Star of Gold', registered in 1979. The plants are fairly strong-growing and relatively compact, and may produce more than one batch of flowers in a year. The flowers are 3-4 cm across, with pointed, somewhat starry segments, and tend to be greenish or yellowish with some spotting. The plants are best grown in pots of coarse mixture and are suited to temperate and subtropical regions.

Dendrobium 'Star Twist'
Cultivar

Nov-Jan

A man-made hybrid between *D. tetragonum* and *D.*

semifuscum, registered in 1981. Plants tend to be slow-growing and have brownish flowers on short racemes. These have a prominent twist in the petals and sepals. This hybrid grows best in subtropical or cool tropical areas, and to over-winter in temperate regions, needs some heat.

Dendrobium 'Sunglow'
Cultivar

Aug-Oct; also sporadic

A man-made hybrid between *D. speciosum* and *D. fleckeri,* registered in 1980. This is a very appealing orchid with widely spreading flowers of warm orange hues. The mid-lobe of the labellum bears masses of white hairs, as in *D. fleckeri,* and red bars and blotches. The plants grow readily and are well suited to temperate and subtropical regions. Plants can be very floriferous and can carry up to 15 flowers on a raceme. May flower almost continuously. This is one of the best native hybrids produced to date.

Dendrobium 'Sun Sprite'
Cultivar

Aug-Oct

This is a man-made hybrid between two natural hybrids (*D.* × *delicatum* and *D.* × *gracillimum*). It was registered in 1975. Plants are strong-growing, with tall pseudobulbs, and carry fairly long racemes of greenish yellow flowers. They can be grown in subtropical or temperate regions and are quite hardy to cold. Hybrids from using a white-flowered form of *D.* × *delicatum* produce strong, erect racemes which carry up to 14 glistening, white flowers.

Dendrobium 'Susan'
Cultivar

Aug-Oct

A man-made hybrid between *D. falcorostrum* and *D. gracilicaule,* which was registered in 1971. This is a popular hybrid with a neat, compact growth habit and racemes of sulphur yellow flowers which are about 2.5 cm across. The racemes tend to be few-flowered (5-10 flowers) and the flowers have a drooping habit, but large plants are attractive when in full bloom. Flowers are highly perfumed. This is an ideal plant for the amateur or enthusiast as it is easily grown in a bush-house or cool glasshouse, or even on a protected veran-dah. Suited to temperate or cool subtropical regions, it is best grown in a pot of coarse mixture.

Dendrobium 'Three Star'
Cultivar

Nov-April

A second generation, man-made hybrid between *D.* × *superbiens* and *D. semifuscum,* registered in 1954. It is a fine orchid for the tropics because it thrives in warm, humid, well lit situations. The pseudobulbs are erect and leafy, and bear strong racemes of flowers which last well when cut. The flowers are about 6 cm across, pale brown with bluish suffusions and twisted perianth segments. For best growth in temperate areas this hybrid needs a heated glasshouse with a minimum temperature of 10°C.

248

Dendrobium 'Sunglow'

D. Cannon

Dendrobium 'Tom Jupp'
Cultivar

Aug-Oct

A second generation, man-made hybrid between *D.* 'Andrew Persson' and *D.* × *delicatum*, registered in 1975. The plants are strong-growing, with fairly long pseudobulbs and thickish leaves. They bear erect racemes with cream to pale pink flowers. This hybrid grows easily in subtropical and temperate regions but is not widely cultivated. It can be grown in a bushhouse or cool glasshouse.

Dendrobium 'Tusco'
Cultivar

Nov-May

A man-made hybrid between *D. nindii* and *D. semifuscum*, which was registered in 1957. It is a free-flowering hybrid that is excellent for the tropics. The pseudobulbs are stout and leafy, and bear strong racemes of flowers 20-40 cm long. The individual flowers are about 6 cm across, with very curly sepals and petals. The colouration varies, but is basically brownish with some blue suffusions and a yellow patch on the labellum. The plants can be grown on slabs or in small pots of coarse material. In temperate regions they need a heated glasshouse with a minimum temperature of 10-15 °C.

Dendrobium 'Verninha'
Cultivar

Nov-March

A man-made hybrid between *D. discolor* and *D. antennatum*, registered in 1961. It is essentially a hybrid for tropical conditions, and the plants are strong-growing and free-flowering. The erect pseudobulbs bear numerous leaves and strong racemes from the upper nodes. The flowers are about 4 cm across and are predominantly pale brown with a greenish white labellum. The lateral petals are erect, and all segments except the labellum are twisted. In temperate regions this hybrid needs a heated glasshouse with a minimum temperature of 10-15 °C. It is best grown in a pot.

Dendrobium 'Virginia Jupp'
Cultivar

Aug-Oct

A man-made hybrid between *D. teretifolium* and *D. linguiforme*, registered in 1969. The plants are much more compact than *D. teretifolium*, although the leaves tend to be more like a shortened, somewhat flattened version of that species than the thumbnail type of *D. linguiforme*. Flowers are white or cream, and are borne on racemes 5-10 cm long. This hybrid has proved to be easy to grow, but it does not always flower freely. It can be grown in a pot, but because it tends to wander, is

249

probably best grown attached to a tree-fern slab. Some variation in the progeny can be obtained by using different forms or varieties of the parents in the cross.

Dendrobium 'Willowie Gold'
Cultivar

Nov-March; also sporadic

A man-made hybrid between *D. discolor* and *D. tetragonum*, registered in 1970. It is an unusual hybrid, of which most characteristics such as the growth habit and flower shape, are like those of *D. discolor*. The pseudobulbs, however, tend to be slightly angular, the flower segments somewhat spidery, and the flowers are produced at odd intervals during the year. The flowers are quite large and golden brownish, with the segments tending to be twisted near the tips. They are borne in racemes 15-30 cm long and last for 2-3 weeks. The plants are much more tolerant of cold than *D. discolor*, but still require a heated glasshouse in temperate regions.

Dendrobium 'Wonga'
Cultivar

Aug-Nov

A second generation, man-made hybrid between *D. speciosum* and *D.* 'Hastings', registered in 1979. It is little-known in cultivation, and is a fairly strong grower, with erect racemes of orange-tinged yellowish flowers which are highly perfumed. The labellum tends to be somewhat hairy on the mid-lobe. This hybrid grows well in a bushhouse or cool glasshouse in subtropical and temperate regions.

Dendrobium 'Yellow Venture'
Cultivar

Aug-Dec

A man-made hybrid which was registered in 1979. It is the result of a cross between *D. speciosum* and *D. adae*. The plants tend to have slender pseudobulbs (thinner than *D. speciosum*) but grow quite strongly. The racemes are erect to arching, and each bears 10-12 fairly crowded flowers which may be cream or yellow. The plants grow easily in a cool glasshouse or bushhouse in subtropical and temperate regions, and are best suited to pot culture.

Dendrobium 'Yondi'
Cultivar

Aug-Oct

A man-made hybrid which was registered in 1982. It is the result of a cross between *D. kingianum* and *D.* 'Eureka'. Plants bear flowers which are about 4 cm across and are a bright cyclamen pink, and have a white, pink-flecked labellum. Plants are reported to be an improvement on the parents and to flower while quite small.

DENDROCNIDE Miq.
(from the Greek *dendros*, a tree; *cnid*, a nettle; alluding to their stinging properties)

Shrubs or **trees** armed with hollow, stinging hairs; **wood** soft and fibrous; **leaves** simple, alternate, glabrous or hairy, petiolate, the margins entire, toothed or lobed; **inflorescence** an axillary panicle; **flowers** unisexual, borne on separate panicles or separate plants; **fruit** a nut; **fruit stalk** becomes fleshy and fruit-like.

A genus of 36 species with 4 described from Australia, and another 2 species from Cape York Peninsula possibly undescribed. The Australian species were previously included in the genus *Laportea* which is now known to be restricted to America. All stem pieces have hollow stinging hairs which contain an intensely irritant and virulent poison. As a group they have little horticultural appeal, yet seem to have a masochistic attraction for enthusiasts.

Dendrocnide cordata (Warb. ex Winkler) Chew
(heart-shaped; in reference to the leaves)

Qld — Stinger

2-4 m x 1-3 m — March-May

Medium **shrub** forming suckering colonies; **stems** very hairy; **leaves** 6-20 cm x 5-10 cm; ovate to cordate, not peltate, dark green, very hairy, the margins coarsely lobed; **panicles** short, axillary; **flowers** small, unisexual, male and female on separate plants; **nut** 0.1-0.2 cm across, each surmounted by a fleshy white stalk.

In Australia this stinger is restricted to rainforests of the Atherton Tableland and Cape York Peninsula; also extends to New Guinea and Timor. It is similar in many respects to *D. moroides*, but the leaves are not peltate. Like that species, its stinging propensity is considerable and the species lacks any horticultural appeal. Propagate from seed or cuttings.

Dendrocnide excelsa (Wedd.) Chew
(lofty; elevated)

Qld, NSW — Giant Stinging Tree

10-38 m x 5-15 m — April-July

Medium to tall, spreading **tree**; **bark** grey, scaly; **timber** soft, fibrous; young parts covered with stiff stinging hairs; **leaves** 15-30 cm x 15-20 cm, broadly ovate, pale green and thin-textured, the margins toothed, on petioles 3-15 cm long; **panicles** axillary, short; **flowers** unisexual, male and female borne on separate trees; **nut** 0.1-0.2 cm across, clustered, each surmounted by a fleshy stalk which may be white, dull pink or mauve.

A large tree, widely distributed in rainforests from Gympie in south-eastern Qld to Bega in south-eastern NSW and often locally common. Plants frequently colonize clearings and rainforest margins. The leaves and young stems are covered with stiff silicaceous hairs which can inflict a nasty sting if handled. The stinging sensation is produced by a virulent toxin and can last for weeks. Even dead leaves can sting. Trees may become partially or completely deciduous during long dry periods. Young plants have large, conspicuously hairy, toothed leaves and can sting more severely than those of mature trees. Naturally, such a large unfriendly plant has very little horticultural appeal. Plants are known from botanical collections (there is a large plant in Melbourne's Royal Botanic Gardens) and in those of enthusiasts. A well drained soil is essential. Fruit are edible and have an acidic taste. The Aborigines used the fibrous root bark to make fishing lines and nets, and the inner bark as a form of cloth. It is also claimed that they used the stinging properties of the hairs as a cure for rheumatics. Propagate from seed which must be sown fresh or cuttings which strike readily.

Dendrocnide moroides (Wedd.) Chew
(resembling the genus *Morus*)
Qld, NSW Mulberry-leaved Stinger
2-4.5 m x 1-3 m April-May
Medium **shrub** forming suckering colonies; **stems** very hairy; **leaves** 6-30 cm x 4-20 cm, broadly ovate to cordate, peltate, dark green to purplish green, very hairy, the margins toothed, the apex pointed; **panicles** short, axillary; **flowers** small, unisexual, male and female on separate plants; **nut** 0.1-0.2 cm across, each surmounted by a fleshy stalk which is pink to purplish.

Widely distributed in rainforests from the Atherton Tableland to the border, and extending to Lismore in NSW. It is a weedy species, usually growing in colonies of sparse, sickly-looking plants. Frequently, a plant consists only of a single stem and a few leaves. The species rapidly colonizes tracks, road verges, rainforest margins and any disturbed soil or clearing. These plants are usually lush and vigorous, often with purplish leaves. This species is the worst of the Australian stingers and is held in a loathing regard by all those who regularly enter rainforests. The very lush plants produce the worst stings, but even insignificant, dead-looking sticks can cause a severe sting. To blunder into a suckering clump of this plant is a nasty experience not to be forgotten, and cases of death have been reported in horses caught in such a predicament. Contact with this plant produces intense stinging sensations which may last for several days, and are associated with swellings in the armpit or groin. Stinging may reactivate periodically over several months, following sweating or contact with water. Slashing clumps produces intense irritation from drifting hairs, to the throat and nasal passages. The fruit are edible but usually contain stinging hairs. Very few people are interested in the cultivation of this species and there is probably more interest in its destruction. Plants are occasionally to be found in botanical collections. Propagate from seed or cuttings which strike readily.

Dendrocnide photinophylla (Kunth.) Chew
(shiny leaves)
Qld, NSW Shiny-leaved Stinging Tree
15-25 m x 5-12 m March-May
Medium to tall, spreading **tree**; **bark** grey, scaly, with prominent pustules; young shoots glabrous, succulent; **leaves** 5-12 cm x 2-6 cm, ovate, dark green and glossy above, dull beneath, thick-textured, the margins entire, the apex drawn out and blunt; **panicles** 5-6 cm long, axillary; **flowers** unisexual, in separate panicles on the tree or on separate trees; **nut** 0.1-0.2 cm across, each surmounted by irregular, fleshy white stalks which are usually clustered.

A large, shiny green tree widely distributed in rainforests from Atherton in north-eastern Qld to Gloucester in NSW. This species is the least offensive of all the stinging trees, the effects of its sting usually lasting only a few hours. Mature plants are frequently

Dendrocnide photinophylla, × ·6

stingless and even the young plants are extremely variable in this respect. The stinging hairs are usually found on succulent stems and leaves. This species has attractive leaves and would make a useful foliage plant for indoors, or an interesting bonsai subject. Because of the variability in the presence of stinging hairs, it should be possible to select and propagate non-stinging forms. The fruits are edible and quite tasty. Plants of this species can be easily grown in well drained soil. They require some protection from direct, hot sun when young, and respond to water and the use of fertilizers. Propagate from seed or cuttings which strike easily.

DENDROLOBIUM Benth.
(from the Greek *dendron*, a tree; *lobos*, a pod or lobe)
Fabaceae
Shrubs; young shoots bearing silky hairs; **stipules** deciduous or persistent; **leaves** trifoliolate; **inflorescence** a crowded axillary raceme or umbel; **flowers** pea-shaped; **fruit** an indehiscent pod, breaking up into segments.

A small genus of 17 species distributed from Indo-Malaysia to northern Australia where there are 2 species. The genus is closely related to *Desmodium* and was formerly included as a section of that genus. The species have limited horticultural appeal and are rarely encountered in cultivation. Propagate from seed which is hard and needs pre-sowing treatment.

Dendrolobium stipatum S. T. Blake
(stalked)
WA, NT
0.3-0.6 m x 0.5-1.5 m July-Aug
Dwarf, hairy **shrub**; young shoots silky-hairy; **stems** in a tuft, arising from a woody base, erect, slender,

251

simple or branched; **leaves** 5-9 cm long, trifoliolate, on petioles 0.7-1.2 cm long; **leaflets** 4-6.5 cm x 1.5-3.3 cm, elliptic to ovate, green above, ashy to glaucous beneath, hairy, the apex blunt or notched; **raceme** short, the flowers crowded; **flowers** about 1.5 cm long, pea-shaped, the petals pink, the standard white in the upper part; **pods** 2-3.5 cm x 0.5 cm, sessile, indehiscent, hairy.

An interesting, small shrub found in open forest of northern areas, often where bushfires occur regularly. It consists of numerous slender stems arising from a perennial, woody rootstock and is unusual in that it begins flowering well before the start of the wet season. Culturally it would be best suited to inland areas or drier tropical regions, and would require well drained soils in sunny or partial-sun situations. Propagate from treated seed and possibly cuttings or transplants.

Dendrolobium umbellatum (L.) Benth.
(flowers in umbels)
Qld, WA, NT
1-4 m x 1-1.5 m   Nov-Jan
Small to medium, bushy **shrub**; young shoots silky-hairy; **leaves** 5-7 cm long, trifoliolate; **leaflets** 3-5 cm x 1.5-3 cm, ovate to oblong, glabrous above, silky-hairy beneath, blunt or acute; **racemes** short, dense, crowded in axillary umbels; **flowers** about 1 cm long, pea-shaped, white, crowded; **standard** broad; **calyx** hairy; **pods** 1-2 cm x 0.5-0.8 cm, indehiscent.

An interesting, widespread shrub which extends to areas of northern Australia where it grows in open forest. Plants are very hardy and decorative when in flower, and could have potential for cultivation in drier tropical zones. Requirements are for well drained soil, in a sunny situation. Propagate from treated seed. Previously known as *Desmodium umbellatum* (L.) DC.

DENHAMIA Meisner
(after Captain D. Denham, an African explorer)
Celastraceae
Shrubs or **trees**; **leaves** alternate, leathery, entire or with wavy or toothed margins; **inflorescence** a cyme or raceme, few-flowered; **flowers** small; **calyx** 5-lobed; **petals** 5; **stamens** 5; **fruit** a capsule; **seeds** with a conspicuous fleshy arillus.

This genus, which is endemic in Australia, consists of 7 species. They are rather unobtrusive plants, rarely encountered in cultivation but nevertheless with some interesting and useful features. They are generally hardy plants, and the seeds are eagerly sought after by birds. Propagation is mainly from seed which must be sown fresh. Cuttings of semi-hardened wood should prove successful for most species.

Denhamia obscura Meisner
(obscure; dull; not remarkable)
WA, NT   Weeping Denhamia
5-10 m x 2-4 m   May-Sept
Tall **shrub** or small **tree**; **bark** splotched with grey or white; **branches** erect or pendulous, slender; **leaves** 5-10 cm x 1-3 cm, oblong or lanceolate, dull green or brownish above, pale beneath, the margins entire or undulate, leathery; **cymes** in the upper axils, some-

times forming a panicle; **flowers** about 0.6 cm across, cream to yellowish; **capsule** 2-2.5 cm long, globular to ovoid, white to yellow, with bright orange seeds, mature July-Oct.

This species occurs in open forest and on rocky escarpments of parts of northern Australia. Has been confused with *D. oleaster*. It is a fairly nondescript tree but with an appealing graceful and pendulous growth form. It has proved to be adaptable and hardy, even under adverse conditions. Trees are long-lived but are generally slow-growing when young. They require a sunny situation, in well drained soil, and are best suited to tropical and subtropical regions. Young plants and sucker-growths have broader, coarsely toothed leaves resembling the English holly. The root bark is bright orange, and useful for bark pictures. It is oily and ignites readily, even when wet, and is thus a very handy tinder material. The wood is yellow and useful for carving. The seeds are attractive to birds. Propagate from seed which must be sown fresh and stem or root cuttings in the wet season.

Denhamia oleaster (Lindley) F. Muell.
(like the genus *Oleaster*)
Qld
5-8 m   Aug-Sept
Tall, slender **shrub** or small **tree**; **bark** grey; **leaves** 5-8 cm x 1.5-2.5 cm, lanceolate, dark green, thick-textured and leathery, pointed or blunt, the margins entire or with a few sharp teeth; **racemes** 2-5 cm long, in the upper axils or terminal; **flowers** about 0.5 cm across, white; **capsule** 1.5-2 cm long, globular, yellowish.

This species is apparently restricted to western Qld, where it occurs in open forest. It has been previously confused with *D. obscura*. A hardy species for cultivation in inland gardens. Plants may be slow-growing but are hardy from an early stage. They require an open, sunny position, in well drained soil. The inner bark is used for the creation of bark pictures. Propagate from seed which must be sown fresh.

Denhamia parvifolia L. S. Smith
(small leaves)
Qld
6-10 m x 3-6 m   Sept-Oct
Small **tree**; **leaves** 0.5-3 cm x 0.4-0.9 cm, elliptic, leathery, dark green and shiny above, paler beneath, the margins bearing a few small teeth; **racemes** 1.5-3.5 cm long, axillary; **flowers** about 0.2 cm long, white to cream; **capsule** 0.6-0.8 cm x 0.5-0.6 cm, ovoid, pale yellow, mature Feb-March.

A little-known species confined to the Burnett District of southern Qld, where it grows on heavy soils along stream banks. It is a hardy small tree, well suited to gardens of inland districts and subtropical regions. Trees are slow-growing, especially when young, and require a sunny situation, in freely draining soil. Propagate from seed which must be sown fresh.

Denhamia pittosporoides F. Muell.
(like the genus *Pittosporum*)
Qld, NSW   Orange Boxwood
3-6 m x 1-3 m   Oct-Nov

Tall, slender **shrub**; **trunk** erect, fluted, often striate; **bark** brown, the underbark green or bright orange; **leaves** 7-13 cm x 1-2 cm, elliptic to linear-lanceolate, simple, dark green and shiny on both surfaces, leathery, when juvenile the margins sharply toothed; **cymes** axillary, short, few-flowered; **flowers** about 0.4 cm across, creamy white; **capsule** 1-1.5 cm across, globular, hard and woody, bright orange, containing 3-5 locules; **seeds** 4-8 per capsule, a bright red aril at one end, mature Feb-June.

A fairly common species, usually occurring in drier rainforests to the west of the divide. Occurs in southern Qld and is rare in northern NSW. It is an interesting tree with a strong resemblance to *Pittosporum phillyreoides*. Plants are slender with narrow, dark leaves and colourful fruit. Birds are attracted to the seeds. Plants are very hardy, and well suited to subtropical and warm temperate regions. Needs very well drained soils. The young plants have broader leaves than those on mature trees, with more prominently toothed margins. They make attractive pot subjects. Propagate from fresh seed or perhaps cuttings.

Denhamia viridissima Bailey & F. Muell.
(very green)
Qld
8-15 m x 2-4 m   Nov-Dec
Small to medium, dense, spreading **tree**; **leaves** 7.5-10 cm x 2.5-4 cm, lanceolate, dark shiny green, drawn out into a pointed apex, the margins entire, veins prominent, often clustered towards the ends of branches; **cymes** or **racemes** terminal; **flowers** 0.5-0.8 cm across, white; **capsule** about 1.5 cm long, fusiform, bluntly 3-angled, yellow; **seeds** 12-18 per capsule, black with a prominent red, fleshy arillus, mature July-Aug.

A small, dense tree restricted to highland rainforests of north-eastern Qld. Little-known in cultivation but could prove to be useful in a somewhat sheltered position. Requires well drained, acid soils and watering during dry periods. Young plants could be useful for indoor decoration. Propagate from seed which must be sown fresh.

Denhamia species
NSW   Mountain Denhamia
3-6 m x 1-2 m   March-April
Medium to tall **shrub**; **bark** grey or brown; young shoots ribbed; **leaves** 2.5-4.5 cm x 1-2 cm, lanceolate, bluntly pointed, simple, the margins entire, dark green and dull above, pale below; **flowers** about 0.4 cm across, cream, solitary in the axils, on slender stalks 0.5-0.8 cm long; **capsule** 1-1.2 cm long, ovoid, pointed, hard and woody, yellow, 3-celled; **seeds** 3-6 per capsule, black with an orange aril, mature April-Aug.

An apparently undescribed species that is common in highland rainforests of mountains along the Bellinger River, north-eastern NSW. Although relatively unknown in cultivation, it is grown by enthusiasts. Requires a moist, shady position and could be a useful shrub for mixing with ferns. The fruit are conspicuous and decorative when ripe. Propagate from seed which should be sown fresh.

Denhamia species
Qld, NSW
5-12 m x 2-5 m   Oct-Dec
Tall **shrub** to small **tree**; **trunk** erect, fluted; **bark** grey-brown, underbark yellow; **leaves** 4-13 cm x 2-4 cm, elliptic to lanceolate, dark green and somewhat shiny, sometimes clustered near the ends of branches, the margins sparsely and bluntly toothed; **cymes** axillary, short, often clustered, on short branchlets or at the base of current season's growth; **flowers** about 0.4 cm across, greenish; **capsule** 0.5-0.8 cm across, ovoid, dull coloured, bearing 3 locules; **seeds** 3-6 per capsule.

An abundant species occurring in rainforests and moister open forests of coastal areas and mountainous districts in south-eastern Qld and north-eastern NSW. Has been confused with *D. pittosporoides*. It is rather a drab plant with few features to recommend it for cultivation. It is, however, quite hardy and bushy. Birds are attracted to the seeds. Young plants have broader leaves with toothed margins. The inner yellow root bark is used in the creation of bark pictures. Propagate from fresh seed or perhaps cuttings.

DENISONIA F. Muell.
(after Sir William T. Denison, Governor of NSW)
Dicrastylidaceae
A monotypic genus, endemic in tropical Australia.

Denisonia ternifolia F. Muell.
(ternate leaves)
NT
0.5-1.5 m x 0.3-1 m   April-May
Dwarf to small **shrub**; **branches** erect, clothed with glandular hairs; **leaves** 2-3.5 cm x 0.3-1 cm, ovate, acute, rigid, in whorls, hairy, strongly fragrant when handled; **flowers** about 1 cm long, pale blue, tubular, with acutely pointed lobes; **calyx** as long as the corolla, hairy; **fruit** about 0.4 cm long, hairy.

A little-known aromatic shrub collected from a restricted region near the Gulf of Carpentaria. It is unknown in cultivation, but could have potential because of its compact habit and aromatic leaves. Would be best suited to tropical or inland regions. Propagate from seed.

DENNSTAEDTIA Bernh.
(after August Wilhelm Dennstedt, 19th-century German botanist)
Dennstaedtiaceae
Terrestrial **ferns**; **rootstock** long-creeping, bearing stiff hairs; **fronds** erect, triangular, finely divided and lacy; **sori** marginal, covered by a cup-shaped indusium.

A genus of about 50 species, one of which is widespread in Australia. Propagate by division or from spore.

Dennstaedtia davallioides (R. Br.) T. Moore
(like the genus *Davallia*)
Qld, NSW, Vic   Lacy Ground Fern
0.5-1.5 m tall
Terrestrial **fern** forming spreading patches; **root-**

253

Dennstaedtiaceae

Dennstaedtia davallioides frond segment, × ·7   pinnule, × 2

stock long-creeping, tough and woody, black; **fronds** erect, broadly triangular, 3-4 times divided, dark green, finely dissected and lacy; **sori** on lobes.

A widespread fern that favours cool, moist, sheltered conditions. Sometimes forms large colonies. It grows readily, but because of its spreading habit, requires a situation in the ground. The fronds are easily damaged by wind and sun, and the fern is best grown in a sheltered, shady situation. It will tolerate quite wet soil, and revels in heavy mulches and a couple of dressings of blood-and-bone annually. The fronds are frequently attacked and severely disfigured by small green caterpillars. Propagate by division of the rhizomes or from spore.

Dennstaedtiaceae Ching

A large, very widespread family of ferns distributed in temperate, subtropical and tropical zones of the world. They are terrestrial ferns with a branching, sub-terranean, hairy rhizome and finely divided lacy fronds. The family includes such ubiquitous genera as *Pteridium*, *Dennstaedtia* and *Hypolepis*. Some are attractive ferns, and a number of exotics are grown as ornamental ferns for pots or in the ground. The family in Australia is represented by 6 genera and about 15 species, some of which are popular in cultivation. Australian genera: *Dennstaedtia*, *Histiopteris*, *Hypolepis*, *Microlepia*, *Oenotrichia* and *Pteridium*.

DENTELLA Forster & G.
(from the Latin *dens*; *dentis*, a tooth; the corolla lobes of some species are minutely toothed)
Rubiaceae

Annual or perennial **herbs**; **stems** spreading, often rooting at the nodes; **leaves** simple, entire, alternate, stipulate; **flowers** tubular, axillary, solitary, sessile or nearly so; **corolla lobes** 5; **calyx** bell-shaped, 5-lobed; **fruit** a capsule.

A small genus of about 10 species distributed from India and Malaysia to northern Australia, where there are 7 species. The Australian species are floriferous small plants but are rarely encountered in cultivation. They are well suited to culture in tropical regions or as pot subjects. Seed is fine and must be sown shallowly under glass. Cuttings strike very quickly and easily.

Dentella browniana Domin
(after Robert Brown)
Qld, NT
prostrate x 0.3-0.5 m   Oct-Feb
Perennial prostrate **herb**; **stems** dying back to a

woody rootstock; **leaves** about 1 cm x 0.1 cm, linear, acute, thickened, bright green and glabrous; **flowers** about.1 cm long, white, tubular, axillary; **capsule** green with white hairs.

A little-known species found along the banks of some northern streams. The plants grow rapidly with the onset of the wet season and die back to a perennial rootstock during the dry. Not known to be in cultivation, but it could be an attractive garden or pot plant for tropical regions. Requires a sunny situation and plenty of water while growing. Propagate from seed or cuttings.

Dentella dioeca Airy-Shaw
(unisexual flowers on separate plants)
NT
prostrate x 0.3-0.5 m   Oct-Feb
Spreading annual **herb**; **stems** prostrate, rooting at the nodes; **leaves** 0.5-1 cm x 0.1-0.2 cm, linear to linear-lanceolate, bright green, glabrous, spreading; **flowers** about 2 cm long and 1 cm wide, tubular, white, axillary, solitary, dioecious; **calyx** about 0.3 cm long, with 5 teeth; **capsule** about 0.2 cm across, globose.

An annual plant that grows around the margins of swamps, lagoons and on flood plains, germinating and flourishing during the wet season. The plants flower in profusion, and the flowers are quite large and showy. Although rarely encountered in cultivation, this showy species has excellent potential as a garden plant for the tropics or as a glasshouse plant in southern Australia. Requires a sunny position and plenty of water while in active growth. Propagate from seed or cuttings which strike easily.

Dentella pulvinata Airy-Shaw
(cushion-shaped)
Qld, SA, NT
prostrate x 0.5-1 m   Jan-May; also sporadic
Perennial **herb** forming a dense, spreading cushion; **leaves** 0.1-0.3 cm x 0.1 cm, linear-elliptical, glabrous, bright green, the margins thickened; **flowers** about 0.6 cm long, tubular, with recurved lobes, hairy inside, withered flowers remain attached; **calyx** bell-shaped, with small stiff bristles; **capsule** about 0.2 cm long, covered with bristly hairs.

An interesting plant that forms dense mats in heavy clay soils subject to flooding, and along stream banks in moist situations. Worth growing for the dense mass of bright green foliage. Suitable as a rockery plant in a sunny situation in temperate regions. Propagate from seed or cuttings which strike readily.

Dentella repens (L.) Forster & G.
(creeping and rooting at the nodes)
Qld, WA, NT
prostrate x 0.5 m   Oct-March
Spreading annual **herb**; **stems** prostrate to suberect, rooting at the nodes; **leaves** 0.6-1 cm x 0.2-0.3 cm, narrow to broadly elliptical or obovate, bright green, glabrous; **flowers** about 1 cm across, tubular, white, axillary, solitary, sessile, 5-lobed; **calyx** tubular, 5-lobed; **capsule** globose, green and somewhat shiny.

Widespread along stream banks, flood plains and

Dentella dioeca D.L. Jones

clay flats of coastal and near-coastal regions of northern Australia. The plants germinate and spread rapidly following heavy rains, finally flowering in profusion before producing seed and dying. Although hardly known in cultivation, this species has excellent potential as a garden plant for tropical and subtropical regions, and also as a flowering pot plant. Requires an open, sunny situation and plenty of water while in active growth. Propagate from seed or cuttings which strike readily.

DEPLANCHEA Vieill.
(after Dr. M. Deplanche, naval surgeon and botanist)
Bignoniaceae

Shrubs or **trees**; young shoots bearing woolly, almost stellate hairs; **leaves** simple, entire, opposite or in whorls; **inflorescence** a crowded, terminal panicle; **flowers** tubular at the base, then with spreading lobes; **calyx** leathery, bell-shaped, with 5 lobes; **corolla lobes** 5; **fruit** a woody capsule; **seeds** flat, with a broad, papery wing.

A small genus of about 9 species with one found in north-eastern Australia. All species are excellent horticultural subjects, but they are rarely met with in cultivation and indeed their requirements remain largely unknown. Propagate from seed which is best sown fresh and lightly covered.

Deplanchea hirsuta (Bailey) van Steenis =
             D. tetraphylla (R. Br.) F. Muell.

Deplanchea tetraphylla (R. Br.) F. Muell.
(leaves in whorls of 4)
Qld
6-12 m x 3-5 m   July-Oct
   Small **tree**, slender or with a substantial canopy; **bark** grey, thick, soft, corky; **branchlets** covered with soft, woolly, yellow-bronze hairs; **leaves** 30-80 cm x 25-40 cm, ovate, blunt, leathery, spreading, crowded at the ends of the branches or in whorls of 4, dark green

and glabrous above, with golden, woolly hairs beneath; **panicle** dense, terminal; **flowers** about 2.5 cm long, bright yellow; **calyx** leathery, with long, pointed lobes; **corolla** tubular, with blunt, spreading or reflexed lobes; **capsule** 5-11 cm x 2-2.5 cm, hard and woody; **seeds** numerous, papery, winged.

   This species has outstanding horticultural potential. It is common in coastal scrubs from Cape York Peninsula to near Townsville, usually in sandy or lateritic soils. The large, leathery, crowded leaves give the plants a very distinctive appearance. Their real attraction lies in the bright yellow flowers which are carried in huge, compact heads at the ends of the branches and are readily seen from a distance. The flowers produce copious quantities of nectar which attracts honey-eating birds and animals. This nectar goes black after a period and somewhat detracts from the beauty of the flowers. Little is known about the cultivation of this very decorative plant. It appears to require acid, sandy, well drained soils and to be best suited to tropical and subtropical regions, in a sunny position. Small plants are very decorative and should be tried for indoor decoration. Propagate from seed which should be very lightly covered with propagation mixture. It is best sown fresh and seems to lose its viability after 3-6 months. Seedlings are very susceptible to damping off. *D. hirsuta* is considered to be merely a form of this species.

DERRIS Lour.
(from the Greek *derros*, a leather covering; in reference to the tough pods)
Fabaceae

   Tall woody **climbers** or rarely **trees**; **stems** twining, glabrous or hairy; **leaves** pinnate, with a terminal leaflet; **leaflets** simple, opposite; **inflorescence** a raceme or panicle; **flowers** pea-shaped, white, yellow, pink, violet or mauve, usually in clusters; **standard** erect or incurved; **wings** spreading; **fruit** an indehiscent, winged pod.

   A large genus of about 80 species widely distributed in tropical regions, with 3 named in Australia and others which are undescribed. Some exotic species are of interest because their roots yield derris powder, used to manufacture insecticides. This material also stupefies fish by blocking the oxygen supply to the gills.

Derris involuta, × ·6

Derris involuta

Plants of this genus are rarely encountered in cultivation. Propagate from seed which has a hard coat and must be treated prior to sowing or from cuttings of semi-hardened wood.

Derris involuta (DC.) Sprague
(rolled inwards; in reference to the margins of the calyx)
Qld, NSW
5-15 m tall   Aug-Jan
Vigorous, woody **climber**; **stems** brown, with pale corky lenticels; young shoots rusty-hairy; **leaves** pinnate, 5-12 cm long, bearing 9-13 leaflets; **leaflets** 3-5 cm x 1-2 cm, ovate, opposite, glossy green; **racemes** 10-25 cm long, axillary; **flowers** 1.2-1.5 cm long, white, pink or mauve, clustered on the raceme; **pods** 4-8 cm x 0.5-0.8 cm, flat, thin, indehiscent, brown, winged.

This vine is distributed along rainforest margins and stream banks from Maryborough in Qld to the Clarence River in north-eastern NSW. It was described from a plant which flowered at Kew Gardens in 1904, 20 years after it had been collected. It is a large, vigorous vine which develops woody trunks up to 8 cm across. Plants may become completely deciduous in dry winters. They are generally very attractive at flowering time and the pods are also decorative in mass. The Aborigines used infusions of roots, leaves and stems of this plant to poison fish in water-holes and small creeks. The species has limited horticultural appeal, but it has been found to be very useful as a groundcover and for stabilizing banks. Plants prefer a sunny situation but can be induced to grow in the shade. They require well drained soil. Propagate from scarified seed or cuttings of semi-mature wood. The species has also been known as *D. scandens*.

Derris koolgibberah Bailey
(Aboriginal name for the Musgrave River)
Qld
10-20 m tall   Oct-Dec
Vigorous, woody **climber**; young shoots densely covered with soft, rusty hairs; **leaves** pinnate, 10-20 cm long, bearing 5-8 leaflets; **leaflets** 8-13 cm x 5-7.5 cm, ovate to broadly oblong, blunt, rusty-hairy on both surfaces; **panicles** up to 50 cm long, axillary or terminal, densely hairy; **flowers** about 0.8 cm long, bright rosy pink, all parts bearing velvety hairs; **pods** 3-5 cm x 1-1.2 cm, oblong, densely hairy.

A little-known climber from lowland rainforests of north-eastern Qld, readily distinguished by the velvety hairs covering most parts. It is rare in nature and relatively unknown in cultivation. It is fairly tropical in its requirements, which are similar to those of *D. trifoliata*. Propagate from scarified seed or cuttings.

Derris trifoliata Lour.
(3-leaved)
Qld, NT
5-15 m tall   Sept-Dec
Vigorous, woody **climber**; young shoots glabrous; **leaves** pinnate, 5-10 cm long, bearing 5-7 leaflets; **leaflets** 3-12 cm x 1.5-2.5 cm, ovate to oblong, leathery, shiny; **racemes** 3-12 cm long, axillary; **flowers** about 1 cm long, white, pink or mauve; **standard** narrow;

keel short; **wings** recurved; **pods** 2.5-4 cm long, oblong, flat.

A common tropical climber found along rainforest margins and stream banks, and often particularly common in coastal communities. Plants may become completely deciduous prior to flowering. The Aborigines used infusions obtained by pounding roots, leaves and stems of this plant to poison fish in water-holes and small creeks. The species has limited horticultural appeal and is mainly tropical in its requirements. It is generally too vigorous for the average home garden, but could perhaps be useful for covering fences etc. It should also be tried for embankment stabilization. A sunny situation in well drained soil is necessary, as well as protection from frost. Propagate from scarified seed or cuttings of semi-mature wood.

DESCHAMPSIA P. Beauv.
(after L. A. Deschamps, a naturalist on d'Entrecasteaux's 18th-century voyage in search of La Perouse)
Poaceae
Perennial **grasses** forming tussocks; **leaves** shorter than the culms, flat, folded or inrolled, ribbed, smooth or scabrous; **culms** erect, smooth or scabrous; **inflorescence** an open panicle.

A genus of about 60 species with a cosmopolitan one extending to subalpine regions of south-eastern Australia. Many species are fodder grasses found on cold, high mountain-tops. Propagate by division or from seed which may need to be stored for 6-12 months before it will germinate.

Deschampsia caespitosa (L.) P. Beauv.
(growing in tufts)
NSW, Vic, Tas, SA   Tufted Hair Grass
0.3-1.2 m tall   Dec-April
Perennial **grass** forming large erect tussocks; **leaves**

Deschampsia caespitosa, × ·2

256

10-60 cm x 0.2-0.5 cm, folded or inrolled, coarse, glabrous, stiff and sharply pointed; **culms** erect, rigid, smooth; **panicle** 10-50 cm x 15-20 cm, sparse and open, green, silvery or purplish, shining; **spikelets** narrow.

A cosmopolitan grass that in Australia is mainly found in highland and subalpine regions. It is believed by some botanists to be introduced and naturalized. It is an extremely attractive grass when in flower, and its shining silvery panicles arouse comment. It can be grown as a container plant, or amongst rocks or small shrubs. It likes a moist position in full sun, and is quite hardy once established. Propagate from seed or by division of the clumps.

DESMODIUM Desv.

(from the Greek *desma*, bond; head-band; *odes*, like; either refers to the necklace-like pods or to the sheath of united stamens)
Fabaceae

Herbs, shrubs, climbers or rarely **trees**; **leaves** pinnately trifoliolate or unifoliolate, reticulate, with small leaf-like appendages at base of leaflets, stipules usually dry, striate; **flowers** pea-shaped, usually small, in terminal racemes or panicles, rarely in axillary umbels or clusters, purple, blue, pink or white; **calyx tube** short, 2 upper lobes more or less united; **standard** oblong to orbicular; **wings** oblong; **keel** obtuse; **upper stamens** free or more or less united with others to form a sheath; **pods** longer than calyx, flat, moniliform.

A large cosmopolitan genus of about 350 species with its main concentration in the tropical regions. About 37 species occur in Australia, and up to 14 are endemic. The majority of these occur in northern Australia, with further representation in eastern NSW, eastern and southern Vic, and also Tas.

They are not common in cultivation. Only a few species are grown by enthusiasts. Plants can be propagated from seed which needs pre-sowing treatment (see Volume 1, page 205). Cuttings of young, vigorous but firm growth produce roots readily.

Desmodium biarticulatum F. Muell. =
Dicerma biarticulatum DC.

Desmodium brachypodum A. Gray
(short peduncles)
Qld, NSW, Vic   Large Tick-trefoil
0.5-1.2 m x 0.5-1 m   Nov-May

Dwarf to small **shrub** or light twining **climber**; **branches** erect, glabrous; **leaves** trifoliolate; **leaflets** 2-6 cm x 1-3 cm, broad-lanceolate to oval, terminal leaflet longest, prominent venation, initially hairy, becoming glabrous; **flowers** pea-shaped, about 0.7 cm long, mauve-purple, borne in erect terminal racemes to 30 cm long; **pods** to 2 cm x 0.4 cm, strongly constricted on one edge, covered with rusty hairs.

Occurs naturally on both sandy and shallow loams. It is not well known in cultivation. Although not a showy plant, it may have some application as an undershrub. Will need relatively well drained soils, with partial to full sun. Propagate from seed or cuttings.

Desmodium brachypodum, × ·5

Desmodium campylocaulon F. Muell. ex Benth.
(bent stem)
Qld, NSW, WA, NT   Creeping Tick-trefoil
0.1-0.2 m x 1-2 m   Sept-Jan

Prostrate or rarely erect, spreading, sometimes stoloniferous **shrub**; **branches** few; **leaves** trifoliolate; **leaflets** 5-10 cm x 0.5-1.5 cm, lanceolate to linear-ovate, terminal leaflet longest, more or less glabrous, rounded or pointed apex, undersurface reticulate; **flowers** pea-shaped, about 0.6 cm long, purple, borne in racemes to 20 cm long; **pods** to 2 cm long, strongly constricted on one edge.

This open, spreading perennial occurs on heavy soils and may be grazed by sheep. Evidently not in cultivation, but may be suitable as a forage plant in semi-arid regions. Should adapt to most soils and may do best in partial or full sun. Could be worth trying as a container plant. May be used as a light groundcover beneath taller plants. Propagate from seed or cuttings.

Desmodium umbellatum (L.) DC. =
Dendrolobium umbellatum (L.) Benth.

Desmodium varians Endl.
(varying)
Qld, NSW, Vic, Tas   Slender Tick-trefoil
0.2-0.5 m tall   most of year

Lightly twining **climber**; **branchlets** slender, usually hairy; **leaves** trifoliolate; **leaflets** to 2.5 cm x 1-1.5 cm, oval to rounded, more or less glabrous; **flowers** pea-shaped, about 0.3 cm long, pink, borne in loose racemes to 12 cm long; **pods** to 2.5 cm long, on long stalks, strongly constricted on one edge.

This species is the most commonly cultivated of the genus. It occurs on a variety of soils, and has a preference for well drained situations. Grows well in semi-shade to partial sun. Has potential as a container plant if used with other species. Frost hardy. Responds well to light pruning. Propagate from seed or cuttings.

DEYEUXIA Clarion ex Beauv.
(after Nicholas Deyeux, 18-19th-century French professor of medicine)
Poaceae

Perennial **grasses** forming clumps or tussocks; **leaves** shorter than or nearly as long as the culms, flat or inrolled, lax or rigid; **culms** erect, smooth or scabrous; **inflorescence** a panicle which may be dense or open and spreading.

A large genus of about 200 species with 29 found in Australia, most of which are endemic. They are commonly known as bent grasses, and are frequent in highland and subalpine situations. Some species have ornamental appeal, but many tend to be coarse and the flowering is brief. Propagate by transplants or division of established clumps or from seed. Seed should be stored for at least 6-12 months before sowing.

Deyeuxia crassiuscula Vick.
(slightly thickened)
Qld, NSW   Bent Grass
20-45 cm tall   Nov-April

Perennial **grass** forming coarse, open clumps; **leaves** 3-12 cm x 0.4-1 cm, flat, bright green, coarse-textured, scabrous; **culms** erect, scabrous; **panicle** 6-10 cm x 1-1.5 cm, dense, spike-like, shiny green tinged with purple, becoming straw coloured with age; **spikelets** turgid.

This is a rather coarse grass with broad leaves that spread in a loose tussock. It is common in subalpine grasslands and is prominent when in flower. It makes a useful container plant and also looks good when planted amongst rocks. It is best suited to areas with a cold winter. Elsewhere, flowering is reduced. Propagate from seed or by division.

Deyeuxia frigida F. Muell. ex Benth.
(cold; frosty)
NSW, Vic   Alpine Bent Grass
0.5-1 m tall   Jan-March

Perennial **grass** forming erect, slender tussocks; **leaves** 15-25 cm x 0.3-0.6 cm, flat, flaccid, glabrous, green; **culms** erect, slender, somewhat scabrous; **panicle** 10-20 cm long, open and loosely spreading, greenish, becoming straw coloured; **awns** prominent, 0.2-0.3 cm long.

A tall grass of cool, moist, somewhat shady situations in highland and subalpine regions. It is very distinctive when in flower, and useful for a cool garden position. Will grow successfully with ferns. Propagate from seed or by division of the clumps.

Deyeuxia imbricata Vick.
(overlapping)
Qld, NSW   Bent Grass
0.6-1 m tall   Aug-Nov; also sporadic

Perennial **grass** forming erect, slender tussocks; **leaves** 10-30 cm x 0.3-0.5 cm, mainly basal, tapered, bright green, slightly roughened; **culms** erect, smooth, glabrous, striate; **panicle** 7-14 cm x 0.5-1.5 cm, dense, spike-like, green or purplish; **spikelets** crowded, small, compressed.

Usually found in open forest, this grass is wide-spread in NSW, but in Qld is confined to mountainous regions near the border. It has no value as a pasture grass but is attractive when in flower, and can be planted amongst small shrubs etc. It does not seem to naturalize readily and become weedy. Propagate from seed or by division.

Deyeuxia quadriseta (Labill.) Benth.
(4 setae)
NSW, Vic, Tas, SA, WA   Reed Bent Grass
0.6-1 m tall   Aug-Dec

Perennial **grass** forming sparse tussocks; **leaves** 10-30 cm x 0.3-0.5 cm, flat or slightly inrolled, dull green; **culms** slender, erect, glabrous; **panicle** 10-15 cm x 1-2.5 cm, crowded, spike-like, bright green; **spikelets** small, crowded.

A widespread grass found in a variety of soils and habitats. It has little forage-value and limited ornamental appeal, although large clumps can be attractive when in full flower. Requires exposure to some sun, and soils of free drainage. Propagate from seed or by division.

DIANELLA Lam. ex Juss.
(after Diana, the Roman goddess of hunting and queen of the woods; *elle*, diminutive suffix)
Liliaceae

Perennial **herbs** forming tufts or spreading colonies;

Deyeuxia imbricata, × ·1

rhizomes subterranean, branched; **roots** wiry or tuberous; **leaves** linear, distichous, the margins entire, serrulate or revolute, sheathing at the base; **leaf-sheath** open or closed, compressed laterally, often keeled; **inflorescence** a panicle; **flowers** white, blue or purple, pedicellate; **perianth segments** 6, spreading; **stamens** 6, the filaments swollen in the upper third; **fruit** a blue, succulent berry containing several black, shiny seeds.

A genus of 25-30 species widely distributed in Asia and the Pacific region. The genus in Australia is under study and the number of species here is uncertain, but it is probably about 15. Studies of the species in north-eastern Qld have shown tremendous genetic variability within a species, making identification by simple procedures very difficult. Variable features include growth habit (whether tufted or colony-forming), leaf dimensions, the presence of aerial growths, and flower colour.

Dianellas are commonly known as flax lilies because the leaves are tough and can be plaited. They are found in the wetter areas of the continent, and some species occur in each state. They grow in a tremendous range of habitats from rainforest to semi-desert areas, and from coastal sand-dunes to inland ranges and tablelands. They are hardy plants, surviving extremes of dryness, temperature and bushfires by virtue of their extensive underground rhizomes and root system. An interesting growth feature is the presence of aerial growths or fans which often appear on the flower panicles.

Dianellas are well suited to cultivation. With their growth habit of tussocks or spreading colonies they are very useful for rockeries, for breaking up lines of shrubbery and as a container plant. Some species such as *D. tasmanica* will grow in shady situations, and mingle well with ferns. Most of the others appreciate sun, and flower best in such an open situation. Once established, the plants are very hardy and long-lived, and require little attention. Flowers are mostly blue and short-lived, but are borne over long periods. Clumps in flower are not generally spectactular but they are very decorative. The flowers are followed by fleshy, shiny berries which are a brilliant blue or purple. Because of the tremendous variation within most of the species, there is a large range of forms available for garden culture, and these can be maintained by vegetative propagation.

Propagation

Dianellas can be propagated from seed or by vegetative techniques. Seed-set is frequently low, with some berries having no seeds at all. Seeds retain their viability for 12-24 months but are best sown fresh. Vegetative propagation can be by division of the clumps which is generally easy. The aerial plantlets usually lack roots and must be potted until sufficiently established to be planted out.

Dianella caerulea Sims
(dark blue)
Qld, NSW, Vic, Tas, WA, NT   Paroo Lily
0.2-1 m x 0.3-2.5 m   Aug-Jan

Tufted perennial **herb** forming spreading patches; **rhizomes** thick, subterranean, much-branched; **leaves** 10-70 cm x 0.5-2.5 cm, distichous, linear, acuminate, dark green, erect, in a fan or tussock, the margins often toothed, a zone of complete fusion between the sheath and blade; **leaf-sheath** overlapping the earlier sheaths and the rhizomes, laterally flattened and acutely keeled; **aerial fans** often present and well developed; **panicle** 30-90 cm long, loose; **flowers** 1-1.6 cm across, pale to dark blue; **stamen filaments** much shorter than the anthers; **berry** 0.8-2 cm long, globular, pale to bright blue or purple.

A widely distributed species, often common in coastal areas on heathlands and even extending to sand-dunes. It is a colony-forming species with some forms growing in rapidly spreading but sparse colonies, while others are more compact. This species is extremely variable throughout its range, and a study in north-eastern Qld has revealed 10 different genetic forms. Some of these forms are excellent for cultivation. Spreading forms from the coast are ideal sand-binders and will withstand salt-laden winds etc. Once established, the plants are very hardy and long-lived. Propagate from seed, by division or from aerial growths.

Dianella ensata (Thunb.) R. J. Henderson — this species does not occur in Australia.

Dianella intermedia Endl.
(intermediate)
NSW (Lord Howe Island)
0.2-0.5 m x 0.5-2.5 m   Dec-Feb

Tufted perennial **herb** forming spreading patches; **rhizomes** slender, wiry, much-branched; **leaves** 10-50 cm x 1-1.5 cm, distichous, linear, acuminate, yellowish green to dark green, the margins entire, in a fan or tussock, a zone of partial fusion between the leaf-blade and sheath; **leaf-sheath** laterally flattened, keeled; **panicle** 20-40 cm long, loose; **flowers** about 1 cm across, pale to dark blue, on stiffly arched pedicels; **stamen filaments** much shorter than the anthers; **berry** about 1 cm long, pale to bright blue.

This species occurs on Norfolk Island and Lord Howe Island. On the latter, it grows on the basalt hills and mountains, where it forms spreading patches. It appears to be hardly tried in cultivation but should grow readily in subtropical and temperate gardens, requiring an open, sunny situation, in well drained soil. Propagate from seed or by division.

Dianella laevis R. Br.
(smooth)
all states   Smooth Flax Lily
0.3-0.8 m x 0.5-1.5 m   Aug-Jan

Tufted perennial **herb** forming compact to shortly spreading clumps; **roots** thickened, tuberous; **rhizome** thick, subterranean, erect or shortly creeping; **leaves** 15-80 cm x 0.5-1.5 cm, distichous, linear, soft-textured, light green, sometimes glaucous, erect in a tussock, the margins entire; **leaf-sheath** open to the base, sheathing the stem, scarcely keeled to rounded; **aerial fans** often present; **panicle** 30-90 cm long, varying from loose to compact; **flowers** 1-2 cm across, usually pale blue; **stamen filaments** much shorter than the anther, which is pale yellow; **berry** 0.8-1.5 cm long, dark blue to purple.

259

Dianella longifolia

A widely distributed flax lily that can usually be recognized by its compact tufts and tuberous root system. It grows in a variety of habitats but is most frequent in moist grassland and open forest. The Aborigines used its tough leaves to plait baskets etc. It is an extremely variable species throughout its range, and a study in north-eastern Qld has revealed 14 different genetic forms. Some of these have excellent horticultural potential. Once established in a permanent position, this species is a hardy and long-lived plant. It is ideal for mingling with small shrubs. Propagate from seed, by division or from aerial growths.

Dianella longifolia R. Br. = *D. laevis* R. Br.

Dianella nemorosa Lam. = *D. caerulea* Sims

Dianella odorata Blume
(fragrant)
Qld   Flax Lily
0.3-0.7 m x 0.3-2.5 m   Aug-Jan
Tufted perennial **herb** forming slowly spreading patches; **rhizomes** thick, subterranean, sparsely branched; **leaves** 10-70 cm x 2-4.5 cm, distichous, linear, acuminate, dark green, erect in a fan or tussock, the interveinal areas translucent, the margins shortly toothed, a zone of partial fusion between the sheath and blade which is Y-shaped in cross-section; **leaf-sheath** overlapping the earlier sheaths and the rhizomes, laterally flattened and acutely keeled; **aerial fans** often present and well developed; **panicle** 30-90 cm long, varying from loose to compact; **flowers** 0.8-1.5 cm across, white, greenish, pale blue or deep blue; **berry** about 1 cm long, oblong, blue to purple.

In Australia this species is apparently restricted to Qld, but there widespread in the tropical lowlands, growing in moist places along stream banks and in rainforests. It is known to consist of at least 5 different genetic forms, one of which has attractively fragrant flowers. The species succeeds well in tropical and subtropical gardens, in a partial-sun situation. Ideal for rockeries or small gardens. Hardy once established. Propagate from seed or by division.

Dianella revoluta R. Br.
(margins rolled back)
Qld, NSW, Vic, Tas, SA, WA   Spreading Flax Lily
0.3-1 m x 0.5-2.5 m   Aug-Jan
Tufted perennial **herb** forming extensive spreading patches; **rhizomes** thick, subterranean, much-branched; **leaves** 10-70 cm x 0.5-1.5 cm, distichous, linear, acuminate, erect in a tussock, dark green, the margins recurved to strongly revolute, stiff; **leaf-sheath** open to partially closed at the base, shortly sheathing, rounded to laterally flattened, keeled; **aerial fans** rarely present; **panicle** 30-90 cm long, loose to very crowded; **flowers** about 1 cm across, pale blue to whitish; **stamen filaments** much shorter than the anthers which are dark brown to black; **berry** 0.8 cm long, oblong, dark blue, shiny.

A very widely distributed species extending from coastal dunes to subalpine regions, and from semi-arid country to moist forests. Plants usually grow in spreading colonies. The species is extremely variable

Dianella tasmanica in fruit   T.L. Blake

throughout its range, and a study in north-eastern Qld has revealed 6 genetic variants. A coastal form from Vic develops very slender, compact tussocks and makes an excellent rockery plant. Spreading flax lily is an ideal garden plant, tolerating poor soils and adversity, yet managing to look decorative. Propagation is by division or from seed.

Dianella tasmanica Lam. ex Juss
(from Tasmania)
NSW, Vic, Tas   Tasman Flax Lily; Blueberry
0.6-2 m x 0.5-2 m   Aug-Feb
Tufted perennial **herb** forming clumps or spreading patches; **rhizomes** thick, subterranean; **leaves** 0.3-1 m x 1.5-4 cm, distichous, linear, keeled midrib, finely toothed margins, Y-shaped in cross-section near the sheathing base; **panicles** loose, on much-branched stem to 1.5 m tall; **flowers** about 1.5 cm across, blue; **stamen filaments** much longer than the anthers; **berry** 1-2 cm long, globular, violet to blue.

This decorative species inhabits cool, damp forests, and often occurs in dense shade. Can form spreading patches, but usually grows as a clump. Has been in cultivation for a long time, and has adapted well to a wide range of conditions. It is suited to moist, acidic soils, doing best in those that are well drained, but it will withstand short periods of waterlogging. Prefers deep shade through to dappled shade, but will tolerate an open, sunny position, which sometimes results in scorching of the leaves. Once established, the plants are hardy and long-lived. Does well in containers. Mixes well with ferns. Propagate from seed or by division of the rhizomes.

260

DIASPASIS R. Br.

(from the Greek *diaspasis*, separation; in reference to the corolla, which is deeply divided with spreading lobes)
Goodeniaceae

An endemic, monotypic genus restricted to south-western WA. It is closely allied to *Scaevola*, which does not have a regularly shaped corolla.

Diaspasis filifolia R. Br.
(thread-like leaves)

WA   Thread-leaved Diaspasis
0.4-0.8 m x 0.3-0.8 m   Sept-Feb

Dwarf, tufting, perennial **herb**; young shoots hairy; **stems** usually erect, slightly branched, becoming glabrous; **leaves** 2-5 cm x 0.1-0.2 cm, linear, terete, young leaves can have a few teeth, usually glabrous when mature; **flowers** about 2 cm across, white or mauve-pink, borne on peduncles in upper axils.

Not well known in cultivation, this species is a dweller of moist, sandy soils in the Darling, Eyre and Roe Districts, and is sometimes found in coastal situations. There is a record of a red-flowered form, but this is not known to be in cultivation. Requirements are for well drained, light to medium soils, with partial to full sun. Plants would probably appreciate watering during dry periods. Suited to container cultivation. Propagate from cuttings taken from young stems.

DICARPIDIUM F. Muell.

(from the Greek *dis*, double; *carpos*, a fruit; *idium*, a small carpel; in reference to the bi-carpellary fruit)
Bombacaceae

A monotypic genus, endemic in northern Australia.

Dicarpidium monoicum F. Muell.
(separate male and female flowers on the same plant)

WA, NT
0.3-1 m x 0.3-0.5 m   Dec-Jan

Dwarf **shrub**; **branches** slender, covered with rigid, stellate hairs; **leaves** 2-3 cm x 0.3-0.5 cm, oblong, dark green, plicate, nearly sessile, densely covered with rigid, stellate hairs, the margins toothed; **flowers** about 0.4 cm across, pinkish, with spathulate petals, 2-3 together in the upper axils; **fruit** of 2 small carpels.

A little-known shrub found in tropical areas extending from near Darwin to the Kimberleys. It has limited horticultural appeal, and is mainly of botanical interest. Would be best suited to a sunny position, in well drained soil in tropical gardens. Propagate from seed and perhaps cuttings.

DICERMA DC. emend Benth.

(refers to the 2 small compartments of the pod)
Fabaceae

Shrubs; young shoots bearing silky hairs; **leaves** trifoliolate; **leaflets** digitate on the end of the petiole; **stipules** persistent, uniting on the stem opposite the petiole; **inflorescence** an elongated, terminal raceme; flowers pea-shaped, small; **wings** scarcely adhering to the keel; **keel** with few obvious appendages; **fruit** an indehiscent pod.

A small genus of 3 species distributed in Burma, New Guinea and northern Australia where there are 1 or 2 species. They are included by many botanists as a section in the genus *Desmodium*. The species have limited appeal for tropical regions. Propagation is from seed which is hard and needs pre-sowing treatment to germinate.

Dicerma biarticulatum DC.
(jointed in 2 places)

Qld, WA, NT
0.3-1 m x 0.5-1 m   Aug-Nov

Dwarf, rigid **shrub**; young shoots silky-hairy; **branches** decumbent to erect; **leaves** trifoliolate; **leaflets** 1-3.5 cm x 0.2-0.4 cm, oblong, rigid, spreading at the end of a short petiole; **stipules** brown; **raceme** 3-9 cm long, terminal; **flowers** about 0.8 cm long, pea-shaped, red or white, crowded; **pods** 1-1.4 cm x 0.2-0.4 cm, sessile, flat, silky-hairy, in large clusters.

A widely distributed species which extends to northern Australia where it grows in open forest, often in rocky areas. It is a hardy shrub with limited potential for cultivation in drier tropical zones. Would require a sunny situation in well drained soils. Clusters of pods are decorative and long-lasting. Propagate from scarified seed and possibly cuttings.

DICHANTHIUM Willm.

(from the Greek *dicha*, in 2 or at variance; *anthos*, a flower; in reference to the lower sessile spikelets of the racemes being different from the rest)
Poaceae

Perennial **grasses** forming upright tussocks; **leaves** flat, usually glaucous, smooth or scabrous; **culms** erect or geniculate, simple or branched, the nodes glabrous or hairy; **inflorescence** a simple raceme, a digitate group of racemes or a panicle; **lower spikelets** on a raceme, either male or sterile, the rest bisexual.

A small genus of tropical grasses comprising 15 species with 10 found in Australia. Two of these are undescribed and another (*D. aristatum* (Poir.) Hubb.) is naturalized and grown as a pasture grass. As a group, they are known as blue grasses. Most species grow in extensive stands in open forest and woodland, and are an important constituent of tropical and inland pastures. Fires are frequent where they grow, destroying the woody, untidy tussocks formed following the previous fire. New leaves are produced rapidly, and in good seasons the tussocks become tall and lush.

Blue grasses, because of their tussock habit and interesting leaf colour, have considerable ornamental appeal. The tussocks tend to become untidy and can be rejuvenated by severe pruning or burning. Propagation is by transplants, division of clumps or from seed which should be stored for 6-12 months before sowing.

Dichanthium affine (R. Br.) A. Camus
(alike)
Qld, NSW, SA, WA, NT   Blue Grass
30-50 cm tall   Oct-Jan
    Perennial **grass** forming tall, slender clumps; **leaves** 10-20 cm x 0.3-0.5 cm, flat, bluish green, striate, hairy; **culms** erect, geniculate, glabrous to hairy; **racemes** 2-4 cm long, 2-6 in a digitate inflorescence, yellowish green suffused with purple.
    A widespread grass found in situations such as woodland and savannah grassland. It is an important pasture species and is botanically very similar to *D. sericeum*, but lacks the conspicuous nodal hairs of that species. It is an attractive grass for a rockery or for mingling amongst small shrubs in an open, sunny position. Propagate from seed or by division of the clumps.

Dichanthium annulatum (Forsskal) Stapf
(ringed or ring-like)
Qld, WA, NT   Sheda Grass
0.5-1 m tall   Nov-Feb; also sporadic
    Perennial **grass** forming erect tussocks; **leaves** 10-30 cm x 0.4-0.6 cm, flat, firm-textured, very glaucous, scabrous, glabrous or hairy; **culms** robust, geniculate, the nodes with a well developed ring of hairs; **racemes** 3-6 cm long, several in a paniculate inflorescence, pale green, often suffused with violet-purple.
    A scattered and fairly infrequent grass of tropical Australia, found in open forest and woodland situations. It is of attractive colouration and worth growing amongst rocks or shrubs. Clumps may get untidy and can be rejuvenated by severe pruning or burning. Propagate by division or from seed.

Dichanthium fecundum S. T. Blake
(fertile; fruitful)
Qld, WA, NT   Curly Blue Grass
0.8-1.4 m tall   Dec-March; also sporadic
    Perennial **grass** forming erect tussocks; **leaves** 10-30 cm x 0.3-0.5 cm, linear, glaucous, sparsely hairy, somewhat scabrous; **culms** erect or geniculate, slender, branched, often covered with white powder; **racemes** 4-6 cm long, solitary or in groups of 2-4, hairy, pale green or purplish.
    A widespread grass of tropical Australia, also extending to New Guinea. It is found in grasslands and woodlands, and is a useful forage species. Tussocks in vigorous growth are very glaucous and attractive. They could be usefully planted amongst rocks or small shrubs, in inland or tropical and subtropical areas. Propagate from seed or by division.

Dichanthium humilius J. Black =
         D. affine (R. Br.) A. Camus

Dichanthium sericeum (R. Br.) A. Camus
(silky-hairy)
all mainland states   Qld Blue Grass
30-80 cm tall   Aug-Jan; also sporadic
    Perennial **grass** forming slender, erect tussocks; **leaves** 8-15 cm x 0.2-0.4 cm, flat, bluish green, often with purple tinges, glabrous or densely covered with white hairs, distributed over the stems and lower parts of the culms; **culms** erect, slender, smooth and glabrous except at the nodes which bear a ring of long white hairs, branched throughout; **racemes** 4-7 cm long, 2-4 arranged in an erect, digitate group; **spikelets** crowded, paired in 2 ranks, awned.
    A very widespread grass that occurs in a variety of habitats, both inland and in near-coastal districts. In some regions such as black soil plains it may be the dominant grass. It is highly nutritious and is eagerly eaten by stock and native animals. It is considered one of the most important native pasture grasses.
    Qld blue grass is highly ornamental when in the vegetative state, but the flowering is rather insignificant. Unthrifty plants lack any appeal but vigorous, lush clumps such as occur following burning and watering are most appealing. The species requires a sunny situation in well drained soil. Propagate from seed or by division of the clumps. Small plantlets produced on the culms can sometimes be induced to form roots.

Dichanthium superciliatum (Hackel) A. Camus =
         D. tenuiculum (Steudel) S. T. Blake

Dichanthium tenuiculum (Steudel) S. T. Blake
(slender culms)
Qld, WA, NT   Tassel Blue Grass
0.8-1 m tall   Dec-March; also sporadic
    Perennial **grass** forming slender tussocks; **leaves** 10-25 cm x 0.5-0.6 cm, flat, glabrous, very scabrous, bluish green to smoky blue; **culms** robust, glabrous, sparsely branched; **racemes** 5-6 cm long, 10-20 in a dense, terminal, digitate inflorescence.

Dichanthium sericeum leaves, × ·45   head, × ·9   fruit, × 4·5

Dichelachne crinita, × ·2

A widespread grass of tropical Australia and New Guinea, found in woodland situations. It responds to fire and is a significant tropical pasture grass. Tussocks in vigorous growth look attractive, and the species could have prospects for cultivation in tropical and subtropical areas. The tasselled inflorescences are an interesting feature. Propagate from seed or by division.

Dichapetalaceae Baillon

A family of dicotyledons consisting of 200 species in 4 genera, most of which are found in tropical regions of the world. They are shrubs, trees or climbers, often pubescent and with simple, stipulate, alternate leaves. Flowers are generally small and not showy, and few species have any horticultural merit. Two species in the genus *Dichapetalum* represent the family in Australia.

DICHAPETALUM Thouars
(from the Greek *dicha*, in 2 or at variance with; *petalos*, petals; in reference to the variation in petals)
Dichapetalaceae

Shrubs or **trees**; **leaves** alternate, entire, stipulate; **inflorescence** a cyme; **flowers** bisexual, regular; **calyx** of 5 segments; **corolla** segments 4-5; **fruit** a drupe.

A large genus of 200 species mainly developed in Africa with 2 species in north-eastern Qld, one endemic. Little is known about their cultural requirements.

Dichapetalum australianum C. White =
D. papuanum (Becc.) Boerl.

Dichapetalum papuanum (Becc.) Boerl.
(from Papua)
Qld
0.5-1.5 m x 0.5-2 m   Jan

Dwarf to small **shrub**; young shoots hairy; **branchlets** angular, bearing lenticels; **leaves** 10-13 cm x 3-4 cm, lanceolate, bright shiny green, apex acuminate, the margins undulate, on petioles 2-3 cm long; **flowers** about 0.5 cm across, greenish, in axillary clusters; **drupe** 1.2 cm x 1 cm, fleshy, 3-lobed, orange-yellow, mature April.

A spreading shrub restricted to rainforests of north-eastern Qld. Although unknown in cultivation, the species is highly ornamental and, because of its small stature, will surely find a place in tropical and subtropical gardens. It would require a protected, shady situation. Propagate from seed or perhaps cuttings.

Dichapetalum timoriense (DC.) Boerl.
(from Timor)
Qld
3-8 m x 1-3 m   Dec-March

Slender, straggly **shrub** or **climber**; **branchlets** bearing soft, rusty or purplish hairs; **leaves** 7-18 cm x 3-10 cm, ovate to obovate or oblong, somewhat leathery or even papery in texture, hairy on the lower surface, the apex shortly pointed; **panicles** 2-4 cm long, in the upper axils, **flowers** about 0.4 cm long, dull, hairy, unisexual or bisexual; **fruit** 1.5-2.5 cm across, globular to pear-shaped, yellow to golden brown, containing 1-3 seeds.

A widespread species, which in Australia is restricted to a few localities in north-eastern Qld. It usually grows as a slender, straggly shrub but will climb if given the opportunity. The fruit and young leaves are edible. The species has limited horticultural appeal but plants are quite decorative when in fruit. Can be grown in tropical regions, in well drained soil in a semi-shady position. Propagate from seed which must be sown fresh and possibly from cuttings.

DICHELACHNE Endl.
(from the Greek *dichelos*, cloven-footed; *achne*, chaff; referring botanically to the glume, the lemma of which is bi-lobed)
Poaceae

Annual or perennial **grasses** forming tussocks; **leaves** flat or inrolled, basal; **culms** erect, unbranched; **inflorescence** a sparse to dense panicle; **spikelets** 1-flowered; **awns** long and prominent.

A small genus of about 6 species restricted to Australia and New Zealand. Four species are found in Australia, 1 of which is undescribed. They are widespread grasses of limited forage-value but with some ornamental appeal. Propagation is by transplants, division of clumps or from seed which should not be sown for 6-12 months after collection.

Dichelachne crinita (L.f.) Hook.f.
(a long mane)
Qld, NSW, Vic,
Tas, SA, WA   Long-hair Plume Grass
0.5-1 m tall   Aug-Dec

Perennial **grass** forming sparse, open tussocks; **leaves** 10-20 cm x 0.3-0.5 cm, flat, green to bluish green, basal and scattered up the culms, glabrous or covered with short hairs; **panicle** 8-20 cm long, dense, slender, bright, verdant, sometimes with purplish tinges, opening with age; **awns** 2-2.5 cm long, crowded and almost concealing the spikelets, erect at first but later spreading and twisting.

A handsome native grass that is widespread in coastal and drier open forests, sometimes extending to highland situations. The clumps are generally sparse but the panicles are very showy when at their peak. A drawback is that the awned seeds catch in clothing and may cause irritation to sensitive skin. This can be avoided by cutting the inflorescences soon after flowering. This grass grows readily and is interesting when planted amongst small shrubs. Propagate by division or from seed.

Dichelachne micrantha (Cav.) Domin
(small flowers)
Qld, NSW, Vic,
Tas, SA, WA   Short-hair Plume Grass
0.3-1.1 m tall   July-Oct

Perennial **grass** forming small, dense tussocks; **leaves** 8-25 cm x 0.3-0.5 cm, flat, dark green, mainly basal, glabrous to scabrous; **culms** slender, erect, glabrous, the nodes generally very dark; **panicle** 8-20 cm long, loose and open, becoming denser with age, with few spikelets; **awns** 1-1.8 cm long, thickened and twisted near the base, abruptly bent.

A widespread grass that occurs naturally in dry woodland and open forest. It has some use as a forage species since it produces leaf growth during winter. It makes an interesting ornamental grass for gardens but is rarely grown. Propagate by division or from seed.

Dichelachne sciurea (R. Br.) Hook. =
D. micrantha (Cav.) Domin

DICHONDRA Forster & G. Forster
(from the Greek dis, double; chondros, a grain; in reference to the double carpels)
Convolvulaceae

Small or prostrate creeping **herbs**; **stems** rooting at nodes; **leaves** stalked, rounded or reniform; **flowers** axillary, small, 5-merous; **carpels** 2, free; **styles** 2; **fruit** composed of 2 fruitlets.

A genus of 4-5 species distributed in the tropical and subtropical regions of the world. One widespread species extends to Australia.

Dichondra repens Forster & G. Forster
(creeping)
all states   Kidney Weed
prostrate x 1-2 m   Sept-Dec

A dense, creeping, perennial **herb**; **stems** rooting at nodes; **leaves** to 4 cm wide, reniform, on long petioles, hairy on both surfaces; **flowers** small, cream-green, solitary, axillary, on peduncles shorter than petioles.

Widely spread throughout the warmer regions of the world. Forms a dense mat of leaves and is sometimes used as an alternative to grass lawns, although it does not withstand constant foot traffic. Very useful as a groundcover for shady, dry situations or between paving, and for other landscape purposes. It can be a problem to eradicate once it is established amongst other plants. Is particularly vigorous in moist soils. Propagate from seed or by division.

Dichondra seed is available commercially. It is probably actually D. micrantha, a non-Australian species.

DICHOPETALUM F. Muell. =
DICHOSCIADIUM Domin

Dichopetalum ranunculaceum F. Muell. =
Dichosciadium ranunculaceum (F. Muell.) Domin

DICHOPOGON Kunth
(from the Greek dichos, double; pogon, a beard; in reference to the 2 appendages on the anthers)
Liliaceae

Perennial tuberous **herbs**; **leaves** narrow, basal, grass-like; **flowers** in loose racemes; **perianth** 6 free-spreading segments, 3 inner ones more or less orbicular, slightly fringed on margins; **anthers** with 2 beard-like appendages; **fruit** a capsule.

Dichopogon strictus, × ·45

Dichosciadium ranunculaceum var. *tasmanicum*   T.L. Blake

An endemic genus of 2 species that occur throughout temperate Australia. They are usually found on sandy to clay-loam soils, in heathland and woodland plant communities. They grow actively during late autumn to spring and then die down over summer. *Dichopogon* is closely allied to *Arthropodium*, which does not have bearded anthers. Both species of *Dichopogon*, however, are included in *Arthropodium* by some botanists. They are delicate perennials that adapt well to garden or container cultivation. The flowers have a pleasant perfume of caramel, chocolate or vanilla, and are borne on light stems that wave in the slightest of breezes. Plants can be scattered informally in gardens to create an attractive natural appearance, or grown in mass in containers.

Propagation is mainly from seed which germinates readily. Transplanting of tubers can also be used with success, provided they are gathered intact. The tubers are on the roots and can be some distance from the rootstock.

Dichopogon fimbriatus R. Br.
(fringed)
Qld, NSW, Vic, SA, WA   Nodding Chocolate Lily
0.2-1 m x 0.2-0.8 m   Sept-Dec

Dwarf, perennial **herb**; **leaves** to 40 cm x about 1 cm, grass-like; **inflorescence** slender, erect, usually undivided; **flowers** to 3 cm diameter, blue to violet, 1-4 together, arising in a bract, drooping, vanilla-scented; **capsule** more or less globular, about 0.8 cm long, on reflexed pedicel.

This species usually occurs naturally on sandy soils, in areas of moderately low rainfall. Likes good drainage, and a warm, relatively sunny location. Multiple plants in pots make an excellent display. Frost tolerant. Responds well to regular light applications of fertilizer. Not as common in cultivation as *D. strictus*. Propagate from seed or by division of the rhizomes.

Dichopogon strictus R. Br.
(erect; upright)
all states   Chocolate Lily
0.2-1 m x 0.2-0.8 m   Sept-Dec

Dwarf, perennial **herb**; **leaves** to 40 cm x up to 1.5 cm, grass-like; **inflorescence** slender, erect, usually branched; **flowers** to 3 cm diameter, blue to violet, rarely white, solitary in a bract, drooping, chocolate to vanilla scent; **capsule** more or less globular, about 0.8 cm long, on erect pedicel.

Widespread and locally common in grassland and open forest. An excellent small herb for growing in rockeries amongst small shrubs, or as a container plant. Once established, the plants are long-lived and frequently become naturalized by self seeding. They like well drained soils, and respond well to regular watering while in growth, and to applications of liquid fertilizers and manures. Propagation is by division of the rhizomes or from seed.

This species can be distinguished from *D. fimbriatus* by its branched inflorescence, solitary flower in each bract, and erect seed capsule.

DICHOSCIADIUM Domin
(from the Greek *dicha*, in 2 ways; *sciadian*, a sunshade)
Apiaceae

An endemic, monotypic genus with 2 varieties. One occurs in the alpine herbfields of NSW and Vic, and the other in Tas.

Dichosciadium ranunculaceum (F. Muell.) Domin
(similar to the genus *Ranunculus*)
NSW, Vic
prostrate x 0.1-0.2 m   Nov-Jan

Dwarf, perennial **herb**, develops fleshy taproot; **leaves** 1.5-3 cm x 1.5-3.5 cm, with long petiole, palmate with 3-5 toothed lobes, shiny, green; **flowers** to 1 cm diameter, white, arranged in an irregular umbel of 3-6 flowers, on a hairy peduncle; **sepals** 5, similar to petals; **petals** 5; **stamens** 5.

A most ornamental herb that can cover large areas in alpine herbfields where the soils are continuously moist. Only recently introduced into cultivation by the

Dichrostachys

National Botanic Gardens, and it is hoped that plants will become more readily available. Needs very well drained soils that are always moist, with dappled shade to partial sun. Likes protection from boulders, logs or other low plants. Has potential as a container plant. Frost tolerant. Propagate from seed.

The var. *tasmanicum* (Hook.f.) Domin has 5-7 lobes on the leaves and 6-12 flowers in the umbel.

DICHROSTACHYS (A.DC.) Wight & Arnold
(from the Greek *dis*, double; *chros*, colour; *stachys*, an ear of corn; in reference to the bi-coloured spikes consisting of bisexual and asexual flowers)
Mimosaceae

Shrubs; **branchlets** or **stipules** modified into spines; **leaves** bipinnate, with a stalked gland between the lower pinnae; **leaflets** small; **inflorescence** a dense cylindrical spike of two types of flower, each a different colour; **lower flowers** asexual, with a rudimentary ovary and prominent staminodes; **upper flowers** bisexual; **petals** 5; **fruit** a pod.

A small genus of 20 species of hardy, rigid shrubs distributed from Africa to Australia where there are 2 species. Their horticultural appeal is limited to interesting flowers and hardiness under adverse conditions. They must be propagated from seed which is hard and needs pre-germination treatment.

Dichrostachys spicata (F. Muell.) Domin
(flowers in spikes)
WA, NT
2-3.5 m x 0.5-1.5 m   Feb-March

Medium **shrub**, much-branched; **branchlets** spiny, 2-5 cm long; **leaves** 3-8 cm long, bipinnate; **pinnae** 1-3 pairs, 2-4 cm long; **pinnules** 0.6-1 cm x 0.1-0.2 cm, 8-12 per pinna, oblong, dull green, glabrous, leathery; **spikes** 2-4 cm long, the lower flowers sterile; **sterile flowers** pink-red in bud, when open displaying prominent, branched white staminodes; **bisexual flowers** about 0.4 cm long, cream to yellowish; **pods** 5-8 cm x 0.3-0.8 cm, twisted, sticky.

An interesting shrub found in dry situations. Although spiny it is a very hardy plant with attractive flowers. It would be best suited to inland or drier tropical regions. Requirements would be for well drained soil, in a sunny position. Propagate from seed which needs treatment to ensure germination.

DICKSONIA L'Herit
(after James Dickson, 18-19th-century British botanist and nurseryman)
Dicksoniaceae

Tree-ferns; **trunk** erect, slender or stout; young fronds bearing hairs; **fronds** spreading in a crown, 3-4 times divided, fine and lacy; **segments** lobed; **sori** rounded, marginal, protected by a 2-valved indusium.

A small genus of about 25 species widely distributed around the world with 3 species endemic in Australia. They are handsome ferns, usually found in moist, sheltered gullies and are very popular for cultivation in suitable gardens and ferneries. Established specimens

Dichrostachys spicata   C.R. Dunlop

generally transplant readily, but as they are protected plants a more suitable method is propagation from spore which germinates readily when fresh.

Dicksonia antarctica Labill.
(from the antarctic regions; southern)
Qld, NSW, Vic, Tas, SA   Soft Tree-fern
0.5-15 m tall

Large **tree-fern**; **trunk** 0.5-15 m x 0.5-1 m, erect, stout, often buttressed at the base, densely covered with brown fibrous roots; **fronds** 2-4.5 m long, spreading in a crown, 3 times divided, lacy, dark green and glossy above, dull beneath; **sori** about 0.1 cm across.

A common tree-fern that often grows in extensive pure stands in moist gullies and cool forests of eastern Australia. The massive, fibrous trunks are an excellent host for epiphytes, particularly small ferns. The fronds of large specimens are produced together in flushes, and present an interesting spectacle. The species is very easy to grow and is an extremely popular garden plant in southern Australia. Once established, the plants are very tolerant of neglect and prove to be surprisingly resistant to dryness. They are best grown in a moist, somewhat sheltered situation but will take a fair amount of exposure to sun, especially if the roots are kept moist. They are an ideal shelter plant for smaller ferns, and also make an excellent tub plant. Large numbers of this tree-fern are sold in nurseries of south-eastern Australia as trunks sawn off at the base and with the fronds trimmed from the top. These grow easily when planted in a suitable position and kept moist. New roots are produced from the base of the trunk, and the plants become self supporting after about 12 months. The first lot of fronds is very susceptible to wind and sun damage, but the next lot is generally much tougher. If planted during dry periods, the trunk should be hosed at least once a day until the weather improves. Established specimens transplant readily; however, if the trunk is sawn off only the upper part will grow. The trunk makes an excellent host for orchids and other ferns. The Aborigines ate the pithy material from near the top of the trunk. This material

266

is rich in starch and was eaten either raw or cooked. Propagate from spore which is best sown while fresh. Young plants are very slow-growing for the first 2-3 years.

Dicksonia herbertii W. Hill
(after Sir Wyndham Herbert, first Premier of Qld)
Qld   Brown-bristle Tree-fern
0.5-5 m tall

Slender **tree-fern**; proliferous plantlets absent from trunk; **trunk** 0.5-5 m x 0.1-0.2 m, erect, slender, dark brown to black, the lower part bearing fibrous roots, the upper part covered in stiff, dull, brownish, spreading, bristly hairs; **fronds** 1-3 m long, coarse, dull green; **sori** about 0.3 cm across, very prominent.

This slender tree-fern is restricted to highland regions of north-eastern Qld where it grows in cool, shady situations, often with the roots in moist to wet soil. It frequently grows in colonies but the plants are not proliferous as in *D. youngiae*. It is an attractive plant, but the numerous stiff bristles may cause a rash if contacted and the plants should not be grown near paths. It grows readily in a moist, sheltered situation and seems well suited to subtropical and temperate regions or highland areas in the tropics. It is relatively fast growing when young, and responds to watering and mulching during dry periods. Propagate from spore which is best sown when fresh.

Dicksonia youngiae C. Moore
(after Lady Young)
Qld, NSW   Red-bristle Tree-fern
0.5-4 m tall

Slender **tree-fern**; **plantlets** developing on trunk; **trunk** 0.5-4 m x 0.1-0.15 m, slender, erect in the upper part, the base often flat on the ground, dark reddish brown, the lower part bearing fibrous roots, the upper part and crown covered in soft, tangled, bright reddish bristly hairs; **fronds** 1-3 m long, coarse, dark shiny green above, dull beneath; **sori** about 0.3 cm across, very prominent.

A slender tree-fern confined to south-eastern Qld and north-eastern NSW, usually growing in moist, sheltered areas on creek banks. It frequently forms colonies, and the plants have a remarkable ability to increase vegetatively by the production of plantlets on the trunk. Trunks of this species lean and fall readily, taking root wherever they touch the ground, with old sections frequently rotting away. Plantlets are produced on erect trunks but seem to be retarded in their development. They grow noticeably when the trunks lean or fall over, eventually becoming independent plants. The crown of each trunk is distinctive, with its mass of red, tangled hairs. It is a very attractive tree-fern, well worth cultivating. It grows readily in a moist, sheltered situation in subtropical and temperate regions. Plants can be quite fast growing and respond to watering, mulching and the application of fertilizers. Bristly hairs on the leaves and trunk may cause a rash if contacted. Propagate from fresh spore or by plantlets.

Dicranopteris linearis   D.L. Jones

Dicksoniaceae Presl

A family of ferns widely distributed in tropical and temperate zones. They are all terrestrial ferns and a significant number are tree-ferns developing a fibrous trunk. The fronds are finely divided and lacy, and the young parts are clothed with hairs. The sori are protected by a reflexed leaf-margin and a true indusium. Species of the tree-fern genera *Cibotium* and *Dicksonia* are popular with fern enthusiasts. In Australia the family consists of 2 genera (*Culcita* and *Dicksonia*) and about 5 species.

DICRANOPTERIS Bernh.
(from the Greek *dicranos*, 2-branched; *pteris*, a fern; in reference to the branching fronds)
Gleicheniaceae

Terrestrial **ferns**; **rootstock** long-creeping, slender, much-branched, bearing stiff, bristly hairs; **fronds** erect, dividing equally in pairs, stiff, spreading, 2-3 times divided, apices of branches protected by branched hairs; **veins** forked 2-5 times; **sori** rounded, lacking indusia.

A small genus of about 10 species with a solitary widespread one extending to Australia. It is a distinctive fern but is rarely encountered in cultivation. Propagate from spore.

Dicranopteris linearis (Burm.f.) Underw.
(linear; in reference to the lobes)
Qld, NSW, WA, NT
0.3-2 m tall

Terrestrial **fern** forming spreading or climbing patches; **rootstock** long-creeping, much-branched, the young parts clothed with bristly hairs; **fronds** 0.3-2 m tall, spreading or climbing, forking one to many times, dark green above, glaucous beneath, a pair of reduced pinnae present at each fork; **segments** spreading at right angles, entire; **sori** rounded, about 0.1 cm across.

A widespread fern found in sunny or semi-protected situations, usually in moist patches of soil but sometimes in surprisingly dry conditions. It is frequently found in clay soils and may colonize banks near roads, railways etc. In open areas the plants show little tendency to climb, but are taller when found amongst other vegetation. It is an easily grown fern and well suited to garden culture. It can be readily established on moist banks such as soaks, clay banks around dams, paths etc. A situation exposed to some sun is most suitable. Plants can be grown in a pot but prove to be very sensitive to drying of the root system, rarely recovering from such an occurrence. Small plants can be transplanted readily but large clumps are very difficult. Propagate from spore which is best sown fresh. Division of the rhizomes is not easy with this species.

The var. *subferruginea* (Hieron.) Nakai differs from the typical form by the persistent, coarse, tangled hairs on the veins. It is chiefly tropical in its distribution.

Dicrastylidaceae J. Drumm. ex Harvey

A family of dicotyledons consisting of about 14 genera and 70 species of small shrubs. As interpreted by some authorities the family is not represented in Australia, and those species previously included in it are now referable to Chloanthaceae (q.v.).

DICRASTYLIS Drumm. ex Harvey
(from the Greek *dicroos*, forked; *stylos*, style; alluding to the deeply 2-branched style)
Chloanthaceae

Dwarf to small **shrubs**, densely covered with hairs; **leaves** simple, sessile or petiolate, decussate or verticillate, entire to slightly toothed, flat to revolute margins, densely hairy; **inflorescence** terminal or axillary cymes; **flowers** tubular, mostly 5-lobed, sometimes irregularly lobed, often hairy outside, densely hairy inside; **calyx** densely woody; **fruit** a capsule.

An endemic genus of 26 species, represented in Qld, NSW, SA, WA and NT. The greatest representation is in WA, where there are 23 species, 18 of which are endemic. They usually occur in low-rainfall regions that have extremely high temperatures.

Dicrastylis is closely allied to *Lachnostachys* and *Physopsis*, which are often known much better by their vernacular name of lamb's tails. *Dicrastylis* does not have its flowers arranged in dense spikes; rather they are loose arrangements.

Very few species have found their way into cultivation, but they have potential due to their orna-mental, hairy foliage and well displayed flowers. They will probably do best in well drained soils, in a warm to hot situation. They are recommended for semi-arid to arid regions. Most species should be frost tolerant and able to withstand extended dry periods. Plants may be short-lived. *Dicrastylis* should also grow well as container plants in temperate Australia.

Some species strike readily from cuttings of firm young growth (eg. *D. lewellinii*, previously known as *D. weddii*). Some of the long-haired species may prove troublesome as they could be more susceptible to fungal attack during humid conditions. Results from seed propagation are not well known.

Dicrastylis brunnea Munir
(deep brown; refers to the hairs that cover most of the plant)
WA
0.3-1 m x 0.3-1 m   Aug-Jan

Dwarf **shrub**; **branches** cylindrical, woody, densely covered by rusty brown hairs; **leaves** 1.5-4 cm x 0.5-1 cm, elliptic-oblong, decussate, more or less sessile, entire, blunt apex, dense yellow-brown hairs on upper surface, often greenish grey hairs below; **flowers** tubular, to 0.9 cm long, white to cream, borne in leafy pyramidal cymes, with purplish yellow or purplish grey hairs; **calyx** densely covered with purplish yellow hairs.

An inhabitant of the dry Eremaean Botanical Province of WA, and probably best suited to cultivation in arid or semi-arid areas. It is most ornamental, with a rusty brown appearance, and its flowers are amongst the largest of the genus. Could succeed as a container plant in temperate regions.

The var. *pedunculata* Munir has its flower-clusters on slender and longer peduncles. Propagate from seed or cuttings.

Closely allied to *D. exsuccosa* and *D. gilesii*, which do not have the rusty brown hairs.

Dicrastylis lewellinii, × ·6

Dicrastylis corymbosa (Endl.) Munir
(in the shape of a corymb)
WA
0.2-0.6 m x 0.3-1.5 m   July-Nov
    Dwarf, spreading, densely woolly **shrub**; **branches**
many, dense covering of white hairs; **leaves** 0.5-1.5 cm
x 0.2-0.5 cm, oblong, decussate, can be verticillate,
sessile, thick, soft, revolute margins, dense covering of
white woolly hairs; **flowers** tubular, about 0.5 cm long,
white, in corymbs of dense, white, woolly cymose
heads; **calyx** densely covered in white hairs.
    A most ornamental dwarf shrub, worthy of
cultivation. It occurs over a wide area in the Avon, Roe,
Eyre, Coolgardie and Austin Districts. Should do best
in very well drained soils, with partial or full sun.
Highly suited as a container plant. Propagate from
seed or cuttings.
    Related to *D. velutina*, which has triangular-ovate
leaves. Previously known as *D. stoechas* Drumm. ex
Harvey.

Dicrastylis exsuccosa (F. Muell.) Druce
(without sap)
WA, NT
0.3-1.5 m x 0.3-1 m   April-Nov
    Dwarf to small **shrub**; **branches** densely covered
with hairs which can be golden yellow; **leaves** 3-10 cm
x 0.7-2.5 cm, lanceolate to ovate-lanceolate, decussate,
petiolate, blunt apex, thick, densely hairy when young,
becoming rough and wrinkled above but retaining
hairs below; **flowers** tubular, about 0.5 cm long,
golden yellow, borne in a pyramidal, terminal panicle
about 7 cm long; **calyx** covered in golden hairs.
    A highly ornamental species from arid regions. It is
worthy of cultivation, but to date has only been grown
by enthusiasts. Probably best suited to arid and semi-
arid regions where good drainage is provided. In temperate areas a hot location may prove suitable. Container cultivation may be a successful alternative. This
is a complex species with quite a few subspecies and
varieties that have differing characteristics of leaves
and hairs. Propagate from seed or cuttings.

Dicrastylis flexuosa (Price) C. Gardner
(flexuose)
WA
0.5-1 m x 0.5-1 m   Aug-Oct
    Dwarf, white-hairy **shrub**; **branches** many; **bran-
chlets** cylindrical, decussate; **leaves** 1-3 cm x 0.5-1 cm,
ovate to oblong-ovate, decussate, sessile, blunt apex,
wrinkled, densely covered in white hairs, margins
crenate; **flowers** tubular, violet-blue, hairy exterior,
glabrous interior, in sessile, decussate cymes terminating the branchlets, 7 flowers per cyme.
    This species is endemic to the Eremaean Botanical
Province of WA. It is best suited to well drained, light
to medium soils, with plenty of sunshine. Recommended for semi-arid regions. Should make an excellent container plant for temperate regions.
Propagate from seed or cuttings.
    Similar to *D. verticillata*, which has small flowers and
verticillate leaves. Similar also to *D. georgei*, which has
petiolate leaves.

Dicrastylis fulva Drumm. ex Harvey
(reddish yellow)
WA
0.6-1.2 m x 0.5-1 m   Sept-Dec
    Dwarf to small **shrub**; **branches** densely covered
with brownish hairs; **leaves** 1-4 cm x 0.5-1 cm, elliptic
to oblong, sessile, decussate, hairy, obtuse, margins
slightly toothed; **flowers** tubular, very small, white,
arranged in terminal corymbose panicles; **calyx**
densely hairy on outside.
    An ornamental species with brownish foliage. It is
from the sandplains north of Perth, with its main
representation around Geraldton. Has had limited
cultivation, and should do best in well drained, light to
medium soils, with partial to full sun. Frost tolerant.
Propagate from seed or cuttings.
    Allied to *D. micrantha*, which has greenish grey hairs
and smaller flowers. Similar also to *D. parvifolia*, which
differs in its smaller and more congested leaves.

Dicrastylis georgei Munir
(after A. S. George, contemporary Australian botanist)
WA
1-1.5 m x 1-1.5 m   June-Sept
    Small hairy **shrub**; **stem** erect, woody; **branchlets**
with dense purple to greyish purple hairs; **leaves** 1.5-
6 cm x 1-3 cm, ovate-subcordate to ovate-rotund,
petiolate, decussate, wrinkled upper surface, reticulate
below, densely hairy all over; **inflorescence** cymose;
cymes sessile, decussate, arranged in distant clusters
along branchlet tips; **flowers** tubular, about 0.5 cm
long, purple or pale pink, hairy.
    A most ornamental species from north-western WA.
Probably best suited to cultivation in tropical regions,
but is worth trying as a container plant in more temperate climates. Requires very well drained soils, with
plenty of sunshine.
    The var. *cuneata* Munir differs, with leaves that are
broadly elliptic, and cuneate towards the base.
    Propagate both forms from seed or cuttings.
    Allied to *D. flexuosa*, which has sessile leaves.

Dicrastylis gilesii F. Muell.
(after Ernest Giles, 19th-century explorer)
WA, NT
1-2.5 m x 1-2.5 m   June-Sept
    Small to medium **shrub**, mainly branching near
base; **branches** densely covered with grey hairs;
branchlets can have purplish grey to greenish grey
hairs; **leaves** 4.5-8.5 cm x 1.5-3 cm, ovate to broadly
lanceolate, petiolate, decussate, dense covering of soft
grey to purple-grey hairs; **inflorescence** cymose;
flowers tubular to about 0.7 cm long, purplish, densely
hairy, borne in pyramidal clusters of up to 15 flowers,
or more in some cases.
    An outstanding species with ornamental hairy
foliage. From the arid interior south-west of Alice
Springs, and far-eastern central WA. It is probably
best suited to semi-arid and arid regions, but may
adapt to container cultivation in temperate areas. Undoubtedly it will need extremely well drained, light to
medium soils, in a warm to hot location.
    The var. *gilesii* forma *densa* Munir is distinguished by
its very compact flower clusters.

The var. *bagotensis* Munir has very long peduncles (to about 6 cm long).

The var. *laxa* Munir differs from other varieties in its loose arrangement of the flower-heads.

Propagate from seed or possibly cuttings.

Dicrastylis lewellinii (F. Muell.) F. Muell.
(after Dr. H. Lewellin)
Qld, NSW, SA, NT
0.3-0.7 m x 0.3-1 m   Aug-Nov

Dwarf **shrub**, often spreading by suckering; **branches** with dense greyish hairs; **leaves** 0.7-1.5 cm x 0.2-0.6 cm, linear to linear-lanceolate, sessile, usually in whorls of 3, recurved margins, densely covered with grey hairs giving a silvery grey appearance; **flowers** tubular, to 0.9 cm long, purplish blue, in clusters that are terminal or arranged in an irregular spike.

This dainty species has been described as having a smoky appearance, due to the greyish, hairy leaves. It often occurs in open sandplains, and needs well drained, light to medium soils, with plenty of sunshine. Plants can sucker over a wide area if allowed to go unchecked. Frost and drought tolerant. An excellent container plant. Propagates readily from cuttings.

Previously known as *D. weddii* Bailey.

Dicrastylis micrantha Munir
(small flowers)
WA
0.6-1.5 m x 0.6-1.5 m   Sept-Nov

Dwarf to small **shrub**; **branches** many, woody, with dense, brownish yellow hairs; **leaves** 1.5-3.5 cm x 0.4-1.5 cm, narrowly elliptic-oblong, decussate, sessile, blunt apex, densely covered in greenish grey hairs; **flowers** tubular, about 0.25 cm long, white, in a flat-topped, corymbose panicle of about 5-6 cm across, profuse.

This decorative dicrastylis occurs mainly north of Geraldton, near the coast. It has had limited cultivation. Requires very well drained soils, with plenty of sunshine. Tolerates extended dry periods and most frosts. Should do well in containers. Propagate from seed or cuttings.

Closely allied to *D. fulva*, which has broader leaves and larger flowers.

Dicrastylis obovata Munir
(ovate leaves, broadest above the middle)
WA
0.6-1.5 m x 1-1.5 m   Oct-Dec

Dwarf to small **shrub**; **stem** erect, woody; **branchlets** covered with short, soft hairs and many leaves; **leaves** 0.7-1.5 cm x 0.3-1 cm, obovate, sessile, decussate, entire, obtuse, covered in short, soft hairs; **inflorescence** a panicle to about 6 cm across; **flowers** tubular, about 0.5 cm long, pale blue or white, many per cluster.

Occurs naturally in the Lake King-Salmon Gums area of south-western WA. Evidently not in cultivation, but is worthy of trial as its flower-heads are quite decorative. Should do best in well drained, light to medium soils, with plenty of sunshine. Tolerant of frost and extended dry periods. Could do well as a con-

tainer plant in temperate regions. Propagate from seed and possibly cuttings.

Has affinities to *D. micrantha* in flower. *D. micrantha* can be distinguished by its oblong leaves.

Dicrastylis parvifolia F. Muell.
(small leaves)
WA
0.3-0.8 m x 0.3-0.8 m   Sept-March

Dwarf **shrub**; **branches** covered with greyish hairs; **branchlets** slender, decussate, with many leaves; **leaves** 0.5-2 cm x up to 0.5 cm, linear to linear-oblong, sessile, decussate, sometimes verticillate, densely hairy, becoming less hairy at maturity, margins slightly recurved, crowded; **flowers** tubular, small, white, unequally 5-lobed, arranged in dense, corymbose panicles, profuse.

A compact shrub that is found in sandplain habitats of the Avon, Coolgardie, Eyre, Irwin and Roe Districts. Its main representation is north-east of Perth. Has had limited cultivation, with success in well drained, sandy soil. Will probably also tolerate clay-loam soils. Needs plenty of sunshine. Tolerant of most frosts. Propagate from seed or cuttings.

Dicrastylis petermannensis Munir
(after the Petermann Ranges)
NT
1-1.5 m x 1-1.5 m   Aug-Oct; also sporadic

Small **shrub**; **branches** densely covered in creamy grey hairs; **leaves** 2.5-6 cm x 0.5-1.5 cm, lanceolate to ovate-lanceolate, decussate, petiolate, entire, dense greenish grey hairs above, dense whitish grey hairs below, somewhat acute apex; **flowers** tubular, about 0.4 cm long, whitish, borne in terminal, cymose heads to 20 cm long, which have a greenish grey to cream appearance due to the dense hairs; **calyx** densely covered in creamy grey hairs.

At present this species is known only from the Petermann Ranges near the WA-SA border. It is best suited to arid locations, but may be successful as a container plant in temperate areas. It has large flower-heads and is certainly worthy of cultivation. Propagate from seed or cuttings.

Dicrastylis sessilifolia Munir
(sessile leaves)
WA
0.5-1 m x 0.5-1.5 m   Sept-Oct

Dwarf **shrub**; **branches** densely covered with greyish white hairs; **leaves** 1.5-3 cm x 0.3-0.7 cm, lanceolate, entire, decussate or also verticillate, sometimes scattered, acute, densely hairy; **flowers** tubular, about 0.5 cm long, white, hairy exterior, borne in pyramidal cymes, with long, slender reddish purple peduncles; **calyx** densely covered with hairs.

An ornamental species from central-south-western WA, near Wiluna, where it grows in sandplain country. Should do best in very well drained, light to medium soils, with plenty of sunshine. May succeed in temperate regions, provided it has a hot, sunny location. Propagate from seed or cuttings.

Dicrastylis stoechas Drumm. ex Harvey =
D. *corymbosa* (Endl.) Munir

Dicrastylis velutina Munir
(velvety)
WA
0.5-1 m x 0.5-1 m   Oct-Feb
   Dwarf **shrub**; **branches** densely covered with
whitish hairs; **leaves** 0.3-2 cm x 0.1-0.8 cm, usually
triangular-ovate, decussate, sessile, wrinkled upper
surface, crenate and revolute margins, dense white
hairs; **flowers** tubular, about 0.5 cm long, white, in
cymose heads to about 2.5 cm diameter; **calyx** with
long white hairs.
   Evidently not in cultivation. This decorative species
occurs over a wide range of south-western WA. It
needs well drained soils, with partial to full sun. Frost
and drought tolerant. Recommended for trial in semi-
arid regions. Propagate from seed or cuttings.
   D. *corymbosa* is similar, but is generally smaller-
growing and has blunt, oblong leaves.

Dicrastylis verticillata Black
(in whorls)
SA
0.5-1 m x 0.5-1.5 m   Oct-Feb
   Dwarf **shrub**; **stem** erect, densely covered in grey
hairs; **branchlets** usually 3 per node, densely hairy;
leaves 0.7-2.5 cm x 0.2-0.5 cm, narrow-linear, in
whorls of 3 or more, sessile, blunt, recurved margins,
hairy wrinkled surface becoming glabrous; **flowers**
tubular, about 0.7 cm long, white, arranged in terminal
cymose racemes, flower-clusters verticillate, 3 flowers
per cyme, profuse.
   A most ornamental species that has had limited
cultivation. It is limited mainly to the Eyre Peninsula.
Prefers well drained soils, in a warm to hot location.
Frost and drought tolerant. Has potential as a container plant. Propagate from seed or cuttings.
   Has affinities to D.*cordifolia* and D. *flexuosa*, which do
not have verticillate leaves.

Dicrastylis weddii Bailey =
D. *lewellinii* (F. Muell.) F. Muell.

DICTYMIA J. Smith
(from the Greek *dictyon*, a net; in reference to the
venation)
Polypodiaceae
   Epiphytic or terrestrial **ferns**; **rootstock** tough,
closely attached to host; **fronds** simple, entire, erect,
thick and leathery; **sori** large, rounded, raised exindusiate.
   A small genus of 4 species, 1 of which is endemic in
Australia. Propagate by division or from spore.

Dictymia brownii (Wikstr.) Copel.
(after Robert Brown, 18-19th-century English
botanist)
Qld, NSW   Strap Fern
0.2-0.5 m tall
   Epiphytic **fern** forming small clumps; **rootstock**
creeping, covered with dark, spreading scales; **fronds**

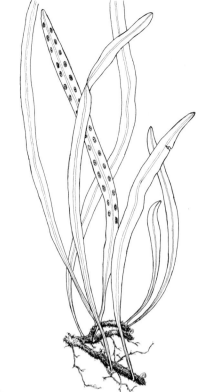

Dictymia brownii, × ·4

stiffly erect, leathery, dark green, the margins entire or
sometimes undulate; **midrib** prominently raised; **sori**
about 0.3 cm across, round or oval, brown.
   A distinctive fern found on trees or rocks, in rainforests or moist situations in open forests. It is easily
grown and is a popular subject for a pot or basket.
Slow-growing until established, but then very hardy
and easy to maintain. Thrives in baskets of tree-fern
fibre. Requires a coarse, acid, epiphyte mixture and
shady, moist conditions. Can be grown in temperate or
tropical regions. Propagate by division of the clumps or
from spore.

DIDISCUS DC. = *TRACHYMENE* Rudge

DIDYMANTHUS Endl.
(from the Greek *didymos*, double; twin; *anthos*, a flower;
referring to the pairs of flowers)
Chenopodiaceae
   A monotypic, endemic genus, confined to WA.

Didymanthus roei Endl.
(after J. S. Roe, first Surveyor-General, WA)
WA
0.1-0.5 m x 1-2 m   sporadic
   Dwarf, spreading, cottony **shrub**; **branches** erect or

spreading, wiry, dense white tomentum; **leaves** about 0.5 cm long, lanceolate, mostly opposite, sessile, entire, covered in white hairs while young, becoming nearly glabrous at maturity; **flowers** bisexual, in pairs; **fruiting perianth** spreading, with 5 winged lobes.

A species with ornamental foliage. It occurs naturally on the edges of saline marshes. Is not well known in cultivation, but could prove to be valuable in the reclamation of saline soils. Would probably do best in relatively well drained, light to medium soils, with plenty of sunshine, in semi-arid or warm to hot coastal regions. Frost tolerant and should withstand extended dry periods. In view of its decorative foliage, it has potential as a container plant. Propagate from seed or cuttings. The woolly tomentum may cause some problems by providing ideal conditions for fungal development. It may be best to propagate from cuttings, using warm conditions with relatively low humidity.

DIDYMOCARPUS Wall.
(from the Greek *didymos*, double; twin; *carpos*, a fruit; the capsule breaks into 2 parts when ripe)
Gesneriaceae

Shrubs or **herbs**; **leaves** opposite, alternate or in a radical rosette; **inflorescence** an axillary cyme; **flowers** small, tubular, widening upwards, divided into 2 lips, the upper lip 2-lobed, the lower 3-lobed, the lobes unequal; **calyx** 5-lobed or toothed, bell-shaped; **stamens** 2 or 4; **fruit** a 2-valved capsule.

A genus of about 120 species with a solitary widespread one extending to high mountains of north-eastern Qld. A number of exotic species are popularly grown as pot plants for glasshouse or indoor decoration; however, the Australian species is not commonly grown. Propagate from seed, by division, transplants or from leaf cuttings.

Didymocarpus kinnearii F. Muell.
(after Robert Kinnear, original collector)
Qld
10-12 cm tall   Nov-May

Perennial **herb** growing in small clumps; **leaves** 6-15 cm x 3-5 cm, lanceolate to ovate or cordate, thin-textured, the upper surface with scattered, septate hairs, the lower surface with silky hairs, the margins serrate, on long, slender petioles, in a spreading radical rosette; **inflorescence** 6-10 cm long, cymose, softly hairy, bearing many small flowers; **flowers** about 0.5 cm long, white, tubular in the basal half, the lobes broad and spreading; **capsule** about 0.6 cm long, narrow, ellipsoid.

A small perennial herb which grows on mossy rocks in highland rainforests of north-eastern Qld. It always grows in moist situations, and the thin-textured leaves have little tolerance of dry conditions. The cultural requirements of this species are similar in many respects to those of African violets. They require cool to warm, moist and humid conditions such as those found on the floor of a bushhouse in tropical or sub-tropical regions, or in a glasshouse in southern Australia. They make ideal pot plants and flower almost continually when grown well. A coarse, well drained mixture rich in humus seems suitable for their culture, and they respond to liquid fertilizers once a fortnight during the warm months. Propagate from seed which is fine and should be sown under glass, from division of the rhizome or from leaf cuttings.

Didymocheton rufum (A. Rich.) Harms =
Dysoxylum rufum (A. Rich.) Benth.

DIDYMOGLOSSUM Desv.
(from the Greek *didymos*, double; twin; *glossa*, a tongue; in reference to the small, alternate fronds)
Hymenophyllaceae

Tiny epiphytic **ferns**; **rootstock** wiry, creeping, bearing fronds at intervals; **fronds** entire, ovate; **sorus** usually solitary, trumpet-shaped.

A small genus of about 20 species with a widespread one having been collected once from north-eastern Qld. Cultivation and propagation are difficult.

Didymoglossum exiguum (Beddome) Copel.
(small; insignificant)
Qld
0.3-0.7 m tall

Tiny epiphytic **fern**; **rootstock** slender, wiry, black, bearing fronds at widely scattered intervals; **fronds** 0.3-0.7 cm long, simple, entire, ovate, the margin near the apex uneven, the fertile fronds more slender and on longer stipes than the sterile ones; **sorus** trumpet-shaped, solitary, apical.

A tiny filmy fern that has been collected only once in Australia, from Mount Bellenden Ker in north-eastern Qld. Its behaviour in cultivation is unknown but it would probably respond to similar conditions as other filmy ferns (see Filmy Ferns).

DIDYMOPLEXIS Griffith
(from the Greek *didymos*, double or paired; *plexis*, plaiting; referring to the column's 2 wings or to the folding made by the union of the petals and dorsal sepal)
Orchidaceae

Small, leafless, saprophytic terrestrial **orchids**; **rhizome** fleshy; **inflorescence** a raceme; **flowers** small, short-lived; **sepals** and **petals** joined in a short tube at the base; **labellum** entire or lobed; **fruit** a capsule.

A genus of about 23 species widely distributed from eastern Africa to Malaysia, with a single widespread species extending to north-eastern Qld. Being a saprophyte, it cannot be grown by any known method.

Didymoplexis pallens Griffith
(pale coloured)
Qld, NT
5-12 cm tall   Nov-Jan

Small, leafless, saprophytic terrestrial **orchid**; **rhizome** fleshy, up to 8 cm long; **inflorescence** 5-12 cm tall, slender, reddish yellow, stem-bracts present, bearing 2-10 erect flowers which are carried on pedicels about 0.5 cm long; **flowers** 0.8-1 cm across, glistening white with a bright orange group of calli on the labellum, cup-shaped, only one open at a time;

Dietes robinsoniana   D.L. Jones

petals and **sepals** broad, the lateral sepals conjoined at the base; **labellum** about 0.5 cm long, triangular or bi-lobed.

A diminutive, delicate terrestrial orchid which flowers rapidly after the first heavy rains of the wet season. In Australia it is restricted to near-coastal lowland areas subject to partial inundation, where the orchid grows on small hummocks. A single flower is produced each day and lasts but a day. Ovary development and seed dispersal occur extremely rapidly, accompanied by elongation of the flower-stem. Being a leafless saprophyte, it cannot be maintained in cultivation by any known method.

DIDYMOTHECA Hook.f. = *GYROSTEMON* Desf.

DIETES Salisb.
(from the Greek *dietes*, biannual; apparently in reference to the leafy shoots and flowering shoots appearing in the same year but in different seasons)
Iridaceae

Rhizomatous perennial **herbs** forming tussocks; **leaves** ensiform to linear, erect and spreading; **inflorescence** an erect, branched, cymose panicle; **flowers** open-petalled, showy, short-lived; **sepals** 3; **petals** 3, similar to sepals; **stamens** 3; **fruit** a capsule.

A small genus of 3 species, 2 in South Africa and one endemic to Lord Howe Island. The African species are common garden plants in Australia and sometimes become locally naturalized. Propagate by division of the clumps or from seed.

Dietes robinsoniana (C. Moore & F. Muell.) Klatt
(after Hercules Robinson)
NSW (Lord Howe Island)   Wedding Lily
1-2 m tall   Oct-Feb

Robust perennial **herb** forming erect tussocks; **rhizome** wiry, branched; **leaves** 0.3-1.8 m x 4-7 cm, ensiform, leathery, stiffly erect in fans, glabrous, dull to shiny green; **inflorescence** an erect, cymose, much-branched panicle to 1.5 m tall; **flowers** 5-8 cm across, white, opening widely, with 6 spreading segments, each flower lasting a day; **capsule** 3-4 cm long, inflated.

Endemic on Lord Howe Island where it is widely distributed in exposed places on cliffs, from sea level to the summits of the mountains. It is a very showy species which captured the early attention of English horticulturists, and has been grown in England as a glasshouse plant for over 100 years. It is an excellent plant for coastal gardens in warm temperate and sub-tropical areas, and can also be grown in containers or as a glasshouse specimen. The plants flower profusely and a single inflorescence may bear up to 100 flowers. Each flower is short-lived, but they are produced continually over a long period. The plant likes an open, sunny aspect, in well drained soil. It responds well to

273

Digitaria

organic manures and mulches, and needs to be kept moist during hot, dry weather. It can be propagated by division of the rhizome or from seed which germinates readily if sown fresh.

Also known as *Iris robinsoniana* C. Moore & F. Muell. and *Moraea robinsoniana* C. Moore & F. Muell.

DIGITARIA Heister ex Fabr.
(from the Latin *digitus*, a finger; in reference to the digitate inflorescences of these grasses)
Poaceae

Annual or perennial **grasses**; **leaves** flat, glabrous or hairy; **culms** erect or digitate; **inflorescences** digitate, of several spike-like racemes, some arranged in whorls; **spikelets** in pairs, glabrous or hairy.

A large genus of nearly 400 species widely distributed in the warmer areas of the world, with about 38 species found in Australia including 8 species naturalized as weeds or pasture plants. Some of the native species form dominant stands in dry inland areas and are valued for the stock feed they provide. Some of the perennial tussock-forming species are attractive when in flower and could be grown in gardens. Propagate by transplants or division of the clumps. They can also be raised from seed which should be stored before sowing for 6-12 months after collection. Sown seed should be covered lightly.

Digitaria ammophila (Benth.) Hughes
(sand-loving)
Qld, NSW, Vic, SA, WA, NT   Spider Grass
30-80 cm tall   Aug-March

Perennial **grass** forming erect tussocks; **leaves** 10-25 cm x 0.3-0.6 cm, flat, softly hairy; **culms** erect or geniculate, slender; **racemes** 8-16 cm long, 10-12 in the inflorescence, the lower ones whorled, spreading; **spikelets** scattered, ovate, covered with long, white silky hairs.

A tough grass widespread in inland districts and often growing in sandy soils. It is a useful pasture grass for drier districts and is quite ornamental when in flower. It likes a well drained soil, in a sunny situation. The clumps should be cut back heavily or burnt after flowering to keep them vigorous. Propagate from seed or by division of the clumps.

Digitaria brownii (Roemer & Schultes) Hughes
(after Robert Brown, 18-19th-century English botanist)
Qld, NSW, Vic, SA, WA, NT   Cotton Panic Grass
12-50 cm tall   all year

Perennial **grass** forming thick, congested tussocks; **leaves** 5-9 cm x 0.3-0.4 cm, flat, glabrous or sprinkled with white hairs, somewhat roughened, with thickened, white, wavy margins; **culms** slender, geniculate; **panicle** 6-11 cm long, consisting of 2-4 digitate, spreading racemes; **spikelets** bearing silky white or purple hairs.

A widespread grass found on sandy to red soils in inland and near-coastal regions, usually in sparse forests. It is often dominant and is a valuable forage species with palatable foliage and stems. It is quite an attractive grass when in flower, and looks appealing

when planted amongst dark rocks. The clumps can be rejuvenated by burning. Propagate by dividing the clumps (which are very tough) or from seed.

Digitaria coenicola (F. Muell.) Hughes
(growing in mud)
Qld, NSW, SA, WA, NT   Spider Grass
15-60 cm tall   all year

Perennial **grass** forming erect tussocks; **leaves** 5-15 cm x 0.3-0.6 cm, flat, linear, densely hairy with velvety hairs on both surfaces, the margins undulate; **culms** erect or geniculate; **racemes** 10-25 cm long, the lower ones in a whorl of 5-10, digitate; **spikelets** in pairs, hairy.

A valuable pasture grass that is widespread throughout the drier inland regions. Its feed is highly palatable and nutritious, and produced with little rain. Spider grass is quite ornamental when in flower, and looks good when planted amongst dark rocks. Propagate from seed or by division of the clumps.

Digitaria divaricatissima (R. Br.) Hughes
(very widely spreading)
Qld, NSW, Vic, SA   Umbrella Grass
15-60 cm tall   all year

Perennial **grass** forming a slender tussock; **leaves** 4-15 cm x 0.3-0.6 cm, flat, tapered to the tip, hairy; **culms** erect or geniculate, slender; **racemes** 15-20 cm long, the lower ones in a whorl of 4-6, spreading stiffly, digitate; **spikelets** in pairs, greenish.

A widespread grass often found in colonies. It has some value as a pasture species and is also quite ornamental when in flower. Can be grown amongst rocks or small shrubs. Propagate from seed or by division of the clumps.

DILLENIA L.
(after John James Dillenius, 17-18th-century German botanist and physician)
Dilleniaceae

Trees; **leaves** alternate, large, with raised parallel veins diverging from the midrib, the petioles often bordered with deciduous wings; **inflorescence** a loose, terminal panicle; **flowers** large, showy; **sepals** 5, spreading; **petals** 5, flimsy; **stamens** numerous; **fruit** compound, of 5-10 carpels.

A genus of about 60 species with one species commonly found in tropical Australia. It is a handsome tree, deserving wide cultivation in tropical regions. Propagate from seed which must be sown fresh.

Dillenia alata (R. Br. ex DC.) Martelli
(winged)
Qld, NT   Red Beech
6-18 m x 5-10 m   Sept-Jan

Small to medium **tree** with a dense canopy; **bark** loose, papery, reddish brown; **leaves** 10-25 cm x 8-12 cm, ovate to orbicular, dark glossy green, thick-textured, slightly roughened, with about 18 prominent veins, the petiole conspicuously winged; **panicle** terminal, loose; **flowers** 6-8 cm across, yellow, lasting 1-2

Dillenia alata   D.L. Jones

days; **sepals** green, cupped; **stamens** numerous; **fruit** of 5-8 glabrous, bright red to crimson carpels which split open to reveal seeds enclosed in a waxy, white aril.

A handsome tree that is widely distributed in near-coastal tropical rainforests and along stream banks in open, sunny situations, with the roots in moist soil. It is prized for such features as the loose, papery bark which is excellent for bark pictures and as a host for orchids, its large glossy leaves, showy yellow flowers and decorative fruits. The red timber is close-grained and easy to work. It is best suited to culture in tropical regions but may have some potential for warm parts of the subtropics. Should make an excellent park or street tree. Also has potential for coastal planting. It likes an open, sunny situation in moist, well drained soils. The plants appreciate regular watering during dry periods, heavy mulching and applications of complete fertilizers during the warm months. Young plants could have potential for indoor decoration or as bonsai specimens. Propagate from seed which must be sown fresh.

Dilleniaceae Salisb.

A family of dicotyledons consisting of about 530 species in 18 genera, mostly developed in the tropics and subtropics but, at least in Australia, with many species also in the temperate zones. They are shrubs, trees or climbers with open-petalled, colourful, short-lived flowers which have a prominent cluster of stamens. The flowers of many species are large and showy, and a few exotics are cultivated in tropical gardens. The family is well represented in Australia by 4 genera and over 90 species. Of these, the genus *Hibbertia* predominates, not only in the number of species but also in its abundance in many plant communities. Many species are grown for their massed production of colourful flowers. Australian genera: *Dillenia*, *Hibbertia*, *Pachynema* and *Tetracera*.

DILLWYNIA Smith
(after Lewis W. Dillwyn, 18-19th-century English botanist)
Fabaceae

Dwarf to medium **shrubs**; **leaves** alternate, simple, linear, terete or somewhat 3-angled, grooved along upper surface, rarely almost flat, with inrolled margins; **stipules** minute or absent; **flowers** pea-shaped, yellow or yellow-and-red, rarely pink, few together in axillary or terminal clusters, rarely solitary; **standard** broader than long; **wings** narrow; **stamens** free; **style** hooked below apex; **fruit** a pod, more or less globular, nearly sessile.

An endemic genus of about 22 species with representation in temperate regions of all states. NSW has the largest number, with a total of 11 different species.

Dillwynias usually inhabit sandheath or dry sclerophyll forest communities. The majority of the species have been tried in cultivation. Many are highly floriferous, providing some unusual colours, and in most cases the flowers are well displayed. They are certainly worthy of greater usage in the landscape.

Most species are fairly adaptable. In general, they like well drained soils, light or heavy in texture. Partial sun is to their liking, but they also grow well in dappled shade or full sun. Pruning after flowering is recommended if dense bushy plants are desired.

Selection of various forms has resulted in the cultivation of some exceptional plants. One example is *D. sericea*, which is extremely variable in colour and the number of flowers produced. A pink form is known to occur in at least two areas of Vic. Other forms with massed flowers along the branchlets are now propagated vegetatively to retain these characteristics.

Scale and the resultant sooty mould which can discolour branches and leaves can be a minor problem. White oil usually controls outbreaks of scale on dillwynias. Some species are susceptible to attack from cinnamon fungus, so it is important that good drainage is provided.

Propagate from seed or cuttings. Being a member of the Fabaceae family, seeds must be collected when mature, but before they are shed from the pods. There is also the need for pre-sowing treatment because of the hard testa (see Volume 1, pages 196-198, Seed Collection, and page 205, Treatment and Techniques to Germinate Difficult Species).

Cuttings of firm new growth produce roots readily. Often, many cuttings can be gained from each slender branchlet if it is in good condition.

Dillwynia acerosa S. Moore
(needle-shaped)
WA
0.3-1 m x 0.5-1 m   Aug-Sept

Dwarf **shrub**; **branches** rigid; **leaves** 0.3-0.4 cm x about 0.1 cm, linear, rigid, margins revolute, initially hairy, becoming glabrous; **flowers** about 0.5 cm long, orange-yellow and purple, axillary and solitary, or in short corymbs.

A species with brightly coloured flowers. Evidently not in cultivation to date. It occurs in inland situations such as the Coolgardie District, and is probably best suited to a warm to hot situation in well drained, light to medium soils. Propagate from seed or cuttings.

Dillwynia acicularis Sieber ex DC.
(needle-like)
NSW
1-5 m x 1-3 m   Aug-Nov
   Small to tall **shrub**; **branchlets** hairy; **leaves** 1-4 cm long, narrow-linear, terete to trigonous, glabrous or slightly hairy, acute apex, straight or curved; **flowers** to 1 cm across, yellow-and-red, in terminal racemes or in upper axils below the terminal raceme; **pods** about 0.6 cm long, inflated.
   Usually occurs in sandstone on the Central Coast and Central Tablelands. Not well known in cultivation. Needs well drained soils, with dappled shade to partial sun. Propagate from seed or cuttings.

Dillwynia capitata J. H. Willis
(in heads)
Vic   Slender Parrot-pea
0.3-0.6 m x 0.3-1 m   Sept-Nov
   Dwarf **shrub**; **branches** slender, minutely hairy; **leaves** to 1 cm long, needle-like, erect, straight, glabrous; **flowers** pea-shaped, small, yellow with red, displayed in densely clustered heads.
   A most ornamental species which occurs in north-eastern Vic, and is well worthy of trial in cultivation. Requires well drained soils, in a location with dappled shade or partial sun. Possibly will tolerate full sun. Has potential for container cultivation. Propagate from seed or cuttings.
   Closely allied to *D. brunioides* from the Sydney sandstone region.

Dillwynia cinerascens R. Br.
(becoming ash grey)
Vic, Tas, SA, WA   Grey Parrot-pea
0.6-2 m x 0.5-1.5 m   July-Nov
   Dwarf to small **shrub**; young shoots slightly hairy; **branchlets** with minute hairs; **leaves** 0.5-2 cm long, filiform, spreading or reflexed, sometimes with recurved apex, becoming glabrous, greyish green at maturity, crowded; **flowers** pea-shaped, about 0.7 cm across, yellow-and-orange, in terminal clusters.
   A showy species that occurs naturally in a wide range of soils and situations. Hardy in cultivation, doing well in relatively well drained soils, with dappled shade to partial sun. Frost tolerant. Can be pruned hard after flowering, to promote bushy growth. Introduced to cultivation in England during 1819, and is grown to a limited extent in USA. Propagate from seed or from cuttings of firm new growth that strike readily.

Dillwynia dillwynioides (Meisn.) Druce
(similar to the genus *Dillwynia*)
WA
1-2 m x 0.6-1.5 m   Aug-Dec
   Small **shrub**; **branches** erect, elongated, glabrous to hairy; **leaves** to 2 cm long, more or less terete, rigid, apex blunt or with small, recurved point, not keeled, usually glabrous; **flowers** pea-shaped, about 1 cm across, yellow-orange with red keel, 1-3 per axil near ends of branchlets.

Dillwynia cinerascens   T.L. Blake

   Very little is known regarding this species. It occurs in the Darling District to the north and south of Perth, where it grows in swampy, peaty sands. In cultivation, requirements should be for relatively well drained soils, in a situation that has dappled shade to partial sun. Propagate from seed or cuttings.

Dillwynia divaricata (Turcz.) Benth.
(forked or spreading)
WA
0.5-1 m x 0.5-1 m   March-April
   Dwarf **shrub**; **branches** much-forked; **branchlets** covered with soft hairs; **leaves** 0.5-2 cm, narrow-linear, usually blunt, glabrous or hairy, scattered but usually crowded near ends of branchlets; **flowers** pea-shaped, about 0.7 cm long, orange-red, solitary or few together, terminal, shorter than leaves; **pods** about 0.6 cm long, inflated.
   A dweller of moist, sandy soils in the Eyre District. Evidently not in cultivation, but worthy of trial due to its orange-red flowers. Probably best suited to well drained but moisture retentive, light soils. Propagate from seed or cuttings.

Dillwynia ericifolia Smith = *D. floribunda* Smith

Dillwynia ericifolia var. **parvifolia** (R. Br.) Benth. =
D. parvifolia R. Br.

Dillwynia ericifolia var. **peduncularis** Benth. =
D. retorta (Wendl.) Druce

Dillwynia ericifolia var. **phylicoides** (Cunn.) Benth. =
D. phylicoides Cunn.

Dillwynia ericifolia var. **tenuifolia** (Sieber ex DC.)
Benth. =   *D. tenuifolia* Sieber ex DC.

Dillwynia floribunda Smith
(abundant flowers)
Qld, NSW
1-2 m x 1-2 m
                                   Aug-Dec; also sporadic

Dillwynia floribunda var. *floribunda*, × ·6

Dillwynia floribunda var. *teretifolia*, × ·6

Small to medium **shrub**; **branchlets** hairy, but can be almost glabrous; **leaves** 0.5-1.5 cm long, more or less flat, straight or slightly curved, tuberculate, erect, crowded; **flowers** pea-shaped, about 1 cm across, yellow or yellow-and-orange, in pairs in upper axils; **pods** to 0.7 cm long, sparsely hairy, inflated.

A showy species that often occurs in the moist soils of heathlands and gullies. Grows in most soil types, but dislikes poorly drained situations. Best suited to dappled shade or partial sun, but will tolerate full sun. Withstands most frosts. Plants can be pruned after flowering. Recorded as introduced to England in 1794. Also cultivated in USA. Propagate from seed or from cuttings which strike readily.

The var. *teretifolia* Blakely differs in always having nearly terete leaves that are channelled above and scarcely tuberculate, and it is a more rigid shrub of 1-3 m high. This variety prefers drier conditions, and was previously known as *D. teretifolia* Sieber ex DC.

Dillwynia glaberrima, × ·6

Dillwynia floribunda var. **sericea** (Cunn.) Benth. =
D. sericea Cunn.

Dillwynia floribunda var. **spinescens** F. Muell. =
D. ramosissima Benth.

Dillwynia glaberrima Smith.
(very smooth)
Qld, NSW, Vic, Tas, SA   Smooth Parrot-pea
1-3 m x 1-2 m   Aug-Nov; also sporadic

Small to medium **shrub**; **branches** erect, not spiny, usually glabrous; **leaves** 0.5-2.5 cm long, more or less terete, glabrous, usually recurved at apex, blunt; **flowers** pea-shaped, about 1 cm across, yellow with red, borne in short or elongated racemes, terminal or axillary; **standard** about twice as broad as high; **peduncles** up to 2 cm long; **pods** to 0.6 cm long, sparsely hairy, inflated.

A common plant of a wide range of habitats and soils. Does well in cultivation, provided it has relatively well drained, light to medium soils, with dappled shade or partial sun, but will also grow in full sun. Frost tolerant. Withstands the harshest of pruning. Can be used as an informal hedge. Grown in England as early as 1800 and is currently cultivated in USA. Propagate from seed or from cuttings which strike readily.

Dillwynia hispida Lindley
(covered with coarse, erect hairs)
NSW, Vic, SA   Red Parrot-pea
0.2-0.6 m x 0.3-1 m   Aug-Nov; also sporadic

Dwarf, spreading to erect **shrub**; **branches** and **branchlets** usually with short hairs; **leaves** to 1 cm long, narrow-linear, more or less terete, smooth or tuberculate, covered with short hairs or rarely glabrous, spreading; **flowers** pea-shaped, to 2 cm across, red to orange-red, borne in well displayed, loose terminal heads.

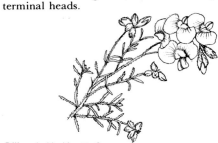

Dillwynia hispida, × ·6

277

Dillwynia juniperina

Dillwynia juniperina, × ·6

An outstandingly ornamental species that can have flowers in varying shades of red. It occurs naturally on both heavy and light soils. In cultivation, it grows well in well drained soils, with dappled shade or partial sun. Tolerant of most frosts and withstands pruning. Excellent plant for containers, including hanging baskets, as the heads of flowers can be pendulous. Propagate from seed or from cuttings which strike readily.

Dillwynia juniperina Lodd.
(similar to the genus *Juniper*)
Qld, NSW, Vic Juniper Pea-bush; Prickly Parrot-pea
1-3 m x 1-2 m   Aug-Dec; also sporadic
   Small to medium, prickly **shrub**; **branchlets** have short hairs; **leaves** to 1.5 cm long, linear, trigonous, spreading to erect, channelled above, usually glabrous, pungent-pointed; **flowers** pea-shaped, about 1 cm across, yellow-and-red, solitary or in terminal or axillary racemes, profuse; **pods** about 0.6 cm long, inflated.
   A decorative prickly species, useful as a refuge plant for native birds, or for directing foot traffic. Grows on a wide range of soils. Prefers good drainage and dappled or partial sun. Recorded as flowering in England during 1820, after introduction in 1818. Propagate from seed or cuttings.

Dillwynia oreodoxa Blakely
(mountain glory)
Vic   Grampians Parrot-pea
1.5-4 m x 1-3 m   Sept-Dec
   Small to medium **shrub**; **branches** usually erect; **leaves** 1-3 cm long, narrow-linear, mainly erect, usually twisted, glabrous, pungent-pointed; **flowers** about 1 cm across, displayed in terminal clusters; **standard** yellow; **wings** yellow; **keel** reddish.
   An upright species that can have an open growth habit. Not well known in cultivation. It requires well drained, light to medium soils, with dappled shade or partial sun. Will possibly tolerate full sun in cool climates. Hardy to frost and snow, and also to extended dry periods. Responds well to light pruning. Propagate from seed or cuttings.
   Similar to *D. glaberrima*, which has a less rigid appearance.

Dillwynia parvifolia R. Br.
(small leaves)
Qld, NSW   Small-leaved Parrot-pea
0.2-1 m x 0.5-1.5 m   Sept-Dec
   Dwarf, spreading **shrub**; **branchlets** hairy; **leaves** to 0.5 cm long, narrow-linear, twisted, glabrous; **flowers** pea-shaped, to 1.5 cm long, yellow-and-red, borne in terminal spikes which are more or less sessile; **pods** to 0.7 cm long.
   In nature, this species is usually found on gravelly, sandy soils. It can be a prostrate or small shrub. Needs similar conditions to those for *D. retorta*. Introduced into England during 1800. Propagate from seed or cuttings.
   Similar to *D. retorta*, which has longer leaves.
   D. ericifolia var. *parvifolia* was the name previously applied to this species.
   The var. *trichopoda* Blakely has flowers arranged on stalked, terminal umbels or corymbs.

Dillwynia phylicoides Cunn.
(similar to the genus *Phylica*)
NSW, Vic
0.2-1.5 m x 1-2 m   Aug-Dec
   Dwarf to small **shrub**; **branches** can be prostrate to erect; **branchlets** with stiff hairs; **leaves** 0.3-1.5 cm, narrow-linear, twisted, usually covered with short rigid hairs, but can be nearly glabrous; **flowers** pea-shaped, about 1.2 cm across, yellow-and-red, solitary or in small terminal or axillary clusters; **pods** about 0.7 cm long, inflated.
   This showy species can be prostrate or erect. It comes from the higher parts of the Blue Mountains, NSW, and their adjacent valleys, and in eastern Vic it occurs in rocky gorges. Not well known in cultivation, but will grow in well drained, light soils, with dappled shade or partial sun. Frost tolerant. Propagate from seed or cuttings.
   D. phylicoides now replaces *D. retorta* var. *phylicoides* and *D. ericifolia* var. *phylicoides*.

Dillwynia prostrata Blakely
(prostrate)
NSW, Vic   Matted Parrot-pea
0.05-0.15 m x 0.5-1.5 m   Sept-Dec
   Dwarf, matting **shrub**; **branches** spreading, hairy; **leaves** to 0.5 cm long, linear, can be broader near the blunt apex, mainly straight, usually glabrous; **flowers** pea-shaped, about 1 cm across, yellow-and-red, solitary or in few-flowered racemes, terminal or in upper axils.
   A subalpine mat plant from south-eastern NSW and north-eastern Vic. Evidently not well known in cultivation, but has potential because of its growth habit. Should grow in well drained, light to medium soils, with dappled shade to partial sun. Suitable for growing amongst taller plants or in containers. Frost tolerant. Propagate from seed or cuttings.

Dillwynia pungens (Sweet) Mackay
(sharply pointed)
WA
0.5-1.5 m x 1-1.5 m   Oct-Dec
   Dwarf to small prickly **shrub**; **branches** often

278

Dillwynia ramosissima, × ·6

glabrous; **leaves** to 0.5 cm long, linear, narrow, broader near the blunt apex, rarely obovate, with inrolled margins, can be minutely tuberculate, usually glabrous; **flowers** pea-shaped, about 1 cm across, yellow-and-red, solitary, terminal or in upper axils; **pods** to 0.7 cm long, inflated, sparsely hairy.

From the Central Coast and Central Tablelands, and the northern part of the South Coast and Southern Tablelands. This dillwynia needs very well drained, light to medium soils, with dappled shade or partial sun. Propagate from seed or cuttings.

Dillwynia retorta (Wendl.) Druce
(twisted)
Qld, NSW
0.5-3 m x 1.5-3 m   Aug-Dec

Dwarf to medium, spreading to erect **shrub**; **branchlets** with many hairs; **leaves** 0.4-1.2 cm long, narrow-linear, twisted, can have minute tubercles, glabrous; **flowers** pea-shaped, to 1.5 cm across, yellow-and-red, borne in terminal or axillary racemes of 1 to many flowers, profuse; **pods** to 0.7 cm long, inflated.

A most ornamental species that inhabits heathland and dry sclerophyll forest. Hardy under a wide range of soils and climatic conditions. Prefers well drained, light to medium soils, with dappled shade or partial sun. Frost tolerant. It usually grows as an open shrub, but will respond well to hard pruning if dense growth is required. Grows well as a container plant. Propagate from seed or from cuttings which strike readily.

D. ericifolia var. *peduncularis* is now considered to be conspecific with *D. retorta*.

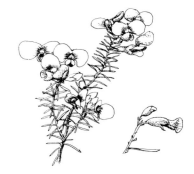

Dillwynia retorta, × ·6

Dillwynia oreodoxa with *Bauera sessiliflora*   T.L. Blake

spreading, can be pendulous; **branchlets** glabrous or slightly hairy; **leaves** to 2 cm long, narrow-linear, more or less terete, straight, rigid, pungent; **flowers** pea-shaped, about 1 cm across, borne in axillary racemes at ends of branches, forming a leafy raceme-like panicle; **standard** yellow; **wings** and **keel** red.

This species was in cultivation in England as early as 1828, under the name of *Eutaxia pungens*. It had the reputation of being an excellent greenhouse plant, cuttings of which formed roots readily. It is not well known in Australia today. As an inhabitant of the southern coastal area, it should grow in temperate regions, provided there is well drained, light to medium soil. Probably best suited to dappled shade or partial sun. Propagate from seed or cuttings.

Dillwynia ramosissima Benth.
(much-branched)
NSW
0.5-1.5 m x 1-2 m   Aug-Nov; also sporadic

Dwarf to small **shrub**; **branchlets** many, spiny,

279

Dillwynia retorta

Dillwynia sericea, × ·65

Dillwynia retorta var. **phylicoides** Cunn. =
D. *phylicoides* Cunn.

Dillwynia rudis Sieber ex DC.
(rough)
NSW
0.5-1.5 m x 1-1.5 m   Sept-Dec
Dwarf to small **shrub**; young shoots hairy;
branchlets develop bristles with age; **leaves** to 1.5 cm
long, narrow-linear, straight or curved, initially hairy,
becoming tuberculate; **flowers** pea-shaped, about
1.5 cm across, crowded in axils near ends of branches;
standard orange and dark red; **wings** orange-and-red;
keel red to reddish brown.
This species has unusually coloured flowers, and is
worthy of being more widely grown in cultivation than
it is at present. An inhabitant of dry heathlands, where
it grows on sandstone and laterites. Should be suited to
most well drained soils, with dappled shade or partial
sun. Propagate from seed or cuttings.

Dillwynia sericea Cunn.
(silky)
Qld, NSW, Vic, Tas, SA   Showy Parrot-pea
0.6-1.5 m x 1-2 m   Aug-Dec
Dwarf to small **shrub**; young shoots with many silky
hairs; **branchlets** usually erect, hairy; **leaves** to 2 cm
long, linear, spreading, usually covered with minute
tubercles, glabrous or hairy, straight or slightly incurved, usually pointed; **flowers** pea-shaped, to 1.5 cm
across, yellow with red, apricot, orange or pink, sessile;
pods about 0.4 cm long, inflated, sparsely hairy.
A most variable species in flower and foliage.
Although not very common in cultivation, quite a few
selected forms with different flower colours are
propagated. A recent introduction with bright pink
flowers is proving adaptable. In general, the species
will grow in most relatively well drained soils, with
dappled shade, partial sun or full sun. It is frost
tolerant and withstands extended dry periods. Excellent for growing beneath tall, established trees.
Plants respond well to pruning, and can be grown successfully in containers. Propagate from seed or from
cuttings which strike readily.
Formerly known as D. *floribunda* var. *sericea*.

Dillwynia stipulifera Blakely
(bearing stipules)
NSW
0.2-0.5 m x 0.5-1 m   Oct-Nov
Dwarf **shrub**; **branchlets** with appressed silky hairs;
leaves to 1.5 cm long, linear, keeled, more or less erect,
smooth, glabrous, straight, pungent; **stipules** slender,
to 0.15 cm long, persistent; **flowers** pea-shaped, about
1 cm across, yellow-and-red, arranged in leafy,
globular heads at ends of branchlets.
Evidently not well known in cultivation, but worthy
of trial in well drained, light to medium soils, with dappled shade or partial sun. It occurs naturally in the
Central Tablelands near Clarence and Wolgan.
Propagate from seed or cuttings.

Dillwynia tenuifolia Sieber ex DC.
(slender leaves)
NSW
0.5-1.5 m x 1-1.5 m   Sept-Feb
Dwarf to small, spreading **shrub**; **branches** slightly
hairy; **leaves** to about 1 cm long, slender, linear, can be
broader at apex, mainly straight, erect or slightly
spreading, clustered, glabrous or with a few hairs,
usually smooth, often ending in a recurved point;
flowers pea-shaped, about 1 cm across, yellow with
red, terminal on short lateral branchlets, usually
solitary; **pods** about 0.5 cm long, inflated.
This is not a well known species. It occurs on the
Cumberland Plateau, and in the lower Blue Mountains, west of Sydney. Should be suited to most well
drained soils, with dappled shade or partial sun.
Propagate from seed or cuttings.

Dillwynia teretifolia Sieber ex DC. =
D. *floribunda* var. *teretifolia* Blakeley

Dillwynia uncinata (Turcz.) J. Black
(hooked)
Vic, SA, WA   Silky Parrot-pea
0.3-1 m x 0.3-1 m   Nov
Dwarf, diffuse to erect **shrub**; young shoots hairy;
branchlets hairy; **leaves** to 1 cm long, terete, thick, obtuse, spreading or often recurved, scattered; **flowers**
pea-shaped, to 1.5 cm across, in short, loose racemes or
clusters, usually terminal, rarely axillary; **standard**
yellow; **wings** and **keel** red.
This species occurs in the Mallee of Vic, in the
southern near-coastal regions of SA, and in the Eyre
District of WA where it can be found in swampy
regions. It is not widely known in cultivation, but
should succeed in relatively well drained soils, with a
warm location. Propagate from seed or cuttings.

DIMOCARPUS Lour.
(from the Greek *dis*, double; *carpos*, a fruit; referring to
the 2-lobed capsules)
Sapindaceae
Shrubs or **trees**; young shoots bearing stellate hairs
in tufts; **leaves** pinnate, subopposite or alternate;
leaflets entire, opposite or alternate; **inflorescence** a
cyme or panicle; **flowers** small, numerous; **calyx** 5-

lobed, hairy; **petals** 5, bearing woolly hairs; **stamens** usually 8; **fruit** a 2-lobed capsule, hairy.

A small, distinctive genus of about 6 species with one endemic in north-eastern Australia. The genus is readily distinguished by its 2-lobed, hairy capsule. They are very rarely encountered in cultivation. Propagate from seed which must be sown fresh.

Dimocarpus australianus Leenh.
(from Australia)
Qld
8-15 m x 3-8 m   Oct-Nov

Small to medium **tree** with a spreading canopy; young shoots yellowish, bearing stellate hairs in tufts; **leaves** 10-18 cm long, pinnate; **leaflets** 2-6, 6.5-9.5 cm x 2-4 cm, elliptic to oblong, thin-textured but leathery, grey-green above, yellowish green to grey beneath, the apex blunt; **panicles** 10-16 cm long, terminal, erect, much-branched, with erect, spreading branches; **flowers** about 0.5 cm across, greenish yellow, densely woolly; **capsule** about 2 cm across, reddish brown, 2-lobed, the surface bearing prominent tubercles, matures Jan.

A rare species known only from a few collections in rainforests of central Cape York Peninsula, but in some areas locally common. It is a handsome tree with a spreading canopy that provides excellent shade. The fruit is similar in many respects to a lychee, and has an edible aril which is esteemed by the population wherever it grows. At this stage plants are rarely cultivated due to lack of propagating material; however, the species should become popular, especially with fruit enthusiasts. It would be best suited to tropical and sub-tropical regions and would probably require a protected position at least when young. Propagate from seed which must be sown fresh.

DIMORPHOCALYX Thwaites
(from the Greek *dis*, double; *morphos*, a shape; *calyx*, the sepals; in reference to the different shaped calyx on male and female flowers)
Euphorbiaceae

Shrubs or **trees**; **branchlets** with prominent lenticels; **leaves** entire, petiolate, alternate; **inflorescence** a cyme, terminal or near the ends of the branches; **flowers** unisexual, showy; **fruit** a capsule.

A small genus of 12 species distributed in Indo-Malaysia, with a single species endemic in north-eastern Qld. The species is unknown in cultivation but has excellent potential for the tropics.

Dimorphocalyx australiensis C. White
(Australian)
Qld
2-4 m x 1-2 m   Jan

Medium **shrub**; young shoots bearing prominent lenticels; **leaves** 7-11 cm x 3-6 cm, elliptical to lanceolate, glabrous, glossy green above, dull and paler beneath, the apex pointed, on petioles 0.5-1.5 cm long; **inflorescence** an elongated, terminal cyme about 3 cm long; **male flowers** about 1.5 cm across, snowy white; **female flowers** unknown; **fruit** a capsule.

This species has been described as a showy shrub

and would certainly be worthy of cultivation. It is known only from rainforest on the Mowbray River in north-eastern Qld. The female flowers and fruit have not been collected. It would require a shady aspect in tropical gardens. Cuttings would be worth trying.

DIMORPHOCHLOA S. T. Blake
(from the Greek *dis*, double; *morphos*, a shape; *chloe*, a grass; probably alluding to the 2 types of inflorescence)
Poaceae

A monotypic grass genus, endemic in Qld and NSW.

Dimorphochloa rigida S. T. Blake
(stiff)
Qld, NSW   Wire Grass
0.5-1 m tall   sporadic

Perennial **grass** forming tall, straggly tussocks or almost thickets; **leaves** 4-7 cm x 0.2-0.3 cm, inrolled, scattered on the culms, tapering from the base, dark green; **leaf-sheaths** striate; **culms** rigid, erect, much-branched at the nodes, wiry, glabrous or hairy; **inflorescence** either terminal or axillary, the axillary one consisting of a solitary spikelet which is self pollinating, the terminal one up to 8 cm long and spike-like.

An unusual grass that grows on poor, rocky sandstone ridges in sparse forest. It is of little ornamental appeal but is of interest to enthusiasts. Plants are very difficult to transplant but can be propagated from the nodal growths.

DIOCLEA Kunth.
(derivation unknown)
Fabaceae

Trees or **climbers**; **stems** lenticellate; **leaves** trifoliolate, alternate; **leaflets** entire, glabrous or hairy; **inflorescence** a panicle or raceme; **flowers** pea-shaped, colourful; **fruit** a woody pod.

A genus of about 50 species, the majority of which are confined to tropical America, one widespread Asian species extending to north-eastern Qld. Propagate from scarified seed.

Dioclea javanica Benth.
(from Java)
Qld
5-20 m tall   Nov-Dec

Vigorous **climber**; young shoots hairy; **stems** ridged, bearing lenticels; **leaves** trifoliolate; **leaflets** 6.5-14 cm x 4-11.5 cm, oblong to elliptical, densely hairy on both surfaces, the apex pointed; **racemes** or **panicles** 10-30 cm long, held erect, in the upper axils; **bracts** conspicuous on young inflorescences; **flowers** about 2 cm long, purple to blue with darker stripes, pea-shaped; **standard** with a yellow patch; **pods** 6.5-16 cm x 2-6.5 cm, oblong, indehiscent, bearing 1-4 seeds; **seeds** 2.5-3.5 cm, round, thick.

A colourful tropical climber which is widespread overseas and extends to north-eastern Qld, where it grows in lowland rainforests. It is perhaps too vigorous to be grown in the home garden but certainly has potential for larger gardens and parks. Best suited to the tropics, and requires well drained soil. Propagate from scarified seed.

Dioclea reflexa J. D. Hook. = *D. javanica* Benth.

DIOSCOREA L.
(after Pedianos Dioscorides, Greek physician who wrote a book on medicinal herbs)
Dioscoreaceae

Climbers with tuberous roots and slender, twining stems; **leaves** alternate, simple, entire, often cordate, petiolate; **aerial tubers** sometimes present in leaf axils; **inflorescence** a spike, solitary or clustered; **flowers** small, dioecious; **fruit** a capsule; **seeds** winged.

A very large genus of 600 species widely distributed in tropical and subtropical regions. Three species occur naturally and a further 2 are naturalized in north-eastern Qld (*D. alata* and *D. reticulata*). Other species are cultivated as garden plants or for their edible tubers (especially *D. sativa* which is sometimes recorded as being native). Propagate by division of the tuberous roots, from aerial bulbils or from seed.

Dioscorea bulbifera L.
(bearing bulbs)

Qld, WA, NT	Aerial Yam
1-3 m tall	Feb-April

Slender, dioecious **climber** with a tuberous root system; **stems** thin, wiry, glabrous; **leaves** 8-15 cm x 6-12 cm, broadly ovate, with a deeply cordate base, bright green, with prominent veins; **aerial bulbils** present in the leaf axils; **male spikes** 2-5 cm long, in clusters; **female spikes** 5-10 cm long, slender, solitary; **capsules** 1-2 cm long, prominently winged at one end.

A common climber of the tropics, found in open forest, along stream banks and also in coastal districts. The plants die back to a perennial root system each year after flowering and fruiting. The tuberous roots were collected by the Aborigines and eaten after soaking and roasting. Aerial yam is an interesting climber for tropical and subtropical regions. It is very useful for mingling amongst shrubbery, and has decorative heart-shaped leaves and fruit. Plants do not smother surrounding plants with strong growth. They are best planted in groups to ensure fruiting, and require a sunny or semi-shady aspect, in well drained soil. Propagation is easily achieved by using the aerial bulbils, which should be half buried in a pot and kept moist until they sprout. They can also be propagated by division of the tubers or from seed.

Dioscorea hastifolia Endl.
(spear-shaped leaves)

WA	Yam
0.5-1.5 m tall	Aug-Oct

Slender, dioecious **climber** with a tuberous root system; **stems** filiform, twining; **leaves** 2-3.5 x 0.5-1.5 cm, linear to linear-lanceolate, the lower ones broadly hastate, bright green, with 3-5 prominent veins; male and female flowers on separate plants, the **male flowers** on spikes 1-3 cm long, the **female flowers**

on racemes about 1 cm long and bearing 2-3 flowers; **capsules** 0.5-1 cm long, prominently winged, pink.

A little-known species of yam which is endemic to WA, where it is distributed in the coastal strip from just south of Perth to near Shark Bay. The plants grow rapidly each year after the autumn rains, flower and fruit and then die back to a tuberous root system. Although little-known in cultivation, this species could add interest to a garden and would never be troublesome. Plants must be grown in groups to ensure fruiting, and should be placed amongst shrubs upon which the stems can twine. A sunny situation, in well drained soil is essential. Propagate by transplants, from seed or by division of the tubers.

Dioscorea transversa R. Br.
(transverse)

Qld, NSW, WA, NT	Native Yam
1-2.5 m tall	Aug-Nov

Slender, dioecious **climber** with a tuberous root system; **stems** thin, wiry, glabrous; **leaves** 5-10 cm x 3-6 cm, cordate or hastate, hard-textured, dark green and shiny, with prominent veins; male and female flowers on separate plants, the **male flowers** on spikes or panicles 3-5 cm long, the **female flowers** on racemes 5-10 cm long; **flowers** about 0.3 cm across, pale green, strongly fragrant; **capsules** 1-1.5 cm long, prominently 3-winged, white, pink or reddish.

A widely distributed climber found in open forest, along stream banks and occasionally even in heathland. The plants die back to a perennial, tuberous root system each year after flowering and fruiting. Clusters of winged capsules are very conspicuous and persist long after the stems have died back. Flowers are highly fragrant on warm days. The thin tubers can be eaten raw or cooked, and were a food for the Aborigines. It has also been reported that Aborigines of the Tully district made a preparation from this species for the treatment of skin cancers. Native yam is a very useful garden climber that adds interest, with its attractive leaves and fruit and fragrant flowers. Its dimensions each year are limited since growth stops once fruiting begins, and the plants do not smother their neighbours with vigorous growth. They are best planted in groups to ensure fruiting, and require a sunny aspect, in well drained soil. They are adaptable to temperate as well as to tropical areas. Propagate by transplants, from seed or by division of the tubers.

Dioscoreaceae R. Br.

A large family of monocotyledons consisting of about 650 species in 6 genera, mostly developed in the tropics, with a few in warm temperate regions. Some are shrubs but most are perennial, herbaceous climbers with the annual stems dying back to a fleshy, tuberous root system. As a group they are known as yams because the starchy tubers of many species are edible, and more than 50 species are cultivated as a food source. Some bear aerial tubers in the leaf axils. In Australia, the family is represented by about 3 species in the genus *Dioscorea*. A few exotics are cultivated for their ornamental features and edible yams, and at least 2 species have become naturalized.

Dioscorea transversa

D.L. Jones

DIOSPYROS L.

(from the Greek *dios*, divine; godlike; *pyros*, wheat; an apparent reference to the fruit of the gods)
Ebenaceae

Shrubs or **trees**; **timber** soft or hard, often dark and close-grained; **leaves** alternate, entire, simple; **flowers** dioecious or bisexual, solitary or in axillary clusters or cymes; **calyx lobes** 3, 4, 5 or 6, enlarging and enclosing the base of the fruit; **corolla lobes** equal to calyx, spreading, glabrous or hairy; **fruit** a berry, becoming soft and fleshy when ripe; **seeds** 1-2 per cell.

A widespread and large genus of more than 500 species with about 15 found in Australia, mainly in Qld and northern tropical areas. A few exotic species are cultivated in Australia for their edible fruit, particularly *D. kaki*, the persimmon from China and Japan. A number of exotic species produce highly valuable woods including the hard black ebony of commerce. The Australian species are all attractive shrubs or trees with dark green foliage and decorative fruits, some of which are edible when soft. They are rarely encountered in cultivation, however, and have the drawback of being slow-growing, especially when they are young. The fruits attract fruit-eating birds such as pigeons and fig-birds. This genus is rather confused in Australia, and is currently being revised. Propagate from seed which must be sown fresh. Some species are known to strike from semi-hardwood cuttings but these may be slow to form roots.

Diospyros australis (R. Br.) Hiern
(southern)
Qld, NSW — Black Plum
4-10 m x 1-3 m — Sept-Oct

Tall **shrub** to small **tree**; young shoots shortly hairy; **leaves** 2.5-8 cm x 1-2.5 cm, elliptical or oval, blunt at the apex, dark glossy green above, pale yellow-green beneath, arranged in 2 regular rows along the branches; **male flowers** 0.5 cm long, stamens 12, yellowish, in dense axillary cymes or clusters; **female flowers** 0.8 cm long, solitary or 2-3 in the axils; **fruit** 1.5-2 cm long, oval, fleshy, black, solitary, enclosed at the base by enlarged calyx, mature Feb-April; **seeds** 0.9 cm long, solitary.

A widely distributed species extending from north-eastern Qld to the Illawarra District of south-central NSW. It is a handsome small tree, readily distinguished by the yellow undersurface of the leaves. The fruit is edible and quite juicy, but must be soft before being eaten. This species is an excellent small tree for a moist protected spot, and grows well with ferns. It requires acid, well drained soils rich in humus, and responds to mulching and regular watering during dry periods. Young plants are somewhat slow-growing and need protection from long periods of direct sun. Propagate from seed which must be sown fresh or from cuttings which may be slow to strike.

Diospyros bundeyana Kosterm.
(from Mount Bundey)
NT
3-4 m x 1-3 m — Nov

Tall **shrub**; young shoots covered with brown hairs; **branchlets** stiff and spreading, sometimes spiny; **leaves** 2.5-7 cm x 2-4 cm, elliptical, papery, dark green, smooth and glabrous above, shortly hairy beneath; male and female **flowers** separate; **fruit** about 2 cm across, brown, with 4 stiffly reflexed, narrow calyx lobes, mature July.

A rarely collected species found on granite hills of western Arnhem Land. It is not known to be in cultivation, but could be of interest for dry tropical and inland regions. Requirements would be for a well drained soil, in a partial-sun situation. Propagate from fresh seed.

Diospyros cargillia F. Muell. =
D. australis (R. Br.) Hiern.

Diospyros compacta (R. Br.) Kosterm.
(compact)
Qld
1-3 m x 1-2 m — Oct

Small to medium **shrub**; young shoots glabrous, somewhat shiny; **leaves** 5-8 cm x 2-3 cm, ovate to oblong, blunt, leathery, with conspicuous reticulate venation, on petioles 1-1.5 cm long; male and female **flowers** separate; **fruit** about 1.2 cm across, globular, solitary in the axils, sessile, the enlarged calyx flat.

A little-known species from north-eastern Qld, found on the mainland and adjacent islands. Appears to have excellent potential as a shrub for cultivation in tropical and subtropical areas but its use is limited by lack of propagating material. Would require well

283

drained soils and a protected situation, at least when young. Propagate from seed which must be sown fresh and possibly from cuttings.

Diospyros cordifolia Roxb.
(heart-shaped leaves)
NT
10-18 m x 3-8 m   April-Aug

Small to medium **tree**; **trunk** with some prickles; young shoots softly hairy; **leaves** 2.5-5 cm x 1.5-3 cm, ovate to ovate-lanceolate, cordate at the base, thinly coriaceous, bright green, the apex blunt or pointed; **female flowers** solitary, the **males** in clusters of three; **flowers** about 0.6 cm across; **fruit** about 1.5 cm across, globular, orange, with expanded calyx lobes.

This species is known from Timor, and in Australia is restricted to NT, where it grows along stream banks in northern parts of the Darwin and Gulf District. It is a handsome tree for tropical and inland regions with a warm to hot climate. The trunk and branches may be prickly. Plants tend to shed leaves during the dry, but this may be offset by watering. Requires well drained soil. Propagate from seed which must be sown fresh.

Diospyros cupulosa (F. Muell.) F. Muell.
(bearing small cups)
Qld
10-15 m x 3-5 m   Oct-Nov

Small to medium **tree**; young shoots sparsely hairy; **leaves** 5-10 cm x 2-4 cm, ovate to oblong, rigid and leathery, dark green and glossy, with prominent reticulate venation, hairy beneath; **female flowers** solitary; **male flowers** in small clusters; **flowers** about 0.6 cm across, yellowish, hairy; **fruit** about 1 cm across, ovoid, the calyx lobes spreading.

Restricted to the Cook and North Kennedy Districts of north-eastern Qld, this species is virtually unknown in cultivation but could have potential in tropical regions. Propagate from fresh seed.

Diospyros fasciculosa F. Muell.
(flowers in clusters)
Qld, NSW
10-20 m x 3-10 m   Sept-Oct

Small to medium **tree**; young shoots glabrous; **leaves** 7-13 cm x 2-4 cm, ovate-oblong or elliptical, blunt, leathery, glossy green above, paler beneath; **female flowers** about 0.6 cm across, yellowish, numerous, in axillary clusters or cymes; **fruit** about 1.2 cm across, globular, fleshy, the base enclosed by the enlarged calyx which has recurved lobes; **seeds** 2-6, shiny.

Widespread in rainforests throughout eastern Qld and north-eastern NSW, this species is occasionally cut as a timber tree, the light coloured wood being suitable for carving. The fruit is reportedly edible but only when soft and juicy. The tree makes an excellent specimen for parks and large gardens in tropical and subtropical regions. It is slow-growing when young and requires protection from direct sun and drying of the root system. An acid, well drained, humus-rich soil is most suitable. Propagate from seed which must be sown fresh.

Diospyros ferrea (Willd.) Bakh.
(rust coloured)
Qld, WA, NT   Native Ebony
6-10 m x 3-5 m   Aug-Dec

Tall **shrub** or small **tree**; **bark** dark, often scaly; young shoots glabrous or hairy; **leaves** 2-10 cm x 1-4 cm, obovate to oblong, dark green, thick and leathery to almost rigid, often shiny; **female flowers** solitary; **male flowers** in small clusters, yellowish, hairy; **fruit** 1-1.5 cm across, ovoid to globular, bright red, containing 1-4 shiny seeds.

A widely distributed and extremely variable species which occurs in a variety of habitats from coastal to inland. All forms have attractive foliage and are decorative when in fruit. They have potential for wider cultivation in tropical and subtropical areas, but are generally slow-growing. They can be grown in well drained soil, in a sunny to semi-shady situation.

There are 4 varieties in Australia.

The var. *geminata* (R. Br.) Bakh. has elliptical to obovoid leaves 3-9 cm x 1.5-3.5 cm, with a blunt apex and shortly stalked fruit in clusters in the axils. These are about 1 cm long, ovoid to globular, bright red, containing 1-4 seeds, the base enclosed by the enlarged, cup-shaped, 3-lobed calyx. It is widely distributed in rainforests between Cairns and Southport, and is an attractive small tree.

The var. *humilis* (R. Br.) Bakh. is a small tree widely distributed in the drier scrubs throughout the coastal and adjacent areas of Qld, north-western WA and NT. It has ovoid to obovoid leaves 2-3 cm x 1-2 cm, with a blunt, rounded apex and solitary fruit in the axils. These are about 1 cm long, ovoid, reddish, containing 2-3 seeds, the base enclosed by the enlarged, 3-lobed calyx. The wood is black and hard, and valued for inlays, turnery and specialized carving. It is a useful small tree which may become deciduous in long dry periods.

The var. *littorea* (R. Br.) Bakh. forma *laurina* (R. Br.) Bakh. is restricted to islands of the North Kennedy District of Qld. It is a small tree with rigid, leathery, oblong leaves 7-10 cm x 2-4 cm, dark green and shiny above, hairy, yellowish flowers and hairy, reddish fruit about 1 cm across.

The var. *reticulata* (R. Br.) Bakh. is distributed on the mainland of north-eastern Qld, and NT, and extends to adjacent islands. It has obovate to oblong leaves 5-8 cm x 2-3 cm, glabrous, blunt, dark green above, with prominent reticulate venation beneath. The flowers are hairy, and the fruit is up to 1.5 cm across and bright red when ripe. It is edible and eagerly sought by fruit-eating birds, especially pigeons.

Propagate all varieties from fresh seed and perhaps cuttings.

Diospyros hebecarpa Cunn. ex Benth.
(hairy fruit)
Qld, NT
7-10 m x 3-6 m   Oct

Small **tree**; young shoots shortly hairy; **leaves** 5-9 cm x 2-3 cm, broadly ovate to oblong, blunt, dark green and shiny above, much paler beneath; male and female **flowers** separate; **fruit** 1.5-2.5 cm across, globular, covered with short hairs; **seeds** 8, flattened.

Restricted to northern areas where it occurs in coastal scrubs and rainforest. It is occasionally cut as a timber tree. The soft, yellow wood is suitable for tool handles and carving. The fruit was used by the Aborigines to poison fish. It is a handsome tree with a dark, compact crown and could be grown in home gardens or parks as a specimen tree. Slow-growing, especially when young, and may need shelter from hot sun. Needs well drained, acid soil. Propagate from seed which must be sown fresh.

Diospyros mabacea (F. Muell.) F. Muell.
(similar to the genus *Maba*)
NSW   Red-fruited Ebony
3-7 m x 2-5 m   Sept-Oct
    Tall **shrub** to small **tree**; **bark** dark, fissured; young shoots finely hairy; **leaves** 6-13 cm x 2-4 cm, oblong to elliptical, thin-textured, dark shiny green, the veins curved, prominent beneath and hairy; male and female **flowers** separate, the parts in 4s, the corolla segments silky-hairy; **fruit** about 2-3 cm across, globular, scarlet, the calyx not enlarged greatly, prominent, the lobes reflexed, mature March.
    A very rare species confined to riverine subtropical rainforests of the Tweed River valley. It has been introduced into cultivation by enthusiasts, and succeeds well in subtropical regions. Plants are very slow-growing and require protection when young. They prefer a deep, organically rich, well drained soil. Propagate from seed which must be sown fresh.

Diospyros maritima Blume
(growing on the seashore)
Qld, WA   Sea Ebony
8-12 m   Nov-Dec
    Small to medium **tree** with a spreading, dense crown; young shoots glabrous; **leaves** 10-25 cm x 3-6 cm, ovate to oblong-elliptical, leathery, dark green and glossy above, dull beneath, drying black; **flowers** about 0.6 cm across, sessile, yellowish, silky-hairy, the females solitary, the males clustered; **fruit** about 2.5 cm across, fleshy, yellow, the base enclosed in the enlarged 4-lobed calyx; **seeds** flattened, shiny, mature July.
    A widespread species from Timor, Indonesia and Samoa, that extends to coastal districts of Cape York Peninsula. It is a handsome tree for coastal regions of the tropics, where it provides excellent shade. It is, however, very slow-growing and cold sensitive, and restricted in its use to the tropics. It needs adequate drainage and a sunny situation. Propagate from seed which must be sown fresh.

Diospyros pentamera (Woolls & F. Muell. ex F. Muell.) Woolls & F. Muell. ex F. Muell.
Qld, NSW   Black Ebony
15-25 m x 5-10 m   Oct-Nov
    Medium slender **tree**; **bark** dark brown to black, scaly; young shoots bearing soft brown hairs; **leaves** 3-8 cm x 1-2.5 cm, elliptical to oblong-lanceolate, leathery, dark green and shiny, slightly yellowish beneath; male and female **flowers** separate on branches or on separate trees; **male flowers** about 0.5 cm

Diplachne fusca, × ·15   head, × 3

long, yellowish, in clusters of 3-5; **calyx** 5-lobed; **petals** 5; **stamens** 15-20; **female flowers** about 0.4 cm long, usually solitary, bearing silky hairs; **fruit** 1.2-1.5 cm across, globular, enclosed at base by enlarged calyx; **seeds** 2-5, angular, mature July-Aug.
    Widespread from north-eastern Qld to Illawarra in south-central NSW, usually in rainforests. Cut as a timber tree as the wood is hard and useful for turnery. This species is relatively unknown in cultivation, but could make an excellent tree for street or park planting. Young plants are relatively slow-growing and may need extra time in the nursery before planting out and may also need to be protected from hot sun. Suitable soils are acid, well drained loams rich in organic matter. Propagate from seed which must be sown fresh.

DIPLACHNE P. Beauv.
(from the Greek *diploos*, double; *achne*, chaff; referring botanically to the glume, the lemma of which is bi-lobed)
Poaceae
    Perennial **grasses** forming tussocks; **leaves** flat or inrolled, glabrous or hairy; **culms** erect or geniculate; **inflorescence** a panicle of simple, spike-like branches; **spikelets** compressed.
    A small genus of about 15 species of grasses mainly distributed in the tropics, with 3 species found in Australia. They grow mainly in wet conditions and have some ornamental appeal. Propagate by transplants, division of the clumps or from seed.

Diplachne fusca (L.) P. Beauv. ex Roemer & Schultes
(brown)
Qld, NSW, Vic, SA, WA, NT   Brown Beetle Grass
20-30 cm tall   March-May; Sept-Oct
  Perennial **grass** forming dense, leafy tussocks; **leaves**
10-22 cm x 0.2-0.4 cm, flat or slightly folded, glabrous,
although somewhat rough, deep green to purplish;
culms erect or geniculate, simple or branched, often
flattened; **panicle** 10-15 cm long, dense, pale green to
purplish, the axes covered with bristles; **spikelets** large,
crowded.
  A widespread grass that grows in or near permanent
water, or in depressions that are periodically inundated, usually on heavy soils. It will also grow on inundated saline soils. It is eagerly eaten by stock but its importance is limited by its sporadic distribution. The
tussocks are decorative when in leaf or flower, and the
species is a worthy garden subject. It can be grown in
soaks or around the margins of dams or ponds, or even
as an emergent aquatic. It is a useful grass to grow in
saline soils. Propagate from seed or by division of the
clump.

Diplachne parviflora (R. Br.) Benth.
(small flowers)
Qld, NSW, SA,
WA, NT   Small-flowered Beetle Grass
0.6-1.2 m tall   June-May
  Perennial **grass** forming dense, tall tussocks; **leaves**
15-30 cm x 0.4-0.8 cm, flat or loosely inrolled, narrow,
pale green, glabrous; **culms** erect, glabrous, smooth
and shiny, or often with a powdery bloom; **panicle** 20-
30 cm x 3-4 cm, open and loose to dense, olive green
becoming straw coloured; **spikelets** solitary, on short,
thin stalks.
  A grass of sporadic occurrence from near-coastal to
inland districts but usually near water or in depressions
subject to seasonal flooding. It is eaten by stock but is
usually only of minor significance as a forage grass. It is
a highly ornamental grass and well worth growing in
gardens for its tall tussocks. It likes plenty of moisture,
in a sunny situation. Propagate from seed or by division
of the clumps.

DIPLARRENA Labill.
(from the Greek *diploos*, double; *arren*, male; refers to
the 2 fertile stamens)
Iridaceae
  Perennial, tussock-forming **herbs** with very short
rhizomes; **leaves** mainly radical, long, flat; **scape** erect,
simple or branched; **flowers** irregular, pedicellate, in a
simple spike or cluster, within 2 sheathing bracts;
perianth divided into 6 segments, 3 outer segments
spreading upwards, 3 inner ones smaller and narrow,
less spreading; **stamens** 3, one without anther; **fruit** a
3-valved capsule.
  An endemic genus of 2 species confined to eastern
Australia. Both are showy and excellent for cultivation.
They are useful clumping species that provide variety
in foliage and form within a garden. The flowers are
delicate and last only a day, but are produced in
profusion over many weeks. This genus has been
known as *Diplarrhena* R. Br. but *Diplarrena* Labill. has
precedence.

Diplarrena latifolia   T.L. Blake

Diplarrena latifolia Benth.
(broad leaves)
Tas
0.3-1 m x 0.5-1.5 m   Sept-Nov
  Tufting, perennial **herb**; **leaves** to 1 m x 2 cm, strap-
like, green, erect, sheathed at the base; **scape** to 1.5 m
long, erect; **flowers** to 6 cm across, white with purple
and yellow markings, produced in succession from the
sheathing bracts and usually each lasting up to 5 days;
outer petals broadly elliptical, almost circular; **capsule**
elongated, 3-celled.
  A most decorative species that has proved hardy in
cultivation, both in Australia and overseas. It does well
in moist, relatively well drained soils, with partial or
full sun. Tolerates dappled shade, but does not flower
as well in this situation. Withstands heavy frosts and
snow-fields. Suitable as a container plant. Propagate
from seed or by division.

Diplarrena moroea Labill.
(similar to the genus *Moraea*)
NSW, Vic, Tas   Butterfly Flag; White Iris
0.3-0.6 m x 0.5-1 m   Oct-Dec
  Tufting, perennial **herb**; **leaves** to 0.6 m x about
1 cm, strap-like, green, sheathing at base; **scape** to
0.6 m long, erect, simple or slightly branched; **flowers**
to 6 cm across, white, white-and-yellow, or sometimes
white with yellow and some purple, honey-scented;
capsule 1.5-2.5 cm long, 3-celled, 3-angled.
  This pleasant member of the iris family is more
widely cultivated than *D. latifolia*. It is an extremely
hardy and adaptable species. Grows in most relatively
well drained soils, with dappled shade, or partial or
full sun. Frost tolerant. An excellent container plant.
Propagate from seed or by division. Often sold as *D.
moraea*. The South African *Moraea iridioides* has been
sold wrongly as *D. moroea*.

DIPLASPIS Hook.f.
(from the Greek *diplasois*, double; in reference to the
fruit composed of paired mericarps)

Apiaceae

Perennial **herbs** with creeping rhizomes; **leaves** radical, in dense tufts, cordate or orbicular, thick; **inflorescence** a simple umbel of many flowers on a simple scape; **petals** ovate, imbricate in bud; **fruit** composed of 2 mericarps.

An endemic genus of 2 species which are mainly confined to high altitudes, eg. the Australian alps and the high mountains of Tas.

They have not been grown to any great extent, and although not extremely decorative, they are useful small plants for bog gardens, and well suited to growing amongst boulders. Propagate from seed or by division.

Diplaspis cordifolia (Hook.) Hook.f.
(heart-shaped leaves)
Tas
0.05-0.1 m x 0.2-0.3 m   Dec-March

Dwarf, perennial **herb** forming a rosette; **leaves** to 6 cm long, with extremely long, hairy petioles; **leaf-blade** cordate-orbicular, margins crenate and revolute, dark green above, purple below; **scape** to 10 cm tall, simple; **flower-heads** about 2 cm across, green and yellowish, more than 20 flowers per umbel.

This herb usually occurs in peaty soils amongst boulders, above 900 m alt. in the western mountains, and near sea level in the west and south-west. It has had very limited cultivation, but grows well as a container plant. In the ground it should do best in well drained, light to medium soils rich in organic material. Needs plenty of sunlight, but should have a cool root system. May grow well in similar conditions to those required by cushion plants. See page 000 for cultivation details. Propagate from seed or by division.

The densely hairy leaves and petioles are the main differences between this species and *D. hydrocotyle*, which is glabrous or slightly hairy.

Diplaspis hydrocotyle (Hook.f.) Hook.
(similar to the genus *Hydrocotyle*)
NSW, Vic, Tas   Stiff Diplaspis
0.02-0.15 m x 0.15-0.3 m   Dec-March

Dwarf, perennial **herb** forming a rosette; **leaves** to 10 cm long, with long, glabrous or slightly hairy petioles; **leaf-blade** cordate, ovate or orbicular, margins entire or crenulate, slightly revolute, shiny, dark green; **scape** to 4 cm long, simple, thick; **flower-heads** small, green and yellowish, usually less than 20 flowers per umbel.

Occurs at high altitudes in the Australian alps and in Tas mountains where it grows in alpine herbfields and bogs. Needs similar conditions to those for *D. cordifolia*. See that species for cultivation notes.

DIPLAZIUM Sw.

(from the Greek *diplazios*, double; in reference to the elongate sori borne in pairs)
Athyriaceae

Terrestrial **ferns**; **rootstock** erect or creeping, often fleshy, bearing dark scales; **fronds** erect or arching, 1-4 times divided, narrow to broadly triangular, sometimes

Diplaspis cordifolia   T.L. Blake

fleshy but usually thin-textured, often brittle; **segments** small to large, the margins entire, lobed or toothed; **sori** elongated, straight or curved, sometimes paired in v's; **indusia** present or absent.

A large genus of about 600 species found in moist areas of the world. Eight species are found in Australia, all confined to the eastern states. They are attractive ferns, usually found in shady, protected situations, often in wet soils. All species grow readily, given suitable conditions. Propagate from spore which is best sown while fresh.

Diplazium accedens Blume =
               Callipteris prolifera (Lam.) Bory

Diplazium assimile (Endl.) Beddome
(related; similar)
Qld, NSW
0.5-1.2 m tall

Medium terrestrial **fern**; **rootstock** tufted, erect, somewhat fleshy, bearing shiny, dark scales; **fronds** 0.5-1.2 m long, erect to arching, narrowly triangular, 2-3 times divided, fine and lacy, very brittle, dark green with dark marks at the main vein junctions; **sori** about 0.1 cm long, crescent-shaped.

A fern of rather delicate appearance, usually found in moist, shady situations but occasionally in drier areas. It grows readily in a large pot or protected garden situation, and will grow quite happily in wet soils. The fronds are very brittle and readily damaged by winds. The species is adaptable and can be grown in tropical or temperate regions. Propagate from spore.

287

Diplazium australe (R. Br.) Wakefield
(southern)
Qld, NSW, Vic, Tas   Austral Lady Fern
0.5-2 m tall

Medium to large terrestrial **fern**; **rootstock** tufted, erect, forming a woody trunk in old plants; **fronds** 0.5-2 m long, semi-erect to arching, broadly triangular, 2-3 times divided, fine and lacy, very brittle, pale green; **sori** about 0.3 cm long, crescent-shaped.

Widespread in cool, shady situations in moist to wet soils. Old plants with a trunk and large fronds could be mistaken for tree-ferns. The species grows readily in a large pot, tub or protected garden situation, in sub-tropical and temperate regions. They become dormant during the cool winter months and grow mainly in spring and summer. Plants must be protected from wind, and are best grown in a shady, moist position. Propagate from spore. This fern is also known as *Allantodea australis* R. Br.

Diplazium cordifolium Blume
(heart-shaped leaves)
Qld
0.3-0.5 m tall

Small terrestrial **fern**; **rootstock** tufted, bearing black scales; **fronds** 0.3-0.5 m long, erect, entire and undivided, linear-cordate, rounded at the base, the apex acuminate, spreading in a tussock, thick-textured, pale green; **sori** 0.3-0.5 cm long, linear, straight or curved.

This fern is widespread through the Pacific region and extends to north-eastern Qld, where it is rare and known only from a few colonies. It grows readily in a pot or shady garden position, in tropical or subtropical regions, but is generally a slow-growing species. Very cold sensitive and needs a heated glasshouse outside the tropics. Propagate from spore.

Diplazium dietrichianum (Luerssen) C. Chr.
(after Amalie Dietrich, 19th-century German collector in Qld)
Qld
0.5-2 m tall

Medium to large terrestrial **fern**; **rootstock** tufted, erect, forming a slender trunk to 0.5 m tall; **fronds** 0.5-2 m long, semi-erect or arching in a tussock, 2-3 times divided, broadly triangular, dark green; **segments** 10-15 cm long, lanceolate, lobed, the basal lobes longest; **sori** 0.3-0.6 cm long, elongated, often in v's.

A large fern found in moist, sunny areas of north-eastern Qld, often in wet soils. It is rarely encountered in cultivation and seems well suited to subtropical regions. Too large for a pot, and best grown in a protected, moist situation in the ground. Propagate from spore.

Diplazium dilatatum Blume
(widened into a blade)
Qld, NSW
0.5-2 m tall

Medium to large terrestrial **fern**; **rootstock** tufted, erect, forming a slender trunk to 0.5 m tall; **fronds** 0.5-2 m long, semi-erect or arching in a tussock, 2-3 times

divided, broadly triangular, dark green; **segments** 10-15 cm long, lanceolate, lobed, narrow at the base, the central lobes longest; **sori** 0.4-0.8 cm long, elongated, often in v's.

A handsome fern found in moist, protected situations from north-eastern Qld to north-eastern NSW, and extending from the coast to high elevations in the tropics. It is an excellent plant for sheltering smaller ferns. The sori make an interesting and delicate pattern on the undersurface of the fronds. It is easily grown in a sheltered garden position, and appreciates moist soil and regular mulching. Suited to tropical, subtropical and temperate regions. Propagate from spore.

Diplazium japonicum Milde =
　　　　　　Lunathyrium japonicum (Thunb.) Kurata

Diplazium melanochlamys (Hook.) Moore
(black stems)
NSW (Lord Howe Island)
0.5-1.5 m tall

Medium **fern**; **rootstock** erect, forming a small trunk, black, fleshy; **stipes** fleshy, black; **fronds** 0.5-1.5 m x 0.3-0.8 m, 3-pinnate, coarsely divided, dark green, glossy; **segments** thin-textured, prominent black patches on rhachis; **sori** 0.5-1 cm long, solitary or in v's.

This fern is confined to Lord Howe Island, but there is widespread and common in moist, shady positions. In cultivation, it is primarily a collectors' item but it has proved to be very adaptable, succeeding in temperate and subtropical regions. It requires a moist, shady situation, in acid, well drained soil. Propagate from spore.

Diplazium proliferum Thouars =
　　　　　　Callipteris prolifera (Lam.) Bory

Diplazium species
Qld
0.5-1.5 m tall

Medium terrestrial **fern**; **rootstock** tufted, erect, forming a short, dark woody trunk in old plants; **fronds** 0.5-1.5 m long, semi-erect to arching, broadly triangular, dark green, 2-3 times divided, fine and lacy, very brittle, with purplish marks at the main vein junctions; **sori** about 0.2 cm long, linear.

A tall, lacy fern found in dense highland rainforests of north-eastern Qld. Very handsome when well grown, and suited to temperate and subtropical regions or highland areas of the tropics. Fairly cold tolerant but requires protection from frost. Fronds are brittle and must be sheltered from wind. Prefers a moist, shady situation. Propagate from spore.

Diplazium sylvaticum (Bory) Sw.
(growing in forests)
Qld
0.3-0.6 m tall

Small to medium terrestrial **fern**; **rootstock** slow-creeping, fleshy, bearing numerous brown scales; **fronds** 0.3-0.6 m long, erect, spreading in a tussock,

288

Diplocaulobium glabrum   B. Gray

once-divided, bright green, somewhat fleshy and brittle, producing plantlets from near the apex; **segments** 6-15 cm long, linear-lanceolate, shallowly lobed; **sori** 0.2-0.4 cm long, linear, often in v's.

Widespread in cool, moist areas along stream banks. Little-known in cultivation but easily grown in a large pot or protected situation in the ground. Somewhat cold sensitive and needs a glasshouse outside the tropics. The fronds are brittle and easily damaged by wind. Propagate from spore and from the plantlets produced on the fronds.

DIPLOCAULOBIUM (H. G. Reichb.) Kraenzlin
(from the Greek *diploos*, double; *caulos*, a stem; *bios*, life)
Orchidaceae

Epiphytic **orchids**; **rhizomes** much-branched; **pseudobulbs** short, erect, smooth or furrowed, often angled, swollen, crowded; **leaves** solitary, terminal on each pseudobulb; **flowers** short-lived, borne in succession, opening widely.

A large genus of more than 70 species widely distributed in the Pacific region, with 2 species found in north-eastern Qld. They are very similar in many respects to the genus *Dendrobium*, but the basal part of the labellum is not joined to the column to form a closed spur. A number of exotic species are frequent in orchid collections but the native species are grown mainly by enthusiasts. Propagate by division or from seed sown under sterile conditions.

Diplocaulobium glabrum (J. J. Smith) Kraenzlin
(glabrous)
Qld
                                        sporadic

Epiphytic **orchid** forming spreading patches; **rhizome** 0.5 cm thick, much-branched, with prominent bracts; **pseudobulbs** 2-5 cm x 1-1.5 cm, ovoid to obovoid, smooth, yellow-green; **leaves** 2.5-8 cm x 1-1.5 cm, solitary at the top of each pseudobulb, ovate, leathery, emarginate; **flowers** about

2.5 cm across, white and pale yellow, opening widely, short-lived, solitary; **labellum** curved.

A hardy orchid which extends from New Guinea, down eastern Cape York Peninsula to the Whitfield Range just south of Cairns. It grows in dense, congested patches, always on trees, in well lit situations such as along rainforest margins. The bulbs and leaves are usually a characteristic yellowish green, as if bleached. The flowers open widely and are quite attractive but are short-lived, usually lasting about half a day. Flowers are produced in succession, and a large plant may be quite floriferous. The species grows easily on a slab of cork or weathered hardwood, or in a pot of very coarse mixture. In the tropics it can be grown with the protection of a bushhouse, while in temperate zones it needs a heated glasshouse with a minimum temperature of 10-15°C. Plants must be kept dry in the winter, and need plenty of light to flower. Propagate by division of the clumps or from seed.

Diplocaulobium masonii (Rupp) Dockr.
(after W. Mason, first collector of the species)
Qld
                                            Nov

Epiphytic **orchid** forming spreading patches; **rhizomes** wiry, much-branched; **pseudobulbs** 2-4 cm x 0.6-0.8 cm, ovoid, quadrangular, sometimes winged; **leaves** 1.5-3 cm x 0.5-1.2 cm, ovate, thinly coriaceous; **flowers** about 3 cm across, white and yellowish, opening widely, short-lived; **labellum** curved, with brown marks on the inside.

A rare species that has not been relocated since its original discovery on trees on the edge of a mangrove swamp near Cape Tribulation. Cultural requirements are probably as for *D. glabrum*.

DIPLOCYCLOS (Endl.) Post & Kuntze corr. Jeffrey
(from the Greek *diploos*, double; *cyclos*, a circle; a probable reference to the tendrils)
Cucurbitaceae

Climbers with weak stems and tendrils; **leaves** palmate or lobed; **inflorescence** an axillary umbel; **flowers** unisexual, the males and females separate or in the same cluster; **fruit** a succulent, many-seeded berry.

A small genus of 3 species with a widely distributed one extending to Australia. The species has poisonous fruits and little horticultural appeal. It can be propagated from seed or cuttings.

Diplocyclos palmatus (L.) C. Jeffrey
(palmate)
Qld, NSW, WA, NT   Striped Cucumber
1-5 m tall   Nov-Jan

Slender **climber**; **stems** weak, slender; **leaves** 3-5 cm x 4-10 cm, deeply palmate, bright green, thin-textured, nauseous, the lobes lanceolate, angular, the margins entire or toothed; **tendrils** 1-2 branched; **flowers** small, greenish, unisexual, in short axillary clusters; **berry** 2-3 cm long, ovoid, yellow or red with conspicuous white stripes.

A slender climber widely distributed in tropical Asia and Africa, and extending to Australia where it is com-

289

Diploglottis

Diplocyclos palmatus, × ·6

mon in northern areas and down the east coast to north-eastern NSW. The fruits are the only decorative feature of the species but these are reputedly poisonous, so it has little horticultural appeal. The leaves and stems give off a nauseous smell if handled. Plants are readily grown in a sunny to partially shaded situation, in well drained soil. Commonly grown in USA as *Bryonopsis laciniosa.* Propagate from seed or cuttings.

DIPLOGLOTTIS Hook.f.

(from the Greek *diploos,* double; *glottis,* a tongue; the inner scale of petals is divided into two)
Sapindaceae

Erect **shrubs** or **trees** with a spreading crown; **leaves** large, pinnate, covered with velvety or scurfy hairs; **inflorescence** an axillary or terminal panicle; **flowers** small, numerous; **calyx** 5-lobed; **petals** 4-6; **stamens** 8, unequal; **fruit** a globular, somewhat fleshy capsule; **seeds** large, bearing a fleshy arillus.

This genus, which is endemic in Australia, consists of 8 species. All are handsome trees with a spreading crown of large pinnate leaves. They are excellent as shade trees or for sheltering delicate plants. They are well suited to cultivation in eastern Australia and deserve to be more widely grown. The seeds of all species are enclosed by a fleshy aril which is acid and can be used to make jams or drinks. They have considerable potential as street or avenue trees, or as park specimens. Propagate from seed which must be sown fresh.

Diploglottis australis Radlk. =
D. cunninghamii (Hook.) Hook.f.

Diploglottis bracteata Leenh.
(prominent bracts)
Qld
6-10 m x 3-5 m   Aug-Sept
Small **tree** with a sparse, spreading crown; young parts bearing short, pale, appressed hairs; **leaves** 20-50 cm long, pinnate; **leaflets** 8-14, 5.5-20 cm x 2-6.5 cm, elliptical to oblong, glabrous and dark green above, scurfy beneath, somewhat leathery; **panicles** 15-30 cm long, axillary, much-branched; **flowers** 0.4-0.6 cm across, brownish-hairy; **capsule** 2.8-3.6 cm

across, somewhat globose, 3-lobed, leathery, wrinkled.

This handsome species is restricted to rainforests of the Atherton Tableland, usually at high elevations. It is frequently found as a remnant tree along roadsides and in cleared paddocks, tolerating full sunshine. It has proved to be hardy in subtropical areas, and may even be successful in temperate regions. Seedlings tend to be slow-growing in their first year, and require shade and moisture while small. Well drained soils rich in organic matter are suitable for the growth of this species. Propagate from seed which must be sown fresh.

Diploglottis campbellii Cheel
(after R. A. Campbell, original collector)
Qld, NSW   Small-leaved Tamarind
10-18 cm x 3-5 m   Nov-March
Small to medium **tree** with a spreading crown; **bark** grey-brown, fissured; young shoots angular, bearing soft, woolly, fawn hair; **leaves** 10-30 cm long, pinnate, spreading; **leaflets** 4-8, 7-15 cm x 2-6 cm, broadly lanceolate, glabrous, dull green, entire, drawn out into a blunt point, the base unequal; **panicles** 5-16 cm long, much-branched; **flowers** 0.4-0.6 cm across, creamy brown, hairy, crowded, fragrant; **capsule** 4-6 cm across, 3-lobed, small, hard, yellowish brown; **seeds** 3, about 2 cm across, round, hard, enclosed by a red, juicy aril, mature Feb-March.

An uncommon rainforest tree restricted to a few localities in south-eastern Qld and north-eastern NSW. It is a very handsome species with an attractive, spreading, shady crown. It is grown to some extent in Brisbane suburbs. The fruit is very decorative and the juicy red aril is refreshingly acid. It can be eaten raw or made into drinks and jellies.

Young plants require protection from hot sun, but once established grow rapidly. They are well suited to subtropical conditions. Propagate from seed which must be sown fresh.

Diploglottis cunninghamii (Hook.) Hook.f.
(after Alan Cunningham, 19th-century botanist)
Qld, NSW   Native Tamarind
10-20 m x 3-8 m   Sept-Nov
Small to medium **tree** with a spreading crown; **trunk** fluted at base; **bark** grey; young shoots angular, densely clothed with velvety brown hairs; **leaves** 30-80 cm long, pinnate; **leaflets** 8-12, 10-30 cm x 2-6 cm, elliptic to lanceolate, spreading, dark green and somewhat hairy above, densely clothed with velvety brown hairs beneath, blunt; **panicles** large, terminal, rusty-hairy; **flowers** about 0.4 cm across, yellow to brown, hairy, 2-3 lobed; **seeds** 3, large, enclosed by an orange-yellow, fleshy aril, mature Nov-Jan.

Widespread in rainforest from south-eastern Qld to Illawarra in southern NSW. A very common and distinctive tree that is popular in cultivation. Young plants are extremely handsome, with large, coarse, hairy leaves, but unfortunately do not take kindly to indoor conditions. They are also very sensitive to drying out and sunburn, and require some protection until about 1.5 m tall. In suitable conditions they are very fast growing. They can be grown in tropical, subtropical and temperate regions, and prefer an acid,

Diploglottis cunninghamii new growth   D.L. Jones

well drained soil rich in organic material. They respond to artificial watering and light dressings of fertilizer. The wood is pale brown and hard, but has limited use for indoor work. The jelly-like aril surrounding the seeds is very acid but refreshing when chewed. It can also be used in drinks or to make jams. Propagate from seed which has a limited life and must be sown fresh. Seeds are often difficult to collect because they are avidly eaten by birds and fruit bats. Previously known as *D. australis* Radlk.

Diploglottis diphyllostegia (F. Muell.) Bailey
(2 types of leaves)
Qld
10-20 m x 3-10 m   May-June
Small to medium **tree** with a spreading crown; **bark** grey; young shoots angular, sparsely clothed with brown hairs; **leaves** 20-40 cm long, pinnate; **leaflets** 2-9, 8-15 cm x 2-6 cm, elliptic to lanceolate, spreading, dark green and slightly hairy above, sparsely hairy beneath, blunt; **panicles** 10-30 cm long, axillary or nearly terminal; **flowers** about 0.4 cm across, creamy brown, with 5-6 petals; **capsule** 1-1.7 cm across, yellowish or orange, hairy, 2-3 lobed; **seeds** 2-3, large, enclosed by a yellow, fleshy aril, mature Sept-Dec.
Restricted to rainforests of north-eastern Qld, but there fairly widespread and common above 800 m alt. An attractive tree resembling *D. cunninghamii* and with similar cultural requirements. The fleshy arils are edible and make an attractive jelly or drink. The trees grow well in subtropical areas, and there are mature specimens in Brisbane. Propagate from seed which must be sown fresh. The species can be distinguished from *D. cunninghamii* by its smaller, less hairy, thinner-textured leaves.

Diploglottis harpullioides S. T. Reynolds
(similar to the genus *Harpullia*)
Qld
3-5 m x 1-3 m   Nov-Feb
Tall **shrub**; young shoots striate, bearing appressed hairs; **leaves** 20-58 cm long, pinnate; **leaflets** 2-8, 8-

25 cm x 3.5-8 cm, elliptical to oblong, abruptly pointed at the apex, dull green and glabrous; **panicles** 1-2.5 cm long, axillary, branching from near the base; **flowers** about 0.5 cm across, hairy; **capsule** 3-4 cm across, brownish, usually 2-lobed, the seeds enclosed in a fleshy red aril.
A small, relatively sparse species restricted to rainforests of north-eastern Qld between near Cooktown and Babinda. It is relatively unknown in cultivation but could have potential for gardens and containers because of its size. For cultivation it requires well drained, moist soils, in a shaded situation. Propagate from seed which must be sown soon after collection.

Diploglottis macrantha L. S. Smith ex S. T. Reynolds
(large anthers)
Qld
2-5 m x 1-2 m   July-Sept
Medium to tall **shrub** with a single trunk; young shoots densely clothed with dark brown hairs; **leaves** 36-80 cm long, pinnate; **leaflets** 6-10, 8-25 cm x 6-8.5 cm, oblong to ovate, the apex acuminate, dark green and shortly hairy above, rusty-hairy beneath; **panicles** 6-13 cm long, slender, axillary; **flowers** about 0.8 cm across, cream to yellowish, densely hairy; **capsules** 1.5-2 cm across, obovoid, yellow with rusty hairs; **seeds** brown, enclosed by an orange-red aril.
A little-known dwarf species confined to rainforests of the ranges of central and northern Cape York Peninsula. It is virtually unknown in cultivation but has appeal by virtue of its large, pinnate leaves, and would probably make a handsome container plant. Best suited to a shady garden situation, in tropical and subtropical regions. Propagate from seed which must be sown fresh.

Diploglottis pedleyi S. T. Reynolds
(after Les Pedley, contemporary Qld botanist)
Qld
7-9 m x 2-4 m   Aug-Oct
Small **tree** with a sparse, spreading crown, often multi-stemmed; young shoots furry, densely covered with long, rusty red hairs; **leaves** 50-80 cm long, pinnate; **leaflets** 10-24, 10-30 cm x 3-6 cm, narrowly oblong to elliptic, glabrous, leathery, dark green; **panicles** 6-10 cm long, axillary, usually clustered near the ends of branchlets, sparsely branched; **flowers** up to 1 cm across, cream, glabrous, fragrant; **capsule** 3.5-5.5 cm across, usually 1-lobed, leathery; **seeds** 3-4 cm long, nearly enclosed by a fleshy red aril.
A little-known species restricted to coastal rainforests of north-eastern Qld in the Babinda-Innisfail region. It is virtually unknown in cultivation and would be best suited to tropical or warm subtropical regions. It is a striking plant with flushes of deep rusty red, hairy new growth and would be well worth growing in containers. Propagate from seed which must be sown fresh.

Diploglottis smithii S. T. Reynolds
(after L. S. Smith, former Qld botanist)
Qld
7-15 m x 3-8 m   Aug-Oct
Small to medium **tree** with a spreading crown;

291

young shoots densely furry with rusty hairs; **leaves** 30-45 cm long, pinnate; **leaflets** 8-12, 6-20 cm x 2.5-7 cm, elliptical, generally narrowing abruptly at each end, glabrous and dark shiny green above, dull and shortly hairy beneath, somewhat leathery; **panicles** 10-30 cm long, axillary or in terminal clusters; **flowers** about 0.5 cm across, somewhat hairy; **capsule** 1.4-1.8 cm across, ellipsoid, yellowish with a white or greyish bloom; **seeds** large, brown, enclosed in a red, juicy aril.

A locally common species found in coastal rainforests of north-eastern Qld, from south of Cooktown to near Innisfail. Juvenile plants have a markedly different appearance to mature trees, having large, dark green simple leaves to 35 cm x 10 cm. The species is little-known in cultivation, being best suited to tropical regions or warm situations in the subtropics. It requires shade and protection when young, and a well drained soil rich in organic matter. Seedlings have proved to be slow-growing in the subtropics. Propagate from seed which must be sown fresh.

DIPLOLAENA R. Br.

(from the Greek *diploos*, double; *chlaina*, a cloak; refers to the double row of bracts that envelop the flower-heads)
Rutaceae

Small to medium **shrubs** with stellate tomentum; **leaves** alternate, petiolate, entire; **flower-heads** terminal, pendent, shortly pedunculate or nearly sessile; **flowers** sessile, in dense heads, surrounded by rows of broad bracts; **calyx** none; **petals** 5, small, narrow; **stamens** 10, prominent; **styles** united in single long style with 5-lobed stigma; **fruit** 2-valved cocci.

An endemic genus of 6 species which are restricted to south-western WA. It seems that with pending botanical revision this number of species will increase. Most species are found on or near the coast, often in calcareous soils. *D. microcephala*, including its varieties, is the most widespread, occurring from Shark Bay to east of Esperance, and venturing further inland than the other species.

All are decorative, with their pendent flower-heads, and grow as bushy shrubs. General requirements are for very well drained soils, and they prefer the protection of other shrubs. They do, however, grow in the open if protection for their root systems is provided. Excellent as container plants. All respond well to pruning. Tip pruning from when plants are young is recommended. Cut flowers have long-lasting qualities. Plants generally do not require fertilizers, but tolerate light applications of a slow-release type. Scale may attack some plants but control with white oil is usually adequate.

To date, plants have not been readily available due to lack of propagating material. It is hoped that this will be rectified in the future, because diplolaenas deserve wider recognition as decorative plants.

Propagation

Some success has been achieved from seed sown in autumn and germinating within 3-8 weeks, but seed is not always readily available. Cuttings of firm young growth usually form roots within 6-12 weeks at 20-25 °C. Attack by *Botrytis* mould can be damaging to cuttings as the hairy stems and moist, humid conditions are perfect for its spread. Adequate ventilation of propagation areas will usually prevent *Botrytis* attacks. Cuttings have a tendency to drop leaves, and these should be removed immediately to prevent *Botrytis* development.

Diplolaena andrewsii Ostenf.

(after C. R. P. Andrews, 20th-century educator and botanist)
WA
0.6-1.5 m x 1-1.5 m   Aug-Sept

Dwarf to small **shrub**; **branches** densely hairy; **leaves** 1.5-3 cm long, broadly ovate, thin, sparsely hairy above, densely hairy below, apex rounded, margins not prominently recurved; **flower-heads** 1-1.5 cm across, pendent, stamens pale red, exserted.

This species occurs on the Darling Scarp, east of Perth. It has the smallest flower-heads of the genus. At this stage it does not seem to be in cultivation. Should be suited to well drained soils, with semi-shade to partial sun, but will possibly tolerate full sun in southern Australia. Propagate from seed or cuttings.

Diplolaena angustifolia Hook.

(narrow leaves)
WA   Native Rose; Yanchep Rose
1-1.5 m x 1-1.5 m   July-Oct

Small bushy **shrub**; **branches** densely hairy, often with rusty tonings; **leaves** 2-5 cm x 0.2-0.6 cm, linear, obtuse, margins revolute, glabrous above, white tomentum below; **flower-heads** to 3.5 cm across, pendent; **stamens** to 3 cm long, crimson to pale orange.

This outstanding species has large, pendent flower-heads. It occurs on limestone and sand from Dongara to Perth, where it often grows as an undershrub to taller shrubs and eucalypts. Needs very well drained, light to medium soils, with semi-shade to partial sun. Tolerates alkaline soils. Ideal container plant. Can suffer frost damage. Plants can be propagated from seed, but best results are obtained from cuttings.

This species is easily distinguished, as all the others have much broader leaves.

Diplolaena angustifolia, × ·65

Diplolaena microcephala var. *velutina*   W.R. Elliot

Diplolaena dampieri Desf.

(after William Dampier, 17th-century English adventurer and explorer)
WA   Southern Diplolaena
1-2 m x 1-2 m   July-Oct
Small bushy **shrub**; **branches** densely hairy; **leaves** to 4 cm long, oblong-obovate, obtuse, dark green and glabrous above, whitish and densely hairy below, aromatic; **flower-heads** about 2-2.5 cm long, pendent; **stamens** red or yellow; **bracts** densely hairy, white to pale green.

One of the most widely cultivated species. It has proved adaptable to most well drained soils, and tolerates semi-shade to open sun. It likes protection from other plants around its root area. Occurs naturally in coastal dune areas north of Cape Leeuwin in the south-west. Withstands fairly heavy pruning. Tolerant of light to medium frosts. Introduced to England during 1837. This species grows well in California, USA. Propagate from cuttings and possibly seed.

Diplolaena drummondii Benth. =
D. microcephala Bartling var. *drummondii* Benth.

Diplolaena ferruginea Paul G. Wilson

(rust coloured)
WA
0.5-0.7 m x 0.5-1 m   Sept-Oct
Dwarf **shrub**; **branches** densely hairy; **leaves** to 2 cm x 1 cm, broadly elliptic, glandular and glabrous or glabrescent above, dense cream to rusty hairs below; **flower-heads** about 2 cm wide, pendent; **outer bracts** narrow-triangular, with rusty hairs; **stamens** crimson.

A recently described ornamental species from the Mount Leseur and Cockleshell Gully regions. It will need very well drained, light to medium soils, with dappled shade to partial sun. Worthy of trial as a container plant. Should tolerate calcareous soils. Propagate from cuttings.

Diplolaena grandiflora Desf.

(large flowers)
WA
1-3 m x 1.5-3 m   July-Oct
Small to medium, spreading **shrub**; **branches** rigid, spreading, densely hairy; **leaves** 2-5 cm long, ovate to broadly oblong, obtuse, dark green above, densely hairy on both surfaces; **flower-heads** to about 4 cm across, pendent; **bracts** green, hairy; **stamens** pink to red.

This species has the largest flower-heads of the genus, and it is common on the red sand of the Irwin District. Also inhabits calcareous soils near the coast. Requires very well drained, light to medium soils and, as do other species, it prefers dappled shade. It will, however, grow in situations where its upper foliage receives plenty of sunshine. Has proved difficult to establish in eastern Australia, where it probably needs a warm, draught-free location. Propagate from cuttings and also seed if available.

Diplolaena microcephala Bartling

(small heads)
WA   Lesser Diplolaena
1-2 m x 1-2 m   July-Sept
Small bushy **shrub**; rusty hairs on new growth; **branches** densely hairy; **leaves** 1-5 cm x up to 1.5 cm, broadly elliptic to obovoid, obtuse, thick, hairy above, densely hairy below; **flower-heads** about 2.5 cm across, pendent; **bracts** pale green; **stamens** green to reddish, very hairy.

A widespread species from Shark Bay to east of Esperance, where it grows in a wide range of habitats. Not common in cultivation. It requires similar conditions to those for the other species. Tolerates light frosts. Propagate from cuttings and possibly seed.

The var. *drummondii* Benth. has thin, flat leaves and very long stamens, while the var. *velutina* Paul G. Wilson has thick, hairy leaves and dense rusty tomentum on the bracts.

DIPLOPELTIS Endl.

(from the Greek *diploos*, double; *pelte*, a small shield; refers to the conspicuous gland in the flower)
Sapindaceae
Perennial **herbs** or **shrubs**, sometimes suckering;

Diplopeltis eriocarpa

branches usually hairy; **leaves** scattered, entire, lobed or divided; **flowers** monoecious, in terminal racemose panicles, pink, blue or white, male and female flowers similar except for stamens and carpels; **sepals** 5; **petals** 4, rarely 5, clawed; **fruit** composed of 3 cocci.

An endemic genus of 5 species, 4 of which are found mainly near the coast in WA. The other is from the Hamersley Range in the north-west, and extends eastwards to NT, near the Qld border.

They have had limited cultivation but show potential, with delicate flowers that are produced over a long period. They should have application for areas that receive low rainfall with medium to high temperatures for most of the year.

The male and female flowers on the same plant open at different times to ensure cross pollination.

Diplopeltis have the common name of Pepperflowers, which refers to the fruit being similar to the shape of pepper fruits.

Plants are readily propagated from cuttings of firm young growth. Details regarding seed germination are not known.

Diplopeltis eriocarpa (Benth.) Hemsley
(woolly fruit)
WA   Hairy Pepperflower
0.5-1 m x 0.5-1 m   July-Oct

Dwarf **shrub**; **branches** erect or spreading, dense covering of long hairs; **leaves** 1.5-5 cm x up to 2.5 cm, deeply lobed to the midrib, margins often incurved, more hairs below than above; **flowers** about 1.5 cm across, pink, profuse, borne in relatively tight panicles; **petals** usually 4, similar in size, lower 2 spread horizontally.

A very showy species from the north-western coast. It is suited to most well drained soils, including alkaline types. Needs a warm to hot situation. Has potential as a container plant in temperate climates. Propagate from cuttings.

Diplopeltis huegelii Endl.
(after Karl von Hugel, 19th-century Austrian botanist)
WA   Pepperflower
0.5-2 m x 0.5-2 m   Aug-Nov; also sporadic

Dwarf to small **shrub**; **branches** erect or spreading, sparse; **branchlets** hairy; **leaves** 1.5-4 cm x 0.5-1.5 cm, most variable, more or less sessile, oblong to narrow-cuneate, can have linear, oblong or cuneate lobes, hairs mainly on margins, stipular lobes often present, upper leaves usually entire; **flowers** to 1.5 cm across, white and pale pink, well displayed in loosely arranged panicles, petals similar in size; **cocci** about 0.4 cm long.

This dainty species has had limited cultivation. It seems to grow best in very well drained soils, with plenty of sunshine. Will probably tolerate dappled shade in warm to hot locations. Needs regular pruning to maintain bushy growth. Tolerates alkaline soils. Suitable for container cultivation. Propagate from seed.

The var. *subintegra* A. S. George differs in its smaller leaves, which are entire or shortly lobed and nearly glabrous.

Diplopeltis intermedia A. S. George
(intermediate)
WA
0.5-1 m x 1-2 m   Sept-Nov; also sporadic

Dwarf, spreading **shrub**; **branches** many; **branchlets** densely hairy; **leaves** about 1 cm x 0.5 cm, oblong to obovate to cuneate, deeply divided or lobed towards tip, stipular lobes near base, hairy, margins incurved, slightly sticky; **flowers** about 1 cm across, pink or white, borne in loose panicles, petals similar in size; **cocci** about 0.3 cm long.

A species with potential because of its dense, spreading habit. Should do well in soils that have good drainage, and with plenty of sunshine. Worthy of experimentation as a container plant.

The var. *incana* A. S. George is known as the grey pepperflower, and has flat leaves, densely covered with hairs, giving a grey appearance.

Propagate both varieties from cuttings, although the var. *incana* may be prone to attack of grey mould.

This species is intermediate between *D. eriocarpa* and *D. huegelii*.

Diplopeltis petiolaris F. Muell. ex Benth.
(with petioles)
WA   Pepperflower
1-2 m x 1-2 m   March-Oct; also sporadic

Small **shrub**, erect or spreading, can sucker from underground stems; **branches** hairy, resinous; **leaves** 1-7 cm x 0.5-4 cm, obovate-spathulate, petiolate, margins irregularly crenate except near the base where they are entire, hairs above and below, sticky; **flowers** about 1.5 cm across, pink to blue, well displayed

Diplopeltis huegelii, × ·6

294

beyond the foliage, in terminal panicles; **cocci** about 0.4 cm long, hairy.

From the Kalbarri region, this species needs very well drained soils, with partial or full sun. Could prove useful for soil erosion control in coastal regions, due to its suckering habit. Tolerant of alkaline soils. Has potential as a container plant. Propagate from cuttings and possibly seed.

Diplopeltis stuartii F. Muell.
(after John McDouall Stuart, 19th-century Australian explorer)
WA, NT   Desert Pepperflower
0.1-0.3 m x 0.5-1 m   July-Sept
Dwarf, trailing, perennial **herb**; **branches** hairy; **leaves** 1-2.5 cm x 0.5-1.5 cm, oblong to narrow-cuneate, often lobed, more or less sessile, apex truncate, upper surface becoming glabrous; **flowers** to 1 cm across, white and pink or rarely pale pink, borne in loose panicles; **petals** in dissimilar pairs, the upper pair broad, lower pair narrow; **cocci** about 0.3 cm diameter.

An inhabitant of sandy or gravelly soils, extending from Wittenoom in the west to near the Qld border in NT. It is not known whether the species is in cultivation, but it is certainly worth trying in tropical and subtropical regions. May make a pleasing container plant in temperate areas. Should tolerate light to medium frosts.

The var. *glandulosa* A. S. George differs in having only glandular hairs, longer leaves and larger fruit.

Propagate both forms from cuttings.

DIPLOPOGON R. Br.

(from the Greek *diploos*, double; *pogon*, a beard; in reference to the lobes of the lemma)
Poaceae
A monotypic grass genus, endemic to south-western WA.

Diplopogon setaceus R. Br.
(bristly)
WA   Bristle Grass
30-60 cm tall   Oct-Dec
Perennial **grass** forming slender, upright tussocks; **leaves** 5-10 cm long, inrolled, flexuose or recurved, erect, forming a rush-like clump; **culms** erect, wiry; **panicle** 1-2 cm long, dense, ovoid, dark green, with numerous bristle-like awns; **spikelets** solitary.

An interesting grass endemic to the South-West Botanical Province, where it grows in moist, sandy soil amongst rushes and sedges. Little is known of its cultural requirements; however, it makes an unusual pot plant and could be grown beside dams, ponds etc in temperate regions. Propagate by division of the clumps or from seed.

DIPLOPTERYGIUM (Diels) Nakai

(from the Greek *diploos*, double; *pterygos*, a wing; in reference to the forked fronds which resemble birds' wings)
Gleicheniaceae
Terrestrial **ferns**; **rootstock** wiry, creeping, bran-

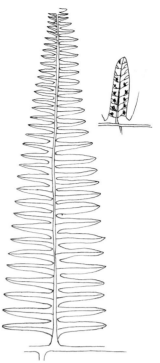

Diplopterygium longissimum pinna, × ·3   pinnule, × 1·8

ched; **fronds** erect, forking regularly and climbing by these divisions, twice-divided; **segments** lobed, dormant buds protected by stipule-like fronds and scales; **sori** small, of few sporangia.

A small genus of about 20 species, a widespread one extending to north-eastern Qld. It is rarely encountered in cultivation. Propagate from spore which must be sown fresh.

Diplopterygium longissimum Nakai
(very long)
Qld   Giant Scrambling Fern
2-6 m tall
Terrestrial **fern** forming large scrambling clumps; **rootstock** long-creeping, much-branched, wiry; **fronds** erect, forking repeatedly and climbing, twice-divided, dark green above, glaucous beneath, the young parts protected by scales and stellate hairs; **segments** linear, oblique; **sori** very small.

A coarse, scrambling fern that forms large tangled clumps along stream banks, road embankments and forest margins, usually in clay soils. It is a colonizing plant, growing rapidly on exposed earth but quickly dying out when overshadowed by forest. The species is not easy to grow and greatly resents disturbance. Potted plants are sensitive to drying of the root system, and rarely recover from it. It is most successful if grown in tropical regions, on a moist clay bank exposed to some sun. Propagate from spore. Root division is extremely difficult.

295

DIPLOSPORA DC.
(from the Greek *diploos*, double; *spora*, a seed or ovule; referring to the 2 or more ovules in each carpel)
Rubiaceae

Shrubs or **trees**; **leaves** entire, opposite or whorled; **stipules** present; **inflorescence** a small raceme or cyme; **flowers** small, bisexual; **calyx** 4-toothed; **corolla** tubular, with 4 spreading lobes; **fruit** a globular berry.

A genus of about 25 species found in China, India and Malaysia, with 3 species endemic to Australia. They are fairly nondescript plants and are rarely seen in cultivation. Propagate from seed which has a limited viability and must be sown fresh.

Diplospora australis Benth.
(southern)
Qld

6-8 m x 1-3 m   Sept-Oct

Tall **shrub** or small **tree**; **branchlets** slender, glabrous; **leaves** 5-8 cm x 1.5-3 cm, lanceolate to ovate, thick and leathery, dark shiny green, drawn out into a long blunt point; **racemes** 2-4 cm long, terminal; **flowers** about 0.4 cm long, greenish; **berry** rounded, blackish.

This species is restricted to the rainforests of Cape York Peninsula. It is a small tree with attractive shiny leaves, and could have potential for ornamental planting. Would require a protected situation in tropical or subtropical regions. Propagate from seed which should be sown fresh.

Diplospora cameronii C. White
(after M. A. Cameron, original collector)
Qld

5-10 m x 3-5 m   Oct

Tall **shrub** to small, spreading **tree**; young shoots yellowish green; **leaves** 5-7 cm x 1.5-2 cm, lanceolate to oblanceolate, dark glossy green above, dull and pale green beneath; **stipules** triangular; **cymes** terminal or in the upper axils, 3-4 cm long; **flowers** small, white, sessile; **corolla** tubular, 0.2 cm long; **fruit** 0.8 cm long, ovoid, bilocular, mature Dec.

A rare species of sporadic occurrence in south-eastern Qld. It is an attractive tree with shiny foliage, and is fairly decorative in flower. Although little-known in cultivation, the species is of interest to enthusiasts. Requirements are probably as for *D. ixoroides*. Propagate from seed which must be sown fresh.

Diplospora ixoroides F. Muell.
(resembling the genus *Ixora*)
Qld

3-5 m x 1-3 m   Oct

Tall, slender **shrub**; **stems** forking, glabrous; **leaves** 2.5-4 cm x 0.8-2 cm, ovate to oblong, thick and leathery, dark green and glossy, narrowed to a blunt apex; **racemes** 2-3 cm long, in terminal clusters, each one with 3-5 flowers; **flowers** about 0.3 cm long, greenish white, sparsely hairy; **berry** about 0.5 cm across, ovoid, black, bearing 3-4 shiny brown seeds.

Widely distributed in rainforests between Cairns and Rockhampton. An attractive shrub which should find a place in enthusiasts' collections, even if only for its leathery, shiny leaves. Plants require well drained soils, and respond to mulching and watering during dry periods. They prefer a shady situation. As well as being garden subjects they could have potential for indoor decoration. Propagate from seed which must be sown fresh.

DIPODIUM R. Br.
(from the Greek *dis*, double; *pous*, a foot; in reference to the paired stipes of the pollinia)
Orchidaceae

Terrestrial **orchids**, either leafless saprophytes or with leaves, or climbing, leafy epiphytes with stems of indeterminate length; **roots** of the terrestrial species are thick and fleshy, those of the epiphytes wiry; **leaves** if present distichous, linear, channelled, acuminate, abscissing near the base; **leaf-sheath** adherent; **inflorescence** a raceme, in the leafless species stout, erect, bearing sheathing bracts; **flowers** large, showy, colourful, opening widely; **labellum** with a hairy patch; **fruit** a capsule.

A genus of 22 species of orchids distributed from Malaysia through various islands to Australia where there are 5 species. Some species are leafless saprophytes that are not amenable to cultivation. The epiphytic species can be cultivated under similar conditions to those for other epiphytic orchids, while the leafy terrestrial species can be grown in pots of a well drained soil mixture. Propagate from seed or, in the case of the epiphytes, by careful division of the stems. Seed must be sown under sterile conditions.

Dipodium ensifolium F. Muell.
(sword-shaped leaves)
Qld

0.1-1 m tall   Oct-Jan

Terrestrial **orchid** forming sparse clumps; **stems** 1-4, thin, tough, erect or straggling, leafy throughout; **leaves** 5-20 cm x 1-1.5 cm, linear, channelled, erect or curved, bright green, the apex acuminate, sheathing at the base; **raceme** 15-55 cm long, erect, axillary, bearing 2-20 flowers on pedicels 1-2 cm long; **flowers** 2.5-4 cm across, pink to mauve with deep purple spots and blotches, opening widely.

Endemic in north-eastern Qld, where it is found in the coastal lowlands and adjacent highland regions between the Herbert and Endeavour Rivers. It occurs up to 1200 m alt. on the Atherton Tableland, in open forest in gravelly, clayey soils, while in lowland situations it grows in open forest or grassland, often in sandy soils. It commonly grows in areas subject to bushfires, and the growth above-ground is destroyed every few years, although one occasionally sees clumps of lanky, untouched growths. The flowers are colourful and quite attractive. The species grows readily in a pot but is somewhat difficult to flower, especially in the southern states. It can be grown in well drained, loamy soil but a coarse, open mix is safer since the roots rot readily if the drainage is impeded. In tropical areas this orchid can be grown in a bushhouse, while in cold temperate regions a heated glasshouse is necessary. Watering should be aimed at keeping the compost moist but not overwet, and the plants should be placed where

Dipodium ensifolium

B. Gray

they receive plenty of light. Liquid fertilizers are of benefit during the growing-season. Large clumps can be divided with care but this is not recommended. Most propagation is from seed.

Dipodium hamiltonianum Bailey
(after James Hamilton)
Qld, NSW, Vic
30-80 cm   Nov-Feb

Leafless, saprophytic terrestrial **orchid**; **inflorescence** 30-80 cm tall, slender, usually yellowish green, stem-bracts blunt, bearing 3-25 flowers in a sparse raceme in the upper one-third; **pedicels** 1.5-2.8 cm long; **flowers** 3-6 cm across, greenish yellow with numerous red spots, lax or semi-pendent, the perianth segments not curling.

Widely distributed and often growing with *D. punctatum*, but much rarer than that species. Flowers are of a very attractive colouration. Being a leafless saprophyte, the results in cultivation are identical to those outlined under *D. punctatum*.

Dipodium pandanum Bailey
(with growth resembling a *Pandanus*)
Qld
climber   Aug-Oct

Terrestrial or climbing **orchid**; **stems** wiry, rarely branched, usually twisting spirally; **roots** developing from the nodes; **leaves** 20-50 cm x 4-6 cm, linear-acuminate, distichous, spreading or recurved, stiff, bright green, sheathing at the base; **raceme** 20-60 cm long, usually bending at right angles to the stem, axillary, 1-5 per plant, bearing 5-25 flowers on pedicels 1-2 cm long; **flowers** 3-4 cm across, creamy white with red spots and blotches, opening widely.

This species is found in New Guinea and north-eastern Qld, where it is restricted to the Iron and McIlwraith Ranges of Cape York Peninsula. It is a most unusual orchid which climbs trees in a spiral manner, and occasionally the stems may reach 8 m long. They are very brittle, however, and the fallen pieces grow over the ground until they contact an object on which to climb. As the stems grow forward, the leaves at the base die but the stems remain alive for long periods and are covered by persistent brown leaf-bases. The species can be grown readily but may be difficult to flower, especially in temperate regions. Plants are very sensitive to cold, and in southern areas must be grown in a heated glasshouse with a minimum temperature of 12°C. They can be grown on a large slab of tree-fern fibre, and are best left undisturbed. Small pieces can be started by inserting the base in a pot of coarse mixture and placing the upper part of the stem in contact with a tree-fern totem. For flowering, the plants must receive plenty of light. Propagate from seed or sections of stems.

297

Dipodium punctatum (Smith) R. Br.
(spotted)
Qld, NSW, Vic, Tas, SA, NT   Hyacinth Orchid
30-90 cm tall   Nov-Feb

Leafless, saprophytic terrestrial **orchid**; **roots** thick and fleshy, much-branched; **inflorescence** 30-90 cm tall, sturdy, usually deep purplish red, stem-bracts acute, bearing 6-50 flowers in a crowded raceme in the upper quarter; **pedicels** about 1.5 cm long; **flowers** 2.5-4 cm across, pale to deep pink, mauve or deep reddish, usually with large spots and blotches, and perianth segments frequently curling back at the tips.

A widely distributed and often common orchid which usually grows in open forest. It is recorded that the Aborigines collected and ate the fleshy roots. Being a leafless saprophyte, it cannot be maintained in cultivation by any known method. Many plants are dug up each year and transferred to gardens and, while some people claim success, the plants usually die after perhaps flowering for a year or two, as their accumulated reserves are exhausted. Propagation from seed appears to be impossible. Some people have claimed success after sowing seed on heaps of sawdust or bush litter, in close proximity to established eucalypts.

Dipodium stenocheilum O. Schwarz
(a narrow labellum)
Qld, NT
30-90 cm tall   Nov-Jan

Leafless, saprophytic terrestrial **orchid**; **inflorescence** 30-90 cm tall, stout, usually pale green, stem-bracts stout, bearing 3-30 flowers in a sparse, elongated raceme in the upper one-third; **pedicels** 2-3 cm long; **flowers** 5-8 cm across, white with a few spots, lax or semi-pendent, the perianth segments curling.

A little-known species from open forests of NT, and possibly other tropical regions. It is a handsome species but, being a leafless saprophyte, it cannot be cultivated.

DIPTERACANTHUS Nees emend Bremek.
(from the Greek *dis*, double; *pteron*, a wing; differing from the genus *Acanthus* by the capsule being flattened at the base)
Acanthaceae

Perennial **herbs** or rarely **shrubs**; **branches** erect, prostrate or decumbent, hairy, often scabrous; **leaves** simple, petiolate, in opposite pairs, sometimes fleshy, hairy; **flowers** tubular, the upper part divided into 5 spreading lobes; **calyx** 5-lobed, hairy; **stamens** 4; **fruit** a capsule.

A genus of about 10 species, 3 found in Australia, mainly in northern and inland regions. They are small hairy perennials with useful features for cultivation. Propagate from seed which should be sown shallowly or stem cuttings which strike very easily.

Dipteracanthus bracteata R. Br.
(prominent bracts)
Qld, NT
0.1-0.3 m x 0.1-0.3 m   sporadic all year

Dwarf, perennial **herb**; **stems** erect, slender, undivided, hairy; **leaves** 1-5 cm x 0.3-0.7 cm, obovate to oblong, hairy, somewhat thickened, in opposite pairs; **flowers** 1-3 cm long, lilac to blue, tubular, prominent bracts enclosing the sepals; **corolla lobes** with crisped margins; **capsules** about 1.5 cm long, flattened near the base.

Widespread on heavy soils, this perennial is rarely encountered in cultivation. It appears to have potential for tropical, subtropical and inland regions, and could be used in rockeries or as a small garden plant. It would probably flower almost continuously with regular watering. Propagate from seed or cuttings.

Dipteracanthus corynothecus (F. Muell. ex Benth.) Bremek.
(club-shaped capsule)
Qld, NSW, NT
0.1-0.15 m x 0.1-0.2 m   sporadic all year

Dwarf, perennial **herb**; **stems** decumbent or erect, roughened, hairy; **leaves** 0.5-2.5 cm x 0.3-0.5 cm, ovate, thickened, scabrous, hairy, in opposite pairs on the stems; **flowers** 1-1.5 cm long, tubular, lilac or white, delicate, with 5 rounded lobes, solitary in the upper axils; **capsule** about 1.2 cm long, club-shaped.

An attractive hairy perennial herb, widespread in heavy soils and frequently common in brigalow country. It is rarely grown but could prove to be a useful small plant for subtropical and inland areas. Very hardy and will flower for long periods, especially if watered. Needs a sunny position in freely draining soil and could be a good rockery plant. Propagate from seed or cuttings which strike easily. The var. *grandiflorus* Bremek. represents a large-flowered form from NT.

Dipteridaceae Seward & Dale

A small family of ferns consisting of about 8 species in the genus *Dipteris*. They are terrestrial or lithophytic ferns with large flabellate fronds which are divided into equal halves. The young parts are clothed with bristle-like scales. A solitary widespread species extends to north-eastern Qld.

Dipteracanthus corynothecus, × ·75

Dipteris conjugata   B. Gray

DIPTERIS Reinw.

(from the Greek *di*, two; *pteron*, a wing; in reference to the fronds divided into halves resembling wings)
Dipteridaceae

Terrestrial **ferns**; **rootstock** creeping, bearing stiff bristles; **fronds** erect, terminating a long stalk, fan-like, consisting of spreading halves, each irregularly and deeply lobed; **sori** very small, lacking indusia.

A small genus of about 6 species with a solitary widespread species extending to north-eastern Qld. Although they are very decorative ferns of distinctive appearance, they are extremely difficult to cultivate. Propagation from spore may be worth trying.

Dipteris conjugata Reinw.

(paired; as in the halves of the frond)
Qld
0.5-2 m tall

Terrestrial **fern** forming erect clumps; **rootstock** creeping, stout, covered with stiff bristles; **stalks** erect, woody, up to 2 m tall; **fronds** terminal, divided into fan-shaped, spreading halves, each deeply divided into lobes; **sori** small, numerous over the lower surface.

A rare fern found in coastal rainforests of north-eastern Qld, usually in exposed patches of moist clay. It is extremely attractive, but unfortunately has often proved to be difficult to grow. Transplanted specimens, whether large plants or sporelings, often linger for a while before dying, although they have grown strongly in some instances. Those removed with a large clump of soil have a better chance of survival. Propagation from spore may hold some of the answers.

DISCARIA Hook.

(from the Greek *discos*, a disc; in reference to the prominent disc on the flowers)
Rhamnaceae

Rigid, spiny, much-branched **shrubs**; **branchlets** thorny, opposite; **leaves** simple, entire, opposite, often absent; **flowers** bisexual, small, in axillary clusters or racemes; **calyx** 4-5 lobed, bell-shaped or tubular; **petals** cucullate, small or absent; **stamens** 4 or 5; **ovary** subtended by a large fleshy disc; **fruit** a capsule.

A small genus of about 10 species with a solitary one in Australia. They have very little horticultural appeal. Propagate from seed which is small.

Discaria pubescens (Brongn.) Druce

(slightly hairy; downy)
Qld, NSW, Vic, Tas   Australian Anchor Plant
0.3-1 m x 0.5-1 m   Sept-Nov

Dwarf, prickly, densely branched **shrub**; young stems hairy; **stems** erect, with axillary spreading spines 1-2.5 cm long; **leaves** 1-1.5 cm x 0.2-0.4 cm, oblong or cuneate, blunt, sparse, often only on young shoots; **flowers** about 0.5 cm across, cream or white, numerous, in dense, compact axillary racemes; **sepals** prominent; **petals** minute; **capsule** about 0.5 cm across, 3-lobed, the disc-like base persistent.

A prickly shrub, scattered along stream banks and stony ridges in dry open forest. It lacks wide horticultural appeal but is an interesting subject when in flower, and makes a useful small refuge plant for birds, or a traffic barrier. It requires well drained soil in a hot, sunny position. Propagate from seed or possibly cuttings.

DISCHIDIA R. Br.

(from the Greek *dis*, two; *chidon*, a crown; in reference to the divided corona segments)
Asclepiadaceae

Epiphytic or lithophytic **herbs**; **stems** slender, sometimes twining, rooting at the nodes; **leaves** simple, alternate or in opposite pairs, fleshy, sometimes specialized as pitchers; **inflorescence** a compact, axillary umbel; **flowers** small, tubular to urn-shaped; **corolla lobes** 5, spreading; **corona** of 5 segments, bifid at the tip; **anthers** containing 2 pollinia each; **fruit** a follicle; **seeds** with a tuft of hairs.

An interesting genus of about 80 species with 3 found in north-eastern Qld. A fourth species, *D. timorensis* Decne., has been erroneously recorded. They are specialized plants, similar in many respects to the genus *Hoya*. They have a very intricate pollination mechanism, and the flowers are visited by few species of insect. They are interesting plants for bushhouse or glasshouse culture but are grown mainly by enthusiasts. They are best grown in a coarse mixture of materials such as charcoal, weathered hardwood, pinebark, leaf mould and moss, but can also be mounted on slabs of fibrous material such as tree-fern fibre or elkhorn peat. All species are very cold sensitive and are best suited to warm regions. Propagation is readily carried out by division of the clumps. Small sections of stem should be laid on sphagnum moss and held in warm, humid conditions until established. Seed is rarely collected and must be sown soon after collection as it rapidly loses viability.

Dischidia major (Vahl) Merr.
(large)
Qld   Rattle Skulls
                                        Aug-Nov
Epiphyte forming spreading clumps with climbing strands; **stems** fairly thick, becoming glabrous, twining; **roots** adventitious, arising from the nodes, growing into the pitchers; **leaves** of 2 types, flat and pitcher; **flat leaves** 1-3.5 cm x 1-2 cm, orbicular, flat, pale green, borne 10-15 cm apart on twining stems; **pitchers** 6-12 cm x 2-5 cm, erect or pendulous, gibbous at the base, waxy, pale yellow-green, hollow, purplish red inside, produced in clusters 1-3 cm apart on non-twining stems; **umbels** axillary; **flowers** 0.6-0.8 cm x 0.3-0.4 cm, pale yellow or striped with green, tubular; **follicle** 3-5 cm x 0.5 cm, linear, curved, yellow-green, pendulous.

A widespread and conspicuous plant which extends to lowland areas of north-eastern Qld, where it grows on fibrous-barked trees in open but humid situations such as along stream banks or in swamps. The conspicuous pitchers are an ornamental feature and function as a water reservoir for the plant. They are usually well stocked with the plant's roots and are also inhabited by ants. The plant can be readily grown in tropical regions, but in southern Australia it is difficult to cultivate and requires warm, humid conditions with a minimum winter temperature of 15°C. It can be grown in a pot or basket of coarse mixture topped with moss, or on a slab of tree-fern or elkhorn fibre. Propagate from seed which is rarely collected or by division of the clumps. Twining stems with flat leaves are most successful as cuttings, striking readily if held in warm, moist conditions. Clusters of pitchers are best left undivided. The species was previously well known in Australia as *D. rafflesiana* Wall.

Dischidia nummularia R. Br.
(coin-shaped)
Qld   Button Orchid
                                Oct-Dec; also sporadic
Epiphyte forming spreading clumps with pendulous or climbing strands; **stems** slender, glabrous; **roots** adventitious, arising from the nodes; **leaves** 0.5-1.5 cm x 0.8-1.2 cm, rounded or ovate, flattish, in pairs, thick and fleshy, mealy grey to white; **umbel** simple, compact, of 1-5 flowers; **flowers** 0.2-0.3 cm x 0.2-0.3 cm, white, tubular, with reflexed lobes, a ring of sparse hairs inside; **follicles** 2-2.5 cm x 0.3-0.5 cm, linear, pale green or yellow, pendulous, the apex curved.

A widespread and common plant which extends to lowland areas of north-eastern Qld, where it grows on rough-barked trees, in open but humid situations such as in swamps or along stream banks. The strings of mealy, round leaves give rise to the common name; however, the species is not an orchid. It grows readily in tropical regions and is a popular subject for glasshouse or bushhouse culture. It can also be established on suitable garden trees in warm tropical areas. In cooler regions it requires the protection of a heated glasshouse over winter. Plants can be mounted on slabs of tree-fern or elkhorn peat, or grown in a container of coarse epiphyte orchid mixture. It makes an attractive subject for a hanging basket. The species resents

shade, and should be grown in a well lit situation. Once established, the clumps are tolerant of dryness, and very easy to maintain. Propagate from seed or by division of the clumps. Small stem-sections can be induced to grow by placing on fresh, growing sphagnum moss.

Dischidia ovata Benth.
(ovate)
Qld
                                        Aug-Nov
Epiphyte forming spreading clumps with pendulous or climbing strands; **stems** slender, glabrous; **roots** adventitious, arising from the nodes; **leaves** 2-3 cm x 1-1.8 cm, broadly ovate, thick and fleshy, green or reddish with conspicuous white lines giving a variegated appearance, on slender petioles about 0.6 cm long; **umbels** axillary, bearing 3-8 flowers; **flowers** about 0.6 cm long, tubular, inflated at the base, yellowish, stained with red, a dense ring of silky white hairs in the throat; **follicle** 2.5-4 cm x 0.4 cm, linear, terete, yellowish.

An attractive epiphyte which is apparently endemic in north-eastern Australia. It has a neat appearance and the variegated leaves are most decorative. In shady situations, they are green with conspicuous silvery white variegations, whereas when exposed to the sun

Dischidia nummularia   D.L. Jones

Dischidia major growing in a hanging basket   D.L. Jones

Diselma archeri Hook.f.
(after William H. Archer, 19th-century Tas botanist)
Tas   Cheshunt Pine
2-6 m x 1-4 m

Dioecious, small to tall **shrub**; **branches** initially erect, spreading with maturity; **leaves** very small, scale-like, opposite, decussate, appressed, imbricate, keeled; male and female cones on separate plants; **male cones** small and solitary, terminal, with crimson pollen sacs; **female cones** consist of 2 pairs of scales, with only upper pair fertile; **seeds** small, winged, 2 per cone.

A conifer that has been in cultivation for many years, grown mainly by enthusiasts. It occurs naturally at altitudes of about 1000-1400 m, in moist soils. It is relatively slow-growing, and is best suited to cool temperate regions. Does best in moist, well drained soils, in a cool location. Withstands frost and snow. Appreciates summer watering. Grows well as a container plant. Is cultivated overseas.

Propagation can be from seed which is dispersed from the cones as soon as it is ripe. Plants are also propagated readily from cuttings, using firm but vigorous tip growth. Roots usually form within 3-5 months.

DISPHYMA N. E. Br.
(from the Greek *dis*, double; *phyma*, a tumour or tubercle)
Aizoaceae

Creeping perennial **herbs**; **branches** often producing adventitious roots at nodes; **leaves** succulent, opposite; **flowers** solitary, on long, erect pedicel; **styles** 5; **fruit** a capsule with 5 valves.

A genus of 3 species, 1 endemic to Australia and the other 2 restricted to New Zealand.

Disphyma australe (Aiton) J. Black =
                    D. crassifolium (L.) L. Bolus

Disphyma blackii Chinn. =
                    D. crassifolium (L.) L. Bolus

Disphyma clavellatum (Haw.) Chinn. =
                    D. crassifolium (L.) L. Bolus

they assume deep reddish tonings and the variegation is less noticeable. The species grows on trees or rocks, in humid lowland situations. It is a prized decorative plant for bushhouse or glasshouse culture, and is very easy to grow. It is extremely cold sensitive and best suited to tropical and coastal subtropical regions, although it is readily grown in a heated glasshouse in temperate Australia. The plants prefer humid conditions and can be grown in a coarse mixture topped with moss, or on slabs of tree-fern fibre or elkhorn peat. The reddish colour can be maintained in the leaves if the plants are grown in a well lit situation. Propagate from seed which is rarely available or by division of the clumps.

Dischidia rafflesiana Wall. = *D. major* (Vahl) Merr.

DISELMA Hook.f.
(from the Greek *di*, two; *selma*, upper deck; alluding to the 2 fertile scales)
Cupressaceae

An endemic, monotypic genus restricted to Tas.

Diselma archeri, × 2·4

301

Disphyma crassifolium

Disphyma crassifolium (L.) L. Bolus
(thick leaves)
Qld, NSW, Vic, Tas,
SA, WA   Rounded Noon-flower
prostrate x 1-2 m   Oct-Feb; also sporadic

Spreading perennial **herb**; **branches** rooting at nodes; **leaves** 0.5-5 cm x 0.4-1 cm, almost cylindrical, opposite, acute or rounded apex, usually glabrous, varying shades of green through to purple and red; **flowers** daisy-like, 2-5 cm across, pink to magenta with white at base; **pedicel** to 10 cm long, erect; **capsule** to 1.2 cm diameter, 5-angled.

A common plant of the seaside, on rocky cliffs or in saline soils. It also inhabits saline areas up to 500 km inland. Most reliable in cultivation, provided it receives plenty of sun, although it will tolerate dappled shade. Does well in most soils. Frost and drought tolerant. Useful for embankments. Especially suited to saline soils. Propagate from stem or leaf cuttings which root very readily. It can also be grown from seed which is expelled when ripe.

Disphyma crassifolium   A. Gray

DISSILIARIA F. Muell. ex Benth.
(from the Latin *dissilio*, to fly apart; in reference to the splitting capsules)
Euphorbiaceae

Large **shrubs** to large **trees**; **stems** with prominent, corky pustules; **leaves** simple, entire or with toothed margins, glabrous, petiolate, opposite or in 3s; **inflorescence** a cyme, borne near the ends of branches; **flowers** small, unisexual, each of 6 perianth lobes arranged in 2 series; **fruit** a capsule.

A small genus of 3 species, all endemic in Australia. They are rainforest trees and are rarely seen in cultivation. Propagate from seed which has a short viability and must be sown soon after collection.

Dissiliaria baloghioides F. Muell. ex Benth.
(resembling the genus *Baloghia*)
Qld   Lancewood
15-25 m x 3-10 m   March

Medium to tall **tree**; **bark** dark brown, flaky, with pustules; **leaves** 4-10 cm x 1.5-3 cm, ovate, glossy green above, paler beneath, narrowed to a blunt point, often in 3s; **cymes** terminal, or near the ends of branches; **flowers** about 0.5 cm across, unisexual, in separate clusters; **capsule** about 2 cm across, globular, covered with soft, downy hairs, splitting when ripe.

Restricted to the rainforests of Qld, but there widely distributed from Cape York Peninsula to near Brisbane. It is cut as a timber tree, the wood hard, close-grained and durable in the ground. The Aborigines used the wood for a variety of purposes including spear-shafts and nullahs. Young plants are attractive, with glossy leaves and could have application for indoor decoration. Trees generally grow too large for home garden use but could make a very attractive addition to parks and municipal gardens. They are slow-growing in their early stages and need protection from over-exposure to hot sun. Once established, they are quite hardy. Freely draining soils are essential. Propagate from seed which must be sown fresh.

Dissiliaria laxinervis Airy-Shaw
(lax venation)
Qld
15-25 m x 5-10 m   Jan

Small to medium **tree**; **branchlets** with small elliptical pustules; **leaves** 10-18 cm x 5-7 cm, broadly elliptical, leathery, dark green and glossy, the apex drawn out and blunt, the margins somewhat undulate; **flowers** unknown; **capsule** about 1.5 cm across, yellow-green, with rusty hairs, grouped 1-4 in the leaf axils, mature June.

A relatively unknown species, restricted to rainforests of the Cook District in north-eastern Qld. It is a handsome tree that could be suitable for parks in tropical and perhaps subtropical areas. Propagate from seed which must be sown fresh.

Dissiliaria muelleri Baillon ex Benth.
(after F. von Mueller, first Government Botanist, Vic)
Qld
5-12 m x 3-6 m   May-Aug

Tall **shrub** or small **tree**; **branchlets** with small corky pustules; **leaves** 2-6 cm x 1-3 cm, ovate, dark green, the apex blunt, the margins irregularly toothed; **cymes** terminal or near the ends of branches; **flowers** about 0.5 cm across, unisexual, in separate clusters; **capsule** about 1.2 cm across, globular, furrowed, corky, splitting when ripe.

This species occurs in dry rainforests around Rockhampton, and extends south to near Gympie. It is a fairly nondescript plant with limited horticultural appeal. Requires well drained soil and is quite hardy once established. Young plants are slow-growing and need protection from direct sun. Propagate from seed which must be sown fresh.

DISSOCARPUS F. Muell.
(from the Greek *dissos*, double; in pairs; *carpos*, fruit; in reference to the arrangement of the fruit)

Chenopodiaceae

Small, densely hairy **shrubs**; **leaves** narrow, alternate, usually crowded; **flowers** bisexual, small, sessile, axillary, 2-15 in globular clusters; **perianth** with 5 small lobes; **stamens** 5; **styles** 2-3; **fruit** a nut, enclosed in the perianth which becomes tubular and develops very short spines or tubercles.

An endemic genus of 2 species inhabiting the semi-arid to arid regions of the continent. Recent botanical revision has reinstated the generic name. Previously the species were included under *Bassia* All.

Dissocarpus biflorus (R. Br.) F. Muell.
(2-flowered)
Qld, NSW, Vic, SA, NT
0.5-1 m x 0.5-1.5 m   most of year
Dwarf **shrub**; **branches** with dense grey hairs; **leaves** 0.5-1 cm long, narrow-linear, soft, hairy, grey; **flowers** about 0.2 cm long, in axillary pairs, united at the base; **fruit** cylindrical, spineless or 1-2 minute spines, woolly.

This widespread species is highly suited to arid and semi-arid climates. It is spineless or bears only minute spines, and has wider application in cultivation than *D. paradoxus* or species of the closely related genera of *Eriochiton* and *Sclerolaena*. It also has decorative foliage. Will adapt to most dry soils, with full sun, but also tolerates partial sun in hot climates. Propagate from seed or cuttings.

There are the following varieties:

var. *cephalocarpus* (F. Muell.) A. J. Scott — which can have 3-8 fruits in a head;

var. *villosus* (Ising) A. J. Scott — from SA, which has long hairs on the fruits.

Previously known as *Bassia biflora* (R. Br.) F. Muell.

Dissocarpus paradoxus (R. Br.) F. Muell.
(unusual; strange)
Qld, NSW, Vic, SA, WA, NT
0.3-0.5 m x 0.5-1.5 m   most of year
Dwarf, spreading **shrub**; **branches** woolly, greyish; **leaves** 0.5-1.5 cm x up to 0.3 cm, flat, soft, densely hairy, greyish; **flowers** small, 6-16, united in dense axillary cluster; **fruit** globular, about 1 cm diameter, covered in dense, woolly hairs, 2-5 spines for each flower-head, just protruding through the wool.

Cultivation requirements are similar to those for *D. biflorus*. Propagate from seed or cuttings.

The var. *latifolius* (J. Black) Ulbr. from Qld, NSW and SA differs in having broader (0.5-0.8 cm), slightly wedge-shaped leaves.

Previously known as *Bassia paradoxa* (R. Br.) F. Muell.

DISTICHLIS Raf.
(from the Greek *distichos*, in 2 rows; in reference to the leaves)
Poaceae

Perennial **grasses** with long-creeping rhizomes; **leaves** in opposite rows, rigid, pointed; **inflorescence** a raceme or spike; **spikelets** few, compressed, dioecious.

A genus of 12 species of grasses, one of which is found on saline soils of south-eastern Australia. It has value in the stabilization and reclamation of salted soils. Propagate by transplants.

Distichlis distichophylla (Labill.) Fassett
(leaves in 2 rows)
Vic, Tas, SA   Australian Salt Grass
10-30 cm tall   Oct-Dec; also sporadic
Perennial **grass** forming spreading, stoloniferous clumps; **stems** creeping or ascending, rigid; **leaves** 2.5-5 cm long, thin, rigid, pointed, in two opposite rows, crowded, glabrous; **raceme** 1-1.5 cm long, terminal, bearing 2-5 spikelets; **spikelets** flattened, straw coloured, unisexual.

A coarse, prickly grass that forms thick clumps on saline marshes and saltpans in coastal and inland districts. It has no commercial appeal whatsoever but is an extremely important grass for stabilizing salted soils, and may be used by soil conservation authorities. It transplants readily in clumps and may also be established by sprigging small pieces.

DISTICHOSTEMON F. Muell.
(from the Greek *distichos*, in 2 rows; *stemon*, stamen; referring to the 2 series of stamens)
Sapindaceae

Shrubs; young shoots woolly; **leaves** simple, alternate, entire; **inflorescence** an interrupted spike or panicle; **flowers** small, lacking petals; **sepals** 5-8; **stamens** more than 20, in 2 or more ranks; **fruit** a winged capsule.

A small genus of 4-6 species, endemic in northern Australia. They are closely related to *Dodonaea* but with a different inflorescence and floral arrangement. Propagate from seed which has limited viability or from cuttings.

Distichostemon filamentosus S. Moore
(thread-like)
Qld, WA, NT
0.3-1.5 m x 0.3-1 m   April-May
Small **shrub**; young shoots softly hairy; **leaves** 2.5-7 cm x 1-2.5 cm, oblong to obovate, thin-textured, olive green to greyish, the apex acute, the margins coarsely toothed and hairy; **flowers** small, greenish, in terminal clusters or on racemes 2.5-7 cm long; **sepals** 5

Distichlis distichophylla   T.L. Blake

303

Distichostemon hispidulus

or 6; **stamens** numerous; **capsule** shortly hairy, green with purple wings.

A little-known species which grows in open forest and rocky areas of northern Australia. It is a hardy, fairly attractive shrub which could have potential for cultivation in drier inland zones. Requirements are a well drained soil in a sunny situation. Propagate from seed or cuttings.

Distichostemon hispidulus (Endl.) Baillon
(slightly covered with coarse, erect hairs)
Qld, WA, NT
3-6 m x 1-2.5 m   Nov-Feb
Medium to tall **shrub**; young shoots softly hairy; **leaves** 2.5-9 cm x 1-3 cm, oblong, blunt, greyish green, with soft, velvety hairs on both surfaces, the margins entire or sometimes toothed; **racemes** 2.5-8 cm long, in the upper leaf axils, sometimes paniculate; **flowers** small, greenish, with numerous crowded stamens; **capsule** about 1 cm long, 3-angled, winged, softly hairy.

A shrub found in inland and near-coastal communities of northern Australia and adjacent islands. It extends to exposed situations near the sea and can also grow in peaty soils subject to periodic inundation. Features useful for cultivation include attractive woolly leaves and the ability to grow in inland and coastal regions of the tropics. The species could be useful for screening purposes and perhaps also as a pot plant. Propagate from seed and possibly cuttings. The species has also been known as *D. phyllopterus* F. Muell.

Distichostemon phyllopterus F. Muell. =
D. hispidulus (Endl.) Baillon

DIURIS Smith
(from the Greek *dis*, double; *oura*, a tail; in reference to the paired lateral sepals)
Orchidaceae

Terrestrial **orchids**; **tubers** ovoid or elongated; **leaves** basal, linear, channelled, flat or erect and tussock-like, straight or twisted; **inflorescence** a loose raceme; **flowers** colourful, opening widely; **lateral petals** erect and ear-like; **lateral sepals** elongated and pendent; **labellum** 3-lobed; **fruit** a capsule.

A genus of about 25 species of terrestrial orchids, most of which are endemic to Australia, but 1 is found in Timor. The exact number of species in the genus is uncertain because many species hybridize freely, resulting in a bewildering range of forms, a number of which have been named as species. As a group, these orchids are frequently called 'doubletails', in reference to the prominent, paired lateral sepals, but many other vernacular names have also been applied. *Diuris* species usually grow in colonies, and occur in a variety of habitats. They are a familiar and colourful group of terrestrial orchids, and it is natural that they should become popular in cultivation. They are generally easy subjects to grow in pots, in bushhouses or cool glasshouses. A suitable soil mix must be very freely draining and should contain about one third of good leaf mould or hardwood shavings. The plants need plenty of water while they are in active growth, and when dormant over the summer they should be kept dry. Most species

increase their tuber number by vegetative reproduction and need to be repotted every 2-3 years. Repotting is carried out over the summer while the plants are dormant. The tubers are placed 3-4 cm deep in the new potting mix and kept dry until the new shoot appears above-ground. While in active growth, the orchids respond to applications of weak animal manures or liquid fertilizers. Flower-spikes of some tall species may become top-heavy and should be staked to avoid damage to the tubers. Plants may be severely attacked by slugs and snails, and should be protected by baits at all times. *Diuris* species can also be established in natural garden situations such as under eucalypts that occur naturally in the area. These introductions will only be successful in the absence of slugs and snails. Propagation of diuris is from seed which must be sown aseptically in tubes or flasks of nutrient media. Some species have been successfully propagated by sprinkling seed around the base of parent plants. Germination of many species of *Diuris* has been successfully carried out by research workers at the National Botanic Gardens, Canberra.

Diuris abbreviata F. Muell. ex Benth.
(abbreviated; shortened)
Qld, NSW
15-45 cm tall   Aug-Nov
Terrestrial **orchid**; **leaves** 12-20 cm x 0.2-0.4 cm, 1-3 per plant, linear, channelled; **flower-stem** 15-45 cm tall, bearing 3-8 flowers in a loose raceme; **flowers** about 2 cm across, yellow to orange, blotched with brown; **labellum** with small lateral lobes, the mid-lobe much longer than wide.

This species is found in north-eastern and south-eastern Qld, and north-eastern NSW as far south as Newcastle. It grows in open forest and also coastal heathlands, usually in clumps or small colonies. It is an easy and rewarding species to grow, and flowers well in cultivation. Vegetative increase is slow. Propagate from seed or by natural increase.

Diuris aequalis F. Muell. ex Fitzg.
(uniform in size)
NSW
10-30 cm tall   Sept-Nov
Terrestrial **orchid**; **leaves** 10-20 cm x 0.2-0.4 cm, 2 per plant, linear, channelled; **flower-stem** 10-30 cm tall, bearing 2-4 flowers; **flowers** about 2 cm across, orange; **labellum** with lateral lobes larger than the mid-lobe.

An imperfectly understood species from central and south-eastern NSW. Behaviour in cultivation is largely unknown, but probably similar to other species. Propagate from seed or by natural increase.

Diuris althoferi Rupp
(after G. W. Althofer, original collector)
NSW
15-35 cm tall   Sept-Oct
Terrestrial **orchid**; **leaves** 15-20 cm x 0.2-0.3 cm, 2-3 per plant, linear, channelled; **flower-stem** 15-35 cm tall, bearing 3-7 flowers in a loose raceme; **flowers** 2-3 cm across, clear lemon yellow with dark blotches;

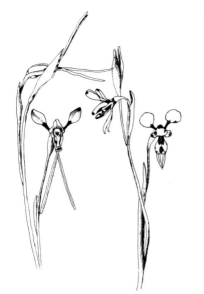

left *Diuris sheaffiana*, × ·6   right *Diuris laxiflora*, × ·6

labellum with lateral lobes about half as long as the mid-lobe.

An imperfectly understood species known only from the Mudgee district of NSW. It may be a form of *D. platichila* or a hybrid from that species. It grows and flowers readily in cultivation. Propagate from seed or by natural increase.

Diuris aurea Smith
(golden yellow)
Qld, NSW   Golden Donkey Orchid
15-50 cm tall   April-Nov
    Terrestrial **orchid**; **leaves** 10-25 cm x 0.3-0.6 cm, 1-2 per plant, linear, channelled; **flower-stem** 15-50 cm tall, bearing 2-5 flowers in a loose raceme; **flowers** 2.5-4.5 cm across, deep golden yellow to orange, sometimes with a few brown markings; **labellum** large and prominent, the lateral lobes large, the mid-lobe very broad.
    A widely distributed and often locally common species that extends from north-eastern Qld to near Sydney. It usually grows in open forest, often in colonies. In cultivation it is a very easy species to grow and is rewarding because it flowers freely. The flowers are large and colourful, and a potful is most attractive. Tall spikes may need staking as they tend to become top-heavy. Introduced to England during 1810. Vegetative increase is slow. Tubers can become quite large. Propagate from seed or by natural increase.

Diuris brevifolia R. Rogers
(short leaves)
SA
15-40 cm tall   Nov-Dec
    Terrestrial **orchid**; **leaves** 7-12 cm x 0.2-0.4 cm, 4-8 per plant, erect, linear; **flower-stem** 15-40 cm tall, bearing 1-5 flowers in a loose raceme; **flowers** 2.5-4 cm across, bright yellow with a prominent dark blotch on

either side of the dorsal sepal; **labellum** with very short lateral lobes and a long, pointed mid-lobe.
    Restricted to SA, where it is distributed in southern areas, often in sandy soils near the coast. It grows readily in cultivation but tends to be shy of flowering. Propagate from seed or by natural increase. Some authorities include this species as a variety of *D. sulphurea* R. Br. [var. *brevifolia* (R. Rogers) J. Weber & R. Bates].

Diuris brevissima Fitzg. ex Nicholls =
                              D. maculata Smith

Diuris carinata Lindley = *D. laxiflora* Lindley

Diuris colemaniae Rupp = *D. tricolor* Fitzg.

Diuris emarginata R. Br.
(notched at the tip)
WA   Tall Donkey Orchid
20-40 cm tall   Sept-Dec
    Terrestrial **orchid**; **leaves** 15-30 cm x 0.2-0.4 cm, 2-3 per plant, linear, channelled, the tips flexuose or spiralling; **flower-stem** 20-40 cm tall, stout, bearing 3-4 long stem-bracts, 2-7 flowers in a loose raceme; **flowers** 2.5-3.5 cm across, yellow with brown markings.
    Widely distributed in the coastal areas of the south-west, this species grows in moist, sandy soils, usually in scattered colonies. It is a very easy species to grow in cultivation but can be difficult to flower. It increases vegetatively, and the clumps should be repotted every couple of years.
    The var. *pauciflora* (R. Br.) A. S. George is a slender variety with fewer and smaller flowers. It is co-extensive with the typical variety and behaves similarly in cultivation. Both forms increase naturally and can also be propagated from seed.

Diuris laevis Fitzg.
(smooth)
WA   Nanny-goat Orchid
15-35 cm tall   Sept-Oct
    Terrestrial **orchid**; **leaves** 8-10 cm x 0.1-0.2 cm, 3-7 per plant, narrow-linear, erect, spirally twisted; **flower-stem** 15-35 cm tall, wiry, with several loosely sheathing bracts, bearing 1-5 flowers; **flowers** 2.5-3.5 cm across, pale yellow with brown blotchings; **labellum** with small, fringed lateral lobes and a broad, shovel-shaped mid-lobe.
    Restricted to the south-west, where it grows on sandy or peaty soils of near-coastal districts, often in moist situations. Can be more difficult to grow than other species, and tends to be shy of flowering. Standing the base of the pot in a saucer of water during their growing-season definitely aids in growth. They are very slow to increase vegetatively. Propagate from seed or by natural increase.

Diuris laxiflora Lindley
(loose flowers)
WA   Bee Orchid
10-45 cm tall   Sept-Nov
    Terrestrial **orchid**; **leaves** 5-15 cm x 0.1-0.2 cm, 2-3

305

Diuris longifolia

per plant, narrow; **flower-stem** 10-45 cm tall, bearing leaf-like stem-bracts, 1-4 flowers at the apex; **flowers** about 2.5 cm across, yellow with dark red-brown blotches; **labellum** with small lateral lobes and a rounded mid-lobe.

A common species of the south-west, where it grows in wet, sandy to peaty soils. It grows readily in cultivation but is somewhat slow to increase vegetatively, and may be shy of flowering. It appreciates peat moss in the potting mix, and the base of the pot standing in water during the growing-season. Propagate from seed or natural increase. *D. carinata* Lindley is a synonym of this species.

Diuris longifolia R. Br.
(long leaves)
NSW. Vic, Tas, SA, WA   Donkey Orchid;
                         Wallflower Orchid
10-50 cm tall   July-Nov
Terrestrial **orchid**; **leaves** 7-25 cm x 0.4-0.8 cm, 1-3 per plant, slender, linear, deeply channelled; **flower-stems** 10-50 cm tall, sturdy, bearing 1-8 flowers in a loose raceme; **flowers** 2.5-5 cm across, variable in colour from clear yellow to yellow-and-brown with mauve suffusions; **labellum** with lateral lobes as long as the mid-lobe.

A very common species distributed over much of temperate Australia, and mainly found in open forest and heathland. It grows in large, often congested colonies, and flowering is markedly stimulated by bushfires. The flowers are very colourful and always attract interest. In cultivation, the species grows readily and increases rapidly by vegetative means. Unfortunately the plants are often very shy of flowering and need some stimulus. Clumps should be repotted every 2 years. Because of the variability in the species and the rapid vegetative propagation, it is possible to select and maintain different colour forms. Propagate from seed or by natural increase.

Diuris longifolia, × ·5

Diuris maculata Smith
(spotted)
NSW, Vic, Tas, SA   Leopard Orchid
15-35 cm tall   July-Nov
Terrestrial **orchid**; **leaves** 15-25 cm x 0.3-0.5 cm, 2-3 per plant, linear, channelled; **flower-stem** 15-35 cm tall, bearing 2-8 flowers in a flexuose raceme; **flowers** 1.5-3 cm across, yellow with dark brown spots and blotches, often heavily blotched; **labellum** with lateral lobes as long as the broad mid-lobe.

A very widespread and common orchid distributed in a variety of habitats, and usually growing in scattered colonies. It is extremely variable in colour, vigour and features of the flowers. It grows and flowers readily in cultivation and a potful in flower is highly ornamental. Vegetative increase is fairly slow. It is possible to select and maintain different growth and colour forms in cultivation. This orchid hybridizes freely with many other species, resulting in bewildering hybrid forms. Introduced to England during 1823. Propagate from seed or by natural increase.

Diuris × palachila R. Rogers
(broad lip)
NSW, Vic, Tas, SA
10-40 cm tall   Aug-Nov
Terrestrial **orchid**; **leaves** 8-17 cm x 0.2-0.4 cm, 2-8 per plant, erect, linear; **flower-stem** 10-40 cm tall, bearing 1-5 flowers on long, slender pedicels; **flowers** 2-3.5 cm across, yellow with brown spots and blotches; **labellum** with lateral lobes about half as long as the broad mid-lobe, and with markedly toothed margins.

This species is a natural hybrid between *D. maculata* and *D. pedunculata*. It is usually found in open, forested areas close to the parents and is an extremely variable taxon. It grows and flowers very easily, and is an excellent plant for cultivation. Propagate by natural increase.

Diuris pallens Benth.
(pale coloured)
NSW
10-15 cm tall   Aug-Nov
Slender terrestrial **orchid**; **leaves** 6-10 cm x 0.2-0.4 cm, 2-7 per plant, narrow-linear, semi-erect; **flower-stem** 10-15 cm tall, wiry, bearing 1-3 flowers; **flowers** 1.5-2 cm across, pale yellow, with drooping sepals and petals; **labellum** with very small lateral lobes, the mid-lobe large and ovate-lanceolate.

A small, easily overlooked species distributed in north-eastern NSW, in moist grasslands. It is generally similar to *D. pedunculata*, but in cultivation tends to be weak and slow-growing. Propagate from seed or by natural increase.

Diuris palustris Lindley
(growing in swamps)
Vic, SA   Swamp Diuris
8-15 cm tall   July-Dec
Terrestrial **orchid**; **leaves** 5-8 cm x 0.1-0.2 cm, 8-10 per plant, erect in a tussock, dark green, linear, twisted; **flower-stem** 8-15 cm tall, dark reddish, flexuose, bearing 1-4 flowers; **flowers** about 1.5 cm across, nodding, pale yellow, heavily blotched with red-brown,

Diuris punctata var. *punctata* forma *blakneyae*   E. Rotherham

especially on the reverse of the petals and sepals; **labellum** with lateral lobes as long as the mid-lobe.

A fairly uncommon species usually found in moist depressions, in small scattered colonies. The flowers have a delightful spicy perfume that is released on warm days. It grows easily in a pot of peaty mix with the base of the pot immersed in a saucer of water during the growing-season. Plants are generally very slow to increase vegetatively. Propagate from seed or by natural increase.

Diuris pedunculata R. Br.
(with a peduncle)
NSW, Vic, Tas, SA   Golden Moth; Snake Orchid
15-40 cm tall   Aug-Nov
Terrestrial **orchid**; **leaves** 10-20 cm x 0.2-0.3 cm, 5-7 per plant, narrow-linear, erect; **flower-stem** 15-40 cm tall, slender, bearing 1-5 flowers on long, slender pedicels; **flowers** 2.5-6.5 cm across, clear canary yellow with a few brown striations, generally nodding; **labellum** with prominently toothed lateral lobes which are much shorter than the ovate-lanceolate mid-lobe.

A widespread and very common orchid, usually growing in colonies. It often grows in semi-moist depressions and is frequent in grasslands. It extends to subalpine regions where it flowers Jan-Feb. It is a most rewarding species to grow, flowering freely and increasing in numbers by vegetative propagation. The species is extremely variable, and selected forms can be maintained in cultivation by natural division. Very large-flowered forms are particularly showy. Plants from subalpine situations tend to have rounded sepals and petals. Plants from the basalt plain to the west of Melbourne often have deep orange flowers. The species hybridizes freely with other *Diuris* species. Propagate from seed or by natural increase.

Diuris platichila Fitzg.
(broad lip)
Qld, NSW
20-60 cm tall   Aug-Oct
Terrestrial **orchid**; **leaves** 15-30 cm x 0.2-0.5 cm, 1-3 per plant, linear, channelled; **flower-stem** 20-60 cm tall, bearing 2-7 flowers on long pedicels; **flowers** 2.5-3 cm across, clear lemon yellow, with a few dark blotches on the labellum and dorsal sepal; **labellum** with long lateral lobes and a very broad mid-lobe.

A fairly common species found in open forests of south-eastern Qld and north-eastern NSW, frequently in hilly or mountainous districts. It grows and flowers well in a pot, and is a very decorative species. Vegetative increase is slow. Flower-spikes need staking. Propagate from seed or by natural increase.

Diuris punctata Smith
(spotted)
Qld, NSW, Vic, SA   Purple Donkey Orchid
20-60 cm tall   Aug-Nov
Terrestrial **orchid**; **leaves** 15-30 cm x 0.2-0.5 cm, 1-4 per plant, linear, channelled; **flower-stem** 20-60 cm tall, stout, bearing 2-10 flowers; **flowers** 3-8 cm across, variable in colour from white, through mauve to deep purple, rarely yellow; **labellum** with small lateral lobes and a broad, shovel-shaped mid-lobe.

This species, the most attractive member of the genus, is widespread and sometimes locally common, usually growing in colonies in moist, grassy areas. It grows strongly in cultivation and flowers freely. A well grown potful is a very attractive sight. Tall spikes may need staking to prevent tuber damage. *D. punctata* is an extremely variable species with many named forms and varieties, most of which have been tried in cultivation.

The typical form is restricted to the central coast of NSW, while var. *punctata* forma *blakneyae* Bailey is the most widespread and colourful form, extending from Qld to SA. It is also the most commonly grown.

The var. *alba* (R. Br.) Ewart & B. Rees is widely distributed in near-coastal regions of Qld and NSW. It has white to lilac coloured flowers with pointed petals.

The var. *albo-violacea* Rupp ex Dockr. is a stout, compact form restricted to the basalt plain near Melbourne, and now in danger of extinction. Its flowers are purplish white and have a delightful spicy fragrance. It has been raised successfully from seed by Mark Clements of the National Botanic Gardens, Canberra.

The var. *longissima* Benth. has long, slender lateral sepals and is distributed from south-eastern Qld to northern Vic.

The var. *minor* Benth. and var. *parvipetala* Dockr. are small forms from south-eastern Qld and north-eastern NSW.

All forms can be propagated from seed or by natural increase.

Diuris purdiei Diels
(after A. Purdie, 20th-century WA botanical collector)
WA
12-45 cm tall   Oct-Nov
Terrestrial **orchid**; **leaves** 8-10 cm x 0.2-0.4 cm, 5-

307

Diuris sulphurea   E. Rotherham

10 per plant, in an erect tussock, linear, twisted; **flower-stem** 10-45 cm tall, stout, bearing 1-5 flowers; **flowers** about 3 cm across, yellow with purplish striations and blotches; **labellum** with a very broad mid-lobe and lateral lobes about half as long.

Restricted to the south-west, where it grows in sandy soils near the coast, and on granite outcrops. It grows readily in cultivation but can be a bit shy of flowering. Natural increase is slow. Propagate from seed.

Diuris setacea R. Br.
(bristly)
WA   Bristly Donkey Orchid
10-20 cm tall   Sept-Nov
Terrestrial **orchid**; **leaves** 5-10 cm x 0.1-0.2 cm, 4-8 per plant, narrow and bristle-like, erect; **flower-stem** 10-20 cm tall, wiry, bearing 1-2 flowers; **flowers** about 2.5 cm across, yellow with a few reddish markings; **labellum** with lateral lobes much longer than the mid-lobe.

An attractive, small species restricted to the Albany district where it grows in peaty soils. It grows readily in cultivation and seems easy to flower. Propagate from seed or by natural increase.

Diuris sheaffiana Fitzg.
(after G. H. Sheaffe, early District Surveyor, Forbes, NSW)
Qld, NSW
20-40 cm tall   Sept-Oct
Terrestrial **orchid**; **leaves** 3-8 cm x 0.1-0.3 cm, 2-3 per plant, erect or lax, dark green, linear; **flower-stem** 20-40 cm tall, bearing 3-6 flowers; **flowers** 2.5-3 cm across, deep orange, sometimes speckled with red, especially on the dorsal sepal; **labellum** with short lateral lobes, often blotched with purple.

A little-known species from south-eastern Qld, and north-eastern and central-western NSW, usually growing in open forest and grassland. It grows readily in a pot of well drained mixture, and flowers freely. Plants may be slow to increase vegetatively. Propagate from seed or by natural increase.

Diuris sulphurea R. Br.
(sulphur yellow flowers)
Qld, NSW, Vic, Tas, SA   Tiger Orchid
25-75 cm tall   Aug-Nov
Terrestrial **orchid**; **leaves** 18-50 cm x 0.3-0.5 cm, 1-3 per plant, linear, channelled; **flower-stem** 25-75 cm tall, bearing 3-6 flowers; **flowers** 4-5.5 cm across, bright sulphur yellow with dark brown blotches, 2 characteristic blotches on the dorsal sepal; **labellum** with lateral lobes about half as long as the mid-lobe.

A widely distributed orchid found in scattered colonies in open forest. It is a very colourful and decorative species, and is usually the latest of the genus to flower in an area. Flowering is greatly stimulated by bushfires. It grows readily in a pot but tends to reproduce freely and be shy of flowering. Propagation is from seed or by natural increase.

Diuris tricolor Fitzg.
(3-coloured)
Qld, NSW
20-40 cm tall   Sept-Oct
Terrestrial **orchid**; **leaves** 10-20 cm x 0.3-0.8 cm, 1-3 per plant, channelled; **flower-stem** 20-40 cm tall, slender, bearing 2-7 flowers; **flowers** about 3 cm across, orange-yellow with a few purplish markings; **labellum** with small lateral lobes and a broad mid-lobe.

A fairly uncommon species found in hilly districts, in open forest of south-eastern Qld and north-eastern NSW. It grows readily in cultivation and is showy when in flower. Natural increase is slow. Propagate from seed.

Diuris venosa Rupp
(veined)
NSW
10-40 cm tall   Nov-Jan
Terrestrial **orchid**; **leaves** 5-12 cm x 0.2-0.3 cm, 3-5 per plant, linear, erect, channelled; **flower-stem** 10-40 cm tall, bearing 1-4 flowers; **flowers** about 2.5 cm across, lilac to mauve with conspicuous dark striations; **labellum** with short, blunt lateral lobes and a prominent, heart-shaped mid-lobe.

A decorative species restricted to the Barrington Tops, NSW, where it grows in scattered colonies in

Diuris longifolia × *pedunculata*   J. Fanning

moist, peaty soil. It is not an easy species to maintain in cultivation, and it tends to die following any setback. It needs a peaty soil mixture, and the base of the pot should be stood in a saucer of water while the plant is in active growth. The soil mix should also be kept moist while the tubers are dormant. Natural increase is very slow. Propagate from seed.

DIURIS CULTIVARS and HYBRIDS

Diuris hybrid (*D. longifolia* × *D. pedunculata*)
A natural hybrid that occurs sporadically where the parents grow together. It is a vigorous plant with large, striking flowers. These may be up to 8 cm across, and have conspicuous elongated petals. They are of a golden yellow colouration and are sterile. Plants grow readily in cultivation but tend to be shy of flowering.

Diuris hybrid (*D. longifolia* × *D. sulphurea*)
A rare natural hybrid that occurs occasionally where the parents grow together. It flowers only after bush-fires, and is rarely collected. The flowers are about 4 cm across, and are yellow with dark blotches and wallflower suffusions. The flowers are sterile. Plants grow readily in cultivation but are very difficult to flower.

Diuris hybrid (*D. maculata* × *D. palustris*)
A rare natural hybrid that occurs occasionally where the parents grow together. The flowers are about 2.5 cm across and are yellow with many dark blotches and markings. They tend to be nodding, and are fragrant. This hybrid grows and flowers quite readily in cultivation.

Diuris hybrid (*D. maculata* × *D. pedunculata*)
A common natural hybrid that occurs frequently where the parents grow together. The hybrids are fertile, and backcross with the parents to create puzzling hybrid forms. Some of these hybrids are quite showy, and adapt well to cultivation. Selected forms can be allowed to increase naturally until a pot of the one form is built up. Plants of these hybrids flower quite freely.

Diuris hybrid (*D. palustris* × *D. pedunculata*)
A very rare natural hybrid that occurs sporadically where the parents grow together. The flowers are about 2.5 cm across and are yellow with a few dark striations. The flowers tend to be held stiffly erect. The leaves are an upright tussock of 6-8, twisted in the manner of *D. palustris*. The flowers are sterile. Plants tend to be difficult to grow in cultivation and except for the first couple of years after collection, are shy of flowering. The hybrid has been named as *D.* × *fastidiosa* R. Rogers.

Diuris hybrid (*D. pedunculata* × *D. punctata*)
A natural hybrid that occurs sporadically where the parents grow together. It has large flowers (to 6 cm across) of an unusual pale orange colour, with elongated petals. The plants are sterile. They grow fairly readily in cultivation but tend to be shy of flowering. Natural increase is slow.

Diuris 'Pioneer'
This is the first man-made hybrid of an Australian terrestrial orchid to be registered as a cultivar. It is the result of a cross between *Diuris maculata* and *D. longifolia*. The same cross occurs naturally but is of very sporadic occurrence. It is a vigorous grower with a flower-stem to 40 cm tall with large (to 4 cm across), colourful flowers of yellow blotched with reddish brown. Plants are sterile and must be propagated vegetatively.

DODONAEA Miller
(after Rembert Dodoens, 16th-century Flemish botanist)
Sapindaceae

Shrubs or small **trees**; **branchlets** often angular, can be terete, ribbed or flattened, usually glandular, glabrous or hairy; **leaves** simple or pinnate, scattered or spirally arranged, opposite in one species, usually sticky; **flowers** small, unisexual or bisexual, axillary or terminal; **sepals** 3-7; **petals** none; **stamens** 4-10; **fruit** a capsule with 2 to 6 angles or wings.

The genus *Dodonaea* occurs principally in Australia but also extends to the tropical regions of America, Africa and Asia. The genus is currently undergoing botanical revision, and consists of 53 named species and about 8 undescribed species, as well as varieties and subspecies. All species are included in this volume.

Representatives occur in all states and are to be found in a wide range of climates, habitats and soils. The majority are found in semi-arid to arid zones, with a few species extending to the higher-rainfall fringes of the continent. They are rare in the tropics and absent from subalpine communities. They commonly occur in various types of woodland and forests, growing in a diverse range of soils. Plants commonly grow in rocky situations, but are also found on sand, sands with a clay subsoil, sandy loams and clay-loams. Frequently, the soils where they grow are rich in lateritic or granitic gravels. Some species occur on soils overlying limestone, eg. *D. humilis* and *D. tepperi*.

As a group, dodonaeas have the common name of hop-bushes, gained from the decorative fruiting capsules which were used by the early settlers as a substitute for European hops in the brewing of beer.

Dodonaeas have much to offer cultivation as the species are generally hardy to a wide range of climates, and adapt to a variety of soils. In general, they do best in relatively well drained soils, but some withstand extended periods of waterlogging, eg. *D. procumbens*. Most species seem to be tolerant of strong winds, and some withstand fairly exposed coastal conditions, eg. *D. baueri* and *D. ceratocarpa*.

Their growth habit varies from spreading prostrate or mound-forming plants, to small trees, with many forms of intermediate shrubs. The bark on taller species can develop interesting papery textures.

Many species have decorative fern-like foliage, while the ornamental aspect of the coloured fruits is the strong point of others. Plants can be covered with reddish 'hops', eg. *D. lobulata*, *D. microzyga* and *D. tenuifolia*.

Flowers of dodonaeas are generally insignificant and can be bisexual or unisexual. The male and female flowers are usually on separate plants. It is unusual for plants to have only bisexual flowers. Only plants with female or bisexual flowers produce capsules, and it is not necessary for them to be fertilized for the capsules to develop (see Volume 1, page 199, Parthenocarpy). Such capsules will not contain viable seeds and will frequently ripen earlier on the plant than the fertile capsules.

Some members of the genus have been cultivated for many years. *Dodonaea triquetra* was introduced to England during 1790, and *D. filiformis* about 1830.

There is only a limited number of species regularly grown. One of the most common is an unnamed species with affinities to *D. tenuifolia*. It has been sold as *D. adenophora* and *D. microzyga*. Others grown widely include *D. boroniifolia*, *D. lobulata* and *D. viscosa*.

Many species have aromatic foliage, with the aroma apparent on a sunny day or immediately after rain. Some people find the aroma pleasant, but others think it objectionable. A common dodonaea in cultivation is the purple-leaved plant, usually available as *D. viscosa* 'Purpurea'. This did not originate from Australia, but was a selection made in the South Island of New Zealand.

Quite a few of the lesser-known species, eg. *D. hirsuta* and *D. petiolaris*, have ornamental qualities in growth habit, foliage and fruits, and deserve to be cultivated. Really, all species have some characteristics that make them appealing for cultivation. Although some may not necessarily appeal in nature, in cultivation they frequently develop better growth and produce many hops.

After rain quite a few species attain a whitish colouration on the foliage, eg. *D. multijuga* and *D. species aff. tenuifolia* (Qld, NSW). It is not a disease, but evidently is due to exudation of resinous material which dries to form a whitish powder.

In cultivation, the major disadvantage with some dodonaeas (usually those that have extremely sticky branchlets and leaves) is that they are readily attacked by scale. Early identification of such attacks and spraying with white oil is generally sufficient to combat the problem. Foliage and branches can become covered with sooty mould if the scale is not killed early in its development. Chewing-pests do not present a threat to foliage growth. There may be the odd attack by stem-borers on some species.

Most dodonaeas respond favourably to the use of slow-release fertilizers, but growth is usually satisfactory without them. Pruning is one of the most important factors. Tip pruning of young plants is highly recommended. Judicious, regular pruning can help to promote good growth, and the picking of branches when in fruit is a good method which also provides material for interior decoration. Pruning cuts should not be made too far down on old wood, as sometimes regrowth will not eventuate.

Propagation

Dodonaeas can be propagated from seed or cuttings. Seed is easy to collect. As the capsules reach maturity they are gathered and can be readily broken apart to obtain the small blackish seed. The seed has a hard outer coat which can make germination difficult. Simple treatment by soaking the seeds in very hot water prior to sowing promotes quicker germination, and results in a better percentage of seedlings. Germination is usually fairly rapid for most species. Seedlings often appear after 14 days, but they can take up to 7 weeks to germinate.

Good results with cuttings have been obtained with quite a few species. Generally, young growth that is just starting to become slightly woody gives better results than soft tip growth. There is, however, a need for more experimentation with a larger number of species.

Dodonaea adenophora Miq.
(bearing glands)
WA
0.3-2 m x 0.5-3 m   June-Aug

Dwarf to medium, multi-stemmed **shrub**; **branches** mainly spreading, many; **branchlets** terete to slightly angular, viscous, reddish brown; **leaves** pinnate to about 1 cm long; **leaflets** 3-7, to 0.5 cm long, usually opposite, viscous; **flowers** male and female usually on separate plants, axillary; **capsules** 4-winged, to 0.8 cm x 1 cm, initially green, maturing to red, reddish brown or purplish brown.

This ornamental and hardy species occurs over a wide part of southern WA. It is found in the semi-arid and arid mallee scrublands and open woodland, where it grows on granitic sands and red sandy loams. Has had limited cultivation. Best suited to well drained soils, with partial or full sun. Frost and drought tolerant. Propagate from seed or cuttings.

D. adenophora has affinities to *D. microzyga*, which has larger leathery capsules.

For many years an unnamed eastern Australian species has been sold under the name of *D. adenophora*. It is quite distinct, with larger leaves that have up to 22 leaflets. This is one of the most commonly cultivated species. See *D. species aff. tenuifolia* (Qld, NSW).

Dodonaea amblyophylla Diels
(blunt leaf)
WA
0.5-2 m x 1-2 m   March-June; also sporadic
    Dwarf to small **shrub**; **branches** erect; **branchlets**
terete, slightly hairy; **leaves** 2-4.5 cm x 0.4-1 cm,
simple, sessile, linear to oblanceolate, bright green,
thick, leathery, entire, apex acute or blunt, glabrous;
flowers unisexual or bisexual, 4-6 in clusters at ends
of branchlets; **capsules** 3-4 winged, to 1.2 cm x 1.7 cm,
glabrous, wings prominent, maturing dark brown to
purplish black during Sept-Oct.
    This species occurs mainly in the goldfields areas of
WA, where it grows on sand or sandy loam. It is also
found near the coast in the south-eastern corner of the
state. This large-fruited species is not common in cultivation. It requires very well drained, light to medium
soils, with partial or full sun. Tolerates extended dry
periods and most frosts. Propagate from seed or
cuttings.
    D. amblyophylla has affinities to *D. bursariifolia*, which
is easily distinguished when in fruit as its capsules have
very narrow wings.

Dodonaea angustifolia L.f. see under *D. viscosa* ssp. C

Dodonaea angustissima DC.
                        see under *D. viscosa* ssp. D

Dodonaea aptera Miq.
(without wings)
WA
1-3.5 m x 1-2.5 m   March-July
    Small to medium **shrub**; young growth viscous;
branches many; **branchlets** glabrous, angular or
furrowed, can be flattened; **leaves** 2-6 cm x 1-3 cm,
simple, elliptic to obovate, petiolate, entire, glabrous,
olive green above, paler below; **flowers** in terminal
clusters, male and female on separate plants; **capsules**
slightly 3-4 angled, to 0.6 cm x 0.7 cm, without wings.
    A recent introduction to cultivation, this species is
readily distinguished by its broad leaves and wingless
hops. It is recorded from the coastal region of the
Darling District, where it grows in alkaline soils. It is
recommended as a coastal plant. Best suited to well
drained, light to medium soils, with partial to full sun,
but will tolerate dappled shade. A vigorous groundcovering form is proving very adaptable; it will be interesting to see whether it retains its low growth. Propagate from seed or cuttings.

Dodonaea attenuata Cunn. see under *D. viscosa* ssp. D

Dodonaea baueri Endl.
(after Ferdinand and Frederick Bauer, 19th-century
botanical artists)
SA
0.2-1 m x 0.5-1.5 m   .   Dec-April; also sporadic
    Dwarf to small **shrub**; **branches** intricate, rigid,
glabrous; **branchlets** angular, glandular, with dense
short hairs; **leaves** 0.6-1.8 cm x 0.3-1 cm, simple,
oblong to orbicular, slightly sinuate, irregularly
toothed, dark green above, pale green below; **flowers**
male and female on separate plants, usually solitary,

Dodonaea boroniifolia, × ·7

on short axillary peduncles; **capsules** 3-5 angled, more
or less wingless, to 0.5 cm x 1 cm, glandular, maturing
to dark red to brown.
    This species is endemic to southern SA (including
the Flinders Ranges). It usually occurs on rocky sites,
ranging from inland to coastal situations, where it can
be exposed to harsh climatic conditions. Has had only
very limited cultivation, but should grow well in most
well drained soils. A warm to hot location, with partial
to full sun is recommended. The species is worth trying
in alkaline soils. Propagate from seed or cuttings.

Dodonaea boroniifolia G. Don
(leaves similar to those of the genus *Boronia*)
Qld, NSW, Vic   Fern-leaf Hop-bush;
                                 Hairy Hop-bush
0.5-2 m x 0.7-2 m   May-Dec
    Dwarf to small, spreading **shrub**; **branches**
spreading; **branchlets** usually sticky, terete or
sometimes angular, hairy; **leaves** pinnate, to 4 cm long;
leaflets 6-14, to 0.8 cm long, opposite or irregularly
alternate, narrow-obovate to oblong, shiny, viscous,
dark green, entire except for toothed apex; **flowers**
male and female on separate plants, rarely bisexual, in
axillary clusters; **capsules** 4-winged, to 2 cm x 2 cm,
maturing to pink or purplish red, Nov-April.
    This dodonaea occurs from western Vic to Charters
Towers, Qld. It is found in a wide range of soils, and
tolerates various climatic conditions, growing in plant
communities from woodlands through to heathland. It
is very decorative, and well known in cultivation.
Adapts to most well drained soils. Grows in dappled
shade to full sun. Frost hardy. Tolerates extended dry
periods. Some forms have capsules that are a delicate
green for many months before gaining purple
colouration. Responds well to pruning, as branches
can become quite long and leggy. Propagate from seed
or cuttings.

Dodonaea aff. **boroniifolia**
Qld
0.5-1.5 m x 0.8-2 m   May-June
    Dwarf to small, spreading **shrub**; **branches** arching;
branchlets terete, slightly hairy; **leaves** pinnate, 0.5-
1 cm long, **rhachis** slightly winged; **leaflets** to 0.7 cm x
0.3 cm, 4-8, opposite, obovate, olive green, viscous,
margin entire, apex 3-6 toothed; **flowers** male and

311

female on separate plants, in clusters at ends of branchlets; **capsules** 4-winged, 1.1-1.6 cm x 0.9-1.4 cm, viscous, glabrous, bright red to purple, mature June-Sept.

The distribution of this species is confined to the North Kennedy District, where it occurs in open forests, usually growing on soils derived from sandstone. It is probably best suited to cultivation in tropical and subtropical regions, but is worthy of experimentation in temperate climates. Needs well drained, light to medium soils, with partial to full sun. Pruning may be required to promote bushy growth. Propagate from seed or cuttings.

D. boroniifolia can be distinguished from this species by its axillary flowers.

Dodonaea bursariifolia F. Muell.
(leaves similar to those of the genus *Bursaria*)
NSW, Vic, SA, WA   Low Hop-bush;
      Small Hop-bush
0.5-1.5 m x 0.7-1.5 m   Aug-Dec

Dwarf to small **shrub**; **branches** spreading, crowded; **branchlets** terete to angular, usually hairy; **leaves** 0.8-3.5 cm x 0.3-1.5 cm, simple, obovate-cuneate, entire, obtuse, leathery, glabrous, bright green; **flowers** unisexual, rarely bisexual, 2-3 together at ends of branchlets; **capsules** 3-4 angled, to 1 cm across, glandular, maturing brown to yellow, wing very narrow.

An inhabitant of sandy loam soils in mallee communities. It is not cultivated to any great extent. Its main application should be in semi-arid regions. Will prefer well drained, light to medium soils and plenty of sunshine, although it will probably tolerate partial sun. Should grow well in alkaline soils. The fruits can take up to 12 months to mature and there is the opportunity for selection of good fruiting forms. Propagate from seed or cuttings.

Dodonaea caespitosa Diels
(tufted)
WA
0.2-0.5 m x 0.5-1.5 m   Feb-March

Dwarf, compact **shrub**; **branches** rigid, entangled, crowded; **branchlets** angular or furrowed, red, viscous, slightly hairy; **leaves** 0.2-1 cm x up to 0.1 cm, simple, sessile, linear to almost terete, bright green, thick, viscous, glabrous or slightly hairy, margin revolute, apex blunt; **flowers** male and female on separate plants, solitary, sessile in axils; **capsules** 3-4 angled, about 0.5 cm x up to 1 cm, with horn-like appendages, scattered, glandular, viscous, glabrous, dark red to brown, mature Oct-Nov.

A pleasant clumping species from south-western WA. It occurs in a wide range of soils in semi-arid mallee scrub and mallee heath communities. Is also found in saline soils. Evidently not well known in cultivation, but it should prove adaptable to most soil types. Will need relatively good drainage and a warm location, with partial to full sun. Could be useful for saline soils, and has potential as a container plant. Propagate from seed or cuttings.

D. caespitosa has affinities to *D. divaricata*, which has broader leaves that can be toothed.

Dodonaea ceratocarpa   P. Armstrong

Dodonaea camfieldii Maiden & Betche
(after J. H. Camfield, botanical collector)
NSW
prostrate-0.5 m x 1-2 m   Nov-Feb

Dwarf, spreading **shrub**; **branches** prostrate to procumbent, often producing roots at nodes; **branchlets** ascending, winged, slightly hairy; **leaves** 1-3.5 cm x 0-3.1 cm, simple, linear to oblong, often appear stem-clasping, undulate, leathery, margin entire to irregularly toothed, revolute, viscous, usually glabrous; **flowers** male and female on separate plants, rarely bisexual, in terminal clusters; **capsules** 4-winged, to 1.4 cm x 1.6 cm, glandular, glabrous to slightly hairy, brown to purplish, mature Oct-Nov.

This species should find wider applications in cultivation. It occurs mainly on Hawkesbury sandstones, in dry sclerophyll forest or woodland. Should adapt to varied soil and climatic conditions. Probably best suited to situations with dappled shade to partial sun. May tolerate full sun. Hardy to most frost. Has potential as a container plant. With its characteristic of producing roots at nodes, it could be useful for soil erosion control on a small scale, eg. for embankments and large sloping garden beds. Propagate from seed or cuttings.

Dodonaea ceratocarpa Endl.
(horned fruits)
WA
0.6-2.5 m x 1-2.5 m   Aug-Feb

Dwarf to medium **shrub**; **branches** spreading, angled; **branchlets** angular to almost winged, usually glabrous; **leaves** 1.5-4.6 cm x 0.3-1.2 cm, simple, oblanceolate to obovate, entire and revolute, sometimes with 2-3 teeth near apex, slightly leathery, dark green; **flowers** male and female on separate plants, in terminal clusters; **capsules** 3-4 angled, to 0.8 cm x 1 cm, with small horn-like appendages on upper outer edges, light brown to purple-brown at maturity.

This species has ornamental foliage and fruits. It occurs on or near the coast of southern WA, as well as slightly inland. It grows in a wide range of soil and climatic conditions. Is best suited to very well drained, light to medium soils, with partial to full sun. The coastal form tolerates coastal exposure and is very hardy to strong winds. Both forms are suitable for cultivation in slightly alkaline soils. Responds well to pruning. Has potential as a container plant. Propagate from seed or cuttings.

Dodonaea concinna Benth.
(elegant; pretty)
WA
0.5-1.5 m x 0.6-1.5 m   Aug-Nov
   Dwarf to small, compact **shrub**; **branches** many, slender, spreading; **branchlets** angular to slightly ribbed, slightly hairy; **leaves** pinnate, about 2 cm long; **leaflets** to 1 cm x 0.1 cm, 4-12, opposite or alternate, linear, concave or channelled above, crowded, olive green; **flowers** male and female on separate plants, axillary; **capsules** 3-4 winged, to 1 cm x 1.3 cm, dark red, purple or black, mature Aug-Feb.
   A hardy and decorative fern-leaved species that occurs in the Eyre and Roe Districts. It grows in sandy loam soils, often containing lateritic gravel. Should be adaptable to most well drained, light to medium soils, and may also tolerate heavy soils. A position with partial to full sun usually means better fruit-production, although plants will also grow in dappled shade. Frost and drought hardy. Selected forms have potential as container plants. Capsules can take up to 12 months to reach maturity. Propagate from seed or cuttings.

Dodonaea coriacea (Ewart & Davies) McGillivray
(leathery)
Qld, WA, NT
0.5-2 m x 1-2 m   March-July
   Dwarf to small, spreading **shrub**; **branches** often entangled; **branchlets** angular to terete, light orange, viscous, glandular, slightly hairy; **leaves** 1.4-3 cm x 0.4-1 cm, simple, sessile, oblanceolate to obovate, viscous, slightly hairy or glabrous, margin entire or toothed, apex truncate and toothed; **flowers** bisexual, rarely unisexual, in clusters at ends of branchlets; **capsules** 3-winged, 1.2-1.9 cm x 1.4-1.8 cm, prominent wings, viscous, glabrous or slightly hairy, creamy yellow, sometimes with purple, mature July-Oct.
   This ornamental species is generally confined to the arid regions of Australia receiving less than 500 mm annual rainfall. It occurs in deep red sands and on rocky hills. Has had only limited cultivation, and should do best in semi-arid to arid areas. Must have very well drained soils, with a warm to hot location. May have potential as a container plant in temperate regions. Propagate from seed or cuttings.
   Previously known as *D. peduncularis* Lindley var. *coriacea* Ewart & Davies.

Dodonaea cuneata Smith see under *D. viscosa* ssp. E

Dodonaea cuneata var. **rigida** Benth.
                     see under *D. viscosa* ssp. G

Dodonaea divaricata Benth.
(spreading)
WA
0.2-0.5 m x 0.5-1.5 m   Aug-Oct
   Dwarf, spreading **shrub**; **branches** rigid, entangled, crowded; **branchlets** terete to slightly furrowed, dark red to brown, glandular, slightly hairy; **leaves** 0.3-0.8 cm x 0.1-0.3 cm, simple, linear to obovate, olive green, thick, viscous, slightly hairy, margin revolute, entire or 1-4 irregular teeth, apex acute; **flowers** male and female on separate plants, solitary, axillary or at ends of branchlets; **capsules** 3-angled, 0.3-0.6 cm x 0.4-0.6 cm, with short horn-like appendages, glandular, glabrous and dark red-brown, mature Dec-Jan.
   D. divaricata usually occurs in the mixed, open mallee scrub of the Irwin and Avon Districts. It grows on heavy soils that can have a high proportion of stone and gravel. Should prove to be adaptable to a wide range of soil types, and should cope with soils that have fair to poor drainage. Needs partial or full sun. Has value as a groundcovering species, and may prove to be a good container plant. Propagate from seed or cuttings.

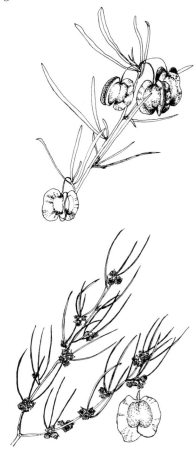

Dodonaea filifolia (2 forms), × ·6

Dodonaea ericaefolia

D. caespitosa is similar, but has narrower and almost terete leaves.

D. pinifolia can be distinguished by its larger leaves, mainly 4-angled fruits and by the flowers, which are in cymes.

Dodonaea ericaefolia G. Don = *D. filiformis* Link

Dodonaea ericoides Miq.
(resembles the genus *Erica*)
WA
0.3-1 m x 0.3-1 m   April-Dec; also sporadic
Dwarf **shrub**; **branches** opposite, slender, entangled, usually erect; **branchlets** angular or furrowed, red, usually very hairy; **leaves** 0.2-1.5 cm x 0.1-0.2 cm, simple, opposite, linear to narrow-oblanceolate, sessile, olive green, sometimes purplish, rigid, usually hairy, margin strongly revolute, entire or with 2-4 teeth, apex acute; **flowers** male and female on separate plants, solitary at ends of branchlets; **capsules** 4-angled, inflated, very hairy, with grey tomentum, mature mainly Sept-Nov.

This uncommon ornamental species inhabits coastal regions, and is found up to 100 km inland, from about 90 km north of Perth to Geraldton. It usually occurs on rocky sites, or in soils that have high laterite content. *D. ericoides* is readily identified, as it is the only species with opposite leaves. Evidently not well known in cultivation, it is best suited to relatively well drained soils. Will tolerate heavy soils. Prefers partial or full sun, but will tolerate dappled shade in semi-arid regions. Has potential as a container plant. Propagate from seed or cuttings.

Dodonaea filifolia Hook.
(thread-like leaves)
Qld, NSW   Thread-leaf Hop-bush
1-5 m x 1-3 m   June-Aug
Small to tall **shrub**; **branches** usually erect, slender, often arching; **branchlets** angular or terete, glabrous; **leaves** 2-8 cm x about 0.3 cm, simple, narrow-linear; **flowers** lime green, male and female on separate plants, borne in clusters at ends of branchlets; **capsules** 3-4 winged, to 1.3 cm x 1.3 cm, prominent wings, yellowish, deepening to bright red to purple-brown when mature during Sept-Dec.

This species grows as a slightly open shrub, and is not common in cultivation. It is, however, quite ornamental, with a profuse display of hops. It occurs naturally in gravels and loams of sandstone plateaux and hillsides in the Great Dividing Range. Hardy in most soils, provided drainage is adequate. Likes plenty of sunshine. Frost tolerant. Propagate from seed or cuttings.

Dodonaea species aff. *filifolia* (Qld, NSW) differs in having hairy branches and shorter, falcate leaves. *D.* species aff. *filifolia* (WA) has rigid, glandular, linear leaves.

Dodonaea aff. filifolia (Qld, NSW)
Qld, NSW
1.5-2.5 m x 1-2 m   March-April
Small to medium, erect **shrub**; **branches** slender;

branchlets terete to slightly angular, slightly hairy; **leaves** 2.5-6 cm x 0.1 cm, simple, sessile, linear to sub-filiform, falcate, olive green, viscous, glabrous, margin entire, apex blunt; **flowers** male and female on separate plants, in clusters at ends of branchlets; **capsules** 4-winged, 1-1.2 cm x 1.1-1.4 cm, prominent wings, glandular, slightly hairy, reddish brown, mature Aug-Dec.

This species has been included as a form of *D. filifolia*, which differs in its longer, straight leaves and glabrous branchlets. Is generally confined to the Great Dividing Range from the Stanthorpe-Wallangarra area in south-eastern Qld, to a southern limit near Sydney. Occurs in dry sclerophyll forest, on granite or sandstone. Needs very well drained soils, with partial or full sun. Tolerant of light to medium frosts and extended dry periods. Pruning may be needed to promote bushy growth. Propagate from seed or cuttings.

Dodonaea aff. filifolia (WA)
WA
1-2 m x 1-2 m   sporadic
Small **shrub**; **branches** ascending, rigid, not usually crowded; **branchlets** terete to slightly angular, glabrous; **leaves** 3.5-10.5 cm x 0.1 cm, simple, sessile, terete or 4-ribbed, erect, rigid, viscous, glabrous, apex acute to pungent; **flowers** male and female on separate plants, in axillary clusters; **capsules** 3-4 winged, 1-1.5 cm x 1.4-1.7 cm, glandular, glabrous, red or reddish brown, mature Sept-Nov.

A species of arid to semi-arid regions, where it grows on sand-dunes or lateritic rises, sandy creek-beds and in red sandy loam. Best suited to well drained soils, in areas of low rainfall. Needs plenty of sunshine for most of the year. It may grow as a container plant in temperate regions. Should tolerate most frosts. Propagate from seed or cuttings.

This species is different from *D. filifolia*, which has soft, linear leaves and terminal clusters of flowers.

Dodonaea hexandra, × ·65   fruit, × 2

314

Dodonaea filiformis Link
(in the form of a thread)
Tas
1-2.5 m x 1-2 m   Sept-Oct
   Small to medium, multi-stemmed **shrub**; **branches** crowded; **branchlets** terete or slightly angular, slightly hairy; **leaves** 1.2-2.5 cm x up to 0.1 cm, simple, linear, often falcate, bright green, viscous, crowded; **flowers** male and female on separate plants, in clusters at ends of branchlets; **capsules** 3-winged, about 1 cm x up to 1.7 cm, glabrous, dark red, mature Dec-Jan.
   This Tas endemic species is found in the eastern half of the island, and occurs in dry sclerophyll forest. It grows as a very dense shrub, and some forms selected for their display of hops are most showy. Seems to adapt to a wide range of relatively well drained soils. Will grow in dappled shade through to full sun. Frost tolerant. Excellent screening plant and does well in containers. Responds favourably to pruning. Propagate from seed or cuttings.
   Previously known as *D. ericaefolia* G. Don, and is sometimes sold as *D. ericifolia*.

Dodonaea hackettiana W. Fitzg.
(after Sir J. W. Hackett, first Chancellor, University of WA)
WA   Hackett's Hop-bush; Perth Hop-bush
2-5 m x 2-4 m   July-Oct
   Medium to tall **shrub**; **branches** erect to spreading, with rows of white to brown hairs; **branchlets** ribbed or angular, with 2-3 rows of white hairs; **leaves** to 6 cm x 0.9 cm, simple, alternate, lanceolate to narrow-elliptic, margin entire, apex acute, bright green; **flowers** male and female on separate plants, in terminal clusters; **capsules** 3-winged, to 1.3 cm x 1.7 cm, wings prominent, glabrous, orange-brown, mature Oct-Jan.
   This species is confined to alkaline, sandy soils or limestone near Perth. It is now recognized as rare in nature, and needs to be conserved. Cultivation and propagation can help to ensure that this species does not become extinct. Best grown in well drained, light to medium soils, and should be suitable also for alkaline soils. Withstands dappled shade through to full sun. Has some potential as a screening and wind-break plant. Propagate from seed or cuttings.

Dodonaea hexandra F. Muell.
(6 stamens)
Vic?, SA, WA?
0.3-0.6 m x 0.5-1.5 m   May-July
   Dwarf, spreading **shrub**; **branches** slender, intricate; **branchlets** angular and furrowed, usually hairy, reddish; **leaves** 0.6-3.5 cm x about 0.1 cm, simple, linear, channelled below, viscous, bright green, revolute margins, glabrous; **flowers** male and female on separate plants, 1-2 per axil or terminal; **capsules** 3-angled, about 0.7 cm x 0.7 cm, inflated, with 3 minute wings on upper edge of each angle, glabrous, brown, mature Sept-Nov.
   This species usually occurs on alkaline soils as an undershrub in mallee communities. It has been grown mainly by enthusiasts, and it has potential for low-rainfall regions and calcareous soils. Needs relatively well

Dodonaea hirsuta, × ·6

drained soils and partial to full sun. Frost and drought hardy. Propagate from seed or cuttings. It is doubtful whether this species occurs in Vic and WA; further searches are required to verify.
   This species occurs naturally with *D. humilis*. Hybrid plants have been found with intermediate characteristics. Such hybrids may be in cultivation if they have been propagated from cuttings. To date, fruits of these hybrids have not been found to contain viable seed.

Dodonaea hirsuta (Maiden & Betche) Maiden & Betche
(long hairs)
Qld, NSW
0.5-2 m x 0.5-1.5 m   June-Sept
   Dwarf to small **shrub**; **branches** spreading, crowded; **branchlets** terete, with many long hairs; **leaves** 0.5-0.6 cm x 0.3-0.5 cm, simple, triangular to obovate, dark green above, paler below, rough, with long hairs, margin revolute, apex truncate and 3-5 toothed; **flowers** male and female on separate plants, 1 per axil; **capsules** 3-4 winged, 1.1-1.6 cm x 1.2-1.7 cm, hairy, orange-brown to deep red at maturity.
   D. hirsuta distribution is restricted to the Stanthorpe-Wallangarra region of south-eastern Qld, and north-eastern NSW. It occurs on the slopes and hillsides, growing in granite or sandstone-derived soils. Although a decorative species when in fruit, it is not well known in cultivation. It is best suited to very well drained soils, with partial or full sun. Has potential as a container plant. Propagate from seed or cuttings.
   Is similar to *D. peduncularis* in flower and capsule, but that species has longer leaves.

Dodonaea humifusa Miq.
(spread on the ground)
WA
prostrate x 1-2 m   March-June; also sporadic
   Dense, mat-like dwarf **shrub**; **branches** prostrate, many, often rooting at nodes; **branchlets** angular, glabrous or slightly hairy, slightly viscous; **leaves** 1-4 cm x 0.2-0.8 cm, oblanceolate to obovate, rarely linear, flat, leathery, crowded, margins entire or 1-4 toothed at apex, glabrous or slightly hairy on midrib and margins; **flowers** male and female on separate plants, rarely bisexual, usually 3-4 in clusters at ends of branchlets;

315

Dodonaea humilis

Dodonaea humifusa, × ·6

capsules 3-5 angled, 0.4-0.9 cm x 0.5-0.8 cm, wings very small, initially light green, maturing to light brown to pinkish brown.

This species occurs in open woodland to the north and north-east of the Stirling Range. It grows on sandy loam which can contain lateritic gravel and have a clay subsoil. It is very hardy, and although not particularly ornamental in fruit, it certainly has potential for many applications in the landscape and horticulture. Its dense coverage and ability to self layer make it suitable for soil erosion control on embankments, and for use as a good living mulch. Suited to most soils, with dappled shade to full sun. Initially, plants may be slow-growing. Withstands frost, extended dry periods and water-logging. Grows well in containers. Propagate from seed, cuttings which strike readily or layers.

Dodonaea humilis Endl.
(low)
SA
0.1-1 m x 1-1.5 m   Nov-March
Dwarf, spreading, compact **shrub**; young growth viscous; **branches** many, minutely hairy to glabrous, procumbent; **leaves** pinnate, 1-5 cm long; **leaflets** 0.3-1 cm long, 2-14 or sometimes more, obovate to cuneate, opposite, rigid, 3-7 teeth at apex; **flowers** male and female on separate plants, in short terminal clusters; **capsules** 4-lobed, 0.5-0.7 cm x up to 0.6 cm, covered in rusty-red hairs, reddish brown, mature Aug-Dec.

An inhabitant of mallee scrub, both inland and near the coast, where it usually grows on limestone or in sands overlying limestone. It withstands extended dry periods. Its low, compact growth habit offers potential in cultivation. Has been grown only to a limited extent to date. Should do well in well drained, light soils, including calcareous types. Prefers partial or full sun. Worthy of experimentation in coastal situations. Suited to container cultivation. Propagate from seed or cuttings.

Dodonaea aff. **humilis**
WA
0.2-0.6 m x 0.5-1 m   Aug-Sept
Dwarf multi-stemmed **shrub**; **branches** erect; **branchlets** angular or ribbed, very hairy; **leaves** pinnate, to 2 cm long; **rhachis** winged; **leaflets** to 0.2 cm x 0.2 cm, 8-18, oblong, opposite or alternate, dark green, viscous, hairy, strongly recurved margins; **flowers** rarely bisexual, male and female on separate plants, 1-2 per axil; **capsules** 2-valved, to 0.8 cm x 0.6 cm, inflated, hairy, reddish brown, mature Nov-Dec.

A unique species of the genus because of its 2-valved capsules. It is very restricted in nature, being confined

to two sites in the Roe District. It occurs on shallow, sandy clays, clay-loams that contain lateritic gravel and on alkaline, red clay-loams. Has had only very limited cultivation to date. Should grow in a wide range of soils. Needs a warm location, with partial to full sun. Frost and drought tolerant. Propagate from seed or cuttings.
Differs from *D. humilis,* which has 4-lobed capsules.

Dodonaea inaequifolia Turcz.
(unequal leaves)
WA
1-5 m x 1-3 m   April-June
Small to tall **shrub; branches** slender, spreading; **branchlets** angular, viscous, glandular, glabrous; **leaves** pinnate, to 6 cm long; **leaflets** to 2.5 cm long, 17-23, opposite or irregularly alternate, linear, channelled above, obtuse, leathery, thick, viscous, usually hairy; **flowers** male and female on separate plants, in axillary clusters; **capsule** 3-4 winged, to 0.8 cm x 1.5 cm, narrow, spreading wings, glabrous, dark red to reddish brown, mature Aug-Dec.

A widespread species in south-western WA, where it usually grows on or near granite outcrops, or in lateritic soils. Has had limited cultivation, and may have potential for use as a light screen plant. Suited to well drained, light to medium soils, with dappled shade to full sun. May need light pruning while young to promote lateral growth. Propagate from seed or cuttings.
This species has affinities to *D. tenuifolia,* which has longer leaves, glabrous leaflets and a 4-winged capsule.

Dodonaea lanceolata F. Muell.
(lanceolate)
Qld, WA, NT
1-4 m x 1-3 m   throughout the year
Small to tall **shrub; branches** slender, spreading; **branchlets** angular or flattened, glabrous; **leaves** 3-15 cm x 1-3 cm, simple, elliptic, long petiole, margin thickened, entire or slightly wavy, shiny, glabrous, bright to dull olive green; **flowers** bisexual or unisexual, male and female can be on same plant, in terminal panicles; **capsules** 3-winged, up to 2 cm x 2 cm, glabrous, initially green, maturing bright red or purple-brown to brown.

An erect shrub that occurs in a number of habitats including open woodland, rocky hills and near creeks and rivers in central and northern Australia. Uncommon in cultivation, yet seed has been offered for sale for many years. Should do best in arid or semi-arid and tropical climates, in well drained soils. Experimentation needed in temperate regions. Propagate from seed or cuttings.

A form that occurs inland along the Great Dividing Range, on or near the eastern coast (nearly its entire length) and on offshore islands, has subsessile leaves. It will become a separate variety in the forthcoming revision.

Dodonaea larreoides Turcz.
(resembles the genus *Larrea*)
WA
1-4 m x 0.8-3 m   Feb-March
Small to medium **shrub; branches** rigid, spreading;

316

branchlets angular, viscous, glandular, glabrous; **leaves** pinnate, 3-6.5 cm long; **rhachis** winged; **leaflets** to 0.7 cm x 0.3 cm, 17-31, opposite or sometimes alternate, oblong to obovate, dark green above, paler below, glabrous or slightly hairy on the entire margin, apex often 2-3 toothed; **flowers** male and female on separate plants, in axillary clusters; **capsules** 3-4 winged, to 0.9 cm x 1.3 cm, viscous, glabrous, red to purplish brown, mature Aug-Dec.

The distribution of this species is in the semi-arid areas north of Perth to north of Geraldton. It is found mainly slightly inland on a variety of soils including calcareous types, and usually occurs in mallee scrub or woodlands. *D. larreoides* is not amongst the most ornamental members of the genus, but is well worth growing in arid to semi-arid regions, especially on alkaline soils. It needs relatively well drained soil. Tolerant of most frost. Propagate from seed or cuttings.

This species has affinities to *D. inaequifolia*, which differs by having much narrower leaflets and broader capsules.

Dodonaea lobulata F. Muell.
(small lobes)
NSW, SA, WA   Lobed-leaf Hop-bush
1-3 m x 1-2 m   May-July
Small to medium **shrub**; **branches** many, slender, spreading; **branchlets** angled to slightly ribbed, viscous, glabrous or slightly hairy; **leaves** 1.5-5.5 cm x up to 0.5 cm, simple, narrow, flat to concave above, margins with prominent, small rounded lobes, shiny, often viscous; **flowers** male and female on separate plants, 2-3 in axillary clusters; **capsules** 3-4 winged, 1-1.5 cm x 1.4-2.1 cm, pendulous, on prominent stalks, glabrous, pink to red-brown, mature July-Feb.

This species is widespread in semi-arid regions, where it occurs on rocky hills, slopes and ridges of various types. In the WA goldfields it grows on sandy loam or clay soils that have a high gravel or rock content. Female plants can be very showy when in full fruit. It is best suited to warm to hot climates, and will grow in most well drained soils. Tolerant of slightly alkaline soils. Withstands frost and drought. Has potential as a container plant in temperate regions. Is susceptible to sooty mould, which is controlled by spraying scale with white oil. Propagate from seed or cuttings.

Dodonaea macrossanii, × ·6

Dodonaea macrossanii F. Muell. & Scortech.
(after J. M. Macrossan)
Qld, NSW
0.2-0.4 m x 0.5-1.5 m   May-Aug
Dwarf, spreading **shrub**; **branches** prostrate; **branchlets** terete or angled, densely hairy; **leaves** pinnate, to 0.8 cm long; **leaflets** 2-6, to about 0.3 cm long, opposite, angular-obovate to obovate, 2-3 toothed apex, thick, hairy; **flowers** male and female on separate plants, solitary in axils; **capsules** 3-4 lobed, to 0.4 cm x 0.6 cm, with long hairs, reddish brown, mature Sept-Nov.

This groundcovering species occurs in the Darling Downs Region of south-eastern Qld and on the western slopes of the Great Dividing Range, just south of the Qld-NSW border, where it grows in sandy soils. It is not well known in cultivation at present. Grows well in most well drained soils, and is highly suited to sandy and gravelly soils. Will tolerate dappled shade through to full sun. Hardy to frosts and extended dry periods. Propagate from seed or cuttings.

Dodonaea megazyga F. Muell. ex Benth.
(large leaflets)
Qld, NSW
2-5 m x 1.5-3 m   Aug-March
Medium to tall **shrub**; **branches** more or less erect, slender; **branchlets** angled, slightly hairy; **leaves** pinnate, 8-25 cm long; **leaflets** 1.7-4.5 cm x 0.4-0.8 cm, 13-31, irregularly alternate, sometimes opposite, lanceolate, flat, dark green above, paler below, margin entire or revolute, apex acute; **flowers** male and female on separate plants, borne in axillary panicles; **capsules** 3-4 winged, 0.8-1.1 cm x 1.4-2.3 cm, often profuse, borne on slender stalks, glabrous, reddish brown, mature Sept-March.

The distribution of this species is from just north of the border in south-eastern Qld, extending south to around Sydney. It is commonly found on sandstone in dry sclerophyll forest and on edges of rainforest. It is an erect species with ornamental foliage. Prefers well drained, light to medium soils, with dappled shade to partial sun, but should tolerate full sun in temperate regions. Hardy to most frosts. Should be useful as a light screening plant. Propagate from seed or cuttings.

Dodonaea microzyga F. Muell.
(small leaflets)
Qld, NSW, SA, WA, NT   Brilliant Hop-bush
0.5-1.5 m x 0.8-2 m   May-July
Dwarf to small, spreading **shrub**; **branches** many, rigid; **branchlets** terete or slightly angular, dark red-brown, viscous, glandular, glabrous or slightly hairy; **leaves** pinnate, 0.4-1.2 cm long, with terminal leaflet; **leaflets** 3-11, to 0.8 cm x 0.3 cm, oblanceolate to obovate, viscous, with prominent glands, apex obtuse or rounded; **flowers** male and female on separate plants, 1-2 per axil, profuse; **capsules** 3-4 winged, 1-1.4 cm x 1-1.8 cm, glabrous or slightly hairy, viscous, red to reddish brown or purple, mature Aug-Dec.

A most ornamental species, as its vernacular name suggests. It occurs over a large area of semi-arid Australia. Adapts well to cultivation, where it needs relatively well drained, light to medium soil. Does best

Dodonaea multijuga

in full sun, but tolerates partial sun, although it usually renders less colourful, less profuse fruits. Frost hardy, and responds well to pruning. Has potential for cultivation in large containers. Propagate from seed or cuttings.

The form which occurs in WA is distinct from the populations in other states. It has larger leaves, and toothed or notched leaflets, and will be raised to varietal status in the forthcoming revision.

Dodonaea multijuga G. Don
(many leaflets)
Qld, NSW
1-2.5 m x 1-3 m   Aug-Nov; also sporadic
Small to medium **shrub**; **branches** erect to spreading, densely covered with hairs; **branchlets** angular, glandular, densely covered with hairs; **leaves** pinnate, 2-7 cm x 0.3-1.3 cm; **rhachis** slightly winged; **leaflets** to 0.8 cm x 0.4 cm, 10-30 or more, oblong to obovate, opposite or alternate, margins revolute, apex toothed, or notched and reflexed; **flowers** male and female on separate plants, in many-flowered axillary panicles; **capsules** 3-4 winged, to 1.6 cm x 2 cm, slightly inflated, initially and remaining lime green, or maturing to purple-and-green, or red to reddish brown.

There are some ornamental-fruited forms of this species. One is a spreading shrub of about 1 m x 2 m, which has bright light green fruits. Others are more upright and produce reddish or purplish fruits. Its distribution is from the Moreton District in south-eastern Qld, to NSW where it occurs in the Northern Tablelands, Central Tablelands and Central Coast Regions. It usually occurs in sandy soil or sandstone, but has proved adaptable by growing well on heavy soil types in cultivation. Is suited to a situation with dappled shade through to full sun. Withstands frost and extended dry periods. Responds well to light pruning. Capsules usually take 6-8 months to mature. Propagate from seed or cuttings.

Dodonaea nematoidea Sherff
see under *D. viscosa* ssp. B

Dodonaea oxyptera F. Muell.
(sharp wings)
Qld, WA, NT
1-3 m x 1-2 m   Jan-Feb
Small to medium **shrub**; **branches** slender, spreading; **branchlets** terete, angled or slightly ribbed, densely hairy, sometimes woolly; **leaves** pinnate, 0.4-4 cm long; **leaflets** 0.4-1.5 cm x up to 0.4 cm, 4-12, oblong to oblanceolate or obovate, dark green above, paler below, blunt apex, margins recurved; **flowers** rarely bisexual, with male and female flowers on separate plants, 1-2 per axil; **capsules** 4-winged, to 0.8 cm x 1.2 cm, densely hairy, reddish brown, mature May-Aug.

From the tropical regions of northern Australia, such as the islands of the Gulf of Carpentaria, the dry rocky hills of Arnhem Land and to the Kimberleys in the west.

Recommended for tropical regions, as it does not occur naturally any further south than latitude 18°S.

Dodonaea microzyga   T.L. Blake

Suited to very well drained soils, with a situation that receives plenty of sunshine. Pruning should promote bushy growth, as in nature it is often scraggy. Propagate from seed or cuttings.

Dodonaea pachyneura F. Muell.
(thick nerves)
WA
1.5-4 m x 2-3 m   March-July
Small to medium **shrub**; **branches** erect; **branchlets** terete to slightly angular, glandular, glabrous to slightly hairy; **leaves** 2.5-8 cm x 0.2-0.7 cm, simple, linear or oblanceolate, leathery, thick, glabrous or slightly hairy along midrib and margin, entire or slightly toothed; **flowers** male and female on separate plants, in axillary clusters; **capsules** 3-4 winged, to 0.9 cm x 1.6 cm, glandular, viscous, glabrous, red to purple-brown, mature Aug-Sept.

The range of this species is from Meekatharra in the south to the Hamersley and Robinson Ranges in the north. It occurs on rocky hillsides, ironstone ridges or red, sandy and stony soils. Not well known in cultivation, but requires very well drained soils, with partial or full sun. It is a species with ornamental fruits. Propagate from seed or cuttings.

Dodonaea peduncularis Lindley
(with peduncles)
Qld, NSW   Stalked Hop-bush
0.5-2 m x 1-2.5 m   June-Oct
Dwarf to small **shrub**; **branches** many; **branchlets** nearly terete to angular, slightly hairy; **leaves** 0.5-2 cm x 0.2-0.5 cm, simple, oblanceolate to narrow-cuneate, olive green, leathery, margin entire, apex usually 3-4 toothed, glabrous; **flowers** male and female on

318

separate plants, borne in clusters at ends of branchlets; **capsules** 4-winged, 0.7-1.3 cm x 0.8-1.3 cm, prominent wings, glabrous, initially cream, maturing to dark red to brown during Oct-Dec.

D. peduncularis occurs over a wide range of the western slopes of the Great Dividing Range, where it is usually confined to sandy soils and sandstone hills. Evidently not well known in cultivation. Should be suited to semi-arid areas, as it likes plenty of sunshine and well drained soils. Experimentation is required in temperate regions. Propagate from seed or cuttings.

Dodonaea peduncularis var. **coriacea** Ewart & Davies
= *D. coriacea* (Ewart & Davies) McGillivray

Dodonaea petiolaris F. Muell.
(with petioles)
Qld, NSW, WA, NT   Balloon Hop-bush;
                    Green Hop-bush
1-3 m x 1-2.5 m   March-July; also sporadic
Small to medium **shrub**; **branches** erect, often sparse; **branchlets** angular, viscous, sometimes slightly hairy; **leaves** 3.5-8 cm x 1-2.5 cm, simple, petiolate, lanceolate to ovate, entire to sinuate, concave, shiny, often sticky, bright green; **flowers** male and female on separate plants, in terminal clusters; **capsules** 3-winged, to 4 cm x 3 cm, glabrous, prominent venation, inflated, narrow wings, yellow or pink to purple-red, mature Aug-Nov.

Dodonaea multijuga green-fruited form   W.R. Elliot

The female plants of this species can be very decorative, with large bladder-like fruits which may be on the plants for the greater part of the year. In nature, the species inhabits rocky hills and ridges, and gibber plains, in shallow soils often with a clay subsoil. Little is known of its performance in cultivation, but it certainly has potential. Should do well in most well drained soils, in a warm to hot situation. Would tolerate semi-shade in arid to semi-arid regions. Frost hardy. Propagate from seed or cuttings.

Dodonaea physocarpa F. Muell.
(inflated fruit)
Qld, WA, NT
1-2 m x 0.6-2 m   Oct-Jan
Small, multi-stemmed, spreading **shrub**; **branches** many, spreading; **branchlets** angular, ribbed, hairy; **leaves** pinnate, 2-5 cm long; **rhachis** deeply channelled above; **leaflets** 1-2 cm x 0.2-0.5 cm, 6-12, opposite or alternate, oblong to obovate, light green, scabrous, viscous, hairy, margins entire or slightly wavy, apex acute, usually with 2-3 teeth; **flowers** rarely bisexual, male and female on separate plants, in axillary or terminal clusters; **capsules** 4-6 winged, to 2.5 cm x 2.5 cm, inflated, wings narrow, slightly hairy or glabrous, pink to reddish brown, mature Feb-Aug.

Female plants of this species are very showy when in fruit. It is restricted to tropical regions, where plants occur on stony soil in shrubland or low woodland. Best suited to tropical and subtropical areas. Needs very well drained soils and a position where it receives plenty of sunshine. In temperate regions, it is worth trying as a container plant. Propagate from seed or cuttings.

Dodonaea pinifolia Miq.
(leaves similar to those of the genus *Pinus*)
WA
0.2-1.5 m x 1-2 m   Dec-March
Dwarf to small, spreading **shrub**; **branches** intricate; **branchlets** angled or ribbed, dark red-brown, viscous, slightly hairy; **leaves** 0.8-3.5 cm x 0.1-0.3 cm, simple, sessile, linear to terete, or oblong to obovate, viscous, usually glabrous, margin revolute, entire or with 1-4 teeth; **flowers** rarely bisexual, male and female on separate plants, 1-3 per axil; **capsules** 3-4 angled, 0.5-1.2 cm x 0.5-2 cm, inflated, with short to long horn-like appendages at wing-tips, glabrous, light to pinkish brown, mature Aug-Nov.

A widespread species of south-western WA, where it usually occurs in rocky areas, sandy or gravelly soils or sandy loam. It is found in various plant communities from heathland through to dry sclerophyll forest. Not well known in cultivation at present. It should prove adaptable to a wide range of soils and climatic conditions. Probably best suited to well drained, light to medium soils, with partial or full sun. Frost and drought tolerant. Propagate from seed or cuttings.

Dodonaea pinnata Smith
(pinnate)
NSW
0.5-1.5 m x 1-2 m   Aug-Oct
Dwarf to small **shrub**; **branches** spreading; **branchlets** terete or slightly angular, with long hairs; **leaves**

Dodonaea platyptera

pinnate, 2-4.5 cm long; **rhachis** channelled above; **leaflets** 0.4-1 cm x about 0.2 cm, 10-26, opposite or alternate, narrow-obovate to ovate, olive green above, paler below, viscous, densely hairy, margins entire, revolute; **flowers** male and female on separate plants, borne in terminal clusters; **capsules** 4-winged, to about 2 cm x 2 cm, wings prominent, viscous, glabrous or few hairs on wings, reddish brown to purple, mature Nov-Feb.

This species is confined to the Central Coast Region, where it occurs on sandstone or sandy soil in dry sclerophyll forest. Evidently not well known in cultivation, it needs very well drained, light to medium soils, with dappled shade to partial sun. Should tolerate full sun in southern Australia. Suited to container cultivation. Propagate from seed or cuttings.

Dodonaea platyptera F. Muell.
(broad wings)
Qld, WA, NT
2-10 m x 1.5-5 m   Feb-March; also sporadic

Medium or tall **shrub** to small **tree**; **branches** erect; **branchlets** terete or angular, usually hairy; **leaves** 4.5-10 cm x 1-3.3 cm, simple, petiolate, narrow to broad-elliptic, viscous, glabrous or slightly hairy on margin and midrib, margin entire, apex usually acute; **flowers** male and female on separate plants, in axillary and terminal clusters; **capsules** 3-4 winged, 0.7-1.5 cm x 2-3 cm, prominent long wings, glabrous, light brown, mature June-Oct.

A tropical species that occurs north of latitude 18°S. It grows in close proximity to water, eg. creeks, rivers and estuaries, and is usually found on sandy soils. Not well known in cultivation, but has potential for tropical and subtropical regions. It is suitable for coastal exposure, and will grow well in alkaline soils. Needs well drained, light to medium soils. Should grow in semi-shade through to full sun. Propagate from seed or cuttings.

Dodonaea polyandra Merrill & Perry
(many stamens)
Qld
1.5-8 m x 1-4 m   May-Oct

Small to tall **shrub** or small **tree**; **branches** spreading; **branchlets** acutely angled or flattened, viscous, glabrous; **leaves** 6-12 cm x 1.6-4.2 cm, simple, petiolate, narrow-elliptic to obovate, dark green above, paler beneath, viscous, glabrous, margins entire, slightly wavy, apex pointed; **flowers** male and female on separate plants, in clusters at ends of branchlets; **capsules** 2-3 winged, 1.6-3 cm x 1.5-3 cm, glabrous, purplish to purplish brown, mature Oct-Dec.

D. polyandra is confined mainly to the east and north coasts, and offshore islands of Cape York Peninsula. It also occurs in the western part of New Guinea. Is usually on rainforest margins, but is also found in scrublands and woodlands. Is often near river or creek banks, in gravelly or sandy soils. Evidently not well known in cultivation, and is best suited to tropical and subtropical regions. Should do best in moist but well drained, light to medium soils, in a position with dappled shade or partial sun. Propagate from seed or cuttings.

Dodonaea polyzyga F. Muell.
(many yokes)
WA, NT
2-3 m x 2-3 m   May-Aug

Medium, spreading **shrub**; **branches** spreading; **branchlets** terete or angled, viscous, densely hairy; **leaves** pinnate, 9-18 cm long; **rhachis** narrow; **leaflets** 1-2.2 cm x up to 0.6 cm, 28-46, opposite or alternate, oblong to oblong-lanceolate, olive green above, paler below, viscous, sparsely hairy, margin entire, apex pointed; **flowers** unisexual or bisexual, in large terminal clusters; **capsules** 3-winged, 1.2-1.7 cm x 2-2.7 cm, viscous, sparsely hairy, pink to red-brown, mature May-Aug.

The distribution of this species is confined to the Kimberleys in north-eastern WA, and the Victoria River area in NT. It usually grows on rocky slopes. It has decorative ferny foliage and large fruits, which should lead to its increased cultivation in tropical regions. It may survive as a container plant in sub-tropical regions. Needs very well drained soils, in a location that receives plenty of sunshine. Tolerates alkaline soils. May need regular pruning while young, to promote bushy growth. Propagate from seed or cuttings.

D. polyzyga has some affinities to *D. megazyga*, which differs in its larger leaflets and smaller capsules.

Dodonaea procumbens F. Muell.
(procumbent)
NSW, Vic, SA   Trailing Hop-bush
prostrate-0.6 m x 1-3 m   Oct-Feb

Dwarf, spreading **shrub**; young growth slightly viscous; **branches** many, often rooting at nodes, glabrous, prostrate to procumbent; **branchlets** angular or flattened, slightly hairy; **leaves** 1-3 cm x up to 0.9 cm, simple, angular, obovate to oblanceolate, light to dark green, viscous, pointed, apex usually 1-4 toothed but can be entire, rarely glabrous, crowded; **flowers** rarely bisexual, male and female on separate plants, 1 or 2 in clusters terminating branchlets; **capsules** 3-4 winged, about 1 cm x 1 cm, glabrous, initially green, maturing to reddish to dark purple during Oct-Feb.

An extremely versatile species. It occurs naturally in a wide range of soils and climatic conditions. Is an excellent groundcover due to its dense foliage. Grows in most soils, and tolerates shady to sunny situations. Withstands periodic inundation and extended dry periods. Frost tolerant. Various forms have been selected. Some are virtually prostrate, while others are more shrubby in appearance. Has potential for wide landscaping usage. Excellent for embankments. Propagate from seed or from cuttings which strike readily.

Dodonaea ptarmicaefolia Turcz.
(leaves similar to those of the genus *Ptarmica*)
WA
1-4 m x 1-2.5 m   April-July

Small to tall **shrub**; **branches** erect and spreading; **branchlets** angular or ribbed, viscous, glabrous; **leaves** 2.5-6 cm x 0.2-0.6 cm, linear to linear-lanceolate, petiolate, bright green, leathery, viscous, margin serrated for upper two-thirds of leaf, glabrous, apex acute and recurved; **flowers** male and female on separate plants, borne in short axillary clusters; **capsules** 3-4 winged, to about 1 cm x 1.6 cm, prominent wings, glandular, initially green, maturing to purplish red to reddish brown during Oct-Jan.

This species occurs in the Avon, Roe and Eyre Districts, from Tammin in the west to Peak Charles in the east, and south to the coast. It is generally found on sandy loams in mallee scrub communities. Should prove adaptable to a wide range of conditions, but is not well known in cultivation. Needs relatively well drained soils, with partial to full sun. Withstands frost and extended dry periods. Propagate from seed or cuttings.

The var. *subintegra* Benth. will be placed as a subspecies of *D. viscosa* in the forthcoming revision.

Dodonaea rhombifolia Wakef.
(rhomboid-shaped leaves)
NSW, Vic   Broad-leaf Hop-bush
0.6-2.5 m x 1-2 m   Aug-Dec

Dwarf to medium **shrub**; **branches** erect; **branchlets** acutely angled, usually whitish, viscous; **leaves** 5-9 cm x 1-3 cm, simple, oblanceolate to narrow-ovate, petiolate, entire, thickish, dull grey-green above, margin entire, revolute, fairly crowded; **flowers** male and female on separate plants, in axillary clusters; **capsules** 4-winged, to 1.3 cm x 2.5 cm, wings large, initially green to light brown, then reddish, mature Sept-Feb.

A fairly bushy species, uncommon in nature. It occurs from the Warrumbungle Mountains in the north, to north-eastern Vic. It grows on mountains in rocky situations, or in stony gullies and creek-beds. Not well known in cultivation, but the greyish green foliage and large hops should ensure that it becomes better known. Needs well drained, light to medium soils, with dappled shade to partial sun. Will probably tolerate full sun. Frost hardy. Propagate from seed or cuttings.

D. rhombifolia has affinities to *D. truncatiales*, which has narrower leaves and shorter fruits.

Dodonaea rupicola C. White
(of rocky areas)
Qld   Velvet Hop-bush
0.6-1.5 m x 0.6-1.5 m   Aug-Nov

Dwarf to small **shrub**; **branches** spreading; **branchlets** terete to slightly angular, many long hairs; **leaves** pinnate, 2-4.5 cm long; **rhachis** winged; **leaflets** 0.4-1 cm x 0.2-0.4 cm, 10-18, opposite, oblong to oblanceolate, dark to olive green above, paler below, rough, densely hairy, margin entire, recurved or revolute, apex often toothed; **flowers** male and female on separate plants, in terminal multi-flowered clusters; **capsules** 4-winged, to 0.9 cm x 1.5 cm, densely hairy, red-brown, mature Aug-Nov.

This leafy, decorative species occurs only in the Glasshouse Mountains, where it inhabits open forest, growing on shallow soils. Certainly has potential for cultivation. Highly recommended for tropical and subtropical areas. Experimentation is needed in temperate regions. Should do best in well drained soils rich in organic material, in dappled shade or partial sun. Has potential as a container plant. Propagate from seed or cuttings.

Dodonaea serratifolia McGillivray
(serrated leaves)
NSW
1-1.5 m x 0.6-1 m   Sept-Oct

Small erect **shrub**; **branches** erect; **branchlets** angular to flattened; **leaves** 3.5-7.5 cm x 0.7-1.2 cm, simple, petiolate, narrow-elliptic, dark green above, paler below, glabrous, margin with irregular small teeth, apex acute or blunt; **flowers** bisexual and unisexual, with male and female on separate plants, in clusters at ends of branchlets; **capsules** 3-4 winged, 1.4-1.8 cm x 1.3-1.5 cm, glabrous, light brown to pinkish purple, mature Sept-Nov.

This species is very restricted in nature, occurring only in the Northern Tablelands east of Glen Innes, where it grows on granitic soils in dry sclerophyll forest. Not well known in cultivation, but should do best in very well drained soils, with dappled shade to partial sun. It may tolerate full sun. Hardy to most frosts and withstands extended dry periods. Propagate from seed or cuttings.

Dodonaea stenophylla F. Muell.
(slender leaves)
Qld, NSW, NT
1-4 m x 1-2.5 m   sporadic

Small to medium **shrub**; new growth viscous; **branches** crowded, erect; **branchlets** angular to flat, glabrous, viscous; **leaves** 4-12 cm long, simple, narrow-linear, rigid, margins entire and thickened, revolute, glabrous, viscous, apex acute; **flowers** male and female on separate plants, in axillary clusters, or rarely terminal; **capsules** 3-4 winged, to about 1 cm x 1.5 cm, long wings, glabrous, light brown to reddish brown, mature throughout the year.

D. stenophylla has its widest distribution in Qld, where it is found mainly in forests of the Great Dividing Range. It also occurs in the New England area of NSW, and there is an isolated occurrence near Daly Waters in NT. It grows on a wide range of soil types, and should prove adaptable in cultivation. To date it is mainly grown by enthusiasts. Best suited to well drained soils, with partial to full sun. Tolerates moderate frosts and extended dry periods. Will grow in alkaline soils. In nature, this is often a straggly shrub, but regular pruning while young should produce bushy growth. Propagate from seed or cuttings.

Dodonaea stenozyga F. Muell.
(narrow yoke)
Vic, SA, WA
0.6-1.5 m x 1-1.5 m   Aug-Dec

Dwarf to small **shrub**; young growth viscous; **branches** many, erect; **branchlets** very slender, angular, glabrous; **leaves** pinnate, sometimes with a terminal leaflet, 0.9-4 cm long; **leaflets** 1-2.5 cm x up to 0.2 cm, 2-10, distant, linear-terete, bright green, glabrous, viscous, channelled above; **flowers** male and female on separate plants, 1-3 per axil; **capsules** 4-winged, to 1.5 cm x 1.7 cm, prominent wings, viscous, glabrous, purple-brown, reddish or black, mature Sept-Feb.

This species occurs in the drier parts of temperate Australia, where it is found on a wide range of soil types, growing amongst mallee scrub or in open wood-

Dodonaea tenuifolia

land. It is not well known in cultivation. Has potential for use as a low shrub in regions of low rainfall, and it may grow in a north-facing position, such as against a wall, in southern Australia. Needs good drainage. Should adapt to most soils including calcareous types. Hardy to frost and drought. Propagate from seed or cuttings.

Dodonaea tenuifolia Lindley
(slender leaves)
Qld
1-3 m x 1-2 m

Feathery Hop-bush
March-April

Small to medium **shrub**; **branches** many, dense; **branchlets** flattened, angled, glabrous; **leaves** pinnate, 5.5-12 cm long; **leaflets** 0.9-3 cm x up to 0.2 cm, 9-25, linear, distal ones opposite, basal ones alternate, viscous, leathery, margin entire; **flowers** male and female on separate plants, in multi-flowered axillary clusters; **capsules** 4-winged, 0.5-1 cm x 1-2 cm, brownish purple to red, mature Aug-Dec.

This species has had only limited cultivation, and is certainly worthy of wider planting. It requires well drained, light to medium soils, and plenty of sunshine. Should be frost hardy. Has potential also as a container plant. Propagate from seed or cuttings.

Has some affinities to *D. microzyga*, which has fewer leaflets and much larger hops.

Dodonaea aff. tenuifolia (Qld, NSW)
Qld, NSW
1-3.5 m x 1-3 m

Dec-May

Small to medium **shrub**; **branches** slender, spreading; **branchlets** angular to terete, viscous, glandular; **leaves** pinnate, 2-4.5 cm long; **rhachis** rarely winged; **leaflets** 0.5-1.6 cm x up to 0.2 cm, 8-22, opposite or alternate, linear to narrow-oblong, glabrous to slightly hairy, entire to slightly wavy, apex blunt; **flowers** male and female on separate plants, in axillary clusters; **capsules** 3-4 winged, 1-1.3 cm x 1-1.5 cm, glandular, glabrous, red to reddish brown, mature May-Nov.

This showy species occurs in south-eastern Qld and north-eastern NSW, where it grows in open forest and woodlands. Is hardy and adapts to a wide range of soils, provided there is adequate drainage. Likes partial or full sun. Frost and drought tolerant. A form with leaves to 5.5 cm long, and toothed leaflets, has been in cultivation for many years under the wrong names of *D. microzyga* and *D. adenophora*. It is very decorative when its fruits reach maturity, the bushes becoming red. Adapts to most soils including relatively well drained clay-loams. Is an excellent screening plant. It will be given subspecific rank in the botanical revision. Propagate from seed or cuttings. Propagation should be from cuttings, to retain characteristics of selected forms.

This species is distinguishable from *D. tenuifolia* because it has prominently angular branchlets and long capsule wings. *D. species aff. tenuifolia* from SA has much shorter leaves and smaller capsule wings.

Dodonaea aff. tenuifolia (Qld, NSW)   T.L. Blake

Dodonaea aff. tenuifolia (SA)
SA
1-2 m x 1-2 m

Feb-April

Small **shrub**; **branches** spreading, crowded; **branchlets** angular to slightly ribbed, viscous, glandular, glabrous; **leaves** pinnate, 0.8-2 cm long; **rhachis** narrowly winged; **leaflets** 0.3-0.8 cm x about 0.1 cm, 10-16, opposite, linear, bright green, viscous, glands on undersurface only, glabrous, margin entire, apex sometimes 1-2 toothed; **flowers** rarely bisexual, male and female on separate plants, axillary; **capsules** 3-4 winged, 0.7 cm x 1-1.5 cm, viscous, glandular, sparsely hairy, pink to reddish brown, mature Sept-Nov.

This species is not well known. It occurs in the south-eastern region, north of St. Vincent's Gulf, and across to near Renmark in the east. This region is semi-arid, and in cultivation plants will need very well drained soils, with plenty of sunshine. Should tolerate most frosts. In temperate regions it should do well as a container plant. Propagate from seed or cuttings.

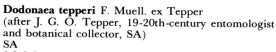

Dodonaea aff. *tenuifolia* (Qld, NSW), × ·6

Dodonaea triangularis, × ·6

Dodonaea tepperi F. Muell. ex Tepper
(after J. G. O. Tepper, 19-20th-century entomologist and botanical collector, SA)
SA
0.3-0.6 m x 0.8-1.5 m   Dec-March
　　Dwarf, spreading **shrub**; **branches** entangled, crowded; **branchlets** angular, slightly hairy; **leaves** 0.8-1.4 cm x up to 0.3 cm, simple, subsessile, linear, oblong to narrow-elliptic, viscous, glabrous, margin revolute, entire to slightly wavy, apex acute or blunt; **flowers** rarely bisexual, male and female on separate plants, 1 per axil; **capsules** 3-4 angled, about 0.5 cm x 0.5 cm, wingless, glabrous, dark red-brown with orange markings, mature Aug-Nov.
　　This is an extremely rare species which apparently occurs only in one locality, east of Adelaide. All efforts should be made to conserve it in nature, and propagation could aid in its survival. The plants could be used in a re-establishment programme. It is not common in cultivation and should do best in fairly well drained soils. Is suited to growing on alkaline soils as it occurs naturally on grey sandy clay overlying limestone. Best planted where it will receive plenty of sunshine. Propagate from seed or cuttings.

Dodonaea triangularis Lindley
(triangular)
Qld, NSW
1.5-3.5 m x 1-2.5 m   Feb-June; also sporadic
　　Small to medium **shrub**; **branches** usually erect; **branchlets** angular, often flattened, usually hairy; **leaves** 1.5-6 cm x 0.6-3 cm, simple, obovate to oblanceolate, dull green, leathery, usually hairy, margin entire, thickened and revolute, apex usually 3-5 lobed; **flowers** rarely bisexual, male and female on separate plants, in axillary and terminal clusters; **capsules** 3-4 angled, to about 1 cm x 1.2 cm, wings small, usually hairy, red to dark brown, mature Sept-Dec.
　　A species from south-eastern Qld and the upper Hunter Valley region of NSW. It is usually found in soil derived from sandstone, in the Great Dividing Range. It is not well known in cultivation, but seed has

been offered for sale over many years. Should grow well in soils with good drainage. Will tolerate dappled shade through to full sun. May require pruning to promote bushy growth. Propagate from seed or cuttings.

Dodonaea trifida F. Muell.
(3-forked)
WA
0.5-1.5 m x 0.5-1.5 m   Sept-Dec
　　Dwarf to small **shrub**; **branches** spreading, densely hairy; **branchlets** slightly pendent, terete or nearly so; **leaves** 0.7-1.2 cm x up to 1 cm, simple, sessile, obovate to triangular, 3-toothed or 3-lobed above the middle, hairy on margin and midrib, or glabrous; **flowers** male and female on separate plants, many in terminal clusters; **capsules** 3-angled, to 1 cm x 1 cm, slightly hairy, dark red to brown with purple tonings, mature Nov-Feb.
　　A rare species in nature, it is mainly confined to coastal regions from Albany to Kundip, north of Hopetoun. It occurs as an undershrub on sandy loams or gravelly soils, mainly on hillsides. This ornamental dodonaea is uncommon in cultivation. Should grow in most well drained soils, in situations of dappled shade through to full sun. Propagate from seed or cuttings.

Dodonaea triquetra Wendl.
(3 angles)
Qld, NSW, Vic   Large-leaf Hop-bush
1.5-4 m x 1.5-3 m   Jan-April; also sporadic
　　Small to medium **shrub**; **branches** erect to spreading, slender, can be reddish; **branchlets** angular or flattened, glabrous; **leaves** 5-12 cm x 1-5 cm, simple, narrow to broad-elliptic, can be lanceolate to ovate, glabrous, margin entire or wavy, apex acuminate to acute, prominent central nerve; **flowers** rarely bisexual, male and female on separate plants, in clusters at ends of branchlets; **capsules** 3-winged, 1-1.6 cm x 1-1.5 cm, glabrous, brown to purple with yellow wings, mature mainly Sept-Jan.
　　Predominantly a coastal species, but also occurring

323

Dodonaea truncatiales

Dodonaea triquetra, × ·6

on the eastern slopes of the Great Dividing Range. Its distribution is from Bundaberg, Qld, to Orbost, Vic. Usually found on sand or sandstone in dry and wet sclerophyll forests, often colonizing disturbed areas. Adapts well to cultivation. Best suited to well drained, light to medium soils, with dappled shade to partial sun. Responds well to light pruning. Could be useful as a light screening plant, but usually short-lived in tropical and subtropical regions. Propagate from seed or cuttings.

Dodonaea truncatiales F. Muell.
(truncate)
NSW, Vic

Angular Hop-bush;
Propeller Hop-bush

1-3 m x 1-2 m   Aug-Dec

Small to medium **shrub**; **branches** erect; **branchlets** angular, ribbed, glandular, initially hairy, becoming glabrous; **leaves** 5.5-13.5 cm x 0.5-1.5 cm, simple, sessile, erect, narrow-elliptic, can be oblanceolate or linear, dark green above, viscous, glabrous, margin entire or with small teeth, apex acute, central nerve prominent; **flowers** rarely bisexual, male and female on separate plants, in axillary clusters; **capsules** 3-4 winged, to 0.8 cm x 2.5 cm, glandular, glabrous or slightly hairy, light brown or red to purple-brown at maturity.

This species is usually found in dry sclerophyll forests on sandstone. In NSW it occurs in the Central Coast, South Coast and Central Tableland Regions, while in Vic it is known from far East Gippsland. The interesting propeller-like hops are attractive. Should adapt to most well drained soils. It is suited to dappled shade or partial sun. Hardy to at least light frosts. May require pruning to promote bushy growth. Propagate from seed or cuttings. Has been sold under the incorrect spelling *D. truncaliales.*

The var. *heterophylla* Maiden & Betche will be raised to specific level in the new revision. See *D.* aff. *truncatiales.*

Dodonaea aff. **truncatiales**
Qld, NSW
1-3 m x 1-2.5 m   Aug-Dec

Small to medium **shrub**; **branches** erect, crowded; **branchlets** angular, ribbed to flattened, glandular, glabrous; **leaves** usually simple, sometimes pinnate; **simple leaves** 3.5-8.5 cm x 0.2-0.5 cm, sessile, linear to linear-lanceolate, dull green, viscous, glabrous, margin entire and revolute; **lateral leaflets** 0.6-3 cm x 0.2-0.3 cm, 1-10, opposite to alternate, glabrous; **flowers** male and female on separate plants, in axillary clusters; **capsules** 4-winged, to 0.7 cm x 2.5 cm, wings narrow and long, glabrous, brown at maturity.

This species is found mainly on the western slopes of the Great Dividing Range, where it grows on stony ridges or in sandy loam. On occasions, it grows on the red soil plains or in open forest. It differs from other species in having both simple and pinnate leaves. It is quite ornamental when in fruit, but is not familiar in cultivation. Best suited to semi-arid areas, but could prove adaptable to subtropical and temperate climates. Needs well drained soils, with partial or full sun. May prove useful as a low screening plant. Propagate from seed or cuttings.

Also known as *D. truncatiales* var. *heterophylla* Maiden & Betche.

Dodonaea viscosa trunk detail   D.L. Jones

324

Dodonaea viscosa   D.L. Jones

Dodonaea viscosa Jacq.
(sticky)
all states   Hop-bush
1-5 m x 1-4 m   Aug-Nov

In Australia this species is a complex that is currently recognized as having 7 forms which warrant sub-specific rank. Some of these forms are undescribed, while others are named species. Because of the forth-coming revision, these subspecies are included and are designated by: *D. viscosa* ssp. A, *D. viscosa* ssp. B etc. Intergradation occurs between these forms, but in general they can be distinguished by their leaf shape and dimensions.

D. viscosa extends over a wide area of Australia, in all states and on Lord Howe Island. It can grow in exposed coastal situations through to the arid regions of Central Australia. The species is widespread overseas, occurring in New Guinea as well as in the tropical regions of America, Africa and Asia.

Dodonaea viscosa ssp. A (Qld, NSW)

A spreading **shrub** of 1-3 m x 1-3 m; **branches** many, spreading; **branchlets** angular to flattened, glabrous; **leaves** 7-13 cm x 2-4 cm, elliptic or rarely obovate-elliptic, petiolate, margin irregularly wavy, glabrous; **flowers** bisexual, male and female on separate plants, in terminal clusters; **capsules** 2-3 winged, to 2.3 cm x 2.5 cm, yellow, light brown, dark reddish brown or pink to purple at maturity.

In Australia this subspecies occurs mainly in coastal north-eastern Qld and neighbouring offshore islands, with isolated occurrences in north-eastern NSW. It is also found in New Guinea and the tropical regions of America, Africa and Asia. It grows on sandy soils and sand-dunes near the sea. Needs very well drained, light to medium soils, with partial or full sun. Suited to coastal exposure and will develop into a dense shrub. Propagate from seed or cuttings.

Dodonaea viscosa ssp. B (Qld, NSW)
Also known as *D. nematoidea* Sherff

Grows to 2-6 m x 1.5-4 m; **branchlets** are usually glabrous; **leaves** 7-17 cm x 1.5-2.5 cm, lanceolate to narrow-elliptic, margin entire or slightly wavy, apex acute; **capsules** 3-4 winged, to 2.8 cm x 2.8 cm.

This is found mainly to the east of the Great Dividing Range. Its distribution is from Cairns in the north to Grafton and Lord Howe Island in the south. It grows in wet sclerophyll forest or woodlands, usually occurring on rocky slopes and hills, but also on sandy soils. In cultivation, it will need well drained, light to medium soils, in a situation that has dappled shade to partial sun. May tolerate full sun. Possibly needs supplementary watering in dry months. Propagate from seed or cuttings. Also occurs in tropical America, Africa and Asia.

Dodonaea viscosa ssp. C (Qld, NSW, Vic)
Also known as *D. angustifolia* L.f. or *D. viscosa* var. *linearis*.

Dense **shrub** of 1.5-5 m x 1.5-4 m; **branchlets** usually glabrous; **leaves** 7.5-16 cm x 0.5-1 cm, linear-lanceolate, tapering to apex and base, viscous, margin entire or slightly wavy; **capsules** 3-4 winged, to 2.8 cm x 2.8 cm.

The distribution in Australia of this ssp. is from the Darling Downs Region in south-eastern Qld, south through eastern NSW until it just reaches East Gippsland in Vic. Is commonly found in dry sclerophyll forest or woodland. Needs similar cultivation requirements to those for *D. viscosa* ssp. B, but will withstand drier conditions once established. Propagate from seed or cuttings.

Also occurs in tropical America, Africa and Asia.

Dodonaea viscosa ssp. D (Qld, NSW, Vic, SA, WA, NT)
Also known as *D. angustissima* DC. and *D. attenuata* Cunn.

A multi-stemmed **shrub** of 2-4 m x 2-4 m; **branchlets** glabrous; **leaves** 3-9.5 cm x 0.1-0.6 cm, linear to narrow-oblong, sessile, margin wavy to slightly toothed, apex acute to blunt; **capsules** 3-4 winged, to 2.8 cm x 2.8 cm.

Dodonaea viscosa

Dodonaea viscosa ssp. D, × ·5

This dense subspecies occurs in the semi-arid and arid regions, where it is found on sandy soil in rocky situations. Is fairly adaptable to differing soil types, but must have relatively good drainage. Will grow in dappled shade, but prefers partial or full sun. Frost and drought tolerant. Due to its multi-stemmed growth habit, it is ideally suited as a hedge and windbreak plant for dry areas with sandy soils. Is also capable of regenerating rapidly, and has become a naturalized weed in some pastoral areas. Withstands regular pruning. Propagate from seed or cuttings.

Dodonaea viscosa ssp. E (Qld, NSW, Vic, SA)
Also known as *D. cuneata* Sm.

An open to dense, spreading **shrub** to 1-3 m x 1-3 m; **branchlets** glabrous to slightly hairy; **leaves** 1.2-4 cm x 0.4-1.2 cm, triangular to narrow-obovate, cuneate, can be viscous, margin entire or slightly wavy, apex truncate or blunt, can be 2-3 toothed; **capsules** 3-4 winged, to 2.8 cm x 2.8 cm, can be blackish brown at maturity.

An ornamental shrubby subspecies that occurs in a variety of soils, plant associations and climatic conditions. Adapts well to cultivation, growing in most soil types, provided there is adequate drainage. Prefers partial or full sun. Frost and drought tolerant. Responds well to light pruning. Propagate from seed or cuttings. Readily distinguished from other forms by its small and often triangular leaves.

Dodonaea viscosa ssp. F (Qld, NSW, SA, WA, NT)
An erect to spreading **shrub** to 1.5-4 m x 1-3 m; **branchlets** glabrous; **leaves** 3-8 cm x 1-2.5 cm, spathulate to obovate, viscous, margin entire to slightly wavy, apex usually rounded, mucronate; **capsules** to 2.8 cm x 2.8 cm.

Generally restricted to arid regions, where it is usually found in rocky situations, both on hills and along creeks. It has had very limited cultivation, and should be best suited to very well drained, light to medium soils, with a hot, sunny aspect. Should respond to hard pruning, as it is recorded that this ssp. coppices readily after bushfires. May have application as a low hedge or windbreak in semi-arid to arid regions. Propagate from seed or cuttings.

Dodonaea viscosa ssp. G (Qld, NSW, Vic, Tas, SA, WA)
Also known as *D. spatulata* Sm., *D. cuneata* var. *rigida* Benth. and *D. viscosa* var. *arborescens* (Hook.) Sherff.

An erect to spreading **shrub** to 1.5-4 m x 1.5-3 m; **branchlets** glabrous to slightly hairy; **leaves** 2.5-8.5 cm x 0.6-1.6 cm, obovate to spathulate, viscous, margin entire to slightly wavy, or with small teeth, apex usually blunt or rounded, sometimes mucronate; **capsules** 3-4 winged, to 2.8 cm x 2.8 cm.

A widespread subspecies that ranges from temperate to semi-arid regions, where it generally occurs in sandy loams. This is the best-known ssp. of the complex, and is often sold just as *D. viscosa*. It is hardy and adaptable. Will grow in most well drained soils, with a warm to hot situation. Tolerates dappled shade, but prefers partial or full sun. Tolerant of frost, extended dry periods and moderate coastal exposure. Withstands hard pruning, even coppicing, once plants are established. Propagate from seed or cuttings.

It is possible that this ssp. is also in cultivation as a large-leaf form of *D. cuneata*, which originated in the Grampians, Vic.

Dodonaea species (Gawler Ranges)
SA
0.2-1 m x 0.5-1.5 m   Feb-March
Dwarf, spreading **shrub**; **branches** entwined, spreading; **branchlets** angular, viscous, glandular, slightly hairy; **leaves** 0.7-1.8 cm x 0.3-0.6 cm, simple, oblong to narrow-elliptic, olive green, viscous, glandular, glabrous to hairy, margin entire or with small teeth, apex usually truncate, with 2-3 teeth; **flowers** small, male and female on separate plants, 1-2 per axil; **capsules** 4-winged, 1-1.3 cm x 0.9-1.4 cm, viscous, usually glabrous, purple-red, mature Sept-Nov.

The granite rocks of the Gawler Ranges provide the habitat for this decorative species. It may prove hardy in temperate zones, as it occurs with species such as *Grevillea aspera* and *Halgania cyanea*, which grow well in such areas. It will need very well drained soils and a location that receives plenty of sunshine. Frost and

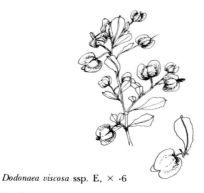

Dodonaea viscosa ssp. E, × ·6

326

drought tolerant. Has potential as a container plant. Propagate from seed or cuttings.

This species has affinities to *D. viscosa* ssp. E, which differs in having clusters of flowers, and glabrous foliage.

DOLICHANDRONE (Fenzl) Seemann
(from the Greek *dolichos*, long; *andros*, man; the 4 fertile stamens are longer than the staminode)
Bignoniaceae

Shrubs or small **trees**; **leaves** simple or pinnate, scattered or in whorls; **inflorescence** a terminal raceme; **flowers** bisexual, tubular, the lobes nearly equal and spreading; **stamens** 4, in pairs, a fifth reduced to a staminode; **fruit** a narrow capsule; **seeds** broadly winged on each side.

A small genus of 9 species with 3 found in tropical Australia. They are attractive shrubs, worth cultivating in tropical regions. Propagate from seed.

Dolichandrone alternifolia (R. Br.) Seemann
(alternate leaves)
Qld
3-5 m x 1-2.5 m   Oct-Jan

Medium to tall **shrub**; **bark** grey, roughened; **leaves** 2-6 cm x 0.5-2 cm, simple, ovate to broadly lanceolate, thick and leathery, dark green, the apex drawn out and acuminate, alternate; **racemes** 2-6 cm long, terminal; **flowers** 2.5-3 cm long, white to cream, tubular; **capsule** 15-30 cm x 0.3-0.5 cm, terete or somewhat flattened.

A little-known species found in forests and heathlands of north-eastern Qld. As it is slender and has attractive flowers it could have potential for cultivation in tropical and subtropical regions. The cultural requirements would be similar to those of *D. heterophylla*. Propagate from seed.

Dolichandrone filiformis (DC.) Fenzl
(thread-like)
Qld, WA, NT
3-8 m x 1-3 m   Dec-March

Medium to tall, slender **shrub** or small **tree**; **bark** brown, rough; **leaves** opposite or alternate, pinnate; **leaflets** 2-4 pairs, 10-25 cm x 0.2-0.4 cm, linear, terete, bright green; **racemes** compact, terminal; **flowers** 2.5-3 cm long, white or pale yellow, tubular, with lobes 1-1.3 cm long; **capsule** 20-35 cm long, terete or somewhat flattened.

A slender shrub confined to tropical districts, where it grows in open forests and woodlands. It is an attractive shrub with excellent horticultural potential for tropical and subtropical areas. It requires a sunny situation in freely draining soils. Propagate from seed or cuttings.

Dolichandrone heterophylla (R. Br.) F. Muell.
(variable leaves)
Qld, WA, NT   Lemonwood
2-6 m x 1-2.5 m   Oct-Jan; April-June

Medium to tall, slender **shrub** or small **tree**; **bark** tough, furrowed, dark brown; **leaves** variable, simple or pinnate, in whorls of 3, crowded on young shoots; **simple leaves** 2.5-12 cm x 0.1-0.4 cm, lanceolate, thick and leathery, grey-green; **pinnate leaves** with 3-13 leaflets; **leaflets** 2.5-10 cm x 0.1-0.4 cm, linear to lanceolate, leathery; **racemes** compact, on long peduncles, terminal; **flowers** 2.5-4 cm long, white, tubular, highly fragrant, the short terminal lobes with crisped margins; **capsule** 10-30 cm x 1-1.5 cm, terete or somewhat flattened; **seeds** about 2.8 cm long, prominently winged.

An attractive shrub widely distributed across northern Australia, growing in open situations, usually on deep sandy soils and extending to the coast on sand-dunes. Horticultural features include the thick, furrowed bark, interesting growth habit and showy, fragrant flowers. It is an excellent shrub for tropical regions or inland areas with a hot, dry climate. Its requirements are for sandy, well drained soil, in a sunny position. Young plants could make interesting subjects for bonsai culture. Propagate from seed which retains its viability for a couple of years. Cuttings would be worth trying.

Dolichos biflorus L. =
Macrotyloma uniflorum (Lam.) Verdc.

DONATIA Forster & G. Forster
(after Vitaliana Donati, 18th-century Italian naturalist)
Donatiaceae

Perennial, evergreen, cushion-like **herbs**; **leaves** small, densely overlapping; **flowers** small, solitary, sessile, mostly terminal; **petals** 5-10, free; **stamens** 2-3; **filaments** free; **fruit** an indehiscent capsule.

The genus *Donatia* is composed of 2 species. One occurs in Tas and New Zealand, and the other is restricted to subantarctic South America. For cultivation and propagation information in regard to this genus, see Cushion Plants, page 135.

Donatia novae-zelandiae Forster & G. Forster
(from New Zealand)
Tas   Cushion Plant
prostrate x 0.5-1 m   Oct-Jan

Dwarf, perennial **herb** forming a hard cushion-like mat of foliage; **branches** with adventitious roots; **leaves** about 0.5 cm long, linear-subulate, sessile, erect and

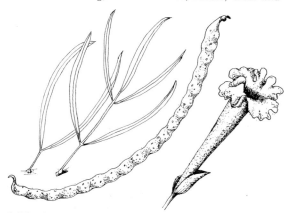

Dolichandrone heterophylla, × ·6

327

appressed, bright green; **flowers** about 0.6 cm diameter, white, terminal, solitary, scattered over the plant.

This interesting species frequents mountain peaks, in wet and exposed areas where snow persists for several months of the year. Not commonly cultivated, but certainly has potential for alpine gardens, or as a container plant. For detailed cultivation requirements, see Cushion Plants, page 135. Propagate from cuttings.

Donatiaceae (Mildbr.) Skottsb.

A small monogeneric family of 2 species distributed in Tas, New Zealand and southern parts of South America. They are dicotyledons found in subalpine situations, and have a dense cushion-like growth habit, and free petals and stamens. Plants are rarely cultivated and are mainly of interest to enthusiasts. They are closely related to the genus *Stylidium* and are sometimes included as a subfamily (Donatioideae) in the family Stylidiaceae.

Donatia novae-zelandiae   T.L. Blake

DOODIA R. Br.

(after Samuel Doody, 17th-century curator, Chelsea Physics Garden)
Blechnaceae

Small terrestrial **ferns**; **rootstock** erect or creeping, sometimes producing runners; **fronds** erect, pinnate or pinnately lobed, generally of harsh texture, alike or dimorphic; **fertile segments** linear; **sori** round, oblong or elongated, in 1-2 rows, bearing indusia.

A small genus of about 15 species but the exact number is uncertain due to variability and overlap of forms. Six species are known from Australia; however, all species can be difficult to identify, and appear to be variable. They are excellent ferns for cultivation and deserve to be much more widely grown. Propagate by division or from spore which is best sown while fresh.

Doodia aspera R. Br.

(roughened)
Qld, NSW, Vic   Prickly Rasp Fern
0.15-0.4 m tall

Small terrestrial **fern** forming spreading patches; **rootstock** creeping, black, densely covered with black scales, producing long stolons; **fronds** 0.15-0.4 m tall, erect, lanceolate, pinnatifid, roughened and harsh-textured, pale to dark green, sometimes yellowish, non-dimorphic; **segments** spreading, margins toothed; **sori** round, in 2-4 rows.

This handsome fern is usually found in colonies in open forest, often in fairly sunny situations, but it also grows in shaded rainforest. It is an excellent fern for garden culture, tolerating a wide range of soil types, and positions from shade to nearly full sun. Plants build up into small colonies by underground stolons, and the new fronds are usually of decorative bright pink or red shadings. It grows equally well in tropical or temperate regions and is quite frost hardy. It can also be established under large trees, tolerating the root competition without serious setback and it is sufficiently versatile to succeed as a basket subject. It requires well drained soil. Propagate by division of the clumps or from spore.

Doodia caudata (Cav.) R. Br.

(with a tail)
Qld, NSW, Vic, Tas   Small Rasp Fern
0.1-0.3 m tall

Small terrestrial **fern** growing in discrete clumps; **rootstock** tufted, erect, bearing numerous woody frond-bases; **fronds** 10-30 cm long, semi-erect or pendent, narrow, pinnate, dark green and somewhat rough, dimorphic; **segments** straight or upcurved, narrow, the terminal pinnae united into a tail; **sori** rounded or elongated, becoming confluent.

A common and widespread fern of eastern Australia, growing in a range of habitats such as moist earth-banks, exposed rock crevices, stream banks in rainforest and dry scree slopes. It is an excellent fern for culture, succeeding in a pot, basket or suitable position in the garden. It will withstand a fair amount of sunshine but seems to grow best where it receives morning sun or filtered sun. It mixes well with rocks, and looks good in a rockery. An acid, well drained soil produces satisfactory growth which can be improved by regular mulching and light applications of fertilizer.

This fern is extremely variable and some of its forms are very difficult to identify. The var. *laminosa* F. Muell. is a very distinctive form which has the upper two-thirds of the frond united into a tail, with the lower one-third lobed. The sori are elongated and are often carried parallel to the midrib. It makes an excellent pot plant and is found sporadically in NSW and Qld. Another distinctive frond-type is found on forma *triloba* F. Muell., which is a form of var. *caudata*. Its fronds have an undivided tail at the apex, and its basal pinnae are attractively lobed. Its rhizome usually bears numerous woody stipe-bases, and the plants grow in rocky slopes of the ranges of south-central Qld. All forms can be propagated from spore or offsets.

Doodia heterophylla (Bailey) Domin

(bearing fronds of more than one kind)
Qld
0.2-0.45 m tall

Small terrestrial **fern** growing in discrete clumps;

Doodia heterophylla   D.L. Jones

rootstock tufted, erect or creeping, bearing numerous woody stipes; **fronds** 0.2-0.45 cm tall, erect, stiff and somewhat roughened, pinnatifid, dark green, markedly dimorphic, the sterile fronds shorter and spreading, the fertile fronds erect, with narrow segments; **segments** spreading, curved, the lower ones reduced to lobes; **sori** elongated, becoming confluent.

A little-known fern found in open forests of south-eastern Qld, often amongst rocks. It is an attractive, neat fern for a pot or protected rockery, but is rarely grown. It grows quite rapidly in suitable conditions, and is fairly easy to raise from spore which is the best method of propagation.

Doodia maxima J. Smith
(largest)
Qld   Giant Rasp Fern
0.3-1 m tall

Medium sized terrestrial **fern** forming slowly spreading clumps; **rootstock** tufted, forming a short, erect trunk, producing stolons; **fronds** 0.3-1 m tall, erect, stiff and harsh-textured, dark green, broadly lanceolate, markedly dimorphic; **segments** 6-10 cm long, spreading, linear, with serrate margins, the fertile ones about half as wide as the sterile; **sori** elongated, in rows close to midrib.

This species, the largest of the genus, is a very handsome fern eagerly sought after by collectors, but it is rarely collected, and there is speculation that it may be a natural hybrid. The plants form robust, attractive clumps and are very hardy once established, steadily developing into a spreading clump by underground stolons. Propagation from spore appears to be difficult and division seems to be the most reliable method.

Doodia media R. Br.
(central; middle)
Qld, NSW, Vic, Tas   Common Rasp Fern
0.3-0.6 m tall

Small terrestrial **fern** forming spreading clumps; **rootstock** creeping, producing underground stolons; **fronds** 30-60 cm tall, erect, harsh-textured, dark green, pinnate, not dimorphic; **segments** spreading; **sori** in 2 rows, becoming confluent with age.

A widespread and common fern found in a variety of situations throughout eastern Australia, often in colonies. It is a hardy fern, readily growing in any partially protected garden situation. Responds to mulches and regular light dressings of fertilizer. Will withstand considerable sunshine, especially if regularly watered or grown in moist soil. The new fronds are often an attractive purplish pink. The species is somewhat variable and some forms may be difficult to identify. The ssp. *australis* B. S. Parris is a larger form, with some of the lobes adnate at the base. It is confined to north-eastern Qld. All forms can be propagated by division or from spore.

Doodia squarrosa Col.
(rough or scurfy)
Qld
0.3-0.9 m tall

Small to medium terrestrial **fern**; **rootstock** creeping; **fronds** 30-90 cm long, erect, narrow to broad-lanceolate in outline, dark green and harsh-textured, pinnate, not dimorphic; **segments** spreading, linear, slender, with a long, drawn out apex.

Localized and uncommon in south-eastern Qld, this species grows around the margins of wet soil, usually in small colonies. It is an attractive fern but is rare in cultivation. It grows easily in a pot or in a semi-shady situation in the ground. Propagate from spore or by division.

DORYANTHES Corr. Serr.
(from the Greek *dory*, a spear; *anthos*, a flower; refers to the long, narrow flower-stem)
Agavaceae

Tall perennial **herbs**; **leaves** lanceolate, very long, in a large tussock; **flower-stem** tall, stout, simple, with short leaves or bracts; **flowers** large, red, in short spikes, arranged in terminal globular or oblong heads; **perianth** spreading, with 6 nearly equal deciduous segments; **stamens** 6; **fruit** a capsule.

An endemic genus comprising 2 species. Some botanists consider this genus to have one species plus varieties. It is confined to coastal regions of south-eastern Qld, and northern and central NSW. Plants usually occur in forests or on rocky slopes, in well drained soils. Hillsides can be spotted with the prominent bright green foliage. Attractive in flower, with the flower-heads well displayed above the decorative foliage.

Doryanthes are popular in cultivation. *D. excelsa* was introduced to gardens in England during 1800, and is the most widely grown species at present. Generally, plants are slow-growing and take 5-10 years before the first flowering stems are produced. In general, they

329

Doryanthes excelsa

prefer well drained soils with some organic material incorporated. They usually respond well to light applications of fertilizer. Adaptable to partial or full sun, and will tolerate dappled shade. Highly suited as container plants; however, containers need to be large enough to allow for mature development. The flowerheads are readily damaged by frost, yet the foliage is tolerant. Established clumps, many years of age, can be successfully transplanted.

Propagate from seed which germinates readily without pre-sowing treatment. Clumps can be divided. Suckers are sometimes formed at the perimeter of a plant, and these can be removed while small, to be potted on until ready for planting.

Doryanthes excelsa Corr. Serr.
(tall; noble)
NSW — Flame Lily; Giant Lily; Gymea Lily; Illawarra Lily
1-1.5 m x 1.5-3 m — Aug-Feb
Tall perennial **herb**; **leaves** to 1.5 m x 10 cm, lanceolate, erect, radical, entire, bright green; **scape** 2-5 m tall, with scattered small leaves; **flowers** about 10 cm long, dark red, arranged in a large, terminal, globular head about 30 cm across; **capsule** to 10 cm long, black to dark brown.

This showy species is common in its habitat of the sandstone areas around Sydney. It is usually found in light forest or in open areas. Very amenable to cultivation. See genus description for cultivation notes. A white-flowered form was recorded as occurring in nature during the 19th century. There is no evidence to show that this form was cultivated.

Doryanthes excelsa var. **palmeri** (W. Hill) Bailey
see under *D. palmeri* W. Hill ex Benth.

Doryanthes guilfoylei Bailey
see under *D. palmeri* W. Hill ex Benth.

Doryanthes palmeri W. Hill ex Benth.
(after Dr. E. Palmer, botanical explorer)
Qld, NSW — Spear Lily
1-3 m x 1.5-6 m — Sept-Nov
Tall perennial **herb**; **leaves** to 3 m x 20 cm, lanceolate, erect, radical, entire, bright green; **scape** 2-5 m tall, with many small leaves; **flowers** about 6 cm long, reddish brown, arranged in a terminal elongated head to 1 m long; **capsule** to 7 cm long, dull greenish brown.

Not as common in cultivation as *D. excelsa*, this species is quite spectacular, with its long terminal flower-heads. It adapts well in cultivation. Grows and flowers successfully as far south as Melbourne. For cultivation details, see genus description.

This is also known by some authors as *D. excelsa* var. *palmeri* (W. Hill) Bailey. *D. guilfoylei* Bailey, sometimes known as *D. excelsa* var. *guilfoylei* (Bailey) Bailey, is also a variant of *D. palmeri*. It can have scapes up to 5 m tall and leaves to 3 m long x up to 20 cm wide.

Doryanthes excelsa   W.R. Elliot

DORYOPTERIS J. Smith
(from the Greek *dory*, a spear; *pteris*, a fern; in reference to the frond-shape of some species)
Sinopteridaceae

Small terrestrial **ferns**; **rootstock** tufted or creeping; **fronds** erect, palmatifid to cordate, lobed, somewhat leathery, sometimes bearing plantlets; **sori** in a continuous marginal band.

This genus consists of about 35 species, 2 of which are found in north-eastern Qld. They are attractive ferns but are somewhat difficult of culture, and are seen only in enthusiasts' collections. Propagate from spore.

Doryopteris concolor (Langsd & Fisch.) Kuhn
(uniform color)
Qld
0.05-0.15 m tall
Small terrestrial **fern**; **rootstock** short-creeping; **fronds** erect, tufted, on shiny, black, winged stipes, palmatifid, dark green, dimorphic; **lobes** of sterile fronds much broader than those on fertile fronds; **sori** marginal, brown.

This fern is quite drought resistant and is usually found in shady but dry locations such as under shrubs or dry rock-ledges in open forest. The fronds curl and shrivel during long dry periods but refreshen and regreen following rain. The species is somewhat tricky to grow, especially in areas with a cool, moist winter climate. It prefers warm conditions and low humidity, and seems best suited to warm, dry areas. It can be grown in the ground or in a pot of well drained, acid mixture. It should not be overpotted, and fairly large plants can be maintained in a relatively small pot. Propagate from spore.

Doryopteris ludens (Wall ex Hook.) J. Smith
(shining; polished)
Qld
0.15-0.35 m tall

Small terrestrial **fern**; **rootstock** creeping, somewhat thick and wiry; **fronds** erect, widely spaced, on long black stipes, palmatifid, with 5 unequal lobes, the lowest longest, dark green, strongly dimorphic; **lobes** of fertile fronds longer and much narrower than sterile fronds; **sori** marginal, brown.

A widespread fern that is very rare in Australia, known from only a couple of localities on Cape York Peninsula. It usually grows in rocky situations and is quite drought resistant. Culturally its requirements are similar to *D. concolor*. It is very cold sensitive and requires a heated glasshouse outside the tropics. Propagate from spore.

Doryopteris concolor   B. Gray

DORYPHORA Endl.

(from the Greek *dory*, a spear; *phoreus*, a carrier; referring to the pointed tips of the anthers)
Atherospermataceae (alternative *Monimiaceae*)

Trees; young shoots bearing soft hairs; **bark** fragrant, pustular or flaky; **leaves** simple, entire, opposite, with toothed margins; **flowers** solitary or in clusters of 2-4 at the ends of stout peduncles, hairy; **sepals** and **petals** alike; **staminodes** alternating with stamens; **fruit** an enlarged perianth, splitting to release hairy fruitlets (carpels).

A small, unique genus of 2 species endemic in eastern Australia. Both are large timber trees with characteristically fragrant leaves, bark and sapwood. They are rarely encountered in cultivation. Propagate from seed which is best sown when fresh.

Doryphora aromatica (Bailey) L. S. Smith
(spicy; aromatic)
Qld   Grey Sassafras
20-30 m x 5-12 m   Jan-April

Medium to tall **tree**; **trunk** straight, cylindrical; **bark** grey, with fairly large pustules; young shoots clothed with fine grey hairs; **leaves** 5-15 cm x 1-2 cm, elliptical, elongated in outline, dark green above, paler beneath, the apex long and drawn out, the margins toothed, fragrant when crushed; **peduncles** 1-2.5 cm long, in clusters of 2-4 in the upper axils; **flowers** about 1 cm across, white; **fruit** 1-2.5 cm long, tubular or club-shaped, splitting to release up to 6 fruitlets, each of which bears a plume of fine hairs about 2.5 cm long at one end, mature Oct-Jan.

Restricted to rainforests of north-eastern Qld, where it is widespread in both lowland and highland districts. All parts of the plant have a distinct and spicy fragrance. The timber is pale yellow and used in cabinet making and joinery. The tree is little-known in cultivation, and generally grows much too large for home gardens. It could make an interesting park tree for tropical and subtropical regions. Requirements are for deep, well drained, acid soils, and the plants may need protection from direct sun when small. They would probably benefit from mulches and artificial watering during dry periods. Propagate from seed which must be sown fresh.

Doryphora sassafras Endl.
(after the aromatic American genus *Sassafras*)
Qld, NSW   Yellow Sassafras
20-30 m x 5-10 m   May-Oct

Medium to large **tree** with a compact crown; **trunk** straight, cylindrical; **bark** grey, finely scaly; young shoots soft and silky; **leaves** 7-10 cm x 1-2 cm, elliptical, opposite, simple, dark green above, paler beneath, the apex blunt, the margins coarsely toothed, fragrant when crushed; **flowers** 2-3 cm across, white, bearing silky hairs, borne on short racemes, often in groups of 3; **perianth lobes** 6; **stamens** 6, with 6 staminodes; **fruit** 0.6-2 cm long, ovoid, brown, splitting on one side to expose fruitlets covered with hairs about 1 cm long, mature March-Aug.

A widespread species extending from south-eastern Qld, throughout eastern NSW to near the Vic border. It is found in moist forests and rainforests, and is most frequent at high elevations, where it is sometimes the dominant tree of the forest. The timber is yellow, and valued for floors, cabinet work and turnery. The tree is planted on a limited scale in reafforestation projects. All parts of the plant have a distinct, sweet fragrance. The tree is handsome but is rarely encountered in cultivation as it grows much too big for home gardens. It could have uses in parks or for street planting. It is best suited to districts with a cool climate, and the plants succeed well in mountainous areas. Deep, well-drained, humus-rich soils produce rapid growth,

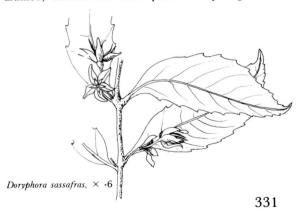

Doryphora sassafras, × ·6

331

although the species has proved adaptable in its tolerance of climate and soil type. Young plants make unusual pot subjects and could be used for indoor decoration. Propagate from seed which should be sown fresh.

DRACAENA Vand. ex L.
(from the Greek *drakaina*, a female dragon)
Agavaceae

Shrubs, **trees** or leaning **climbers**; **trunk** and **branches** marked with annular scars; **leaves** linear, simple, scattered, crowded near the end of the stems, the base stem-clasping; **inflorescence** a terminal panicle; **flowers** tubular at the base, with 6 spreading lobes; **stamens** 6; **fruit** a succulent berry.

A large genus of about 150 species with a solitary widespread species extending to northern Australia. Many exotic species and hybrids are commonly grown in tropical and subtropical gardens for their colourful leaves. They also make handsome pot plants and are excellent for decoration indoors or in glasshouses.

Dracaena angustifolia Roxb.
(narrow leaves)
Qld, NT
3-8 m x 1-2 m   Dec-March

Woody **climber** or slender, tall, suckering **shrub** or small **tree**; **stems** slender, branched, pithy; **leaves** 12-40 cm x 1-3 cm, linear, spreading, dark green, the apex acute, crowded near the apex of the stem; **panicle** 15-40 cm long, terminal, much-branched; **flowers** about 1.5 cm long, greenish white, tubular at the base, the lobes linear, spreading; **fruit** 1-1.5 cm across, brownish, pulpy, containing 1-3 large seeds.

A widespread species which extends to northern Australia; it is widely distributed in north-eastern Qld, the Torres Strait islands and the Top End. It grows in sheltered rainforests and is often locally common. The stems tend to be weak and thread their way through surrounding shrubs and trees which act as a support. The plants grow readily but are not commonly cultivated. They are very sensitive to cold, especially frosts. They can be grown in tropical gardens, where they require a shady position, or as glasshouse plants in temperate zones. They can be propagated from seed which has a short life and must be sown fresh, from cuttings which strike readily or by transplanting basal suckers.

Dracaenaceae Salisb.
A family of monocotyledons consisting of about 2 genera and 150 species scattered through the tropics. Genera in this family are included by most authors in the tribe Dracaeneae of the family Agavaceae.

DRACOPHYLLUM Labill.
(from the Greek *draco*, a dragon; *phyllon*, a leaf; in reference to the similarity of the leaves to those of the Dragon Tree *Dracaena draco*)
Epacridaceae

Dwarf **shrubs** to small **trees**; **leaves** resemble those of monocotyledons, crowded at branchlet tips, the bases overlapping and sheathing, narrow, often rigid and concave, prominent annular scar after leaf-drop; **flowers** in terminal or lateral panicles, racemes or spikes, rarely solitary; **calyx** of 5 persistent sepals; **corolla** cylindrical or bell-like, with 5 spreading lobes; **stamens** 5; **fruit** a capsule.

A genus of about 30 species. There are 5 species in Australia, and representation also in New Zealand, New Caledonia and the Antarctic. They usually inhabit moist mountain gullies and exposed subalpine regions. They are slow-growing and long-lived.

Foliage is decorative, and some species have large terminal panicles of flowers which are certainly ornamental. They are not common in cultivation, although *D. secundum* was introduced to England in 1823. This species responds well to pruning, and it is highly likely that others will too. Plants are well suited to container cultivation. Propagation is best from fresh seed using the bog method (see Volume 1, page 204). Propagation from stem cuttings has not been very successful to date. Experimentation with vigorous young growth that is just starting to become firm may have some potential.

Dracophyllum fitzgeraldii F. Muell.
(after R. D. Fitzgerald, 19th-century botanist and Deputy Surveyor-General, NSW)
NSW (Lord Howe Island)
3-12 m x 3-6 m   July-Nov

Medium **shrub** to small palm-like **tree**; **leaves** to 30 cm long, flat, sword-like, broadest at base, margins initially serrated, fine striation; **flowers** tubular, inflated, about 0.6 cm long, white to pale pink, arranged in congested terminal panicles; **bracts** deciduous.

This is the largest species in the genus and it is endemic to Lord Howe Island. It inhabits moist, sheltered gullies or exposed slopes and peaks where the gathering of cloud is a regular occurrence. Mature specimens are very old, and it is expected that under cultivation plants may take quite a few years to reach even 3 m in height. The species is not well known in cultivation. It needs very well drained soils that are moist for most of the year. Preference is probably for shade, dappled shade or partial sun rather than full sun.

Has potential as a large container plant. Propagate from fresh seed.

Dracophyllum milliganii Hook.f.
(after Dr. Joseph Milligan, 19th-century surgeon and naturalist, Tas)
Tas
0.5-4 m x 0.5-2 m   Dec-Jan

Small to medium, erect **shrub**; young shoots pinkish; **branches** few, only at base; **leaves** 15-90 cm long, usually in a terminal crown at end of branch, broad, sheathing base, tapering to a long point, margins minutely serrulate; **flowers** small, tubular, inflated, white but sometimes pale pink, borne in long panicles to 40 cm long, each flower-head subtended by pinkish and green bracts; **capsule** small.

An inhabitant of moist, sheltered gullies, mountain

Dracophyllum milliganii   T.L. Blake

slopes or ridges. Not well known in cultivation. Should be suited to cold temperate climates. Needs very well drained, but moisture retentive, light to medium soils. Frost and snow tolerant. Worthy of trial as a container plant. Propagate from fresh seed.

Dracophyllum minimum F. Muell.

(smallest)

Tas   Cushion Plant
prostrate x 0.3-1 m   Nov-Jan; also sporadic

A compact, dwarf, cushion-like **shrub**; **branches** many, short, erect; **leaves** to 0.5 cm long, narrow, sheathing base, imbricate, slightly concave, tapering to an acute point, rigid, margins slightly serrulate; **flowers** small, tubular, white, solitary, terminal, sessile, surrounded by leaves; **fruit** initially crimson, becoming brown and hard when ripe.

One of the quaint cushion plants that are common at altitudes above 1200 m. It grows in association with plants of similar character. Cultivation experience with this species is very limited. It requires very well drained, light to medium soils that have ample organic material to retain moisture. At the same time they require plenty of light. For further cultivation details, see Cushion Plants, page 135. Propagate from seed, cuttings or by division.

This species is very similar to *Abrotanella forsteroides*, which has small daisy-like flowers.

Dracophyllum sayeri F. Muell.

(after W. Sayer, original collector)

Qld
2-4 m x 2-4 m   July; also sporadic

Medium, spreading **shrub**; **branches** often horizontal, often intertwined; **leaves** to 45 cm x 2 cm, crowded at ends of branches, sheathed at base, tapering to a long, narrow point, concave, margin more or less smooth; **flowers** small, tubular, white to pink, profuse, borne in terminal panicles, prominent bracts to 7.5 cm long, white and pink; **capsule** small, broader than long.

This species has ornamental foliage, and is confined to high elevations in north-eastern Qld, where it enjoys moist, humid conditions throughout the year. Has had limited cultivation, both in the ground and as a container plant. It needs well drained, light soils in a situation of semi-shade to dappled shade. Propagate from seed or cuttings.

Dracophyllum secundum R. Br.

(arranged on one side)

NSW
0.2-1.5 m x 0.5-1.5 m   July-Dec

Dwarf to small **shrub**; **branches** spreading or erect; **leaves** 5-16 cm x 0.5-0.8 cm, linear-lanceolate, spreading, concave, sheathing at base, margins serrulate, glabrous; **flowers** tubular, to 1 cm long, white to pink, borne in narrow terminal panicles or loose racemes to 15 cm long; **capsule** small, globular.

An inhabitant of rocky cliffs and gullies in the sandstone regions around Sydney, and in the Blue Mountains. Also withstands coastal exposure on rock faces. Despite its limited cultivation it is proving adaptable. Grows in light to medium, well drained to relatively well drained soils. Prefers some root protection in full

Dracophyllum fitzgeraldii   D.L. Jones

sun. Withstands limited dry periods. An excellent container plant. The flowers can be arranged in spike formation around the stems. Propagate from seed. Cuttings are difficult and usually very slow to form roots, if they do at all.

DRAKAEA Lindley
(after Miss Drake, 19th-century English botanical illustrator)
Orchidaceae

Slender terrestrial **orchids**; **tubers** rounded, subterranean; **leaves** solitary, basal, fleshy, usually ovate; **flower-stem** erect, wiry, bearing a solitary flower; **flowers** of unusual or bizarre shape; **sepals** and **petals** greatly reduced; **labellum** peltate, on an irritable stalk, the lamina ovate, lobed, the surface glabrous or bearing glands or hairs; **fruit** a capsule.

A genus of 4 species, all endemic to the south-west of WA. They are unusual terrestrial orchids which die back to a dormant subterranean tuber over the dry summer months, and grow again with the onset of rains in the autumn, flowering in the spring. All species have been tried in cultivation and the response has been generally disappointing. Growers specializing in terrestrial orchids have limited success. Frequently, leaves will appear in the first couple of years following collection but they are difficult to maintain, and generally fade away. Best results have been obtained using mixtures based on sandy loam derived from coastal heathland, and combining this with coarse sand and eucalypt shavings. The mixture should be allowed to dry out completely when the tubers are dormant over summer. The successful raising of the species from seed may provide the answer to their cultivation.

Drakaea elastica Lindley
(elastic; returning to its original position)
WA   Praying Virgin
10-20 cm tall   Sept-Oct

Slender terrestrial **orchid**; **leaf** 1-1.5 cm x 0.3-

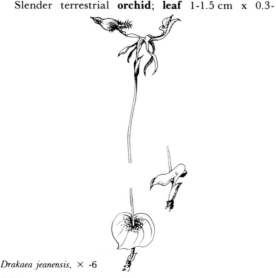

Drakaea jeanensis, × ·6

0.5 cm, ovate to cordate, thick-textured and fleshy, dark green above, whitish beneath; **flower-stem** 10-20 cm tall, wiry, purplish, bearing a solitary flower; **flower** of very unusual shape; **sepals** and **petals** 0.5-0.8 cm long, very slender, reddish green, reflexed; **labellum** about 1 cm long, hammer-shaped, purplish, with a warty surface, peltate, on a stalk about 0.5 cm long.

An unusual orchid, widespread throughout the south-west in coastal heaths and open forest. For response in cultivation, see the genus heading. *D. fitzgeraldii* Schltr. is considered to be a synonym of this species.

Drakaea fitzgeraldii Schltr. = *D. elastica* Lindley

Drakaea glyptodon Fitzg.
(an extinct South American animal of unusual shape)
WA   King-in-his-carriage
15-30 cm tall   Sept-Oct

Similar in most respects to *D. elastica*, but the **labellum lamina** is dark red-brown, with prominent glands on the short lobe, and hairs on the basal part of the main lobe, the apex smooth and glabrous.

An unusual orchid found on sandy soils of the coastal plain of the lower south-west. For response in cultivation, see the genus heading.

Drakaea jeanensis R. Rogers
(after Jean S. Rogers, original collector of the species)
WA   Hammer Orchid
15-30 cm tall   Oct-Nov

Similar in most respects to *D. elastica*, but the **labellum lamina** has congested glands and branched hairs on the short lobe, and the main lobe is smooth and glabrous.

An unusual orchid confined to sandy soils of the coastal plain of the south-west. For response in cultivation, see the genus heading.

Drakaea species
WA   Warty Hammer Orchid
10-30 cm tall   Sept

Similar in most respects to *D. elastica*, but the **leaf** is more prominently rounded and the **labellum lamina** has congested warty growths on the short lobe, and hairs and warts on the main lobe which has a prominently upturned tip.

An undescribed species found in heathland communities, on sandy soils of the south-west. For response in cultivation, see the genus heading.

DRAPETES Dansk ex Lam.
(from the Greek *drapetes*, runaway slave or fugitive)
Thymelaeceae

Dwarf **shrubs**; **leaves** alternate, imbricate; **flowers** bisexual or unisexual, sessile or subsessile, in small terminal heads; **receptacle** funnel-shaped, petaloid; **sepals** 4, 1-2 globular yellow glands at base of tube; **stamens** 4, joined to the throat, alternating with lobes; **fruit** a small ovoid drupe.

This small genus has 1 endemic representative in Tas, and others in Malaysia, New Zealand and southern South America.

Drosera species, × ·7

1. D. burmanni,   2. D. erythrorhiza,   3. D. dichrosepala,   4. D. bulbosa,   5. D. omissa,   6. D. occidentalis,   7. D. petiolaris

Drapetes tasmanica Hook.f.
(from Tasmania)
Tas
0.1 m x 0.3-1 m   Dec-Feb

Dwarf, self layering, tufting perennial **shrub**; **branches** slender, somewhat fleshy, often intertwined; **leaves** about 0.3 cm long, linear, erect, imbricate, yellowish green, hairy near apex; **flowers** tubular, small, white, hairy, 4-6 together in terminal heads; **fruit** a small drupe.

This carpeting, self layering species frequents mountain summits and plateaux in the north and west, where it usually grows in moist, sheltered locations. Not well known in cultivation. It requires very well drained, light to medium soils with organic material to prevent drying out. Suited to semi-shade or partial sun. Frost and snow tolerant. Has potential for gardens or containers. Propagate from cuttings or layers.

DRIMYS Forster & G. Forster — see under
BUBBIA Tieghem and *TASMANNIA* R. Br.

DROSERA L.
(from the Greek *droseros*, dewy; in reference to the prominent glandular hairs imparting a dewy appearance)
Droseraceae

Annual or perennial **herbs**; **basal rosette** present or absent; **rootstock** branching or tuberous, ascending; **stems** present or absent, simple or branched, erect or climbing; **leaves** simple, petiolate, the **lamina** often modified, bearing sessile and stalked glandular hairs; **inflorescence** a raceme or cyme; **flowers** open-petalled, mostly showy, usually bearing 5 petals but ranging from 4-8; **fruit** a capsule.

A large genus of over 100 species distributed in temperate and tropical parts of the world. They are very well developed in Australia where there are about 54 named species and many more awaiting description. Forty-two species are known from the south-west of WA alone, and the rest are distributed in the eastern states and in tropical areas. Droseras as a group are known as sundews and are found in very poor, acid soils which are wet for at least part of the year. Some grow in or around the margins of permanent bogs and swamps. They are a popular group of plants for cultivation.

Droseras are carnivorous and trap insects and small animals on specialized glandular hairs found mostly on the margins and upper surface of the leaf lamina. These hairs, also known as tentacles, are sensitive to very light pressure, which results in an inward and downward movement of the tentacles, and bears any trapped insect on to the leaf surface. Any struggle by the insect sensitizes surrounding tentacles so that the insect eventually becomes firmly trapped. Enzymes released by the glandular heads of the tentacles break down the insect's protein, and the soluble parts are absorbed by the leaf. Frequently, this breakdown is aided by bacterial activity. Thus the trapping and digestion of small animals by the leaves aids in the plant's

335

nutrition, but it is not essential as the plant can live without this supplementary source. The extra materials obtained from digestion are probably of significance in flower and seed production.

Sundews are a variable group within which there are a number of different growth habits enabling species to be grouped together. This is a convenient split since it reflects to some extent their behaviour in cultivation.

'Pygmy Sundews' — about 17 species of miniature plants with tiny dense rosettes and prominent stipules which are a drought resistant mechanism. They are indigenous to Australia and some have quite large, colourful flowers. They grow fairly readily in cultivation. They increase by gemmae, which are small vegetative buds formed on the petioles, are splashed off by the autumn rains and grow to form a new plant. Despite their tiny size, the pygmy sundews are very drought resistant and easily survive a long, hot summer.

'Rosetted Sundews' — these may or may not have an underground tuber but all possess a basal rosette of leaves. Some species within this group are difficult to maintain in cultivation. Those with a tuber die back to this organ in early summer and remain dormant over the hot, dry part of the year until the weather cools and the rains come. This tuber has a fibrous, channelled attachment to the soil surface, through which each new shoot is produced. This fibrous attachment is important and should not be disturbed. If it is disturbed the tuber should be placed close to the soil surface, and it will eventually find its own level again. New tubers are produced each year to replace the old, and some species produce extra ones, and so increase in number by this method. This method of reproduction is remarkably similar to that of some Australian terrestrial orchids which have a symbiotic relationship with a soil fungus. Perhaps the difficulty of maintaining some of these sundews in cultivation is because they have a similar relationship. The addition of eucalypt sawdust or shavings to the mixture may prove beneficial.

'Climbing and Erect Sundews' — small to very large species with erect or climbing stems using their sticky leaves for support. Some have a basal rosette but most have the lower leaves reduced to scales. The majority of species have subterranean tubers. As a group, they are difficult to maintain in cultivation, the usually good response in the first year being followed by a decline.

'Fan-leaved Sundews' — a group of 3 species, all from WA. They have a basal rosette and one or more erect leafy stems. As a group, they are very difficult to grow. (*D. platypoda, D. ramellosa* and *D. stolonifera.*)

'Narrow-leaved Sundews' — 3 species with erect, linear to strap-shaped leaves. These are generally easy in cultivation. (*D. arcturi, D. binata* and *D. indica.*)

'Proliferous Sundews' — a unique species from Qld, which proliferates from the end of the raceme. Grows readily. (*D. prolifera.*)

'Membranous-leaved Sundews' — 2 tropical species with upright rootstocks and rosettes of pale green, thin-textured leaves and small, inconspicuous flowers. Both species grow readily (*D. adelae* and *D. schizandra*). Cultivation

Drosera species, × ·55

1. D. macrantha ssp. *macrantha,*   *2. D. huegelii,*   *3. D. heterophylla*

As a general rule, sundews resent lime and fertilizers of any description. They like an acid soil mix and are best grown in a fairly small, shallow pot. Peat moss, either by itself or mixed with equal parts of coarse sand, is usually a satisfactory medium, although some species like live, growing sphagnum moss. Sundews like a well lit aspect, and are best watered from below by capillarity. The easiest way is to stand the base of the pot in a container or tray of water. For those species that go dormant, the tray should be emptied as the plants wither, and they should be allowed to dry out until signs of active growth reappear. Some species can be successfully grown in terraria (*D. binata* and *D. prolifera*) and in hanging baskets (*D. binata*). If grown indoors, droseras are susceptible to damage from aerosol sprays.

Some droseras may adapt to cultivation in an in-ground situation as bog garden plants or in moist soils. Experimentation is needed with a large range of species to enable evaluation of their performances. *D. arcturi, D. binata* and *D. peltata* may prove to be amongst the easiest species to cultivate in such a way. A method

Drosera arcturi

T.L. Blake

used to establish alpine cushion plants may have application to droseras (see Cushion Plants, page 135).

Propagation

Most sundews can be propagated from seed and some can be propagated vegetatively from leaf and stem cuttings. Seeds are best sprinkled around the parent plant but they can also be sown on an empty pot of media. The best time for sowing is in spring. Seedlings can be pricked out with care, and are best if handled in a clump of soil with as little disturbance as possible. Leaf cuttings can be successful for some species. The whole leaf and petiole should be cut off with a razor blade, as close to the stem as possible. Healthy mature leaves are used, and after cutting are laid face up on a bed of live sphagnum moss. They must be laid flat so that they are in full contact with the moss, and the whole pot should be sealed in a plastic bag or terrarium and placed in a shady position. After 2-3 months, small plants appear and these can be pricked out when large enough to handle. Species that can be propagated in this way include *D. adelae, D. binata, D. schizandra* and *D. spathulata. D. binata* is so easy that its leaves can be cut into sections about 2.5 cm long. Species with erect rootstocks, such as *D. adelae, D. binata, D. indica* and *D. schizandra*, can be propagated by cutting the stem into 2 cm lengths and treating these as for leaf cuttings. The thick roots of some species can be removed and treated in a similar manner, eg. *D. binata* and *D. hamiltonii*.

Drosera adelae F. Muell.
(after Adelae de L'Arbre)
Qld   Lance-leaved Sundew
                                                Nov-Jan
Small perennial **herb**; **rootstock** erect, bearing dead leaves at the base; **leaves** 10-20 cm long, narrow-lanceolate, erect, thin-textured and soft, dull green, smooth beneath with bristles on the upper surface and margins, the veins prominent, the **petioles** woolly; **inflorescence** 20-30 cm tall, woolly, curved near the top, bearing several flowers; **flowers** about 0.6 cm across, greenish.

An unusual sundew that grows around springs and swamps in the near-coastal ranges of north-eastern Qld. It is readily recognized by the rosette of lanceolate leaves. It grows easily in cultivation, but as it is cold sensitive it needs a heated glasshouse. A pot of growing sphagnum moss is recommended as the growing medium; however, peat moss is a good substitute. The mixture must be kept moist at all times and the plants given a fair amount of light. It can be propagated from seed or leaf cuttings which strike readily.

Drosera andersoniana W. Fitzg. ex Ewart & J. White
(after Anderson, the original collector)
WA   Sturdy Sundew
                                                Aug-Sept
Small, erect perennial **herb**; **rootstock** tuberous; **rosette** basal, sparse; **stem** erect, fleshy, up to 25 cm

337

tall; **radical leaves** about 1.5 cm long, spathulate, sparse; **stem-leaves** 2-3 cm long, in 3s, the **lamina** round, the **petioles** slender and thread-like; **inflorescence** terminal, of loose, open clusters; **flowers** about 1.5 cm across, white, pale pink or red, the sepals hairy.

A stout, erect sundew from east of Perth, where it grows in colonies in gravelly soil. It grows readily in a small pot of peaty soil with the base of the pot immersed in water. Plants become dormant during summer, and reappear in the autumn. Propagate from seed.

Drosera androsacea Diels = *D. parvula* Planchon

Drosera arcturi Hook.
(arctic)
NSW, Vic, Tas   Alpine Sundew
                                                   Nov-Jan
Small perennial **herb**; **rootstock** erect, branched; **rosette** basal, sparse; **leaves** 2-7 cm x 0.3-0.5 cm, oblong to linear, blunt, incurved to concave, reddish, tapering to a short **petiole**; **inflorescence** 3-10 cm long, erect, thick and fleshy, bearing a single flower; **flowers** about 1 cm across, white, held erect.

Restricted to subalpine regions above approximately 1200 m alt. but there locally common in sphagnum bogs, along the margins of streams and in wet moss pockets. It grows readily in cultivation but must be kept moist at all times as the plants may die if they dry out. Live sphagnum moss can be used as a culture medium, but may need to be trimmed so that it does not overgrow the sundew. A pot of shredded peat moss is also successful and the base of the pot should be kept immersed in water. Propagate from seed, by division of the clumps or from pieces of root or leaf cuttings.

Drosera auriculata Backh. ex Planchon
(shaped like an ear; with auricles)
Qld, NSW, Vic, Tas, SA
10-80 cm tall   Aug-Dec
Slender, erect, perennial **herb**; **rootstock** tuberous; **rosette** usually absent, if present small and sparse; **stems** 10-70 cm tall, erect, simple or sparsely branched; **radical leaves** 0.4-1.2 cm long, spathulate, dark green; **stem-leaves** 0.5-1.5 cm long, alternate or clustered, the **lamina** orbicular, peltate, pale green, the **petiole** filiform; **inflorescence** 3-10 cm long, terminal, bearing 2-8 flowers; **flowers** 1-1.5 cm across, white or pale pink.

This species is most common in south-eastern Australia and is absent from the tropics; in Qld it is restricted to the south-east. It grows in scattered colonies in a wide variety of habitats and soil types and frequently with *D. peltata*. Like that species it can be grown in non-limed infertile garden soil or in a pot of peaty soil mixture. The base of the pot should be immersed in water while the plants are in active growth from April-Dec. When the plants are dormant the potting mixture should be allowed to dry out. Seedlings can be raised by sprinkling seed on the soil surface around the parent plants during spring. The species is easily confused with *D. peltata* but has glabrous calyx lobes.

Drosera banksii R. Br.
(after Sir Joseph Banks)
Qld, NT   Sundew
                                                  May-June
Tiny perennial **herb**; **rootstock** non-tuberous; **rosette** absent; **stem** 3-10 cm tall, erect, rather weak, with dense red hairs near the top; **leaves** 0.5-1 cm long, scattered up the stem, the cupped **lamina** 0.1-0.2 cm across, the **petioles** thread-like; **inflorescence** terminal, short, bearing few flowers; **flowers** about 0.6 cm across, white.

A little-known tropical species that grows in sandy soils subject to periodic inundation. Its behaviour in cultivation is unknown.

Drosera barbigera Planchon
(having a beard)
WA   Drummond's Sundew
                                                  Sept-Oct
Small perennial **herb**; **rosette** basal, fairly sparse, sometimes more than 5 cm across, held above-ground on a stalk which is covered with old growth; **leaves** 1-1.5 cm long, widening towards the narrow **lamina** which is not more than 0.2 cm wide; **inflorescence** 5-10 cm tall, flexuose, brilliant red, bearing 3-10 flowers; **flowers** about 2 cm across, brick red to orange, with a black centre.

A very colourful pygmy sundew which grows in colonies, on gravelly soil under stunted woodland to the east of Perth. The flowers are large and full by comparison with many other species, and resemble a poppy. When dormant the growth-bud is protected by stipules and withered dead leaves. Cultivation and propagation are probably as for *D. pygmaea*. It was previously known as *D. drummondii* Lehm.

Drosera binata Labill.
(forked; branched)
Qld, NSW, Vic, SA, Tas   Forked Sundew
20-80 cm tall   Sept-April; also sporadic
Erect perennial **herb**; **rootstock** erect or creeping, branched; **leaves** 5-60 cm long, erect, linear, forked one or more times, pale green to reddish, the **petiole** smooth, 3-40 cm tall, the forked **lamina** 5-15 cm long, a few extra-large glandular hairs at the top of each branch; **inflorescence** up to 80 cm long, erect, smooth, branching in the upper part, bearing many flowers; **flowers** about 2.5 cm across, white or pink, on short pedicels.

A widely distributed sundew which commonly forms colonies on wet, peaty soil, usually in heathland. It extends north as far as Fraser Island. The leaves unroll like those of young ferns. This species is probably the easiest to grow, and most rewarding of all sundews. Plants can be grown in a mixture of peat/sand or live sphagnum moss, and are best planted in a broad, shallow container. The base of the container should be permanently immersed in water.

Forked sundew is a variable species and there are numerous forms which have been selected for cultivation and are propagated vegetatively.

'T Form' — deep green to reddish leaves fork once; stops growth and becomes dormant in winter. It is the

Drosera bulbosa   P. Armstrong

This miniature, erect sundew grows on wet, sandy flats of the coastal plain near Perth. It is of interest mainly to carnivorous-plant enthusiasts. It probably requires cultural conditions similar to those for *D. menziesii*.

Drosera bulbosa Hook.
(with a bulb)
WA   Red-leaved Sundew
                                              April-Aug
Small perennial **herb**; **rootstock** tuberous; **rosette** basal, rounded, flat, dark red or purple, 2-4 cm across; **leaves** 1-2.5 cm long, spathulate, narrowing to **petioles**, up to 2.5 cm wide, rounded at the apex; **inflorescence** 2-3 cm long, glandular, bearing a single flower; **flowers** about 2.5 cm across, white, finely dotted with black.
An autumn-winter-flowering sundew that grows in sandy, peaty soil around the margins of swamps near Perth and in the south-west. It is an attractive species that requires similar conditions to those for *D. whittakeri*.

Drosera burmanni Vahl
(after Johannes Burmann, 18th-century Dutch botanist)
Qld, NSW, WA, NT   Tropical Sundew
                                             March-July
Small annual or perennial **herb**; **rosette** basal, fairly dense, 3-5 cm across, green to purplish red; **leaves** 1.5-2.5 cm x 0.5-0.8 cm, spathulate, with long glandular hairs, narrowing gradually to a **petiole**; **inflorescence** 6-15 cm tall, erect, curved near the apex, bearing 3-10 flowers; **flowers** about 0.8 cm across, white, with conspicuous reddish sepals.
A widespread and often locally common non-tuberous sundew that is most abundant in tropical regions. It extends as far south as north-eastern NSW. It grows in open forest and heathland, and seems to prefer depressions subject to periodic inundation. Some forms are annuals, others are perennials. It can be grown in cultivation but it tends to be short-lived, and as it is readily raised from seed it is probably best treated as an annual. A pot of peat/sand mix stood in water is satisfactory. Seed should be sown around the parent plants in spring.

Drosera dichrosepala Turcz.
(bi-coloured sepals)
WA   Rusty Sundew
                                              Oct-Nov
Tiny perennial **herb**; **rosette** fairly sparse, with a conspicuous central bud; **leaves** about 0.7 cm long, the **petiole** narrow, then suddenly expanded into the **lamina** which is elliptical; **inflorescence** 2-4 cm long, bearing 2-7 flowers; **flowers** about 2 cm across, white, the petals narrow.
A little-known species that grows in scattered colonies, on gravelly soils beneath stunted heath of the Eyre District. Cultivation and propagation are probably as for *D. pygmaea*.

Drosera drummondii Lehm. = *D. barbigera* Planchon

most common form in cultivation and was introduced into England over 100 years ago.
'Dichotoma Small Form' — leaves twice-divided, pale green, slender, becomes dormant in winter.
'Dichotoma Giant Form' — leaves twice-divided, yellowish green, thick-textured and vigorous, becomes dormant in winter.
'Multifida' — leaves divided 3-4 times, dark red tentacles, grows throughout the year. One of the best forms. A pink-flowered clone of this much-divided sundew is also known, as well as one that may be divided from 7 to 13 times. This latter form is native to Stradbroke Island, and is a very attractive pot or basket plant. It is sometimes known by the name 'Extrema'.
All forms of *D. binata* can be readily propagated from seed or leaf and stem cuttings.

Drosera bulbigena Morrison
(reproducing by bulbs)
WA   Midget Sundew
                                                  Sept
Small, erect, perennial **herb**; **rootstock** tuberous; **rosette** absent; **stems** 3-8 cm tall, erect, flexuose, leafy; **leaves** about 0.8 cm long, scattered, the **lamina** about 0.3 cm across, half-round, the **petioles** thread-like; **flowers** about 1.2 cm across, white, about 6 borne at the apex of the stem.

Drosera erythrorhiza   P. Armstrong

Drosera erythrorhiza Lindley
(red roots)
WA   Red Ink Sundew
                                                March-July
Small perennial **herb**; **rootstock** tuberous; **rosette** basal, dense, green or bronze, 4-8 cm across, rounded; **leaves** 3-5 cm long and up to 2.5 cm wide, almost round and with little distinct petiole; **inflorescence** 3-8 cm long, erect, branched in the upper half, bearing 10-30 flowers; **flowers** about 1.5 cm across, white or pale yellow.

This sundew has the unusual habit of the inflorescence appearing with the rosette of leaves as they emerge following the autumn rains. The buds are well developed on emergence, and the flowers open before the rosette is fully expanded. After flowering, the rosette continues expansion until it is a handsome and often colourful growth in its own right, of up to 10 cm across. The subterranean tubers and roots of this species are bright red, and are often used to make coloured dye or ink. The plant is widespread in the South-West Botanical Province. It grows in sandy heathlands and often forms extensive colonies. Flowering is promoted in the year following bushfires. This species grows readily for a couple of years in cultivation, but is difficult to maintain and dies out. It needs a pot or saucer of sand/peat mixture, and must be kept moist while in active growth and dry while dormant.

Drosera fimbriata De Buhr
(fringed)
WA   Sundew
5-10 cm tall   Oct
Small perennial **herb**; **rootstock** tuberous; **rosette** absent; **stem** 5-10 cm long, erect, simple; **leaves** dimorphic, the lower ones 0.4-0.6 cm long, linear, whorled, the upper ones 0.8-2.3 cm long, the **lamina** orbicular, the **petiole** thread-like; **inflorescence** a terminal raceme bearing 5-15 flowers; **flowers** 1-1.5 cm across, white.

A rare species known only from deep, white sand between Many Peaks and Jerramungup. It is of interest to enthusiasts but its behaviour in cultivation appears to be unknown. Propagate from seed.

Drosera gigantea Lindley
(exceptionally large)
WA
0.5-1 cm tall   Aug-Dec
Erect climbing **herb**; **rootstock** tuberous; **rosette** absent; **stem** 0.5-1 cm tall, erect, climbing, stout, sometimes zigzagged, branching, the branches spreading through surrounding vegetation, green or reddish; **leaves** 1-2 cm long, the **lamina** shield-shaped, with 2 long, pointed lobes, the **petiole** filiform; **inflorescence** an open terminal cluster; **flowers** about 1.6 cm across, white.

An erect sundew from coastal districts of the South-West Botanical Province, where it often grows in dense masses in wet, sandy soil and in peaty swamps. It flowers profusely and can be grown in a pot of peat/sand mix, with the base of the pot immersed in water while the plants are in growth. When dormant they should be allowed to dry out. Propagate from seed sprinkled around the parent plant.

Drosera glanduligera Lehm.
(glandular hairs)
NSW, Vic, Tas, SA, WA   Scarlet Sundew
                                                 Aug-Oct
Tiny perennial **herb**; **rosette** basal, crowded, bright green, up to 3 cm across, flat; **leaves** 1-1.5 cm long, pale green, the **petiole** slender, the **lamina** elliptical, about 0.5 cm across; **inflorescence** 2-10 cm tall, erect, thickened near the middle, hairy, bearing 15-20 flowers; **flowers** about 1 cm across, bright orange fading to pink.

A widespread species found on moist banks and damp heathlands of temperate regions, usually in colonies. It is a little gem of a plant, and makes an attractive and interesting specimen for a shallow pot or saucer. Needs an acid, peaty soil mix and responds best if kept continually moist. Propagate from seed or leaf cuttings.

Drosera graniticola N. Marchant
(growing in granite)
WA   Sundew
10-20 cm tall   Aug-Sept
Small perennial **herb**; **rootstock** tuberous; **rosette** absent; **stem** 10-20 cm tall, unbranched, erect; **leaves** 1-1.5 cm long, solitary or in groups of 2 or 3, the **lamina** about 0.2 cm across, on slender or sometimes

Drosera macrantha   W.R. Elliot

winged **petioles**; **inflorescence** a terminal raceme bearing 5-12 flowers; **flowers** 0.8-1.2 cm across, white, the petals spathulate.

A recently described species which grows in the soil pockets of granitic outcrops in the South-West Botanical Province. It is an attractive species that has been seldom grown in cultivation. It probably has similar requirements to those of *D. gigantea*. Propagate from seed.

Drosera hamiltonii C. R. P. Andrews
(after A. G. Hamilton, botanical collector)
WA   Rosy Sundew
                                        Nov-Dec
Small perennial **herb**; **rootstock** non-tuberous; **rosette** basal, about 4 cm across; **leaves** dark red, 1-2 cm long, the **petiole** narrow, gradually expanding to the **lamina**; **inflorescence** 20-30 cm tall, hairy, erect, bearing 5-12 flowers; **flowers** about 2 cm across, bright pink.

A little-known sundew that grows in peaty, swampy soils of the South-West Botanical Province. It is a very ornamental species with colourful flowers borne well above the rosette. It can be grown in a pot or saucer of peaty mixture, with the base of the pot immersed in water. Plants should not be allowed to dry out but are best if kept moist all year. Propagate from seed.

Drosera heterophylla Lindley
(different types of leaf)
WA   Swamp Rainbow
                                        June-Aug
Small, erect, perennial **herb**; **rootstock** tuberous; **rosette** absent; **stem** 10-30 cm tall, erect, somewhat flexuose, wiry; **lower leaves** reduced to scale-like bracts about 1 cm long; **upper leaves** insectivorous, about 1.5 cm long, the kidney-shaped **lamina** about 0.3 cm across, the **petiole** filiform; **flowers** about 3 cm across, white or pale pink, with 8 narrow petals, on long pedicels at the end of the stem.

This showy, erect sundew grows in shallow water in low areas subject to partial inundation, in the south-west. Cultivation and propagation are probably as for *D. menziesii*.

Drosera huegelii Endl.
(after Baron Carl von Hugel, Austrian traveller and naturalist)
WA   Bold Sundew
                                        June-Sept
Small, erect, perennial **herb**; **rootstock** tuberous; **rosette** absent; **stem** 30-60 cm tall, erect, flexuose, stiff and wiry; **leaves** few and scattered, usually up to 6 per stem, about 3 cm long, the **lamina** about 0.8 cm across, round, deeply cupped, the **petiole** thin and thread-like; **inflorescence** terminal, loosely branched; **flowers** about 2.5 cm across, white or pinkish.

This slender, erect sundew grows in wet sands and clays of the South-West Botanical Province. Cultivation and propagation are as for *D. menziesii*.

Drosera indica L.
(Indian)
Qld, NSW, Vic, SA, WA, NT   Indian Sundew
                                        May-Aug
Small annual or perennial **herb**; **rootstock** erect to scrambling, 3-50 cm tall, bearing dead leaves at base; **leaves** 5-12 cm long, filiform to linear, yellowish green, scattered along the stem, the margins often inrolled, the **petiole** sometimes flattened and glabrous; **inflorescence** 10-30 cm long, axillary in the upper leaves, bearing 3-20 flowers; **flowers** about 2 cm across, white, pink, mauve, purple or orange, the sepals bearing tiny glandular hairs.

A widely distributed Asian species that extends to Australia, where it is most common in the tropics. In hot, dry inland areas it is an annual, and is found along seasonal watercourses. It can be grown in a pot or saucer of peat kept continually moist by immersing the bottom of the container in water. It grows best in warm, humid conditions and is propagated from seed or stem cuttings. A number of forms exist, varying in size and flower colour.

Drosera leucoblasta Benth.
(a pale or white bud)
WA
                                        Sept-Oct
Tiny perennial **herb**; **rosette** basal, very tight, about 2 cm across, the central bud prominently silvery, and conical; **leaves** about 1 cm long, the **petiole** narrow and flattened, the **lamina** expanded suddenly, about 0.3 cm across, round; **inflorescence** 5-8 cm long, flexuose, bearing 2-6 flowers; **flowers** 1.5-2.4 cm across, white, pale pink or orange.

A most attractive sundew which is widespread and often common on sandy soils in heathland and open forest. When dormant the central bud is protected by the elaborate silvery stipules. Cultivation and propagation are probably as for *D. pygmaea*. *D. miniata* Diels is now included in this species.

Drosera lovellae Bailey = *D. spathulata* Labill.

Drosera macrantha Endl.
(large flowers)
Vic, Tas, SA, WA   Bridal Rainbow Sundew;
                                       Climbing Sundew
40-120 cm tall   June-Oct

341

Small, climbing, perennial **herb**; **rootstock** tuberous; **rosette** absent; **stems** to 150 cm tall, weak, climbing; **leaves** 7-9 cm long, the lamina about 0.8 cm across, rounded, cupped, yellowish green, the petiole slender and thread-like; **inflorescence** terminal, branched, bearing 5-30 flowers; **flowers** 2.5-3 cm across, white or pink, sweetly scented.

A very attractive and common sundew widely distributed in the South-West Botanical Province, the stems climbing or trailing through surrounding vegetation. Unfortunately it is fairly difficult to cultivate, although it can be established in poor, infertile garden soils or in a pot of sand, peat moss and sawdust. The plants should be kept moist while in active growth and allowed to dry out over summer. Propagate from seed.

The typical form has shallowly fringed ovate sepals and is restricted to WA. The ssp. *planchonii* (J. D. Hook ex Planchon) N. Marchant has broadly ovate sepals which are deeply fringed. It is distributed in SA, Vic and Tas, and does not grow as tall as the typical form. It grows in a variety of soils from sandy loams to stony clays. The stems grow actively from autumn to spring and then die back to a pearly white subterranean tuber. The subspecies seems easier to grow than the typical form. It can be grown in a pot of sand, peat moss and old sawdust or infertile garden soil. The stems need staking, and the plants should be kept moist while in growth and allowed to dry out over summer. Propagate from seed.

Drosera macrophylla Lindley
(large leaves)
WA   Sundew
                                      June-Sept
Small perennial **herb**; **rootstock** tuberous; **rosette** basal, dense, flat, 5-6 cm across, green; **leaves** 2-4 cm long, often as wide, rounded at the apex, tapering to the base; **inflorescence** erect, 2-6 cm tall, bearing 1 to several flowers; **flowers** about 2.5 cm across, intensely white.

A handsome free-flowering sundew that grows in extensive and sometimes congested colonies, on sandy soils in open forest of the South-West Botanical Province. Numerous flowers are produced from several scapes which arise from the centre of the rosette. This is a delightful sundew, but unfortunately it is difficult to grow. It prefers sand/peat mixture in a small pot, and must be kept moist while in active growth and dry while dormant.

Drosera marchantii De Buhr
(after Dr. N. Marchant, contemporary botanist, WA)
WA   Sundew
10-50 cm tall   Aug-Oct
Small, weak, perennial **herb**; **rootstock** tuberous; **rosette** absent; **stem** 10-40 cm tall, erect, simple or branched; **leaves** 10-20, scattered up the stem, 0.8-1 cm long, the **lamina** about 0.3 cm across, orbicular, peltate, shallowly cupped, the **petiole** thin and thread-like; **inflorescence** a short terminal panicle bearing 2-10 flowers; **flowers** 2.5-3 cm across, pink.

A species restricted to a small area of south-western WA, where it grows in moist, sandy soils or laterite. Behaviour in cultivation is probably similar to that of *D. menziesii*.

Drosera menziesii R. Br.
(after Archibald Menzies, naval surgeon who collected in WA)
WA   Pink Rainbow
10-35 cm tall   Aug-Oct
Small, erect, perennial **herb**; **rootstock** tuberous; **rosette** absent; **stems** erect, 10-35 cm long, flexuose; **leaves** about 3 cm long, in groups of 3, the **lamina** about 0.4 cm across, rounded, concave, reddish, the **petiole** slender and thread-like; **inflorescence** terminal, clustered; **flowers** about 2.5 cm across, pink to dark red, with broad petals.

An erect, rather variable sundew which is widespread in the South-West Botanical Province, growing in moist sands and sandy clay-loams. It is a showy species and can be grown in a pot of sand and peat moss, with the base of the pot immersed in water while the plants are growing. When the plants are dormant the soil mix should be kept dry. The tubers should not be disturbed. Propagation is by sprinkling seed around the parent plants.

The typical form has red, pink or purple flowers and the sepals fringed throughout, while ssp. *thysanosepala* (Diels) N. Marchant has white flowers, the sepals of which are fringed on the tips.

Drosera microphylla Endl.
(small leaves)
WA   Purple Rainbow
                                             Sept
Small, erect, perennial **herb**; **rootstock** tuberous; **rosette** absent; **stem** erect, flexuose, climbing; **leaves** 2-3 cm long, solitary or in groups of 3, the rounded, cupped **lamina** about 0.3 cm across, the **petiole** filiform; **inflorescence** terminal, 1 to several per plant, each bearing a solitary flower; **flowers** up to 2 cm across, dominated by the large, shiny green sepals, the petals pink or purplish, occasionally deep red or white.

Widely distributed in the South-West Botanical Province, this erect species grows in wet soils on sandplains and granite outcrops. The flowers are short-lived and in cool, moist weather may self pollinate without opening. A superior form is the var. *macropetala* Diels, which is taller and has larger flowers. Cultivation and propagation are as for *D. menziesii*.

Drosera neesii ssp. *borealis*   P. Armstrong

Drosera neesii ssp. *neesii*   P. Armstrong

green, the **lamina** about 0.3 cm across, shield-shaped, with 2 pointed lobes; **flowers** in a terminal cluster on each branch, about 1 cm across, white.

A slender sundew which grows in scattered colonies amongst thick reeds in peaty swamps of the South-West Botanical Province. It is a very free-flowering species but appears to be little-tried in cultivation. It probably requires similar conditions to those for *D. gigantea* and *D. modesta*.

Drosera neesii Lehm.
(after C.G.D. Nees von Esenbeck, 19th-century German botanist)
WA   Jewel Rainbow
                           Sept-Oct

Erect or climbing, perennial **herb**; **rootstock** tuberous; **rosette** absent; **stems** 15-50 cm tall, erect, flexuose, sometimes climbing; **leaves** 3-5 cm long, in groups of 3, shiny golden-green, the **lamina** shield-shaped, cupped, with 2 pointed lobes, the **petioles** thread-like; **inflorescence** a close terminal cluster; **flowers** about 3 cm across, light to deep pink, occasionally yellow, the sepals fringed.

Areas subject to periodic inundation are the favoured habitat of this sundew, which is found in coastal regions of the South-West Botanical Province. Plants may be erect or climbing, and often grow in protected situations. Growth and propagation are as for *D. modesta*. The species is somewhat variable. The typical form has golden-green leaves and is found mainly in coastal localities, while the ssp. *borealis* N. Marchant has deep red leaves and extends some distance inland.

Drosera miniata Diels = *D. leucoblasta* Benth.

Drosera modesta Diels
(modest; unassuming)
WA   Sundew
                              Oct

Small, erect, climbing **herb**; **rootstock** tuberous; **rosette** absent; **stems** 0.5-1 cm tall, erect, climbing, flexuose, yellowish green; **leaves** about 3 cm long, in 3s, yellowish green, the **lamina** shield-shaped, with 2 pointed lobes, the **petiole** filiform; **inflorescence** an open terminal cluster of about 12 flowers; **flowers** about 1.5 cm across, white, the sepals with fringed margins.

This is one of the few sundews to grow in cool, moist, sheltered situations such as gullies and along stream banks. It is a climbing species found in the south-west. It can be grown in a pot of sand/peat mixture, with the bottom of the pot immersed in water while the plant is in growth. It likes cool conditions. When dormant the mix is allowed to dry out and not watered until growth begins again. Propagate by sprinkling seed around the parent in the spring.

Drosera myriantha Planchon
(numerous flowers)
WA   Sundew
15-35 cm tall   March-Dec

Slender, erect, perennial **herb**; **rootstock** tuberous; **rosette** absent; **stem** erect, 15-35 cm tall, sparsely branching in the upper half; **leaves** about 1.5 cm long, in groups of 3 in the lower half, singly on the branches,

Drosera nitidula Planchon
(shining)
WA
                              Nov-Dec

Tiny perennial **herb**; **rosette** basal, tight, often less than 1 cm across; **leaves** about 0.4 cm long, the **petiole** expanding to the **lamina** which is about 0.1 cm across and round; **inflorescence** 2-4 cm tall, bearing 4-6 flowers; **flowers** about 0.4 cm across, white, the sepals folding inwards after flowering.

A fairly common species of wet, sandy heathlands near Perth. It grows easily in cultivation, but it is mainly cherished by carnivorous-plant enthusiasts. It can be grown in a small pot of sandy soil/peat mixture, with the base of the pot immersed in water. Seed will germinate if sprinkled around the base of the plant.

Drosera occidentalis Morrison
(western)
WA   Western Sundew
                              Nov-Dec

Tiny perennial **herb**; **rosette** basal, rather sparse, often less than 1 cm across; **leaves** about 0.5 cm long, the **lamina** only about 0.1 cm across, rounded and concave; **inflorescence** 2-3 cm tall, bearing a single flower; **flowers** about 0.3 cm across, white.

A little-known, tiny species found on moist, sandy flats around swampy areas of the South-West Botanical Province. Cultivation and propagation are probably as for *D. pygmaea*.

343

Drosera omissa Diels
(abandoned; disregarded)
WA   Bright Sundew
                                                Jan-Feb
Tiny perennial **herb**; **rosette** dense, bright gold-green, the old withered leaves persisting for some time; **leaves** about 1.2 cm long, the **petiole** expanding to the **lamina** which is about 0.5 cm across, very hairy; **inflorescence** 5-7 cm tall, erect, bearing several flowers; **flowers** about 0.5 cm across, white.

A little-known species found in localized wet soaks in the South-West Botanical Province. Cultivation and propagation are probably as for *D. pygmaea*.

Drosera paleacea DC.
(chaffy stipules)
WA   Dwarf Sundew
                                                Sept-Dec
Tiny perennial **herb**; **rosette** basal, very dense and compressed, about 1.5 cm across, the central bud shiny; **leaves** about 0.6 cm long, the **petiole** narrow, the rounded **lamina** about 0.2 cm across; **inflorescence** 3-6 cm tall, erect, closely packed, with 20-30 flowers down one side, the apex curved; **flowers** 0.3-0.5 cm across, white.

A decorative, tiny sundew that grows in sandy soil around swamps near Perth. It is mainly of interest to carnivorous-plant enthusiasts and can be grown in a small pot of sandy soil/peat mixture, with the base of the pot immersed in water. Seedlings can be raised by sprinkling the seed around the mature plants.

Drosera pallida Lindley
(pale)
WA   Pale Rainbow
                                                Aug-Sept
Small, climbing, perennial **herb**; **rootstock** tuberous; **rosette** absent; **stems** 0.3-1 cm tall, climbing, flexuose; **leaves** about 6 cm long, in 3s, the rounded, cupped **lamina** about 0.4 cm across, the **petiole** filiform; **lower leaves** scale-like; **inflorescence** an open, terminal cluster; **flowers** about 2 cm across, white.

A robust climbing sundew that grows in moist heaths surrounding freshwater lakes of the South-West Botanical Province. Response in cultivation is largely unknown but it would probably need similar conditions to those for *D. menziesii*.

Drosera parvula Planchon
(very small)
WA   Cone Sundew
                                                Sept-Oct
Tiny perennial **herb**; **rosette** basal, tight, the central bud cone-shaped; **leaves** about 0.8 cm long, the **lamina** about 0.3 cm across, round; **inflorescence** 2-6 cm tall, erect, bearing several flowers; **flowers** about 1 cm across, white.

A fairly common species found in heathland and under stunted woodland in sandy soils of the South-West Botanical Province. It often forms extensive colonies on white sands and although the flowers are small they are borne in profusion. When the plant is dormant its centre is protected by the stipules and dead leaves from previous seasons. Cultivation and propagation are probably as for *D pygmaea*. This species was previously known as *D. androsacea* Diels.

Drosera peltata Thunb.
(peltate leaves)
Qld, NSW, Vic, Tas, SA, WA   Pale Sundew
10-50 cm tall   Aug-Dec
Slender, erect, perennial **herb**; **rootstock** tuberous; **rosette** present or absent, basal, sparse; **stems** 10-30 cm tall, erect, simple or branched; **radical leaves** 0.4-1.5 cm long, spathulate, pale green; **stem-leaves** 0.4-1.5 cm long, peltate, often in groups of 2-6, the **lamina** orbicular, pale green, the **petiole** filiform; **inflorescence** 3-10 cm long, terminal, bearing 2-8 flowers; **flowers** 1-1.5 cm across, white or pale pink.

A widely distributed Asian species which extends to Australia, where it is widespread in tropical and temperate regions of eastern Australia and on some granite outcrops near Coolgardie in WA. In the tropics it extends as far north as Cooktown, and in temperate regions it is common as far south as Tas. It grows in scattered colonies in a wide variety of habitats and soil types. It makes an interesting pot plant and can be established in a non-limed infertile garden soil. It grows well in a pot of sand/peat moss or sandy loam/peat moss/sawdust. The base of the pot should be immersed in water while the plant is in active growth, usually from April-Dec. When the plant is dormant the mixture should be allowed to dry out. At no time should fertilizer or lime be added to the mixture. Seedlings can be raised by sprinkling seed on the soil surface around the parent plants. The best time to sow seed is spring. This species is frequently confused with *D. auriculata*, which has been treated as a subspecies of *D. peltata* by some authors. The two species can usually be separated by the hairy calyx lobes of *D. peltata*. In south-eastern Australia *D. auriculata* has a dark coloured appearance. Seedlings of the 2 species are quite different.

Drosera peltata ssp. **auriculata** (Backh. ex Planchon)
Conn = *D. auriculata* Backh. ex Planchon

Drosera petiolaris R. Br.
(conspicuous petioles)
Qld, WA, NT   Woolly Sundew
                                                July-Dec; also sporadic
Small perennial **herb**; **rootstock** erect, densely covered with old leaves; **rosette** basal, sparse to dense, woolly, up to 10 cm across; **leaves** 3-5 cm long, spreading, curved or arched, green to reddish, with conspicuous woolly hairs, the **lamina** about 0.4 cm across, round, the **petiole** up to 0.3 cm wide and very prominent; **inflorescence** 5-30 cm tall, erect, curved near the top, bearing numerous flowers in a 1-sided raceme; **flowers** about 1.5 cm across, white or pink to purplish.

A widely distributed and common tropical sundew that grows as scattered individuals in open forest, heathland and even rainforest margins, usually in moist depressions. It can be grown in a pot of sand/peat mix, with the base of the pot immersed in water. It prefers to be kept continually moist, and likes a sunny position. Seedlings can be started by scattering seed around the parents. Leaf cuttings can also be successful.

Drosera planchonii J. D. Hook. ex Planchon =
   D. macrantha ssp. *planchonii* (J. D. Hook. ex Planchon)
                                             N. Marchant

Drosera platypoda Turcz.
(broad feet)
WA   Fan-leaved Sundew
                                           Sept-Oct
   Small perennial **herb**; **rootstock** tuberous; **rosette**
basal, sparse or absent; **stem** erect, to 25 cm tall, leafy,
simple or branched; **stem-leaves** about 1 cm x 0.5 cm,
fan-shaped, **inflorescence** short, simple or branched,
arising from the rosette or apex of the leafy stem;
flowers about 2 cm across, white, numerous.
   A tall, slender sundew found in sandy and gravelly
soils of the South-West Botanical Province. Flowering
takes place mainly after bushfires. The species appears
to be rather difficult to grow.

Drosera platystigma Lehm.
(a broad, flat stigma)
WA   Sundew
                                             Oct-Nov
   Tiny perennial **herb**; **rosette** dense, rounded, about
1.2 cm across, the central bud shiny green; **leaves**
about 0.6 cm long, the **petiole** thickened, the **lamina**
about 0.2 cm across, reddish, round; **inflorescence** 4-
6 cm tall, hairy, dark, bearing 3-5 flowers; **flowers**
about 1.2 cm across, red, shiny, with 3 prominent dark
veins in each petal.
   A pretty little sundew found in colonies in sandy and
gravelly soil of the South-West Botanical Province.
Should make an attractive pot plant, but chiefly of interest to carnivorous-plant enthusiasts. Propagation
and cultivation are probably as for *D. pygmaea*. *D.
sewelliae* Diels is now included in this species.

Drosera prolifera C. White
(proliferous; producing plantlets)
Qld   Trailing Sundew
                                          all year
   Small, spreading, perennial **herb**; **rootstock** weak;
rosette basal, sparse; **leaves** 3-5 cm long, erect to
spreading, pale green, the **lamina** distinctly kidney-
shaped, the **petiole** smooth and slender; **inflorescence**
10-18 cm long, weak and trailing, bearing 4-6 scattered
flowers, a vegetative bud near the apex becomes proliferous; **flowers** about 0.8 cm across, greenish cream.
   An interesting tropical species restricted to moist,
sheltered areas on the summit of Thornton's Peak in
north-eastern Qld. The plants grow in colonies, and
spread by the proliferation of plantlets on the inflorescence. This species grows very easily in cultivation
and quickly builds up into a spreading potful. It can be
grown in a pot or saucer of sphagnum moss or peat,
and must be kept moist at all times. In cool areas the
plants need the protection of a glasshouse. Propagate
by separation of rooted plantlets or division of clumps.

Drosera pulchella Lehm.
(beautiful)
WA
                                             Nov-Jan
   Tiny perennial **herb**; **rosette** basal, dense, bright

green, up to 2 cm across; **leaves** about 0.9 cm long, the
petiole very narrow, expanding suddenly to the **lamina**
which is about 0.3 cm across and round; **inflorescence**
about 5 cm tall, bearing up to 4 flowers; **flowers** 1-
1.2 cm across, pink with dark veins near the base, or
rarely red.
   A pretty little sundew that grows around springs and
soaks, and also in swamps of the South-West Botanical
Province. It can be grown in a small pot of sandy soil
but must be kept moist at all times. Propagate from
seed or leaf cuttings.

Drosera pycnoblasta Diels
(a densely packed bud)
WA   Pearly Sundew
                                           Sept-Nov
   Tiny perennial **herb**; **rosette** basal, tight, the central
bud white and shiny; **leaves** about 0.4 cm long, the
lamina only about 0.1 cm across, round; **inflorescence**
2-3 cm tall, bearing 6-8 flowers; **flowers** about 0.5 cm
across, white.
   A tiny sundew found on heathland, on moist to dry
sandy soils, often in white sands and usually in large
colonies. Although the flowers are small they are borne
in profusion. When dormant the growing bud is
protected by the stipules and previous season's
withered leaves. Cultivation and propagation are
probably as for *D. pygmaea*.

Drosera pygmaea DC.
(tiny; dwarf)
Qld, NSW, Vic, Tas, SA, WA   Pygmy Sundew
                                           June-Jan
   Tiny perennial **herb**; **rosette** basal, tight, the central
bud surrounded by shiny stipules; **leaves** about 0.2 cm
across, on **petioles** 0.8 cm long; **inflorescence** erect, up
to 2.5 cm tall, bearing a single flower; **flowers** about
0.4 cm across, white, with 4 petals.
   This tiny plant is found in stunted heathland,
usually growing in moist, sandy soil. It is mainly of interest to carnivorous-plant enthusiasts, and can be
grown in a small pot of sandy soil/peat mixture, with
the base of the pot immersed in water. It can be propagated from seed or leaf cuttings.

Drosera radicans N. Marchant
(rooting from stems)
WA   Sundew
7-18 cm tall   Aug-Sept
   Small perennial **herb**; **rootstock** tuberous; **rosette**
absent; **stem** 7-18 cm long, erect or trailing, the lower
nodes often bearing red, adventitious roots; **leaves** 0.6-
0.9 cm long, the **lamina** about 0.3 cm across, the
petiole thin and thread-like; **inflorescence** a terminal
raceme, bearing 5-15 flowers; **flowers** 0.7-0.9 cm
across, white.
   A recently described species found between Kalbarri
and Geraldton in south-western WA, where it grows in
seasonally inundated clay soils. Little is known of its
performance in cultivation, but it probably requires
conditions similar to those for *D. menziesii*. Propagate
from seed or stem cuttings.

Drosera ramellosa

Drosera ramellosa Lehm.
(much-branched)
WA   Branched Sundew
                                              Aug

Small perennial **herb**; **rootstock** tuberous; **rosette** basal, sparse, inconspicuous, pale yellowish green, a few leafy stems 1-10 cm tall arise from the rosette; **radical leaves** about 0.8 cm long, spathulate; **stem-leaves** about 1 cm long; **inflorescence** 2-3 cm long, arising from the rosette or leafy stem, branched, lengthening after flowering; **flowers** about 1.6 cm across, white.

A little-known, easily overlooked sundew found on sandy soil near Perth and other regions of the South-West Botanical Province. It appears to be rather difficult to maintain in cultivation, the usual response being to disappear the year following collection.

Drosera schizandra Diels
(split anthers)
Qld   Notched Sundew
                                              Nov-Jan

Small perennial **herb**; **rootstock** erect, branched; **leaves** 6-10 cm x 4-5 cm, cuneate, held erect in a rosette, thin-textured, pale green, the lower surface smooth, the upper surface and margins bristly, a slight constriction marking the junction of petiole and lamina; **inflorescence** 5-8 cm long, erect, bearing up to 10 flowers; **flowers** about 0.5 cm long, greenish white, not expanding widely.

An unusual tropical sundew which is restricted to boggy areas and the margins of springs on Mount Bartle Frere in north-eastern Qld, above 1500 m alt. It can be grown easily, but for best results likes warm, humid conditions such as in a heated glasshouse. It can be grown in live sphagnum moss or peat moss, and the mixture should be kept moist at all times by standing the pot in water. Propagate from seed or leaf cuttings. The leaves should be cut as low as possible near the rhizome.

Drosera scorpioides Planchon
(curved like a scorpion's tail)
WA   Shaggy Sundew
                                              Sept-Nov

Tiny perennial **herb**; **rosette** basal, sparse, held erect from the soil, the base covered with the shaggy remains of previous growth; **leaves** 1.4-1.6 cm long, the **petioles** erect, slender, the **lamina** about 0.5 cm long, oval; **inflorescence** 2-6 cm tall, erect, the apex curved, crowded with up to 30 flowers, densely woolly; **flowers** about 1 cm across, white.

This tiny, decorative species grows in wet, sandy or gravelly soil of the Eyre District. It is of interest mainly to carnivorous-plant enthusiasts, and probably requires conditions similar to those for *D. paleacea*.

Drosera sewelliae Diels = *D. platystigma* Lehm.

Drosera spathulata Labill.
(spoon-shaped leaves)
Qld, NSW, Vic, Tas   Spoon-leaf Sundew
                                              sporadic all year
Small perennial **herb**; **rootstock** tuberous; **rosette**

basal, about 2 cm across, reddish, somewhat dense and crowded; **leaves** about 1 cm long, spoon-shaped, the **lamina** expanding to about 0.5 cm across; **inflorescence** 5-20 cm tall, erect, glandular, woolly, bearing 5-15 flowers; **flowers** about 1 cm across, white, pink or red, with hairy sepals.

A very widely distributed sundew that extends from Asia to the east coast of Australia, and New Zealand, where it grows in and around swamps, soaks etc. It is quite an attractive species and is popular with carnivorous-plant enthusiasts. A small pot of peat/sandy soil mix is suitable for its culture and the plants should be kept continually moist. Clumps increase naturally by vegetative reproduction. *D. lovellae* Bailey is now included in this species.

Drosera stolonifera Endl.
(bearing stolons)
WA   Leafy Sundew
4-18 cm tall   July-Sept
Small perennial **herb**; **rootstock** tuberous; **rosette** basal, sparse, bearing a few leaves, several leafy stems arising from the basal rosette; **radical leaves** 0.5-2.5 cm long, spathulate, green or reddish; **stem-leaves** fan-shaped, in whorls of 3-4, varying from stout and green to slender and purplish; **inflorescence** 3-6 cm long, much-branched, usually arising from the rosette,

Drosera stricticaulis   B. Crafter

346

Drosera subhirtella   T.L. Blake

but occasionally from a leafy stem; **flowers** about 1.6 cm across, white.

A variable species found in the South-West Botanical Province in open forest and woodland, usually in colonies. It is an attractive species but appears to be rather difficult to maintain in cultivation. Some forms have very sticky leaves that are capable of catching large insects such as butterflies. In all, 4 subspecies are known. The typical form has numerous stem-leaves, and the leaves and stems are red to dark green. It occurs on the sandy soils of the coastal plain.

The ssp. *compacta* N. Marchant has a very compact rosette of yellowish red leaves and few stem-leaves. It is found in the far south-west, growing in sand or laterite.

The ssp. *rupicola* N. Marchant is an interesting variant which only occurs on granite outcrops, growing in dense colonies. It has attractive lime green leaves, each with a small lamina (0.5-0.8 cm across).

The ssp. *humilis* (Planchon) N. Marchant has yellow-green leaves, each with a large lamina 2-4 cm long.

Propagate from seed.

Drosera stricticaulis (Diels) O. Sarg.
(erect, straight stems)
WA   Sundew
                                                         Oct
Small, erect, perennial **herb**; **rootstock** tuberous; **rosette** absent; **stem** to 40 cm tall, erect, self supporting; **lower leaves** about 1 cm long, linear, **upper leaves** 1-2 cm long, in groups of 3, gold-green, the **lamina** about 0.5 cm across, the **petiole** thread-like; **inflorescences** short, several at the apex of the stem; **flowers** 2.5-3 cm across, rosy pink.

A showy, erect sundew found in places subject to periodic inundation. Its behaviour in cultivation is uncertain, but it probably should be treated as for *D. macrantha*. Propagate from seed.

Drosera subhirtella Planchon
(somewhat hairy)
WA   Aug-Sept
Small, climbing, perennial **herb**; **rootstock** tuberous; **rosette** absent; **stem** to 50 cm tall, flexuose, trailing or climbing, reddish, with a few black, glandular hairs; **leaves** 2-6 cm long, the **lamina** about 0.3 cm across, reddish, round, the **petiole** slender and thread-like; **inflorescence** terminal, branched, the flowers clustered; **flowers** about 2.5 cm across, shiny yellow.

A showy sundew that grows in gravelly soils under open forest of the South-West Botanical Province. Its behaviour in cultivation is uncertain, but it should probably be treated the same as *D. macrantha*. Propagate from seed. The typical form has hairy sepals, while the ssp. *moorei* (Diels) N. Marchant has glabrous sepals.

Drosera subtilis N. Marchant
(fine; thin)
WA, NT   Sundew
3-18 cm tall   Feb-March
Small annual or perennial **herb**; **rootstock** non-tuberous; **rosette** absent; **stem** 3-18 cm long, simple or sparsely branched, the lower part developing adventitious roots; **leaves** 0.4-1.4 cm long, the **lamina** about 0.15 cm across, orbicular, the **petiole** thread-like; **inflorescence** a 1-sided raceme bearing 10-30 flowers; **flowers** 0.6-0.8 cm across, white or pale yellow.

A recently described species known only from a couple of localities in north-western WA and NT. Nothing is known about its behaviour in cultivation. Propagate from seed.

Drosera sulphurea Lehm. =
D. neesii Lehm. ssp. *neesii*

Drosera thysanosepala Diels =
D. *menziesii* ssp. *thysanosepala* (Diels) N. Marchant

Drosera whittakeri Planchon
(after Whittaker, the original collector)
Vic, SA   Scented Sundew
                                                   July-Oct
Small perennial **herb**; **rootstock** bulbous; **rosette** basal, rounded, flat, 2-4 cm across, bright green to reddish; **leaves** 1-2.5 cm long, spathulate, narrowing to **petioles**; **inflorescence** 2-4 cm long, erect, fairly thick, bearing a single flower; **flower** 2.5-3 cm across, white, fragrant, after flowering peduncles recurve below the leaves.

A widely spread and familiar species that grows in a variety of habitats including heathland and open forest. It may grow as scattered individuals, but more often occurs in dense colonies. It is a very pretty plant and is excellent for culture in a shallow pot or saucer, although it is sometimes difficult to maintain. It likes

347

an acid, peaty soil mixture and the plants need to be kept moist while in active growth. During summer they die back to an elongated subterranean tuber, and at this stage must be kept dry until they reappear. Plants can be established in the garden in temperate regions but only in non-limed, infertile soil. The flowers, which are large, showy and fragrant, are produced on a succession of single-flowered scapes which emerge from the centre of the rosette.

The var. *praefolia* (Tepper) J. Black has the unusual habit of producing the flowers before the leaves appear above-ground. It is native to SA. Propagate from seed or by a vegetative increase in the number of tubers produced.

Drosera zonaria Planchon
(girdled; zoned)
WA

Painted Sundew
April-May

Small perennial **herb**; **rootstock** tuberous; **rosette** basal, dense, round, reddish, 4-5 cm across; **leaves** 1-2.5 cm long, spathulate, rounded, green or yellowish, the margins fringed with bright red; **inflorescence** 3-4 cm tall, erect, branched in the upper part, bearing 10-20 flowers; **flowers** 0.6-0.8 cm across, white or pale yellow, sweetly scented.

A widespread and common sundew that always grows in deep sands of the South-West Botanical Province. It forms extensive, often dense colonies, and has rarely been found in flower as this phenomenon seems to be triggered by a particular combination of summer fires and autumn rains. As with *D. erythrorhiza*, the inflorescence emerges with the leaves in the autumn. Cultivation notes as for *D. erythrorhiza*.

Droseraceae Salisb.
(Sundew Family)

A family of dicotyledons consisting of 105 described species in 4 genera. Three of the genera are monotypic, while the fourth, *Drosera*, comprises over 100 species distributed in temperate and tropical parts of the world. All members of the family are annual or perennial herbs which trap insects and other small animals by various mechanisms. The proteins of the trapped animals are then broken down by enzymes, and the soluble chemicals are absorbed by the leaves. This trapping is a supplementary source of nutrition for the plant but is not essential for its survival. Plants of this family often grow in very infertile soils, frequently in swamps. One, *Aldrovanda vesciculosa*, is aquatic. They are of tremendous interest to plant enthusiasts, and many native and exotic species are cultivated for their ornamental and interest value. Australian genera: *Aldrovanda* and *Drosera*.

DRUMMONDITA Harvey
(after James Drummond, first Government Botanist, WA, and his brother Thomas, botanical and zoological collector, northern USA)
Rutaceae

Small **shrubs**; **leaves** heath-like, alternate; **flowers** tubular, solitary, terminal; **sepals** 5; **petals** 5, erect, imbricate in bud, papery in texture; **stamens** 10, only

Drosera whittakeri growing in moss beds   T.L. Blake

those opposite petals bear anthers; **fruit** 5 kidney-shaped cocci.

A small endemic genus of 4 species with the interesting distribution of 3 in WA and the other collected in the Gilbert River area in northern Qld. This species, *D. calida* (F. Muell.) Paul G. Wilson, has not been relocated since the original collection. For many years some species were included in the genus *Philotheca* Rudge. The main difference between these two genera is that *Philotheca* has soft petals and all the stamens are fertile, while *Drummondita* has papery petals and only the stamens opposite the petals bear anthers.

Drummondita species are not well known in cultivation at present, as they are mainly grown by enthusiasts, yet they have potential because of their small, brightly coloured flowers and long flowering period. They occur naturally in sandy and gravelly soils in regions that have a relatively low rainfall. In general, they need well drained soils, and in most cases they tolerate partial to full sun. They make excellent rockery plants, and could prove successful for containers in temperate regions.

Plants are best propagated from cuttings of firm new growth, although they can be slow to form roots. Application of a rooting hormone is usually beneficial.

Drummondita ericoides Harvey
(similar to the genus *Erica*)
WA
0.3-1 m x 0.3-0.6 m   Aug-Nov

Dwarf **shrub**; **branches** usually erect; **leaves** to 1 cm x 0.1 cm, linear, semi-terete, channelled above, crowded, apex blunt, terminating with a gland; **flowers** tubular, about 1.5 cm long, terminal, solitary, erect, almost sessile, yellowish green or white, with prominent violet stamens.

From the Irwin District, where it grows in stony soils. It was originally collected near Moresby's Range, north-east of Geraldton. Must have a warm to hot position, in very well drained soils. Tolerant of at least light frosts. Propagate from cuttings.

Previously known as *Philotheca ericoides* (Harvey) F.

Muell. It is allied to *D. miniata*, which is larger in most respects and has orange-red flowers.

Drummondita hassellii (F. Muell.) Paul G. Wilson
(after Albert Y. Hassell, 19-20th-century botanical collector and politician, WA)
WA
0.3-0.5 m x 0.3-0.6 m   June-Dec

Dwarf **shrub**; **branches** usually erect; **leaves** 0.3-1.2 cm long, semi-terete to obovoid, smooth, crowded, ending in an acute mucro; **flowers** tubular, to 2 cm long, yellow or red, terminal, solitary, erect, almost sessile.

This species is probably the best-known in cultivation. It is widespread in southern WA, usually growing in lateritic soils, in quite open situations. Best suited to very well drained, light to medium soils, with partial or full sun. Frost tolerant. Withstands light pruning. Has potential as a container plant. Propagate from cuttings.

The variety *longifolia* Paul G. Wilson has also found its way into cultivation. It differs from the above in its longer leaves (to 1.7 cm long) with a warty surface. The flowers are yellow and can be up to 2.5 cm long.

Previously known as *Philotheca hassellii* F. Muell.

Drummondita miniata (C. Gardner) Paul G. Wilson
(red)
WA
1.5-2.5 m x 1-2 m   June-Aug

Small to medium **shrub**; **branches** many, spreading; **leaves** 1-1.5 cm long, thick, linear, glandular to warty, initially with grey tomentum, glabrous only at maturity, crowded; **flowers** tubular, to 2.7 cm long, orange-red petals with violet anthers, terminal, solitary or up to group of 3, with short peduncles.

The most outstanding species in the genus, with brightly coloured flowers. It occurs in the Austin District, on laterite. Little is known regarding its performance in cultivation. Should do well in a warm to hot location with well drained, light to medium soils. Suited as a rockery plant but tends to be short-lived. Successful as a container plant. Frost tolerant. Propagate from cuttings.

Previously known as *Philotheca miniata* C. Gardner.

Drummondita hassellii, × ·6

DRYADODAPHNE S. Moore
(from the Greek *dryas*, a wood nymph; *daphne*, the Greek name for the Bay Laurel (*Laurus nobilis*); apparent references to an oak and the bay laurel)
Monimiaceae

A monotypic genus found in New Guinea and north-eastern Qld.

Dryadodaphne trachyphloia Schodde
(rough bark)
Qld
12-25 m x 5-10 m   Sept-Oct

Medium **tree**, columnar in shape; **bark** greyish, orange-brown in the fissures; **branchlets** brittle; **leaves** 5-20 cm x 1-2.5 cm, elliptical, dark green above, paler beneath, the apex long and drawn out, the margins toothed; **petiole** channelled on the upper surface; **cymes** at least 5-flowered, axillary; **flowers** about 1 cm across, the petals purplish brown; **fruit** about 2 cm long, club-shaped, splitting into fruitlets.

A majestic tree from New Guinea and north-eastern Qld where it is known only from the rainforests of Mount Molloy, Mount Lewis and Mount Spurgeon. The tree is virtually unknown in cultivation, but it appears to have excellent potential for parks in tropical and perhaps subtropical regions. Requirements are for deep, well drained, acid soils, and the plants may need protection from hot sun when small. Propagate from seed which must be sown fresh.

DRYANDRA R. Br.
(after Jonas Dryander [1748-1810], Swedish curator of Sir Joseph Banks' botanical collections)
Proteaceae

Dwarf to tall **shrubs** and small **trees**; new growth soft and densely hairy; **leaves** alternate, rarely entire, usually with prickly teeth, or pinnatifid, with many small regular lobes or segments, upper surface usually glabrous, without veins, lower surface can have whitish tomentum and prominent venation; **flowers** sessile, in pairs, arranged in dense, terminal or lateral heads, surrounded by imbricate scale-like bracts, often with a ring of floral leaves similar to stem-leaves; **receptacle** flat or convex, densely hairy; **perianth** usually yellow, rarely glabrous; **fruit** woody follicles, often hairy, concealed by spent flowers.

A genus of outstanding horticultural potential, restricted to south-western WA. There are about 50 named species and at least a further 50 yet to be described.

Fossil collections have revealed that *Dryandra* was a genus with very wide distribution in Europe and the northern hemisphere prior to man's occupation of the Earth. The species in WA are remnants of what was once a very large genus.

Dryandra is closely related to *Banksia*. The main differences are that *Dryandra* has flower-heads with a flat or convex receptacle, and they are usually surrounded by overlapping, scale-like bracts which can be most decorative, as in *D. proteoides*, *D. quercifolia* and *D. tenuifolia*. Also, the seed follicles are readily removed

Dryandra

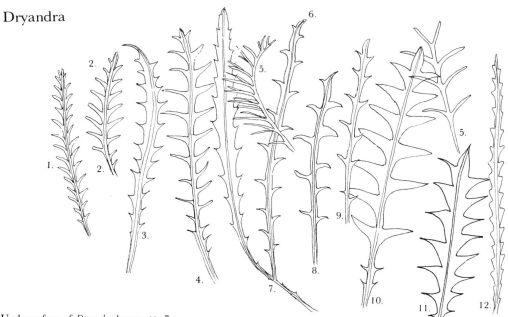

Undersurface of *Dryandra* leaves, × ·7

1. D. fraseri,   2. D. serratuloides,   3. D. vestita,   4. D. cirsioides,   5. D. ashbyi (2 forms),   6. D. horrida,

7. D. cynaroides,   8. D. erythrocephala,   9. D. seneciifolia,   10. D. armata,   11. D. arborea,   12. D. polycephala

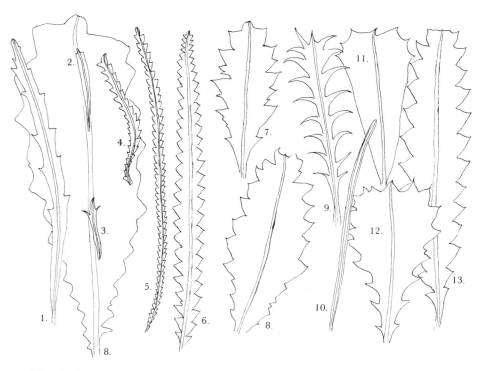

Undersurface of *Dryandra* leaves, × ·7

1. D. carduacea,   2. D. tridentata,   3. D. carlinoides,   4. D. kippistiana,   5. D. baxteri,   6. D. serra,

7. D. quercifolia,   8. D. praemorsa (2 forms),   9. D. falcata,   10. D. speciosa,   11. D. sessilis,   12. D. cuneata,   13. D. concinna

from the spent flower-heads of *Dryandra* species, whereas in *Banksia* they are firmly attached to a woody cone.

Dryandras usually occur in well drained soils that have a high percentage of lateritic gravel incorporated with sand. They also grow in deep sands, and in some cases in granitic areas. It is most unusual for them to occur in situations where they are subject to periodic inundation.

There is quite a range of growth habits in the genus *Dryandra*. The majority of species grow as small shrubs. *D. formosa*, *D. hewardiana* and *D. sessilis* can reach the stature of a tall shrub or even a small tree. Many of the dwarf, clumping and spreading species have been sought after and tried in cultivation, eg. *D. drummondii*, *D. fraseri* and *D. nivea*. *D. pteridifolia* can form large mounds of 1 m x 2-3 m.

Although many species have very decorative flower-heads, eg. *D. praemorsa* and *D. nobilis*, the first aspect of this genus to make an impression is usually the variation in foliage. Some appear fern-like, eg. *D. drummondii* and *D. nivea*. The leaves can be soft or rigid, and colour varies from grey-green to deep green. Other species such as *D. quercifolia* have leaves reminiscent of the well known holly (*Ilex*). In some cases there are species with long, serrated leaves, and some of these produce soft new growth ranging in colour from buff to red.

Some dryandras have become well known because they are commonly sold as cut flowers. In addition to their decorative characteristics they have long-lasting qualities, eg. *D. formosa*, *D. praemorsa* and *D. quercifolia*.

Not only are fresh flower-heads used for floral decoration, but species such as *D. proteoides* with its ornamental bracts, are excellent for use when fully dried. The leaves can also be dried and used in floral decoration.

The flowers can be arranged in large terminal heads or in small heads terminating short lateral branches. The predominant colour is yellow; however, there are many shades from white to orange, and at least one species has purplish tonings. There is also some variation within single species, eg. *D. praemorsa* and *D. quercifolia*, in which a form with pink tonings is very attractive.

The clumping species do not display their flower-heads to advantage, and generally they are partially or fully hidden by the foliage, which is very ornamental in itself.

Most dryandras are excellent for attracting native birds. Both honey-eaters and insect-eaters are able to feed, and are regular visitors to flowering plants.

Cultivation

Although many species have been cultivated since they were first discovered (about 26 species were grown in England during the 19th century), it cannot be stated categorically that *Dryandra* is a hardy genus. Some species have adapted very well, eg. *D. drummondii*, *D. formosa*, *D. fraseri* , *D. nivea*, *D. praemorsa*, *D. pteridifolia*, *D. quercifolia*, and *D. tenuifolia*, but others need special conditions if they are to survive and become desirable plants.

The first prerequisite for success is very good drainage. Dryandras seem to adapt to most soil types, provided drainage is adequate. Most species are susceptible to cinnamon fungus (see Volume 1, page 168), and one of the most practical ways to combat this is to make sure drainage is more than adequate.

In heavy soil locations, one fairly simple way to help improve drainage is to gain a greater depth of top-soil by adding extra soil to growing areas, or by creating mounds. It is recommended that local soil be used, and this should be thoroughly incorporated with the existing top-soil. The addition of gypsum at 1.5 kg/m² is usually successful in creating better drainage in heavy soils (see Volume 1, pages 65-69).

Most species prefer plenty of sunshine, but some are also tolerant of a partial-sun situation. In temperate regions, a location sheltered from cold winds and receiving a large amount of sunshine is usually required for successful cultivation.

Dryandras prefer to have a fairly stable soil temperature and moisture content. The use of ground-covering plants in their vicinity is usually beneficial to their continued growth, as these prevent the sun from heating the soil to extremes, and also help by drawing excess moisture from the soil. In nature, dryandras usually dominate any association with other plants. Most seem to be frost tolerant, and some are capable of withstanding light snowfalls. Many of the species grow extremely well as container plants, and some of the more difficult species, eg. *D. ferruginea*, *D. polycephala* and *D. speciosa*, are worthy of cultivation in this manner.

Maintenance

Watering. As with most members of the Proteaceae family, dryandras withstand extended dry periods once they are established. They seem to be more drought tolerant than *Banksia* species. Young plants may need watering through their first summer. Summer watering should be done during cool spells of weather (see Volume 1, page 168). Deep soaking is imperative, as light applications only promote surface roots, which means plants will continue to need regular watering.

Fertilizing is not usually needed once plants are established. Most species occur naturally in soils that are low in nutrients, and therefore are capable of growing well without any added nutrients.

Small doses at the time of planting can be useful in promoting growth. Nearly all dryandras are susceptible to excess phosphorus. It is therefore best to use fertilizers with a low phosphorus level.

Regular use of fertilizers on dryandras tends to promote top growth at the expense of root growth, and usually means flower production is also delayed.

Yellowing of leaves from the margins can be caused by lack of nitrogen or an iron deficiency. This can be corrected by applications of nitrogenous fertilizers or iron chelates.

Pruning. The best method of obtaining bushy plants is to tip prune while the plants are young. Most shrubby species respond well to regular light pruning once established. Species grown for cut flower production, e.g. *D. formosa*, withstand fairly severe pruning once well established. Dwarf clumping species appreciate pruning

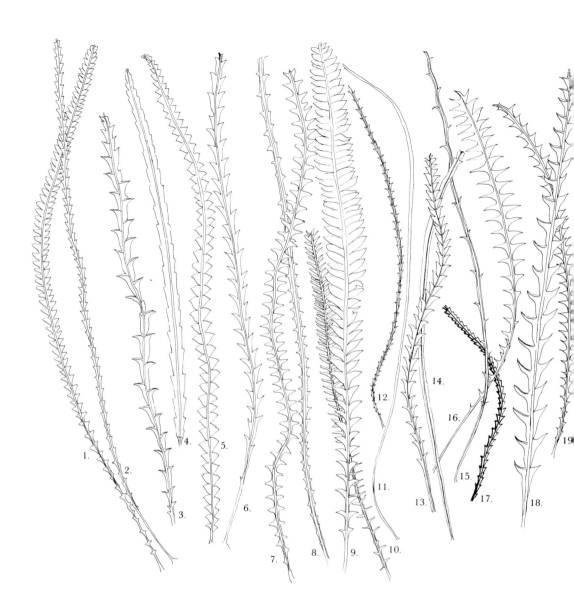

Undersurface of *Dryandra* leaves, × ·55

1. D. nivea, 2. D. nivea form, *3. D. proteoides.*

4. D. squarrosa, 5. D. mucronulata, 6. D. longifolia.

7. D. obtusa, 8. D. foliosissima, 9. D. brownii

10. D. tortifolia, 11. D. subulata, 12. D. pulchella.

13. D. tenuifolia form, *14. D. tenuifolia* form, *15. D. comosa.*

16. D. nana, 17. D. shuttleworthiana, 18. D. hewardiana.

19. D. stuposa

when young. Large pruning wounds should be sealed with a suitable material.

Propagation

Dryandras can be grown from seed or cuttings. To date, seed has been scarce due to pests which eat the embryo, thus thwarting the collection of good seed. An easy way to see whether seeds are likely to be in good condition is to cut the seed-heads with secateurs, from the base of the receptacle. When cut, if there is borer waste, it is most unlikely that the seed will be viable (for extraction of seed, see Volume 1, page 200). Another method of seed extraction is to carefully cut away two-thirds of the follicle along the junction of the two valves. Nail clippers or other suitable implement can be used. Usually the seed can then be easily removed.

Some species, such as *D. formosa*, *D. praemorsa* and *D. sessilis*, release their ripe seed while the seed-heads are still on the plant.

Seed should be sown in a well drained, sterilized mixture, such as 3 parts coarse sand, 1 part peat moss and 1 part soil, then covered to its own depth with the mix (see Volume 1, page 193). The container should be kept moist until germination occurs, which is usually within 21-90 days. Germination is best when cycling temperatures occur, ie. variation between day and night temperatures. Autumn and spring are usually the best times to sow seed. It is recommended that young seedlings be potted on as soon as they can be handled. This can be at the cotyledon stage. Less damage to the fibrous root system is likely to occur at this stage.

At the seedling stage, young plants can be attacked by fungal diseases such as damping off. They should be checked regularly, and may need to be treated immediately with a fungicide if an attack is under way. It is important to place seedling trays in a well ventilated position. This may help to overcome fungal attacks, which are more likely to occur in still, humid conditions.

Dryandras can also be grown from cuttings; however, some difficulty has been experienced with this method. The species with branchlets that are more or less free of hairs have been found to produce roots more readily than those that are hairy. Cuttings can take quite a long time to produce roots. It is best to use cuttings of firm new growth. Softwood cuttings have been found to be satisfactory during winter. Application of hormone rooting powder may be beneficial in promoting a strong root system.

It has been suggested that leaf-node cuttings from clumping species should be tried. *D. calophylla* has been propagated by root division, and this method may have application with similar species.

Grafting. Some work has been undertaken using *Banksia* species as stock, because all *Dryandra* species are thought to be susceptible to cinnamon fungus. As there are some excellent forms that do not produce large numbers of seeds, there is the need to develop successful grafting methods. Initial results with seedling grafts are promising, but there is still much to be done in this field.

Dryandra arborea C. Gardner
(tree-like)
WA   Yilgarn Dryandra
4-6 m x 3-4 m   May-Oct
Tall **shrub**; bark black, thick, rough; **branches** spreading; **branchlets** short; **leaves** 3-5 cm long, oblanceolate, petiolate, deeply lobed, with pungent points, reticulate beneath, glaucous to green, dense; **flower-heads** 4-5 cm across, terminal, bright yellow.

This species occurs further inland than any other member of the genus. It grows on shallow soils with a hard clay subsoil, on ironstone hills north of Southern Cross and Coolgardie. Not commonly encountered in cultivation. Is best suited to inland regions with low rainfall, but plants have grown well in high-rainfall regions. Adapts to a variety of soils from light through to heavy-textured, provided they drain very well. Needs plenty of sunshine, but will tolerate dappled shade in semi-arid areas. In temperate climates should do best in north-facing location. Frost tolerant. Propagate from seed.

Dryandra arctotidis R. Br.
(after the genus *Arctotis*)
WA
prostrate x 2-3 m   Aug-Nov
Dwarf, climbing, spreading **shrub**; **branches** prostrate, usually below ground level; **leaves** 7.5-14 cm x 1 cm, dark green to green, divided to midrib, lobes to 0.8 cm long, prickly, revolute margins, white undersurface; **flower-heads** about 2.5 cm across, cup-like, yellowish green, terminal, long brown bracts, surrounded by leaves.

Not well known in cultivation, although recorded as being grown in England by 1830. This groundcovering species with fern-like leaves inhabits rocky, loamy clay or clay soils, where it grows amongst other plants. Good drainage is essential. Best suited to dappled shade or partial sun, but will tolerate full sun. Frost tolerant. Propagate from seed or cuttings of firm young tip growth. The closely allied *D. brownii* Meisner has longer bluish green leaves.

Dryandra arctotidis var. **tortifolia** (Kippist ex Meisner) Benth. = *D. tortifolia* Kippist ex Meisner

Dryandra armata R. Br.
(thorns or spines)
WA
1-1.5 m x 1-1.5 m   Aug-Sept
Small prickly **shrub**; **branches** many, densely hairy when young; **leaves** to 8 cm long, lobes nearly to midrib, pungent tips, rigid, flat or undulate, undersurface with reticulate venation and sometimes tomentose; **flower-heads** 2.5-4 cm across, yellow, terminal, surrounded by floral leaves which are longer than flowers; **bracts** hairy or becoming glabrous; **perianth** 2.5-3 cm long, hairy.

One of the more widely dispersed species, occurring from north of Geraldton to Cape Arid, east of Esperance. Usually occurs as an understorey plant, on lateritic gravel or sand/laterite mixture. Requires well drained soils, and tolerates dappled shade to full sun. Withstands light frosts. Has a tendency to retain dead leaves on lower branches. The species was introduced

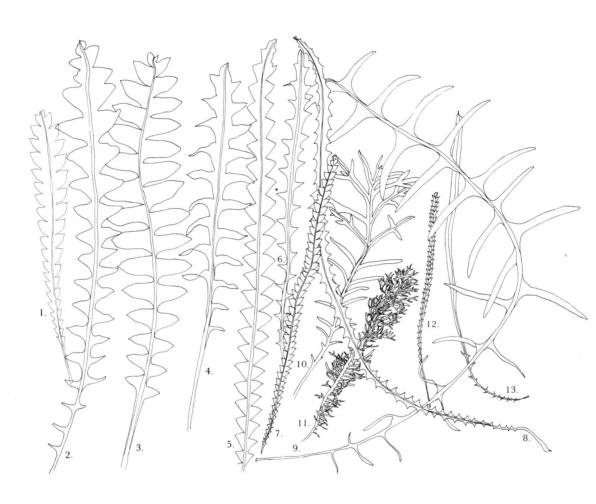

Undersurface of *Dryandra* leaves, × ·5

1. D. foliolata,   2. D. calophylla,   3. D. drummondii,   4. D. ferruginea,   5. D. nobilis,   6. D. conferta,

7. D. formosa,   8. D. plumosa,   9. D. pteridifolia,   10. D. preissii,   11. D. bipinnatifida,   12. D. sclerophylla,

13. D. subpinnatifida

354

to England by 1803. Propagate from seed or cuttings of firm young tip growth.

Dryandra ashbyi B. L. Burtt
(after Edwin Ashby, 20th-century botanical collector)
WA
0.5-2.5 m x 1.5-2 m   May-Aug
   Dwarf to medium **shrub**; **branches** rigid, glabrous; **branchlets** often yellow; **leaves** 4-8 cm x 1.5-2.5 cm, irregular, narrow, pungent lobes, bluish to dark green and glabrous above, whitish tomentum below, margins revolute; **flower-heads** 2-4 cm across, greenish yellow with black bracts, terminal on very short branchlets, appearing sessile, surrounded by floral leaves.
   A species from the Geraldton region, where it grows in gravelly, sandy soils. Little-known in cultivation. Best suited to very well drained soils. Prefers partial or full sun. Frost and drought tolerant. Has potential as a container plant. May need pruning to promote bushy growth. Propagate from seed or cuttings.
   Closely allied to *D. fraseri* which has greyish branchlets that can be very hairy, and longer leaves with closely spaced lobes.

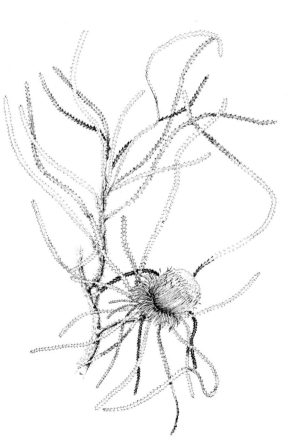

Dryandra baxteri, × ·45

Dryandra baxteri R. Br.
(after William Baxter, 19th-century English gardener and botanical collector)
WA
1-2 m x 1-2 m   Sept-Oct
   Small bushy **shrub**; **branches** many, densely hairy; **leaves** to about 30 cm long x about 0.5 cm, divided to midrib into many small triangular segments which are usually reflexed near tips, margins recurved to revolute, light green above and whitish below; **flower-heads** to 5 cm across, pale yellow, sometimes with light purplish tonings, borne in axils, often hidden by long floral leaves.
   A species with ornamental foliage, from the Stirling Range where it grows in gravelly sands. It was originally introduced to England in 1824. Needs very well drained soils, with partial to full sun. Limited knowledge of its reaction to cultivation suggests that it certainly has potential. Develops as a bushy shrub. Suitable for large containers. Tolerant of light to medium frosts. Propagate from seed or cuttings.

Dryandra bipinnatifida R. Br.
(once-divided and with deeply lobed pinnae)
WA
prostrate x 1-2 m
   Dwarf, slow-growing **shrub**; **branches** underground, densely hairy; **leaves** 15–30 cm long, pinnate, usually erect, with acute linear segments which can be entire or again divided; **flower-heads** about 6 cm long, pale yellow, borne on branches at ground level, not densely surrounded by floral leaves; **bracts** brown, very long, hairy, often becoming glabrous.
   One of the prostrate species that has interesting foliage and displays its flower-heads well. Occurs in lateritic gravels or sands, on or near the coast from south of Perth to south of Geraldton. Has had limited cultivation. Is suited to very well drained soils, in a warm situation. Tolerates dappled shade to full sun. Excellent potential as a container plant. The species was introduced into England by 1840. Propagate from seed or cuttings.
   Sometimes mistaken for *D. preissii*, which has larger, coarser leaves that are much less deeply lobed than those of *D. bipinnatifida*.

Dryandra calophylla R. Br.
(beautiful leaves)
WA
prostrate x 1-3 m   Sept-Nov
   Dwarf, spreading **shrub**; **branches** prostrate, usually below ground level, covered in deciduous hairy bracts when young; **leaves** 15-40 cm x about 3-5 cm, dark green, upright, divided nearly to midrib, forming large, rigid triangular lobes with pungent tips, small, acute lobes near base, margins recurved, undersurface white; **flower-heads** about 3 cm across, dull mauve to reddish pink, sometimes brownish, borne at the ends of the covered branches on the outskirts of the plant, profuse.
   A most ornamental groundcovering species. It occurs on deep and shallow soils with a hard clay subsoil in the south-western corner of WA. Not well known in

Dryandra carduacea

cultivation. Appreciates well drained, light to medium soils, with dappled shade to full sun. Tolerant of extended dry periods and most frosts. An excellent container plant. Was grown in England by 1830. Propagate from seed. Undamaged seed is difficult to obtain because the flower-heads are on the ground, and the seed is often eaten. Experimentation with cuttings is required, as they have so far proved to be only moderately successful as a method of propagation.

This species has been confused with *D. drummondii*, which is readily distinguished by its clumping habit of growth, and bluish green leaves.

Dryandra carduacea Lindley
(resembling a thistle)

WA — Pingle
2-4 m x 1.5-3 m — June-Sept

Medium **shrub**; **branches** upright, long, from near base of plant, slightly hairy or glabrous; **leaves** 4-8 cm x 1-1.5 cm, dark green, linear, cuneate to lanceolate, prickly-toothed margins, white undersurface, very small prickly lobes at leaf-base; **flower-heads** to 4 cm across, yellow, surrounded by floral leaves, borne terminally on short lateral branchlets along the branches; **seed** shed when ripe, about April-May.

Although not widely grown, this showy species has proved fairly adaptable. It usually occurs in lateritic soils, with overhead shelter provided by tall trees. It is found both north and south of Perth. Best suited to well drained soils, with partial to full sun. Tolerant of frost and extended dry periods. Can become straggly, but responds well to pruning. An excellent producer of nectar, and useful as a cut flower which also dries well and can be used for years. The species was introduced to England about 1840. Propagate from seed or cuttings of firm young growth which usually strike readily.

Dryandra carlinoides Meisner
(similar to the genus *Carlina*)

WA
0.6-1.5 m x 1-1.5 m — Sept-Nov

Dwarf to small, dense **shrub**; **branches** usually erect; **branchlets** often nearly verticillate; **leaves** to 3 cm long x about 0.5 cm, green, linear or lanceolate, entire or with few scattered teeth, convex, slightly rigid and prickly, margins recurved, undersurface whitish; **flower-heads** about 3 cm across, pinkish green to cream, terminal, profuse.

A free-flowering species from the Irwin and Darling Districts, where it occurs amongst other shrubs, on shallow, gravelly soils, or on sands with a clay subsoil. Has had only limited cultivation, but certainly has potential for wider use. Needs very good drainage, and should be suited to light to heavy soils. Tolerates dappled shade to full sun. Withstands extended dry periods and most frosts. Suitable for cultivation as a container plant. An excellent cut flower. Propagate from seed and possibly cuttings.

Dryandra cirsioides Meisner
(similar to the genus *Cirsium*)

WA
1-1.5 cm x 1-1.5 cm — July-Oct

Small prickly **shrub**; **branches** thick, densely hairy,

Dryandra carlinoides   W.R. Elliot

usually upright; **leaves** 7-10 cm x 2-3 cm, dark green, long, narrow, spreading lobes, rigid, prickly, margins revolute, often whitish undersurface; **flower-heads** 4-5 cm across, bright yellow, terminal on short branchlets, giving a clustered appearance near ends of branches, profuse.

A very prickly species that forms dense thickets in its natural habitat. It occurs in the southern sand plain east of Esperance, and grows in coarse granitic loam or gravel. Requirements are for well drained, light to medium soils, with plenty of sun, although it will tolerate dappled shade. Frost and drought tolerant. Will probably grow in calcareous soils, as some of the soils in its natural habitat are slightly alkaline. Propagate from seed. Some success has been gained from cuttings, although root formation is likely to be slow.

Sometimes difficult to distinguish from *D. armata*, which usually has smaller flower-heads and closely spaced lobes with recurved margins.

Dryandra comosa Meisner
(hairy tufts)

WA
2-3 m x 3-4 m — Aug-Oct

Small to medium, dense **shrub**; **branches** many, to ground level, slightly hairy; **leaves** 15-30 cm x up to 0.5 cm, rigid, flexuose, with small, distant, prickly teeth or lobes, wide yellowish midrib, undersurface hairy; **flower-heads** 3-4 cm across, yellow, axillary, with a few long floral leaves, on older wood; **outer bracts** prominent, ovate, with short points, brown; **inner bracts** lanceolate to linear.

Evidently not well known in cultivation. This species occurs in the Avon District, where it is restricted to the

Wongan Hills region, growing on the peaks and slopes of laterite hills. Should do best in well drained, light to medium soils, with plenty of sunshine, although will probably tolerate dappled shade. Frost tolerant. Propagate from seed and possibly cuttings, although the hairy stems may present problems. Now gazetted as a rare species.

Dryandra concinna R. Br.
(elegant; pretty)
WA
1-3 m x 1-1.5 m   Sept-Nov

Small to medium, upright **shrub**; **branches** hairy; **leaves** 5-15 cm x up to 2.5 cm, green, with triangular lobes halfway to midrib, ending in short, pungent mucro, margins recurved, reticulate above, whitish tomentum below; **flower-heads** about 2 cm across, yellowish green, axillary, surrounded by short floral leaves, profuse, on very short branchlets; **bracts** oblong to oblong-linear, obtuse and densely hairy.

Restricted to the Stirling Range, where it occurs on the upper slopes and summits. Grows in rocky loam or clay-loam soils. Not well known in cultivation. Needs well drained soils, and does best in partial or full sun. Frost tolerant. Suitable as a low windbreak. Potential as a container plant because of its upright habit. Propagate from seed and possibly cuttings.

This species has affinities to *D. serra*, which has narrower leaves with small lobes.

Dryandra conferta Benth.
(crowded)
WA
1-2 m x 1-2 m   July-Oct

Small bushy **shrub**; **branches** short; **leaves** 7.5-20 cm x 0.8-1.5 cm, crowded, with pungent-pointed, distant lobes more than halfway to the midrib, rigid margins, strongly recurved, shiny dark green above, whitish tomentum below; **flower-heads** 2.5-3.5 cm across, bright yellow, styles reflexed, axillary, surroun-

ded by floral leaves, profuse, often crowded along the branches; **bracts** many, narrow-lanceolate to linear.

Regarded as a most attractive species, yet it is not well known in cultivation. The typical form occurs in the Mogumber region, where it is recorded as growing in heath communities, on gravelly soils. Best suited to very well drained, light to medium soils, with partial or full sun. Should tolerate most frosts and extended dry periods. Has some use as a cut flower. Propagate from seed or possibly cuttings. Future revision will recognize 6–7 species from this complex.

Dryandra cuneata R. Br.
(wedge-shaped)
WA   Wedge-leaved Dryandra
1.5-3 m x 1-2 m   May-Oct

Small to medium, upright or spreading **shrub**; **branches** rigid, usually hairy; **leaves** 4-7.5 cm x up to 4 cm, light green, obovate to oblong-cuneate, serrated, undulate margins, pungent, rigid, reticulate, pale green undersurface; **flower-heads** to 4.5 cm across, bright greenish yellow, terminal on short branchlets near ends of branches, profuse; **bracts** densely hairy, surround flower-heads.

A common plant of coastal heaths between Albany and Esperance. It is also found as far inland as Lake Grace. Usually grows on deep, white sand, but is also found on gravelly soils. Although introduced into England during 1803, it has had only limited cultivation in Australia. Reports suggest it may prove to be quite adaptable. Best suited to well drained, light to medium soils. Appreciates plenty of sunshine, but will tolerate dappled shade, where it may become straggly. Responds well to pruning, and is good as a cut flower. Has potential for use as a container plant. Tolerant of light to medium frosts. Propagate from seed or possibly cuttings.

Dryandra cynaroides C. Gardner
(similar to the genus *Cynara*)
WA
1-1.5 m x 1-2 m   Oct-Feb

Dwarf to small **shrub**; **branches** erect, few, densely hairy when young; **leaves** 5-7 cm x about 1.5 cm, linear-lanceolate, nearly whorled, with distant, pungent lobes, margins recurved, broad midrib, glabrous, smooth, with prominent nerves above, white tomentum below; **flower-heads** 4-5 cm x 4-5 cm, orange-yellow to yellow, terminal on short lateral branchlets, surrounded by floral leaves; **outer bracts** very narrow, with long hairs.

Evidently not in cultivation, this species occurs on lateritic sands in the Avon District. Should do well in well drained, light to medium soils, with partial to full sun. Tolerant of extended dry periods and light to medium frosts. Propagate from seed and possibly cuttings, although the hairs may cause a few problems.

In comparison with *D. vestita*, *D. cynaroides* has shorter, narrower and thinner leaves; also hairier flower-heads and bracts.

Dryandra dorrienii Domin = *D. falcata* R. Br.

Dryandra erythrocephala, × ·55

357

Dryandra drummondii Meisner
(after James Drummond, first Government Botanist,
WA)
WA
0.5-1.3 m x 1-1.5 m   Jan-May; also sporadic
     Dwarf, clumping **shrub**; young shoots often velvety
and brownish to red; **branches** short, covered in rusty
hairs; **leaves** to 90 cm x 8 cm, fern-like, often flexuose,
dull blue-green, long, entire petiole with rusty hairs at
base, upper half divided to midrib, lobes elongated-
triangular, dull green, with prominent reticulate
venation below; **flower-heads** about 5 cm across, light
brownish yellow often with purplish red styles, terminal,
borne amongst the foliage; **bracts** covered in rusty hairs.
     A very ornamental and adaptable species which oc-
curs naturally amongst shrubs and trees, on gravelly
loam. Grows in most soils that have good drainage.
Tolerates dappled shade through to full sun. Frost
tolerant. An excellent container plant. Propagate from
seed (see comments for *D. calophylla*). Some success has
been achieved with cuttings but further experimentation is required. The lack of suitable stems and the
dense covering of hairs can create some problems with
propagation from cuttings.
     This species was incorrectly sold under the name of
D. calophylla for many years. The two differ markedly in
foliage and growth habit. *D. calophylla* grows like a
prostrate banksia, its leaves are upright, and grey-green
above and white below.

Dryandra erythrocephala C. Gardner
(red head; refers to the colour of the flower-heads)
WA
1-2 m x 1-1.5 m   Dec-March; also sporadic
     Small, usually upright, dense **shrub**; **branches**
upright; **leaves** 5-12 cm x about 1.5 cm, dull green,
linear-lanceolate, nearly whorled, rigid, with distant,
pungent, narrow lobes which are sometimes opposite,
margins revolute, hairy base; **flower-heads** about 2 cm
across and 4 cm long, cream with dull purplish brown,
terminal on short lateral branchlets, profuse; **bracts**
long, narrow, hairy.
     Although different from most of the other *Dryandra*
species due to its flower colour and summer flowering,
this species is still not widely cultivated. It occurs on
sandy or gravelly sandy soils, amongst low shrubs in
the south, and extends inland to the Coolgardie District. Attractive both in foliage and flower. Should do
best in very well drained, light to medium soils, with
partial or full sun. Tolerant of extended dry periods
and most frosts. Worth trying as a container plant.
Propagate from seed or possibly cuttings which are
usually slow to strike.

Dryandra falcata R. Br.
(sickle-shaped)
WA   Prickly Dryandra
1-3 m x 1-2 m   Sept-Nov
     Small to medium, prickly, dense **shrub**; **branches**
few, with many short and long hairs when young;
leaves 3-10 cm long, bright green, more or less
cuneate, rigid, with pungent lobes nearly to midrib,
shiny and glabrous above, light green below; **flower-
heads** 4-5 cm across, bright yellow, terminal, surroun-

Dryandra falcata   T.L. Blake

ded by floral leaves which are usually longer than
flowers; **bracts** 1-2 cm long, narrow, with long hairs.
     A showy species that is common in heathland com-
munities from the Stirling Range to Esperance. It occurs on sandy or gravelly soils. Has had limited cultivation. Originally introduced to England in 1824.
Requires very well drained, light to medium soils, with
plenty of sunshine. Probably tolerates dappled shade.
Propagate from seed. Some success has been gained
from cuttings but more experimentation is needed.

Dryandra favosa Lindley = *D. armata* R. Br.

Dryandra ferruginea Kippist ex Meisner
(rust coloured)
WA
0.3-2 m x 0.6-2 m   Aug-Nov
     Dwarf to small **shrub**; **branches** can be covered
in the low-growing form; **leaves** 15-45 cm x up to
3.5 cm, green, rigid, erect, scattered, pungent,
triangular lobes on upper part, extending more than
halfway to midrib, margins revolute, slightly rough
above, rusty tomentum, with prominent reticulation
below; **flower-heads** 5-9 cm across, yellow, with
prominent light orange-brown bracts which are as long
as flowers, protea-like, terminal on short branchlets

358

that are usually on old wood, may be partially hidden by leaves.

A most ornamental and complex species with 6-7 forms. The lower-growing forms usually occur on sandplains amongst other shrubs, whereas the tall forms grow on dry, gravelly ridges in central-southern WA. Best suited to very well drained, light to medium soils, with partial or full sun. Frost tolerant. Withstands fires, and hard pruning could be beneficial to promote new growth. Has potential as a container plant. Propagate from seed. Some success has been achieved with cuttings.

D. ferruginea has affinities to *D. proteoides*, which has narrower leaves with distant lobes.

Dryandra floribunda R. Br. =
                D. sessilis (Knight) Domin

Dryandra foliolata R. Br.
(with leaflets)
WA
1-2.5 m x 0.6-2.5 m   Oct-Dec
Small to medium, prickly **shrub**; **branches** densely hairy when young; **leaves** 7.5-18 cm x 1-2.5 cm, green, erect, deeply serrated almost to midrib, slightly prickly, recurved margins, reticulate above, tomentose below; **flower-heads** about 2 cm across, light cream with dull red at base, purplish red styles sessile, axillary, surrounded by spreading floral leaves, often crowded along branches, mainly on older wood, usually partly hidden by the leaves; **bracts** densely tomentose, rusty brown.

This species occurs above 600 m alt. in the Stirling Range. It grows in rocky loam, often in exposed locations. Has proved to be fairly adaptable in cultivation. Best suited to well drained, light to medium soils. Withstands a fair amount of moisture during winter and spring. Prefers partial or full sun. Frost tolerant. The foliage is ornamental, and pruning promotes bushy growth if desired. An excellent container plant. Propagate from seed or possibly cuttings.

Dryandra ferruginea   T.L. Blake

Dryandra foliosissima C. Gardner
(abounding in leaves)
WA
1.5-2.5 m x 1.5-2 m   June-Aug
Small to medium, bushy **shrub**; **branches** many; **branchlets** many, short, spreading; **leaves** to 35 cm x 1 cm, linear, spreading or recurved, crowded and often entangled, with many short, distant, triangular lobes extending halfway to midrib, margins recurved, green and glabrous above, reticulate and hairy below; **flower-heads** to 4 cm across, pale yellow, terminal on branches and lateral branchlets, surrounded by floral leaves; **outer bracts** linear-subulate; **inner bracts** cream and reddish brown, densely hairy.

This dryandra is not well known, but it is a species with very decorative foliage. It occurs in gravelly soils in the southern sandplains. Needs very well drained, light to medium soils. Will grow in dappled shade to full sun. Tolerates light to medium frosts and extended dry periods. Excellent for large containers. Propagate from seed or cuttings of firm young growth.

This species is allied to *D. mucronulata*, which has flat leaves.

Dryandra formosa R. Br.
(beautiful form)
WA   Showy Dryandra
3-8 m x 2-5 m   Sept-Nov
Medium **shrub** to small **tree**; **branches** many, usually upright, hairy especially when young; **leaves** 5-20 cm x up to 1 cm, green, regularly divided to midrib by many triangular to falcate lobes, margins recurved, not rigid, undersurface with tomentum; **flower-heads** to 10 cm across, bright yellow-orange, terminal on branches and short lateral branchlets, profuse, surrounded by floral leaves which are longer than flowers; **seed** usually dispersed when ripe.

One of the best-known and most commonly grown dryandras. Originally introduced to England in 1803. Renowned as a cut flower because of its lasting qualities. Occurs on stony or peaty soils near Albany, and east to the Stirling Range. Fairly adaptable in cultivation. Does best in very well drained, light to medium soils, with partial sun. Also tolerates dappled shade and full sun. Success has been gained by growing *D. formosa* near the base of established trees, where the moisture content of the soil is fairly constant.

Reports also state that the species is able to cope with watering during extended dry periods. Will withstand some coastal exposure. Tolerant of most frosts. Responds well to hard pruning, as experienced in the cut flower trade. Propagate from seed or cuttings.

Dryandra fraseri R. Br.
(after Charles Fraser, first Superintendent, Sydney Botanic Gardens)
WA
0.3-1 m x 1-2 m   May-Sept
Dwarf, spreading **shrub**; **branches** initially hairy, erect or spreading; **branchlets** usually hairy; **leaves** 5-10 cm x about 1 cm, divided to midrib, spreading, often recurved, green to glaucous, lobes narrow, prickly,

with recurved margin, whitish undersurface; **flower-heads** 4.5-6.5 cm across, pale to bright yellow ageing to orange, pinkish in bud, terminal on branches or short lateral branchlets, often appearing axillary, surrounded by floral leaves longer than flowers; **bracts** fairly inconspicuous.

A very common species from north of Geraldton to the Stirling Range in the south. Usually occurs in gravelly soils. There are at least 2 distinct forms. One is more or less prostrate and the other erect. Also, some plants can have glaucous leaves which are very ornamental. The species has proved to be one of the most adaptable in cultivation and is grown as a cut flower. Requires well drained soils, and will grow in dappled shade to full sun. Flowers much better in a sunny situation. Tolerant of frost, extended dry periods and some coastal exposure. Is excellent value as a container plant. Can suffer from dieback in the central area of the plant, but regular pruning will overcome this problem. Introduced to England about 1840. Propagate from seed or cuttings of firm young growth which strike readily.

D. ashbyi differs from this species in having shorter leaves, and flower heads that are without pinkish tonings.

Dryandra hewardiana Meisner
(after Robert Heward, 19th-century English botanist)
WA
2-5 m x 3-6 m   July-Oct
Small to tall, erect to spreading **shrub**; **branches** usually erect, can be very long, initially hairy, usually becoming glabrous; **leaves** 10-30 cm x 1-1.5 cm, spreading, distant, prickly lobes more than halfway to midrib, ending in pungent points, dark green and shiny above, whitish below, recurved margins; **flower-heads** 3-4 cm across, lemon yellow, axillary, surrounded by a few long floral leaves, profuse, on very short branchlets along branches.

Readily recognized by its flower-heads along the branches, this species has adapted well to cultivation. In nature, it occurs near the coast or slightly inland, in the regions of Dongara and Moora. Usually grows in stony or gravelly soils. Does best in relatively well drained soils, with plenty of sunshine, although it will tolerate dappled shade in hot, dry climates. Withstands frost and extended dry periods. Pruning promotes bushy growth if desired, and plants can be used for informal hedges. Useful as a cut flower. Propagate from seed which is produced in large quantity or from cuttings of firm young growth.

D. patens Benth. is considered conspecific with *D. hewardiana*.

Dryandra horrida Meisner
(very prickly or thorny)
WA   Prickly Dryandra
0.5-1.5 m x 1-1.5 m   May-Aug
Dwarf **shrub**; **branches** densely hairy, with brownish tomentum; **leaves** 8-16 cm x up to 1 cm, narrow, distant, prickly, narrow lobes more than halfway to midrib, margin revolute, undersurface whitish; **flower-heads** 2-2.5 cm x about 3.5 cm long, bright yellow and yellow-orange, on old wood, surrounded by

a few long floral leaves, axillary; **bracts** prominent, with long, brown plumose hairs.

Not well known in cultivation. This prickly-foliaged species occurs in the Avon District, where it grows in sandy soils or lateritic clays, with casuarina scrub. Should be best suited to very well drained, light to medium soils, with partial or full sun. Propagate from seed or possibly cuttings.

Dryandra kippistiana Meisner
(after Richard Kippist, 19th-century English botanist)
WA
0.6-1.5 m x 0.6-1.5 m   Aug-Oct
Dwarf to small, erect or spreading **shrub**; **branches** many, hairy; **leaves** 2.5-6 cm x 0.5-1 cm, green, divided to the midrib, many closely spaced, narrow, prickly segments with revolute margins, whitish undersurface; **flower-heads** 4-5 cm across, light yellow to pink-and-yellow, profuse, terminal or axillary, with few floral leaves.

A most outstanding species that can be literally covered with flower-heads on the upper foliage. From the Avon District, where it commonly occurs on gravel ridges with a clay subsoil, growing amongst other plants. Should prove to be adaptable. Suited to well drained, light to heavy soils, with partial or full sun. Tolerant of most frosts. Withstands heavy pruning, and is an excellent cut flower. Has potential as a con-

Dryandra hewardiana   T.L. Blake

Dryandra kippistiana

B. Crafter

tainer plant. Propagate from seed or cuttings of firm young growth.

Superficially similar to some forms of *D. polycephala*, which have longer leaves, with more widely spaced segments. Flower-heads are usually smaller.

Dryandra longifolia R. Br.
(long leaves)
WA
1-2.5 m x 1-3 m   June-Aug

Small to medium, spreading **shrub**; young shoots hairy, with bronze tonings; **branches** densely hairy; **leaves** 15-35 cm x 1-1.5 cm, with lanceolate or triangular lobes more than halfway to midrib, margins revolute, undersurface with whitish hairs; **flower-heads** to 3.5 cm wide, yellow, terminal on short branches, surrounded by long floral leaves.

This species was originally introduced to England, where it was greatly admired, in 1805. Not very well known in cultivation at present. It occurs in the Eyre District, in granitic areas. Should grow in very well drained, light to medium soils, with partial or full sun, and will probably tolerate dappled shade. Withstands light to medium frosts. Has potential as a container plant. Propagate from seed. The hairy branches may present problems in propagation from cuttings but it would be worthy of trial.

Dryandra mucronulata R. Br.
(small sharp points)
WA   Sword-fish Dryandra
1-2.5 m x 3-4 m   April-July; also sporadic

Small dense **shrub**; **branches** densely hairy, many; **leaves** to 30 cm x 1.5 cm, divided more than halfway to midrib into triangular lobes, margins flat, undersurface tomentose; **flower-heads** about 4 cm across, greenish yellow to golden yellow, with prominent, hairy, brown bracts, terminal or axillary, often on short branchlets along branches, surrounded by floral leaves.

An ornamental-foliaged species that occurs where the Avon, Eyre, Darling and Roe Districts are in close proximity to each other. It grows on the plains through to the peaks of the Stirling Range. Has had limited cultivation, although plants were grown in England by 1824. Needs well drained, light to medium soils, with partial to full sun. Probably also tolerates dappled shade. Withstands light to medium frosts. Propagate from seed or cuttings of firm young growth. The hairy stems may pose some problems.

Dryandra nana Meisner
(dwarf)
WA   Dwarf Dryandra
0.3 m x 0.6-1.5 m   Sept-Nov

361

Dryandra nivea

Dwarf, clumping or creeping **shrub**; **branchlets** usually short and erect; **leaves** 5-12 cm x about 0.4 cm, almost divided to midrib, pungent lobes spreading to 0.7 cm long, margins more or less flat; **flower-heads** about 4 cm x 2.5 cm, yellow, cup-like, with extremely long styles, surrounded by floral leaves 5-10 cm long; **bracts** extend about half the length of flower-heads.

This decorative species is restricted to the Hill River region in the sandplains north of Perth. Its requirements are for very well drained, light to medium soils, with partial to full sun. Will tolerate dappled shade in warm to hot climates. Frost tolerant. Ideal container plant. Propagate from seed.

Similar to *D. nivea*, which has much shorter styles and leaf-segments.

Dryandra nivea (Labill.) R. Br.
(snow white)
WA — Couch Honeypot
0.1-1 m x 0.6-3 m — July-Oct

Dwarf, clumping or spreading **shrub**; **branches** erect or spreading; **leaves** 15-40 cm x 0.5-0.8 cm, fern-like, with pungent lobes, can be divided to midrib, glabrous and dark green above, whitish tomentum below, margins strongly recurved; **flower-heads** to 2.5 cm x 4 cm, somewhat cup-like, brownish yellow with pinkish perianth and greenish style, prominent brown, hairy bracts, terminal on short branchlets, surrounded by long floral leaves, often hidden by foliage.

A widespread and variable species, of which there are at least six distinct forms. One is low and has creeping branches, while another grows as a clump with dense foliage. Occurs over a wide area of south-western WA, where it grows in sandy and gravelly soils. Adapts well to cultivation, provided it is grown in a well drained situation. Will grow in dappled shade through to full sun. Frost and drought tolerant. An excellent container plant. Propagate from seed or cuttings. Was originally introduced to cultivation in England in 1803.

Allied to *D. nana*, which has much longer styles and leaf-segments.

Dryandra nobilis Lindley
(stately)
WA — Golden Dryandra; Great Dryandra
1.5-4 m x 2-4 m — June-Oct

Medium, bushy **shrub**; **branches** erect, hairy; **leaves** to 32 cm x 1.5 cm, spreading to erect, concave, divided to midrib, lobes triangular, pungent, margins recurved, green above, whitish with prominent reticulation below, new growth densely hairy; **flower-heads** 5-7 cm across, orange-yellow, surrounded by floral leaves, more or less sessile, borne along branches.

One of the most outstanding dryandras. It inhabits gravelly soils in the Avon District. Suited to most well drained soils and does well in heavy soils. Prefers plenty of sunshine, but in semi-arid regions it will withstand shaded conditions. Drought and frost tolerant. Pruning while young is recommended, particularly if it is to be used as a screening plant, as old plants can become open and straggly. Suited to large containers. Propagate from seed or cuttings which have been used with limited success.

Dryandra obtusa R. Br.
(blunt)
WA — Shining Honeypot
0.1-0.3 m x 1-3 m — June-Oct

Dwarf, spreading **shrub**; **branches** horizontal, usually just below ground level, tips very hairy and blunt; **leaves** to 30 cm x 2 cm, fern-like, erect, rigid, often divided to midrib, spreading, obtuse lobes, shiny dark green leaves above, whitish tomentum below, margins strongly recurved; **flower-heads** 3-5 cm across, terminal, usually around edge of plant, protruding above the soil, prominent, long brown to reddish bracts for length of flower-head.

An interesting and ornamental but slow-growing species from the sandheaths of the Eyre and Roe Districts. Plants can be surrounded by a ring of flower-heads during a good season. Needs very well drained, light to medium soils, with partial to full sun. Will tolerate growing as an undershrub in semi-arid climates. Frost and drought tolerant. Worthy of experimentation as a container plant. Cultivated in England as early as 1803. Propagate from seed. Cuttings may prove quite a challenge.

Dryandra patens Benth. = *D. hewardiana* Meisner

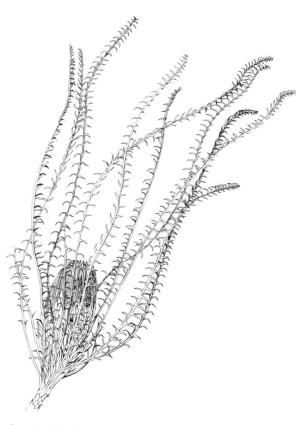

Dryandra nivea form, × ·4

362

Dryandra nivea form

T.L. Blake

Dryandra polycephala, × ·4

Dryandra plumosa R. Br.
(feathery; plumed)
WA
1-2 m x 1.5-3 m   Aug-Oct

Small dense **shrub**; young growth covered in yellowish bronze hairs; **branches** hairy; **branchlets** with rusty hairs; **leaves** to 35 cm x 1-2 cm, mainly spreading, divided halfway to midrib, triangular lobes with small recurved point at apex, green above, whitish tomentum below, margins recurved; **flower-heads** to 4 cm x 4 cm, pale yellow, sessile, axillary; **outer bracts** tomentose; **inner bracts** long, plumose, more or less hidden by the dense floral leaves.

This species has extremely ornamental foliage, the whole plant often adopting a dense, pendulous appearance. Its new growth is eye-catching. It occurs in sandy, gravelly soils in the Eyre District. Has adapted well to cultivation, in light to medium soils with good drainage. Suited to partial or full sun, and will possibly tolerate dappled shade in semi-arid climates. Introduced to cultivation in England during 1803. Propagate from seed. Revision will recognize another species.

At one stage this species was grown under the wrong name of *D. longifolia*.

Dryandra polycephala Benth.
(many heads)
WA   Many-headed Dryandra
1-3 m x 1-2 m   Aug-Nov

Small to medium **shrub**; **branches** many, slender, more or less glabrous; **leaves** 5-20 cm x about 0.6 cm, spreading, rigid, concave, divided halfway to the midrib, distant, triangular, pungent lobes, bright green, margins recurved; **flower-heads** to 4 cm across, bright yellow, profuse, terminal on short branchlets, with leaves 2-5 cm long.

An outstanding species that is commonly available as a cut flower, both fresh and dried. It grows naturally on sandy gravel, often as an understorey plant in the Darling Ranges, and in the Irwin District. Has a preference for very well drained soils and partial or full

sun. Is grown successfully in sandy soils. Withstands hard pruning. Tolerant of drought and most frosts. An excellent container plant. Propagate from seed or cuttings. Some experimentation with seedling grafts has been successful using *Banksia integrifolia* and *B. spinulosa* as the stock.

Closely allied to *D. kippistiana*, which has narrow leaf-segments, much closer together than those of *D. polycephala*.

Dryandra praemorsa Meisner
(as though bitten off)
WA   Cut-leaf Dryandra;
                    Urchin Dryandra
2-4 m x 2-4 m   July-Oct

Medium **shrub**; **branches** densely hairy; **leaves** 5-13 cm x 2-6 cm, obovate-oblong, cuneate, truncate, undulate, prickly-toothed or lobed, dark green above, white tomentum below, margins recurved; **flower-**

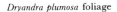

Dryandra plumosa foliage

W.R. Elliot

363

Dryandra preissii

Dryandra cirsioides   T.L. Blake

heads to 3 cm wide, orange-yellow to yellow, terminal on short branchlets, surrounded by floral leaves.

Not very common in cultivation, this is an interesting species from the south-west, where it grows in gravelly soils, often where trees provide a light canopy of foliage. Best suited to well drained, light to medium soils, with partial sun. Tolerant of most frosts. Ideally suited as a container plant. Propagate from seed and it is worth trying cuttings.

This species can be mistaken for *D. bipinnatifida* at the seedling stage. Some seed sold as *D. preissii* is actually that of *D. pteridifolia*.

Dryandra proteoides Lindley
(similar to the genus *Protea*)
WA   King Dryandra
1-2.5 m x 1-2 m   July-Sept
Small to medium **shrub**; **branches** usually erect; **leaves** to 27 cm x about 1.5 cm, erect to spreading, fairly rigid, nearly divided to midrib, spreading, triangular, distant lobes with pungent apex, green above, dense white tomentum below, margins recurved; **flower-heads** to 10 cm across, yellow and brown, borne on very short branchlets on old wood; **outer bracts** dark brown, short and broad; **inner bracts** to 6 cm x 0.7 cm, light brown, longer than flowers, ciliate apex.

This species produces the largest flower-heads of the genus, but they are not displayed to their full advantage. It occurs naturally in gravelly soils of the Avon District, from Northam to Narrogin. Adapts to a wide range of well drained soils, with partial to full sun. Tolerates dappled shade in warm climates. Withstands drought and frost. Excellent for large containers. Responds well to light pruning. Introduced to cultivation in England during 1840. Propagate from seed; there has been limited success with cuttings.

Has some similarities to *D. ferruginea*, which has broader leaves and many triangular lobes.

Dryandra pteridifolia R. Br.
(fern-leaved)   Fern-leaf Dryandra;
WA   Tangled Honeypot
0.3-1.5 m x 1-2.5 m   April-June; Sept-Jan
Dwarf to small, clumping or creeping **shrub**; young growth hairy and often with reddish pink tonings; **branches** densely hairy; **leaves** to 70 cm x 12 cm, erect to spreading, very long petiole, divided to midrib, spreading triangular lobes ending in sharp mucro, green above, whitish below, with prominent reticulation, revolute margins; **flower-heads** 6-8 cm across, pale red to pale yellow, terminal on short stems, surrounded by the long leaves; **bracts** long, broad, hairy.

One of the hardier species that has adapted well to cultivation. It usually occurs naturally in sandy soils of the Irwin, Avon, Roe and Eyre Districts. There are many forms, one of which is spreading with prostrate branches, and another, which is the more commonly grown, is an impenetrable mound. The prostrate form has smaller leaves and flowers. Needs well drained soils, and tolerates dappled shade through to full sun. Frost tolerant. Suited to container cultivation. Introduced to England in 1823. Propagate from seed.

heads 6-8 cm across, yellow or pink with yellow, terminal on short branchlets, profuse.

A commonly cultivated, showy species from the gravelly soils of the Darling District, where it is restricted to jarrah forest. It is one of the most adaptable of the dryandras. Prefers well drained soils, with dappled shade or partial sun, but tolerates full sun. Grown commercially for cut flowers, although it does not always develop long stems. The form that has flowers with pink tonings is eagerly sought by growers. Useful as a screening plant. Plants can become woody and sparse-foliaged after 5-7 years. To overcome this, it is best to tip prune plants regularly. This promotes lateral growth, giving a bushy appearance. Cutting of flowering stems for decoration also assists. Introduced to England in 1848. Propagate from seed or cuttings.

Similar to *D. quercifolia*, which differs with its very prominent bracts and stiffer, smaller leaves.

Dryandra preissii Meisner
(after Ludwig Preiss, 19th-century German botanist)
WA
0.2-0.3 m x 0.5-1 m   Sept-Nov
Dwarf, spreading **shrub**; **branches** procumbent; **leaves** 10-22 cm x 5-8 cm, fern-like, divided to midrib, spreading lobes to 4 cm long, dark green above, rusty tomentum and prominent reticulation below; **flower-**

364

Dryandra proteoides   T.L. Blake

above, reticulate, with white tomentum below; **flower-heads** to 2.5 cm x 2 cm, pale yellow, terminal on short branchlets, surrounded by floral leaves.

From the southern Avon District, where it occurs in gravelly soils. It is not a well known species, either in nature or in cultivation. Should be suited to very well drained soils, in a situation that receives plenty of sunshine. Tolerant of frost. Propagate from seed or cuttings.

D. purdieana has affinities to *D. armata* and *D. cirsioides*. Further study is needed on this complex group of species in order to establish constant differences.

Dryandra quercifolia Meisner
(leaves similar to those of the genus *Quercus*)
WA   Oak-leaf Dryandra
2-4 m x 1.5-3 m   April-Nov; also sporadic
Medium **shrub**; new growth often covered in bronze hairs; **branches** densely hairy; **leaves** to 7.5 cm x 4 cm, obovate-oblong to oblong-cuneate, undulate, rigid, irregular short lobes with pungent mucro, glabrous when mature, often shiny above, reticulate, green below; **flower-heads** to 8 cm across, yellow to greenish yellow or rarely with pink, profuse, terminal, surrounded by floral leaves longer than flowers; prominent brown **bracts**.

An outstanding cut flower species as the decorative heads are borne on long branchlets, and last well. In nature, it occurs on a variety of gravelly soils, and has adapted to a wide range of soil types with good drainage in cultivation. Best suited to a location with partial or full sun. A form with pink-and-yellow flowers was recently introduced to cultivation. Withstands hard pruning. Tolerant of drought and frost. Propagate from seed or cuttings of firm young growth which strike readily.

Dryandra pulchella Meisner
(beautiful)
WA
1-2 m x 1-2 m   Oct-Jan
Small **shrub**; **branches** many, ascending, slightly hairy; **leaves** to 20 cm x up to 0.5 cm, greyish green, crowded, very small, spreading lobes with mucro, revolute margins; **flower-heads** to 2.5 cm across, greenish yellow, with long, distinctly curled styles, appearing axillary, on very short branchlets, surrounded by many long floral leaves which have hairy petioles.

This species is endemic to the Wongan Hills region, where it occurs in mallee scrubland, on lateritic, gravelly soils. Evidently not well known in cultivation. Should be suited to well drained, light to medium soil, with partial or full sun. Its ornamental foliage gives it potential as a container plant. Propagate from seed or cuttings. Now gazetted as a rare species.

Dryandra purdieana Diels
(after Alexander Purdie, former Director of Technical Education, WA)
WA
1-2 m x 1-1.5 m   July-Sept
Small **shrub**; **branchlets** hairy while young, with grey bark at maturity; **leaves** 5-10 cm x 2.5-3 cm, lobed nearly to midrib, pungent, triangular lobes, glabrous

Dryandra runcinata Meisner =
                            D. ferruginea Kippist ex Meisner

Dryandra sclerophylla Meisner
(hard leaves)
WA
0.5-1.5 m x 0.5-1.5 m   Sept-Nov
Dwarf to small, multi-stemmed **shrub**; **branches** few; **leaves** to 15 cm x 0.4-0.8 cm, rigid, often clustered near ends of branchlets, many small, spreading segments with pungent tips, margins revolute; **flower-heads** about 2.5 cm across, yellow, axillary or terminal, surrounded by long floral leaves; **bracts** awned, densely hairy.

This species may not be in cultivation. It occurs on lateritic soils in the Irwin District. Should do best in well drained, light to medium soils, with partial to full sun. Has potential as a container plant. Propagate from seed. Experimentation with cuttings is needed.

Dryandra seneciifolia R. Br.
(leaves similar to those of the genus *Senecio*)
WA
0.5-1 m x 1-1.5 m   July-Nov
Small **shrub**; **branches** erect, hairy; **leaves** to 12 cm x 1 cm, spreading and sometimes reflexed, crowded,

365

Dryandra sessilis   T.L. Blake

usually divided halfway to midrib, distant, short, narrow, pungent lobes, dark green above, white tomentum below; **flower-heads** about 2 cm across, purplish red and yellow, profuse, borne on very short branchlets, appearing sessile, often hidden by floral bracts.

Although introduced to cultivation in England and Europe during 1840, this species is not commonly grown. It occurs in sandy, gravelly soils in the southwest. Requires very well drained, light to medium soils, with partial to full sun. Withstands pruning if required. Propagate from seed or cuttings.

Dryandra serra R. Br.
(toothed)
WA
1.5-5 m x 1.5-3 m   Oct-Nov
Small to tall **shrub**; **branches** erect, more or less glabrous; **branchlets** slender; **leaves** 5-20 cm x 1-2 cm, divided halfway to midrib, many triangular teeth with pungent tips, flat, green and reticulate above, white tomentum below; **flower-heads** small, 1.5-2 cm across, pale yellow, on short branchlets, surrounded by spreading floral leaves.

This species is not well known in cultivation, although it was introduced to England prior to 1835. It occurs in the Stirling Range, where it often grows in lateritic gravels. Requires very well drained soils, with partial or full sun. May become spindly, and pruning while young, especially at planting time, should promote bushy growth. Continued tip pruning or more severe pruning is recommended if a bushy plant is desired. Unpruned plants are useful for narrow areas. Propagate from seed or cuttings.

Has affinities to *D. formosa*, which has leaves which are divided to the midrib.

Dryandra serratuloides Meisner
(similar to the genus *Serratula*)
WA
1-1.5 m x 0.5-1 m   July-Sept
Small **shrub**; **branches** erect, densely hairy; **leaves** to 8 cm x 2-3 cm, lobed nearly to midrib, lobes linear, variable in width, and longer midway along leaf, pungent; **flower-heads** to 3 cm x 2 cm, pale yellow with brown bracts, borne on older wood, surrounded

by long and hairy floral leaves that have widely spaced lobes.

An uncommon species from the Irwin and Avon Districts, where it occurs on gravelly soils. Not well known in cultivation. Best suited to very well drained soils, with partial or full sun. Should tolerate extended dry periods and light to medium frosts. Has potential as a container plant. Propagate from seed or possibly from cuttings.

There is some confusion regarding this species. Plants grown from seed and sold as *D. obtusa* have been identified as *D. serratuloides*. *D. obtusa* is distinguishable by its longer leaves.

Dryandra sessilis (Knight) Domin
(sessile)
WA   Parrot Bush
2-6 m x 1.5-3.5 m   July-Nov
Medium to tall **shrub** or small **tree**; young growth covered in silky hairs; **branches** erect; **leaves** to 5 cm x 3 cm, cuneate, many, more or less sessile, scattered prickly lobes, light greyish green, glabrous on both surfaces; **flower-heads** about 3.5 cm across and 4 cm long, pale greenish yellow to cream, profuse, terminal, surrounded by floral leaves which are longer than flowers.

Well known in cultivation, this species is the most common of dryandras in nature. It occurs in a wide range of habitats, from jarrah forest to coastal limestone. There are at least three distinct forms. The one described is the so-called normal form, which is renowned for the nectar-producing quality of the flowers. The form with smaller leaves is very inferior in this regard. The var. *magna* has large leaves and is known only from Cape Naturaliste. Although preferred cultivation requirements are well drained soils, it has proved hardy in a variety of soils, with partial to full sun. Frost tolerant. Pruning while young promotes bushy growth. As this species sheds its seeds when ripe, regular observation is needed to collect seed. Introduced to England during 1803. Propagate from seed or cuttings.

Dryandra shuttleworthiana Meisner
(after Robert J. Shuttleworth, 19th-century English botanist)
WA
0.5-1 m x 0.5-1 m   June-Sept
Dwarf **shrub**; **branches** densely hairy; **leaves** 5-12 cm x 0.4-0.7 cm, fern-like, rigid, almost divided to midrib, with many small triangular and blunt lobes, dark green above, whitish tomentum below; **flower-heads** to 6 cm across, purplish violet with very hairy orange-brown bracts, pendent, appearing sessile along the branches, without floral leaves.

An interesting species with pendent, shaggy flower-heads. From the Irwin and Avon Districts, where it grows in sandy or gravelly soils. Uncommon in cultivation due to lack of propagating material. Requires very well drained situations, with partial to full sun. Frost tolerant. Has potential as a container plant. Propagate from seed or cuttings.

Closely allied to *D. speciosa*, which has similar flower-heads but differs in its entire leaves.

Dryandra tenuifolia prostrate form   T.L. Blake

Dryandra speciosa Meisner
(showy or handsome)
WA   Shaggy Dryandra
0.6-1.5 m x 1-1.5 m   May-Oct
Dwarf to small **shrub**; **branches** more or less erect, hairy; **leaves** to 8 cm x about 0.2 cm, narrow-linear, entire, slightly recurved tip ending in rigid point; **flower-heads** to 7 cm across, reddish or pale orange with narrow, brown to grey bracts, pendent, terminal on short branchlets, usually only a few floral leaves.
A much sought after species from the sandy and gravelly soils of heathland areas. Its pendent flower-heads have feather-like bracts which tend to hide the colourful flowers. Needs very well drained, light to medium soil, with partial to full sun. Sometimes slow-growing and sparse in cultivation. Recommended as a container plant. Frost tolerant. Withstands light pruning. Propagate from seed and cuttings of firm young growth which strike readily.
Closely allied to *D. shuttleworthiana*, which has lobed leaves and very similar flowers which are distinct for the genus.

Dryandra squarrosa R. Br.
(overlapping and bent backwards)
WA
1.5-2.5 m x 1-1.5 m   Aug-Oct
Small to medium, erect **shrub**; new growth hairy; **branches** usually glabrous; **lower leaves** 10-25 cm x about 0.6 cm; **upper leaves** about 5-12 cm long, regularly toothed about halfway to midrib, each tooth ending in a pungent mucro, flat or undulate, shiny green above, white tomentum below; **flower-heads** to 2 cm x 3.5 cm, yellow, axillary, surrounded by a few floral leaves.
An uncommon species that was originally collected near King George Sound in 1828. Some success has been gained by growing it in well drained, heavy soils where it receives plenty of sunshine. Should also grow satisfactorily in well drained, light to medium soils. Has potential as a container plant. Propagate from seed or cuttings.

Dryandra stuposa Lindley
(tufts of long, matted hair)
WA
2-3 m x 1.5-3 m   Feb-Aug
Small to medium, bushy **shrub**; **branches** with long rusty hairs; **leaves** 7.5-20 cm x 0.6-1.5 cm, lobed almost to midrib, lobes triangular and pungent, widely spaced, dark bluish green above, paler below; **flower-heads** to 5 cm across, yellow ageing to orange or orange-brown, terminal, often on short branchlets; **bracts** relatively inconspicuous.
A showy species from the Darling District. It occurs in open forest on sandy or lateritic gravels, often forming dense thickets. Has proved difficult to maintain in cultivation. Needs very well drained soils, with partial or full sun. Has potential for cultivation in large containers. Propagate from seed or cuttings.
Is similar to *D. nobilis*, which has larger leaves and flower-heads.

Dryandra subpinnatifida C. Gardner
(lower part of leaves is pinnatifid)
WA
1-2 m x 1-2 m   Aug-Oct
Small **shrub**; **branches** many, erect, glabrous; **branchlets** short, spreading, crowded with leaves; **leaves** to 25 cm x 0.5 cm, lobed at base or with fine spines, becoming entire except for a few lobes near the pungent apex, green and glabrous above, white tomentum below, margins recurved; **flower-heads** about 2 cm across and 4 cm long, pale yellow, on short branchlets on older wood, surrounded by floral leaves with spreading, fine-awned lobes.
An unusual species that is not well known in cultivation. It occurs on lateritic soils in the Avon and Roe Districts. Prefers very well drained, light to medium soils, with partial to full sun, but will probably tolerate dappled shade. Propagate from seed. It is also worth experimenting with cuttings.

Dryandra subulata C. Gardner
(awl-shaped)
WA   Awled Honeypot
0.2-0.4 m x 0.3-1 m   Sept-Oct
Dwarf, clumping **shrub**; **branches** spreading, initially hairy, becoming glabrous; **leaves** to 30 cm x 0.2 cm, entire, grass-like, erect, not rigid, dark green above, whitish tomentum below, margins revolute; **flower-heads** to about 2.5 cm across, pinkish brown, terminal on short branchlets, surrounded by stiff, narrow bracts, surrounded by a few floral leaves or floral leaves absent.
An uncommon, tufting species from the Irwin District, where it occurs on sandy soils in heathland. Has had very limited cultivation to date. Requires very well drained soils, with partial to full sun. Should be an excellent container plant. Hardy to light frost. Propagate from seed. Experimentation with cuttings is worthwhile.

Dryandra tenuifolia R. Br.
(slender leaves)
WA
0.1-2.5 m x 1-5 m   Aug-Oct

367

Dwarf, creeping or small to medium, upright **shrub**; **branches** nearly glabrous; **leaves** 5-30 cm x 0.5-0.8 cm, lower one-third usually entire, upper two-thirds divided to midrib, with falcate-triangular lobes which have pungent apex, dark green above, whitish tomentum below, margins revolute; **flower-heads** to 5 cm across, pale yellow with pink to brown flowers, surrounded by brown bracts which are longer than the flowers, profuse, borne on very short branchlets, produced along the branches and amongst the leaves.

An adaptable and ornamental species with at least two distinct forms. The prostrate form has leaves with fewer lobes than the shrubby form. Both forms occur in the south-west, where they grow in sandy soils with a clay subsoil. Grows well in most well drained soils, doing best in partial or full sun. Will tolerate dappled shade. The prostrate form is an excellent groundcover and does well amongst taller plants. Frost tolerant. Responds very well to pruning. Recorded as introduced to England in 1803. Propagate from seed or from cuttings of firm young growth which strike readily.

Dryandra tortifolia Kippist ex Meisner
(twisted leaves)
WA
0.3-0.5 m x 0.5-1 m   Aug-Sept
Dwarf, clumping **shrub**; **leaves** to 30 cm x about 1 cm, erect, fern-like, rigid, divided to midrib, many narrow, pungent lobes, dark green above, whitish tomentum below, margins recurved, concave in cross-section; **flower-heads** to 3.5 cm across and 6.5 cm long, yellow, on very short branchlets; **bracts** brown, about half as long as flowers, surrounded and often hidden by long floral leaves.

A species with ornamental foliage and growth habit. It is not well known in cultivation. Occurs in grey sandy soil, in heathland north and south of Perth. Should do best in well drained, light to medium soils, with plenty of sunshine. Has potential as a container plant. Tolerant of most frosts. Propagate from seed or cuttings.

Allied to *D. arctotidis*, and was known as *D. arctotidis* var. *tortifolia* (Kippist ex Meisner) Benth.

Dryandra tridentata Meisner
(3 teeth or prongs)
WA   Yellow Honeypot
0.3-0.5 m x 0.5-1 m   Sept-Oct
Dwarf **shrub** with a lignotuber; **branches** erect, to 50 cm tall; **leaves** 2-4 cm x 0.4-1 cm, broad-linear to linear-cuneate, usually 3-toothed at apex, otherwise entire, tapering to petiole, green above, prominent reticulation below; **flower-heads** 4-5 cm across, yellow, just protruding through the soil.

An interesting species from the Avon and Irwin Districts, where it grows on sandy soils in the heathlands. Has had limited cultivation. Appreciates very well drained, light soils. May adapt to heavier soil types. Prefers partial or full sun. Should respond well to pruning if plants develop lignotubers in cultivation. Has potential as a container plant. Propagate from seed or cuttings.

Dryandra vestita Kippist ex Meisner
(clothed)
WA
0.3-1 m x 0.5-2 m   Aug-Oct
Sprawling dwarf **shrub**; new growth with brownish hairs; **branches** densely hairy; **branchlets** often woolly; **leaves** to 1.5 cm x 1.5 cm, linear to linear-cuneate, lobed halfway to midrib, ending in pungent points, green above, reticulate below, often with rusty tomentum, margins recurved; **flower-heads** 2.5-3.5 cm across and about 4.5 cm long, yellow, with narrow plumose bracts, borne on short branchlets, surrounded by floral leaves.

Although this decorative species is in cultivation, it is not well known, and has proved to be a very slow grower. It is from the Avon District, where it usually grows in sandy, lateritic soils in open heathland, or in association with casuarinas. Requirements are very well drained, light to medium soils, with partial to full sun. Frost and drought tolerant. Potential as a container plant. Propagate from seed. Experimentation with cuttings is worthwhile.

Drymaria filiformis Benth. =
              Stellaria filiformis (Benth.) Mattf.

DRYMOANTHUS Nicholls
(from the Greek *drymos*, forest; *anthos*, a flower; in reference to the forest habitat of the species)
Orchidaceae
Epiphytic **orchids** of monopodial growth; **stems** short, simple or branched, when old clothed with persistent leaf-bases; **leaves** few, distichous, simple, entire, often falcate; **inflorescence** an axillary raceme; **flowers** small, dull coloured; **fruit** a capsule.

A small genus of 2 species, one endemic in north-eastern Qld, the other in New Zealand. They are small epiphytes of uncommon occurrence and are rare in cultivation. Propagate from seed which must be sown on sterilized media.

Drymoanthus minutus Nicholls
(small)
Qld   Dec-Feb
Small epiphytic **orchid**; **stems** 1-4 cm long, solitary

Drymoanthus minutus   B. Gray

Drymophila moorei   D.L. Jones

Perennial **herbs**; **rhizome** creeping, fleshy; **stems** simple or sparsely branched; **leaves** simple, entire, sheathing at the base, spreading in 2 opposite rows; **flowers** 6-petalled, on recurved stalks; **styles** 3; **fruit** a berry.

An endemic genus of 2 species.

Drymophila cyanocarpa R. Br.
(blue fruit)
NSW, Vic, Tas   Turquoise Berry
0.2-0.4 m x 0.1-0.2 m   Nov-Jan

Dwarf, perennial **herb**; **stem** arching, leafy, simple or branched; **leaves** 2.5-8 cm long, lanceolate to oblong, more or less sessile, prominent longitudinal nerves, glabrous, shiny; **flowers** about 1.5 cm diameter, white, borne on spreading to recurved pedicels, striate petals; **berry** globular to oblong, about 1 cm diameter, turquoise-blue.

A most ornamental lily, with bright turquoise-blue berries. It frequents moist, shady forests. Not common in cultivation but should be suited to moist, well drained, light to medium soils, with full to partial shade. It may tolerate sun for limited periods. Can be grown as a container plant. Mixes well with ferns. Frost tolerant. Propagate from seed or by division of rhizomes.

Drymophila moorei Baker
(after Charles Moore, Superintendent of Sydney Botanic Gardens, late 19th century)
Qld, NSW
0.2-0.4 m x 0.1-0.2 m   Nov-Jan

Dwarf, perennial **herb**; **leaves** 2.5-8 cm long, ovate to oblanceolate; **flowers** about 1.5 cm diameter, white to pink, borne on spreading to recurved pedicels; **berry** globular to oblong, about 1-2 cm diameter, yellow to orange.

From gullies in the near-coastal forests of northern NSW, along rivers such as the Clarence, Macleay and Hastings, and south-eastern Qld. It requires similar growing conditions to those for *D. cyanocarpa*. Does well as a container plant. May be damaged by frost but suckers from underground. Propagate from seed or by division of rhizomes.

Differs from *D. cyanocarpa* in its broader leaves and yellowish berries.

or branched; **leaves** 0.5-2.5 cm x 0.2-0.7 cm, 2-5 per plant, oblong to elliptical, falcate, dark green; **racemes** 0.4-1 cm long, axillary, 1-2 per plant, bearing 2-7 flowers on short pedicels; **flowers** about 0.25 cm across, greenish with a white labellum; **petals** and **sepals** about 0.25 cm long, spathulate; **labellum** about 0.22 cm long, oblong.

This diminutive orchid is endemic in north-eastern Qld, between the Burdekin and Mulgrave Rivers. It is found in coastal lowlands and the adjacent highlands, on rainforest trees or bottlebrushes, growing in streams or along banks. Floristically this orchid has no appeal, but it is grown by orchid collectors. It is best attached to a piece of cork, paperbark or weathered hardwood and held in humid conditions, but where it receives adequate air movement.

Once established, the plants are quite hardy and easy to maintain, but they must not be overwatered or held in dark, stagnant conditions. In southern areas they must be grown in a heated glasshouse with a minimum temperature of 5°C. In tropical areas a bushhouse is suitable. Propagate from seed which must be sown under sterile conditions.

DRYMOPHILA R. Br.
(from the Greek *drymos*, wood or forest; *philos*, loving; refers to its habitat of moist, shady forests)
Liliaceae

DRYNARIA (Bory) J. Smith
(from the Greek *dryas*, a wood nymph or *drynad*, to whom the oak was sacred; the sterile fronds of these ferns resemble oak leaves)
Polypodiaceae

Epiphytic **ferns**, rarely terrestrial; **rootstock** stout, fleshy, creeping, branched, covered with scales; **fronds** of 2 types produced alternately; **sterile fronds** short, papery, lobed; **fertile fronds** erect, pinnate or lobed; **sori** rounded, exindusiate.

A distinctive genus of ferns comprising about 20 species, 3 of which are found in northern Australia. They are very popular subjects for cultivation and make attractive basket plants. Propagate by division of the rhizomes or from spore.

Drynaria quercifolia (L.) J. Smith
(leaves like those of the oak genus *Quercus*)
Qld, WA, NT
0.3-1 m tall

Epiphytic or lithophytic **fern** forming large, spreading clumps; **rootstock** thick, fleshy, covered by soft brown scales; **sterile fronds** 20-40 cm x 10-30 cm, erect, shallowly lobed, rounded at the base, brown and . papery; **fertile fronds** 0.3-1 m long, pinnately lobed, dark green and tough, erect; **sori** rounded, in 2 regular rows.

A widely distributed and common fern found on trees or rocks of northern Australia, usually in fairly sunny, humid situations. In dry areas the plants become completely deciduous and grow new fronds with the onset of the wet season. The species sometimes grows in large, spreading colonies, especially in sandstone crevices. It is a very easy and rewarding fern to grow, succeeding best in tropical and subtropical regions. It can be grown as a garden plant amongst rocks, attached to trees or in a pot or basket. It requires a coarse, acid, epiphyte mixture, and the plants should be allowed to dry out in cool winter weather. Outside the tropics it needs a heated glasshouse over winter, and the plants must be kept dry until the weather warms up. The Aborigines of NT ate the rhizomes in times of food shortage. Propagate by division of the clumps or from spore which is best sown fresh.

Dryopteris poecilophlebia, × ·15

sori, × 1·2

Drynaria rigidula (Swartz) Beddome
(somewhat rigid)
Qld, NSW   Basket Fern
0.3-1.5 m tall

Epiphytic, lithophytic or occasionally terrestrial **fern**; **rootstock** thick, fleshy; **sterile fronds** 10-36 cm x 5-10 cm, lanceolate, erect, lobed, stiff, brown and papery; **fertile fronds** 0.3-1.5 m long, erect or arching, pinnate; **segments** 5-20 cm long, linear-lanceolate, stalked, with minutely toothed margins; **sori** rounded, in a single row.

Widespread throughout eastern Qld and northeastern NSW, in moist forests and rainforests, sometimes forming huge clumps. It is readily grown in tropical and subtropical regions, but in temperate areas usually requires the winter protection of a glasshouse where it can be allowed to dry out. Can be grown in a pot, basket or on a slab of tree-fern trunk. In the tropics it is often seen as a garden plant on rocks or trees, and may become weedy in glasshouses, bushhouses and ferneries. It is a very drought tolerant fern, and in cold winters the plants are best kept dormant and allowed to dry out. A couple of handsome cultivars were found in the wild in the early part of this century and are highly prized by enthusiasts. Of these, the cv. 'Whitei' is the best, having broad pinnae with deeply and irregularly lobed margins. The pinnae of the cv. 'Vidgenii' are also irregularly lobed, but are much longer and narrower than those of cv. 'Whitei'. Both cultivars make exceptional basket plants. The cultivars are propagated by division or from tissue culture, the species by division or from spore.

Drynaria sparsisora (Desv.) Moore
(scattered sori)
Qld
0.3-1 m tall

Epiphytic or lithophytic **fern** forming spreading clumps; **rootstock** thick, fleshy, bearing stiff, spreading scales; **sterile fronds** 10-30 cm x 10-15 cm, erect, broadest towards the middle, shallowly and irregularly lobed, brown and papery; **fertile fronds** 0.3-1 m long, pinnately lobed, dark green, erect; **sori** rounded, irregularly scattered.

A handsome fern that grows on rocks or trees along streams and rock outcrops, frequently in rainforest. It is not common in cultivation, but grows readily if treated as for *D. quercifolia*. Propagate by division or from spore.

DRYOPOA Vick.
(from the Greek *drys*, an oak; *poa*, a grass)
Poaceae

A monotypic genus, endemic in south-eastern Australia.

Dryopoa dives (F. Muell.) Vick.
NSW, Vic, Tas   Giant Mountain Grass
1-4.5 m tall   Oct-March

Biennial **grass** forming very large, loose tussocks; **leaves** 20-150 cm x 1.5-2.5 cm, flat, arching, bright green, often with a prominent white midrib and lesser veins; **culms** up to 2.5 cm thick, robust, erect; **panicle** 15-45 cm x 15-20 cm, very loose and spreading, straw coloured; **spikelets** about 0.8 cm long; **lemma** awnless, with 5 prominent dorsal ribs.

A large grass which is common in cool, moist mountain forests, but not extending into subalpine regions. It is a colonizing species, frequently found on disturbed earth, alongside roads and tracks and on burnt ground. It germinates and grows rapidly, usually forming a tussock in the first year and flowering in the second year, before dying, although vigorous plants may flower in the first year. It is a graceful grass with considerable ornamental appeal. It likes shady, moist conditions and can be grown beside streams or amongst ferns. Young plants transplant readily. Plants may become naturalized in a garden, but are never a problem as they are easily removed. Seed can be germinated in moist soil where the plants are wanted, or the seedlings can be transplanted. Propagate by transplants or from seed which seems to germinate readily soon after collection.

Previously known as *Festuca dives* F. Muell.

DRYOPTERIS Adans.
(from the Greek *drys*, an oak; *pteris*, a fern; some species are ferns growing in oak woods)
Aspidiaceae

Terrestrial **ferns**; **rootstock** erect or creeping, the apex densely covered with scales; **stipes** erect, grooved on the upper surface; **fronds** 2-4 times divided, glabrous, coarse to finely divided; **sori** rounded, bearing indusia.

A large genus of about 150 species with a solitary widespread one extending to north-eastern Qld. A further species is awaiting transference to another genus. A number of exotic species is commonly grown by fern enthusiasts. Propagate from spore which is best sown while fresh.

Dryopteris poecilophlebia (Hook.) C. Chr.
(coloured veins)
Qld
0.3-1 m tall

Terrestrial **fern** forming slowly spreading clumps; **rootstock** creeping, hard and black, knobby; **fronds** 0.3-1 m tall, erect or arching, pinnate, coarse and dull green; **segments** 8-16 cm x 2-4 cm, lanceolate to oblong, with prominent veins, the margins toothed, drawn out at the apex; **sori** very small, rounded, scattered.

A coarse fern found in rainforests of north-eastern Qld, usually in scattered colonies. It grows easily in a pot of well drained, acid soil mix fortified with leaf mould, or in a shady position in the ground. Somewhat frost sensitive and needs protection in areas with cold winters. Propagate by division of the rhizomes or from spore. This fern will be transferred to another genus in the future.

Dryopteris sparsa (D. Don) O. Kuntze
(scattered)
Qld
0.3-0.8 m tall

Terrestrial **fern** forming discrete clumps; **rootstock**

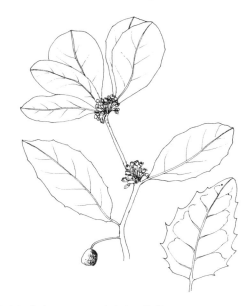

Drypetes lasiogyna var. *australasica*, × ·3

erect, tufted, bearing large papery scales; **stipes** erect, brown, polished, in a tuft; **fronds** 0.3-0.8 m tall, broadly triangular, 2-3 times divided, dull green, thin-textured and brittle; **sori** rounded, covered by an indusium.

An attractive fern, widespread through the Pacific region but rare in Australia where it is restricted to a couple of mountains in north-eastern Qld. It grows in cool, shady situations, always in rainforest. This fern makes a handsome pot plant and also grows easily in a cool, shady, moist position in the garden. Can be grown in a suitable position outside in temperate regions. Very susceptible to wind damage. Propagate from spore which is best sown while fresh.

DRYPETES Vahl.
(from the Greek *dryppa*, an olive fruit; referring to the similarity of the fruits)
Euphorbiaceae

Shrubs or **trees**; **leaves** alternate, entire, leathery, petiolate; **flowers** dioecious, axillary, small, solitary or in clusters or racemes; **male flowers** of 4-5 petals, 4 to numerous stamens; **female flowers** large, 4-5 petals; **fruit** an indehiscent drupe; **seed** solitary, furrowed on one side.

A large genus of about 200 species widely distributed throughout Asia and Africa, with 3 species endemic in eastern Australia and another possibly endemic to Lord Howe Island. Although they are handsome and relatively common trees, they are rarely encountered in cultivation. Propagate from seed which for best results should be sown soon after ripening.

Drypetes australasica (J. Mueller) Pax & K. Hoffm. = *D. lasiogyna* (F. Muell.) Pax & Hoffm. var. *australasica* (J. Mueller) Airy-Shaw

371

Drypetes lasiogyna

Drypetes lasiogyna var. *australasica*   D.L. Jones

Drypetes lasiogyna (F. Muell.) Pax & Hoffm.
(a hairy ovary)
Qld, NSW, WA, NT   Grey Bark; Yellow Tulip
8-20 m x 3-8 m   Oct-Dec
   Small to medium **tree**; **trunk** straight, often flanged
at the base; **bark** grey to brown, scaly, shedding in
flakes, giving a mottled appearance; **leaves** 2-10 cm x
1-3 cm, elliptical, bright green and glossy on both surfaces, the margins wavy, often toothed; **juvenile leaves**
sharply toothed and holly-like; **flowers** about 0.4 cm
across, yellowish, solitary or in small axillary clusters
or racemes, unisexual, each sex on separate trees;
female flowers usually solitary, on longer stalks; **drupe**
1-2 cm long, oval to oblong, bright red and shiny, succulent, the flesh yellow, matures Feb-April.
   Widespread in coastal rainforests of NT and north-
western WA, and distributed from north-eastern Qld
to north-eastern NSW (also Lord Howe Island), extending well inland in a couple of localities, usually in
dry rainforests. It is cut as a timber tree, and the wood
has excellent qualities for carving and use in engraving
blocks. The fruit is recorded as being edible but no
details are known.
   The plants make attractive small trees, suitable for
gardens in temperate and tropical regions. They have a
dense, dark green canopy useful for shade and shelter,
and are decorative when in fruit, although male and
female trees are needed. They are well suited to rainforest gardens, and require shelter from direct hot sun
when small. They succeed in well drained, acid soils,
and respond to mulches and light applications of fertilizer. Young plants could make attractive bonsai subjects. Propagate from seed which must be sown fresh.
   The type form (var. *lasiogyna*) is distinguishable by
its hairy ovary and fruit, and is found in WA, NT and on
the Atherton Tableland in north-eastern Qld. This form
is extremely variable in the size and shape of its leaves
and fruit. The var. *australasica* (J. Mueller) Airy-Shaw,
with glabrous fruit, is widely distributed in Qld and
north-eastern NSW, but has not been found in NT.

Drypetes subcubica (J. J. Smith) Pax & Hoffm.
(nearly cubic)
Qld
8-12 m x 2-5 m   Nov-Dec
   Small, slender **tree**; **bark** grey, smooth; **leaves** 8-
20 cm x 2-6 cm, ovate to elliptical, dark green,
leathery, the margins sometimes toothed, the apex
drawn out into a long point; **flowers** about 0.6 cm
across, creamy green, in axillary clusters, unisexual;
drupe about 2 cm long, nearly cubic in shape, with 4
longitudinal grooves.
   This species, which is found in a couple of In-
donesian islands, also occurs in a limited area of rainforest north of Cairns. It is little-known in cultivation,
but with its decorative large leaves, it could have
potential for tropical gardens. Plants would be well
suited to rainforest gardens, and require protection
from direct sun when small. Young plants could make
attractive pot plants. Propagate from seed which must
be sown fresh.

DUBOISIA R. Br.
(after Charles du Bois, London merchant who had a
botanical garden at Mitcham, Surrey, 17-18th
centuries)
Solanaceae
   Shrubs or **trees**; **bark** corky; **leaves** simple, alternate, the margins entire; **inflorescence** a cyme or
panicle; **flowers** small, tubular, with flared corolla
segments; **corolla lobes** 5; **sepals** 5; **fruit** a fleshy berry.
   A small genus of 3 species, 2 of which are endemic in
Australia, the third extending to New Caledonia. They
are of interest because their leaves contain important
alkaloids used in drug manufacture. Plants are grown
commercially in Qld for the alkaloid extraction, and
wild plants are also harvested. Seed germination of all
species is erratic. Cuttings strike fairly readily.

Duboisia hopwoodii (F. Muell.) F. Muell.
(after Henry Hopwood)
Qld, NSW, SA, WA, NT   Pituri
2-5 m x 1-3 m   Aug-Jan
   Medium to tall **shrub**; **bark** corky, brownish yellow
to purplish; **branchlets** slender, the tips drooping;
leaves 7-15 cm x 1.5-3 cm, linear to lanceolate, dark
green; **cymes** terminal; **flowers** to 2 cm x about 0.5 cm,
bell-shaped, white with purple stripes, the tips of the
petals spreading; **berry** about 0.6 cm across, globular,
black, fleshy.
   A widely distributed and common plant found in the
arid regions, growing on freely drained, sandy or
gravelly soils. The leaves are highly toxic to stock and
camels. The Aborigines chewed the powdered leaves
and twigs because they produce a narcotizing effect
and assuage hunger. The name 'Pituri' was used by the
Aborigines of the Mulligan River in western Qld for
the chewing concoction and this name was then widely
adopted by the early white settlers, who also chewed
plugs of the material. The plant was of importance to
the Aborigines, and it was used as an article of trade.
Also, small water-holes were infused with leaves and
branches of the plant, to stupefy and disorient emus,
making them easy prey for hunters. Pituri plants may
sucker freely, especially if the parent bush is destroyed.
The species grows readily in cultivation, and is well
suited to gardens in inland areas. Plants withstand cutting and can be a useful screen. They require a sunny

aspect, in well drained soil. Propagate from seed, cuttings or by transplants. Seed is difficult and slow to germinate. Root cuttings could be well worth trying.

Duboisia leichhardtii (F. Muell.) F. Muell.

(after L. Leichhardt, early explorer)

Qld, ?NSW	Corkwood
2-6 m x 1-3 m	July-Nov

Medium to tall **shrub**; **bark** corky, grey to yellowish brown; **leaves** 5-12 cm x 1-2 cm, lanceolate, pale green, glabrous, thin-textured, the apex pointed; **panicles** terminal, much-branched; **flowers** about 1.2 cm x 0.6 cm, white, sometimes tinted with mauve, bell-shaped, with flared tips, each tapered to a fine point; **berry** about 0.6 cm long, oval, black, fleshy.

This species is restricted to central and southwestern Qld. It has also been reported for the Far Western Plains of NSW, but these records are apparently erroneous. The species is chiefly found west of the Dividing Range. It usually grows in fertile soils, in association with *Callitris* spp. The leaves are poisonous to stock, but they are generally unpalatable. They contain important alkaloids similar to those of *D. myoporoides*, but often in higher concentrations, and wild and cultivated populations are harvested for their extraction. A hybrid with *D. myoporoides* is important in this respect. Commercial plantations of *D. leichhardtii* have been established using transplanted seedlings, and these may be harvested mechanically or by hand. Plants grow readily in a sunny aspect, in well drained soil, and respond to the use of fertilizers. They can be propagated from cuttings which may be difficult to strike or from seed which germinates erratically.

Duboisia myoporoides R. Br.

(like the genus *Myoporum*)

Qld, NSW	Corkwood
3-12 m x 2-5 m	July-Oct; also sporadic

Duboisia myoporoides   D.L. Jones

Tall **shrub** or small **tree**; **bark** corky, pale brown to grey, fissured; **leaves** 5-12 cm x 1-2.5 cm, lanceolate, pale to bright green, glabrous, shiny, thin-textured, the apex blunt; **panicles** terminal, much-branched; **flowers** about 0.6 cm across, white with purplish markings, bell-shaped, with flared tips, each blunt; **berry** about 0.6 cm long, globular, black, fleshy, mature Oct-Dec.

Widely distributed from Cape York Peninsula to north-eastern NSW, in coastal districts and the adjacent ranges and tablelands. The species also occurs in New Caledonia and New Guinea. It is usually a component of open forest, but is also very frequent in coastal communities, and may also occur on rainforest margins. Plants are quite attractive when in flower, and may flower nearly continuously. They are extremely fast growing when young, often achieving 5 m in 18 months. They are also quite long-lived. The leaves are poisonous to stock, and cases of human poisoning (non-fatal) are also known. The leaves contain valuable alkaloids (hyoscyamine, nor-hyosciamine, scopolamine and atropine) which are used in the production of sedatives and ophthalmics. Their effects were known as early as the 1850s. Assays of dried leaves show up to 7% alkaloids. These drugs gained importance in World War II (scopolamine prevents seasickness), when large quantities of leaves were harvested for their extraction. Since then this species has increased in importance as a drug-source, and the leaves are harvested from wild populations and cultivated plantations, and exported to Europe. There is interest in the selection and propagation of superior clones, as there is variation in the drug content. A hybrid between this species and *D. leichhardtii* is grown in the south Burnett area of Qld. Commercial plantations are harvested mechanically. The Aborigines knew of this plant and drank a concoction of the sap of the trunk, which induced a stupor. The plant is easy to grow, requiring well drained soils in a sunny situation. It withstands regular pruning, and responds to the application of fertilizers. Propagate from seed which should be sown fresh from the fruit or from cuttings of semi-hardwood which may be slow to strike. Seeds may also be slow and erratic to germinate.

Intergeneric hybrids are known between this species and *Cyphanthera albicans* in areas of NSW where the two grow together.

Duboisia hybrids

Because of their importance for drug production, populations of *Duboisia* have been sampled to obtain high-yielding clones. A hybrid between *D. myoporoides* and *D. leichhardtii* was found to be particularly high-yielding, and is now propagated vegetatively and grown in plantations in the south Burnett District of Qld. The plants are grown in rows and may be harvested mechanically or by hand. As they are trimmed regularly, they become compact and bushy. After harvest, the stems are allowed to wilt in the shade, and the leaves are stripped, packed and exported to Europe for drug extraction. Plants are generally short-lived in this intensive cultivation system, and must be replaced after 5-7 years. Propagate from cuttings of semi-hardened wood under mist and with rooting hormones.

DUNBARIA Wight & Arnold
(after M. Dunbar, 18-19th-century Scottish academic)
Fabaceae

Climbers or **trailers**; **stems** twining, tomentose; **leaves** trifoliolate; **leaflets** with resinous dots; **flowers** pea-shaped, small, usually yellow, solitary or in axillary racemes; **fruit** a linear, flat, dehiscent pod, not indented between the seeds.

A genus of about 25 species, well developed in tropical Asia, with 2 species found in northern Australia. Propagate from seed which is hard and needs treatment to germinate.

Dunbaria conspersa Benth.
(scattered or thickly covered)
Qld, NT
3-5 m tall   Aug-Nov
Slender **climber**; **stems** twining, bearing fine hairs, with resinous dots; **leaves** trifoliolate; **leaflets** 1-2.5 cm x 0.5-1 cm, rhomboidal, the terminal one 3-lobed, with wavy margins; **flowers** about 0.5 cm long, yellow, solitary or in pairs in the leaf axils; **pods** 2.5-3.5 cm long, straight or falcate, glabrous or slightly hairy.

A widespread species which extends to the coastal districts of tropical Qld, and NT and adjacent islands. Plants are never showy in flower but the species is an interesting slender climber that could be grown in tropical and subtropical gardens. It requires well drained soil. Propagate from scarified seed.

Dunbaria singuliflora F. Muell.
(solitary flowers)
Qld, NT
prostrate-2 m tall   Jan-April
Slender herbaceous **climber** dying back to a thickened perennial storage root; **stems** slender, prostrate or climbing; **leaves** trifoliolate; **leaflets** 1-2 cm x 0.5-0.8 cm, ovoid, the terminal one lobed; **flowers** about 0.5 cm long, yellowish, pea-shaped, solitary in the axils; **pods** about 2 cm long, straight, hairy.

This species is found in northern areas, where it grows in open forest, flowering and fruiting during the wet season and dying back to a perennial rootstock in the dry. It has limited horticultural appeal, with small dull coloured flowers. Will grow in inland areas or drier tropical regions, in a sunny situation. Propagate from scarified seed or by transplants.

DURANDEA Planchon
(after J. F. Durande, 18th-century French botanist)
Linaceae

Climbers; **stems** hairy; **tendrils** coiled or recurved, arising from sterile inflorescences; **leaves** alternate, simple, entire or with serrate margins, stipulate; **inflorescence** cymose or paniculate; **flowers** open-petalled, showy, short-lived; **sepals** 5; **petals** 5; **fruit** a drupe.

A genus of 15 species distributed in Melanesia with a solitary species in north-eastern Qld. Propagate from seed which must be sown fresh or from cuttings.

Durandea jenkinsii (F. Muell.)
(after W. S. Jenkins)
Qld
5-10 m tall   Nov-Jan
Medium **climber**; shoots glaucous, angular; **leaves** 10-18 cm x 4-6 cm, narrow-lanceolate to ovate, thick and leathery, dark green, glossy on both surfaces, the margins crenulate; **tendrils** stout, arising at the base of new growth; **panicles** in the upper leaf axils; **flowers** about 0.8 cm across, bright yellow, with a prominent bunch of stamens; **drupe** about 1 cm across, globular.

An attractive climber found in lowland rainforests of north-eastern Qld, often near the coast. It is recorded that the Aborigines used the recurved tendrils as fish hooks. Plants are very showy in flower, and the species is deserving of wider cultivation in tropical and subtropical areas. It is a strong climber and is ideal for covering unsightly objects. It can also be trained up existing trees. The flowers are short-lived but produced in mass. Plants require a well drained soil in a semi-shady situation. Propagate from seed or cuttings of semi-ripened wood. The species was previously known as *Hugonia jenkinsii* F. Muell.

Duriala villosa (F. Muell.) Ulbrich =
Maireana enchylaenoides (F. Muell.) Paul G. Wilson

DYSOPHYLLA Blume
(from the Greek *dysodes*, fetid; evil-smelling; *phyllon*, a leaf; in reference to the strong-smelling leaves)
Lamiaceae

Herbs; **leaves** opposite or whorled; **flowers** small, in dense cylindrical spikes; **calyx** equally 5-toothed; **corolla** with short tube, the limb of 4 equal or nearly equal lobes; **stamens** 4; **anthers** terminal, 1-celled; **fruit** a nut.

This tropical genus has only 1 representative in Australia. It is closely related to *Mentha*.

Dysophylla verticillata Benth.
(leaves in whorls)
Qld, WA, NT
0.3-0.6 m x 0.5-1 m   May-Oct; also sporadic
Dwarf, perennial **herb**; **stems** glabrous or pubescent, decumbent at base, becoming upright; **leaves** 2-8 cm long, linear-lanceolate, verticillate, sessile, entire, margins recurved, strong-smelling; **flowers** very small, white, borne in cylindrical spikes to 5 cm or more long.

Confined to northern Australia where it inhabits low-lying areas such as the edges of water-holes. On occasions it grows in shallow water. It is probably best suited to tropical and subtropical regions, and should grow in moist soils, in partial to full sun. The leaves have a strong, almost obnoxious fragrance. Worthy of experimentation in southern states. Propagate from cuttings.

DYSOXYLUM Blume
(from the Greek *dysodes*, fetid; evil-smelling; *xylon*, wood; referring to the onion-like odour of the wood of some species)
Meliaceae

Shrubs or more usually **trees**; **bark** and **wood** often with unpleasant or fetid odour; **leaves** pinnate, spreading; **leaflets** simple, entire, opposite or alternate; **inflorescence** a raceme or panicle; **flowers** tubular at the base; **calyx** 4-5 lobed or divided into 4-5 sepals; **petals** 4-5, free or joined to a staminal tube, the lobes spreading; **stamens** 8-10, the filaments joined into a tube; **fruit** a capsule, 1-5 celled, splitting when ripe.

A large genus of about 200 species widely distributed through India, Malaysia and the Pacific region, with about 10 species found in Australia. They are all found in NSW or Qld, 1 species extending to NT. They commonly grow in rainforests, but a couple of species also extend to moist situations in open forests. Most of the Australian species are endemic, and a couple are important timber trees yielding valuable timber. All species are decorative and have many features to recommend them for cultivation; however, some grow too large for the average home garden but still could find a use for park planting. When grown in an open situation they develop a spreading crown and provide excellent shade. The flowers are small but the fruit is often decorative. The plants are easily grown in eastern Australia, but need some protection when young. Propagate from seed which deteriorates and must be sown soon after collection.

Dysoxylum arborescens (Blume) Miq.
(tree-like)
Qld
6-10 m x 3-8 m   Nov-Dec

Tall **shrub** or small **tree**; young shoots glabrous; **bark** of branches dark red; **leaves** 12-20 cm long, pinnate; **leaflets** 14-18, 7-15 cm x 2.5-5 cm, oblong to elliptical, unequal at the base, the apex bluntly toothed, dark green and glabrous on both surfaces; **panicle** 6-10 cm long, glabrous; **flowers** about 1 cm across, yellowish, membranous, the calyx bearing 5 sharp lobes; **petals** 5; **capsule** 1.5 cm long, hairy, 5-celled; **seeds** 5 per capsule, about 0.8 cm long.

A widespread species found in South-East Asia and various Pacific islands, and extending to rainforests of north-eastern Qld. It is virtually unknown in cultivation, but appears to have strong potential as a garden or park plant for tropical and subtropical regions. Plants would probably require acid, well drained soils, and some protection when small. Propagate from seed which should be sown while fresh.

Dysoxylum cerebriforme Bailey
(shaped and marked like the brain)
Qld
6-10 m x 3-6 m   Dec-March

Tall **shrub** or small **tree**; young shoots bearing rusty hairs; **leaves** 15-30 cm long, pinnate; **leaflets** 7-9, 3-10 cm x 2.5-6 cm, lanceolate, dark green above, paler and hairy beneath, with white hairs, rounded at the base, the points drawn out; **racemes** or panicles 6-10 cm long, in the upper axils; **flowers** about 1 cm long and 2 cm across when open; **capsule** 2.5 cm long, pear-shaped, 5-celled, brownish; **seeds** 5, orange-red, prominently wrinkled, densely covered with a mixture of short orange and white hairs, mature Feb.

Dysoxylum oppositifolium   D.L. Jones

A little-known species restricted to lowland rainforests of north-eastern Qld. It is virtually unknown in cultivation, but appears to have excellent potential as a small shade tree for tropical gardens. Would probably require acid, well drained, humus-rich soil. Propagate from seed which must be sown fresh.

Dysoxylum densivestitum C. White & Francis = D. schiffneri F. Muell.

Dysoxylum fraseranum Benth.
(after Charles Fraser, first Superintendent of Sydney Botanic Gardens)
Qld, NSW   Rosewood
12-25 m x 3-8 m   April-Dec

Medium to large **tree** with a dense, spreading crown; **bark** light brown, scaly; young shoots hairy; **leaves** 15-25 cm long, alternate, pinnate; **leaflets** 6-12, 5-10 cm x 2-4 cm, narrow to broadly elliptical, often slightly falcate, dark glossy green above, dull and paler beneath, numerous oil dots, the apex blunt; **panicles** axillary, in the upper leaf axils; **flowers** 0.8-1 cm across, cream, white or pale mauve, fragrant; **petals** 4-5; **capsule** 2-4 cm across, globular to pear-shaped, 3-4 celled, pink to rusty brown; **seeds** 6-8 per capsule, 0.8 cm long, red, mature April-Dec.

A common rainforest tree, especially in highland districts, extending from south-eastern Qld to Wyong near Sydney. It is a very attractive tree that should find widespread application for large gardens or parks, but generally grows too big for home gardens. The crown is spreading and dense, and supplies excellent shade. Trees in rainforests are tall and slender, but those planted in the open develop a spreading or rounded crown. The species is easily grown on well drained, acid soils, and performs best in soils rich in organic matter. Seedlings require some protection from direct summer sun when small. Established plants are quite hardy, and tolerant of dry periods. Small plants may need some protection from frosts. The timber is reddish and has a delicate perfume like that of roses. It is highly regarded and is excellent for carving and cabinet work. Propagate from seed which must be sown while fresh.

Dysoxylum klanderi F. Muell.
(after Dr. Klander)
Qld
12-15 m x 5-12 m   March-July
   Medium **tree** with a spreading canopy; young shoots glabrous; **leaves** 8-25 cm long, pinnate; **leaflets** 4-8, 10-15 cm x 3-5 cm, ovate-lanceolate, opposite, glossy dark green on both surfaces, the apex acuminate; **panicles** 30-60 cm long, pendulous, fairly narrow, the branches bearing small clusters of flowers near the end; **flowers** about 1.5 cm across, greenish yellow, hairy; **petals** 4; **capsule** about 2.5 cm across, globular to pear-shaped, 4-celled, brownish, shortly hairy; **seeds** 4 per capsule, ovate, 3-sided, mature Dec.
   Restricted to rainforests of north-eastern Qld, but there fairly widespread and often locally common. It is a handsome tree, perhaps too large for a home garden but suitable for parks and possibly as a street tree. It requires a protected situation in acid, well drained soil, and is best suited to coastal areas of tropical or subtropical regions. Propagate from seed which must be sown fresh.

Dysoxylum latifolium Benth.
(broad leaves)
Qld
15-20 m x 5-8 m   Oct-Nov
   Medium, spreading **tree**; young shoots hairy; **leaves** 10-15 cm long, pinnate; **leaflets** 7-10 cm x 1-3 cm, 4-5 per leaf, ovate to oblong, fairly thick-textured, dark green; **racemes** 10-12 cm long, clustered; **flowers** about 1.2 cm across, cream-brown, hairy; **petals** 4; **capsule** about 2 cm across, globular, 2-celled, brown; **seeds** 1-4 per capsule.
   A little-known species restricted to rainforests of north-eastern Qld. Virtually unknown in cultivation but could have potential as a park tree in tropical and subtropical areas. Propagate from seed which must be sown fresh.

Dysoxylum muelleri Benth.
(after Baron F. von Mueller, first Government Botanist, Vic)
Qld, NSW   Red Bean
15-25 m x 3-10 m   Jan-July
   Medium to large **tree**; **bark** grey, somewhat scaly; young shoots hairy; **leaves** 30-60 cm long, pinnate, stiff and spreading; **leaflets** 11-21, 6-15 cm x 1.5-3 cm, ovate to lanceolate, green above, paler and often hairy beneath, the base very unequal, the apex pointed; **panicles** 10-25 cm long, in the upper leaf axils; **flowers** 1-1.5 cm across, cream, numerous, hairy, tubular at the base; **petals** 4; **capsule** 1.5-2 cm across, globular, 2-5 celled, brown, hairy, wrinkled; **seeds** 2-10 per capsule, about 1 cm long, red, mature Nov-Feb.
   A widespread tree found in rainforests from Cairns to the Bellinger River in north-eastern NSW. Because of its size it is best suited to acreage planting, large gardens or parks, and could have potential as a street tree. It develops a dense, spreading canopy and provides excellent shade. The plants require protection and moisture when young, but are easy to maintain once growing strongly. They can be successfully grown as far south as Melbourne, but need protection from

Dysoxylum pachyphyllum   T.L. Blake

frosts when young. The timber is a deep reddish brown and is excellent for carving, indoor fittings and joinery. The bark and sapwood smell strongly of onions. Propagate from seed which must be sown fresh.

Dysoxylum oppositifolium F. Muell. ex DC.
(opposite leaves)
Qld, NT   Pink Mahogany
8-18 m x 3-10 m   July-Nov
   Small to medium **tree** with a spreading canopy; **bark** grey-brown, smooth; young shoots densely clothed with short hairs; **leaves** 15-25 cm long, pinnate; **leaflets** 4-10, 6-12 cm x 2-5 cm, ovate-lanceolate to lanceolate, dark green above, paler beneath, glabrous or slightly hairy, drawn out in a pointed apex; **racemes** 1-5 cm long, narrow and spike-like, in the upper leaf axils; **flowers** about 1 cm across, yellowish, finely hairy; **petals** 4; **capsule** 2.5-3 cm across, 4-6 celled, yellowish brown, hairy; **seeds** up to 4 per capsule, about 1.3 cm long, red, mature Oct-Nov.
   A little-known species found in rainforests of north-eastern Qld, and the Top End. It is a handsome tree with a spreading shady canopy, and is very decorative when in fruit. It may grow too large for the average home garden, although it is very slow-growing,

especially in the early stages. Young plants require protection from direct sun.

The species appears to grow well in moist to wet soils, and is best suited to tropical or subtropical regions. Propagate from seed which must be sown fresh.

Dysoxylum pachyphyllum Hemsley
(thick leaves)
NSW (Lord Howe Island)
6-18 m x 3-8 m   Oct

Small to medium, sparse **tree**; young shoots glabrous; **leaves** 30-40 cm long, pinnate; **leaflets** 8-12, 4-15 cm x 3-5 cm, oblong to elliptical, dark green and glabrous on both surfaces; **panicle** 6-15 cm long, glabrous, few-flowered; **flowers** 0.3-0.4 cm across, yellowish brown; **petals** 5, thick-textured; **stamens** 8; **capsule** 3-4 cm long, leathery, brown, 3-celled; **seeds** with a thick aril, mature Oct-Dec.

An attractive tree with a spreading canopy. Endemic to Lord Howe Island, where it grows in rainforests. It is virtually unknown in cultivation, but appears to have potential for parks in subtropical and warm temperate regions. Plants would probably require well drained soils and protection when young. Could make an excellent bonsai subject. Propagate from seed which should be sown while fresh.

Dysoxylum pettigrewianum Bailey
(after W. Pettigrew)
Qld   Spurwood
20-30 m x 10-15 m   Jan

Tall **tree** with a spreading canopy; **stem** prominently buttressed; **bark** dark brown, with numerous pustules; young shoots densely clothed with brown hairs; **leaves** 8-15 cm long, pinnate; **leaflets** 7-13, 6-20 cm x 2-8 cm, narrow-elliptical to ovoid, drawn out into a long, blunt apex, dark green, shortly hairy, rounded at the base; **spikes** 5-8 cm long, hairy, axillary; **flowers** about 1.2 cm long, yellowish, densely hairy, bell-shaped at the base; **petals** 4-5; **capsule** about 2.5 cm across, pear-shaped, 4-celled, glabrous, with 4 prominent, acute ridges; **seeds** up to 4 per capsule, about 2 cm long, red, mature Nov.

A large tree of coastal and highland rainforests of north-eastern Qld. It is reportedly deciduous during dry periods in the winter. It attains dimensions that are much too large for the average home garden, but could be useful for planting in parks and large gardens. It develops a dense, spreading canopy that provides excellent shade and shelter. Best suited to tropical and subtropical regions in an acid, well drained soil rich in organic matter. Young plants require shading until about 1 m tall. Propagate from seed which must be sown fresh.

Dysoxylum rufum (A. Rich.) Benth.
(reddish)
Qld, NSW   Hairy Rosewood
10-18 m x 3-8 m   Jan-March

Small to medium **tree** with a spreading crown; **bark** grey, wrinkled; young shoots densely covered with rusty hairs; **leaves** 30-60 cm long, pinnate, alternate; **leaflets** 8-19, 9-20 cm x 1-3 cm, lanceolate to broadly lanceolate, entire, grey-green above, rusty-hairy beneath, unequal at the base, the apex drawn out into a fine point; **panicles** or racemes axillary; **flowers** 0.8-1 cm long, tubular, white, fragrant, becoming overpowering; **capsule** 1.5-2 cm across, globular, densely covered with rigid, yellow-brown hairs, matures May-Sept.

A widespread tree found in moist forests and rainforests of eastern Australia, from the Atherton Tableland to Port Macquarie. Its spreading canopy provides excellent shade and shelter for delicate plants such as ferns. The species has excellent attributes and deserves to be widely planted as a street tree in parks and public gardens for shade and its pleasing appearance. It grows readily in most acid, well drained soils, and responds to artificial watering, fertilizers and mulches. Young plants may require some protection from direct sun, and frost may kill the growing point. The species can be grown successfully in temperate regions, where its ultimate size is much reduced. The flowers have a heady perfume that becomes sickening as they age. The hairs on the fruit are irritating when handled. The timber has an onion-like odour. Propagate from seed which must be sown fresh. This species is sometimes classified as *Didymochoton rufum* (A. Rich.) Harms.

Dysoxylum schiffneri F. Muell.
(after Dr. R. Schiffner)
Qld   Yellow Mahogany
15-25 m x 5-10 m   Aug-Sept

Medium to tall **tree** with a spreading crown; **bark** grey-brown, smooth; young shoots bearing silky hairs; **leaves** 10-18 cm long, pinnate; **leaflets** 5-15 cm x 2-6 cm, ovate to lanceolate, opposite, thin-textured, papery, dark green; **racemes** 5-10 cm long, 2-3 together in the upper axils; **flowers** about 1.2 cm across, white, silky-hairy; **petals** 4; **capsule** about 2.5 cm across, globular, 4-celled, brown, becoming hairless; **seeds** 4 per capsule, turgid, brown, mature Nov-Dec.

This species occurs in lowland and highland rainforests of north-eastern Qld, extending to about 1400 m alt. on Mount Bellenden Ker. It has potential for planting as a specimen tree in parks, or as a street or avenue tree. Tropical and subtropical regions are most suitable for its culture, and it needs a position exposed to semi-shade or dappled sun, in acid, well drained soil. Propagate from seed which must be sown fresh.

Dysoxylum sericoflorum C. White =
                              D. cerebriforme Bailey

DYSPHANIA R. Br.
(from the Greek *dysphanes*, obscure; scarcely visible; in reference to the minute flowers)
Dysphaniaceae

Annual **herbs**; **leaves** alternate, fleshy, hairy, aromatic; **flowers** very small, unisexual or bisexual; **female flowers** in short clusters on a spike-like arrangement; **perianth segments** usually 2-4, hooded, sometimes inflated.

Dysphania rhadinostachya

Dysphania simulans, × ·6

An endemic genus of 9 species generally occurring in low-rainfall regions, eg. Central Australia. Not common in cultivation, but certainly has some potential in arid and semi-arid areas.

Dysphania rhadinostachya (F. Muell.) A. J. Scott
(slender flower-spike)
Qld, SA, WA, NT
0.1-0.5 m x 0.5-1 m   July-Sept
   Annual dwarf **herb**; **stems** erect or ascending, glandular, hairy; **leaves** to about 3 cm long, ovate, lobed, hairy; **flowers** small, globular clusters arranged in axillary or terminal spikes, simple or branched, very short, to about 12 cm long, fragrant.
   An annual, best suited to semi-arid areas. It grows quickly in most soils, but needs full sun for best results. Useful for preventing erosion in recently cleared areas. Has decorative seed-spikes to about 15 cm long. Propagate from seed.
   Previously known as *Chenopodium rhadinostachyum* F. Muell.

Dysphania simulans Tate
(similar)
Qld, NSW, SA, WA, NT
0.1-0.3 m x 0.2-0.3 m   sporadic
   Dwarf, erect annual **herb**; **stems** decumbent to ascending, hairy; **leaves** 1-2 cm long, elliptic, thin, 2-4 pairs of lobes, hairy; **flowers** small, arranged in dense, narrow spike to 25 cm long; **perianth segments** 3.
   A dweller of saline soils and areas prone to flooding. From inland regions. May prove useful in the reclamation projects for saline soils. Direct sowing of seed should be the most suitable method. Application of fertilizers would help to promote fast growth.

Dysphaniaceae (Pax) Pax
   A small endemic family of dicotyledons consisting of 9 species in the genus *Dysphania*. They are small perennial herbs, closely related to the genus *Chenopodium*, and are included in the family Chenopodiaceae by some authorities.

378

E

Ebenaceae Gurke

A family of dicotyledons consisting of about 500 species in 2 or 3 genera principally developed in tropical rainforests of the Indomalesian region, with a few extending into temperate zones. They are mostly small to medium sized trees with alternate, simple leaves and a fleshy fruit that is astringent unless very ripe. Some species are important timber trees and produce the black, hard wood known as ebony. A number have edible fruit, the most significant of which is the Chinese or Japanese Persimmon (*Diospyros kaki*). The family is represented in Australia by about 15 species in the genus *Diospyros*.

ECDEIOCOLEA (F. Muell.) Shaw
(from the Greek *ecdeia*, falling short; *colcos*, a sheath; in reference to the deciduous scale-leaves)
Ecdeiocoleaceae

An endemic monotypic genus confined to south-western WA.

Ecdeiocolea monostachya F. Muell.
(single flower-spike)
WA
1 m x 1 m   Aug-Oct
A large, tufting, perennial **herb**; **stems** to 1 m tall, simple, without basal leaves but with imbricate sheathing scales at base, and one sheathing scale below spike; **inflorescence** a terminal spike of unisexual flowers, cream; **male flowers** stamens 3, filaments free; **female flowers** staminodia 3.

A most ornamental rush-like species. It occurs over a wide area from north-east to north of Perth, where it is found both in inland and coastal situations. It grows on a variety of soils, including various sands, lateritic gravels and sandy clays. Recorded as horticulturally desirable. Should grow well in a wide variety of soils which are moist for extended periods, and in a position where the plant receives partial or full sun. Experimentation as a bog garden plant is warranted. Propagate from seed.

Ecdeiocoleaceae Cutler & Airy-Shaw

A monogeneric family of monocotyledons containing a solitary species (*Ecdeiocolea monostachya*) which is confined to WA. The species is a large perennial herb with many floral features similar to those of Restionaceae, and the growth habit of Xyridaceae. The species is rarely cultivated.

ECTROSIA R. Br.
(from the Greek *ectrosis*, a miscarriage; in reference to the asexual or unisexual flowers in the spikelets)
Poaceae

Annual or perennial **grasses**, perennials forming tussocks, the annuals slender; **leaves** linear, flat or rolled; **culms** erect, slender; **inflorescence** a terminal panicle; **spikelets** solitary, laterally flattened; **lemmas** awned.

An endemic genus of about 12 species of grasses. They are mostly annuals with little forage or ornamental value; however, one perennial species has potential for the tropics. Propagate from seed which may need storage for up to 12 months before sowing or by division of the clumps.

Ectrosia schultzii Benth.
(after Frederick Schultze, naturalist, NT)
Qld, WA, NT
25-60 cm tall   Nov-Feb
Perennial **grass** forming upright tussocks; **leaves** 10-23 cm x 0.3-1 cm, linear, rigid, flat or rolled; **culms** 25-60 cm long, erect, simple, slender; **panicle** 5-17 cm x 3-12 cm, dense, green to purplish; **spikelet** 0.7-0.9 cm long, narrow; **glumes** lanceolate, green to purple.

Widely distributed in forests of northern Australia, this grass forms an attractive tussock that is particularly appealing in good seasons following fires. It has scope for garden planting in tropical and adjacent inland regions. Requirements are for a well drained soil in a sunny position. Tussocks can be rejuvenated by burning or cutting back. Propagate from seed or by division of the clumps.

EHRETIA L.
(after D.G. Ehret, 18th-century German botanical illustrator)
Boraginaceae (alternative *Ehretiaceae*)

Tall **shrubs** or **trees**; **leaves** alternate, simple, entire or the margins toothed; **inflorescence** of axillary or terminal panicles or cymes; **flowers** small, white; **calyx** with 5 lobes; **corolla** with a short tube and 5 spreading lobes; **fruit** a lobed drupe.

A genus of about 50 species with about 6 from Australia. They are shrubs or trees with large clusters of small colourful fruit. Some species are cut for their timber. Only one Australian species is grown and that on a limited scale. Propagate from seed which for best results must be sown fresh.

Ehretia acuminata R. Br.
(ending in a short, sharp point)
Qld, NSW   Koda
10-25 m x 5-12 m   Sept-Nov

Medium to tall **tree**; **stem** deeply channelled; **bark** light grey, fissured; young shoots light green, with prominent white lenticels; **leaves** 7-15 cm x 3-6 cm, ovoid to elliptical, thin-textured, dark green above, paler beneath, the margins coarsely toothed; **panicles** 5-20 cm long, terminal or in the upper axils; **flowers** about 0.5 cm across, white, sweetly scented; **fruit** 0.4-0.6 cm across, globular, orange, splitting when ripe, mature March-April.

This tree is widely spread through countries such as India, Japan and the Philippines. In Australia it is distributed from north-eastern Qld to south-central NSW. It grows in rainforests and moist forests, frequently along stream banks. The trees become particularly noticeable when fruiting, as the whole canopy becomes covered with clusters of orange fruit. These are attractive to fruit-eating birds. In the rainforests, the trees can be recognized by the channelled trunk covered with light, fissured bark. They are cut for timber which is used for furniture and joinery. Trees may become deciduous in dry times. The ripe fruit is edible and has a sweetish taste. Koda is a fast growing and useful tree, suitable for gardens and acreage planting. It is an excellent shelter species for ferns and more delicate rainforest plants. It requires well drained soil and responds favourably to moisture and fertilizers. Propagate from seed which has a short viability and is best sown fresh.

Ehretia membranifolia R. Br.
(very thin-textured foliage)
Qld, NSW
5-10 m x 3-5 m   Oct-Nov

Tall **shrub** or small **tree**; **branchlets** slender, glabrous; **leaves** 3-6 cm x 1-3 cm, oblong to oblanceolate, thin-textured, light green, the margins entire, the apex blunt; **cymes** 2-4 cm long, loose, in the upper axils; **flowers** about 0.4 cm across, white, the tubular base about 0.2 cm long; **fruit** about 0.4 cm across, bright red, 4-celled.

A species of protected situations along stream banks and in dry rainforests. It is distributed between Rockhampton and northern NSW, most frequently on the western slopes of the Great Dividing Range. Plants become deciduous during dry periods. Little is known about the behaviour of this species in cultivation, but it could be an attractive and useful shrub for drier inland areas. Would require well drained soils. Propagate from fresh seed.

Ehretia saligna R. Br.
(willow-like)
Qld, WA, NT   Peach Wood
3-6 m x 2-4 m   April-May; Sept-Oct

Tall **shrub** or small **tree**; young shoots bright green; **leaves** 5-13 cm x 2-4 cm, linear to linear-lanceolate, thin-textured, bright green above, glaucous beneath, the margins entire, the apex drawn out into a fine point; **cymes** 2-5 cm long, axillary; **flowers** about

0.4 cm across, white, the tubular base about 0.2 cm long; **fruit** about 0.4 cm across, reddish, 4-celled.

Distributed across northern Australia and adjacent islands, this species grows along stream banks. Plants become deciduous during the dry season. The wood is orange, attractively patterned and can be used for furniture construction. Although unknown in cultivation, this species could be an attractive shrub for inland and drier tropical regions. Would need a sunny aspect in well drained soil. Propagate from fresh seed.

Ehretia species (Leichhardt)
Qld
1.5-2 m x 0.5-1.5 m   Aug

Small dense **shrub**; young shoots coppery green; **leaves** 2-3 cm x 1-1.5 cm, ovate to obovate, hard-textured, scabrous, brownish green, the margins bearing a few white spines; **panicle** about 1 cm long, dense, few-flowered, terminal; **flowers** about 0.2 cm across, white; **drupe** 0.2-0.3 cm across, globose, ridged, reddish, hard.

An apparently undescribed species which is confined to brigalow scrub of the Leichhardt District. Its flowers and fruit are too small to be considered decorative but it is a very hardy, rigid shrub which could find a place in inland gardens. Requirements would be for well drained soil in a sunny situation. Propagate from seed.

Ehretiaceae Lindley
A family of dicotyledons consisting of 13 genera and 400 species mostly found in tropical and subtropical regions. They are mainly shrubs or trees with a few herbs that have alternate, simple leaves and cymose inflorescences. The fruit is a hard, dry drupe. Some species are cut as timber trees. The family is poorly represented in Australia and is frequently included in Boraginaceae. Australian genera: *Cordia, Ehretia* and *Halgania*.

Ehretia membranifolia, × ·5

EICHLERAGO Carrick
(after Dr. Hj. Eichler, former Curator of Herbarium Australiense, CSIRO, ACT)
Lamiaceae
A monotypic endemic genus restricted to the Murchison River, WA. This genus is closely allied to *Prostanthera*, which has a larger middle lobe in the lower corolla lip, while *Eichlerago* has 3 equal lobes.

Eichlerago tysoniana Carrick
(after Isaac Tyson, 19th-century botanical collector)
WA
1-2 m x 1-2 m   unknown
Small **shrub**; young growth hairy; **branches** hairy, often intertwined, reddish brown; **branchlets** densely hairy, brown; **leaves** 0.2-0.4 cm x 0.2-0.4 cm, decussate, elliptic to orbicular, hairy, margin entire or rarely irregularly lobed, flat or sometimes wavy; **flowers** tubular, to about 1 cm long, white streaked with purple, exterior glabrous, lobes equal, blunt, upper 2, lower 3, solitary, axillary on short pedicels along branchlets; **stamens** not exserted; **calyx** 2-lipped, hairy exterior; **fruit** globular, about 0.3 cm diameter.
This species has evidently not been seen in nature since the last collection in 1898. Efforts have since been made to explore the Mount Narryer area in the upper Murchison River region, but climatic conditions have prevented botanists reaching their destination. From limited information available, it would appear to be a decorative species worthy of cultivation. Propagation of such a rare and isolated species would ensure that there is a bank of plant material in cultivation and may enable it to be saved from extinction.
It should be suited to warm to hot climates with low rainfall. Will probably grow best in well drained soils. Cultivation as a container plant would be worth trying in temperate regions. Propagate from seed and probably cuttings.

Elaeagnus triflora, × ·35

ELACHANTHUS F. Muell.
(from the Greek *elachys*, short; small; *anthos*, a flower; in reference to the shortness of the florets)
Asteraceae
Although there has been reference to this genus comprising up to 3 species, it is most likely a monotypic endemic genus.

Elachanthus pusillus F. Muell.
(very small)
NSW, Vic, SA, WA   Elachanth
0.05-0.1 m x 0.1-0.2 m   Aug-Sept
Dwarf, erect or branched annual **herb**; **branches** hairy; **leaves** 0.3-1 cm long, narrow-linear, alternate, hairy; **flower-heads** about 0.5 cm across, greenish yellow, terminal; **outer flowers** female; **inner flowers** bisexual but sterile; **fruit** an achene.
This small annual dwells in semi-arid regions. Little-known in cultivation but may have some potential as a container plant. Should do best in very well drained soil in a sunny location. Will probably tolerate extended dry periods and most frosts. Propagate from seed or cuttings.

Elaeagnaceae Juss.
A small family of dicotyledons consisting of about 50 species in 3 genera chiefly found in the northern hemisphere. They are intricately branched shrubs or climbers with opposite or alternate leathery leaves. All parts are covered with silvery or golden scales, especially while young. The flowers lack petals and are carried in racemes. The fruit is a mealy drupe. The family is represented in Australia by a solitary species of *Elaeagnus* which is also found overseas.

ELAEAGNUS L.
(probably after Dioscorides' name for the Wild Olive *Elaeagnos*)
Elaeagnaceae
Shrubs, **trees** or **climbers**; **leaves** and young stems covered with stellate hairs or peltate scales; **leaves** alternate, entire; **inflorescence** an abbreviated spike or cyme, axillary; **flowers** bisexual or unisexual, tubular at the base, with 2 or 4 spreading lobes; **fruit** a drupe.
A genus of about 45 species from Europe, Asia and North America with 1 species extending to Australia. A couple of exotic species are grown as garden plants in temperate regions but the Australian species is rarely cultivated. Propagate from fresh seed or cuttings.

Elaeagnus latifolia L. = *E. triflora* Roxb.

Elaeagnus triflora Roxb.
(flowers in threes)
Qld
2-4 m x 3-5 m   Aug-Nov
Straggly **shrub** or **subclimber**; **stems** erect, leaning on surrounding vegetation or straggly, or climbing vigorously; young shoots bearing scurfy silvery scales; **leaves** 5-15 cm x 3-5 cm, ovate to ovate-lanceolate or elliptical, dull green above, the underside with golden or rusty scales; **spikes** or clusters axillary or terminal

383

on short shoots; **flowers** about 0.5 cm across, white, with 4 spreading petals; **fruit** 1.5-2 cm long, oblong, red, pendulous, mature Dec-March.

An Asian species which extends to Australia. It is widely spread throughout eastern Qld from the north to the south, and grows in lowlands and on adjacent ranges and tablelands. It is usually found along stream banks and rainforest margins but can also grow in quite open situations. In such positions it adopts a bushy habit and throws out long, straggly shoots. In shady situations or when in contact with surrounding vegetation, these shoots can climb vigorously. The fruit is edible but rather dry and mealy tasting. The species has limited garden appeal but could be of interest to fruit enthusiasts. Decorative features include the red fruit and the silvery new growth. Plants occupy a fairly large area and are best grown in an open situation where they can be trained to form a bush. Needs well drained soil and is suitable for tropical, subtropical and warm temperate regions. Propagate from seed which for best results should be sown fresh or from cuttings.

Elaeocarpaceae DC.

A family of dicotyledons consisting of about 350 species in 12 genera. They are trees or shrubs, mostly found in the tropical and subtropical areas of Indo-malaysia with a significant development in South America, New Guinea and north-eastern Australia. They have simple, stipular leaves which may be opposite or alternate in a spiral, and showy cup-shaped flowers with fringed or lacerated petal margins. The fruit is a capsule or colourful drupe. The seeds of several species are edible. In Australia the family comprises 43 species in 5 genera, most of which are found in rainforests or moist sclerophyll forests. Many species are popular in cultivation. Australian genera: *Aceratium, Aristotelia, Elaeocarpus, Peripentadenia* and *Sloanea*.

ELAEOCARPUS L.

(from the Greek *elaia*, the olive tree; *carpos*, fruit; a reference to the apparent similarity of the fruit of some species to olives)
Elaeocarpaceae

Tall **shrubs** or **trees**; **leaves** alternate, spirally arranged, occasionally opposite, simple, entire or with toothed margins; **inflorescence** an axillary raceme, sometimes a panicle; **flowers** bisexual, usually bell-shaped, often showy; **sepals** 4-5, overlapping; **petals** 4-5, broad, entire or with the tips delicately fringed or lobed; **stamens** more than 30; **fruit** a drupe, with a hard inner layer.

A large genus of 200 species widely distributed in tropical areas of Asia, Malaysia, Indonesia and New Guinea with 28 species in Australia, including about 8 which are undescribed. The genus is somewhat confused in Australia and is at present under revision. The vast majority of the Australian species are found in the rainforests of north-eastern Qld, while a solitary species is widespread across tropical Australia, and one extends south into Vic and the Bass Strait islands of Tas. Most species occur in rainforests, and a couple of hardier types are found in open forest and heathlands.

All species have horticultural merit by virtue of their spreading shady canopies, delicate flowers and colourful fruit. The smaller-growing species are ideal for home gardens, while the larger types are suitable for parks and acreage planting. They succeed best in organically rich soils and respond to heavy watering during dry periods. Birds are attracted to the flowers and fruit of many species. In some the old leaves turn brilliant red to scarlet before falling. Propagate from seed or cuttings. Seed is generally notoriously difficult to germinate in any quantity, the usual response being a low percentage of seedlings out of a large batch sown. Cracking the hard outer coat prior to sowing produces some improvement, as does fermentation in plastic bags and storage in moist peat moss for 6-12 months. Further experimentation is needed to develop a reliable seed germination technique. Cuttings may be slow to strike but are generally reliable.

Elaeocarpus angustifolius Blume
(narrow leaves)
Qld, NSW, NT   Blue Fig; Blue Quandong
15-30 m x 8-15 m   March-June; Dec

Medium to tall **tree** with an open, graceful crown; **bark** light grey; young shoots pale green; **leaves** 7-18 cm x 3.5-5 cm, elliptical, dark green and glossy above, paler beneath, thin-textured, prominent hairy domatia beneath, the margins finely toothed, old leaves turn bright red to scarlet; **racemes** 5-15 cm long, arising on leafless parts of stems; **flowers** 1.2-1.5 cm long, greenish white, on one side of the raceme, the petals finely fringed; **drupe** 2-3 cm across, globular, bright shiny blue, mature Aug-Jan.

A widespread and often locally common tree found in rainforests and along stream banks from north-eastern Qld to the Nambucca River in NSW. It is cut as a timber tree for its pale wood which is very useful for a wide variety of purposes, and in great demand. The fruit has a thin layer of edible flesh around the hard stone. This can be eaten fresh (although it is rather tasteless), and was also mixed with water by the Aborigines to make an edible paste. The seeds have a hard, attractively pitted coat and can be used as Chinese checkers or for necklaces etc. They have religious significance in Asian countries. Blue quandong is a very attractive tree with a graceful open, spreading canopy. It is ideal for parks and acreage planting and can be grown in tropical, subtropical and temperate areas. The blue fruit is borne in mass and is highly decorative. Another interesting feature are the clusters of old leaves which turn brilliant red or scarlet before falling. Trees require well drained soils and like plenty of moisture. When young they prefer some protection from direct sun. In good conditions plants are very fast growing. They respond to applications of fertilizers or manures. Seed is difficult to germinate in any quantity, with usually only sporadic seedlings appearing out of large batches. Cracking the hard seed-coat gives some improvement. Fermenting the seeds and storing in plastic bags of moist peat moss also promotes germination. The species can also be grown from cuttings. *E. angustifolius* has also been well known as *E. grandis* F. Muell. and *E. sphaericus* (Gaetn.)Schumann.

Elaeocarpus costatus fruits   D.L. Jones

Elaeocarpus arnhemicus F. Muell.
(from Arnhem Land)
Qld, NT
6-15 m x 3-5 m   Oct-Dec

Tall **shrub** or medium, spreading **tree**; **bark** pale grey, smooth; young shoots hairy, the stems reddish; **leaves** 3-8 cm x 3-4 cm, oblong to ovate, blunt, dark green above, dull beneath, the margins crenate, the apex blunt; **racemes** 2-8 cm long, in the upper axils; **flowers** about 0.5 cm long, greenish white, fringed; **drupe** 1.2-1.5 cm long, ovoid, bright blue, mature Nov-Jan.

Distributed along stream banks of northern Australia and extending down Cape York Peninsula to the North Kennedy District in regions with a distinct dry season. It often grows in rocky areas and is a very ornamental species, particularly when in fruit. Ideally suited to gardens in tropical and subtropical areas. The fruit is relished by birds, especially pigeons. Likes a sunny situation in well drained soil. Appears to be fairly frost hardy. Propagate from seed which is difficult to germinate in any quantity or from cuttings.

Elaeocarpus baeuerlenii Maiden & R. T. Baker =
E. kirtonii F. Muell. ex Bailey

Elaeocarpus bancroftii F. Muell. & Bailey
(after T. L. Bancroft)
Qld   Ebony Heart;
Johnstone River Almond
15-30 m x 8-15 m   March-June

Medium to tall **tree** with a graceful spreading canopy; **bark** dark grey, scaly; young shoots clothed with fine silky hairs; **leaves** 7-15 cm x 2-6 cm, oblanceolate to lanceolate, dark green and glossy above, paler beneath, the margins entire, the apex blunt; **racemes** 4-8 cm long, on leafless parts of stems; **flowers** about 1.5 cm long, white, bell-shaped, the margins of the petals lobed; **drupe** 2.5-3.5 cm across, globular, greenish black, mature July-Oct.

A large tree of the rainforests of north-eastern Qld in lowland and highland areas. It is cut as a timber tree, and the heavy, dark brown wood is used for floors and scantling. The kernel of the fruit is edible, with a pleasant nutty flavour, but is surrounded by a very hard shell. The tree grows too large for the average home garden, although in southern areas its dimensions will be reduced by the climate. Grows very successfully in Brisbane. An ideal park tree. Requires well drained soils, with plenty of moisture during the growing-season. Seedlings are quite fast growing. Propagate from seed which is difficult to germinate (see *E. angustifolius*).

Elaeocarpus coorangooloo J. F. Bailey & C. White
(an Aboriginal name)
Qld   Brown Quandong
10-15 m x 5-8 m   April-Aug

Medium **tree** with a spreading crown, twigs with conspicuous lenticels; **leaves** 7-10 cm x 3-6 cm, ovate-lanceolate, dark green and glossy above, paler beneath, prominently veined, the margins with small blunt teeth, the apex blunt to rounded, old leaves turn scarlet before falling; **racemes** 5-8 cm long, crowded in the upper parts of the branchlets; **flowers** about 0.4 cm long, cream, fringed or lobed, hairy; **drupe** 1-1.6 cm across, globose, blue, mature March-April.

An uncommon species restricted to rainforests of north-eastern Qld including parts of the Atherton Tableland. Trees can be very floriferous. They probably grow too large for home gardens but could have uses in parks and large gardens or for acreage planting. Requires well drained, organically rich soils and is best suited to tropical and subtropical regions. Propagation from seed may be difficult (see *E. angustifolius*).

Elaeocarpus costatus M. Taylor
(fluted; ribbed; in reference to the seed)
NSW (Lord Howe Island)
4-6 m x 1-3 m   Jan-March

Tall **shrub** or small **tree**; **bark** brown, roughened; young shoots silky-hairy; **leaves** 5-8 cm x 2.5-3.5 cm, oblong to ovate-lanceolate, crowded towards the ends of branchlets, leathery, the margins irregularly toothed, the apex blunt or pointed, the main veins on the underside prominent; **racemes** 1.5-3 cm long, crowded at the base of previous year's growth, few-flowered; **flowers** about 1 cm long, nodding, creamy white, hairy, the margins fringed; **drupe** 1.5-2 cm x 1-1.5 cm, ovoid, bright blue, mature June-Aug.

This species is endemic to Lord Howe Island, where it is fairly uncommon and found mainly on the mountains. It is a very decorative species of modest size, with attractive flowers and colourful fruit. Although it has good prospects at this stage, it does not appear to have been introduced into cultivation. Would be a suitable garden plant for temperate and subtropical regions in a semi-protected situation in well drained soil. Propagate from seed which is difficult to germinate and possibly from cuttings.

Elaeocarpus cyaneus Ait. ex Sims =
E. reticulatus Smith

385

Elaeocarpus eumundi Bailey
(from Eumundi, south-east Qld)
Qld, NSW
15-25 m x 5-12 m   Nov-Dec

Medium to tall **tree** with a dense crown; **bark** grey, smooth; young shoots hairy; **leaves** 7-13 cm x 3-5 cm, lanceolate to oblanceolate, glossy green above, paler beneath, crowded towards the ends of branchlets, entire or the margins of the distal part of the leaf toothed, the apex drawn out into a blunt point, on petioles 2-5 cm long; **racemes** 4-6 cm long, on leafless branches; **flowers** about 1 cm long, deep cream, sweetly scented, the narrow petals conspicuously fringed; **drupe** 1.4-1.8 cm long, ovoid, dark blue, mature May-June.

A widespread species found in rainforests from Cape York Peninsula to north-eastern NSW. It is cut as a timber tree, with the brown, closely grained timber being used for cabinet work and joinery. The species is a handsome tree and, although its final size in the rainforests is very large, experience has shown that plants are very slow-growing and could be used as garden subjects. They have attractive foliage and are showy when in flower and fruit. They require well drained soil and respond to mulching, fertilizing and watering during dry periods. Propagate from seed which is difficult to germinate and perhaps from cuttings.

Elaeocarpus ferruginiflorus C. White
(rusty-hairy flowers)
Qld   Quandong
6-10 m x 3-6 m   Dec-Jan

Small **tree**; **trunk** gnarled and twisted; young shoots bearing rusty hairs; **leaves** 4-5.5 cm x 2.5-3 cm, elliptical to oblong, bright green, the venation prominent, the undersurface of young leaves bearing rusty, woolly hairs which are easily rubbed off, the margins with very large blunt teeth; **racemes** 4-7 cm long, in the upper axils; **flowers** about 0.5 cm long, cream, fringed, the outer side of the petals covered with brown hair; **drupe** 1.5-1.8 cm long, ovoid to ellipsoid, pale blue-green, mature Feb-March.

This species is found in stunted rainforest of north-eastern Qld, usually at high elevations. Although not known to be in cultivation, its size, habit and flower and fruit details indicate good potential for home garden culture in subtropical and perhaps temperate regions. Would require well drained soils and probably a protected situation. Propagate from seed which may be difficult to germinate. Cuttings would be worth trying.

Elaeocarpus foveolatus F. Muell.
(foveoles or small pits on the seed or leaf)
Qld   White Quandong
15-25 m x 5-12 m   Oct-Dec

Medium to tall **tree** with a spreading canopy; **bark** light brown; young shoots bearing silky hairs; **leaves** 5-10 cm x 1-4.5 cm, lanceolate to oblanceolate, dark green and glossy above, paler beneath, with prominent veins, the margins toothed, the apex pointed; **racemes** 3-5 cm long, in the upper axils; **flowers** about 1.2 cm across, white, cup-shaped, the petals and calyx lobes

silky-hairy; **drupe** 1.5-2.5 cm long, oval, blue, mature Nov-Jan.

A large tree of the rainforests of north-eastern Qld in the lowlands and adjacent ranges and tablelands. It is cut as a timber tree, and the pale wood is used for cabinet work and in racing sculls. The crown of some trees of this species appears brown when viewed from the ground. The species appears to be rarely grown but has features which indicate potential for parks and large gardens. Suitable for tropical and subtropical climates. Seedlings from highland areas, such as the Atherton Tableland, could probably be grown in temperate regions, where the climate would reduce their ultimate size. Requires well drained soils, and plants need plenty of water until established. Propagation from seed is difficult (see *E. angustifolius*).

Elaeocarpus grahamii F. Muell.
(after Dr. George Graham)
Qld
15-25 m x 8-12 m   Oct-Dec

Medium to tall **tree**; **bark** grey; young shoots silky-hairy; **leaves** 10-15 cm x 2.5-4 cm, ovate-lanceolate, papery-textured, dark green above, dull beneath, the venation prominent, the veins on the underside bearing silky hairs; **racemes** 7-12 cm long, numerous, in the upper axils; **flowers** about 0.6 cm long, white, on very long pedicels (1.5 cm), the petals finely fringed; **drupe** about 1 cm long, obovoid, bright blue, mature Dec-Jan.

Restricted to lowland rainforests of north-eastern Qld. The species is virtually unknown in cultivation but could be a useful tree for parks or acreage planting in tropical and subtropical regions. Requirements and propagation as for *E. bancroftii*.

Elaeocarpus grandis F. Muell. = *E. angustifolius* Blume

Elaeocarpus holopetalus F. Muell.
(entire petals)
NSW, Vic   Black Olive-berry
10-20 m x 3-10 m   Dec

Small to medium **tree**; **bark** dark grey-brown; young shoots hairy; **leaves** 3-7 cm x 1.5-2.5 cm, oblong to narrow-lanceolate, dark green and glabrous above, densely hairy beneath, the margins sharply toothed; **racemes** 1-2 cm long, in the upper axils; **flowers** about 0.5 cm long, white; **calyx** clothed with yellow hairs; **petals** not fringed; **drupe** about 0.9 cm long, oblong to ovoid, black with a bluish bloom, mature March-Aug.

In the southern part of its range this species grows in cool, moist fern gullies in wet sclerophyll forest, while in the northern part of its range it commonly occurs in beech forests at high elevations. It ranges from East Gippsland in Vic to central NSW. It is a very handsome tree with a dense crown, but is not particularly showy either in flower or fruit. The species is unusual for the genus in that the margins of the petals are not fringed. Plants can be grown readily in the garden but do best in moist soil in a shady situation. It is an excellent species for growing with ferns. Best suited to temperate regions. Propagate from seed which is difficult to germinate or from cuttings.

Elaeocarpus angustifolius fruits

D.L. Jones

Elaeocarpus johnsonii F. Muell.
(after William Johnson, analytical chemist)

Qld   Kuranda Quandong
10-18 m x 5-10 m   Nov-Dec

Small to medium **tree**; young shoots pink, covered with rusty hairs; **leaves** 9-12.5 cm x 1.5-2.5 cm, obovate to oblong, dark green and glabrous above, bearing dark brown hairs beneath, the margins very hairy and finely toothed; **racemes** about 4 cm long; **flowers** about 0.5 cm long, cream, hairy; **drupe** about 5 cm x 2 cm, ovoid, dark blue, with a waxy bloom and brown dots, mature Feb-March.

A tree of the rainforests of the Atherton Tableland. The fruit is large and colourful, and it is recorded that the Aborigines ate the kernel. Trees probably grow too large for the average home garden but they should be tried in temperate regions, where their size may be reduced by the climate. They should prove to be an excellent tree for parks in tropical and subtropical regions. Propagate from seed which may be difficult (see *E. angustifolius*).

Elaeocarpus kirtonii F. Muell. ex Bailey
(after W. Kirton, original collector)

Qld, NSW   Silver Quandong
15-25 m x 5-15 m   Jan-March

Medium to large **tree**; **bark** pale grey, smooth; young shoots softly hairy; **leaves** 9-18 cm x 3-5 cm, elliptical to lanceolate, dark green and shiny on both surfaces, the veins very prominent, the margins prominently toothed; **racemes** 8-15 cm long, axillary, often on leafless parts; **flowers** about 1 cm long, white, sweetly scented, the petals fringed; **drupe** 1-1.3 cm long, ovoid, pale blue, mature Nov-Jan.

A species of sporadic distribution in mountain ranges from Eungella in central Qld to the Dorrigo Plateau. As well as occurring well inland (eg. Bunya Mountains), the species occurs in near-coastal rainforests. It is cut as a timber tree as its brown soft wood is suitable for a wide range of purposes, and is excellent for bending. The species probably grows too large for the average home garden but would be excellent for parks and acreage planting. Requirements are for well drained soil in a partially protected situation. Plants respond to mulching, fertilizers and water during dry periods. Propagate from seed which may be difficult to germinate.

Elaeocarpus largiflorens C. White
(large flowers)

Qld   Tropical Quandong
12-20 m x 5-12 m   Jan-March

Small to medium **tree**; **bark** brown, pustular; young shoots densely clothed with brown hairs; **leaves** 8-15 cm x 4-6 cm, elliptical to ovoid, dark green and glossy above, paler beneath, the margins entire or finely toothed, the apex pointed; **racemes** 4-8 cm long, in the upper axils; **flowers** about 1.2 cm across, white, each petal split into 2-6 lobes; **drupe** about 2 cm long, ovoid, blue, mature Dec-Jan.

Distributed in the lowland and highland rainforests of north-eastern Qld between Mount Spec and Mount Spurgeon. It is occasionally cut as a timber tree but the light brown wood has limited use. The sap is reported as being irritant. The species is rarely cultivated but would seem to be worthy of trial in tropical and subtropical regions. Requirements for cultivation and propagation would probably be similar to those of *E. bancroftii*.

Elaeocarpus longifolius C. Moore =
           E. kirtonii F. Muell. ex Bailey

Elaeocarpus longipetiolatus C. White =
           E. michaelii C. White

Elaeocarpus michaelii C. White
(after Norman Michael, original collector)

Qld
10-15 m x 5-8 m   June-July

Small to medium **tree**; **bark** grey; **leaves** 8-25 cm x 3-6 cm, elliptical to lanceolate, glabrous, thin-textured, bright green, the margins with widely spaced small teeth; **racemes** 4-6.3 cm long, on the older parts of the branchlets; **flowers** about 1 cm long, white, fringed, the sepals and pedicels bearing silky hairs; **drupe** about 1.3 cm across, globose, bright blue, mature Dec-Feb.

A little-known species from the lowland rainforests of north-eastern Qld in the North Kennedy District. Appears to have excellent potential for parks or larger gardens in tropical and perhaps subtropical areas. Propagate from seed which may be difficult to germinate. This species has also been known as *E. longipetiolatus* C. White.

387

Elaeocarpus obovatus

Elaeocarpus obovatus G. Don
(obovate leaves)
Qld, NSW   Blueberry Ash
15-25 m x 5-12 m   Sept-Nov

Medium to tall **tree**; **bark** grey, with corky pustules; young shoots bearing silky hairs; **leaves** 5-9 cm x 2-3.5 cm, elliptical to obovate, dark green and shiny above, dull beneath, the margins of the upper two-thirds irregularly toothed; **racemes** 6-12 cm long, axillary or from leafless stems; **flowers** about 0.4 cm long, cream to white, unusual scent, the petals fringed; **drupe** 0.6-1.2 cm long, oval to globular, deep blue, mature Jan-March.

A widespread species found in moist and dry rainforests, moist open forests and coastal scrubs from Proserpine in Qld to Wyong, south of Sydney. It is cut as a timber tree, the tough white timber being suitable for a wide variety of uses. Despite its ultimate large size in rainforests, this is a very useful plant for many purposes. It withstands moderate coastal exposure and will tolerate brackish water around its roots. Its growth habit is dense, making it ideal for screening or shelter, and the flowers and fruit are ornamental. Although it grows best in a well drained, organically rich loam, it will tolerate a wide variety of soils including those subject to waterlogging. Young plants are fairly fast growing and respond to the use of fertilizers. Can be grown in temperate regions. Propagate from seed which is difficult to germinate or from cuttings.

Elaeocarpus reticulatus Smith
(net-like venation)
Qld, NSW, Vic, Tas   Blueberry Ash;
                       Blue Olive-berry
8-15 m x 3-5 m   Oct-Jan

Tall **shrub** or small **tree**; **bark** brown, with vertical cracks; young shoots pink-red; **leaves** 5-11 cm x 2-3.5 cm, oblong to narrow-lanceolate, dark glossy green above, paler beneath, the venation conspicuous, the margins shallowly toothed, the apex pointed, old leaves turn bright red to scarlet; **racemes** 2-10 cm long, axillary; **flowers** 0.6-0.9 cm long, white or pink, bell-shaped, unusually scented, the petals prominently fringed; **drupe** 0.8-1.3 cm long, oval to globular, dark shiny blue, mature May-Oct.

A widespread species which extends from Fraser Island in Qld to eastern Vic, and Flinders Island in Tas. It is a very common species found in many different habitats including dry rainforest and coastal scrubs. It is undoubtedly the most popular and widely grown member of the genus, being prized for its delicate, showy flowers and brilliant blue fruit which is held on the plants for a long period. Although most plants are white-flowered, pink forms are also known, these varying from pale to deep rosy pink. It makes an excellent garden plant as it is adaptable and easily managed. It can be grown in a variety of soils, and in positions from full shade to full sun. Once established, plants are quite hardy to dryness. They respond to pruning and can be trimmed into a hedge. Suited to subtropical and temperate regions, and they are hardy enough to be grown in inland towns, provided they are watered. Also tolerate moderate coastal exposure.

388

The cv. 'Prima Donna' is an attractive pink-flowered form grown mainly in south-eastern Qld.

Propagate from seed which is very difficult to germinate or from cuttings which strike fairly easily.

Elaeocarpus ruminatus F. Muell.
(folded as in the surface of the rumen)
Qld   Brown Quandong
15-25 m x 5-12 m   Nov-Dec

Medium to tall **tree** with a spreading canopy; **bark** grey-brown; young shoots clothed with silky hairs; **leaves** 6-12 cm x 2-5 cm, elliptical, dark green and glossy above, paler beneath, the margins toothed, the apex abruptly narrowed, old leaves turning red to scarlet; **racemes** 5-13 cm long, axillary, crowded towards the ends of branchlets; **flowers** about 0.6 cm long, greenish white, cup-shaped, silky-hairy; **drupe** about 1.5 cm across, globose, blue, mature Dec-Jan.

A tree of the rainforests of north-eastern Qld in the Cook, North Kennedy and South Kennedy Districts. It is occasionally cut as a timber tree for its pale wood which is used for cabinet making. Adventitious roots are frequently noticeable near the base of the trunk. The species appears to be rarely grown but could be a handsome tree for parks, large gardens and acreage planting in tropical and subtropical regions. Requires well drained soils and plenty of moisture. Propagate as for *E. angustifolius*.

Elaeocarpus reticulatus, × ·5   flowers   fruit

Elaeocarpus obovatus　　　D.L. Jones

Elaeocarpus reticulatus　　　D.L. Jones

Elaeocarpus sericopetalus F. Muell.
(silky-hairy petals)

Qld	Northern Quandong
15-25 m x 5-12 m	Dec-Jan

Medium to tall **tree** with a spreading canopy; **trunk** often with coppice shoots; **bark** dark brown, with pustules; young shoots pale green, finely hairy; **leaves** 6-10 cm x 1.5-2 cm, lanceolate, dark green and glossy above, paler beneath, the margins with blunt teeth, old leaves turning bright red to scarlet; **racemes** 4-6 cm long, axillary or on leafless parts of stems; **flowers** about 0.6 cm long, white, bell-shaped, the petals clothed with silky hairs; **drupe** 1.5-2 cm long, oval, blue, mature Nov-Dec.

A tree of the rainforests of north-eastern Qld in the Cook and North Kennedy Districts. It is sometimes cut for its light brown timber which is fairly soft. The species is virtually unknown in cultivation but could be useful for parks or large gardens, and acreage planting in tropical and subtropical regions. Requirements and propagation as for *E. bancroftii*.

Elaeocarpus sphaericus (Gaertn.)Schumann =
E. angustifolius Blume

Elaeocarpus stellaris L. Smith
(star-like; in reference to the shape of the cross-section of the fruit)

Qld	
12-18 m x 5-10 m	Oct-Nov

Small to medium **tree**; **bark** brownish; **branchlets** with short, appressed hairs; **leaves** 10-18 cm x 4.5-8 cm, dark green and glabrous, somewhat thick-textured, elliptical to obovate, the apex shortly acuminate, the margins toothed, somewhat decurved; **racemes** 1-3 cm long, bearing 3-6 flowers, sometimes branched; **flowers** about 0.5 cm long, white, fringed; **drupe** 5-6 cm x 4-5 cm, broadly ovoid, blue, with 5 large vertical lobes or wings, mature Dec-Jan.

A species of restricted distribution, known only from rainforests along a couple of creeks near Innisfail in north-eastern Qld. The large fruits are decorative and are generally eaten by rats. The species is apparently not in cultivation but could make a handsome tree for parks in tropical and perhaps subtropical regions. Leaves on seedlings may be up to 30 cm x 10 cm. Seed may be difficult to germinate (see *E. angustifolius*).

ELAEODENDRON J. F. Jacq. ex Jacq.
(from the Greek *elaeos*, an olive; *dendron*, a tree; in reference to the resemblance of the fruit to an olive)
Celastraceae

Shrubs or small **trees**; **leaves** opposite or alternate, simple, entire or crenate; **inflorescence** an axillary cyme, often clustered; **flowers** small, unisexual or bisexual; **calyx** 3-5 parts; **petals** 3-5, spreading; **fruit** a succulent or dry drupe, 1, 2 or 3-celled; **seed** solitary.

A small genus of about 17 species with 2 found on eastern mainland Australia and another on Lord Howe Island. They are generally decorative plants, but are rarely seen even though they adapt readily to cultivation. They can be propagated from seed which has a limited viability and is best sown fresh. Some species can also be struck from cuttings of half-ripened wood.

Elaeodendron australe

Elaeodendron australe Vent.
(southern)
Qld, NSW   Red Olive-berry
5-10 m x 3-5 m   Aug-Nov

Tall **shrub** or small **tree**; **bark** dark brown, wrinkled; young shoots bright green and shiny; **leaves** 4-11 cm x 2-4 cm, dark green and shiny above, paler beneath, thick and leathery, the margins toothed or scalloped, the apex blunt; **cymes** 2-5 cm long, axillary; **flowers** 0.4-0.6 cm across, greenish, male and female separate; **drupe** 1.2-1.6 cm long, ovoid, bright red or orange, mature March-June.

A widespread small tree found in dry rainforests and moister open forests between Ayr in north-eastern Qld and Kiama in southern NSW. It is particularly common in some coastal communities. Plants adapt very well in cultivation and the species makes an ideal garden subject. The fruits are extremely decorative and hang for many weeks, contrasting pleasantly with the foliage. Plants fruit while quite young. The species can be grown in tropical, subtropical and temperate regions. It requires well drained soil in a sunny or partially shady situation. Plants can be quite fast growing and respond to water, mulches and fertilizers.

The var. *angustifolium* Benth. is a sparser-growing form with dark green, narrow leaves (1-2 cm wide) which are rarely toothed. It is found in much drier regions than the typical form, even occurring west of the divide. It occurs in Qld and NSW. Propagate both forms from seed which must be sown fresh or from cuttings which may be slow to strike.

Elaeodendron curtipendulum Endl.
(shortly pendulous)
NSW (Lord Howe Island)   Tamana
4-6 m x 1-3 m   March-April

Tall **shrub** or small **tree**; **bark** greyish brown, ribbed; **leaves** 4-6.5 cm x 2.5-3.5 cm, elliptical to obovate, thick and leathery, often reflexed, the margins sparsely crenate, somewhat glaucous, shiny above, pale beneath, the apex blunt; **panicles** much-branched, in the upper axils; **flowers** unisexual, about 0.5 cm across, yellowish green; **male flowers** clustered; **female flowers** in spreading cymes; **drupe** about 2 cm long, ovoid, leathery, bright bluish green, finely dotted.

This species is endemic to Lord Howe Island, where it is widespread and fairly common. It can be a handsome species with a dense crown but frequently tends to be straggly. It has good potential for cultivation but at this stage little is known about its performance. Would be suitable for temperate regions in a semi-protected situation in well drained soil. Propagate from seed which is best sown fresh.

Elaeodendron melanocarpum F. Muell.
(black fruit)
Qld   Black Olive Plum
5-10 m x 3-5 m   Oct-Nov

Tall **shrub** or small **tree**; young shoots bright green, glabrous; **leaves** 4-10 cm x 2-4 cm, dark green and shiny above, paler beneath, thick and leathery, the margins scalloped, the apex blunt; **cymes** 2-5 cm long, axillary; **flowers** about 0.3 cm across, greenish, male

Elaeodendron australe   D.L. Jones

and female separate; **drupe** 1.5-1.8 cm long, ovoid to globular, bright shiny black.

Distributed in drier rainforests from Cairns to the MacPherson Ranges on the border. Trees closely resemble *E. australe* and can be distinguished only by the smaller flowers and shiny black fruit. This species is not as popular in cultivation as *E. australe*, even though its habit and foliage are so similar. It grows readily in a semi-protected position in well drained soil. Plants may be more frost tender than *E. australe*. Propagate from fresh seed and perhaps cuttings. Seed of this species has proved to be difficult to germinate.

Elaeodendron microcarpum C. White & Francis =
Pleurostylis opposita (Wall.) Alston

Elaphoglossaceae (Hert. & Ching)Pichi.-Serm.
A family of ferns comprising the solitary genus *Elaphoglossum*. This family is included in Lomariopsidaceae by most authors.

ELAPHOGLOSSUM Schott ex J. Smith
(from the Greek *elaphos*, a stag; *glossa*, a tongue; in reference to the shape of the fronds of some species)
Lomariopsidaceae

Epiphytic **ferns**; **rootstock** creeping, fleshy, clothed with papery scales; **fronds** simple, stalked, entire or with undulate margins, dimorphic; **fertile fronds** smaller than sterile fronds, the whole of the lower surface covered with sporangia.

A large genus of about 400 species mostly found in tropical regions with 2 species in north-eastern Qld.

They are interesting ferns but are rarely encountered in cultivation. Propagate by division of the clumps or from spore which may be difficult to raise successfully.

Elaphoglossum callifolium (Blume)T. Moore
(leaves like those of the genus *Calla*)
Qld
15-50 cm tall
Epiphytic **fern** forming small to medium clumps; **rootstock** shortly creeping, thick, fleshy; **scales** thin, brown, papery; **fronds** 15-50 cm x 4-6 cm, lanceolate, erect, shiny green above, dull beneath, the margins wavy; **fertile fronds** covered with black sori on the lower surface.

This fern is widely distributed through the Pacific region, and in Australia is confined to a few highland areas of north-eastern Qld, where it grows on boulders and trees. Plants can be cultivated readily in a pot or basket of open, coarse mixture and must be held in shady, humid conditions where there is adequate air movement. In southern Australia they may need the protection of a glasshouse. Propagate by division of the clumps or from spore.

Elaphoglossum conforme Sw.
             see under *Elaphoglossum* species

Elaphoglossum species
Qld
5-20 cm tall
Epiphytic **fern** forming small clumps; **rootstock** shortly creeping, fleshy; **scales** light brown, papery; **fronds** 5-20 cm x 1-4 cm, lanceolate, erect, shiny green above, dull beneath, the margins wavy; **fertile fronds** covered with black, crystalline sori on the lower surface.

A common fern of north-eastern Qld, extending

Elaphoglossum species, × ·6

from lowland to highland areas and usually growing on trees or rocks in rainforest. Plants resent disturbance and can be somewhat difficult to maintain in cultivation. They require a coarse epiphyte mixture and prefer a small pot to a large one. They can also be attached to a slab of tree-fern fibre or elkhorn peat. Shade and humidity are essential for their successful establishment and culture. Propagate from spore. This species was previously confused with *E. conforme* Sw., a species not native to Australia.

Elatinaceae Dum.
A small family of dicotyledons consisting of about 40 species in 2 genera widely distributed in tropical and temperate parts of the world. They are annual or perennial herbs or shrubs with a few aquatics which can adapt to a terrestrial environment. Few species have any horticultural merit. In Australia the family consists of about 6 species in the genera *Bergia* and *Elatine*.

ELATINE L.
(from the Greek *elatine*, a fir tree; an apparent reference to the leaves resembling those of the fir tree)
Elatinaceae
Small glabrous **herbs**, either aquatic or growing on mud; **stems** rooting at the nodes; **leaves** simple, entire, opposite or in whorls, stipulate; **flowers** bisexual, small, usually solitary, in the leaf axils; **flowers** 2, 3 or 4-partite; **fruit** a dehiscent capsule.

A small genus of 20 species with a solitary species widespread in Australia. They have very limited horticultural appeal. Propagate from pieces of stem or seed sown on mud.

Elaeodendron curtipendulum   D.L. Jones

Elatine gratioloides Cunn.
(resembling the genus *Gratiola*)
all states   Waterwort
prostrate x 0.3-2 m   Nov-March; also sporadic

Small annual **herb** growing as an aquatic or rooting in mud; **stems** 3-20 cm long, brittle, rooting at the nodes; **leaves** 0.2-2.0 cm x 0.1-0.4 cm, ovate to elliptic, opposite, bright green, with minute marginal glands; **flowers** about 0.2 cm across, greenish to pinkish, solitary, in the axils; **capsule** 0.15-0.25 cm across, globular, flattened, splitting when ripe.

A small plant which grows as a submerged aquatic in still water or as a mat plant on mud. Although short-lived, the species can be a useful aquatic for an aquarium. Requires strong light and may be fast growing. Can also be grown as a bog plant and in these conditions usually naturalizes readily. Plants can be propagated easily from stem pieces or from seed sown on mud (see Aquarium Plants and Aquatics, Volume 2, pages 216-217).

ELATOSTEMA Gaudich.
(from the Greek *elatos*, elastic; *stemon*, a stamen; an allusion to the stamens springing up)
Urticaceae

Slender or shrubby perennial **herbs**; **stems** fleshy, rooting at the base, simple or branched; **leaves** simple, alternate or opposite, distichous, asymmetrical, toothed; **flowers** very small, unisexual, in axillary, unisexual heads which have 4-5 outer bracts; **male flowers** of 4-5 segments; **female flowers** of 3-4 segments; **fruit** a tiny nut.

A large genus of about 200 species mostly found in the tropics with 2 species in Australia. They are very similar to nettles but lack any stinging hairs. The species have limited appeal and are grown only by enthusiasts.

Propagate from stem cuttings which root very easily or from the minute seed which must be sown on the surface of a pot of moist media.

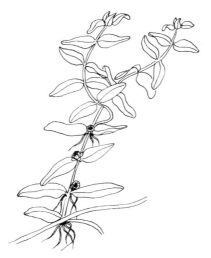

Elatine gratioloides, × 1·5

Elatostemma reticulatum var. *grande*   D.L. Jones

Elatostema reticulatum Wedd.
(netted veins)
Qld, NSW   Rainforest Spinach
0.3-0.6 m x 0.3-0.45 m   Sept-Nov

Perennial **herb** forming coarse clumps; **stems** 30-60 cm tall, bright green and fleshy, rooting at the base, often arched below the tip; **leaves** 7-15 cm x 3.5-6 cm, lanceolate to ovate-elliptical, falcate, dark green, rough-textured, scattered up the stems, the margins coarsely toothed; **male flower-heads** 1-2.5 cm across, somewhat flattened, green; **female flower-heads** 0.6-0.8 cm across, globular, greenish; **flowers** about 0.2 cm across; **nuts** about 0.1 cm long.

A coarse herb which forms clumps along streams in rainforests, often growing in the stream bed itself. It may be recognized by the fleshy stems with the asymmetric leaves arranged along them in one plane. It is widespread, extending from north-eastern Qld to north-eastern NSW, and is often locally common. The young stems and leaves can be eaten either raw or after cooking, and are nutritious and tasty. The species can be grown easily but demands cool, shady, humid conditions and an abundance of water. It can be grown with ferns or as a pot plant. Potted specimens make excellent indoor plants, tolerating quite dark conditions, but must be kept moist. Can also be used in a large terrarium. Likes an open, friable potting mixture. In suitable conditions seedlings appear readily. Old plants can be rejuvenated by cutting back and/or repotting.

The species consists of 2 varieties. The typical form is found on the mainland, while var. *grande* Benth. is endemic on Lord Howe Island. The latter has much larger and broader leaves (to 20 cm x 8 cm) and the flower-heads are borne on long peduncles.

Both forms can be propagated from transplants, seed sown on the top of moist media or from cuttings which strike readily.

Elatostema stipitatum Wedd.
(stalked)
Qld, NSW
0.1-0.3 m x 0.1-0.2 m   Sept-Dec

Perennial **herb** forming weak, slender clumps; **stems**

10-30 cm long, diffuse or prostrate, sparsely branched, fleshy, rooting at the base; **leaves** 2-5 cm x 0.5-2.5 cm, oblong to ovate-lanceolate, sessile, dark green, thin-textured, the margins deeply toothed, the lower margin embracing the stem as an auricle; **male flower-heads** 0.4-0.6 cm across, prominently stalked; **female flower-heads** 0.1-0.2 cm across, prominently stalked; **flowers** about 0.1 cm across.

A slender herb found in moist, shady situations in rainforests, usually on boulders beside streams. It is restricted to north-eastern NSW and extreme south-eastern Qld. It can be grown readily in a shady, moist situation but is rather an insignificant species with little appeal. Can be a useful terrarium plant. Propagate from seed or cuttings which strike easily.

ELATTOSTACHYS Radlk.

(from the Greek *elaterios*, driving away; *stachys*, a spike; an obscure derivation)
Sapindaceae

Trees with paper-thin bark; new growth hairy; **leaves** pinnate, alternate; **leaflets** simple, opposite or alternate, entire or toothed, venation prominent; **inflorescence** an axillary cyme, raceme or panicle; **flowers** small, often hairy, unisexual or bisexual; **sepals** prominent; **petals** minute; **fruit** a 3-4 celled capsule, splitting when ripe; **seeds** large, 1 per cell, with a small basal aril.

A small genus of about 14 species with 2-4 found in eastern Australia. They are small trees with some ornamental features, but are rarely grown except by enthusiasts. Propagate from seed which must be sown fresh.

Elattostachys nervosa (F. Muell.) Radlk.

(prominent veins)
Qld, NSW   Beetroot Tree
8-15 m x 5-10 m   April-May; Sept-Nov

Small to medium **tree**; young shoots bronze, clothed with rusty hairs; **bark** grey, smooth, very thin; **leaves** 10-15 cm long, pinnate, alternate; **leaflets** 7-25 cm x 2.5-6 cm, 3-6 per leaf, lanceolate, often falcate, dark green and somewhat shiny above, paler and dull beneath, venation prominent; **racemes** 3-5 cm long, axillary, clothed with pale brown hairs; **flowers** 0.3-0.4 cm across, brownish; **capsule** 1.2-1.8 cm across, pinkish red to red, globular, 3-celled, pink inside; **seeds** about 0.6 cm long, brown to black, with a short red aril, mature July-Oct.

Usually a small tree with a spreading canopy, this species often grows in the drier types of rainforest, and also in moist areas of open forest. It is distributed between Gympie in Qld and Barrington in northern NSW. Birds feed on the seeds. It is quite an ornamental species that could find a place in gardens, but at this stage seems to be grown mainly by enthusiasts of rainforest plants. Plants are rather sensitive to poor drainage and must be grown in very well drained soils. They are usually quite slow-growing but respond to the use of fertilizers and mulches. They appear to be quite hardy to moderate frosts and can be grown in sub-tropical and temperate regions. Propagate from seed which has short viability and must be sown soon after collection. Frequently, a large percentage of the capsules are parthenocarpic and never open.

Elattostachys xylocarpa (Cunn. ex F. Muell.)Radlk.

(thick, hard fruit)
Qld, NSW   White Tamarind
10-20 m x 5-10 m   Dec-April

Small to medium **tree**; young shoots deep purplish red, clothed with rusty hairs; **bark** dark brown, thin and papery; **leaves** 6-12 cm long, pinnate, alternate; **leaflets** 5-8 cm x 1.5-3 cm, elliptic, smooth and somewhat shiny green above, dull and hairy beneath, the margins coarsely toothed, venation prominent beneath; **racemes** or **panicles** 3-5 cm long, axillary, red; **flowers** about 0.3 cm across, brownish, woolly; **capsule** 1.5-2 cm across, globose, grey-brown, 3-4 celled, wine coloured inside; **seeds** about 0.8 cm long, glossy black, with a small aril, mature April-May.

Widespread in dry rainforests between Rock-hampton and the Clarence River. The wood is yellowish and very tough, and could be used for tool-handles. Birds eat the seeds. This is an ornamental small tree with a spreading canopy of attractive leaves. New flushes of growth are particularly colourful. Although rarely grown, it has potential for wider cultivation in parks and private gardens. Plants are, unfortunately, rather slow-growing, especially when young. They need well drained soil and will tolerate exposure to sun from an early age. Seedlings make decorative pot plants but are not successful indoors. Suitable for subtropical and warm temperate regions. Propagate from seed which has a short viability and must be sown soon after collection. Frequently, a large percentage of the capsules are parthenocarpic and never open.

ELEOCHARIS R. Br.

(from the Greek *helos*, a marsh; *charis*, grace; favour; alluding to their preference for a marshy habitat)
Cyperaceae

Annual or perennial **herbs**, chiefly aquatic; **rhizome** creeping, much-branched; **leaves** reduced to sheathing bracts; **culms** erect or lax, wiry or spongy, hollow, often septate, the apex blunt and bearing sterile scales or an inflorescence; **inflorescence** a compressed, terminal, many-flowered spikelet; **glumes** spirally arranged, overlapping; **flowers** bisexual; **perianth** of bristles; **fruit** a nut, crowned by an enlarged style-base.

A large genus of about 160 species with about 28 in Australia, 11 of which are endemic. They are mostly aquatics growing in permanent water or in heavy soils subject to periodic inundation. They have limited horticultural appeal, but can be planted around the margins of dams and ponds to act as protection for water-birds and fish, and to improve the aesthetic appearance. Propagation can be from seed which must be sown in a pot with the base standing in water and the top covered by glass. Clumps can be readily divided and planted direct into their final position. The best time for division is in the spring and early summer (see Aquatics, Volume 2, page 217).

Eleocharis acuta R. Br.

(a short, sharp point)
Qld, NSW, Vic, Tas, SA, WA   Common Spike-rush
0.3-0.9 m tall   Nov-April

Perennial **aquatic**; **culms** 30-90 cm x 0.2-0.3 cm,

erect, terete, in distant tufts, 3-sided, striate; **sheaths** purplish, striate; **spike** 1.5-3 cm x 0.3-0.7 cm, linear, sharply pointed apex, dark brown or variegated; **glumes** brown, blunt; **nut** obovate, shiny, yellowish brown.

A widely spread species found in swamps, lagoons and along the margins of sluggish watercourses. It can be established readily in a dam or pond with a muddy bottom, but may spread rapidly. Very useful as a shelter for water-birds and fish. Propagate by division.

Eleocharis cylindrostachys Boeck
(cylindrical spikes)
Qld, NSW   Spike-rush
0.3-0.5 m tall   Nov-April

Perennial **aquatic**; **culms** 30-50 cm tall, tufted, rounded, striate; **sheath** membranous, striate; **spike** 1-2 cm x 0.2-0.3 cm, linear, cylindrical, blunt, pale coloured; **glumes** dense; **nut** about 0.15 cm long, turgid, golden brown, obovate.

A widely distributed aquatic that forms spreading colonies in shallow (up to 40 cm deep) water. It could have potential for cultivation in ponds and around the margins of dams as a protective plant for fish. Propagate by division.

Eleocharis difformis S. T. Blake
(of unusual structure)
Qld   Spike-rush
0.1-0.3 m tall   Nov-March

Perennial **aquatic**; **culms** 10-30 cm x 0.7 cm, somewhat rounded to 4-angled, green, in tufts; **spike** 5-12 cm x 0.13-0.15 cm, linear, pointed, green, erect, few-flowered; **glumes** obovate, blunt; **nut** about 0.15 cm long, obovate, turgid, pale coloured.

This aquatic grows in marshy ground or in water about 60 cm deep. It is known only from Moreton and Stradbroke Islands near Brisbane. Its behaviour in cultivation seems to be unknown but is probably similar to that of species such as *E. sphacelata*. Propagate by division.

Eleocharis dulcis (Burm.f.) Trin. ex Hersch.
(sweet tasting)
Qld, NSW, NT   Tall Spike-rush; Water-chestnut
0.3-1 m tall   Nov-March

Perennial **aquatic** spreading by long, slender stolons which bear tubers about 0.7 cm across; **culms** 0.3-1 m x 0.3-0.8 cm, tufted, erect, cylindrical, septate; **spike** 2.5-5 cm x 0.5-1 cm, cylindrical, blunt, pale coloured; **glumes** oblong, striate, dense; **nut** about 0.2 cm long, obovate, shiny, yellowish.

This species is widely distributed in swamps throughout eastern Qld and is also found in NT. It is an extremely variable species, frequently having thin, wiry culms. The tubers on the rhizomes are edible and were eaten by the Aborigines, either raw or after baking. They are reportedly very tasty and are also nutritious. In Asian countries a selected form of this species is widely cultivated and the tubers are harvested and used in cooking. These tubers are known as water-chestnuts. In tropical Qld the Aborigines wove the culms of this species into mats. The species can be readily grown as an aquatic or a bog plant in sub-

tropical and tropical regions. Propagate by division or from the tubers.

Eleocharis equisetina Presl
(resembling the horsetail genus *Equisetum*)
Qld, NSW   Spike-rush
0.3-1 m tall   Nov-March

Perennial **aquatic**; **culms** 0.3-1 m x 1-3 cm, tufted, erect, rigid, terete, shiny, striate; **sheaths** purple-brown, lax; **spike** 2-4 cm x 0.3-0.4 cm, cylindrical, pale coloured; **glumes** elliptic, incurved, dense; **nut** about 0.2 cm long, turgid, ribbed, golden brown.

A widely distributed aquatic found in swamps from north-eastern Qld to north-eastern NSW. Its behaviour in cultivation seems to be largely unknown but is probably similar to that of species such as *E. sphacelata*. For ponds and dams in tropical, subtropical and temperate regions. Propagate by division.

Eleocharis fistulosa (Poir) Link
(tubular; hollow throughout)
Qld   Spike-rush
0.3-0.6 m tall   Nov-March

Perennial **aquatic**; **culms** 30-60 cm x 0.3-0.4 cm, erect, spongy, acutely 3-angled, hollow; **spike** 2-3.5 cm x 0.3-0.4 cm, cylindrical, pale coloured; **glumes** appressed, densely clustered; **nut** about 0.15 cm long, obovate, turgid, shiny, straw coloured.

This is a widespread species which in Australia is confined to north-eastern Qld, where it grows in wet depressions. It could be a useful aquatic for dams and ponds in tropical and northern inland regions. Propagate by division.

Eleocharis nuda C. B. Clarke
(naked; unadorned)
Qld, ?NSW   Spike-rush
0.1-0.3 m tall   Nov-March

Perennial **aquatic**; **culms** 10-30 cm x 0.2 cm, terete or 3-angled, wiry, green, prominently striate; **sheaths** purplish, striate; **spike** 1.5-3 cm x 0.3-0.5 cm, cylindrical, greenish, pointed; **glumes** ovate, striate, appressed; **nut** obovate to pear-shaped, shiny, brown, ribbed.

This species grows in *Melaleuca* swamps of north-eastern Qld and it could be a useful aquatic for ponds or dams in tropical areas. Doubtfully recorded for NSW. Propagate by division.

Eleocharis pallens S. T. Blake
(pale coloured)
Qld, NSW, Vic, SA, WA, NT   Pale Spike-rush
30-50 cm tall   Nov-April

Perennial **aquatic**; **culms** 30-50 cm x 0.1-0.3 cm, densely tufted, erect, slender, smooth or with 9-10 furrows; **sheaths** membranous, appressed, striate; **spike** 1-2 cm x 0.2 cm, linear, cylindrical, sharply pointed, pale brown or tawny; **glumes** ovate, pointed; **nut** about 0.15 cm long, obovate, turgid.

A widely distributed aquatic of the interior regions of Australia, growing in permanent lagoons, channels and water-holes. Plants provide excellent cover for young fish, and the species could be useful around the margins of ponds and dams. Propagate by division.

Eleocharis sphacelata

D.L. Jones

Eleocharis philippinensis Svenson
(from the Philippines)
Qld, NSW, WA   Spike-rush
0.3-1 m tall   Nov-March
    Perennial **aquatic**; **culms** 0.3-1 m x 0.3 cm, erect or lax, acutely 4-5 angled, hollow, without septa; **sheaths** dark purple; **spike** 4-6 cm x 0.3 cm, linear, sharply pointed, greenish; **glumes** ovate, blunt, green; **nut** about 0.15 cm long, obovate, club-shaped, dark brown, deeply pitted.
    In Australia this species is widespread in tropical and subtropical Qld, and is also found in WA and on the north coast of NSW. It grows in swamps and can be easily recognized by the soft, acutely and unequally 4-angled culms. It is a spreading species, forming dense clumps and may have some potential as an aquatic for tropical and subtropical regions. Propagate by division.

Eleocharis plana S. T. Blake
(flat)
Qld, NSW, SA   Spike-rush
0.3-0.8 m tall   Nov-April
    Perennial **aquatic**; **culms** 30-80 cm x 0.2-0.4 cm, erect, striate, finely furrowed; **sheaths** striate, truncate; **spike** 1-1.5 cm x 0.2 cm, linear, cylindrical, pale brown; **glumes** ovate, pointed; **nut** about 0.15 cm long, obovate, turgid, shiny.
    An aquatic distributed in eastern Australia from central Qld to central NSW. It appears to be unknown in cultivation but could have some application for ponds, dams or water gardens in subtropical and temperate regions. Propagate by division.

Eleocharis pusilla R. Br.
(small; tiny)
Qld, NSW, Vic, Tas, SA   Small Spike-rush
0.02-0.25 m tall   Nov-March
    Small perennial **aquatic**; **culms** 2-25 cm x 0.5 cm, tufted, erect or arching, wiry; **sheaths** loose, roughened; **spike** 0.2-0.7 cm x 0.15 cm, ovate to lanceolate, sharply pointed; **glume** ovate, blunt; **nut** about 0.1 cm long, shiny, 3-sided.
    A small aquatic which grows around the margins of swamps and in shallow depressions, often in black soil areas. It can be grown readily around the margins of dams or ponds or as a bog plant, and is usually easy to control. Propagate by division.

Eleocharis sphacelata R. Br.
(withered; dead-looking)
Qld, NSW, Vic, Tas, SA, NT   Tall Spike-rush
0.5-2 m tall   Nov-April
    Robust perennial, emergent **aquatic**; **rhizome** woody, thick, bearing 2 rows of culms; **culms** 0.5-2 m x 0.5-1.1 cm, cylindrical, erect, hollow, septate, the apex blunt and bearing sterile scales or an inflorescence; **spike** 3-6.5 cm x 0.4-1.3 cm, cylindrical, acute, brown;

395

Eleocharis spiralis

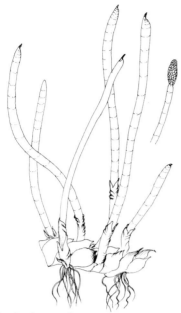

Eleocharis sphacelata, × ·2

glumes numerous, densely packed, striate; **nut** about 0.2 cm long, tawny yellow, turgid.

A common aquatic which may grow in small clumps or as large stands in fresh water. The rhizomes root in the mud and the culms are emergent. The species is readily grazed by stock. It is a vigorous species which can be introduced into shallow dams or ponds to provide protection for fish, and to add a pleasing aspect. It can, however, spread rapidly during the summer months and may need to be kept in check. Propagate by division.

Eleocharis spiralis (Rottb.)R. & S.
(in a spiral)
Qld   Spike-rush
0.3-0.5 m tall   Nov-March

Perennial **aquatic**; **culms** 30-50 cm x 0.2-0.35 cm, densely clustered, erect, bluntly 3-angled in the lower part, acute towards the apex; **spike** 1.5-3.5 cm x 0.4-0.6 cm, dense, cylindrical, pale coloured, the apex pointed; **glumes** wedge-shaped, densely packed, leathery, striate; **nut** about 0.17 cm long, elliptic, glossy, straw coloured.

A widespread Asian species which in Australia is restricted to north-eastern Qld, where it grows in swamps of fresh and brackish water. It could be a useful aquatic for tropical regions. Propagate by division.

ELEPHANTOPUS L.
(from the Greek *elephas*; *elephantos*, elephant; *pous*, a foot; in reference to the similarity of the roots' layout to an elephant's footprint)
Asteraceae

Perennial, hairy **herbs**; **leaves** sessile, entire or den-tate; **flower-heads** few-flowered, surrounded by leaf-like bracts; **corolla** tubular, 5-lobed, often more deeply cleft on one side; **fruit** an achene.

A genus of about 32 species with 25 in the Americas and the others in Africa and Asia. One species extends to Australia, and it can become a troublesome weed in tropical areas.

Elephantopus scaber L.
(rough)
Qld, NT
0.1-1 m x 0.1-0.6 m

Dwarf, rigid, perennial **herb**; **rootstock** creeping; **stems** 1-3, arising from base, terete, smooth or slightly ribbed, warty, densely hairy, whitish; **radical leaves** 5-40 cm x 1-6 cm, form a rosette, spathulate, most variable, base semi-sheathing, slightly hairy to glabrous and warty above, densely hairy below, apex blunt; **stem-leaves** 3-15 cm x 0.2-3 cm, narrow-oblong to obovate, on flowering branches, semi-sheathing on non-flowering branches; **flower-heads** blue, mauve or purple, 20-50, forming terminal, hemispherical, compound heads about 2.5 cm across.

This species is widespread in the tropics of Asia, as well as in the tropical regions of north and north-eastern Qld and NT. It was originally collected by Banks and Solander at Endeavour River, Qld.

It is regarded as a troublesome weed in Qld, and it should not be cultivated if there is the likelihood of it spreading. Best suited to tropical and subtropical regions, in well drained, acid soils. Tolerates dappled shade and partial or full sun. Propagate from seed or cuttings.

The liquid from boiled roots and leaves is used in India for the treatment of bowel and stomach disorders, while in some parts of Asia it is used as a diuretic.

ELIONURUS Humb. & Bonpl. ex Willd.
(from the Greek *elyein*, roll; *oura*, a tail; in reference to the waving or twisting ears)
Poaceae

Annual or perennial **grasses**; **leaves** linear, basal or ascending the culms, flat or rolled; **culms** erect, branched sparsely from the upper axils; **inflorescence** a simple raceme; **spikelets** in pairs, one spikelet sessile and bisexual, the other stalked and sterile.

A small genus of 25 species of tropical and subtropical grasses with a solitary species in northern Australia. It is rarely encountered in cultivation and has limited horticultural appeal. Propagate from seed which may need storage for up to 12 months before sowing or by division of the clumps.

Elionurus citreus (R. Br.)Munro ex Benth.
(with a lemon scent)
Qld, NT   Lemon-scented Grass
0.5-1 m tall   Oct-Dec

Perennial **grass** forming slender clumps; **leaves** 20-40 cm x 0.2-0.3 cm, linear, flat, glabrous, striate, flexuose, basal and scattered on the culms, lemon-scented when crushed; **culms** 50-100 cm tall, erect, glabrous, smooth, branched from the upper nodes;

inflorescence a series of racemes, 3-12 cm long; **spikelets** in pairs, each dissimilar.

Distributed mainly in northern Australia (although it just reaches south-eastern Qld), this species is often locally common in the sandy soils and gravels of stream beds and also in near-coastal sand-dunes. It is a slender grass, with limited horticultural appeal other than the attractive lemon scent of its leaves when crushed. It can be grown as a garden or container plant in tropical and subtropical regions. Propagate from seed or by division of the clumps.

ELYTHRANTHERA (Endl.) A. S. George

(from the Greek *elytron*, a cover; *anthera*, an anther; in reference to the prominent column wings which hide the anther)
Orchidaceae

Small terrestrial **orchids**; **tubers** rounded, enclosed in a brown, papery sheath, subterranean; **leaves** solitary, basal, glandular, hairy; **flower-stems** erect, wiry, glandular, hairy; **flowers** small, hand-like, opening widely, brightly coloured and glossy; **labellum** small, inconspicuous, with 2 prominent, hinged basal calli; **column** with prominent hooded wings; **fruit** a capsule.

A small genus of 2 species endemic to south-western WA. They are colourful terrestrial orchids with glossy flowers which look like they have been enamelled, and give rise to the popular name of enamel orchids. They grow actively in autumn-winter, flower in spring and die back to the subterranean tuber over summer.

They have proved difficult to cultivate, and growers specializing in terrestrial orchids are the most successful. The usual response of plants is to grow for a couple of years but to gradually become smaller and fade away. Generally, plants are very shy of flowering in cultivation. Fair results have been obtained using mixtures based on sandy loams derived from coastal heathlands, in combination with coarse sand and broken down leaf mould or eucalypt shavings. Repotting may hasten the decline of the plants. Watering should be regular while they have above-ground growth, and the soil mixture should be allowed to dry out while they are dormant. Species may be propagated successfully by sprinkling seed around the base of established plants.

Elythranthera brunonis (Endl.) A. S. George
(after Robert Brown, English botanist of the 18-19th centuries)
WA   Purple Enamel Orchid
Aug-Dec

Terrestrial **orchid**; **leaf** 2.5-8 cm x 0.4-0.6 cm, narrow-lanceolate, dark green, purplish towards the base, covered with short glandular hairs; **flower-stem** 15-30 cm tall, wiry, bearing 1-3 flowers; **flowers** 2.5-4 cm across, dark purple and glossy on the inside, paler and blotched on the outside; **labellum** small, with 2 long basal calli.

This species is restricted to the south-west but there it is a very widespread and often locally common orchid. The very colourful, glossy flowers attract immediate attention. Its performance in cultivation is variable, transplants often surviving for years and even increasing in number. The species rarely seems to thrive, however, and is extremely shy of flowering. Seedlings can be raised by scattering seed on the surface of the pot.

Elythranthera emarginata (Lindley) A. S. George
(notched)
WA   Pink Enamel Orchid
Aug-Dec

Terrestrial **orchid**; **leaf** 4-10 cm x 0.5-1 cm, lanceolate, erect, dull green, purplish towards the base, covered with soft glandular hairs; **flower-stem** 12-30 cm tall, wiry, bearing 1-4 flowers; **flowers** 3-4 cm across, bright pink and glossy on the inside, paler and spotted on the outside; **labellum** small, notched at the apex, with 2 long, linear basal calli.

This species, which is restricted to the south-west, grows mostly on gravelly soils in open forest, but is also to be found on peaty soils around the margins of swamps. Its performance in cultivation is similar to that of *E. brunonis* but it is more sensitive to disturbance and tends to die out readily.

EMBELIA Burm.

(after the local name of a species in Ceylon)
Myrsinaceae

Straggly **shrubs** or **climbers**, rarely trees; **leaves** simple and entire or toothed, alternate; **petiole** margined to glandular; **inflorescence** a raceme or panicle, axillary or terminal; **flowers** small, bisexual; **calyx** 4-5 lobed; **petals** 4-5, free; **stamens** 4-5; **fruit** a small, globose, 1-2 seeded drupe.

A widespread genus of about 130 species distributed in tropical and subtropical regions with 2, possibly 3,

Elythranthera emarginata, × ·5

Embelia australiana

species endemic in eastern Australia. They are rarely encountered in cultivation. Propagate from seed which has limited viability and must be sown fresh.

Embelia australiana Benth. & J. D. Hook.
(Australian)
Qld, NSW
5-15 m tall   Sept-Nov

Tall woody **climber**; young shoots glabrous; **branchlets** usually zigzagged; **leaves** 3-8 cm x 1-4 cm, elliptical to obovate, thin-textured but leathery, glossy green above, dull beneath, the apex blunt or pointed; **racemes** 1.5-2.5 cm long, axillary; **flowers** about 0.4 cm across, greenish white; **petals** 4; **drupe** about 1.6 cm across, globular, bright red, very hard, mature March-May.

An endemic climber distributed in rainforests from north-eastern Qld to north-eastern NSW. It is a strong-growing species and its major ornamental feature is the bright red fruit. It can be grown from tropical to temperate regions but is perhaps too vigorous for the average home garden. Could be suitable for large municipal gardens. Requires well drained soil. Propagate from seed which is best sown fresh.

Embelia flueckigeri F. Muell.
(after Dr. Frederick Flueckiger)
Qld
2-5 m x 1-3 m   Oct

Straggly **shrub** or **subclimber**; young shoots glabrous; **petioles** distinctly corrugated; **leaves** 6-10 cm x 2-3.5 cm, elliptic, somewhat thick-textured, glossy green on both surfaces, the margins entire; **panicles** short, much-branched, terminal, densely clothed with brown hairs; **flowers** about 0.5 cm across, greenish white; **calyx** 5-lobed, hairy; **stamens** 5, protruding; **drupe** about 1.5 cm across, globular, reddish.

A little-known species restricted to lowland rainforests of north-eastern Qld. May be shrubby in the open but tends to straggle if amongst other plants. Has limited horticultural appeal. Best suited to tropical and warm subtropical conditions, in well drained soil. Propagate from seed which must be sown fresh.

Embelia australiana, × ·6

EMBLINGIA F. Muell.
(after Thomas Embling, 19th-century Melbourne physician)
Emblingiaceae

A monotypic endemic genus confined to on or near the western coast, north of Perth in WA.

Emblingia calceoliflora F. Muell.
(flower like a small shoe)
WA
prostrate x 0.6-1.5 m   Aug-Oct; also sporadic

Dwarf, spreading, perennial **shrub**; **branches** with stiff hairs; **leaves** 2.5-4 cm x 1-2 cm, lanceolate to elliptical, mostly opposite, narrowed to a petiole, acute apex, stiff; **flowers** tubular, about 1.2 cm long, orange or greenish cream to white, hairy, on short axillary stalks; **capsule** to about 1 cm broad.

An ornamental species that occurs in the Carnarvon and Irwin Districts, from near the Exmouth Gulf, and extends southward to north of Moora. It grows on red or white sand over limestone. Plants in the northern regions have orange flowers, while those to the south have greenish cream, often with pink and yellow tonings. Has had only limited cultivation, but shows potential. Needs very well drained soils and plenty of sunshine. Tolerant of light to medium frosts and extended dry periods. Suited to alkaline soils. Worth trying as a container plant. Propagate from seed or cuttings which may prove successful.

Emblingiaceae (Pax) Airy-Shaw

A monogeneric family consisting of the monotypic genus *Emblingia* which is endemic to Australia. The species is regarded as a relict and is sometimes included in the family Capparaceae even though it has many features divergent from that family.

EMBOTHRIUM J. & G. Forster — for Australian species previously included in this genus see *OREOCALLIS.*

EMMENOSPERMA F. Muell.
(from the Greek *emmeno,* I abide in or I cleave to; *sperma,* a seed; in reference to the way the seeds remain attached after the fruit falls away)
Rhamnaceae

Trees; **leaves** opposite, glabrous, simple, entire; **inflorescence** an axillary cyme or panicle; **flowers** small, pedicellate; **calyx** prominent, bell-shaped, with 5 lobes; **petals** minute, alternating with the calyx lobes; **stamens** 5; **fruit** 2-celled, with one seed in each cell.

A small genus of 3 species, one in eastern Australia, one in WA and the other in New Caledonia. The eastern Australian species is a handsome tree rarely seen in cultivation. Propagate from seed which may be difficult to germinate.

Emmenosperma alphitonioides F. Muell.
(resembling the genus *Alphitonia*)
Qld, NSW   Bonewood; Yellow Ash
15-30 m x 5-15 m   Sept-Oct; March

Medium to tall **tree** with a dense, spreading crown;

Enchylaena tomentosa   W.R. Elliot

bark grey, slightly wrinkled, bright yellow when cut; young shoots bright green; **leaves** 3.5-10 cm x 1.5-4 cm, ovoid to elliptical, dark glossy green above, paler beneath, the apex blunt; **panicles** 3-6 cm long, much-branched, terminal or in the upper axils; **flowers** about 0.4 cm across, white, somewhat bell-shaped; **fruit** about 0.9 cm across, 2-celled, the outer part shedding and exposing shiny bright red seeds, mature March-Nov.

A widespread tree found in rainforests from Cape York Peninsula to the Illawarra region of southern NSW. It is cut as a timber tree, with the pale yellow-brown wood being used for tool-handles, boat building and scantling. It is a very handsome tree that probably becomes too large for the average home garden, but it is eminently suited to parks and acreage planting. Plants can be fast growing and, although they require some shade when young, once established they are generally hardy to sun. A well drained, organically rich soil produces best growth, but the species will adapt to poorer well drained soils. Plants respond to the application of mulches, fertilizers and water during dry periods. Suited to tropical, subtropical and temperate regions. Propagate from seed which may be erratic in its germination, some batches germinating freely and without difficulty.

EMPODISMA L. Johnson & Cutler
(from the Greek *empodisma*, an obstacle or hindrance; in reference to the masses of wiry stems)
Restionaceae

Perennial **herbs** with rhizomes; **stems** wiry, usually intertwined, appear leafless; **flowers** dioecious, in axillary spikelets; **female spikelets** axillary, single-flowered; **male spikelets** several-flowered; **male** and **female perianth** with membranous tepals; **staminodes** absent in female flowers; **fruit** a pale, smooth, indehiscent nut.

A genus of 2 species. One is endemic to south-west WA, and the other is widespread in south-eastern Australia and in New Zealand. Both are rarely cultivated. Propagate from seed and possibly by division of the rhizomes.

Empodisma gracillimum (F. Muell.) L. Johnson & Cutler
(very slender)
WA
0.3-1 m x 1-2 m   July-Nov

Semi-climbing perennial **herb**; **rhizomes** unbranched; **stems** slender, many, usually interlocked, leafless except for reflexed **sheathing scales**, upper scales appressed, lower scales lax; **flowers** small, brown, arranged in spikelets, male and female on separate plants; **nut** small.

A fairly common herb from the South-West Botanical Province. It can cover large areas in its swampy habitat. Not common in cultivation. Suitable for moist to wet soils that receive some sunshine. Should do well on the edge of a shallow pool with a clay base, or as bog garden plant (see Aquatics, Volume 2, page 217). Is worth experimenting with in pots or hanging baskets. Propagate from seed.

Previously known as *Calorophus gracillimus* F. Muell. and *Hypolaena gracillima* (F. Muell.) Benth.

Empodisma minus (Hook.f.) L. Johnson & Cutler
(smaller)
Qld, NSW, Vic, Tas, SA
0.5-2 m x 0.5-2 m   July-Nov

Semi-climbing perennial **herb**; **rhizome** branched; **stems** slender, many, often flexuose, leafless except for reflexed **sheathing scales** which are loose; **flowers** small, brown, arranged in spikelets, male and female on separate plants; **female spikelets** sessile or with short stalks; **nut** small.

A widespread species, common in wet areas. Has similar requirements to *E. gracillimum*. Previously known as *Calorophus minus* Hook.f. The simplest way to distinguish between the species is by the rhizomes. Those of *E. minus* are branched, while those of *E. gracillimum* are not.

ENCEPHALARTOS Lehm. — for Australian species formerly placed in this genus see *MACROZAMIA*.

ENCHYLAENA R. Br.
(from the Greek *enchylos*, succulent; *laina*, a cloak; in reference to the perianth enlarging and becoming fleshy)
Chenopodiaceae

Dwarf to small **shrubs**; **branches** slender, striate, hairy; **leaves** alternate, slender; **flowers** small, solitary, axillary; **perianth** 5-lobed; **stamens** 5; **fruiting perianth** semi-globular, fleshy, enclosing the fruit.

This endemic genus is currently undergoing botanical revision. Until recently it was regarded as monotypic, but is now known to comprise several species. *Enchylaena* is represented in all states except Tas. It is common in semi-arid and arid regions. Plants are readily propagated from seed or cuttings.

Enchylaena tomentosa R. Br.
(with short, soft hairs)
Qld, NSW, Vic, SA, WA, NT   Barrier Salt-bush;
                                   Ruby Salt-bush
0.3-1 m x 0.5-1.5 m   throughout year

399

Dwarf, spreading **shrub**; **branches** woody, slender, hairy, often straggly; **leaves** 0.8-2 cm x 0.1-0.2 cm, linear, terete, succulent, green to bluish green, hairy; **flowers** small, solitary, axillary; **fruiting perianth** about 0.5 cm diameter, initially yellow, maturing to dark red, can be profuse.

This attractive species with succulent foliage produces berries that change colour as they mature, providing an unusual ornamental feature. It is widespread in semi-arid and arid zones, and usually occurs on sandy soils that may be calcareous. It is also found in coastal regions and on a variety of soils, some of which can be slightly saline. Has excellent potential for areas of low rainfall. Grows well in most soils that have good drainage, including those of poor quality. Withstands some salinity. Tolerates dappled shade in arid climates, but does best in a situation with partial to full sun. Succeeds in temperate zones. Drought and frost tolerant. Responds well to pruning. An excellent container plant. Propagate from seed which germinates readily or from cuttings.

ENDIANDRA R. Br.

(from the Greek *endo*, within; inside; *andros*, a man; a reference to the fertile stamens of the inner whorl) *Lauraceae*

Trees; **leaves** alternate, simple, glabrous or hairy; **inflorescence** a cyme or panicle, in the upper axils; **flowers** bisexual, opening widely or closed and turbinate, or bell-shaped; **perianth** segments 6; **stamens** in 2 whorls, the outer whorl sterile and reduced to a ring, or absent, the inner 3 fertile; **fruit** a drupe.

A genus of about 100 species distributed through the Pacific region from Malaysia to Australia, where there are 36 species including about 12 which are undescribed. All the Australian species appear to be endemic. They occur in the rainforests of tropical and subtropical eastern Australia, and proliferate in north-eastern Qld. Their insignificant flowers and sporadic fruit production limit their horticultural appeal, but most species have attractive leaves and a dense canopy. They are of interest to rainforest enthusiasts and could be suitable as park trees. Some species make attractive pot plants and could be tried for indoor decoration. Propagate from seed which has a short viable period and is best sown fresh.

Endiandra acuminata C. White & Francis
(drawn into a long, sharp point)
Qld   Brown Walnut
15-25 m x 8-12 m   March-June; Aug; Oct
Medium to large **tree**; young shoots covered with rusty hairs; **leaves** 5-9 cm x 2.5-4 cm, elliptical or ovoid, dark glossy green above, dull beneath, with a short, blunt point; **panicles** 2-5 cm long, in the upper axils; **flowers** 0.6-0.8 cm across, brownish, softly hairy; **drupe** 2-2.5 cm long, narrow-ovate, black, mature Oct-Dec.

A large tree of the coastal rainforests of north-eastern Qld to the north and south of Cairns. It is cut for its brown timber which is useful for cabinet making and plywood. Trees grow too large to be recommended for the average home garden but would make a handsome addition to a park or large block. They succeed

best in tropical and warmer subtropical regions, in a well drained soil. Propagate from seed which must be sown fresh. Also known as *E. subtriplinervis* C. White & Francis.

Endiandra anthopophagora Domin
(eaten by man)
Qld
15-25 m x 5-15 m   Feb-March
Medium to tall **tree**; young shoots bearing rusty hairs; **leaves** 14-18 cm x 5-6 cm, ovate-lanceolate, thick and papery in texture, bright green and shiny above, paler beneath; **cymes** 3-5 cm long, in the upper axils; **flowers** about 0.2 cm across, globular, pinkish; **drupe** 3-4 cm across, pinkish red, somewhat rounded, mature Jan-Feb.

A rare tree confined to lowland rainforests near Cairns. It is unknown in cultivation but is of interest botanically, and the species could have potential for planting in parks in tropical regions. The derivation of the epithet indicates that the fruit may have been eaten by the Aborigines. Propagate from seed which must be sown fresh.

Endiandra compressa C. White
(somewhat flattened)
Qld, NSW   Queensland Greenheart
10-25 m x 3-8 m   Nov-Dec
Medium to large **tree**; **bark** light grey to whitish, smooth; young shoots pale green; **leaves** 8-18 cm x 2-3 cm, elliptical or lanceolate, dark glossy green above, pale glossy green beneath, the apex blunt; **panicles** about 1.5 cm long, axillary; **flowers** about 0.3 cm across, creamy yellow; **drupe** 3-5 cm across, bluish black, rounded but flattened laterally, mature Jan.

Widely distributed from north-eastern Qld to north-eastern NSW but very rare in the latter state and known only from a single locality. It is found in rainforests and is occasionally cut as a timber tree for its hard, strong wood. It is a large tree with an attractive spreading canopy, insignificant flowers and large fruit which are most noticeable when they are on the ground. Horticulturally, its uses are limited because of its size; however, it should be an excellent tree for parks and large public gardens. Would be worth trying in temperate regions. Young plants require some protection from hot sun. Well drained soil is essential and the plants respond to mulching, slow-release fertilizers or animal manures, and watering during dry periods. Propagate from seed which must be sown fresh.

Endiandra cowleyana Bailey
(after E. Cowley)
Qld   Northern Rose Walnut
10-20 m x 5-12 m   Jan-March
Small to medium **tree**; **bark** smooth, grey; young growth hairy; **leaves** 5-7 cm x 2-3 cm, ovate-lanceolate, dark green above, paler beneath, the apex drawn out and blunt; **panicles** 5-8 cm long, in the upper axils, shortly hairy; **flowers** about 0.2 cm across, greenish, hairy on the outside; **drupe** 3-4 cm x 1.5-2 cm, ovoid, black, mature July-Oct.

Restricted to north-eastern Qld but there wide-

400

spread in highland and lowland rainforests. It is cut as a timber tree, the grey wood being light and easy to work. Trees grow too large for the average home garden but could be planted as specimens in parks. Best suited to tropical and subtropical regions. Young plants require protection from direct sun for the first 12-18 months. Suitable soils must drain freely and best growth is achieved in loams rich in organic matter. Propagate from seed which must be sown fresh.

Endiandra crassiflora C. White & Francis
(thick, fleshy flowers)

| Qld, NSW | Dorrigo Maple; Dorrigo Walnut |
| 8-20 m x 3-6 m | Nov-March |

Small to medium **tree**; **bark** grey, with numerous pits; young shoots densely covered with soft, golden hairs; **leaves** 5-10 cm x 3-4.5 cm, elliptical to ovate, bright green, glossy and somewhat crinkled on the upper surface, the lower surface grey; **racemes** 2-4 cm long, axillary; **flowers** about 0.4 cm across, greenish pink to reddish, opening widely, the segments thick and fleshy; **drupe** 1.5-2.5 cm long, oval, blue-black, mature Sept-Dec.

Distributed in rainforests from the MacPherson Range of south-eastern Qld to near Wauchope in NSW, usually in highland situations. It has pale pink, close-grained timber but the trees are not commonly cut. Trees have an attractive spreading canopy and would have potential for parks or large gardens in subtropical and temperate regions. They require a well drained, organically rich soil and respond to mulching, fertilizers and watering during dry periods. Propagate from fresh seed.

Endiandra dichrophylla F. Muell.
(leaves of 2 colours)

| Qld | Brown Walnut |
| 10-18 m x 3-8 m | Nov-Jan |

Small to medium **tree**; **bark** grey to brown, wrinkled; young shoots silky-hairy; **leaves** 5-15 cm x 3-6 cm, ovate-lanceolate, dark green and glossy above, shiny pale brown beneath, the apex shortly pointed; **panicles** 4-7 cm long, slightly hairy, in the upper axils; **flowers** about 0.2 cm across, greenish white; **drupe** 2-3 cm long, elliptical, black.

Fairly widely distributed in the rainforests of north-eastern Qld, between the Atherton Tableland and Mount Spec. The timber is nearly white and is occasionally used for general purposes. Trees are rarely grown but have potential for parks or acreage planting. Requires well drained soils and responds to water during dry periods. Best suited to tropical and subtropical regions. Propagate from seed which must be sown fresh.

Endiandra discolor Benth.
(two unlike colours)

| Qld, NSW | Rose Walnut |
| 10-25 m x 5-12 m | Oct-Nov |

Medium to large **tree** with a spreading canopy; **bark** brown, roughened by craters; young shoots finely hairy; **leaves** 6-10 cm x 2-3 cm, elliptical to ovate, bright shiny green above, grey beneath, with prominent domatia on the underside; **panicles** 3-5 cm long, in the

upper axils; **flowers** about 0.2 cm across, bell-shaped, greenish; **drupe** 2-2.5 cm long, oval, black, mature March.

Widely distributed in rainforests from Tully in north-eastern Qld to near Gosford in NSW, frequently growing near streams. The timber is hard, dull pink and is used for plywood, veneers, flooring and joinery. The trees grow too large for the average home garden but should be excellent park trees for tropical, subtropical and temperate regions. Requirements are for well drained soil in a semi-shady position. Young plants need protection from direct sun. Propagate from fresh seed.

Endiandra exostomonea F. Muell. ex Bailey
(exserted stamens)

| Qld | |
| 10-20 m x 5-8 m | March-June; Aug; Oct |

Small to medium **tree**; young shoots hairy; **leaves** 8-13 cm x 2.5-4 cm, ovate-lanceolate, papery in texture, dark green and glossy above, duller and shiny beneath, the veins hairy; **panicles** 3-6 cm long, numerous, in the upper axils; **flowers** about 0.2 cm across, brownish; **drupe** 4-5 cm long, ovoid to cylindrical, yellow, mature July-Sept.

This species occurs in rainforests of north-eastern Qld. It has been confused with *E. virens* but the two are quite distinct and their ranges do not overlap. *E. virens* can be distinguished by its narrow, dark green leaves. Plants of *E. exostomonea* are rarely cultivated; however, with its attractive foliage the species could find a place in parks and gardens in tropical and subtropical regions. Seedlings require protection from the sun for the first 12-18 months. Soils of free drainage are essential. Propagate from fresh seed.

Endiandra glauca R. Br.
(bluish)

| Qld | Brown Walnut |
| 6-10 m x 3-5 m | Dec-June |

Tall **shrub** or small **tree**; **bark** greyish; young shoots bearing short, rusty hairs; **leaves** 7-14 cm x 3-6 cm, elliptical to oblong, bright green and shiny above, white or glaucous beneath, the veins rusty-hairy, the apex acuminate; **panicles** 4-8 cm long, rusty-hairy, in the upper axils; **flowers** about 0.4 cm across, bell-shaped; **drupe** 1-1.5 cm long, oval, black, mature June-Sept.

A small tree of rainforests extending from Cape York Peninsula to Mount Spec near Townsville. It has ornamental foliage and attractive new growth, and appears to have good potential for tropical and subtropical gardens. A shady situation in well drained loamy soil is most suitable for its culture. Young plants are decorative in a pot and would be worth trying for indoor decoration. Propagate from seed which must be sown fresh.

Endiandra globosa Maiden & Betche
(nearly spherical)

| Qld, NSW | Ball Nut; Black Walnut |
| 10-20 m x 5-10 m | Oct-Dec |

Small to medium **tree** with a spreading canopy; **bark** grey to brown, finely scaly; young shoots light green,

Endiandra hayesii

hairy; **leaves** 10-15 cm x 4-6 cm, lanceolate to elliptical, dark green and glossy above, paler beneath, the main veins prominent, the apex drawn out and blunt; **racemes** or **panicles** 2-8 cm long, much-branched, in the upper axils; **flowers** 0.3 cm across, white, bell-shaped; **fruit** 4-6 cm across, shiny blue-black, mature April-May.

A species of fairly restricted distribution, confined to extreme south-eastern Qld and north-eastern NSW in the Brunswick and Tweed River valleys. It grows in rainforests, often on alluvial flood plains and also in open forest in stony soils. Trees have a spreading canopy and attractive glossy leaves. They are particularly noticeable during flushes of new growth, which is light green and contrasts with the mature foliage. The fruit is large and interesting but, because of its dull colouration, it is hardly noticeable while on the tree. Despite its ultimate size, this is a handsome tree worthy of wide cultivation. It makes an excellent shade tree and would be suitable for gardens and parks. Young plants can be fast growing in a suitable position. Preference is for a well drained, organically rich loam but the species has proved adaptable to a variety of well drained soils. It is hardy in temperate regions, withstanding moderate frosts. Propagate from seed which loses its viability within a month of ripening.

Endiandra globosa fruit

D.L. Jones

Endiandra hayesii Kosterm.
(after H. C. Hayes, original collector)

Qld, NSW	Rusty Rose Walnut
8-20 m x 3-8 m	Oct-Nov

Small to medium, often crooked **tree**; **bark** grey-brown, scaly; young shoots pale green, covered with pale brown hairs; **leaves** 6-13 cm x 4-8 cm, ovoid, dull and furry above, brown and bearing pale hairs beneath, tapering to a long, drawn out, blunt point; **panicles** 3-6 cm long, in the upper axils, few-flowered; **flowers** about 0.4 cm across, pale green; **drupe** 2.5-2.8 cm long, oblong, black, mature Aug and March.

This species is restricted to cool, sheltered gullies of lowland rainforests between the Clarence River in north-eastern NSW and Burleigh Heads in south-eastern Qld. Flushes of new growth are pale coloured and furry. The species is unknown in cultivation due to lack of propagating material and it is rarely collected in fruit. Probably requires well drained, acid soils and protection from sun when small. Propagate from seed which must be sown fresh. Cuttings would be worth trying.

Endiandra hypotephra F. Muell.
(underside of leaf is grey)

Qld	Blue Walnut
15-20 m x 5-12 m	Feb-June

Medium **tree**; young shoots densely covered with soft rusty hairs; **leaves** 7-14 cm x 3-8 cm, ovate to broadly elliptic, thick and leathery, dark green and glabrous above, pale or white on the underside, with brown hairs on the main veins, the apex long and drawn out; **panicles** or **cymes** 4-10 cm long, dense, in the upper axils; **flowers** about 0.8 cm across, greenish; **drupe** 2-2.5 cm long, oblong to cylindrical, blue-black, mature June-Sept.

A handsome tree found in the rainforests of north-eastern Qld. Trees have a spreading canopy and attractive foliage, and could be grown in parks or large gardens in tropical and subtropical regions. Young plants require protection from hot sun. Well drained soil is essential. Propagate from seed which must be sown fresh.

Endiandra insignis Bailey = *E. pubens* Meisner

Endiandra introrsa C. White
(turned inward; a reference to the anthers)

NSW	Dorrigo Plum; Red Plum
12-30 m x 5-12 m	Oct-Dec

Medium to large **tree** with a dense crown; **bark** reddish brown, with numerous pustules; young shoots softly hairy; **leaves** 5-8 cm x 2-3 cm, ovate to elliptical, smooth and green above, covered with grey, waxy bloom beneath, the apex drawn out into a long, blunt point; **racemes** 1-2.5 cm long, curved, axillary, the peduncle reddish; **flowers** about 0.3 cm across, greenish to yellow, bell-shaped; **drupe** 4-7 cm across, initially yellow or red, then black, glossy, mature Feb-March.

A large handsome tree commonly found in rainforests at high elevations, and restricted to north-eastern NSW. It has excellent potential as a shade tree for parks or large gardens, and would succeed in subtropical and temperate regions. Plants grow fairly rapidly in the first few years and prefer protection from direct sun while small. They need well drained soil, and respond to mulches and fertilizers. Propagate from seed which must be sown fresh.

Endiandra longipedicellata C. White & Francis
(long pedicels)
Qld
10-15 m x 5-8 m   June-Oct
Small to medium **tree**; young shoots densely covered with rusty hairs; **branchlets** flattened or 4-angled; **leaves** 5-10 cm x 2.5-4.5 cm, elliptical to lanceolate, the apex acuminate, glabrous, green above, covered with rusty hairs beneath; **panicles** 5-10 cm long, in the upper axils; **flowers** about 0.2 cm across, narrowly bell-shaped, hairy; **drupe** 4-5 cm x 2.5 cm, ovoid to oblong, dark blue to blue-black, with a glaucous bloom, mature June-Oct.
A little-known species restricted to rainforests of north-eastern Qld. It has potential for parks and large gardens in tropical and subtropical regions but little is known about its behaviour in cultivation. Propagate from seed which must be sown fresh.

Endiandra lowiana Bailey =
E. virens F. Muell. ex Meisner

Endiandra microneura C. White
(minute veins)
Qld
10-20 m x 5-10 m   March-April
Medium **tree**; **bark** grey; young shoots angular,

Endiandra muelleri leaves, flowers and fruit, × ·5

hairy; **leaves** 10-17 cm x 4-6 cm, elliptical, stiff and somewhat leathery, dark green above, paler beneath, the apex pointed; **panicles** 4-8 cm long, in the upper axils; **flowers** about 0.15 cm across, bell-shaped, white; **drupe** 7-8 cm x 3.5-4 cm, oblong, yellow, mature Nov-Dec.
Restricted to lowland rainforests of north-eastern Qld, sometimes in rather swampy conditions. It appears to be unknown in cultivation, but could have potential for parks or acreage planting in tropical and perhaps warm subtropical regions. May be able to tolerate soils with slow or impeded drainage. Propagate from fresh seed. The species was previously known as *E. reticulata* C. White.

Endiandra montana C. White
(growing on the mountains)
Qld   Brown Walnut
4-8 m x 3-5 m   Nov-Dec
Tall **shrub** or small **tree**; young shoots with minute hairs; **leaves** 8-11 cm x 2.5-4 cm, elliptical, somewhat leathery, shiny above, dull beneath, the apex bluntly pointed; **panicles** short, in the upper leaf axils, few-flowered; **flowers** about 0.4 cm across, cream, turbinate; **drupe** 3-4 cm x 2.5-3 cm, ovoid, flattened on each side, yellow, mature July-Oct.
An interesting species confined to a couple of mountain-tops of north-eastern Qld. It grows as a gnarled shrub or small tree and, although it is hardly known in cultivation, it could have tremendous potential for gardens in tropical and subtropical regions. Requires a well drained soil and plants may need some protection when young. Propagate from seed which must be sown fresh.

Endiandra muelleri Meisner
(after F. von Mueller, first Government Botanist, Vic)
Qld, NSW   Mueller's Walnut
15-20 m x 5-12 m   Nov-Jan
Medium **tree** with a spreading canopy; **bark** brown, with large, loose flakes; young shoots pale green to

Endiandra introrsa   D.L. Jones

pink, finely hairy; **leaves** 5-9 cm x 2-4 cm, ovate-lance-olate to elliptical, glossy green on both surfaces, the apex blunt; **panicles** 3-5 cm long, in the upper axils; **flowers** about 0.3 cm across, greenish or reddish, globular or bell-shaped; **drupe** 1.5-2.5 cm long, oblong to oval, black, mature April-May.

A widely distributed rainforest species extending from central Qld to Comboyne in NSW. Plants also occur in moist eucalypt forest. It is a handsome tree with a fairly dense crown of dark green leaves. Flushes of new growth are pinkish red and attractive. Flowers are insignificant and the fruit is hardly noticeable. In its final size, the species is too large for the average home garden but plants are slow-growing and would be suitable for many years. Can be grown in tropical, sub-tropical and temperate regions. Propagate from seed which must be sown fresh.

Endiandra palmerstonii (Bailey) C. White & Francis
(after Christie Palmerston)
Qld   Queensland Walnut
20-25 m x 8-15 m   March-May

Large **tree** with a spreading canopy; **bark** grey-brown, wrinkled; young shoots clothed with fine brown hairs; **leaves** 10-15 cm x 4-7 cm, ovoid to elliptical, dark green and shiny above, paler beneath, the veins hairy; **panicles** 4-8 cm long, in the upper axils; **flowers** about 0.3 cm across, greenish; **drupe** 4-6 cm across, globular, pinkish green to bright red (often falling while green), ribbed, mature Nov-March.

A large tree of north-eastern Qld, growing in lowland and highland rainforests. It is an important timber tree, its dark brown wood being used for cabinet making, veneers etc. The seeds are about 2.5 cm across and are enclosed in a hard shell. Aborigines ate the seeds after roasting, pounding and immersing them in running water. In nature rats feed on the seed and leave the gnawed shells beneath the tree. Queensland walnut is a very handsome tree but its extreme size prohibits its use in home gardens. It would, however, be an excellent tree for parks, large gardens and acreage planting. Best suited to tropical and sub-tropical regions. It should be tried in temperate regions where the climate may reduce its size to manageable proportions. Seedlings are very fast growing and require protection from direct sun for the first 12-18 months. A well drained, organically rich soil produces best growth but the plants are quite adaptable. Propagate from seed which must be sown fresh.

Endiandra pubens Meisner
(clothed in downy hairs)
Qld, NSW   Hairy Walnut
10-25 m x 6-12 m   March-May; Nov

Small to medium or tall **tree** with a spreading canopy; **bark** light grey to whitish; young shoots reddish, densely covered with rusty hairs; **leaves** 7-20 cm x 3-8 cm, ovoid, the upper surface bright glossy green and smooth, the lower surface densely covered with rusty hairs; **panicles** 2-4 cm long, in the upper axils; **flowers** about 0.3 cm across, whitish, bell-shaped; **drupe** 4-8 cm across, globular, pale green becoming deep red, mature Oct-Feb.

A widely distributed species extending from north-eastern Qld to north-eastern NSW in the region of the Bellingen River. It is always found in rainforest, often in very sheltered, moist situations. It can develop into a handsome tree with a spreading, dense canopy and attractive flushes of new growth. The fruit is large and colourful but, unfortunately, it is borne sporadically and not in abundance, although there seems to be variation between trees in this characteristic. Plants have proved to be very slow-growing, but seedlings are attractive. They need protection when young and respond to the use of mulches and fertilizers. Propagate from seed which must be sown fresh. Superior fruiting trees should also be tried from cuttings.

Endiandra reticulata C. White =
E. microneura C. White

Endiandra sankeyana Bailey
(after J. R. Sankey)
Qld   Sankey's Walnut
15-25 m x 5-12 m   Oct-Nov

Medium **tree**; young shoots clothed with pale brown hairs; **leaves** 6-14 cm x 4-7 cm, elliptical to lanceolate, dark green and glabrous above, paler beneath, the main veins reddish brown with velvety hairs; **panicles** 4-8 cm long, in the upper axils; **flowers** about 0.2 cm across, yellow; **drupe** 2.5-3 cm across, globose, blue-black, sometimes flattened, the calyx star-like, mature June-Nov.

Restricted to rainforests of north-eastern Qld. Probably too large for home gardens but could be used in parks and for acreage planting. Well drained soil is essential. Propagate from seed which must be sown fresh.

Endiandra sieberi Nees
(after F. W. Sieber, botanical collector from Prague)
Qld, NSW   Corkwood; Pink Walnut
10-25 m x 5-10 m   June-Oct

Medium to tall **tree** with a sparse canopy; **bark** corky, deeply fissured, grey to reddish brown; young shoots pale green; **leaves** 5-10 cm x 3-5 cm, elliptical to lanceolate, dull green on both surfaces, the margin pale yellow or translucent; **panicles** 4-6 cm long, in the upper axils; **flowers** about 0.4 cm across, white; **drupe** 1.5-2.5 cm long, oval to oblong, purplish black, shiny, mature March-Nov.

A widespread tree in coastal and near-coastal areas from central Qld to Jervis Bay in south-central NSW. It usually grows on sandy soils and may be found in rainforest or drier, open situations amongst casuarinas. The timber is hard, close-grained and suitable for tool-handles and cabinet work. Although not a good shade tree, the plants have a distinctive character and decorative bark, and are deserving of cultivation. They are good for coastal zones but they will not withstand severe salt-laden winds. Seedlings will take exposure to sun when quite small and have a reasonable growth rate. They can be damaged by frosts. Potted plants could probably be trained into interesting bonsai specimens. The fruit is eaten by the larger fruit pigeons. Propagate from seed which must be sown fresh.

Endiandra sieberi   D.L. Jones

A species of lowland rainforests and moist eucalypt forests, extending from south-eastern Qld to Comboyne in NSW. It has good potential for gardens because of its small stature, long, narrow dark green leaves and large fruit which are, unfortunately, not borne in mass. Plants have proved to be fairly slow-growing. They require well drained soil and the application of mulches and water during the growing-season. Frosts may cause severe damage to young plants. Propagate from seed which must be sown fresh.

ENNEAPOGON Desv.
(from the Greek *ennea*, nine; *pogon*, a beard; in reference to the nine plumose awns which characterize this genus)
Poaceae

Annual or perennial **grasses**, often glandular-hairy; perennials forming tussocks, annuals slender; **leaves** linear, flat or rolled, glabrous or hairy; **culms** erect, slender; **inflorescence** a dense, spike-like panicle, usually terminal on the culm but sometimes also axillary; **spikelets** with a single floret; **lemma** with an apical fringe of 9 short, soft plumose awns.

A genus of about 30 species of grasses mostly developed in tropical regions of Africa and Australia. In the latter there are 19 species, mostly endemic. Some species are quite decorative and have potential for cultivation. Propagate from seed which may need storage for up to 12 months before sowing or by division of the clumps.

Enneapogon lindleyanus (Domin) C. E. Hubb.
(after John Lindley, 18-19th-century professor of botany at University College, London)
Qld, SA, WA, NT   Wiry Nineawn
20-45 cm tall

Perennial **grass** forming compact tussocks; **leaves** 3-8 cm x 0.1-0.2 cm, linear, glabrous, smooth, stiffly spreading, usually tightly rolled; **culms** 20-45 cm tall, wiry, erect; **panicle** 1-1.5 cm x 0.5-0.8 cm, globular to oblong, dense and compact, purplish, held well above the foliage; **spikelets** small, with a short awn.

An attractive, tough grass usually found on rocky, shallow soils. In cultivation plants develop a compact tussock and flower freely. Would be a very useful grass for a rockery. Requires well drained soil in a sunny aspect. Propagate from seed or by division.

Enneapogon nigricans (R. Br.) Beauv.
(blackish)
Qld, NSW, Vic, SA, WA   Nigger-head
20-45 cm tall

Perennial **grass** with glandular hairs; **leaves** 7-14 cm x 0.2-0.3 cm, hairy, linear, usually inrolled; **culms** 20-45 cm tall, wiry; **panicle** 3-7 cm x 1-1.5 cm, oblong to narrow-cylindrical, dark purple to blackish, dense; **lemma** with prominent, stiff awns.

A widespread and often locally common grass, usually noticed in drier inland regions. It is an interesting species with an ornamental dark coloured, dense inflorescence. It can be grown in a pot or general garden situation, but likes an open aspect and tends to die out if crowded by other plants. Growth is slow and the

---

Endiandra subtriplinervis C. White & Francis =
    E. acuminata C. White & Francis

Endiandra tooram Bailey
(an Aboriginal name)
Qld   Tooram Walnut
5-12 m x 3-5 m   Dec-Feb

Tall **shrub** or small **tree**; young shoots slender, densely covered with rusty brown hairs; **leaves** 10-15 cm x 2.5-3.5 cm, oblong, glabrous and green above, dull and thinly hairy beneath; **panicles** 3-6 cm long, in the upper axils; **flowers** 0.2-0.3 cm across, creamy green; **drupe** 2.5-3.5 cm across, orbicular, dorsally flattened, dark purple, fleshy, mature Sept-Nov.

Found in rainforests of north-eastern Qld. The fruit is fleshy, and it is recorded that it was eaten by the Aborigines; however, it is unknown whether any prior preparation was used. The species has potential as a garden plant for tropical and subtropical regions. It prefers a shady to semi-shady situation and grows best in organically rich, loamy soil. Plants are rather slow-growing. They respond to mulches, fertilizers or manures, and watering during dry periods. Young plants should be tried for indoor decoration. Propagate from seed which must be sown fresh.

Endiandra virens F. Muell. ex Meisner
(green)
Qld, NSW   White Apple
5-10 m x 3-6 m   March-April

Tall **shrub** to small **tree**; **bark** grey, with soft, vertical, corky ridges; young shoots light green, hairy; **leaves** 6-15 cm x 2-3 cm, narrow-lanceolate to oblong, dark green and glossy above, paler beneath; **panicles** 3-6 cm long, in the upper axils; **flowers** about 0.3 cm across, greenish, bell-shaped; **drupe** 6-10 cm across, globular, yellow to bright orange or red, a thin layer of yellow pulp around the large seed, mature May-July.

Enneapogon oblongus

Enneapogon nigricans, × ·2

tussocks are small and sparse. Propagate from seed or by division.

Enneapogon oblongus N. Burb.
(oblong)
Qld, SA, WA, NT   Purple-head Nineawn
30-60 cm tall

Perennial **grass** forming sparse tussocks; **leaves** 8-20 cm x 0.2-0.4 cm, linear, hairy, bluish green, flattened or loosely rolled; **culms** 30-60 cm tall, stout, erect; **panicle** 1-3 cm x 0.8-1.2 cm, oblong, compact and dense, purplish, held well above the foliage; **spikelets** small, crowded.

A widespread grass of inland regions, usually growing on shallow, stony soils. It is virtually unknown in cultivation but it could have potential for a rockery in a sunny position. It apparently grows readily and flowers freely if watered regularly. Propagate from seed or by division.

Enneapogon purpurascens (R. Br.) Beauv.
(purplish)
Qld, NT
30-50 cm tall

Perennial **grass** forming sparse, short clumps; **leaves** 8-15 cm x 0.2-0.4 cm, linear, flat or inrolled, bearing glandular hairs; **culms** 30-50 cm tall, slender, wiry, erect; **panicle** 4-7 cm x 1-1.5 cm, oblong to narrow-cylindrical, purplish, dense; **lemma** with prominent awns.

A grass of open woodlands, usually in rocky sites or on stony soil. Flowering plants are most attractive but the species tends to be weak-growing. Worth trying in a sunny rockery in drier tropical, subtropical and inland regions. Propagate by division or from seed.

ENTADA Adans.
(a name from southern India)
Caesalpiniaceae

Strong woody **climbers**; **stems** slender; **leaves** bipinnate, the terminal pair of pinnae modified into tendrils; **pinnae** opposite; **pinnules** simple, entire; **inflorescence** a spike or raceme; **flowers** small; **petals** 5; **calyx** 5-toothed; **stamens** 10; **fruit** a large woody pod.

A genus of about 30 species of vigorous climbers with 2 widespread species extending to parts of northern Australia, and another endemic. The species have limited horticultural appeal because of their size, the chief feature of interest being the huge woody pods. Propagate from seed which has a very hard coat that must be sliced open or broken to allow germination to take place.

Entada phaseoloides (L.)Merrill
(resembling the genus *Phaseolus*)
Qld, NT   Matchbox Bean
10-20 m tall   Oct-Dec

Robust woody **climber**; young shoots shortly hairy; **leaves** bipinnate, consisting of a petiole 5-15 cm long, 1-2 pairs of pinnae and 2 terminal tendrils; **pinnae** 10-15 cm long, each bearing 8-10 leaflets; **leaflets** 5-12 cm x 3-6 cm, oblong to obovate, dark green, the apex blunt or notched; **racemes** 15-30 cm long; **flowers** about 3 cm long, creamy white becoming brownish, crowded, with a strong musty perfume; **pods** 60-120 cm x 8-12 cm, contracted between the seeds, brown, hard; **seeds** about 5 cm across, round and thick, blackish.

A widespread tropical climber, the seeds of which are dispersed by the ocean currents. In Australia it is restricted to the tropics, extending south as far as Mackay. It is most common in coastal scrubs but extends some distance inland, growing in rainforests and communities along stream banks. The huge woody pods which hang in clusters are the most distinctive feature. The Aborigines used the bark and seeds as a fish poison and also ate the seeds after roasting or baking, grinding to a pulp and immersing in running water for 10-12 hours. The early settlers hollowed out sections of the pods to make waterproof containers for their matches. The species grows far too vigorously for a home garden but is of interest for botanical collections and perhaps acreage planting. In good conditions plants grow very strongly and the initial growth from the seed can be quite amazing. The species is very cold sensitive and suitable only for tropical regions and perhaps coastal areas of the subtropics. It needs well drained soils and a strong tree for support. Propagate from seed which has a very hard coat which must be sliced prior to sowing.

Entada pursaetha DC.
(an overseas name)
Qld
20-30 m tall   Oct-Dec

Large, vigorous **climber**; **leaves** bipinnate, of 1-2 pairs of pinnae, the rhachis ending in a forked tendril; **pinnae** 7-12 cm long, with 6-10 leaflets; **leaflets** 2.5-9 cm x 1-1.4 cm, elliptical to obovate, bright green, the apex notched; **spikes** 7-23 cm long, axillary or clustered on lateral branches; **flowers** about 0.8 cm long, cream to greenish yellow; **pods** 0.5-2 m x 7-15 cm, woody, straight; **seeds** 5-5.7 cm x 3.5-5 cm, thick, brown.

A widespread vine found in Africa and Asia, and extending to north-eastern Qld where it grows in riverine rainforests. It grows into a huge liane, and is far too big for cultivation except perhaps in acreage planting or for botanical collections. Plants are very cold sensitive and the species is suitable only for tropical regions. Requires well drained soil and a strong support. In suitable conditions plants can grow with astonishing speed. Propagate from the large seed which needs slicing or filing prior to sowing.

Entada scandens Benth. = *E. phaseoloides* (L.)Merr.

ENTEROPOGON Nees
(from the Greek *enteros*, a gut; *pogon*, a beard; an apparent reference to the ciliate ligule)
Poaceae

Perennial **grasses** forming tussocks; **leaves** flat, glabrous; **culms** unbranched; **inflorescence** a digitate panicle; **spikes** narrow, rigid, spreading; **spikelets** compressed, with one bisexual floret and 1-2 asexual florets; **lemmas** awned.

A small genus of about 9 species of grasses with 4 found in Australia. One species is grown on a limited scale. Propagate from seed which may need storage for up to 12 months before sowing or by division of the clumps.

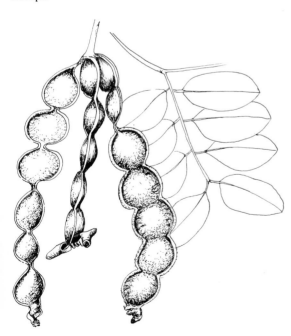

Entada phaseoloides leaves and fruit, × ·06

Enteropogon acicularis (Lindley) Lazarides
(stiff, slender and pointed)
Qld, NSW, Vic, SA, WA, NT   Spider Grass;
                                   Umbrella Grass
0.4-1 m tall   Aug-Dec

Perennial **grass** forming tussocks; **leaves** 5-20 cm x 0.2-0.4 cm, linear, acuminate, flat, often curled in the tussock; **culms** 40-100 cm long, erect, tough; **inflorescence** of 2-14 spikes radiating from the end of the culm; **spikes** 4-20 cm long, rigid; **spikelets** compressed, purplish; **lemmas** awned.

A widespread useful forage grass found in a variety of open situations in drier regions, but growing strongly in moist depressions and along channels etc. Plants are quite ornamental when in flower and are an attractive addition to a rockery or garden. They require a sunny situation in well drained soil. Tussocks can be rejuvenated by burning or cutting back. Propagate from seed or by division.

ENTOLASIA Stapf
(from the Greek *entos*, inside; within; *lasios*, shaggy or woolly; possibly a reference to the hairy lemmas)
Poaceae

Perennial **grasses** forming straggly tufts; **leaves** flat, basal and scattered up the stems; **culms** erect, wiry, freely branched; **inflorescence** a panicle or spike; **spikelets** deciduous at maturity; **lemmas** bearing silky hairs.

A small genus of about 5 species with 3 found in Australia. They are straggly grasses with little appeal except for enthusiasts. Propagate from seed which may need storage before sowing and perhaps from aerial nodes.

Entolasia stricta (R. Br.) Hughes
(stiff and upright)
Qld, NSW
20-80 cm tall   Aug-Oct; also sporadic

Perennial **grass** forming loose, straggling tufts; **leaves** 0.5-10 cm x 0.1-0.7 cm, flat or incurved, basal and scattered up the culms; **culms** 20-80 cm tall, erect or straggly, slender and wiry, branched freely; **panicle** 2-10 cm long, narrow, terminal; **spikelets** green or purplish.

A straggly grass with very limited ornamental appeal. It is widespread in open forest, on shallow sandy or stony soils. Plants resent disturbance but the species can be propagated by aerial sections, some of which form roots.

Epacridaceae R. Br.
(Heath Family)

A family of dicotyledons consisting of about 400 species in 30 genera, mostly developed in Australia and Indomalaysia. They parallel the development of the family Ericaceae and are mainly small shrubs, plus a few small trees. They frequently grow in infertile sandy soils which may be boggy. When they dominate the vegetation, the community is known as heathland. The leaves are usually small and rigid, and the flowers are tubular, often showy and nectariferous. The family is extremely well developed in Australia with some 340

Epacris

species in 28 genera, the vast majority of which are endemic. These are mainly to be found in temperate areas, where they occur in a variety of vegetation types. They are particularly well developed in coastal heathlands and sandplain communities. A unique group has developed in the highland and subalpine regions of NSW, Vic and Tas. The family is much less conspicuous in tropical regions and nearly absent in the arid zones. One species, *Trochocarpa laurina*, is a small tree in rainforests. Many of the smaller heaths have attractive colourful flowers which produce an impressive display, frequently in the winter months before the majority of wildflowers are out. These heaths are popular subjects for garden or pot culture. One species, *Epacris impressa*, is the floral emblem of Vic.

The family is divided into 2 subfamilies, one of which has 6 tribes, all represented in Australia. The placement of the genus *Wittsteinia* is controversial and may be included in this family or Ericaceae. Main Australian genera grown: *Acrotriche, Andersonia, Archeria, Astroloma, Brachyloma, Choristemon, Coleanthera, Conostephium, Cosmelia, Cyathodes, Dracophyllum, Epacris, Leucopogon, Lissanthe, Lysinema, Melichrus, Monotoca, Needhamia, Oligarrhena, Pentachondra, Prionotes, Richea, Rupicola, Sphenotoma, Sprengelia, Styphelia, Trochocarpa, Wittsteinia* and *Woollsia*.

EPACRIS Cav.

(from the Greek *epi*, upon; *acris*, a summit; refers to elevated habitats of some species)
Epacridaceae

Heath-like **shrubs**; **leaves** sessile or petiolate, jointed, alternate; **flowers** long and tubular or campanulate, solitary, axillary, on short peduncles or almost sessile, in upper axils or along branches; **corolla** 5-lobed, white, pink or red; **style** inserted in a central tubular depression; **sepals** and **bracts** overlapping on the outside of the corolla; **fruit** a capsule, splitting along the midrib of the carpels to release very fine seed.

A genus of approximately 40 species. Their representation in Australia is restricted to the eastern and south-eastern regions, with distribution in south-eastern Qld, NSW, Vic, Tas and SA, where a total of about 35 species occur. Epacris also occur in New Caledonia and New Zealand.

In Australia they grow naturally in a diverse range of habitats. They are found from near sea level to varying elevations including Mount Kosciusko, Australia's tallest mountain. Epacris occur on many soil types from moist swampy sands through to dry, gravelly and rocky hillsides. The vernacular name of heath is commonly used as many species are amongst the predominant plants of coastal and alpine heathland communities. The tubular flowers undoubtedly reminded the early settlers of heather.

Epacris have always been popular garden plants because of their massive floral displays. Many species were introduced to England in the early 1800s, eg. *E. longiflora, E. obtusifolia, E. pulchella* and *E. purpurascens*. Many hybrids using forms of *E. impressa* were introduced, but these have long since disappeared from cultivation.

All species are worthy of cultivation. Some such as

E. longiflora are adaptable to a wide range of conditions, whereas *E. impressa* has proved to be short-lived in some situations. A form known as *E. impressa* 'Bega', with orange-red flowers which appear for extended periods, is one of the hardiest forms of this species grown to date. Nurserymen and enthusiasts are selecting interesting and hardy forms of *E. impressa*, and hope that plants will prove to be more reliable in cultivation in the future.

In general, most epacris grow best in relatively well drained soil, whether it is sandy or heavy loam. They appreciate an overhead canopy that provides dappled shade, but will tolerate a fair amount of sun. They can withstand very wet soils, but prefer moist soils for most of the year.

All species are tolerant of frosts, although soft young growth may be damaged by heavy frosts. Plants respond very well to hard pruning after flowering, as some can become straggly in cultivation. Tip pruning from the time of planting helps to promote bushy growth.

Epacris are suited to cultivation in containers. If grown in this way, regular pruning is recommended.

Most species react favourably to light applications of slow-release fertilizer, but this is not usually required once plants are established.

The most common pest is scale, with resultant sooty mould giving the branches a dull blackish appearance. Applications of white oil should control scale. On some occasions plants may be subject to attacks by leaf-eating caterpillars or wood borers.
Propagation

Plants can be propagated from seed or cuttings.

Seed is not usually available commercially. To gather the fine seed, the ripening capsules need to be kept under regular observation and collected just when beginning to split in the warm weather. They should then be placed in a paper bag or dry, open container. When the seed is released it can be stored in an airtight container.

There is some evidence that the seed has a dormancy period and that best results are obtained if it is stored in a cool, dark place for 3-6 months before sowing. Further experimentation is needed in this area.

Plants are not commonly grown from seed, but good

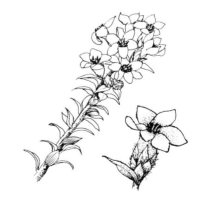

Epacris barbata, × ·75   flower, × 2·2

success can be gained by using the bog method or capillary beds. See Volume 1, page 204 for further information.

The seed is very fine, and must be spread thinly over the surface of the propagation mixture. If the medium is kept constantly moist, germination should occur within 10-20 weeks.

During the 18th century in England, freshly ripened seed was successfully germinated on a sandy medium under a bell jar. In Vic seed has germinated extremely well in containers that have stock plants of *Epacris impressa* var. *grandiflora* growing in them. The containers were watered regularly by drip irrigation, thus keeping the coarse granite sand mulch virtually constantly moist. The result was a lawn-like mass of seedlings.

Plants are more commonly propagated from stem cuttings. Difficulty is experienced with cuttings that are too woody. The best type of material to use is young growth that has just become firm. Branchlets that appear too soft and floppy are usually better than the harder wood. The application of rooting hormones is helpful in promoting stronger root systems.

Epacris acuminata Benth.
(a long, sharp point)
Tas
0.5-1 m x 0.3-0.6 m   Aug-Nov

Dwarf **shrub**; **branches** many, usually upright; **branchlets** short and erect; **leaves** 0.5-1 cm long, broadly ovate, subsessile, concave, erect, pungent apex, base rounded or cordate; **flowers** tubular, with spreading lobes, to 0.8 cm across, white, solitary, in upper axils, forming short leafy spikes; **sepals** and **bracts** often pinkish.

Scattered throughout Tas, from sea level to about 1000 m alt. Not well known in cultivation. It is an attractive species, with the combination of pink buds and white flowers. Best suited to moist, well drained, light to medium soils, with dappled shade or partial sun. Frost tolerant. Propagate from seed or cuttings.

Epacris breviflora, × ·65   leaf and flower, × 1·3

Epacris apiculata Cunn.
(short point)
NSW
0.5-2 m x 0.3-1 m   Oct-Dec

Dwarf to small, erect **shrub**; **branches** erect; **branchlets** few; **leaves** to 1 cm x 0.4 cm, cordate-ovate, crowded, pointed apex; **flowers** tubular, with spreading lobes, to 0.8 cm across, white, as long as or shorter than leaves, almost sessile, solitary, in axils along the branches.

A dweller of sandstone regions west of Sydney and in the Blue Mountains. Best suited to moist, well drained light soils, with dappled shade to partial sun. Frost tolerant. Propagate from seed or cuttings.

There has been some confusion regarding a species sold as *E.* species 'Cabramurra'. At one stage it was thought to be a form of *E. apiculata*, but is now recognized as a form of *E. breviflora*.

Epacris barbata Melville
(bearded)
Tas
0.6-1.2 m x 0.5-1 m   Sept-Nov

Dwarf to small **shrub**; **branches** many, usually erect; **branchlets** hairy; **leaves** to 1 cm x 0.5 cm, elliptical-ovate, short and broad stalk, flat, dark green, shiny, pungent apex; **flowers** tubular, with pointed, spreading lobes, about 1.2 cm across, white, solitary, in upper axils, often forming dense terminal heads; **sepals** and **bracts** often tinged red and with prominent, long silky hairs; dark **anthers** slightly protruding.

A very showy, white-flowered epacris. It occurs in the sandy coastal heaths of the Freycinet Peninsula region and has horticultural potential. Best suited to moist, sandy, well drained soils. May adapt to heavier soil types, provided drainage is excellent. A location with dappled shade or partial sun is suitable. May tolerate full sun. Propagate from seed or cuttings.

Epacris bawbawensis Stapf = *E. paludosa* R. Br.

Epacris breviflora Stapf
(short flowers)
Qld, NSW, Vic, Tas   Drumstick Heath
1-2.5 m x 0.5-2 m   Sept-Jan; also sporadic

Small to medium, erect **shrub**; **branches** erect; **branchlets** hairy; **leaves** to 1.5 cm x 0.5-0.8 cm, broadly ovate to elliptic-ovate, ending in pungent tip, spreading or reflexed, midrib prominent, glabrous, margins minutely toothed; **flowers** with spreading lobes, to 1 cm x about 1 cm, white, borne in a terminal leafy spike, sweetly scented.

A dweller of regions beside creeks and swamps, at low and high altitudes. It is not well known in cultivation, although it has been grown for over 20 years as *E.* species 'Cabramurra'. Prefers moist but well drained, light to medium soils that have a high organic content. Will grow in dappled shade through to partial sun. Tolerates full sun in cool temperate climates. Is an excellent container plant but needs regular pruning to maintain bushy growth. Propagate from cuttings of young growth.

Self generation of seedlings has occurred in containers watered by trickle irrigation, so trials from seed

Epacris calvertiana, × ·6   leaf and flower, × 1·2

using the bog or capillary method may be worthwhile (see Volume 1, page 204).

There is some evidence that this species is a hybrid, with *E. petrophila* as one parent.

Plants available in the nursery trade as *E. apiculata* are most likely referable to *E. breviflora*.

Epacris calvertiana F. Muell.
(after Mrs J. S. Calvert, 19th-century botanical artist and collector)
NSW
1-2 m x 0.6-1.5 m   June-Dec

Small, open to upright **shrub**; **branches** erect; **branchlets** brown, hairy; **leaves** 1-2 cm x about 0.3 cm, lanceolate, concave, glabrous, erect to spreading, crowded, ending in a pungent point; **flowers** tubular, to 1.5 cm x 0.2-0.4 cm, cream to white, slightly swollen, in upper axils; **sepals** acute, about half as long as corolla.

This species is from the Blue Mountains and Woronora Plateau. It generally occurs at the base of sandstone cliffs, in protected situations which have moist soil for most of the year. In cultivation it requires very well drained, moist soil, with semi-shade or filtered sunlight. Should tolerate light to medium frosts. Pruning after flowering will promote bushy growth. Propagate from cuttings. The var. *versicolor* Maiden & Betche has white-lobed, pink flowers.

Epacris coriacea Cunn. ex DC.
(leathery)
NSW
0.5-3 m x 1-2 m   Sept-Nov

Dwarf to medium **shrub**; **branches** procumbent to erect; **leaves** 0.4-1 cm long, obovate to broadly ovate, with very short stalk, concave, thick, smooth, blunt

apex; **flowers** campanulate, about 0.3 cm x 0.6 cm, white, borne in axils along the branchlets.

A variable species, with some spreading and some erect forms. It occurs on the Woronora Plateau, where it often inhabits sandstone cliffs. Evidently not well known in cultivation. It requires very well drained soils, and should do best in a situation with dappled shade or partial sun. Frost tolerant. Propagate from cuttings.

Epacris corymbiflora Hook.f.
(flowers in corymbs)
Tas
0.1-0.5 m x 0.3-1 m   Oct-Jan

Dwarf **shrub**; **branches** erect or spreading; **leaves** about 0.4 cm x 0.2 cm, broadly elliptical, subsessile, erect, overlapping, thick, blunt, with sharp incurved tip; **flowers** tubular, with spreading to reflexed lobes, about 0.6 cm long, with prominent brown floral bracts and sepals, borne in terminal corymb.

An outstanding dwarf species which is well represented in wet peaty heaths near the west and south-west coast. Although not well known in cultivation, it shows horticultural potential with its growth habit and well displayed flower-heads. Its requirements are for moist, light to medium soils, rich in organic matter, and a situation with dappled shade to partial sun. Could do very well as a container plant. Pruning is likely to promote bushy growth. Frost hardy. Propagate from cuttings.

Previously known as *E. myrtifolia* var. *corymbiflora* (Hook.f.) Rodway.

Epacris crassifolia R. Br.
(thick leaves)
NSW
0.1-0.5 m x 0.5-1 m   Nov-Jan

Dwarf **shrub**; **branches** slender and trailing or stout and erect; **leaves** 0.6-1.5 cm long, obovate to oblance-olate, blunt, petiolate, margins thickened and sometimes ciliate; **flowers** tubular, variable, to 1 cm long, white, axillary along the branches.

This most variable species is not well known in cultivation. Plants from the Blue Mountains have slender trailing branches and small flowers. This form could be extremely decorative amongst other plants and cascading over boulders.

Those from the coast and near the Blue Mountains have stouter branches and are more upright. Cultivation requirements are for well drained but moist, light to medium soils. Dappled shade to partial sun should help to provide good growth. This species, like many other epacris, has excellent potential for container cultivation. Should be hardy to light to medium frosts. Propagate from cuttings.

Epacris exserta R. Br.
(protruding; in reference to the anthers and stigma)
Tas
0.5-1 m x 0.3-1 m   Sept-Nov

Dwarf **shrub**; **branches** more or less erect, often long and slender; **branchlets** hairy, can be short and crowded; **leaves** 0.6-0.8 cm x up to 0.3 cm, elliptical-lanceolate to obovate, erect to spreading, thick, blunt

410

Epacris corymbiflora   T.L. Blake

diameter, white, solitary, in axils near ends of branchlets, forming leafy clusters.

A small showy species from the Mount Kosciusko region of NSW and the Bogong High Plains area in Vic, where it grows in moist, peaty soils. Has potential for cultivation because of its habit of producing adventitious roots from the branches. Should be suitable for moist but well drained, light soils which are relatively rich in organic matter. Will grow in full sun if root area receives some protection, but dappled shade or partial sun will be more to its liking. Potential as a container plant is worth investigating. Propagate from cuttings of new growth. Aerial layering should also be successful due to the tendency to self layer.

Epacris gunnii Hook.f.
(after Ronald Gunn, 19th-century botanical collector, Tas)
Tas
0.3-1.5 m x 0.3-1 m   Aug-Nov
Dwarf to small **shrub**; **branches** slender, erect; **leaves** 0.3-0.6 cm x 0.2-0.4 cm, broadly ovate, spreading to recurved, cordate base, acuminate, pungent apex, densely crowded; **flowers** tubular, with spreading lobes, about 0.7 cm across, white, subsessile, solitary, in the upper axils of branchlets, forming dense leafy spikes.

This decorative species occurs from sea level to about 1220 m alt. in wet heathlands. Has had limited but successful cultivation. Best suited to moist, light to medium soils, with dappled shade or partial sun. Frost tolerant. Has potential as a container plant. Pruning after flowering produces bushy growth. Propagate from seed or cuttings.

Previously known as *E. microphylla* var. *gunnii* (Hook.f.) Benth.

Epacris hamiltonii Maiden & Betche
(after Alexander G. H. Hamilton, botanist, NSW)
NSW   Hamilton's Heath
0.3-1 m x 0.5-1.5 m   July-Oct; also sporadic
Dwarf **shrub**; **branches** long, spreading, crowded, can be intricate, hairy; **branchlets** with dense covering of soft whitish hairs; **leaves** to 1 cm long, broad-lanceolate, slightly hairy margins; **flowers** campanulate, about 1 cm x 0.4 cm, white, lobes spreading; **style** longer than corolla.

A rare species from near Blackheath in the Blue Mountains, where it grows on sandstone as an undershrub. It may not be in cultivation. Should be suited to moist but well drained, light to medium soils, with dappled shade to partial sun. Frost tolerant. Propagate from cuttings.

Epacris heteronema Labill.
(unequal filaments)
NSW, Vic, Tas
0.3-6 m x 0.3-2 m   Sept-Nov
Dwarf to small or tall **shrub** to small, slender **tree**; **branches** erect; **leaves** 0.6-1.2 cm long, variable, broadly ovate-cordate to narrow-ovate, thick, spreading to erect, broad stalks, margins translucent and finely toothed, pungent apex, crowded; **flowers** tubular, about 0.7 cm long, white, solitary, axillary,

apex, undersurface slightly keeled or 3-5 ribbed; **flowers** campanulate, about 0.7 cm x about same across, white, prominent anthers, axillary, scattered along branches and in leafy clusters at ends of branchlets.

From the north of Tas, where it is usually found growing in cool, moist soils, often on creek and riverbanks. Has had only limited cultivation, but plants have grown well. Requirements are for moist, well drained soils, with dappled shade to partial sun. Prune for bushy growth.

An undescribed variety of the species occurs on the west coast. It differs in its shorter, orbicular leaves and glabrous branchlets.

Propagate both forms from cuttings.

Epacris franklinii Hook.f. = *E. mucronulata* R. Br.

Epacris glacialis (F. Muell.) M. Gray
(of the ice)
NSW, Vic
0.1-0.5 m x 0.5-1 m   Oct-Jan
Dwarf **shrub**; **branches** prostrate to decumbent, often with adventitious roots near the base; **branchlets** becoming glabrous; **leaves** to 0.4 cm x 0.3 cm, rhomboid to obovate-rhomboid, flat or slightly concave, thick, leathery; **flowers** campanulate, about 1 cm

Epacris impressa

Epacris impressa   W.R. Elliot

borne in upper axils, forming a dense terminal head, profuse; **style** shorter than flower.

This species is most variable. It can be a dwarf compact shrub, but on the west coast of Tas it can reach 6 m in height. Such tall plants must be extremely old. It nearly always occurs on swampy ground, whether near sea level or at an alt. of up to about 1200 m.

This species was originally introduced to cultivation in England during 1823. Plants require moist, light to medium soils that are relatively well drained. Will grow in dappled shade to partial sun. Needs root protection if grown in full sun. Should respond well to pruning. Propagate from cuttings.

Has affinities to *E. paludosa*, which has the style protruding from the flowers.

Epacris impressa Labill.
(with impressions)
NSW, Vic, Tas, SA   Common Heath
0.3-2.5 m x 0.2-1 m   Feb-Nov; also sporadic

Dwarf to medium, often spindly **shrub**; **branches** usually erect; **branchlets** usually hairy, slender to thick; **leaves** 0.8-1.5 cm x 0.2-0.6 cm, linear-lanceolate to ovate-lanceolate, sessile, narrowing to a pungent point, hairy or glabrous; **flowers** tubular, to 2 cm x 0.5 cm, white or varying shades of pink and red, sometimes a combination of white and pink, 5 distinct indentations at tube-base, axillary, along branches, forming a long leafy spike, or near ends of branchlets, profuse.

An outstanding species found in coastal heathland and forested regions, where it often grows in extensive sparse colonies and provides an abundance of flowers in a wide array of colours. It is a very popular native plant, and is the Victorian floral emblem. It may flower for a long period, but it is notable for the display produced during the winter months. The species was introduced to cultivation in England, where it was extremely popular, prior to 1826. Many hybrids were raised and named, but have since disappeared. See Epacris Cultivars, page

In general, plants need well drained, light to medium soils, but they will also tolerate heavy soils that do not become·waterlogged. A situation with semi-shade, dappled shade or partial sun is suitable. Plants will grow in full sun, provided the root area receives protection, such as can be given by other plants. Frost tolerant. Will withstand limited periods of dryness once established. Ideally suited to massed groups in informal drifts in the garden. The use of different colour shades offers potential.

Once plants are established, heavy pruning after flowering encourages bushy growth. Very successfully grown as a container plant.

The species is renowned for its attraction to nectar-feeding birds such as honey-eaters and lorikeets.

412

Epacris impressa var. *grandiflora* double-flowered form   W.R. Elliot

Epacris lanuginosa   T.L. Blake

Propagate from seed which must be sown while fresh. The bog and capillary bed methods can give good results (see Volume 1, page 204). Cuttings of firm young growth usually form roots readily.

E. impressa is extremely variable, and selected forms from several regions have been propagated and are available from nurseries. The location is usually added to the name, eg. *E. impressa* 'Bega'. This particular form has orange-red flowers for up to 9 months of the year, and is one of the most adaptable forms for cultivation. There has been some selection of forms on the basis of flowering period, eg. 'Spring Pink' which usually starts flowering during Sept and can last for 2 months.

The var. *grandiflora* Benth. occurs in the Grampians in Vic. It grows in moist, shady situations on sandstone cliffs and in rocky places. It has larger flowers and leaves than the typical form. Forms with double flowers have also been selected for cultivation. Some flowers are composed of up to 10 corollas in a tight arrangement. This variety has proved difficult in cultivation, and plants seem to have a short life span. Soils with extremely good drainage are recommended. Good results have been achieved with container cultivation. Propagate from cuttings.

Epacris lanuginosa Labill.
(very woolly)
NSW, Vic, Tas   Woolly Heath; Woolly-style Heath
0.3-1.5 m x 0.3-1 m   Sept-Feb
   Dwarf to small **shrub**; new growth hairy; **branches** erect, slender, arising from near base of plant; **branchlets** hairy; **leaves** to 1.2 cm x about 0.2 cm, narrow-lanceolate, tapering to a pungent point, crowded, usually imbricate; **flowers** tubular, about 0.8 cm long, white, spreading to erect, prominent spreading lobes, profuse in upper axils.

This showy white-flowered species is a dweller of wet heathlands, from coastal situations to about 100 m alt. Not commonly cultivated, it was first grown in England during the 19th century. It requires moist but well drained soils. Dappled shade or partial sun is suitable.

Will tolerate full sun, provided its root system is protected in some way, eg. by logs, rocks or other plants. Frost tolerant. Prune after flowering. Suited to containers. Propagate from cuttings or seed.

Epacris longiflora Cav.
(long flowers)
Qld, NSW   Fuchsia Heath
0.5-2.5 m x 1-2 m   most of year
   Dwarf to medium **shrub**; **branches** usually spreading, often long if unpruned; **branchlets** hairy; **leaves** 0.5-1.2 cm x 0.3-0.6 cm, ovate-cordate to lanceolate, spreading to reflexed, tapering to a fine point, glabrous; **flowers** tubular, 1.2-4 cm x about 0.4 cm, red with cream or white tips, pendulous along branches.

A most ornamental species, particularly well known on the sandstone regions of the coast and plateau around Sydney, where it grows in the heathlands and dry sclerophyll forests. It is also known from the north coast and a few mountains in the border range. Adapts

413

Epacris marginata

very well to general garden cultivation or to containers. The commonly cultivated form has red bells with white tips, but there are others grown including an all-white form which is known as *E. longiflora* 'White Sport'. The form from the mountains of the Qld-NSW border does not have pendulous flowers; they are held stiffly and borne all around the stems. It is not as attractive as the Sydney form, and is also somewhat more difficult to grow. Plants of the commonly grown form can be left to develop long, sometimes arching branches, from which the flowers hang in row formation. If desired, hard pruning will promote bushy growth, with flowers being produced throughout the plant. Does best in well drained soils, preferring light or medium types, but will grow well in heavy soils, provided the drainage is excellent. Suited to semi-shade, dappled shade or partial sun. There are some reports of it growing well in full shade. Frost tolerant. Cultivated in England by 1803 (see Epacris Cultivars entry, Page 419). Propagate from cuttings using firm but young material for best results.

Epacris marginata Melville
(thickened leaf-margins)
Tas
0.3-1 m x 0.3-0.6 m   Sept-Dec

Dwarf, erect **shrub**; **branches** initially hairy, rough at maturity; **leaves** 0.6-1.2 cm x about 0.3 cm, ovate-cordate, initially erect, becoming spreading to recurved, acuminate, pungent apex, green to somewhat glaucous, crowded, often overlapping; **flowers** tubular, with spreading lobes, about 1 cm across, white, solitary, in upper axils or along branchlets, forming leafy spikes, honey-scented; **style** slightly longer than ovary.

From the Tasman Peninsula in the south of Tas, this species is little-known in cultivation. Should be suited to moist to well drained, light to medium soils, with dappled shade or partial sun. Withstands limited coastal exposure. Propagate from seed or cuttings.

Epacris longiflora   W.R. Elliot

Epacris microphylla R. Br.
(small leaves)
Qld, NSW, Vic, Tas   Coral Heath
0.5-1 m x 0.5-1 m   April-Oct

Dwarf **shrub**; **branches** erect, short or long; **branchlets** hairy; **leaves** to 0.5 cm x about 0.3 cm, ovate-cordate, more or less sessile, very concave, tapering to a fine prickly point, crowded, often imbricate; **flowers** campanulate, about 0.2 cm x 0.4 cm, white, sometimes tinged with pink, buds often pink, profuse in upper axils.

A commonly cultivated species which has proved hardy in a wide range of soil conditions. In nature it inhabits sandy, peaty soils that are subject to waterlogging in winter and then remain moist for most of the year. In coastal heathlands it sometimes grows in extensive colonies. It copes well with heavy soils, provided drainage is adequate. Best planted in a situation that has dappled shade or partial sun; however, it is tolerant of full sun if there is protection for its root area. It likes to grow amongst other small plants. Frost and snow tolerant. Grows well in containers, and responds well to heavy pruning. Propagate from seed or cuttings using firm young growth.

The var. *rhombifolia* L. Fraser & Vick. from the Northern and Central Tablelands of NSW is distinguished by its rhomboid leaves.

Epacris marginata, × ·6   leaf and flower, × 1·8

414

Epacris microphylla   T.L. Blake

Epacris microphylla R. Br. var. **gunnii** (Hook.f.)
Benth = *E. gunnii* Hook.f.

Epacris mucronulata R. Br.
(small sharp points)
Tas
0.6-2 m x 0.6-1.5 m   Oct-Dec
    Dwarf to small **shrub**; **branches** erect, often parallel,
many; **leaves** 0.8-1.2 cm x 0.1-0.2 cm, narrow-elliptical, spreading to recurved, tapering to a fine point,
margins minutely serrulate; **flowers** tubular, about
1 cm long, white, prominent spreading lobes, in upper
axils, on slender stalks.
    An ornamental species that seems to be confined to
stream banks on the west coast of Tas. It may not be in
cultivation, but it has potential because of the well
displayed flowers. Probably will grow best in moist but
well drained, light to medium soils, in a position with
dappled shade or partial sun. Frost tolerant. Propagate
from cuttings of firm young growth.

Epacris muelleri Sonder
(after F. von Mueller, first Government Botanist, Vic)
NSW   Mueller's Heath
0.2-0.5 m x 0.5-1 m   Dec-Feb
    Dwarf **shrub**; **branches** can be slender and
spreading or rigid and erect; **leaves** 0.2-0.5 cm, ovate or

elliptical to lanceolate, almost sessile, thick, blunt;
flowers campanulate, about 0.4 cm long, white, small,
blunt, spreading lobes, borne in upper axils.
    Evidently not in cultivation, this variable species is
very similar to *E. rigida*, which is usually taller and
produces its flowers during Sept-Nov. It occurs in the
Blue Mountains, where it grows in damp, sheltered
locations on sandstone. Will need moist but very well
drained soils, with semi-shade or dappled shade. It
may prove to be tolerant of partial sun. Frost tolerant.
Forms with slender, spreading branches seem to show
potential as container plants. Propagate from cuttings.

Epacris myrtifolia Labill.
(leaves similar to those of the genus *Myrtus*)
Tas   Myrtle Heath
0.3-1 m x 0.3-1 m   Oct-Dec
    Dwarf **shrub**; **branches** erect; **branchlets** hairy;
leaves to 1 cm x 0.5 cm, ovate to orbicular, petiolate,
somewhat imbricate, thick, leathery, apex often blunt
or slightly pointed; **flowers** campanulate, about 0.5 cm
x 0.7 cm, white, prominent spreading lobes, borne in
upper axils; **style** very short, only anthers exserted.
    This species occurs in coastal regions of south and
south-eastern Tas. To date it has had limited cultivation, and results indicate that it is worthy of further
trial. Should adapt to most well drained soils. Would
probably do best in a situation with partial sun. Withstands fairly extreme coastal exposure, and is frost
tolerant. Propagate from cuttings of firm new growth.

Epacris myrtifolia Labill. var. **corymbiflora** (Hook.f.)
Rodway = *E. corymbiflora* Hook.f.

Epacris obtusifolia Smith
(blunt leaves)
Qld, NSW, Vic, Tas, ?SA   Blunt-leaf Heath
1-2 m x 0.6-1.5 m   June-Dec
    Small **shrub**; **branches** few, erect, often slender;

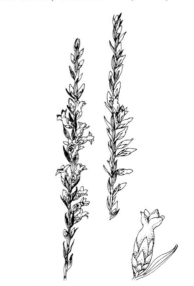

Epacris obtusifolia, × ·55   flower, × 1·6

415

Epacris paludosa

branchlets hairy; **leaves** 0.5-1.2 cm x 0.2-0.3 cm, elliptical to oblanceolate, erect, often imbricate, short and broad petiole, blunt apex; **flowers** tubular, about 0.8 cm long, white to cream, perfumed, axillary, often arranged on one side of branches, strong scent of honey.

An adaptable species that occurs naturally in damp, sandy, peaty soils in heathlands, usually in large colonies. Will grow in most soils, and does not need moisture for most of the year. It prefers dappled shade or partial sun, but will tolerate full sun on the foliage, provided there is root protection. Can keep very narrow. Tip pruning while plants are young is recommended if bushy growth is desired. Also withstands heavy pruning. Introduced to cultivation in England during 1804. Propagate from seed or cuttings of firm young growth which strike readily.

Epacris paludosa R. Br.
(boggy or marshy)
NSW, Vic, Tas   Alpine Heath; Swamp Heath
1-2.5 m x 0.6-2 m   Oct-Jan
Small to medium **shrub**; **branches** many; **branchlets** hairy; **leaves** 0.6-1.2 cm x about 0.3 cm, broadlanceolate, spreading to more or less erect, tapering to a pungent point, thick, prominent veins, margins minutely serrulate, glabrous, crowded; **flowers** tubular, to 0.8 cm x 0.8 cm, white, prominent spreading lobes, borne in small terminal clusters.

In nature this species occurs in mountain forests, subalpine woodlands and alpine heathlands, usually in situations where the soil is sometimes waterlogged, and moist for much of the year. It is found in the Blue Mountains to the north, and in the south it occurs on two peaks on Flinders Island. Cultivation requirements are for moist but well drained, light to medium soils, in a position with dappled shade to partial sun. Withstands heavy pruning. Grows well in containers. Propagate from fresh seed or from cuttings of firm young growth.

Epacris paludosa   T.L. Blake

Epacris petrophila Hook.f.
(rock-loving)
NSW, Vic, Tas   Rock Heath; Snow Heath
0.2-1 m x 0.5-1 m   Dec-March
Dwarf **shrub**; **branches** erect to spreading; **branchlets** slightly hairy, sometimes trailing; **leaves** to 0.5 cm x 0.2 cm, elliptic to obovate-oblong, erect, imbricate, sessile, thick, glabrous, bright green; **flowers** campanulate, about 0.7 cm across, white, borne in axils near ends of branches, forming tight clusters.

A dwarf alpine species which is capable of self layering in nature. Whether this will occur in cultivation is not known. It has potential, with its dense shiny foliage and well displayed flowers. Not well known in cultivation. It appreciates moist, well drained, light to medium soils and a location with dappled shade to partial sun. Frost and snow tolerant. Should grow well in containers. Propagate from seed or from cuttings of firm young growth.

Epacris pulchella Cav.
(beautiful)
Qld, NSW   Coral Heath; Wallum Heath
1-2 m x 1-2 m   Dec-July; Aug-Oct
Small **shrub**; **branches** many, slender; **branchlets** hairy; **leaves** 0.4-0.8 cm long, cordate-ovate, acuminate, glabrous, crowded, nearly imbricate; **flowers** campanulate, about 0.5 cm long, pink but can be white, profuse along branchlets, forming leafy spikes.

A showy species from the coast and further inland on the Blue Mountains in NSW, extending to Fraser Island in Qld. It occurs in heathlands or dry sclerophyll forests, where it grows on sandstone or granite. Hardy in cultivation. Best grown in well

Epacris paludosa, × ·75   flower, × 3

416

drained, light to medium soils, with dappled shade or partial sun. Frost tolerant. Excellent container plant. Some forms may need regular pruning to promote bushy growth. The coastal forms usually flower during March-July, while those from the Blue Mountains bloom during Dec-Feb. Those from Qld may flower in spring and again in autumn. A pink-flowered form with short, broad leaves is popular in cultivation. The flowers often have up to 10 lobes. Propagate from seed or cuttings of firm young growth.

Epacris purpurascens R. Br.
(becoming purple)

NSW	Port Jackson Heath
1-2 m x 0.6-1.5 m	Sept-Nov

Small **shrub**; **branches** erect, long, few; **leaves** 0.8-1.5 cm long, ovate, sometimes cordate at base, tapering to a fine pungent point, concave, spreading or curved upwards, crowded; **flowers** campanulate, to 0.6 cm x 1 cm across, pink, deepening with age, axillary, forming leafy spikes.

A showy species from around Sydney to the Blue Mountains. Found on sandstone and shale, usually where soils are moist for most of the year. It has had limited cultivation in this country, although it was introduced to England during 1803. Well drained, light to medium soils, with dappled shade or partial sun should provide conditions for good growth. Can be pruned hard if desired. Frost tolerant.

The var. *onosmiflora* Maiden & Betche differs in having white flowers. It is found only in the Blue Mountains, where it grows in soils that are wet for extended periods. This variety is more common in cultivation than the typical form.

Propagate both forms from cuttings of firm new growth or from seed.

Epacris reclinata Cunn. ex Benth.
(bent back)

NSW	
0.6-1 m x 0.6-1 m	April-Sept

Dwarf **shrub**; **branches** erect, hairy; **branchlets** often reddish; **leaves** 0.4-0.6 cm x 0.2-0.3 cm, ovate-cordate, dark green, spreading to reflexed, acute, ciliate, shiny above; **flowers** tubular, 0.8-1.5 cm long, pink to red, pendulous, with spreading but not reflexed lobes, borne in upper axils, profuse.

A most ornamental species from moist, sheltered situations on sandstone in the higher areas of the Blue Mountains. A form with reddish pink flowers is the most popular in cultivation. It is well known for its long flowering period. Requires well drained, light to medium soils, with dappled shade through to partial sun. Frost tolerant. Tip pruning while young is beneficial in producing bushy growth. It also responds well to hard pruning after flowering. Grows well amongst other plants and is an excellent container plant. Eagerly visited by honey-eating birds. Propagate from seed or cuttings using firm new growth for best results.

Epacris rigida Sieber ex Sprengel
(rigid; stiff)

NSW	Keeled Heath
0.2-1.5 m x 0.3-1 m	Sept-Nov

Dwarf to small **shrub**; **branches** erect, many; **leaves** to 0.5 cm long, ovate or elliptical to lanceolate, almost sessile, thick, blunt, keeled below; **flowers** campanulate, about 0.3 cm x up to 1 cm, white, with blunt, spreading lobes, on very short stalks, borne in upper axils.

An inhabitant of the Blue Mountains, where it grows in open situations on sandstone. Although recorded as grown in England, it is evidently not in cultivation in Australia. As with many of the species that occur in the Blue Mountains, it is likely to be best suited to moist but well drained soils. A position with dappled shade to partial sun, or full sun with protection provided for the root area, is needed. Propagate from seed or cuttings of firm young growth.

Epacris rigida var. laxa Benth. = *E. muelleri* Sonder

Epacris robusta Benth.
(robust)

NSW	Round-leaf Heath
1-2 m x 0.6-1.5 m	Sept-Nov

Small **shrub**; **branches** erect, rigid, stout; **branchlets** slightly hairy; **leaves** about 0.7 cm long, oval to orbicular, spreading, flat, blunt, thick; **flowers** tubular, to 0.7 cm long, white to cream, spicy-scented, borne in terminal clusters.

Evidently not known in cultivation, this robust shrub occurs amongst granite rocks in south-eastern NSW. It will need very well drained, light to medium soils, and should grow in a situation with dappled shade to partial sun. Propagate from seed or cuttings of firm new growth.

Has some affinities to *E. obtusifolia*, which has its flowers along the branches and has much narrower leaves.

Epacris pulchella form, × ·75   flower, × 3

417

Epacris serpyllifolia R. Br.
(leaves similar to those of the genus *Serpyllum* — 'Wild Thyme')
NSW, Vic, Tas   Thyme Heath
0.2-1.2 m x 0.3-1 m   Sept-Dec
     Dwarf to small **shrub**; **branches** many, either spreading or erect; **branchlets** hairy; **leaves** 0.2-0.6 cm x 0.2-0.3 cm, ovate, pointed, thick, spreading, crowded, dark green but often with reddish tonings during autumn and winter; **flowers** tubular, with spreading lobes, about 0.8 cm x 0.6 cm, white, borne in crowded terminal clusters; **bracts** and **sepals** brownish.
     This variable heath occurs at elevations of over 900 m, in moist, infertile soils. In exposed situations it can form a dense mat but, with shelter, grows as a bushy shrub. Has had limited cultivation in Australia, but is recorded as being grown in England. Should be suitable for moist but well drained, light to medium soils, with dappled shade or partial sun. Withstands pruning, and is frost and snow tolerant. Propagate from seed or cuttings of firm young growth.

Epacris squarrosa Hook.f. =
            E. tasmanica W. M. Curtis

Epacris stuartii Stapf
(after Charles Stuart, 19th-century botanical collector)
Tas   Stuart's Heath
0.3-1 m x 0.3-0.8 m   April-June
     Dwarf **shrub**; **branches** many, mainly erect; **leaves** 0.3-0.8 cm x 0.2-0.5 cm, broadly ovate, mainly spreading, tapering to a fine but firm point, slightly convex, glabrous, can be shiny; **flowers** campanulate, about 0.7 cm x 1 cm, white, prominent spreading lobes, in upper axils or along branches, profuse; **bracts** and **sepals** often pink.
     This species is recorded as being one of the hardiest

Tas endemics. It is very rare in nature, known to occur only on an exposed rocky headland called Southport Bluff, at the entrance to D'Entrecasteaux Channel. It is almost certain that plants will grow taller than 1 m in sheltered areas. Has recently been introduced to cultivation. Requirements are for very well drained soils, with dappled shade to partial sun. If grown in full sun, protection for the root area must be provided. Frost and salt spray tolerant. Propagate from seed or cuttings.

Epacris tasmanica W. M. Curtis
(from Tasmania)
Tas
0.3-1 m x 0.2-0.6 m   June-Oct
     Dwarf, erect **shrub**; **branches** slender, erect, few; **branchlets** few, short; **leaves** to 0.8 cm x 0.3 cm, broadly lanceolate to broadly ovate, sometimes orbicular, more or less flat, erect to spreading, tapering to pungent apex; **flowers** tubular, with spreading lobes, about 1 cm across, solitary, mainly in upper axils; **floral bracts** and **sepals** often reddish.
     This variable but decorative epacris is widespread and abundant in the south and east of Tas. Has adapted well to cultivation in that state. Little is known of its record in other areas. Requirements are for moist, well drained, light to medium soils, with dappled shade or partial sun. Frost tolerant. Responds well to pruning. Has potential as a container plant. Propagate from seed or from cuttings which strike readily.
     Previously known as *E. squarrosa* Hook.f.

Epacris virgata Hook.f.
(twiggy)
Tas   Tamar Heath
0.5-1.5 m x 0.5-1 m   July-Sept
     Dwarf to small **shrub**; **branches** slender, erect, many; **branchlets** more or less glabrous; **leaves** 0.4-

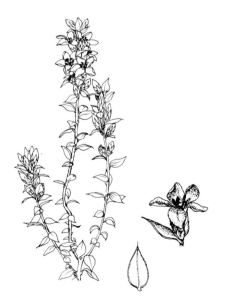

Epacris stuartii, × ·55   leaf and flower, × 1·1   *Epacris tasmanica,* × ·55   leaf and flower, × 1·6

Epacris serpyllifolia

T.L. Blake

1 cm x 0.2-0.3 cm, elliptical-ovate, more or less erect, concave, ending in short, soft point, glabrous; **flowers** campanulate, about 0.8 cm x 0.8 cm, white, slightly spreading lobes, borne in axils along the branches.

Epacris virgata, × ·65

leaf and flower, × 3·2

A very showy white-flowered species that has proved adaptable in cultivation. In nature it is restricted to an area known as Asbestos Hills (also known as Asbestos Range), near Launceston. Grows well in most soils, including fairly heavy clays, but likes moisture for most of the year. Suited to semi-shade through to partial sun. Grows well amongst other fairly dense plants. Frost tolerant. Responds well to hard pruning. Is a good container plant. Propagate from seed or cuttings of firm young growth which strike readily.

EPACRIS CULTIVARS

In England during the middle to late 19th-century epacris cultivars were amongst the most esteemed plants in cultivation. They were very popular as greenhouse subjects and much was written in prominent gardening publications such as 'The Garden', 'Gardener's Chronicle' and 'Floricultural Cabinet' on their selection and cultivation.

Although they were often referred to as epacris hybrids, most were actually selections resulting from cross pollination of selected forms within a species.

By far the most commonly grown were different flower forms of *E. impressa*, which amounted to about 70 named cultivars and some of these are detailed in the following table.

Epacris impressa cultivars considered highly worthy of cultivation in Britain during 19th century:

419

Cultivar	Flowers
E. 'Alba Odorata'	white, highly fragrant
E. 'Ardentissima'	crimson
E. 'Butterfly'	carmine and white
E. 'Candidissima'	white
E. 'Ceraeflorus'	white, dense spikes, 30 cm or more
E. 'Coccinea'	scarlet
E. 'Diadem'*	large, rosy carmine
E. 'Fireball'	deep scarlet
E. 'Fulgens'	red
E. 'Her Majesty'	white
E. 'Hyacinthiflora'	rose
E. 'Ignea'	rose
E. 'Lady Alice Peel'	salmon and white
E. 'Lady Panmure'	white
E. 'Lowi'	red with white tips
E. 'Model'	rose
E. 'Mont Blanc'	white
E. 'Mrs Pym'	rose
E. 'Nivalis'	white
E. 'Princess Beatrice'*	large, pale pink
E. 'Queen Victoria'	white
E. 'Rose Perfection'*	pink
E. 'Rosea'	rose
E. 'Rubella'	purplish rose
E. 'Sunset'	pink to purplish red
E. 'The Bride'	white
E. 'The Premier'	clear rose pink
E. 'Vesta'*	white with pink tips
E. 'Vesuvius'	crimson and scarlet
E. 'Viscountess Hill'	scarlet

*These were introduced in 1884 by Messrs Veitch's Nursery of Chelsea, and were recognized as outstanding selections.

Selections of *E. longiflora*, which was then known as *E. miniata*, were extremely popular. One such selection was awarded the Silver Medal by the London Horticultural Society in 1884, and the report in the 'Floricultural Cabinet' stated it was '. . . considered to be the gem of the plants exhibited in June 1884'.

Other species grown to a lesser extent were *E. microphylla*, *E. pulchella*, and *E. purpurascens* which was renowned for its double-flowered form, then known as *E. onosmaeflora* 'Flore-pleno Nivalis'.

Early this century interest in epacris cultivars began to wane, and at present only a few are known to be cultivated in England.

In Australia many selections have been made, but few have been given names. Evidently, no hybridizing work has eventuated. The most popular are *E. impressa* 'Bega', *E. impressa* var. *grandiflora* 'Double-flowered Form' and *E. impressa* 'Spring Pink', which are all selections from naturally occurring plants.

There is scope for further selection of *E. impressa* forms with different growth habits, flowering times, and flower arrangements, colours and sizes; there is scope also for a deliberately planned hybridization programme within the genus. Species and forms certainly have potential as container plants in a wide range of climates and situations.

EPALTES Cass.
(from the Greek *epalthes*, healing; in reference to the qualities of the root of *E. divaricata*, an Indian species)
Asteraceae

Annual **herbs**; **stems** glabrous or hairy; **leaves** alternate, entire or toothed; **flower-heads** simple, terminal, axillary or in panicles, sessile or with short peduncles, more or less hemispherical; **bracts** unequal, in 3-4 rows; **florets** all tubular; **outer florets** female; **inner florets** male or bisexual; **fruit** an achene.

A cosmopolitan genus of about 17 species. There are 5 endemic species in Australia, in the mainland states.

Epaltes australis Less
(southern)
Qld, NSW, Vic, SA, WA, NT   Spreading Nut-heads
prostrate-0.15 m x 0.3-1 m   Sept-Feb; also sporadic

Dwarf annual **herb**; **branches** spreading, prostrate to ascending; **leaves** 1-6 cm x 0.4-1.5 cm, alternate, cuneate, petiolate, toothed margins, greyish green; **flower-heads** to 0.8 cm across, yellowish brown, hemispherical, nearly sessile or shortly stalked, axillary or lateral; **achenes** to 0.2 cm long, ribbed.

A dweller of moist soils on edges of inland lakes and in temporarily flooded areas. It usually occurs in clayey sands or heavy grey clays which are often alkaline. Has application for planting around edges of dams and pools. Develops into a fairly dense, circular mat. Best suited to situations where it receives plenty of sunshine. Propagate from seed or cuttings.

EPHEMERANTHA P. Hunt & Summerh. = *FLICKINGERIA* A. Hawkes

EPIBLEMA R. Br.
(from the Greek *epiblema*, a cover or patch; an apparent reference to the appendages at the base of the column or perhaps to the column itself which looks like a white patch in the centre of the flower)
Orchidaceae

A monotypic genus endemic to south-western WA.

Epiblema grandiflorum R. Br.
(large flowers)
WA   Babe-in-a-cradle
60 cm tall   Oct-March

Terrestrial **orchid**; **leaf** 12-20 cm x 0.3-0.5 cm, linear, terete, sheathing at the base, erect, solitary, dark green; **flower-stem** 45-60 cm tall, bearing 2-8 flowers in a loose terminal raceme; **flowers** 3-4 cm across, blue to deep mauve, with dark veins and small blotches; **labellum** ovate, with a tuft of linear calli at the base; **column** short, with large, broad, lateral wings.

A unique orchid confined to the south-west, where it grows amongst reeds in swamps and bogs. It flowers sporadically; with abundance in some years, while in others only odd flowering specimens can be found. It has been tried in cultivation and found to be difficult or impossible to maintain.

EPIPHYTES

An epiphyte is a plant which grows on another plant but which does not parasitize that plant. The term is often used loosely to include lithophytes, which are strictly plants which grow on boulders or rocks. The root system of the epiphyte ramifies over the bark of the supporting plant and derives moisture and nutrients for its growth from the bark, litter and animal droppings. Epiphytic growth habits are found in many groups of plants but most frequently in orchids, ferns, rhododendrons, hoyas and members of the family Araceae. Epiphytes are most common where conditions are moist and humid enough to allow their survival. In Australia they reach their best development in moist gullies and rainforests.

Epiphytes are popular in cultivation but, because of their growth habit and prevailing conditions where they occur naturally, they require somewhat specialized conditions for their successful culture. Most species like a fairly humid atmosphere, but with adequate air movement so that conditions do not become stale. This sort of atmosphere is readily maintained in a glasshouse or bushhouse. In suitable subtropical and tropical climates, a wide range of epiphytes can be grown as garden plants attached to or growing up the trunks of suitable trees.

Epiphytes tend to have specialized roots which in nature are often exposed to the atmosphere. These roots rot readily if confined to overwet or oxygen-starved conditions. In cultivation epiphytic plants must be attached to a slab of tree-fern fibre, cork or

Epiblema grandiflorum, × ·45

Epipogium roseum, × ·5

weathered hardwood, over which their roots can wander, or they can be potted in a basket or pot of coarse mixture. Suitable materials for such a mixture include chunks of tree-fern, cork, charcoal, weathered pine-bark, scoria, sand or terracotta. A coarse mixture allows rapid drainage and the entry of oxygen soon after watering. Epiphytes potted in soil generally struggle for survival and their roots rot in the poorly aerated mixture.

EPIPOGIUM Gmelin ex Borkhausen
(from the Greek *epi*, upon; *pogon*, a beard)
Orchidaceae

Terrestrial, saprophytic **orchids**; **rhizome** branched, irregular, rootless, bearing rhizoids; **flower-stem** fleshy; **inflorescence** a raceme; **flowers** pale coloured, short-lived; **fruit** a capsule.

A small genus of 2 species of saprophytic orchids, a widespread one of which extends to Australia. They are impossible to maintain in cultivation by any known method.

Epipogium roseum (D. Don) Lindley
(pale pink)
Qld, NSW
5-60 cm tall   Oct-March

Brittle, pale coloured terrestrial **orchid**; **flower-stem** 5-60 cm tall, fleshy, hollow, the **raceme** nodding until the flowers mature, bearing 3-25 flowers; **flowers** about 0.8 cm across, dingy yellow to pinkish, not opening widely, the segments slender; **labellum** small, cup-shaped.

This widely distributed orchid extends to eastern Australia, where it is found from north-eastern Qld to north-eastern NSW. It grows in colonies in shady forests and is rarely seen. The plants complete their life cycle within a very short period and further growth is confined to subterranean tuber development. Plants are saprophytic and cannot be grown by any known method.

421

ERAGROSTIELLA Bor.
(a diminutive grass similar to the genus *Eragrostis*)
Poaceae

Perennial **grasses** forming spreading clumps or tussocks; **leaves** thin, wispy, flat or inrolled; **inflorescence** an erect spike; **spikelets** awnless.

A small genus of 7 species mainly found in India and Ceylon, with a widespread species extending to north-eastern Qld. This species is a very drought resistant grass with possible breeding potential. Propagate by division or from seed which requires storage prior to sowing.

Eragrostiella bifaria (Vahl.) Bor.
(arranged in 2 rows)
Qld
30-45 cm tall   Dec-March

Slender perennial **grass** forming wispy tussocks; **leaves** 5-25 cm x 0.1-0.3 cm, flat or more often inrolled; **culms** 30-45 cm tall, erect, slender, with few nodes; **spikes** 10-16 cm x 0.5-1 cm, grey, becoming straw coloured; **spikelets** about 1 cm long, leaden grey to straw coloured.

A widespread grass which extends to Australia, where it is restricted to north-eastern Qld and is fairly rare. It occurs in shallow sandy soils over granite and in laterite. It grows in areas subject to long dry spells and is very drought resistant. Studies have shown that it is a true resurrection plant, with apparently dead leaves able to rehydrate and regreen after rain. May have potential in breeding programmes. Can be grown in a pot or rockery. Propagate by division or from stored seed.

ERAGROSTIS Wolf
(from the Greek *eros*, love; *agrostis*, a grass)
Poaceae

Annual or perennial **grasses** forming tussocks or spreading clumps, base of tussock sometimes bulbous and woolly; **leaves** basal or scattered on the culms; **culms** wiry, rarely cane-like; **inflorescence** an open, loose or contracted spike-like panicle; **spikelets** conspicuous, awnless.

A large genus of about 300 species widely distributed in tropical and subtropical regions of the world. About 60 species are found in Australia, most of which are endemic. A few exotic species are common naturalized weeds. As a group they are most familiar in inland regions, growing in low-lying areas where water collects. New growth is often a fresh bright green. Some species are useful pasture grasses but most are of limited significance. A number are very decorative when in flower and are worthy of trial in gardens and containers. Some species tend to be weedy and may naturalize readily. Propagate from seed which may need storage for up to 12 months before sowing or by division of the clumps.

Eragrostis australasica (Steudel) C. E. Hubb.
(Australian)
all mainland states   Bamboo Grass; Cane Grass
1-3 m tall   Aug-Feb

Perennial **grass** forming large spreading clumps;
leaves 10-20 cm x 0.2-0.6 cm, linear, hard and rigid, flat or inrolled, pointed, scattered up the culms; **culms** 1-3 m tall, 0.5-1 cm thick, woody, cane-like, much-branched, with a coating of whitish wax; **panicle** 5-20 cm x 3-15 cm, much-branched, loose, spreading; **spikelets** 0.5-1.2 cm long, linear.

A widespread coarse grass of inland regions, growing in claypans and depressions subject to partial inundation as well as around the margins of swamps and lakes. It could be used in larger ponds and dams as a windbreak and shelter for water-birds and fish. May become a nuisance if allowed to spread unchecked. Can also be grown as a bog plant in moist depressions. Propagate readily by division of the rhizomes.

Eragrostis brownii (Kunth) Nees ex Steudel
(after Robert Brown, English botanist of the 18-19th centuries)
all states   Common Love Grass
10-40 cm tall   Aug-April

Perennial **grass** forming spreading clumps; **leaves** 3-8 cm x 0.2-0.4 cm, bright green, glabrous, flat or inrolled; **culms** 10-40 cm tall, slender, glabrous; **panicles** 2-15 cm x 2-8 cm, loose and open, the lowest branches longest; **spikelets** 0.5-0.9 cm long, flat, purplish.

An attractive widespread grass which occurs in a variety of situations, often common on stream banks and in moist depressions. New flushes of growth are a bright fresh green and are produced rapidly following summer rain or watering. This feature makes it a decorative grass for planting in a rockery or amongst shrubs etc, but it can regenerate readily in moist soils and may become invasive in some areas. Well grown clumps in flower are also decorative. Plants grow readily in a sunny or semi-shady position, and are hardy to frost and dryness, although for best appearance they should be watered. The species has potential as an alternative to lawn in areas that do not have to withstand heavy foot traffic. Plants can be propagated readily by division or from seed.

Eragrostis elongata (Willd.) J. F. Jacq.
(lengthened; drawn-out)
all mainland states   Clustered Love Grass
30-60 m tall   Aug-Dec

Perennial **grass** forming erect, slender tussocks; **leaves** 8-20 cm x 0.2-0.4 cm, glabrous, flat or inrolled, bright green; **culms** 30-60 cm tall, erect, unbranched, smooth; **panicle** 8-16 cm x 1-2 cm, narrow, sparse at the base, clustered towards the apex; **spikelets** about 0.5 cm long, flattened, green to purplish.

A widespread grass which prefers inundated sandy soils. New growth is an attractive fresh green. The slender tussocks of this species mix well with garden shrubs, and plants can be grown in a shrubbery or amongst rocks. They respond to watering. Propagate from seed or by division.

Eragrostis eriopoda Benth.
(a woolly foot)
Qld, NSW, SA, WA, NT   Naked Woollybutt Grass
30-60 cm tall   Aug-Dec

Perennial **grass** forming dense, spreading clumps,

bases of clumps densely woolly; **rhizome** much-branched; **culms** 30-60 cm tall, erect, wiry; **leaves** 3-8 cm x 0.2 cm, narrow, linear, stiffly spreading, rigid; **panicle** 11.5-20 cm x 4-7.5 cm, much-branched, open; **spikelets** 0.6-2.2 cm long, straight or curved, not woolly.

An interesting grass which is widely distributed in semi-arid and arid inland regions, frequently growing on red, clayey soils. It is a very hardy species, surviving severe droughts by virtue of its deep root system. Plants form a dense clump which spreads out in all directions. Eventually the centre dies, leaving a ring-like or U-shaped growth pattern. The leaves are eaten by stock. Clumps are markedly rejuvenated by fires, the new growth being lush and palatable. The species has had limited cultivation as a garden plant but would be worthy of trial in a wide range of climates and soil types. The unusual growth habit is appealing, and when in flower the clumps are quite decorative. Propagate from seed or by division.

Eragrostis laniflora Benth.
(woolly flowers)
Qld, NSW, SA, NT   Hairy Woollybutt Grass
30-50 cm tall   Aug-Dec

Perennial **grass** forming dense, spreading clumps; **rhizome** stout, much-branched, densely woolly; **leaves** 3-8 cm x 0.1 cm, narrow, linear, stiffly spreading, rigid, inrolled; **culms** 30-50 cm tall, erect, wiry; **panicle** 10-20 cm x 5-8 cm, much-branched, sparse; **spikelets** 1-2.5 cm long, woolly, with long soft hairs at the base of each floret.

A grass of drier regions, which grows in deep sandy soils and forms dense clumps. Plants are very decorative when in flower and could have potential in a sunny garden situation. Once established the clumps are very hardy. Propagate from seed or by division.

Eragrostis lanipes C. E. Hubb.
(a woolly foot)
SA, WA   Love Grass
20-50 cm tall   Aug-Dec

Perennial **grass** forming spreading clumps, bases of clumps bulbous, woolly; **leaves** 4-10 cm x 0.2-0.3 cm, flat or inrolled; **culms** 20-50 cm tall, slender, smooth; **panicle** 10-20 cm x 3-8 cm, open and loose, the branches somewhat rigid; **spikelets** about 0.5 cm long, green to purplish.

A little-known grass which is decorative when in flower and could be tried in temperate regions. Would require an open, sunny position and would probably respond to regular watering during dry periods. Propagate from seed or by division of the rhizomes.

Eragrostis microcarpa Vick.
(tiny fruit)
Qld, NSW   Love Grass
30-60 cm tall   Aug-Dec

Perennial **grass** forming spreading clumps, bases of clumps woolly; **leaves** 5-10 cm x 0.2-0.3 cm, linear, hairy on the upper surface; **culms** 30-60 cm tall, slender; **panicle** 15-20 cm x 12-15 cm, pyramid-shaped, open and loose; **spikelets** about 1 cm long, green.

A grass of inland regions, which grows in moist depressions etc. It has some ornamental qualities and could be a useful species for moist soils in an open, sunny situation. Propagate by division of the rhizomes or from seed.

Eragrostis setifolia Nees
(bristle-shaped leaves)
all mainland states   Bristly Love Grass
20-45 cm tall   Aug-Dec

Perennial **grass** forming slowly spreading clumps; **leaves** 4-13 cm x 0.1-0.3 cm, linear, flat or inrolled, rough on the upper surface, the margins sharp and thickened; **culms** 20-45 cm long, slender, wiry, glabrous; **panicle** 7-10 cm x 1-2.5 cm, of separate clusters of spikelets, green; **spikelets** 1-1.5 cm long, flattened.

A wiry grass of heavy soils, often growing on clays which crack when dry. It is most commonly seen in inland regions and remains green for long periods after rain. It is a useful pasture grass. Plants are quite ornamental when in flower and have prospects for garden planting. Requirements are for moist soil in a sunny situation. Propagate from seed or by division.

Eragrostis speciosa (Roem. & Schult.) Steudel
(beautiful)
Qld, NSW, SA, WA, NT   Handsome Love Grass
30-90 cm tall   Aug-Dec

Perennial **grass** forming slender tussocks; **leaves** 5-12 cm x 0.1-0.3 cm, linear, glabrous, inrolled, bright bluish green; **culms** 30-90 cm long, glabrous, bluish green; **panicle** 15-25 cm x 1-2.5 cm, narrow and drawn out, of separate clusters of spikelets, greyish, often drooping; **spikelets** 0.7-3 cm long, narrow, flattened, clustered.

A widely distributed grass of inland areas, usually growing in loose colonies in the alluvial beds of ephemeral streams. It is a very hardy species and after rains looks bright green and fresh. It appears to have potential for garden cultivation in a sunny aspect or even a container. Propagate by division or from seed.

Eragrostis xerophila Domin
(growing in dry places)
Qld, NSW, SA, WA, NT   Knottybutt Never-fail
10-22 cm tall   Aug-Dec

Perennial **grass** forming slowly spreading clumps; **rhizome** stout, knotted, woolly; **leaves** 1-3 cm x 0.3-0.5 cm, linear, rigid, flat, numerous, often recurved; **culms** 10-22 cm tall, sheathed throughout; **panicle** 5-10 cm x 0.4-0.8 cm, of separate clusters of spikelets, green to purplish; **spikelets** 0.5-1.2 cm long, flattened, clustered.

A widely spread grass of inland regions, often growing in low-lying clay soils. It is very drought resistant, surviving long dry periods by means of its thick rhizome. It is tolerant of both saline and calcareous soils and has some use as a pasture grass, withstanding the close grazing of sheep. When in growth it is fairly ornamental and the clumps are always very leafy. It can be grown in a rockery or garden situation amongst low shrubs, and may have excellent prospects in saline soils. Needs a sunny aspect. Propagate from seed or by division.

ERECHTITES Raf. — for Australian species previously included here see *SENECIO* L.

EREMAEA Lindley
(from the Greek *eremaios*, solitary; in reference to the terminal flowers on the branchlets)
Myrtaceae

Dwarf to large **shrubs**, sometimes with lignotubers; **branches** spreading to ascending; **leaves** alternate, narrow and needle-like or broad, usually densely hairy initially, becoming glabrous, often crowded on young shoots; **flowers** single or in clusters of up to about 9, terminal on old growth or along new growth, surrounded by scale-like bracts; **calyx** glabrous or hairy, shiny or dull; **staminal bundles** brightly coloured, orange, pink or violet, rarely white or cream; **fruit** a capsule, more or less globular; **seeds** with conspicuous wings on top and along sides.

Eremaea is an endemic genus of about 17 species. It is currently undergoing botanical revision, which should be finalized within a couple of years. It is restricted to south-western WA, where species occur mainly north of Perth, in sandy soils amongst heathland or low woodland communities. In general, eremaeas occur in areas that have lengthy periods of sunshine. They are subjected to extended dry periods and light frosts, and receive rainfall mainly during winter and early spring.

It is a highly ornamental genus, with the brightly coloured staminal bundles well displayed. All members have potential for cultivation, although at present only *E. beaufortioides* and an unnamed species with affinities to *E. violacea* are grown to any significant extent. Other species that have had limited cultivation include *E. acutifolia*, *E. fimbriata*, *E. pauciflora* and *E. purpurea*.

Most species need an open, warm to hot situation that receives plenty of sunshine, and very well drained, light to medium soils. In general they are sensitive to poor drainage. Some growers have excellent success by raising the level of the garden area, to provide better drainage. See Volume 1, page 65 for further details.

Eremaeas seem to be able to withstand fairly strong winds, but several are susceptible to damage by heavy frosts. Some are useful as low groundcovering shrubs, while others can develop into bushy screening plants, eg. *E. pauciflora* forms.

Many species have potential as popular container plants. In this way a wider range could be grown in temperate zones and in areas with heavy clay-loam soils prone to waterlogging. Difficulty has also been experienced in establishing and maintaining eremaeas in subtropical and tropical zones, and container cultivation could be worth trying in these regions.

Most species respond well to pruning after flowering, and it can be beneficial in promoting bushy growth on leggy, open plants.

Eremaeas are not prone to damage by many pests, but their roots can be subject to attack by cinnamon fungus (*Phytophthora cinnamomi*) in poorly drained soils.
Propagation

Propagation from seed and from cuttings are the two methods commonly used. Grafting has been practised to a limited extent.
Seed. Mature capsules can be collected (about 12 mon-ths after flowering) and placed in a cloth or paper bag. When dry, the seed will be dispersed. It is difficult to obtain seed of some species, eg. *E. fimbriata*, because the orifice of the capsule is too narrow for dispersal. Growers have overcome the problem by grinding the capsules in a meat grinder. Undoubtedly, some seed is damaged by this process but it is usually worthwhile.

Seed can be sown and germinated in the normal manner or by other ways such as the bog method or capillary bed method (see Volume 1, page 204). The seed usually germinates within 2-4 weeks, but it may take over twice that time.
Cuttings. Not all species have been propagated from cuttings, but with those tried, best results are gained when vigorous, firm young growth is used. Some species, eg. *E. beaufortioides*, tend to form extended callus formations which are like thickened roots with a blunt apex. In some cases the callus can develop for 12 months or more before normal roots are formed, or it can continue to develop without forming roots (see Overcallusing, Volume 1, page 234). Examination of the callus tissue by plant pathologists has not yet found the reason for this phenomenon.

Amongst the easier species to propagate from cuttings are *E. purpurea* and the species aff. *violacea* which has large heads of flowers and broad leaves.
Grafting. Some success has been gained using *Kunzea ambigua* as a stock for *E. beaufortioides*. This indicates that the method is worthy of further trial with a wider range of species.

Eremaea acutifolia F. Muell.
(pointed leaves)
WA   Rusty Eremaea
0.5-7 m x 0.8-2.5 m   Aug-Nov
Dwarf, spreading **shrub**; **branches** arched, thick, rigid, hairy; **branchlets** short; **leaves** 0.4-1.2 cm x 0.1 cm, linear, rigid, flat, spreading, crowded, hairy or becoming glabrous, ending in a pungent point; **flowers** about 1 cm across, orange, surrounded by woolly bracts, solitary, forming leafy spikes; **stamens** in bundles, with claws as long as filaments; **fruit** about 1 cm across, broadly cup-shaped, rough, valves partly protruding.

This showy, prickly-leaved species occurs in a restricted area of the Irwin District, where it grows on sandheath. It is not well known in cultivation. Requires very well drained soils in a location that receives maximum sunshine. Tolerates extended dry periods and most frosts. Responds well to pruning. Has potential for use as a groundcover, and should do well as a container plant in temperate zones. Propagate from seed which germinates readily or from cuttings.

There are 3 as yet undescribed species with affinities to *E. acutifolia*. They are worthy of cultivation.

Eremaea aff. acutifolia A
WA
0.25-0.8 m x 1 m   Sept-Nov
Dwarf **shrub** with arching, spreading **branches**, well developed **branchlets** and broad **leaves**. The **flowers** are orange and the large **fruits** have persistent calyx lobes. It occurs near the Arrowsmith River and extends

southwards to the Badgingarra area. Has been wrongly known as *E. acutifolia*.

Eremaea aff. acutifolia B
WA
0.3-1.1 m x 0.5-1 m   Aug-Nov
Dwarf to small compact **shrub** with erect or ascending **branches**, broad-lanceolate **leaves** and abundant orange **flowers**. **Fruits** have 4-7 lobes, but usually 5. Its compact growth habit gives it potential for wide use in gardens or containers. In nature it is distributed from near Eneabba to around Badgingarra.

Eremaea aff. acutifolia C
WA
0.3-0.6 m x 0.6-1.2 m   Sept-Oct
Dwarf **shrub** with fewer, longer **branchlets**, broad-lanceolate **leaves** and dark pink to purplish pink **flowers**. This species is virtually unknown in cultivation. It may grow larger than indicated above. It is found in the area between Eneabba and Jurien.

Eremaea beaufortioides Benth.
(resembling the genus *Beaufortia*)
WA   Round-leaved Eremaea
0.8-3 m x 1-2.5 m   Sept-Dec
Dwarf to small **shrub**; young shoots hairy; **branches** many, arching, spreading, initially hairy, becoming glabrous; **leaves** 0.4-0.6 cm x up to 0.5 cm, ovate, occasionally linear-lanceolate, 3-veined, spreading, recurved, crowded, glabrous when mature; **flowers** about 1 cm across, orange, usually 2-5 per head, terminal, often profuse; **fruit** barrel-shaped, to 0.8 cm across, glabrous.
One of the best-known eremaeas in cultivation. It is from the Irwin and northern Avon Districts, where it grows on sand. Has adapted relatively well to cultivation, growing in a wide range of well drained soils, including clay-loams. Likes a sunny position. In temperate regions it is prone to attack by grey mould during still, humid weather and it therefore does best in an open situation. Some plants may become leggy. Pruning helps promote bushy growth. Tolerant of most frosts and extended dry periods. At least 2 forms are in cultivation. One is low, compact and spreading, with glabrous leaves. The other is more upright, with slightly hairy leaves. Propagate from seed which germinates readily or from cuttings of vigorous firm new growth which will usually produce roots. If mature material is used, rooting can be slow, with the formation of thick, root-like calluses. Sometimes, after 6-12 months, typical roots will form from the callus growth.
An alternative is to graft plants, and success has been achieved using *Kunzea ambigua* as the rootstock.

Eremaea aff. beaufortioides
WA
0.2-1.5 m x 0.8-2 m   Oct-Nov
An erect, compact **shrub** with brilliant purplish pink, well displayed **flowers**. It occurs near the Arrow-

smith River and extends southwards to Cockleshell Gully in the southern Irwin and northern Darling Districts. Virtually unknown in cultivation.

Eremaea brevifolia (Benth.) Domin
(short leaves)
WA
0.3-2.2 m x 0.5-2 m   Aug-Oct
Dwarf to small, spreading **shrub**; young shoots hairy; **branches** many, spreading to ascending; **leaves** to 0.7 cm x 0.7 cm, broadly ovate to obovate, crowded, slightly concave, initially hairy, becoming glabrous; **flowers** to about 0.8 cm across, bright orange, solitary, terminal; **fruit** broadly cup-shaped, 1-1.5 cm across, thick-walled, light brown and rough exterior, valves partly protruding, undulating rim.
This species from the Irwin District grows as a dense shrub on deep sandy soils. It is evidently not known in cultivation. Is not as floriferous as some species but has large flowers. Requires very well drained, light to medium soils, with plenty of sunshine. Should tolerate most frosts and extended dry periods. Propagate from seed or cuttings.
Until recently this species was thought to be undescribed.

Eremaea aff. brevifolia
WA
0.3-1 m x 0.75-1.3 m   Aug-Oct
This species differs in its narrower and strongly recurved **leaves**. Has large orange **flowers**. The **fruits** have spine-like points on the rim. It is not common in cultivation. Tolerates a lightly shaded location, but prefers plenty of sunshine. Seedling plants should develop a lignotuber as they do in nature. This species has a wide distribution, occurring near Leeman in the southern Irwin District and extending southwards to near Pinjarra, south of Perth, in the Darling District.

Eremaea ebracteata F. Muell.
(without bracts)
WA
0.3-1.5 m x 0.5-2.5 m   Sept-Feb
Dwarf, spreading **shrub**; young shoots hairy; **branches** many, hairy; **leaves** 0.6-1 cm x about 0.1 cm, linear, hairy, crowded, apex blunt or pointed; **flowers** about 1 cm across, slightly reddish orange, usually 2-3 per head, terminal, profuse; **fruit** to about 1 cm across, cup-shaped, becoming corky with age, rim sometimes wavy.
This ornamental species usually grows as a dense groundcovering shrub. It occurs on sandheath, such as around Northampton and Kalbarri in the Irwin District. At present it is little-known in cultivation. It must have a very sunny position to grow well, and needs very well drained soils. Hardy to most frosts and drought resistant. In temperate regions it is best grown in a hot situation, eg. near a north or west-facing wall. Suited also to container cultivation. Propagate from seed which is amongst the easiest of the genus to germinate or from cuttings.
Has affinities to *E. purpurea*, which differs in its purplish pink flowers and erect growth habit.

Eremaea aff. **ebracteata**
WA
1-5 m x 3-5 m   Sept-Dec
Densely branched **shrub** with almost terete, linear **leaves** which are initially hairy. The bright orange **flowers** are profuse. The **fruits** are smooth or slightly warty and are about 1 cm across. It is a showy species of the Shark Bay-Kalbarri area in the northern Irwin District. Has potential as a screen plant in semi-arid regions.

Eremaea ericifolia Lindley =
E. pauciflora (Endl.) Druce

Eremaea fimbriata Lindley
(fringed)
WA
0.3-1 m x 0.5-1.5 m   July-Oct
Dwarf, spreading to erect **shrub**; young shoots hairy; **branches** many, initially hairy, becoming glabrous; **leaves** 0.5-1 cm x about 0.2 cm, linear-lanceolate, crowded, margin hairy; **flowers** about 1 cm across, purplish pink with yellow anthers, terminal, usually solitary but can be in pairs; **petals** pink; **calyx** hairy; **fruit** about 1 cm long at maturity, narrow opening.
This is the only pink-flowered *Eremaea* species. It occurs on sand in the Darling District north of Perth, where it grows both inland and in near-coastal situations. Although highly ornamental, it has not been cultivated to a large extent. Must have very well drained soils. Is best suited to sandy types. Does best in plenty of sunshine, but will grow in light shade. Withstands extended dry periods and tolerates most frosts. Has potential as a container plant in temperate zones. Propagate from seed which is often prevented from shedding because of the narrow opening of the fruit. Observation in nature shows that the thin walls of the fruits are sometimes burnt by fire, allowing release of the seed. Germination usually occurs within 8 weeks. Cuttings can be slow to form roots.
Until recently this species was well known as *E. rosea* C. Gardner & A. S. George. The species that was regarded as *E. fimbriata* is undescribed (see *E.* species 'Darling-Irwin').

Eremaea pauciflora (Endl.) Druce
(few-flowered)
WA
0.5-2.5 m x 0.5-2.5 m   Aug-Dec
Small to medium **shrub**; young shoots hairy; **branches** many, hairy; **leaves** to 0.8 cm x 0.1 cm, narrow-linear, erect, margins hairy but can become glabrous, apex blunt; **flowers** about 0.5 cm across, medium to light orange, rarely yellow, surrounded by scale-like bracts, terminal, solitary, few to profuse; **fruit** about 0.8 cm across, globular, smooth, brown.
A widely distributed and variable species from the Avon, Darling, Roe and Coolgardie Districts. It occurs mainly in heathland communities. Although the specific name means few-flowered, some plants can be very colourful during flowering. Selection of floriferous and hardy forms is most desirable. Needs very well drained, light to medium soils, with plenty of sunshine. May require light pruning to promote bushy growth.

Drought and frost tolerant. In temperate zones best growth is usually attained in containers. Propagate from seed which germinates readily or from cuttings which may be slow to form roots.
Previously known as *E. ericifolia* Lindley and *E. pilosa* Lindley.

Eremaea aff. **pauciflora**
WA
2-4 m x 2-4 m   July-Dec
Medium to tall **shrub** with many **branches** and dense foliage. The orange **flowers** have a strong, sweet, spicy scent. It is distinguished from *E. pauciflora* by the small fruit, the calyx tube which is hairy on the lower two-thirds, and the broad calyx lobes which curve inwards. It occurs in the Watheroo-Eneabba region of the southern Irwin District. Should prove useful as a screening plant.

Eremaea pilosa Lindley =
E. pauciflora (Endl.) Druce

Eremaea purpurea C. Gardner
(purple)
WA
0.2-1.3 m x 0.7-1.5 m   Oct-Feb
Dwarf, upright or spreading **shrub**; young growth hairy; **branches** thick, many, often erect; **leaves** to 0.3 cm x 0.1 cm, oblong, flat, crowded, glandular, apex blunt, glabrous when mature; **flowers** about 0.7 cm across, pinkish purple, rarely white, usually solitary, terminal; **fruit** about 0.8 cm across, cup-shaped, smooth or warty, valves just protruding.
This ornamental eremaea occurs north of Perth in the Darling District, in the Gingin-Muchea region, where it grows in sandy soils. It is not well known in cultivation but shows considerable potential. Requires very well drained, light to medium soils and plenty of sunshine. Is hardy to dryness and most frosts. Responds well to light pruning after flowering. Has potential as an upright plant for containers. Propagate from seed or cuttings.
This species is closely allied to *E. ebracteata*, which differs with its reddish orange flowers and spreading growth habit.

Eremaea rosea C. Gardner & A. S. George =
E. fimbriata Lindley

Eremaea violacea F. Muell.
(violet)
WA   Violet Eremaea
0.2-0.5 m x 0.5-2 m   Sept-Jan
Dwarf, spreading **shrub**; young growth hairy; **branches** few; **branchlets** many, crowded, short; **leaves** to 1 cm long x about 0.1 cm, semi-terete, linear, crowded, becoming more or less glabrous, stiff, apex pointed; **flowers** about 0.6 cm across, violet-blue, rarely pale to white or with pink to reddish tones, anthers yellow, usually in groups of 2-7 at ends of branchlets; **fruit** about 0.6 cm across, globular, wide opening, smooth, dark purplish grey, valves protruding.
This extremely ornamental species occurs in the Avon, Darling and Irwin Districts, between the Hill River and Mingenew. It grows on sandy soils of the

426

Eremaea aff. *violacea* B

heathland. Shows high potential for cultivation but as yet is not well known. If grown in an open, sunny position it develops as a near-prostrate plant. It will tolerate dappled shade, although not flowering as well. Needs well drained, light to medium soils. Hardy to light frosts, and withstands extended dry periods. Should make an excellent container plant. Propagate from seed or cuttings.

There are 2 as yet undescribed species with affinities to *E. violacea*. They are in cultivation.

Eremaea aff. **violacea** A
WA
0.3-1.2 m x 0.5-0.7 m   Sept-Dec
This species differs in its initial upright growth habit, which becomes spreading with maturity. It has longer **branchlets** and soft, broadly lanceolate **leaves**. It occurs near Mingenew and extends southwards to Badgingarra. Not common in cultivation, but grows well in partial or full sun.

Eremaea aff. **violacea** B
WA
0.5-2 m x 1-2 m   Oct-Jan
This differs from *E. violacea*, with its stiff, more or less erect **branches** and rigid, lanceolate **leaves** to 1.7 cm x 0.2 cm. The brilliant bluish purple **flowers** with golden anthers are in dense terminal heads of up

to 10 per head. The **fruits** are about 0.8 cm across, glabrous, with prominent calyx lobes. This is an outstanding vigorous species, occurring from the Arrowsmith River to Badgingarra in the south. It grows well in a wide range of situations but does best in an open, sunny position. It can develop long branches, but responds well to pruning if bushy growth is desired. May suffer damage from heavy frosts. For many years this species has been available incorrectly in the eastern states as *E. violacea*.

Eremaea species 'Darling-Irwin'
WA
0.3-1 m x 0.5-1.5 m   Aug-Oct
Dwarf **shrub**; young shoots hairy; **branches** many, mainly erect; **leaves** to about 0.8 cm x 0.4 cm, elliptical, crowded, spreading, with 7 prominent nerves, hairy, apex blunt; **flowers** about 1.2 cm across, orange, solitary, terminal; **fruit** about 1.4 cm across, with prominent calyx lobes, slightly hairy at base.

This undescribed species occurs in the Darling and Irwin Districts, where it grows in open situations on sandy soils. It is very showy when in flower. It is not common in cultivation, and has been grown under the name of *E. fimbriata* for many years. Is best suited to a sunny situation in well drained soils. May need light pruning if bushy growth is desired. Hardy to most frosts and extended dry periods. Worth growing as a

Eremocitrus

container plant in temperate zones. Propagate from seed or cuttings. The same comments apply to cuttings of this species as to those of *E. beaufortioides* and *E. violacea*.

EREMOCITRUS Swingle

(from the Greek *eremos*, desert; lonely places; *citrus*, the Latin name for the citron but applied to related members of the citrus family)
Rutaceae

A monotypic genus endemic to Australia.

Eremocitrus glauca (Lindley) Swingle
(bluish green)

Qld, NSW, SA	Desert Lemon; Native Cumquat; Wild Lime
2-7 m x 1-3 m	Aug-Oct

Medium to tall **shrub** or small **tree**, sometimes forming spiny thickets; **bark** brown, fissured, hard; new shoots angular, finely hairy; **stems** spiny; **leaves** 1-5 cm x 0.2-0.4 cm, narrow-obovate to spathulate, light to dark green, the tip shortly notched; **flowers** 1-1.3 cm across, white to greenish, sweetly scented; **stamens** 10-15; **berry** 1-1.6 cm across, globular, pale yellow, bearing 2-4 grey, furrowed seeds, mature Jan-March.

A hardy plant widespread in semi-arid and arid inland regions, usually growing in heavy clay soils. It may occur in dry woodland or on almost treeless plains. Plants appear to sucker to varying degrees, and grow in clumps of mature individuals which have well developed trunks, or as spiny thickets. Individual mature trees have an attractive shape, with a sturdy slender trunk and a rather narrow drooping crown. In very dry seasons the trees become completely deciduous. The fruit is edible, although sour, and can be used to make a very refreshing drink or a marmalade. A citrus relative adapted to a harsh, dry climate, the species is of interest to Australian and international horticulturists as a potential citrus rootstock and for use in breeding programmes. A selection programme has been carried out by CSIRO and a range of material has been introduced into cultivation. Plants can be grown easily but require a warm climate and soil of unimpeded drainage. They are suited to inland regions and are quite ornamental when well grown. Two drawbacks to its wider cultivation are the thorns and the suckering habit. Initially, plants are slow-growing but once established growth is much faster. They respond to watering during dry periods and to applications of nitrogen-rich fertilizers. Propagation can be by a variety of techniques including seed, stem cuttings which are difficult to strike, root cuttings, suckers or by budding on to citrus rootstocks. Seed retains its viability for a few years.

EREMOPHILA R. Br.

(from the Greek *eremos*, desert; lonely places; *phileo*, to love; in reference to the preferred habitat of many species)
Myoporaceae   Emu-bush; Poverty Bush

Prostrate to tall **shrubs** or small **trees; branches** glabrous, hairy or scaly; **leaves** alternate, opposite or rarely whorled; **flowers** tubular, usually 2-lipped, upper lip 2-4 lobed, lower lip 1-3 lobed, various colours, with spotted or unspotted exterior and interior; **calyx lobes** variable, some enlarging after flowering; **stamens** 4, exserted or not exserted; **fruit** non-splitting, drupe-like or rarely a drupe, glabrous, hairy or scaly, smooth, ribbed or rarely winged.

Eremophila is an endemic genus of about 180 species. Many of these are outstanding ornamentals. At present a revision of the genus is being undertaken, which should be finalized within 2 years. Quite a few species have been described recently, and there are many still undescribed. One species (*E. crassifolia*) crosses intergenerically with *Myoporum platycarpum*. It occurs at Waikerie in SA, and is sometimes cultivated.

The genus is distributed throughout all the mainland states, with most species occurring in semi-arid and arid regions. They grow as undershrubs in low woodlands and mulga scrub, as well as in open situations where they are the main component of the vegetation.

They grow on a variety of soils, including deep sands, rocky and gravelly loams and clay-loams, mostly of neutral to alkaline pH. Many species favour heavier soils which can be subject to inundation for limited periods. The majority grow in regions which have fairly harsh climatic conditions. Day temperatures are usually very high from Sept-March and during this period there is virtually no rainfall. During April-Aug most regions have some frost and a limited rainfall, with warm to hot day temperatures. Those species growing in arid areas can survive periods of 1-2 years or more without rain.

Eremocitrus glauca, × ·55

428

There are many adaptations developed by eremophilas to cope with harsh climates. Several have a lacquer-like appearance on the foliage, caused by an exudate which protects the leaves from strong, drying winds, eg. *E. denticulata* and *E. viscosa*. Other species have a dense covering of silvery hairs which can be felt-like, and which act as a reflectant of the sun's rays and reduce water loss from the stomata. Examples are *E. bowmanii*, *E. leucophylla*, *E. macdonnellii* (Northern Simpson Desert form) and *E. rotundifolia*.

There is considerable variety in growth habit within the genus. Some species are groundcovering plants which can form roots from the nodes on the older branches, eg. *E. biserrata* and *E. serpens*. Many are low-growing shrubs that become rounded in appearance if grown in exposed conditions, eg. *E. glabra*, *E. hillii* and *E. maculata*. Others are large, open shrubs with leggy growth, eg. *E. denticulata* and *E. serrulata*. A small number of species are valued for their upright growth, and these include *E. calorhabdos* and *E. racemosa*. Some develop into small trees and have pendulous branchlets and foliage, eg. *E. longifolia* and *E. santolina*.

A few species are recognized as pastoral weeds in Qld and NSW, and under some circumstances they may be poisonous to stock. These include *E. gilesii*, *E. mitchellii* and *E. sturtii*. They are difficult to control because of their ability to sucker and regrow after fire. In some areas they may become the predominant species, competing with pasture.

A limited number of eremophilas were used by the Aborigines for medicinal purposes. *E. freelingii* was used for the treatment of headaches, *E. longifolia* for colds and *E. gilesii* for body sores.

Some species have limited use as fodder plants, with *E. longifolia* the most commonly used. Other species suitable for fodder include *E. bignoniiflora*, *E. glabra*, *E. latrobei*, *E. mitchellii*, *E. oppositifolia* and *E. polyclada*.

E. mitchellii has been used as a sandalwood substitute. It has sweetly scented, fragrant wood containing about 3% essential oils. It is also an excellent fuel.

Many eremophilas are highly ornamental. Some have large colourful flowers that are often displayed for extended periods. Other species have small flowers that are produced in profusion along the branchlets, creating a spike-like effect. All colours of the spectrum are represented. The most common group is the blue to purple range, and a large number of species have pink tonings. After flowering, some species develop large, eyecatching calyx lobes which can be cream, green, purplish or other colours. Notable examples are *E. cuneifolia*, *E. fraseri*, *E. latrobei*, *E. miniata* and *E. oppositifolia*. Species with large tubular flowers and exserted stamens are commonly visited by honey-eating birds seeking nectar.

Cultivation

The *Eremophila* genus has outstanding potential for cultivation and deserves to be better-known. At present the most commonly grown are *E. glabra* (and its many variants) and *E. maculata*. These have proved adaptable to most soil types and climatic conditions. The majority of other species have been grown by enthusiasts, and much knowledge has been accumulated on their cultivation requirements. Eremophilas also show

potential for increased planting in semi-arid and arid regions, especially on alkaline soils.

Basic cultivation requirements are for good to excellent drainage and a situation that receives plenty of sunshine. Mediterranean climates such as that of Adelaide are very suitable.

Eremophilas do not appreciate long periods of overcast, humid weather, and thus they have proved generally unsuccessful in moist tropical, subtropical and high-rainfall temperate zones.

Once plants are established, artificial watering is not usually required, although in periods of extreme dryness they respond well to deep soakings (see Pests and Diseases later in this entry).

Its capability to grow in soils with a very low moisture content will lead this genus to become significant for areas of low rainfall and high temperatures. Plants are also adaptable to the alkaline soils which are often common in such arid and semi-arid regions.

Most species will tolerate light to medium frosts but some may be damaged or killed by heavy frosts.

Frost tolerance of some Eremophila species

Hardy	Susceptible
E. behriana	E. alternifolia
E. biserrata	E. bowmanii
E. calorhabdos	E. christophori
E. densifolia	E. dalyana
E. drummondii	E. glabra (coastal forms)
E. glabra (inland forms)	E. hillii
E. ionantha	E. laanii
E. longifolia	E. latrobei
E. maculata	E. macdonnellii (some forms)
E. polyclada	E. macgillivrayi
E. scoparia	E. mackinlayi
E. serrulata	E. oldfieldii
E. subfloccosa	E. oppositifolia
E. weldii	E. purpurascens
E. youngii	E. saligna

Those species which have proved fairly reliable in temperate zones include *E. biserrata*, *E. calorhabdos*, *E. decipiens*, *E. densiflora*, *E. denticulata*, *E. dichroantha*, *E. divaricata*, *E. drummondii*, *E. gibbifolia*, *E. glabra*, *E. laanii*, *E. maculata*, *E. polyclada*, *E. serpens*, *E. serrulata* and *E. weldii*.

Some species which are difficult to cultivate in temperate zones can be grown successfully by planting them in built-up soil mounds, which allows for excellent drainage. Those that appreciate this treatment include *E. bowmanii*, *E. clarkei*, *E. granitica*, *E. laanii*, *E. macdonnellii*, *E. macgillivrayi*, *E. pantonii*, *E. racemosa* and *E. youngii*.

Eremophilas can make excellent container plants, thus increasing the range of species which can be grown in areas that do not have suitable conditions for their cultivation in the ground.

In general, eremophilas grow well without applications of fertilizers. Fertilizing can, however, promote healthy vigorous growth which may help plants withstand fungal attack.

Most species respond well to pruning. In those with

an open, leggy growth habit, tip pruning of young plants is recommended as this promotes plenty of lateral shoots. Species that are likely to need tip pruning include *E. calorhabdos* and *E. racemosa*.

Pests are not a major problem with eremophilas. The most common pest is scale, which is more prevalent on species with sticky branches and leaves, eg. *E. decipiens*, *E. denticulata*, *E. drummondii*, *E. resinosa* and *E. serrulata*. Sooty mould may spread rapidly on these species. Control can be effected by applications of white oil, which is best applied during cool conditions. Some caterpillars may disfigure leaf growth, but this is usually of little or no consequence to the continued health of plants. Flea beetles tend to feed near ground level, and may cause damage to young plants during Sept-Nov. They eat young shoots and parts of leaves. Species recorded as being affected include *E. alternifolia*, *E. denticulata* and *E. glabra* forms. Aphids can be a nuisance on some of the species with sticky foliage. They usually do not cause a great deal of damage, but some flowers may be disfigured. Aphids can be kept under control by high-pressure spraying of water from a hose. Some species are subject to root knot nematode attack, which can weaken growth, making plants more susceptible to fungal attacks (see Volume 1, page 161).

There are a few diseases that attack eremophilas, and the most common are the fungi *Botrytis* spp. and *Alternaria* spp. These damage leaf-nodes, leaves and branchlets, and can result in browning of leaves, leaf-drop and blackening of stems. Young plants can die from such attacks, and it is imperative that affected plants be removed or treated as soon as they are noticed.

Attack from *Botrytis* and *Alternaria* is more likely when the soil is moist to waterlogged and climatic conditions are overcast with rain occurring regularly, creating constant high humidity with water on the foliage for lengthy periods of time. Plants are particularly susceptible when grown in sheltered areas.

The best method of avoiding fungal attacks is to grow plants in open areas which allow plenty of air movement and sunshine. Some of the more susceptible species are *E. densiflora*, *E. gibbifolia*, *E. glabra*, *E. hillii*, *E. maculata*, *E. resinosa* and *E. subfloccosa*. It seems that species with broad, crowded leaves and hairy stems are more likely to be adversely affected than those with narrow leaves and glabrous stems.

Cinnamon fungus (*Phytophthora cinnamomi*) may be the cause of death for some species, especially in wet soils. Further research is needed here (see Volume 1, page 168).

Propagation

Eremophilas are propagated from seed, cuttings and to a lesser extent by grafting.

Seed. Only a limited number of species have been grown successfully from seed. Seed germination has proved extremely variable.

Over the years many methods and treatments have been used, including the following:
1. Sowing of freshly collected and untreated seed. There is some evidence to suggest loss of viability with age.
2. Sowing of old seed which has been sunbaked and

subjected to rain. Such seed can sometimes be gathered from below mature plants.
3. Washing seed in water for an extended period in an effort to leach out germination inhibitors prior to sowing.
4. Watering of sown seed with various liquids including (a) iron chelate solution; (b) solutions of metal salts, eg. copper and lead.
5. Extraction of seed from the fruit, and germination on blotting-paper or a similar medium.

The above have been successful to some extent, but do not provide full answers. For example, germination can be sporadic in a single species (9 days to 15 months or more for *E. maculata*). Considerable experimentation is still needed. At Kings Park and Botanic Gardens in WA, about 30 species have been grown from seed, with the first germination occurring within 14-52 days of sowing.

Containers planted with eremophila seed should not be discarded if germination does not take place quickly. Germination of some species has occurred 6 years after sowing.

There is one theory that emus help to break down germination inhibitors as the fruits pass through their digestive systems. Again, further research is necessary in this regard.

Cuttings. This is the most common method used for propagation of eremophilas. Many species produce roots readily from cuttings, but as with seed, some have proved difficult.

Some species produce roots more quickly from firm mature growth than from young growth. Further trials are needed before recommendations can be stated categorically about which type of material should be used. Growths from root suckers usually provide excellent cutting material, with roots forming readily.

Species which have resinous leaves can easily become a sticky mass, especially if the cuttings are collected some time before preparation. One method of separating the cuttings is to dip them in a 50/50 mixture of methylated spirit and water for a very short time. As soon as possible after dipping they should be washed thoroughly in water.

To combat fungal contamination, some propagators rinse the cutting material in a fungicidal mixture prior to preparing the cuttings.

All flowers and buds should be removed from cuttings as they can rot, resulting in infestation of grey mould (*Botrytis cinerea*) which spreads to other plant parts.

Some propagators find treatment of cuttings with a low-strength rooting hormone more beneficial than a medium or strong mixture. Experience will reveal which concentration is best for use under the conditions available.

Once prepared and placed in the propagation structure, the cuttings need to be kept under regular surveillance so that any dead or infected cuttings can be noticed and removed. The leaf-nodes are readily affected, which may mean a rooted cutting has no growth points and is unable to develop new growth. In some cases cuttings may produce sucker growth from below the propagating medium.

Grafting. This method of propagation has been used

on a wide range of species, particularly within the last few years. It has yet to be thoroughly tested, but it certainly has potential for wider use with some species.

Initially, grafting was used with a hardy species (*E. maculata*) as the stock but this did not prove successful. Now *Myoporum insulare*, *M. montanum* and *Eremophila bignoniiflora* are successfully used as the rootstock. This has enabled propagation of species that have proved extremely troublesome from seed or cuttings. Successfully grafted *Eremophila* species:

a) Rootstock — *Myoporum insulare*

E. abietina	E. maitlandii
E. bignoniiflora	E. mitchellii
E. chamaephila	E. pentaptera
E. cuneifolia	E. pterocarpa
E. duttonii	E. pustulata
E. eriocalyx	E. saligna
E. glabra	E. santalina
E. inflata	E. tetraptera
E. interstans	E. virens
E. macdonnellii	E. viscida

b) Rootstock — *Myoporum montanum*

E. denticulata	E. macdonnellii
E. glabra	E. maculata
E. ionantha	E. subfloccosa

c) Rootstock — *Eremophila bignoniiflora*

E. santalina

Grafting offers the benefit that species prone to rotting and lime chlorosis can be grown on a hardy and adaptable rootstock.

Some problems still need to be overcome when grafting eremophilas. Best results will be achieved if vigorous rooted cuttings are used as rootstock. This enables good development of the grafting union and subsequent top growth. Rootstock should not be overwatered during grafting.

The propagation of eremophilas poses many unanswered questions, but further experimentation should provide additional knowledge of the most successful methods. This will enable a greater number of species to become more widely grown.

Eremophila abietina Kraenzlin
(similar to the genus *Abies*)
WA   Spotted Poverty Bush
1-2.5 m x 1-2 m   Aug-Sept
Small to medium **shrub**; **branches** thick, rough, many, very sticky; **leaves** 1.2-2 cm x 0.1 cm, alternate, linear, falcate, dark green, crowded, glabrous; **flowers** tubular, 2-2.5 cm long, white to pale mauve with maroon spots, with short hairs, solitary, on long curved pedicels; **calyx** enlarges after flowering, scarlet or bluish green with cream; **fruit** about 0.7 cm x 0.4 cm.

A showy inhabitant of the Laverton area in the Great Victoria Desert, where it grows on rocky outcrops. Has recently been introduced to cultivation. Will probably prove difficult to grow in other than hot, well drained to dry situations. It may adapt to container cultivation in temperate zones. Tolerant of frost and drought. Propagate from seed or cuttings which are difficult to strike. Grafted successfully on to stock of *Myoporum insulare*.

Eremophila adenotricha (F. Muell. ex Benth.) F. Muell.
(glandular hairs)
WA   Glandular-haired Eremophila
0.6-1.5 m x 0.6-1.2 m   Sept-Nov
Dwarf to small, sticky, hairy **shrub**; **branches** sticky, with rusty glandular hairs; **leaves** 2-3 cm x about 0.1 cm, linear, alternate, crowded, thick and soft, blunt apex, covered with glandular hairs, sticky; **flowers** about 2 cm long, blue, sessile, glabrous exterior; **calyx lobes** hairy, sticky, do not enlarge after flowering; **fruit** ovoid, to 0.5 cm x about 0.3 cm.

Occurs as an erect shrub in the Avon and Coolgardie Districts, where it grows in woodland. Evidently not in cultivation. It will need very well drained soils and a location that receives plenty of sunshine. Drought and frost tolerant. Propagate from seed. Cuttings may prove hard to root because of the dense covering of hairs.

Eremophila alatisepala Chinn.
(winged sepals)
Qld
1.5-2.5 m x 1.5-2.5 m   Aug-Sept
Small to medium, bushy **shrub**; **branches** many, sticky, glabrous; **leaves** 2-6 cm x up to 0.4 cm, linear, crowded, deep green, apex pointed, glandular, sticky, glabrous; **flowers** about 2.5 cm long, cream with pink, axillary, solitary, on pendulous pedicels; **calyx lobes** winged, to 2 cm long, enlarging slightly after flowering; **fruit** to 0.9 cm x 0.6 cm, ribbed, hairy.

A recently named species restricted to the Gregory North and Gregory South Districts. It grows mainly on stony ridges and slopes, and on heavy red clay flats. Its ornamental features make it worthy of cultivation. Will probably adapt to a wide range of well drained soils. Needs a hot, sunny situation. Could be useful as a windbreak in arid or semi-arid regions. Propagate with difficulty from cuttings and possibly seed.

Eremophila alternifolia R. Br.
(alternate leaves)
NSW, SA, WA, NT   Narrow-leaf Fuchsia-bush;
                                           Native Honeysuckle
2-3 m x 2-3 m   Aug-Sept; also sporadic
Small to medium **shrub**; young growth sometimes sticky; **branches** slightly sticky, warty; **leaves** 1-5 cm x about 0.1 cm, linear to subterete, glabrous, apex acute and hooked; **flowers** tubular, 2.5-3 cm long, pink to carmine, rarely white or yellow, spotted interior and glabrous exterior, axillary, solitary, on curved pedicels to 3 cm long, upper lip 4-lobed, lower lip 1-lobed; **stamens** slightly exserted; **calyx lobes** to 1 cm long, purplish, overlapping, broad; **fruit** about 0.6 cm long, ovoid.

Often found on skeletal soils on hills and ranges, and in red loams. It is one of the more commonly cultivated species and has proved to be fairly adaptable. Prefers well drained, heavy soils, with an aspect that is warm to hot for most of the year. Recommended for inland situations, and has potential as an informal hedging plant. Suitable for containers in temperate regions. Hardy to light and medium frosts, and withstands extended dry periods. Regular tip pruning while young

Eremophila angustifolia

Eremophila alternifolia, × ·8

will produce a more compact shrub. Flowers are attractive to honey-eating birds. Propagate from cuttings. Further trials on seed germination are needed.

The var. *latifolia* F. Muell. ex Benth. from SA differs in its thicker, broader leaves which are not hooked. It needs similar conditions to the above.

Eremophila angustifolia (S. Moore) Ostenf. =
E. oldfieldii var. angustifolia S. Moore

Eremophila arachnoides Chinn.
(covered with fine hairs like a spider's web)
SA, WA
2-3 m x 1-2 m   Sept-Oct; also sporadic

Medium, upright, broom-like **shrub**; **branches** ascending, slender, initially white and 4-angled, becoming glabrous and terete, slightly warty; **leaves** 1.5-3.5 cm x about 0.1 cm, narrow-linear, terete to subterete, greyish, opposite, erect or spreading, usually recurved; **flowers** tubular, 1-2.5 cm long, white to mauve, hairy exterior, lobes broad, axillary, solitary or in pairs, on upper parts of branchlets; **stamens** not exserted; **fruit** to 0.7 cm x 0.5 cm, pink, succulent.

A recently described species from the inland near Meekatharra in the Austin District of WA. Evidently not cultivated. It will be best suited to semi-arid and arid regions. Frost tolerant. May grow successfully as a container plant in temperate regions. Propagate from seed or cuttings.

The typical form is confined to WA, while the ssp. *tenera* Chinn. occurs in SA as well. It differs from the above in its branches, which have a more warty appearance.

E. dalyana is very similar, but has dry fruit.

Eremophila battii F. Muell.
(after John D. Batt of Balladonia, WA)
SA, WA, NT
0.3-1 m x 0.5-1 m   Sept-Nov

Dwarf **shrub**; **branches** with white hairs; **leaves** 0.5-

2 cm x up to 0.3 cm, linear-cuneate to linear-spathulate, concave, margin entire or distantly toothed, recurved near tip, hairy; **flowers** tubular, 1.5-3 cm long, blue, blunt lobes, with a few long hairs, axillary, solitary, on short pedicels, on upper branchlets; **stamens** not exserted; **calyx lobes** about 0.7 cm long, hairy; **fruit** about 0.6 cm long.

A Central Australian species which often occurs in spinifex and mulga scrub. It is little-known in cultivation. Has potential for use in dry regions. Succeeds as a container plant in temperate zones. Frost and drought tolerant. Propagate from seed or cuttings.

The var. *major* J. Black is currently not regarded as distinct.

Eremophila behriana (F. Muell.) F. Muell. ex Benth.
(after Dr. Hermann Behr, early botanical collector, SA)
SA
0.2-0.4 m x 0.5-1 m   Sept-Nov

Dwarf **shrub**; **branches** spreading or ascending, rough, hairy; **leaves** 0.8-2.5 cm x 0.3-0.9 cm, obovate-cuneate, rigid, erect, crowded, margins hairy, toothed near apex and often recurved; **flowers** tubular, about 1 cm long, lilac to purple, glabrous exterior, lobes pointed, axillary, solitary, subsessile; **stamens** not exserted; **calyx lobes** short, narrow; **fruit** ovoid, to 0.5 cm long.

Endemic to SA in locations such as York Peninsula, Eyre Peninsula and Kangaroo Island, where it occurs in mallee scrub. Has adapted well to cultivation. Best suited to medium or heavy soils. Appreciates plenty of sunshine. Hardy to frosts and extended dry periods. Responds well to pruning which promotes bushy growth. A prostrate suckering form from Cummins on Eyre Peninsula has potential for wider use. Propagate from seed or cuttings which strike readily.

Eremophila bicolor Chinn. = E. racemosa Endl.

Eremophila bignoniiflora (Benth.) F. Muell.
(flowers similar to those of the genus *Bignonia*)
Qld, NSW, Vic, SA, WA, NT   Bignonia Emu-bush;
Creek Wilga; Eurah;
Gooramurra
2-7 m x 1.5-4 m   May-Nov; also sporadic

Medium to tall, rounded **shrub** or small **tree**; **bark** rough, grey; **branches** glabrous; **branchlets** often pendulous, slightly sticky, glabrous; **leaves** 3-18 cm x up to 1.5 cm, linear-lanceolate to lanceolate, alternate, pale green, glabrous, usually entire, pointed apex; **flowers** tubular, 2-3 cm long, cream, tinged red above, exterior glabrous, green or purplish-spotted interior, lobes rounded and spreading, axillary, solitary, on sticky pedicels to 1.5 cm long, fragrant; **stamens** slightly exserted; **calyx lobes** sticky, broad; **fruit** ovoid, 1.5-2 cm x 1-1.3 cm, more or less succulent before maturity.

A very widespread species of the inland. It usually grows beside waterways, in drainage channels and near lakes, on heavy clay soils that can be subject to flooding. A most adaptable species, ideally suited to heavy soils in semi-arid and arid zones. Does not tolerate highly alkaline soils. It will also grow in temperate regions, provided it receives plenty of sunshine

Hardy to most frosts. Suitable as a screen or low wind-break. Some plants may sucker lightly. Responds well to pruning.

Propagate from seed or cuttings which readily form roots. Root cuttings taken from fairly thick roots can also be successful. This species has been used as a root-stock for grafting of *E. santalina*.

Suspected of poisoning stock in Qld and NSW.

Eremophila biserrata Chinn.

(twice serrate)
WA   Prostrate Eremophila
prostrate x 1.5-3 m   Sept-Nov

Dwarf, spreading **shrub**; **branches** prostrate, self layering at nodes on older wood, glandular, hairy; **leaves** 1.4-3 cm x 0.4-1.1 cm, oblanceolate to spathulate, alternate, petiolate, erect to slightly spreading, crowded, glandular, hairy, margins toothed, apex blunt; **flowers** tubular, to about 3 cm long (including exserted stamens), lime green to yellowish green, with brownish purple on upper lip, exterior hairy, axillary, solitary, erect, on pedicel to 0.5 cm long, often obscured by foliage; **stamens** red; **calyx lobes** to 0.5 cm long, hairy; **fruit** oblong to pear-shaped, about 0.5 cm x 0.4 cm.

A dense, spreading groundcovering species from the Hyden-Forrestiana-Lake King area in the Coolgardie District. It grows on sand or sandy clay soils which can be saline. In cultivation it has adapted well to a wide range of soils including alkaline types. Grows well in open or semi-shaded situations and, because of its ability to self layer, is an excellent stabilizer for small areas of recently disturbed soils. Tolerant of frost and drought. Withstands harsh pruning. Resents foot traffic. The flowers are attractive to honey-eating birds. Propagate from seed, cuttings which root readily or sections of self layered stems. This species has been grown under the unpublished name of *E.* 'versicolor'.

Eremophila bowmanii F. Muell.

(after Edward Bowman, original collector)
Qld, NSW   Bowman's Emu-bush
0.3-2 m x 0.5-2 m   May-Nov; also sporadic

Dwarf to small, spreading or upright **shrub**; **branches** ascending, hairy, brown to reddish brown; **branchlets** warty, densely covered with white hairs; **leaves** 1-5 cm x up to 0.4 cm, alternate, linear to linear-lanceolate, covered with hairs giving a white to glaucous appearance, margin entire and revolute, apex blunt to pointed; **flowers** tubular, to 2.5 cm long, light to dark blue or lavender, spotted interior, glabrous to hairy exterior, lobes blunt, axillary, solitary or rarely paired, on long, hairy pedicels; **stamens** not exserted; **calyx lobes** to 2 cm long, densely hairy; **fruit** ovoid, to about 0.7 cm long.

This species has highly ornamental foliage and flowers. It occurs in inland Qld and NSW, where it grows on the arid shrublands that are dominated by *Acacia* species. It is not common in cultivation. Best planted during warmer months. Does best in well drained soils, with a warm to hot aspect. Tolerates dappled shade but prefers plenty of sun. Frost and drought tolerant. Grows well as a container plant in temperate regions, but can be prone to attack by grey mould during still and humid weather. Propagate from seed or cuttings.

The var. *latifolia* L. S. Smith differs in its lanceolate to orbicular leaves which can be up to 1 cm wide and do not have revolute margins. This variety is also cultivated.

Eremophila brevifolia (A. DC.) F. Muell.

(short leaves)
WA   Spotted Eremophila
1-4 m x 1-4 m   Aug-Dec

Small to medium **shrub**; **branches** glabrous, striate; **leaves** to 1 cm x 1 cm, alternate, ovate to orbicular, glabrous, sessile, glandular on undersurface, margin entire or toothed; **flowers** tubular, to 1.5 cm long, white or pale mauve, exterior glabrous, lobes spreading and blunt, constricted at base, axillary, solitary, in upper branchlets; **stamens** not exserted; **calyx** with glandular spots, glabrous.

An uncommon species in the wild. It occurs in the Avon, Darling and Irwin Districts. Fairly well known in cultivation, and seems adaptable to a wide range of soils, including alkaline types. Needs good drainage. Suited to dappled shade, and partial or full sunshine. Responds well to hard pruning. Hardy to frost and drought. Will grow in semi-exposed coastal situations. May be subject to aphid attack. Propagate from seed or cuttings which readily produce roots.

The var. *flabellifolia* is now considered conspecific.

Eremophila caerulea (S. Moore) Diels

(deep blue)
WA
0.5-1 m x 0.5-1 m   Oct-Nov

Dwarf, dense **shrub**; **branches** spreading or ascending, warty; **branchlets** densely hairy; **leaves** to 1 cm x about 0.1 cm, alternate, narrow-linear, subterete, crowded, ascending, thick, warty, initially hairy, becoming glabrous, apex blunt; **flowers** tubular, about 1.5 cm long, violet, blue or purple with dark spots, exterior slightly hairy, axillary, solitary, near ends of upper branchlets; **stamens** not exserted; **calyx lobes** hairy; **fruit** ovoid, to 0.4 cm x 0.3 cm.

A showy dwarf species from the Coolgardie, Eyre and Roe Districts, where it often occurs in lateritic soils. In cultivation it seems to do well in most types of soil, provided drainage is adequate. Will tolerate dappled shade but prefers partial or full sunshine. Withstands most frost and is drought tolerant. Responds well to pruning. Has high potential as a container plant. Propagate from seed or cuttings which usually form roots readily.

Eremophila calorhabdos Diels

(after the genus *Calorhabdos* or beautiful wand)
WA   Red Rod; Spiked Eremophila
1.5-3 m x 0.5-2 m   July-Feb; also sporadic

Small to medium, upright **shrub**; **branches** erect, few, densely covered with short hairs; **leaves** 1.5-2.5 cm x 0.5-1 cm, ovate-oblong to elliptic-oblong, slightly concave, initially covered in short hairs, becoming glabrous, crowded, margins toothed near apex; **flowers** tubular, to 3 cm (including exserted

Eremophila calycina

Eremophila calorhabdos, × ·7

stamens), pink, red or purplish, exterior glabrous, lobes pointed and recurved, axillary, solitary, on short pedicels, along branches, often forming a leafy spike, profuse; **calyx lobes** pointed, short, glabrous; **fruit** ovoid, about 0.7 cm x 0.5 cm.

A well known species of semi-arid country in the Coolgardie, Eyre and Roe Districts. It usually occurs on sand or loam in the woodlands. Hardy and adaptable in cultivation. Grows in a wide range of soils, but must have good drainage. Will grow in shaded situations, but may develop as a leggy, few-branched plant, whereas if grown in partial or full sun, plants will be bushy. Does well against a wall or fence, in full sun. Pruning from an early stage helps to promote side growth. Tolerant of frost and drought. In temperate climates plants are often subject to attack by grey mould during still, humid weather. This usually starts to develop on dying flowers. Successfully grown as a container plant. The flowers are attractive to honey-eating birds. Subject to attack by grey aphids which can be controlled by pyrethrum. Propagate from seed or cuttings which produce roots readily.

Eremophila calycina S. Moore =
$\qquad$ *E. duttonii* F. Muell.
For plants previously known under *E. calycina* see *E. neglecta* J. Black.

Eremophila chamaephila Diels
(earth-loving)
WA
0.25-0.3 m x 0.25-0.8 m $\qquad$ Sept-Nov
Dwarf, compact **shrub**; **branches** many, spreading; **leaves** to 0.3 cm long, linear-oblong, appressed against branches, glabrous, warty; **flowers** tubular, to 1 cm long, violet to rich purple, exterior glabrous, axillary, solitary, in upper branchlets, lobes pointed; **stamens** scarcely exserted; **calyx lobes** narrow, warty, glabrous; **fruit** ovoid, to 0.3 cm x 0.2 cm.

A dense groundcovering species from low-rainfall areas in the Coolgardie, Eyre and Roe Districts. It is

rare in cultivation but shows potential. Needs well drained soils in a situation of partial or full sunshine. Is tolerant of alkaline soils and withstands semi-exposed coastal conditions. Hardy to frost and extended dry periods. Ideal as a container or rock-garden plant. Propagate from seed or cuttings. Grafted successfully using *Myoporum insulare* as the stock.

Eremophila christophori F. Muell.
(after Christopher Giles)
NT
1-2 m x 1-2 m $\qquad$ May-Sept; also sporadic
Small **shrub**; **branches** erect, sparsely hairy, hidden by leaves; **leaves** 1.5-3 cm x up to 1 cm, elliptic to narrow-lanceolate, alternate, erect, crowded, glandular and hairy or glabrous, margins entire, apex pointed or blunt; **flowers** tubular, about 2 cm long, blue, hairy exterior, lobes pointed, axillary, solitary, subsessile, near ends of branchlets; **stamens** not exserted; **calyx lobes** to 0.7 cm long, narrow; **fruit** narrow-ovoid, to 0.6 cm x 0.2 cm.

A showy Central Australian species that occurs on stony rises in the MacDonnell Ranges. Is mainly cultivated by enthusiasts at present. It can flower continuously. Needs very well drained soils, with partial or full sunshine. Tolerant of most frosts and extended dry periods. Tip pruning when young promotes bushy growth. Propagate from seed or cuttings.

Eremophila clarkei Oldfield & F. Muell.
(after William Clarke, patron of botanical collecting)
SA, WA $\qquad$ Turpentine Bush
1-4 m x 1-3 m $\qquad$ July-Oct; also sporadic
Small to medium **shrub**; **branches** faintly hairy; **leaves** 1.5-5 cm x 0.3-1 cm, alternate, narrow-lanceolate, slightly hairy, margins entire or toothed, apex pointed, sticky; **flowers** tubular, to about 3 cm long, white, mauve, pale blue or pink, hairy exterior,

Eremophila christophori, × ·7

434

solitary, on S-shaped pedicels to 2 cm long, near ends of branchlets; **stamens** not exserted; **calyx lobes** green to purplish, enlarging slightly after flowering; **fruit** ovoid, about 0.8 cm to 0.6 cm.

A most variable and decorative species with differing forms occurring on a wide range of soil types in low-rainfall areas. It is most common on heavy soils. Has had limited cultivation, but shows potential in semi-arid regions. Not highly recommended for planting in temperate zones, where it may do best as a container plant. Tolerant of frost and drought. Pruning is proving beneficial in providing bushy growth. Propagate from seed or cuttings which can be difficult to strike. Plants in cultivation with narrow-toothed leaves and mauve to pink flowers, known as a form of *E. granitica*, are most likely referable to *E. clarkei*.

This species is closely related to and difficult to separate from *E. georgei*, and both are part of a complex with many forms.

Eremophila compacta S. Moore
(compact)
WA
0.5-1.5 m x 1-1.5 m   July-Sept

Dwarf to small compact **shrub**; **branches** have persistent leaf-bases; **leaves** about 1 cm x 0.4 cm, alternate, linear-oblong to oblanceolate, more or less sessile, warty, with dense grey tomentum; **flowers** tubular, to 2 cm long, pale crimson or bluish purple, glabrous, lobes spreading and blunt, solitary, on short pedicels; **stamens** not exserted; **calyx lobes** hairy; **fruit** globular, about 0.9 cm x 0.8 cm.

From the Austin District, where it usually occurs on stony soils. Evidently not very well known in cultivation. It has well displayed flowers and with the grey, hairy foliage it shows ornamental potential. Will need very well drained soils, in a situation that receives plenty of sunshine. Drought and frost tolerant. Propagate from seed or cuttings.

It is suspected that a form of this species has been cultivated under the wrong name of *E. maitlandii* which is very rare in cultivation. This species has some affinities to *E. cordatisepala*, which has sky blue flowers and non-warty leaves.

Eremophila cordatisepala L. S. Smith
(heart-shaped sepals)
Qld, NT
0.3-0.6 m x 0.5-1 m   July-Nov

Dwarf, spreading **shrub**; **branches** many, densely hairy, becoming greyish or yellow with age; **leaves** 0.5-1.5 cm x up to 0.8 cm, alternate, obovate to oblanceolate, crowded, densely hairy, margins entire, apex pointed or blunt; **flowers** tubular, about 2 cm long, sky blue, almost glabrous, axillary, solitary, on short pedicels; **stamens** not exserted; **calyx lobes** enlarging after flowering; **fruit** ovoid, to 1 cm x 0.6 cm, glabrous, usually shiny.

This species occurs on a range of soils including loams and clay-loams which are near limestone outcrops, as well as growing on stony soils in hilly areas. It is extremely rare in cultivation and, to date, has been grown only by eremophila enthusiasts. Is probably best

suited to semi-arid and arid regions, but further trials may prove it hardy also in temperate climates. Should be established during the warmer months. Must have very well drained soils, with plenty of sunshine.

Tolerates alkaline soils, extended dry periods and most frosts. Has potential as a container plant. Propagate from seed or cuttings.

Differs from the allied *E. compacta*, which has reddish flowers and warty leaves.

Eremophila crassifolia F. Muell.
(thick leaves)
Vic, SA   Trim Emu-bush
0.6-1.3 m x 0.6-2.0 m   Aug-Dec

Dwarf to small, spreading **shrub**; **branches** glabrous; **leaves** 0.5-1.5 m, alternate, ovate, subsessile, concave above, thick, rigid, green, glabrous; **flowers** tubular, to 1.2 cm long, blue, occasionally pink or white, glabrous exterior, lobes spreading and pointed, axillary, solitary, subsessile, near ends of branchlets; **stamens** not exserted; **calyx lobes** narrow, margins hairy; **fruit** ovoid, about 0.4 cm long, succulent, glabrous.

A relatively hardy species that occurs in desert areas from the Vic Mallee and extends to Eyre Peninsula in SA. Fairly well known in cultivation, it grows in most well drained soils (including alkaline). Best suited to a sunny situation. Drought and frost tolerant. Responds well to pruning. Propagate from seed or cuttings which produce roots readily.

A natural intergeneric hybrid between this species and *Myoporum platycarpum* occurs in SA, and has had limited cultivation.

Eremophila cuneifolia Kraenzlin
(wedge-shaped leaves)
WA   Pinyuru
0.3-1.5 m x 0.6-1.5 m   July-Oct

Dwarf to small, spreading **shrub**; young growth sticky; **branches** rough and densely hairy; **leaves** 1-2 cm x 0.5-1.5 cm, alternate, wedge-shaped to orbicular, crowded, faintly hairy, sticky, apex often notched; **flowers** tubular, about 2 cm long, pale purple to blue, hairy exterior, axillary, solitary, subsessile, near ends of branchlets; **calyx lobes** enlarge after flowering, 1-1.5 cm long, purplish or rarely whitish; **fruit** ovoid to 0.8 cm x 0.5 cm.

A most ornamental species with well displayed flowers and coloured calyx lobes. It occurs in loamy and rocky soils of arid regions such as the Gibson Desert. Rare in cultivation. Probably best suited to soils with adequate drainage, in positions where there is plenty of sunshine. Worthy of trial as a container plant in temperate zones. Drought and frost tolerant. Propagate from seed or cuttings which are extremely difficult to strike. Grafting on to stock of *Myoporum insulare* has been successful but has not yet been widely tried.

Eremophila dalyana F. Muell.
(after Sir Dominick Daly, 19th-century Governor, SA)
Qld, SA, NT
1-7 m x 1-4 m   May-Sept

Small **shrub** to small **tree**; **branches** covered with

435

silvery scales, warty; **leaves** 2-6 cm x about 0.1 cm, opposite, narrow-linear, somewhat terete, silvery, entire, apex pointed and recurved; **flowers** tubular, to 2.5 cm long, white to lilac or pinkish, constricted near middle, 1-2 slender, curved pedicels per axil; **stamens** not exserted; **calyx lobes** not enlarging after flowering; **fruit** elongated, to 1 cm x about 0.2 cm.

From the stony slopes in arid regions of Central Australia. This species is currently grown only by enthusiasts. Does best in semi-arid regions, with only limited success in temperate zones when ground-grown. Will grow in relatively well drained soils, and needs a warm to hot location. May suffer damage from heavy frosts. Drought tolerant. Responds well to pruning. Tip pruning while young helps promote lateral growth. Has potential as a container plant in temperate regions. Propagate from seed or cuttings.

Eremophila decipiens Ostenf.

(misleading)

SA, WA	Slender Fuchsia
1-2 m x 1-2 m	Aug-Nov; also sporadic

Small **shrub**; young growth sticky; **branches** many, usually sticky, initially hairy, becoming glabrous; **leaves** 2-4 cm x 0.3-1 cm, alternate, linear to lanceolate, flat, initially slightly hairy, becoming glabrous, apex pointed; **flowers** tubular, to 2.5 cm long, red, glabrous, 4 upper lobes, 1 lower lobe, axillary, solitary, on S-shaped pedicels, along branches; **stamens** exserted; **calyx lobes** pointed, glabrous; **fruit** ovoid to globular, to 0.8 cm long, glabrous, succulent.

A widespread species that occurs on a wide range of soils from the Murchison River in WA, and extends to SA. It is one of the most reliable and commonly cultivated species. Seems to adapt well to most soils, provided there is good drainage. It will grow in dappled shade, or partial or full sun. Hardy to frost and extended dry periods. Responds well to pruning. Does well as a container plant. Commonly visited by honey-eating birds when in flower. Propagate from seed or cuttings which usually produce roots readily.

This species is sometimes confused with forms of *E. glabra*, which does not have S-shaped pedicels. *E. maculata* var. *linearifolia* S. Moore is included in this species. In the forthcoming revision it is proposed that the SA forms of this species will be absorbed into *E. glabra*.

Eremophila delisseri F. Muell.

(after Edmund A. Delisser, 19th-century SA squatter and surveyor)

SA, WA	
0.3-1.5 m x 0.6-1.5 m	Aug-Jan

Dwarf to small **shrub**; **branches** densely covered in white hairs, entangled, sometimes warty; **leaves** 0.5-1 cm x 0.3-0.5 cm, mostly opposite, obovate-oblong, subsessile, white or yellowish due to dense covering of stellate hairs, margin entire, apex recurved; **flowers** tubular, 2-2.5 cm long, pale lavender, violet or blue, hairy exterior, lobes blunt, axillary, solitary, sessile, near ends of branchlets; **stamens** not exserted; **calyx lobes** narrow, hairy; **fruit** ovoid, woolly, to 0.6 cm x 0.4 cm.

An eyecatching species with ornamental foliage and flowers. It occurs in arid regions from Ooldea, SA to the Coolgardie District in WA. To date it has been grown mainly by enthusiasts, in low-rainfall regions. Needs very well drained soils and plenty of sunshine. Frost and drought tolerant. Should make a wonderful container or rock-garden plant. Propagate from seed or cuttings which usually produce roots readily.

Eremophila dempsteri F. Muell.

(after Andrew Dempster, one of the original settlers of Esperance, WA)

WA	
2-4 m x 1-2.5 m	Aug-Nov; also sporadic

Medium, upright **shrub**; young growth slightly sticky; **branches** erect, lined, slender; **leaves** 0.6-0.8 cm x about 0.1 cm, alternate, scattered, linear, glabrous, sticky, apex pointed, usually recurved or hooked; **flowers** tubular, about 1.2 cm long, violet to mauve, sweetly scented, exterior hairy, axillary, solitary, on slender pedicels; **stamens** not exserted; **calyx lobes** densely hairy, narrow; **fruit** ovoid, to 0.5 cm x 0.2 cm.

An upright, broom-like species from the Coolgardie, Eyre and Roe Districts, where it grows on sand, loams or lateritic soils. It is ornamental in bud and fruit due to the conspicuous line of hairs on the calyx lobes. Poorly known in cultivation (see below). Suited to most well drained soils. It will grow in dappled shade, but will develop into a stronger shrub if in a sunny location. Frost and drought tolerant. Responds well to pruning. Tip pruning from an early stage helps to promote lateral growth. Propagate from seed or cuttings.

This species has been confused in cultivation with an apparently unnamed species from the Kalgoorlie region, that has affinities to *E. drummondii*. Plants of the latter are commonly cultivated, whereas the true *E. dempsteri* is rarely grown. This unnamed species has a low, spreading growth habit, with longer, flat leaves that are not recurved, and the calyx and waxy purple flowers are glabrous.

Eremophila densifolia F. Muell.

(dense leaves)

WA	
0.1-0.5 m x 1-3 m	July-Dec

Dwarf, spreading **shrub**; **branches** glabrous or hairy; **leaves** to 2 cm x 0.2 cm, narrow-linear to linear-lanceolate or terete, green to purple-green, hairy, sessile, thick, convex below, crowded, margins can be hairy, apex pointed; **flowers** tubular, about 1.2 cm long, purple, violet or blue, glabrous exterior, lobes pointed and spreading, solitary, almost sessile, near ends of branchlets or forming leafy spikes along branches; **stamens** not exsereted; **calyx** hairy; **fruit** ovoid, hairy, to 0.3 cm x 0.2 cm.

A relatively common species from the Coolgardie, Eyre and Roe Districts, where it usually occurs on red-brown earths or lateritic soils. It is well known in cultivation. Prostrate forms are the most commonly grown — one of these has purplish leaves and is becoming popular. Seems to be very adaptable, as it does well in most soils. Tolerates a semi-shaded position but is more vigorous, with a greater display of flowers, if

Eremophila densifolia, × 1   flower, × 2

grown in a position receiving partial or full sun. Suited to coastal planting. Can withstand harsh pruning and is tolerant of extended dry periods and most frosts, although the coastal form can suffer damage in severe frost. Propagate from seed or cuttings which strike well.

An upright species which may be a form of this species or one yet to be described, has been collected.

Eremophila denticulata F. Muell.
(minutely toothed)

WA   Fitzgerald Eremophila
1-2.5 m x 1-3.5 m   Sept-March; also sporadic

Small **shrub**; young growth sticky; **branches** more or less glabrous; **leaves** 2.5-6 cm x up to 1.2 cm, alternate, lanceolate to oblong-elliptical, dark green, petiolate, flat, margins with small teeth, apex pointed, sticky; **flowers** tubular, to 3 cm long, initially yellow, ageing red, constricted just above base, glabrous exterior, lobes pointed, with 4 above and 1 below, solitary, on S-shaped pedicels about 1.5 cm long; **stamens** exserted; **calyx** glabrous; **fruit** ovoid, about 1 cm x 1 cm.

This eremophila is now gazetted as a rare species in nature. It occurs only in a small area south-west of Ravensthorpe, where it grows on loam beside rivers. It is, however, well entrenched in cultivation and has proved most adaptable, growing in a wide variety of soils, provided there is adequate drainage. Will grow in dappled shade but does best where it receives plenty of sunshine. Tolerant of extended dry periods and most frosts. The flowers are excellent for attracting honey-eating birds. Can become leggy, but responds well to harsh pruning. Regular tip pruning is recommended from an early stage to promote lateral growth. May be subject to aphid and scale attack. Propagate from seed or cuttings which produce roots readily.

Eremophila dichroantha Diels
(2-coloured flowers)

WA   Bale-hook Eremophila
1-2 m x 1-2 m   Aug-Dec

Small **shrub**; **branches** very slender, glabrous, can be pendulous; **leaves** about 1 cm x 0.1 cm, opposite, linear to narrow-linear, thin, warty, prominently hooked, glabrous; **flowers** tubular, to 1.2 cm long, blue to violet, scented, slightly inflated, hairy exterior, lobes blunt and spreading, axillary, solitary, on short pedicels, along branchlets, profuse; **stamens** not ex-

serted; **calyx lobes** red-veined, with hairy margins; **fruit** ovoid, about 0.3 cm x 0.15 cm.

A graceful species with slender, often pendulous branches that can be covered with the small bluish flowers. The coloured calyces, although small, are decorative. It grows on sand and lateritic soils in the Coolgardie, Eyre and Roe Districts. Is fairly common in cultivation and has adapted to most well drained soils. Tolerates dappled shade but flowers better in an open, sunny situation. Seems to withstand strong winds. Hardy to extended dry periods and most frosts. Responds well to pruning. Suitable for use as a container plant. Propagate from seed or cuttings which produce roots readily.

Eremophila divaricata F. Muell.
(widely spreading or forked)

NSW, Vic, SA   Spreading Emu-bush
1-2 m x 1.5-3 m   Sept-Feb

Small, spreading **shrub**; **branches** often entwined and ending in spines, glabrous except for new leaf axils; **leaves** 0.5-1.5 cm x up to 0.5 cm, alternate, linear-lanceolate to linear-cuneate, glabrous, margins entire, apex pointed; **flowers** tubular, to 1.5 cm long, blue to lilac, constricted near base, slightly hairy exterior, lobes pointed and spreading, solitary, nearly sessile, along branches; **stamens** not exserted; **calyx lobes** short, fringed margins; **fruit** ovoid, about 0.8 cm long, curved and beaked.

A species of the flood plains, where it usually occurs on soils of medium to heavy texture. Has become relatively well known in cultivation, and grows in a wide range of soil and climatic conditions. Does well in heavy soils. Tolerates semi-shaded conditions but is better suited to a sunny situation. Hardy to extended dry periods and most frosts. Can be pruned harshly, and is a useful low screening plant. Suited to use in large containers. Propagate from seed or cuttings which produce roots quickly.

Eremophila dichroantha, × ·7   flower, × 1·4

Eremophila drummondii F. Muell.
(after James Drummond, first Government Botanist, WA)
WA
0.5-2.5 m x 0.5-1.5 m   July-Nov
   Dwarf to medium, upright **shrub**; young growth sticky; **branches** erect, glabrous, many; **leaves** 2-6 cm x about 0.1 cm, alternate, narrow-linear or terete, erect, glabrous, sticky; **flowers** tubular, to 2 cm long, violet or blue, constricted at base, glabrous exterior, lobes short, axillary, on long slender pedicels, solitary or in pairs, profuse, along branchlets; **stamens** not exserted; **calyx lobes** pointed, glabrous; **fruit** ovoid, to 0.7 cm x 0.4 cm.
   The distribution of this species is in the Avon, Coolgardie and Roe Districts, where it occurs on sandy or lateritic soils in sandheath or woodland. A showy species in flower, but not extremely well known in cultivation. Low-growing and tall forms are grown. Suited to most soils that have adequate drainage. Prefers a sunny situation but is tolerant of dappled shade. Hardy to frost and drought. Responds well to pruning. The low form has potential as a container plant.
   The var. *brevis* S. Moore differs in its much shorter pedicel. The leaves are up to about 2 cm long, and are shiny and sticky.
   Propagate both forms from seed or cuttings which strike readily.
   This species probably includes the apparently undescribed species from Kalgoorlie which has been sold as *E. dempsteri*. It differs in its broader, flat leaves and extended flowering season.

Eremophila aff. *drummondii* 'Kalgoorlie', × ·75

Eremophila duttonii F. Muell.
(after Francis S. Dutton, 19th-century explorer and Premier of SA)
Qld, NSW, SA, WA, NT   Budda;
                              Harlequin Fuchsia-bush
1-4 m x 1-4 m   May-Dec
   Small to medium, dense to sparse **shrub**; young growth sticky; **branches** slightly hairy, rough; **leaves** 2-6 cm x up to 0.5 cm, alternate, linear to narrow-lanceolate, flat, crowded, often only at branchlet-tips, initially hairy, becoming glabrous, margins entire, apex pointed and often hooked; **flowers** tubular, to 3.5 cm long, initially yellow, becoming red with yellow below and inside, glabrous exterior, constricted near base, lobes pointed and slightly reflexed, solitary, on long S-shaped pedicels, in upper axils; **stamens** exserted; **calyx lobes** sticky, enlarge after flowering to 2.5 cm long, green; **fruit** ovoid, to 1 cm x 1 cm, glabrous.
   A common and very showy species of inland plains. It grows as a rounded bush on red sands and loams. Is rare in cultivation and is usually slow-growing. It must have well drained soils, and is probably best suited to loams. Needs maximum sunshine to grow well. Tolerant of alkaline soils, most frosts and drought. Should be useful as a screen and windbreak in arid regions. Honey-eaters gain nectar from the flowers. Doubtful whether it will succeed in temperate zones, but may grow in large containers.
   Propagate from seed or cuttings which can be slow to form roots. Successfully grafted using *Myoporum insulare* as the stock.

Eremophila elderi F. Muell.
(after Sir Thomas Elder, sponsor of 19th-century exploring expeditions)
Qld, SA, WA, NT
0.5-1.5 m x 0.6-1.5 m   July-Oct
   Dwarf to small **shrub**; **branches** hairy; **leaves** 5-8 cm x about 2.5 cm, alternate, ovate to lanceolate, sessile, crowded, densely glandular-hairy, aromatic, margins entire, apex pointed; **flowers** tubular, about 2 cm long, white, lilac or blue with red patches in throat, hairy exterior, lobes pointed, axillary, on long hairy pedicels, solitary or in pairs, near ends of branches; **stamens** not exserted; **calyx lobes** pointed, hairy, do not enlarge after flowering; **fruit** globular to ovoid, 1-1.3 cm long.
   This species has proved difficult to maintain in cultivation. It occurs in the arid interior, where it often grows on stony soils. Seems to require very good drainage and a location that receives plenty of sunshine. Doubtful whether it will succeed in temperate zones, but is worth trying as a container plant. Best established in warmer months. Frost and drought tolerant. Propagate from seed or cuttings which can be difficult to strike. Grafting of this species may increase the life span of cultivated plants.

Eremophila eriocalyx F. Muell.
(woolly calyx)
WA   Desert Pride
1-3 m x 0.6-2.5 m   Aug-Dec
   Small to medium **shrub**; **branches** hairy; **leaves** 0.3-

Eremophila drummondii   W.R. Elliot

2.5 cm x up to 0.2 cm, alternate, linear, hairy, margins revolute, apex pointed; **flowers** tubular, to 3 cm long, cream, pinkish yellow to orange-red or violet, usually hairy exterior, densely hairy interior, lobes blunt, on solitary hairy pedicels, near ends of branchlets, profuse; **stamens** not exserted; **calyx lobes** broad, hairy; **fruit** ovoid, about 0.6 cm long, slightly flattened, ribbed.

An ornamental eremophila from arid WA, where it occurs on clay flats. It is not commonly cultivated, but plants have been successfully established in areas of low rainfall. Needs relatively good drainage and maximum sunshine. Tolerates alkaline soil, most frosts and drought. Plants can become leggy in cultivation, but tip pruning from an early stage will help to promote bushy growth. Grows well in containers.

Eremophila exilifolia, × ·65

Propagate from seed or cuttings which usually strike readily.

For many years this species was known under the illegitimate name of *E.* 'eriobotrya'.

Eremophila exilifolia F. Muell.
(small or thin leaves)
WA
0.5-1 m x 0.8-1.5 m   July-Sept; also sporadic
Dwarf, flat-topped **shrub**; young growth sticky; **branches** hairy, blackish with resin; **leaves** 0.3-0.6 cm x about 0.1 cm, alternate, narrow-linear, crowded, recurved, glabrous, sticky, apex blunt; **flowers** tubular, about 1 cm long, pinkish white, lilac or blue to violet, hairy exterior, lobes pointed, on solitary S-shaped pedicels, along branchlets; **stamens** not exserted; **calyx lobes** slightly hairy, broad; **fruit** ovoid, to 0.6 cm long, glabrous.

Occurs in arid locations such as the Gibson and Great Victoria Deserts, as well as in the Austin District to the west. Rarely cultivated. It should prove difficult to maintain unless it has a very sunny position in well drained soil. Frost and drought tolerant. May adapt to container cultivation. Propagate from seed or cuttings which can be very slow to form roots.

Bees have been observed collecting propolis from the leaves. They use the substance for blocking crevices.

Eremophila exotrachys Kraenzlin =
                                    E. platythamnos Diels

Eremophila falcata Chinn.
(sickle-shaped)
WA
1-2 m   Sept-Oct
Small **shrub**; young growth sticky; **branches** whitish grey, glandular, glabrous; **leaves** 1.5-3.5 cm x up to 0.6 cm, alternate, falcate, erect, margins entire, glabrous, shiny, sticky, apex pointed and hooked; **flowers** tubular, to 2 cm, white to pale purplish pink, exterior slightly hairy, lobes blunt; **stamens** not exserted, pedicels short, 1-4 per axil; **calyx lobes** blunt, slightly hairy, especially on margins; **fruit** ovoid, to 0.6 cm long, beaked.

A recently described species that occurs in arid WA, near Neale Junction in the Great Victoria Desert, and extends north-westwards to the southern region of the Hamersley Range. It occurs in mulga woodland on clay-loams over limestone. Not known in cultivation. Best suited to semi-arid and arid climes. Will need good drainage, with maximum sunshine. Frost and drought tolerant. Propagate from seed or cuttings which have not been successful to date.

This species has some affinities to *E. paisleyi*.

Eremophila foliosissima Kraenzlin
(many leaves)
WA   Poverty Bush
1-2 m x 1-2 m   July-Oct
Small **shrub**; **branches** with white hairs; **leaves** 4-6 cm x about 0.1 cm, narrow-linear to filiform, crowded, covered in long white hairs, margins strongly revolute, apex pointed; **flowers** tubular, about 2 cm

439

long, pale to deep lilac or mauve-blue, exterior glabrous, lobes pointed, solitary, on hairy pedicels, near ends of branchlets; **stamens** not exserted; **calyx lobes** hairy, narrow; **fruit** globular, 0.7 cm x 0.7-1 cm.

An eyecatching species that is an inhabitant of the Austin District. It grows on clay-loams in mulga woodland. Evidently not in cultivation, but will be best suited to semi-arid and arid regions. The soil should be adequately drained, in a sunny location. Frost and drought tolerant. May grow as a container plant in temperate zones. Propagate from seed or cuttings which have so far failed to produce roots. Grafting may have some application if the species is to be grown over a wider range.

Eremophila forrestii F. Muell.
(after Sir John Forrest, 19th-century WA explorer and parliamentarian)
WA
0.3-1 m x 0.6-1.5 m   Aug-Oct
Dwarf **shrub**; **branches** densely hairy, whitish; **leaves** 1.5-5 cm x up to 1 cm, opposite or alternate, obovate-oblong, crowded, thick, soft, margins entire, apex blunt; **flowers** tubular, about 2 cm long, white to violet, exterior slightly hairy, lobes pointed, solitary, on hairy pedicels; **stamens** not exserted; **calyx lobes** hairy, pointed; **fruit** ovoid, to 0.9 cm x 0.6 cm.

A species from arid areas in the Austin District, where it grows on soils derived from greenstone. Evidently not known in cultivation. It has a whitish to yellowish appearance due to the dense covering of hairs. Needs very well drained soils and a position where it can receive maximum sunshine. Drought and frost tolerant. Propagate from seed. Cuttings may prove difficult to strike because the woolly stems and leaves may rot in humid conditions.

Eremophila fraseri F. Muell.
(after Sir Malcolm Fraser, 19th-century WA Surveyor-General)
WA   Burra
1.5-4 m x 1.5-3 m   June-Oct; also sporadic
Small to medium **shrub**; young growth sticky; **branches** rough, glabrous, sticky; **leaves** 1.5-8 cm x 0.5-1.5 cm, alternate, ovate to elliptic, petiolate, glabrous, very sticky, margins wavy, apex pointed; **flowers** tubular, about 4 cm long, brownish red outside, white to pink inside, exterior slightly hairy, lobes pointed, solitary, on long erect pedicels; **stamens** exserted; **calyx lobes** enlarge after flowering, pink to purple; **fruit** ovoid, about 0.8 cm long, glabrous, sticky.

A very sticky species that occurs inland from the Hamersley Range and south-eastwards to near Laverton. It grows on a wide range of soils and is often on open plains or rocky hills in mulga scrub. Extremely rare in cultivation. It should do best in relatively well drained soils, in a hot, sunny situation. Hardy to frost and drought. May be successful as a container plant in temperate zones. Propagate from seed or cuttings. Both methods are difficult.

The Aborigines used the resin from the leaves as a cementing material.

Eremophila freelingii F. Muell.
(after Sir Arthur H. Freeling, 19th-century SA Surveyor-General)
Qld, NSW, SA, WA, NT   Limestone Fuchsia;
                                Rock Fuchsia-bush
1-3 m x 1-3 m   July-Oct; also sporadic
Small to medium **shrub**; young growth hairy and sticky; **branches** hairy, rough; **leaves** 2-8 cm x up to 1 cm, alternate, lanceolate, hairy, sticky, margins entire, apex pointed; **flowers** tubular, to 3 cm long, whitish, lilac, lavender or pale blue, exterior hairy, lobes pointed, 1 or 2 pedicels per axil; **stamens** not exserted; **calyx lobes** hairy, sticky, often overlapping; **fruit** ovoid, to 0.8 cm long, slightly hairy.

This species generally occurs in rocky soils on hills and ridges, and occasionally in mulga scrub. Often found on soils derived from limestone. Cultivated mainly by enthusiasts, who have had success in semi-arid to arid regions, and in temperate zones. Needs a very sunny situation in soils that have excellent drainage. Hardy to drought and light to medium frosts, and is suited to alkaline soils. Tip pruning of plants from an early age helps to maintain bushy growth. Propagate from seed or cuttings which can be slow to produce roots.

Eremophila georgei Diels
(after William J. George, 19th-century WA mine manager and botanical collector)
WA
1-1.5 m x 1-1.5 m   June-Sept
Small bushy **shrub**; **branches** hairy; **leaves** 1.5-2.5 cm x up to 1 cm, alternate, oblong to obovate-oblong, nearly sessile, many white hairs, margins entire or toothed, apex pointed; **flowers** tubular, about 2 cm long, lilac to purple, exterior hairy, lobes pointed, solitary, on long hairy pedicels; **stamens** not exserted; **calyx lobes** purplish, slightly hairy; **fruit** ovoid, about 0.8 cm long.

This beautiful species from the Austin District occurs on red-brown earths in the mulga woodlands. It is rare in cultivation and only grown by enthusiasts to date. Is undoubtedly best suited to arid and semi-arid zones, but it may succeed as a container plant in temperate regions. Must have very well drained soils and a situation that receives plenty of sunshine. Propagate from seed or cuttings which may prove troublesome to strike. Grafting of this species on to a hardy rootstock may enable it to be grown over a wider range.

Along with the closely related *E. clarkei*, this species is part of a complex in which it is hard to separate the varying forms.

Eremophila gibbifolia (F. Muell.) F. Muell.
(humped or swollen leaves)
Vic, SA   Coccid Emu-bush
0.3-1 m x 0.3-1 m   Sept-April; also sporadic
Dwarf twiggy **shrub**; **branches** many, stiff, glabrous; **leaves** to 0.5 cm x 0.1 cm, alternate, linear to linear-oblong, erect, warty, glabrous; **flowers** tubular, to 1.2 cm long, purple or rarely white, exterior glabrous, lobes short, sessile, axillary, along branchlets; **calyx lobes** pointed, small; **fruit** to 0.4 cm long.

This species occurs on heavy soils in mallee scrub-

Eremophila gilesii   T.L. Blake

land, and is one of the most commonly cultivated. The form with purple flowers is the more popular. Has proved most adaptable to a wide range of soils, including alkaline types. Does best in relatively well drained situations, but withstands short periods of waterlogging. Tolerates dappled shade but prefers partial or full sunshine. Hardy to frost and drought. Responds well to pruning. Tip pruning while young is recommended to promote bushy growth. If grown in protected situations it can be subject to attack by grey mould, which results in dieback of branch growth. Is an excellent container plant. Propagate from cuttings which strike readily.

Eremophila gibbosa Chinn.
(humped or swollen)
WA
1-3.5 m x 0.5-1.5 m   April-Oct

Small to medium **shrub**, often suckering from roots; young growth shiny, sticky; **branches** ascending or erect; **branchlets** glabrous, warty, sticky, pale brown; **leaves** 1.5-6.5 cm x 0.5-2.5 cm, alternate, ovate to broad-elliptic, petiolate, sticky, glabrous or with tuft of hairs at apex, margins entire or faintly toothed, apex blunt; **flowers** tubular, about 2.5 cm long, yellowish green, exterior slightly hairy, lobes pointed, solitary, axillary, on long S-shaped pedicels; **stamens** exserted; **calyx lobes** green to purplish; **fruit** globular, about 0.4 cm long, more or less bi-lobed, grey, wrinkled.

From the Coolgardie District, where it occurs in the Fraser Range east of Norseman and extends to the Coolgardie area, usually growing on red clay-loams and on rocky soils. This species has had limited cultivation (under the name of *E. serrulata*). It seems adaptable and grows in most soils. Needs good drainage. Tolerates dappled shade but does better in a sunny situation. Hardy to frost and drought. Can develop as an open plant, but tip pruning from an early age helps promote lateral growth. It is prone to suckering, and damage to the root area can result in a thicket of growth which may be useful as an informal hedge.

Honey-eating birds are regular visitors on flowering plants. Propagate from seed or cuttings which can be slow to form roots. Root suckers have been successfully transplanted.

An allied species is *E. virens*, which has very hairy flowers.

E. serrulata (A. Cunn. ex A. DC.) Druce differs from *E. gibbosa* with its concave, serrated or wavy-margined leaves.

Eremophila gibsonii F. Muell.
(after Alfred Gibson, a member of Giles' 1873-74 expedition)
SA, WA, NT
1-2 m x 1-2 m   Aug-Sept

Small **shrub**; young growth sticky; **branches** glabrous, sticky; **leaves** 2-7 cm x up to 0.2 cm, alternate, linear, glabrous, sticky, margins toothed, apex pointed and often hooked; **flowers** tubular, about 1.5 cm long, white, pale lilac to blue, exterior hairy, lower lobes blunt, 1-2 slender pedicels per axil to about 1.2 cm long; **stamens** not exserted; **calyx lobes** glabrous, green, enlarging after flowering; **fruit** ovoid, to 0.6 cm long, ribbed, sticky.

This species occurs on sand ridges in Central Australia, and extends westwards to the Coolgardie District. Evidently it is little-known in cultivation. Requirements should be for well drained soils, with maximum sun. Best suited to arid and semi-arid climates. Easier to establish during warmer months. Tolerant of frost and drought. Propagate from seed or cuttings which do not produce roots readily.

Eremophila gilesii F. Muell.
(after Ernest Giles, 19th-century explorer)
Qld, NSW, SA, WA, NT   Desert Fuchsia;
Green Turkey-bush
1-1.5 m x 1-2 m   June-Nov; also sporadic

Small **shrub**; young growth sticky; **branches** spreading or ascending, hairy; **leaves** 1-7 cm x up to

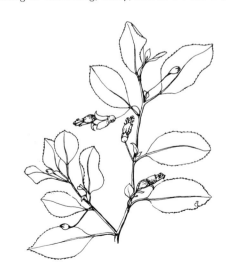

Eremophila gibbosa , × ·45

Eremophila glabra

0.5 cm, alternate, linear or linear-lanceolate, initially channelled above, margins entire, becoming flat and slightly toothed, apex pointed, hairy; **flowers** tubular, to about 3 cm long, pink, pale blue, mauve, lilac or purple, exterior hairy, lobes large, pointed and spreading, 1-2 S-shaped pedicels per axil; **stamens** not exserted; **calyx lobes** narrow, pointed, to 2 cm long, hairy; **fruit** ovoid to globular, to 1.3 cm long, hairy.

A very ornamental species that is widespread throughout the arid regions of Central Australia. It usually occurs on sandy plains and stony ridges, in association with mulga. Ideally suited to cultivation in areas of low rainfall and high temperatures. It requires very well drained soils and maximum sunshine to reach its potential. Best planted in warmer months. Hardy to frost and extended dry periods. Responds well to pruning. Successfully grown as a container plant in temperate regions. Propagate from seed or cuttings.

In NT a further 2 varieties occur: *E. gilesii* var. *argentea* Ewart, has silvery leaves; *E. gilesii* var. *filiforme* Ewart, has extremely narrow leaves.

E. gilesii is closely related to *E. foliosissima*, and the status of many forms is yet to be determined. Collectively these forms are referred to as the *E. gilesii*–*E. foliosissima* complex. In areas such as south-western Qld, *E. gilesii* is commonly regarded as a weed by the agricultural community, because of its effective regeneration. Control measures such as slashing, ploughing and the use of herbicides are successful.

Eremophila glabra (R. Br.) Ostenf.
(glabrous)
Qld, NSW, Vic, SA, WA, NT   Common Emu-bush;
                    Fuchsia-bush; Tar Bush
prostrate-1.5 m x 1-3 m   Aug-March; also sporadic
Dwarf to small **shrub**; young growth often sticky; **branches** faintly to densely hairy giving a whitish appearance; **leaves** 1.5-5 cm x up to 1.2 cm, alternate,

Eremophila glabra 'Murchison River', × ·5

linear-lanceolate to elliptic, flat, glabrous to densely hairy, margins entire or faintly toothed, apex pointed; **flowers** tubular, to 3 cm long, green, yellow, orange or red, exterior glabrous or slightly hairy, constricted near base, lobes pointed, 4 upper, 1 lower, solitary, axillary, on pedicels to 1.5 cm long, profuse; **stamens** exserted; **calyx lobes** short, hairy; **fruit** ovoid to globular, to 1 cm long, more or less succulent.

An extremely variable and complex species which includes many ornamental forms popular in cultivation. In the forthcoming revision some forms may be raised to specific status, while others will become subspecies. Some forms previously thought to be *E. glabra*, or closely related, have already been described as new species, eg. *E. biserrata* and *E. serpens.*

E. glabra is widespread and occurs on a variety of soil types. It is most common on red and brown loams, and calcareous red loams. Has adapted well to cultivation, and it grows successfully on most soils, provided drainage is adequate. Likes plenty of sunshine, but tolerates a position with dappled shade, producing fewer flowers than it otherwise would. Hardy to light or medium frost and drought. Withstands heavy pruning. Plants are prone to attack by grey mould during flowering if climatic conditions are still, humid and overcast or persistently rainy. The ageing flowers are first to succumb, the mould spreading to the stems and leaves. This results in damage or death of branches. It can be controlled by removal of diseased parts by pruning or by use of a suitable fungicide. Many forms are excellent container plants. Some of the more commonly grown forms are:
(a) Prostrate form with yellow flowers, from Mingenew, WA.
(b) Shrubby form with greyish green leaves and red flowers, from the Murray lands.
(c) 'Murchison River', a form with silvery foliage and bright red flowers.

Many others are grown and are usually designated by the area they originated from, eg. Kalgoorlie, Maralinga or Ouyen.

Propagate from seed or cuttings which readily form roots, although buds on the cuttings can die, leaving them barren.

Eremophila glutinosa Chinn.
(sticky)
WA
1-2 m x 1-2 m   May-Sept; also sporadic
Small **shrub**; young growth sticky; **branches** ascending or erect, hairy, sticky; **branchlets** terete, hairy, sticky; **leaves** to about 1 cm x 0.1 cm, alternate, narrow-linear, erect or spreading, crowded, sticky, apex slightly pointed; **flowers** tubular, 1-2.5 cm long, pale to dark lilac, exterior hairy, lobes pointed, solitary, axillary, on short pedicels, along branchlets; **stamens** not exserted; **calyx lobes** unequal, narrow, green to purplish, hairy; **fruit** ovoid, to 0.6 cm long, flattened, ribbed.

A recently described species. It is widespread in the Eremaean Botanical Province, with large populations in the Meekatharra and Wiluna regions. Occurs in rocky situations, as well as in heavy red clay-loams.

A decorative species which produces many flowers over a long period. More or less restricted to the Central Australian region, where it occurs on a very wide range of soils. Erroneously recorded for WA and SA. It is usually found on red earths, but it also frequently occurs on various loams, sands, gravels and stony soils. Grown mainly by enthusiasts to date, but has potential for wider application, especially in arid and semi-arid zones. The soils need to be well drained, and plants grow best in maximum sunshine. Can be damaged or killed by heavy frost. Responds well to pruning. May prove successful as a container plant in temperate zones. Propagate from seed or cuttings.

A natural hybrid with *E. willsii* as the other parent has had limited cultivation.

Eremophila gracilifolia F. Muell.
see *E. longifolia* (R. Br.) F. Muell.

Eremophila granitica S. Moore
(growing on granitic soil)
WA
0.7-2 m x 0.6-1.5 m   July-Oct; also sporadic
Dwarf to small, open **shrub**; young growth sticky; **branches** slender, sticky; **leaves** 2-5 cm x about 0.1 cm, alternate or somewhat opposite, narrow-linear, usually curved, sticky, apex pointed; **flowers** tubular, to 3 cm long, white, mauve or pink, constricted near base, exterior hairy, lobes blunt, 1-2 slender hairy pedicels per axil; **stamens** not exserted; **calyx lobes** about 1 cm long, purplish, overlapping, slightly hairy; **fruit** ovoid, about 0.6 cm x 0.4 cm.

This fine-foliaged species occurs in the Austin and Coolgardie Districts where, despite the specific name, it grows on red loams and sandy soils. Cultivated mainly by enthusiasts to date, but has potential for use in semi-arid to arid areas. The flowers are well displayed, and they can be produced for most of the year. Needs well drained soils. Will grow in heavy clay-loams, provided drainage is adequate. Does best in a sunny situation, but will tolerate shade in areas with high temperatures. Can suffer damage or death from heavy frosts. Propagate from seed or cuttings which can be slow to form roots.

Plants with narrow, toothed leaves and pink to mauve flowers, known as a form of this species, are referable to *E. clarkei*.

Eremophila hastieana W. Fitzg. =
E. leucophylla Benth.

Eremophila hillii E. A. Shaw
(after Ron Hill, 20th-century collector and horticulturist)
SA, WA
0.5-1 m x 1-2 m   Aug-Nov
Dwarf **shrub**; **branches** rough, becoming glabrous; **branchlets** densely hairy; **leaves** 0.7-1.3 cm x 0.4-1 cm, alternate, obovate to orbicular, sessile or shortly petiolate, crowded at ends of branchlets, densely hairy, margins usually slightly lobed, apex rounded; **flowers** tubular, to 3.5 cm long, yellowish, bright red or brick red, exterior hairy, lobes pointed, lower lobes reflexed,

Eremophila glabra 'Kalgoorlie'   W.R. Elliot

Evidently not well known in cultivation, but should be best suited to semi-arid and arid regions. Hardy to frost. Trials are needed in temperate zones, and it may grow successfully in containers. Propagate from seed or cuttings, both of which are difficult.

Eremophila goodwinii F. Muell.
(after Rev. Thomas H. Goodwin, who assisted John Dallachy's collecting expedition in 1888)
Qld, NSW, NT   Purple Fuchsia-bush
1-1.5 m x 1-2 m   all year
Small **shrub**; young growth sticky; **branches** erect to ascending, hairy, sticky; **leaves** 2-7 cm x 0.2-0.6 cm, alternate, linear to linear-lanceolate, flat, green to greyish, hairy, sticky, margins entire or faintly toothed, apex pointed and often recurved; **flowers** tubular, to 2.5 cm long, lilac to bluish purple, exterior hairy, lobes pointed, 1-2 short pedicels per axil; **stamens** not exserted; **calyx lobes** unequal, to 1.5 cm long, hairy; **fruit** globular, to 1 cm long, pointed, hairy.

443

solitary, on short pedicels, in upper axils, profuse; **stamens** exserted; **calyx lobes** to 0.7 cm long, broad, pointed, hairy; **fruit** ovoid, to 0.7 cm x about 0.6 cm, dry.

The greyish leaves and profuse orange-red flowers make this attractive species popular with eremophila enthusiasts. To date it is not widely known in cultivation. It occurs in semi-arid and arid regions such as Ooldea on the Nullarbor Plain, where it grows in rubbly soils with a calcareous subsoil. Does well in cultivation on freely drained soils. Needs plenty of sunshine. Frost and drought tolerant. It develops as a rounded bush if in an open situation. Responds favourably to pruning. Successfully grown as a container plant in temperate zones. Propagate from seed or cuttings which strike readily.

Eremophila homoplastica (S. Moore) C. Gardner
(similar in structure but not from a common source)
WA
0.3-0.8 m x 0.5-1 m   July-Aug; also sporadic
Dwarf dense **shrub**; **branches** wiry, intertwined, hairy; **leaves** to 0.5 cm x about 0.1 cm, oblong, appressed, crowded, warty, furrowed, glabrous, apex pointed; **flowers** tubular, about 1.2 cm long, blue to violet, exterior slightly hairy, lobes rounded, solitary, axillary, on short pedicels, near ends of branchlets; **stamens** not exserted; **calyx lobes** densely hairy; **fruit** ovoid, about 0.5 cm long.

This wiry species is from semi-arid and arid regions in the Austin and Coolgardie Districts, and the Great Victoria Desert. It is rarely encountered in cultivation as it is difficult to maintain. Must have very well drained soils, with maximum sunshine. It is frost tolerant. May succeed as a container plant in temperate zones. Propagate from seed or cuttings. Success with grafting has enabled the species to be grown for longer periods.

Eremophila hughesii F. Muell.
(after Sir Walter W. Hughes, 19th-century founder of University of Adelaide)
WA, NT
1-2 m x 0.6-1.5 m   Aug-Oct
Small **shrub**; **branches** erect or weeping, slender, glabrous; **leaves** 2-6 cm x about 0.1 cm, alternate, linear, thick, glabrous, apex pointed and sometimes hooked; **flowers** tubular, to about 2 cm long, white to lilac-blue, exterior sparsely hairy, not constricted at base, lobes pointed, 1-3 slender glabrous pedicels per axil, near ends of branchlets; **stamens** not exserted; **calyx lobes** to 1 cm long, glabrous or hairy, pointed; **fruit** ovoid, about 0.6 cm long, ribbed, beaked, slightly hairy.

An inhabitant of the Eremaean Botanical Province, where it occurs on rocky hills and sandy spinifex plains. Evidently not in cultivation, but has potential for use in semi-arid zones. Some plants develop pendulous branches, and selection of these forms for cultivation is worthwhile. Will need very well drained soils, in situations with plenty of sunshine. Frost hardy. Propagate from seed or cuttings.

Eremophila imbricata (Benth.) F. Muell.
(overlapping)
WA
0.6-1.3 m x 0.5-1.5 m   June-Sept
Dwarf to small **shrub**; **branches** densely covered with white hairs; **leaves** 1-2.5 cm x 0.6-2 cm, ovate or oblong, crowded, overlapping, thick, whitish, densely hairy, apex blunt; **flowers** tubular, 1.5-2.5 cm long, blue, exterior glabrous, lobes pointed, sessile, in upper axils; **stamens** not exserted; **calyx lobes** to 0.5 cm long, narrow, hairy; **fruit** ovoid, about 0.5 cm long.

This rare species is not well known. It is reported to occur north of Perth, between the Moore and Murchison Rivers, but recently has only been observed near Mount Augustus and Mount Clue stations, north of the Gascoyne River in the Ashburton District, where it grows on stony slopes. Evidently not cultivated, it is a most ornamental species with whitish foliage and blue flowers. Should be best suited to semi-arid and arid regions, but may succeed as a container plant in temperate zones. There is limited knowledge of this species, and in cultivation it may grow larger than indicated above. Because of its rarity, it is imperative that plants be propagated to ensure that a bank of material is available. Propagate from seed or cuttings which may be difficult to strike.

Eremophila inflata C. Gardner
(inflated)
WA
0.6-2 m x 0.6-1.5 m   Nov-Dec
Dwarf to small **shrub**; **branches** slender, glabrous, slightly warty and sticky; **leaves** to 3 cm x 0.4 cm, linear to oblong-lanceolate, petiolate, glabrous, apex more or less blunt; **flowers** tubular, swollen, with small opening, about 1 cm long and 1 cm across, pale violet or pink, exterior hairy, 1 or sometimes 2 pedicels per axil; **stamens** not exserted; **calyx lobes** reflexed after flowering; **fruit** ovoid, about 0.3 cm x 0.25 cm, beaked, hairy.

This is a very distinct species because of its inflated flowers. Occurs in the Avon and Coolgardie Districts. Little-known in cultivation, and only grown by enthusiasts to date. At this stage it appears adaptable. Needs very well drained soils and plenty of sunshine. Hardy to frost and drought. May succeed as a container plant in temperate zones. Propagate from seed or cuttings. *Myoporum insulare* has been used successfully as a grafting stock.

This species will be placed in a monotypic genus as a result of the forthcoming botanical revision.

Eremophila interstans (S. Moore) Diels
(intermediate)
SA, WA
3-6 m x 1-3 m   July-Oct; also sporadic
Erect, broom-like, medium to tall **shrub**; **branches** slender; **leaves** to 1.5 cm x about 0.1 cm, alternate, narrow-linear to almost filiform, silvery scurf on surface, apex pointed and hooked; **flowers** tubular, about 1 cm long, white to lilac, scented, exterior hairy, lobes blunt and spreading, 1-2 pedicels per axil, along ends of branchlets, profuse; **stamens** not exserted; **calyx**

Eremophila latrobei ssp. *latrobei*   T.L. Blake

lobes with densely hairy margins; **fruit** to 0.4 cm x 0.2 cm.

A fairly common shrub of the Coolgardie District, where it grows amongst the understorey of the woodlands, from near the Nullarbor Plain, westward to Norseman and Coolgardie. Also occurs in the Uno Range, SA. To date only cultivated by a few enthusiasts, but it has potential for use in arid and semi-arid regions. Should do well in most soils, provided drainage is adequate and it receives plenty of sunshine. Frost and drought tolerant. Pruning when plants are young should provide bushier growth. Propagate from seed or cuttings which are usually slow to form roots. Successfully grafted using *Myoporum insulare* as the rootstock.

The var. *parviflora* S. Moore will be absorbed with *E. virgata* W. Fitzg. as a subspecies in the forthcoming revision.

Eremophila ionantha Diels
(violet flower)
WA   Violet-flowered Eremophila
0.8-2 m x 0.8-1.5 m   Aug-Nov; also sporadic
Dwarf to small **shrub**; **branches** erect, many; **leaves** 2-4 cm x about 0.1 cm, opposite near ends of branchlets, alternate and opposite on other parts, linear, almost terete, mainly erect, incurved, glabrous, warty; **flowers** tubular, to 2 cm long, pale blue, violet or mauve, exterior glabrous, constricted near base, lobes pointed, solitary, axillary, on curved pedicels, profuse, near ends of branchlets; **calyx lobes** pointed; **fruit** ovoid to acute, about 0.7 cm long, hairy.

An inhabitant of the goldfields region, where it is an undershrub in the woodlands, growing mainly on red

loams. It is a showy, graceful species which has adapted well to cultivation, doing best in well drained, medium to heavy soils, in a position that receives plenty of sunshine. Suited to container planting in temperate zones. Tolerant of most frosts and extended dry periods. Responds well to pruning. Propagate from seed or cuttings which readily form roots.

The var. *brevifolia* Diels has shorter leaves and is in cultivation. Some botanists believe it is not distinct.

Eremophila kochii Ewart = *E. resinosa* F. Muell.

Eremophila laanii F. Muell.
(after van der Laan, 19th-century Dutch medical doctor)
WA
1-3 m x 1-3 m   Aug-Jan; also sporadic
Small to medium, spreading **shrub**; **branches** many, often entangled, can be pendulous, hairy; **leaves** 2-5 cm x up to 0.4 cm, alternate, lanceolate, often reflexed, slightly hairy, apex pointed; **flowers** tubular, to 2 cm long, white, pink or reddish pink, exterior glabrous, lobes blunt, solitary, axillary, profuse; **stamens** usually exserted; **calyx lobes** about 0.5 cm long, hairy; **fruit** globular, about 0.8 cm long.

A showy species which is one of the most adaptable in cultivation. It occurs in the Austin District, growing in heavy soils where the pink and white-flowered forms often grow together. Does well in most soils, provided drainage is adequate. Prefers plenty of sunshine, but tolerates dappled shade. Hardy to frost and extended dry periods. Usually grows to about 1.5 m tall in cultivation, but may develop into a larger shrub in some areas. The white-flowered form may sucker and is usually more vigorous, flowering throughout the year. Can become leggy, but responds well to pruning. Tip pruning from an early age is recommended. Successfully grown as a container plant. Propagate from seed or cuttings which strike readily.

Eremophila lachnocalyx C. Gardner
(soft woolly calyx)
WA
1-2 m x 1-2 m   July-Aug
Small hairy **shrub**; **branches** initially densely covered in white hairs, becoming glabrous; **leaves** about 0.5 cm x up to 0.4 cm, alternate, orbicular to elliptic, often reflexed, crowded, dense covering of grey to white hairs, apex pointed; **flowers** tubular, about 2 cm long, blue-violet, exterior slightly hairy, lobes pointed, not constricted at base, solitary, axillary, near ends of branchlets; **calyx lobes** to 1 cm long, pointed, covered in woolly hairs; **fruit** about 0.6 cm x 0.25 cm, hairy.

This most ornamental species with greyish foliage occurs in the Austin District, on heavy or gravelly loams. It is rare in cultivation. Certainly best suited to arid and semi-arid climates, but may succeed in more temperate climes if grown in a sunny, dry situation. Worth trying as a container plant. Hardy to frost. Propagate from seed. Cuttings may prove troublesome due to the dense felting on the leaves.

Has similarities to *E. mackinlayii*, which has flowers that are constricted near the base, and larger leaves.

Eremophila latrobei F. Muell.
(after Charles J. La Trobe, 19th-century Lieutenant-Governor of Vic)
Qld, NSW, SA, WA, NT   Crimson Turkey-bush;
                                                    Native Fuchsia
1-3 m x 0.6-2 m   June-Oct; also sporadic
   Small to medium **shrub**; **branches** usually erect, hairy to glabrous, warty; **leaves** 1-8 cm x up to 0.6 cm, alternate, linear to oblong, can be filiform, glabrous and green to densely hairy and grey, often warty, margins flat or strongly recurved, apex blunt or pointed; **flowers** tubular, to 3.5 cm long, red to purplish pink, rarely white or yellow, exterior faintly hairy, lobes pointed, slightly constricted near base, solitary, axillary, on pedicels to 1 cm long; **stamens** exserted; **calyx lobes** to 2 cm long, margins hairy, becoming slightly enlarged after flowering; **fruit** ovoid to globular, glossy, to 1 cm long, beaked.
   Widespread through its range, this showy species occurs in a variety of habitats including sandplains, undulating plains with heavy loams, rocky hills and ridges, and mulga communities. There are many differing forms in cultivation, and they are usually designated by the location from which they were originally collected, eg. *E. latrobei* 'Mootwingee'. An attractive form with silvery leaves occurs on soils derived from limestone in north-western Qld and north-eastern NT. Plants do best in areas of low rainfall and high temperature. They need relatively well drained soils. They will grow in situations receiving dappled shade, but prefer partial or full sun. Hardy to extended dry periods and most frosts. They are easier to establish during warm months. Frost damaged plants are usually capable of reshooting. Plants can become leggy, but they respond well to pruning. Tip pruning on young plants is recommended. Successfully grown as a container plant in temperate regions. Propagate from seed or cuttings which may form roots readily but may be difficult to strike.
   The var. *glabra* L. S. Smith occurs throughout the species' range, and differs from the typical form in that it has glabrous leaves and the margins of the calyx lobes are barely hairy.
   The var. *tuberculosa* S. Moore is considered identical to the typical form.

Eremophila lehmanniana (Sonder ex Lehm.) Chinn.
(after J. G. Lehmann, 19th-century German botanist)
WA
0.3-1.5 m x 0.5-2 m   Aug-Oct
   Dwarf to small **shrub**; young growth sticky; **branches** velvety; **leaves** 0.5-1.3 cm x up to 0.6 cm, alternate, ovate to broad-lanceolate, hairy to nearly glabrous, margins entire or toothed; **flowers** tubular, to about 1.8 cm long, white, blue, mauve or violet, exterior glabrous, lobes pointed, solitary, axillary, profuse, on very short pedicels, near ends of branchlets; **stamens** not exserted; **calyx lobes** pointed, glabrous or hairy; **fruit** about 0.4 cm long.
   This species is known from the Avon, Coolgardie, Eyre, Irwin and Roe Districts. It occurs in near-coastal situations as well as inland. Widely cultivated in SA. Should grow in a wide range of soils, provided drainage is adequate. Needs plenty of sunshine. Hardy

to extended dry periods and most frosts. Sometimes subject to grey aphid attack which can be controlled with pyrethrum. Propagate from seed or cuttings.
   The species previously known as *E. woollsiana* F. Muell., and its var. *dentata* Ewart & J. W. White, are now included in *E. lehmanniana*.

Eremophila leonhardiana E. Pritzel =
                                    E. willsii F. Muell.

Eremophila leucophylla Benth.
(white leaves)
SA, WA, NT   Grey Poverty Bush
1-2 m x 1-1.5 m   June-Dec
   Small hairy **shrub**; young growth with yellowish hairs; **branches** densely covered in greyish or yellow hairs; **leaves** 1.5-4.5 cm x up to 1 cm, alternate, narrow to broadly obovate, densely hairy, margins entire and thickened, apex blunt or pointed, ending in a mucro; **flowers** tubular, 2.5-4 cm long, pink, greenish or yellowish pink, pale to deep red or purple, exterior hairy, lobes pointed, solitary, axillary, on pedicels about 0.5 cm long, near ends of branchlets; **stamens** exserted; **calyx lobes** to 1.5 cm long, densely hairy; **fruit** ovoid, to 0.8 cm long, glabrous.
   An eyecatching species with greyish foliage and bright flowers. It is a dweller of mulga and spinifex country. Confined mainly to inland regions, but also occurs near the coast in the Irwin District of WA. At this time it is rare in cultivation. Best suited to arid and semi-arid regions, but has succeeded in temperate areas. Worthy of trial as a container plant. Must be grown in a very well drained situation that receives plenty of sunshine. Should tolerate light to medium frost, but heavy frosts can kill plants. Propagate from seed or cuttings which may rot due to the dense covering of hairs. The following are now considered conspecific: *E. hastieana* W. Fitzg., *E. turtonii* F. Muell. and *E. xanthotricha* F. Muell.

Eremophila linearis Chinn.
(linear; in reference to the leaves)
WA
1-4 m x 0.6-2 m   July-Nov
   Small to tall **shrub**; young growth sticky; **branches** slender, extremely sticky, glabrous, rough; **leaves** to 3 cm x 0.1-0.2 cm, alternate, linear, crowded, glandular, glabrous, sticky, margins entire, apex pointed; **flowers** tubular, to 3 cm long, red above, orange to yellow below, yellow inside, exterior glabrous, lobes pointed, solitary, axillary, on S-shaped glabrous pedicels to 1.5 cm long, near ends of branchlets; **stamens** exserted; **calyx lobes** pointed, glabrous, enlarging slightly after flowering; **fruit** ovoid, to 1.3 cm x 1 cm, grey, beaked.
   A recently described species, widespread in central WA and prominent in the Meekatharra-Wiluna area. It usually grows in red-brown clay-loams which become inundated. Often occurs in mulga woodland or *Eremophila/Cassia* scrubland. Rare in cultivation, it is probably best suited to arid or semi-arid climates. Must have plenty of sunshine. Hardy to frost and drought. Propagate from seed or cuttings which can be difficult to strike.

446

This species was previously included under *E. duttonii*, which occurs mainly in the eastern states, and has broader leaves and hairy, thick branches.

Eremophila linsmithii R. J. Henderson
(after Lindsay S. Smith, 20th-century Qld botanist)
Qld
0.5-2 m x 0.5-2 m   Aug-Dec
Dwarf to small **shrub**; **branches** ascending or spreading from the base, sticky, slightly hairy; **leaves** 3-5 cm x up to 0.3 cm, alternate, linear, spreading, flat, glabrous or faintly hairy, sticky, apex pointed; **flowers** tubular, to 2.5 cm long, white to pale lilac, exterior slightly hairy, lobes pointed, solitary, axillary, on S-shaped pedicels, near ends of branchlets; **stamens** not exserted; **calyx lobes** to 1.5 cm long after flowering, pointed, slightly hairy; **fruit** ovoid, to about 0.7 cm long.

This recently described species occurs in the Warrego and Gregory South Districts of Qld, and may extend into north-western NSW. It grows in 'break-away' country, and is often found on clay-loams. Not common in cultivation, it should do best in areas of low rainfall and high temperatures. Needs a well drained situation, with partial or full sun. Hardy to most frosts and extended dry periods. Worth trying as a container plant in temperate zones. Propagate from seed or cuttings. This species has affinities to *E. gilesii*, which has flowers of a deeper colour and larger fruits. It is also similar to *E. gibsonii*, which has finely toothed leaves.

Eremophila longifolia (R. Br.) F. Muell.
(long leaves)
Qld, NSW, Vic, SA, WA, NT   Berrigan; Emu-bush
2-8 m x 1.5-5 m   July-Dec
Small **shrub** to small **tree**, often suckering; **bark** rough, dark grey; **branches** pendulous, hairy, streaked; **leaves** 3-20 cm x up to 0.8 cm, alternate, linear to linear-lanceolate, dull green, pendulous, hairy or becoming glabrous, margins entire, apex pointed and hooked; **flowers** tubular, to 3 cm long, pink to reddish brown or brick red, exterior hairy, lobes blunt, constricted near base, 1-3 pedicels per axil; **stamens** exserted; **calyx lobes** to 0.8 cm long, triangular, hairy; **fruit** ovoid to globular, to about 1 cm long, glabrous, succulent.

This widely distributed eremophila occurs on most soil types and grows in a wide variety of plant communities. It is sometimes found on heavy clays which are subject to inundation. Does best in semi-arid and arid regions, but will grow in temperate zones. Soils need to be relatively well drained, and plants must receive plenty of sunshine. They withstand climatic extremes very well. Usually plants develop only to about 4 m tall. The tree form is rare in nature. An excellent shade or shelter tree. Disturbance of the roots promotes suckering, which results in dense copses. Plants are capable of regrowth after fire. Useful for soil erosion control, and the foliage is suitable for animal fodder. The Aborigines used bruised leaves of this species for tanning skins. Propagate from seed or stem cuttings which can be very slow to form roots, from root cuttings or by transplanting root suckers.

E. gracilifolia is considered by some botanists to probably be a hybrid between *E. longifolia* and *E. oldfieldii*.

Eremophila macdonnellii F. Muell.
(after Sir Richard G. Macdonnell, 19th-century Governor of SA)
Qld, NSW, SA, NT
0.5-1.8 m x 0.5-5 m   Aug-Feb; also sporadic
Dwarf, spreading **shrub**; **branches** slightly to densely hairy; **leaves** 0.3-2.5 cm x 0.2-1 cm, most variable, linear to ovate, slightly to densely hairy, green to whitish, flat, apex blunt or pointed; **flowers** tubular, to 3.5 cm long, purple, exterior glabrous, flattened, solitary, axillary, on pedicels to 3 cm long; **stamens** not exserted; **calyx** cup-like, with short, pointed lobes; **fruit** more or less globular, to 1.5 cm diameter.

A most ornamental species from Central Australia, where it grows on a range of soils from sands to loams. It is highly variable and comprises many forms. Quite a few of these are cultivated, such as the grey-foliaged form from the northern Simpson Desert. There is also a prostrate form with narrow, hairy green leaves, and 2 forms with broader leaves, which develop into rounded shrubs. All forms require very well drained soils and plenty of sunshine. They do well in containers. Drought and frost tolerant. Although plants may spread, it is unusual for them to reach 5 m across. A mature plant usually reaches about 2 m across. Plants respond well to pruning, and this keeps them to smaller dimensions if required. In southern Australia they can be subject to attack by grey mould during still, overcast, humid conditions. An open position is recommended in such areas. Propagate from seed or cuttings which readily form roots.

There are named varieties of this species but, as their validity is still under consideration, they have not been included here.

Eremophila macgillivrayi J. Black
(after Dr. William D. K. MacGillivray, original collector)
Qld, ?NSW, SA   Dog Bush
1-3 m x 1-3 m   Aug-Oct
Small to medium **shrub**; **branches** thick, corky, brittle, densely covered in minute white to creamy hairs; **leaves** 2-6 cm x about 0.5 cm, alternate, lanceolate to slightly falcate, flat, thick, minutely hairy, appearing waxy, whitish, margins entire, apex pointed; **flowers** tubular, to 2.5 cm long, red, rarely yellow, exterior slightly hairy, lobes pointed, solitary, axillary, on hairy pedicels to 2 cm long; **stamens** slightly exserted; **calyx lobes** pointed, to 1 cm long, hairy; **fruit** ovoid, to 1.6 cm long, hairy.

An extremely ornamental species with whitish foliage and red flowers. It occurs in north-eastern SA and in the Burke and Gregory South Districts in Qld. There is doubt about whether it occurs in NSW. It is rare in cultivation, and is best suited to areas with low rainfall and high temperatures. The soil must be well drained, and the plants need maximum sunshine. Hardy to most frosts and extended dry periods. May succeed as a container plant in temperate zones. Propagate from seed or cuttings which may be slow to form roots.

447

Eremophila mackinlayi F. Muell.
(after John McKinlay, 19th-century explorer)
WA   Desert Pride
1-2 m x 1-2 m   Aug-Dec; also sporadic

Small to medium, hairy **shrub**; young growth covered in yellowish hairs; **branches** with dense covering of white to yellowish hairs; **leaves** 1.5-2 cm x up to 1.2 cm, alternate, obovate to orbicular, petiolate, dense covering of white or yellow hairs, margins wavy, apex blunt; **flowers** tubular, to 3 cm long, blue or purple to violet, exterior hairy, lobes blunt, solitary, axillary, sessile, often bunched, near ends of branchlets; **stamens** not exserted; **calyx lobes** narrow, to 1.2 cm long, hairy; **fruit** ovoid, about 0.8 cm to 0.5 cm, hairy.

This very hairy and decorative species occurs in the Austin and Irwin Districts, where it usually grows in mulga communities. Rare in cultivation and grown mainly by enthusiasts. It has potential for use in arid and semi-arid regions. Needs well drained soils and maximum sunshine to grow to its potential. May reach only 1 m x 1 m. Worth trying as a container plant in temperate regions. Propagate from seed or cuttings which usually form roots readily.

E. strongylophylla F. Muell. is now regarded as conspecific.

Eremophila macmillaniana C. Gardner
(after Sir Robert F. Macmillan, 20th-century Chief Justice of WA)
WA   Grey Turpentine-bush
1.5-2.5 m x 1.5-2.5 m   July-Sept

Small to medium **shrub**; **branches** rigid, erect, with greyish white hairs; **leaves** to 3 cm x 1 cm, spathulate, initially with a dense covering of greyish white hairs, becoming glabrous and sticky, margins entire, apex pointed; **flowers** tubular, to 3.3 cm long, dull red and nearly glabrous exterior, interior yellow, lobes pointed, solitary, axillary or terminal; **stamens** slightly exserted; **calyx lobes** about 2.5 cm long, hairy; **fruit** ovoid to globular, 0.5-0.8 cm x about 0.5 cm.

A dweller of granite soils on 'break-aways', slopes and hills in the Cue-Meekatharra region of the Austin District. This species is not well known in cultivation. It will be best suited to locations with a low rainfall and high temperatures. Should be hardy to most frosts and is drought tolerant. May succeed in temperate areas as a container plant. Propagate from seed or cuttings which have proved extremely difficult to strike.

Eremophila maculata (Ker-Gawler) F. Muell.
(spotted)
Qld, NSW, Vic, SA, WA, NT   Native Fuchsia;
                        Spotted Emu-bush; Spotted Fuchsia
0.5-3 m x 1-3 m   June-Nov; also sporadic

Dwarf to medium, dense **shrub**; **branches** rigid, hairy; **leaves** 0.5-5 cm x up to 1 cm, alternate, oblanceolate, petiolate, glabrous or faintly hairy, green to grey-green, margins entire, apex pointed; **flowers** tubular, to 2.5 cm long, whitish to deep pale pink, red to purplish red, yellow or orange, often with cream or yellow-spotted interior, exterior glabrous, lobes pointed, solitary, axillary, on S-shaped glabrous pedicels to 2.5 cm long, profuse; **stamens** exserted; **calyx lobes** to 1 cm long, pointed, glabrous; **fruit** more or less globular, to 2 cm long, glabrous, green but sometimes purple.

A commonly cultivated ornamental species, renowned for its variation in flower colour and growth habit. In nature it usually occurs on heavy clay-loams which are subjected to periodic inundation. In cultivation plants prefer heavier soil types to sandy soils. Needs relatively good drainage. Will grow in dappled shade but prefers partial or full sun. Hardy to frost and extended dry periods. Responds well to pruning. Tip pruning of young plants helps promote bushy growth. The low-growing, mounding forms are very useful as a living mulch. Each form is commonly known by the name of the locality where collected. Plants are available from nurseries by such names as *E. maculata* 'Morgan', or sometimes they are known by their colour, eg. *E. maculata* 'Aurea'. The species was introduced to England in 1820 and is also currently grown with success in California, USA.

Eremophila mackinlayi, × ·55

Eremophila maculata, × ·55   var. *brevifolia*   var. *maculata*

Eremophila nivea

W.R. Elliot

The var. *brevifolia* Benth., from WA and NT, is a spectacular variety with short, elliptic to orbicular leaves and bright red flowers to 4 cm long. Although cultivated to some extent it is not well known. It does best in a situation with plenty of sunshine.

Plants of this species can be subject to attack by grey mould during still, overcast, humid weather. The spore usually begin growth on spent flowers, and the fungus spreads. If conditions are suitable, dieback of leaves, branches and stems can occur very quickly.

Propagate from seed or cuttings of slightly firm growth which usually produce roots readily.

E. maculata is known to be toxic to sheep and cattle, and is especially so when new growth is produced after rain.

The var. *linearifolia* S. Moore is included under *E. decipiens* Ostenf.

Eremophila maitlandii F. Muell. ex Benth.
(after Maitland Brown, 19-20th-century explorer and botanical collector)
WA   Shark Bay Poverty Bush
1.5-3 m x 1-2.5 m   July-Oct
Small to medium **shrub**; **branches** with dense covering of whitish hairs; **leaves** 3-9 cm x up to 0.5 cm, alternate, linear-lanceolate, covered with silvery hairs, flat, leathery, margins entire, apex pointed; **flowers** tubular, about 2 cm long, pale blue to lilac, exterior slightly hairy, lobes blunt, solitary, axillary, on pedicels about 1.2 cm long, near ends of branchlets; **stamens** not exserted; **calyx lobes** sparsely hairy, becoming papery, pointed, about 1 cm long; **fruit** narrow-ovoid, 0.7-0.9 cm x 0.3-0.6 cm, glabrous.

A common plant from north of Carnarvon, extending to south of Shark Bay. It occurs on sands or loams in open shrublands. Rare in cultivation, as it is best grown in areas of low rainfall and high temperatures. Needs well drained soils, with maximum sunshine. Hardy to most frosts. Pruning should help to promote bushy growth, as plants can become open and leggy. Worth trying as a container plant in temperate zones. Propagate from seed which germinates well or from cuttings.

Plants in cultivation under this name may in fact be *E. compacta*.

Eremophila margarethae S. Moore
(after Margaret Elvire, wife of Sir John Forrest)
WA
0.5-1 m x 0.5-1 m   Aug-Oct
Dwarf **shrub**; young growth yellowish; **branches** hairy, warty; **leaves** 1.5-3 cm x about 0.3 cm, alternate, linear-lanceolate, densely covered in white hairs, margins thickened, entire, apex pointed; **flowers** tubular, to 2 cm long, blue, pale violet or lavender, exterior hairy, lobes blunt, 1-2 very short hairy pedicels per axil, near ends of branchlets; **stamens** not exserted;

449

calyx lobes pointed, with yellowish hairs; **fruit** ovoid, to about 0.7 cm long.

A dweller of semi-arid and arid regions, where it grows in mulga communities. It is rarely encountered in cultivation, but has potential for inland areas. Soils need good drainage and plenty of sunshine. Hardy to most frosts. May succeed as a container plant in temperate zones. Propagate from seed or cuttings.

Some plants grown under this species name in the eastern states are an unnamed species which is described below.

Eremophila merrallii Ewart & J. W. White & B. Wood (after Edwin Merrall, 19-20th-century botanical collector)
WA
0.3-0.5 m x 1 m   Sept-Jan

Dwarf, rounded **shrub**; **branches** hairy; **leaves** 0.5-1 cm x 0.1-0.2 cm, alternate, semi-terete, crowded, grey-green, hairy, smooth above, warty below, margins entire, apex blunt; **flowers** tubular, about 1 cm long, grey-blue, exterior slightly hairy, constricted near base, lobes more or less blunt, solitary, sessile, in upper axils; **stamens** not exserted; **calyx lobes** about 0.3 cm long, pointed, hairy; **fruit** ovoid, to 0.3 cm x 0.2 cm.

The habitat of this species is stony clay-loam beneath *Eucalyptus* species in the region north and south of Southern Cross. It is commonly cultivated in SA, where it has proved quite adaptable. Needs soils with good drainage, and a warm to hot position, but it will tolerate some shade. Hardy to light and medium frosts. Also withstands extended dry periods. Suited for container planting. Propagate from seed or cuttings which usually form roots readily.

This species may become a subspecies of *E. caerulea* in the revision.

Eremophila metallicorum S. Moore
(of the miners)
WA
0.5-1 m x 0.5-1.5 m   Aug-Oct; also sporadic

Small **shrub**; young growth sticky; **branches** slender, woody; **leaves** to 1 cm x 0.2 cm, alternate, linear to oblanceolate, sessile, crowded, scurfy, margins entire, apex pointed; **flowers** tubular, to 2 cm long, blue, exterior hairy, solitary, axillary, on slender hairy pedicels to 1.5 cm long; **stamens** not exserted; **calyx lobes** to about 0.8 cm long, shiny; **fruit** ovoid, about 0.6 cm long, hairy.

This species occurs in the Laverton area of the Austin District. It is related to *E. granitica* but is not regarded as a very ornamental species. Needs a hot position in well drained soils. Should tolerate most frosts and is drought tolerant. Propagate from seed or cuttings.

Eremophila microtheca (F. Muell. ex Benth.) F. Muell.
(refers to the small fruits)
WA   Heath-like Eremophila
0.3-1 m x 0.5-1 m   Aug-Sept

Dwarf, heath-like **shrub**; young growth hairy; **branches** hairy; **leaves** to 0.5 cm x about 0.1 cm, appearing whorled, linear, crowded, initially hairy, becoming glabrous, apex blunt, strong odour; **flowers** tubular, to 1.5 cm long, lilac, exterior faintly scaly, lobes pointed, solitary, axillary, profuse, on very short pedicels, near ends of branchlets; **stamens** not exserted; **calyx lobes** pointed, hairy, to about 0.4 cm long; **fruit** ovoid, about 0.3 cm x 0.2 cm.

This species from the southern Irwin District is rare in nature. It grows in near-coastal situations, on claypans or in sandplain country where plants are most likely to be subjected to inundation during winter. It is little-known in cultivation, and is grown mainly by enthusiasts. Cultivation may be useful in aiding its survival. Should grow in a wide range of soils, and needs plenty of sunshine. Frost tender, damage often occurring to bark while foliage is untouched. Propagate from seed or cuttings.

Eremophila miniata C. Gardner
(red)
WA   Kopi Poverty Bush
1-5 m x 1-3 m   July-Oct; also sporadic

Small to tall **shrub**; young growth sticky; **branches** many, rigid, stout, brittle, rough, hairy; **leaves** 1.5-3.5 cm x up to 0.5 cm, alternate, broad-linear to linear-oblong, clustered, often at ends of branches, hairy, slightly sticky, margins entire, apex pointed; **flowers** tubular, to about 3.5 cm long, red, pinkish, reddish brown or creamy yellow, exterior slightly hairy, spotted, lobes pointed, solitary, axillary, on S-shaped pedicels to 1.5 cm long; **stamens** exserted; **calyx lobes** large, enlarging after flowering, same colour as flowers; **fruit** ovoid, to 0.8 cm long, beaked, glabrous.

A showy eremophila with large flowers and calyx lobes. It is from inland areas, where it usually grows on well drained alkaline soils in the vicinity of salt lakes.

Extremely rare in cultivation, this species is best suited to areas of low rainfall and high temperatures. It should grow in most soils, provided drainage is adequate. Needs maximum sunshine. Tip pruning of young plants may promote bushy growth. Propagate from seed or cuttings which could be slow to form roots.

Eremophila mitchellii Benth.
(after Sir Thomas Mitchell, 19th-century explorer and botanist)
Qld, NSW   Budda; False Sandalwood; Sandalwood
2-9 m x 2-5 m   March-May; Sept-Nov; also sporadic

Medium **shrub** to small **tree**; young growth sticky; **bark** rough, dark brown to nearly black, with oblong pattern; **branchlets** glabrous to slightly hairy; **leaves** to 6 cm x 0.7 cm, alternate, linear-lanceolate, narrowed to base, glabrous, sticky, aromatic, bright green, margins entire, apex pointed and hooked; **flowers** bell-shaped, to 2 cm long, white to pale cream, sweetly scented, exterior glabrous or slightly hairy, lobes blunt, 1-3 sticky curved pedicels per axil, profuse; **stamens** not exserted; **calyx lobes** to 1 cm long, slightly enlarging after flowering, papery; **fruit** ovoid, to 0.6 cm long, hairy, beaked.

A widespread ornamental species in south-western Qld and north-western NSW, where it occurs on a wide range of soil types from sandy loam to heavy clays. It is not commonly cultivated but seems adaptable to most soils. Will grow in dappled shade to full sun. In temperate climates it needs a hot, sunny situation. Capable of suckering if root damage occurs. It is a weed of economic importance in some grazing areas as its foliage is unpalatable to stock, and it is difficult to control its spread, even by ploughing, burning, stick-raking or the use of herbicides. When the scented wood is burnt it emits a pleasant aroma. Propagate from seed, stem cuttings or root cuttings. Stem cuttings are usually very slow to form roots, if they do at all.

Eremophila muellerana C. Gardner
(after F. von Mueller, first Government Botanist, Vic)
WA   Round-leaved Eremophila
1-1.5 m x 1-1.5 m   Aug-Sept
Small **shrub**; **branches** hairy; **leaves** to 1.5 cm x 0.8 cm, alternate, oblong-linear, densely hairy, sulphur yellow, margins entire, apex rounded; **flowers** tubular, to about 3.5 cm long, reddish violet, exterior slightly hairy, lobes blunt, solitary, axillary, on short hairy pedicels; **stamens** not exserted; **calyx lobes** pointed, to 1.5 cm long, with white hairs.

An outstanding species with sulphur yellow foliage and large reddish violet flowers. It occurs in the Austin District, and usually grows amongst mulga. Rarely encountered in cultivation, but it certainly has potential for use in arid and semi-arid regions. Most likely a species which would be difficult to establish in temperate zones, but is worth trial as a container plant. Propagate from seed or cuttings.

Eremophila neglecta J. Black
(neglected)
SA, NT
1.5-2.5 m x 1.5-2.5 m   May-July; also sporadic
Small to medium **shrub** with open habit; **branches** hairy, sticky; **leaves** 3-6 cm x up to 0.7 cm, alternate, lanceolate, glabrous, margins entire, apex pointed; **flowers** tubular, to 3 cm long, cherry red to reddish brown, rarely yellow, exterior faintly hairy, lobes blunt, solitary, axillary, on hairy, sticky pedicels to 2 cm long;

Eremophila neglecta, × ·65

stamens exserted; **calyx lobes** broad, green to red to purple, enlarging after flowering to about 1.5 cm long; **fruit** to 0.7 cm long, flattened, beaked, rough surface.

From Central Australia, where it grows on stony plains with shallow sand over clay. The coloured calyx is a feature of this species. At present it is rare in cultivation, but has potential for wider use in arid and semi-arid zones. May succeed in temperate zones if grown in dryish situations. Needs well drained soils, with plenty of sunshine. Frost and drought tolerant. Propagate from seed or cuttings which are difficult to strike.

Eremophila nivea Chinn.
WA
1-2.5 m x 0.7-1.5 m   Aug-Dec
Small to medium **shrub**; new growth covered in white hairs; **branchlets** densely covered with white hairs; **leaves** about 1.5 cm long, alternate, linear-lanceolate, hairy, greyish, margins recurved, apex pointed; **flowers** tubular, to 2 cm long, lilac, exterior glabrous, constricted near base, lobes blunt, solitary, axillary, profuse, forming leafy spikes, near ends of branchlets; **stamens** enclosed; **calyx lobes** to 1 cm long, densely hairy; **fruit** ovoid, glabrous.

This undescribed ornamental species is a recent introduction to cultivation. It is rare in nature and occurs only in the Three Springs area in the Irwin District, where it grows on sandy soils overlying clay. Although grown mainly by enthusiasts, it has proved adaptable to a range of soil types and climatic conditions. It prefers to grow in heavy soils with adequate drainage, but it will tolerate limited periods of inundation. Does best in plenty of sunshine. Withstands strong winds. Can become leggy but responds well to pruning, which should be started at an early age. Suited to container cultivation. This species appreciates a heavy watering in dry periods, and can respond with a flush of flowers 1-2 weeks later. Propagate from seed or cuttings which strike readily.

This species has been cultivated under the name of *E. margarethae*, which has thick, yellowish grey leaves that appear glabrous.

Eremophila obovata L. S. Smith
(ovate leaves, broadest above the middle)
Qld, SA, WA, NT
0.3-0.6 m x 0.5-1 m   July-Nov; also sporadic
Dwarf, spreading **shrub**; **branches** many, initially densely covered in hairs, often becoming glabrous; **leaves** 0.5-2.2 cm x up to 1 cm, alternate, obovate, spiral, slightly hairy, margins entire, sometimes with few teeth, apex rounded or notched; **flowers** tubular, to 2.2 cm long, blue to pale purple, exterior hairy, lobes blunt, solitary, axillary, on hairy pedicels to 0.6 cm long; **stamens** not exserted; **calyx lobes** to about 1 cm long, pointed, slightly hairy; **fruit** globular, to 0.8 cm diameter, glabrous.

This species is widely distributed in Central Australia, but rare in cultivation. It occurs from the Gregory South District in Qld to the Tanami Desert in WA. Is suited to regions with low rainfall and high temperatures. Must have well drained soils. May succeed as a container plant in temperate zones. Is proving difficult to maintain in cultivation.

The var. *glabriuscula* L. S. Smith from Qld and NT

Eremophila oldfieldii

has glabrous leaves. It has had only limited cultivation to date. Responds well to summer watering. Propagate both forms from seed or cuttings.

Eremophila oldfieldii F. Muell.
(after Augustus Oldfield, 19th-century English botanist and zoologist)
WA
2-4 m x 1.5-4 m   July-Sept
Medium **shrub**; **branches** faintly hairy; **leaves** to 12 cm x 1 cm, alternate, lanceolate, thick, green, glabrous, margins entire, apex pointed; **flowers** tubular, to about 3.5 cm long, red or purplish, exterior glabrous except for pointed lobes, solitary, axillary, on slender, hairy pedicels to 1 cm long; **stamens** exserted; **calyx lobes** to 1.3 cm long, overlapping, pointed, green; **fruit** ovoid, about 0.4 cm long.

A showy, bushy species from the Austin, Coolgardie and Irwin Districts, where it occurs mainly in red loams that are often alkaline. Cultivated mainly by enthusiasts to date, but has potential for areas of low rainfall and high temperatures, where it should be useful as a low screening and windbreak plant. Hardy to most frosts. Propagate from seed or cuttings which can be slow to form roots.

The var. *angustifolia* S. Moore was previously known as *E. angustifolia* (S. Moore) Ostenf. It differs in its narrower leaves and finer-lobed calyx, and is not well known in cultivation. The cherry red flower with its yellow-green calyx is very striking.

Eremophila oppositifolia R. Br.
(opposite leaves)
NSW, Vic, SA, WA   Twin-leaf Emu-bush; Weeooka
1.5-4 m x 1-3 m   June-Dec; also sporadic
Small to medium **shrub**; **branches** greyish due to dense covering of fine hairs; **leaves** 2-10 cm x up to 0.2 cm, opposite, terete to linear, greyish green, apex pointed and usually hooked; **flowers** tubular, to about 3 cm long, cream with pinkish tonings, exterior glabrous, lobes blunt, 1-2 short pedicels per axil, profuse; **stamens** slightly exserted; **calyx lobes** to about 2 cm long, whitish to lime green, blunt; **fruit** oblong, to 0.7 cm long, hairy.

This decorative species is relatively well known in cultivation. It occurs on a wide range of soil types in areas of low rainfall, and adapts to most soils which are fairly well drained. It prefers plenty of sunshine, although it tolerates dappled shade or partial sun in hot climates. Hardy to most frosts and extended dry periods. An additional ornamental feature of this species is the prominent whitish calyx lobes that are retained after flowering. Does well as a container plant. Responds favourably to pruning. Tip pruning of young plants helps to promote early bushy growth.

A further 2 varieties are recognized:
var. *angustifolia* S. Moore, from SA and WA, has greyish green leaves of less than 0.1 cm wide
var. *rubra* C. White & Francis, from Qld and NSW, has broad-lanceolate leaves that are alternate, cream flowers with pink to brown tonings and reddish brown calyx lobes. It is not widely cultivated at present.
Propagate all forms from seed or cuttings which usually produce roots readily.

Eremophila oldfieldii var. *angustifolia*, × ·6

Eremophila ovata Chinn.
(ovate)
NT
0.5-1 m x 0.5-1 m   July-Nov
Dwarf, bushy **shrub**; young growth often brownish yellow; **branchlets** densely hairy; **leaves** to 3.5 cm x 1.7 cm, alternate, ovate to orbicular, slender, petiolate, hairy to glabrous above, densely hairy below, margins flat and entire, apex pointed or blunt; **flowers** tubular, to 2.5 cm long, lilac to scarlet, exterior hairy, lobes blunt, solitary, axillary, sessile, near ends of branchlets; **stamens** not exserted; **calyx lobes** to 1.2 cm long, pointed, hairy; **fruit** globular, to 0.8 cm diameter, hairy.

A recently described species confined to the George Gill and Gardiner Ranges in north-central Australia. It is rare in cultivation, and is best suited to regions with low rainfall and high temperatures. Needs well drained soils. Should be hardy to most frosts. Worthy of trial as a container plant in temperate regions. Propagate from seed or cuttings.

Closely allied to *E. strongylophylla*, which occurs in the Shark Bay area of WA and has leaves with broad, short petioles and slightly wavy margins.

Eremophila pachyphylla Diels
(thick leaves)
WA
1-3 m x 1-2 m   Aug-Nov
Small to medium **shrub**; **branches** glabrous; **leaves** to 2.5 cm x about 0.4 cm, alternate, elliptic-lanceolate, glabrous, margins entire, apex pointed and hooked; **flowers** tubular, to 2 cm long, pale violet, scented, exterior glabrous, lobes blunt, 1-3 per axil, on short pedicels; **stamens** not exserted; **calyx lobes** to 1 cm long, purple, spathulate; **fruit** ovoid-oblong, 0.3-0.4 cm x about 0.2 cm, hairy.

An inhabitant of the Coolgardie and Eyre Districts, where it grows on sandy clays or brown clays in

Eremophila paisleyi, × ·7

Eucalyptus woodland. Is cultivated mainly by enthusiasts at present. The purple calyx adds an ornamental aspect. Best suited to well drained soils, with plenty of sunshine. Hardy to extended dry periods and most frosts. Propagate from seed or cuttings which can be slow to produce roots. Bees are suspected of collecting propolis from the leaves (see *E. exilifolia*). *E. psilocalyx* F. Muell. now has precedence over *E. pachyphylla*.

Eremophila paisleyi F. Muell.
(after J. C. Paisley, mid-19th-century private secretary to Governor of SA)
SA, WA, NT
1-3 m x 0.6-2 m   Aug-Nov
Small to medium **shrub**, variable in form; young growth sticky; **branches** erect, slender, glabrous, warty, sticky; **leaves** 2-5 cm x up to 0.8 cm, alternate, narrow-linear to oblanceolate, glabrous, margins entire, apex pointed and hooked; **flowers** tubular, about 1.5 cm long, white, mauve or bluish, exterior hairy, lobes blunt, scented, in axillary clusters of 2-5, profuse; **stamens** not exserted; **calyx lobes** hairy, obovate after flowering, pinkish; **fruit** ovate, to 0.4 cm long, slightly flattened, beaked, hairy.

An ornamental species that occurs in Central Australia, Great Victoria Desert and in the Austin and Coolgardie Districts of WA. In cultivation it is best suited to areas with low rainfall and high temperatures. Soils need to be well drained. Tolerant of most frosts. Plants can become leggy, but tip pruning from an early age may help to promote bushy growth. Worthy of trial as a container plant in temperate regions. Propagate from seed or cuttings which may be slow to produce roots. All attempts have been unsuccessful so far. Trials with grafting may produce best results.

Some forms tend towards *E. falcata*, which differs in its falcate leaves and whitish grey branches.

Eremophila pantonii F. Muell.
(after Joseph A. Panton, 19-20th-century public servant and botanical collector)
WA   Broombush
1-4 m x 0.6-2 m   Aug-Dec
Small to medium **shrub**; **branches** erect, glabrous, warty; **leaves** 1-2.5 cm x about 0.2 cm, alternate, linear, folded lengthwise, glabrous to hairy, warty, margins entire, apex pointed and hooked; **flowers** tubular, to about 2 cm long, lavender to violet, exterior hairy, lobes pointed, constricted near base, solitary, axillary, sessile, near ends of branchlets, profuse; **stamens** not exserted; **calyx lobes** small, hairy, overlapping; **fruit** ovoid, to 0.8 cm x 0.4 cm, hairy.

One of the broombushes, so called because the long, erect, slender branches end with clusters of branchlets and leaves. It occurs in the Austin and Coolgardie Districts, where it grows mainly on red, sandy loams. Has been in cultivation for some time, and is proving adaptable to a wide range of well drained soils. Does best in areas with low rainfall and high temperatures, but also grows successfully as a container plant in temperate regions. Tip pruning from planting helps to promote bushy growth. The planting of 2-3 plants close together is used successfully to create a dense screen. Hardy to light and medium frosts, and extended dry periods. Propagate from seed or cuttings which can be slow to form roots.

Eremophila pantonii, × ·6

453

Eremophila parvifolia J. Black
(small leaves)
SA, WA
0.6-1.5 m x 0.6-1.5 m   Sept-Oct; also sporadic

Dwarf to small **shrub**; **branches** slender, glabrous, often entangled; **leaves** to 0.2 cm x 0.2 cm, alternate, ovate to orbicular, thick, glabrous, apex pointed; **flowers** tubular, about 1 cm long, pale lilac to purple, exterior glabrous, lobes blunt, constricted near base, solitary, axillary, along ends of branchlets, profuse; **stamens** not exserted; **calyx lobes** pointed, hairy margins; **fruit** globular, about 0.3 cm diameter, glabrous.

This species, which has minute foliage, occurs on the coastal plains from near Fowlers Bay, SA to Eucla, WA, and is also found in the Coolgardie District of WA. It is grown mainly by enthusiasts. Needs well drained soils and plenty of sunshine, although it withstands some shade in areas which regularly have high temperatures. Tolerant of calcareous soils, frost and extended dry periods. Propagate from seed or cuttings which strike readily.

Eremophila pentaptera J. Black
(5-winged)
SA
0.2-0.5 m x 0.2-0.5 m   Aug-Nov

Dwarf **shrub**; **bark** rough, corky; **branches** erect, greyish white; **leaves** 1-3.5 cm x up to 1 cm, alternate, oblanceolate, succulent, glabrous except for the entire margins, apex blunt; **flowers** tubular, to 3.5 cm long, violet to blue-purple, exterior glabrous, lobes rounded, solitary, rarely 2 per axil, on 5-winged warty and glabrous pedicel; **stamens** not exserted; **calyx lobes** to 1.2 cm long, pointed, glabrous except for margins; **fruit** oblong, about 0.6 cm long, ribbed.

An ornamental species which has proved difficult to establish in cultivation. It occurs in semi-arid to arid regions of north-eastern SA, where it is limited to a few sites on flats near rivers. This species is suspected to be short-lived. In nature it establishes very quickly, growing rapidly during wet years and declining when normal conditions return. Flood debris has been found through bushes. Best suited to areas with low rainfall and high temperatures. Tolerant of most frosts. Makes an ornamental container plant. Propagate from seed or cuttings which are usually quick to produce roots. This species has been successfully grafted using *Myoporum insulare* as the rootstock.

Eremophila phillipsii F. Muell.
(after George B. Phillips, 19th-century WA public servant)
WA
1-1.5 m x 1-1.5 m   Sept-Nov

Small **shrub**; **branches** glabrous, sticky; **leaves** to 4 cm x about 0.1 cm, alternate, linear, glabrous, strong odour, apex pointed; **flowers** tubular, to 1.5 cm long, violet, exterior hairy, lobes blunt, solitary, axillary, on long, slender pedicels, near ends of branchlets; **stamens** not exserted; **calyx lobes** pointed, to about 0.4 cm long, densely hairy; **fruit** about 0.4 cm x 0.2 cm, compressed.

From the Coolgardie, Eyre and Roe Districts, where it occurs on a range of soils. It is not well known in cultivation, possibly because it is regarded as difficult to maintain and because of the strong leaf odour, which is considered offensive by some people. Needs well drained soils and plenty of sunshine. Tolerant of extended dry periods and most frosts. Propagate from seed or cuttings which have proved very difficult to strike.

Eremophila platycalyx F. Muell.
(broad calyx)
WA   Granite Poverty Bush
1-3 m x 1-2 m   Aug-Nov

Small to medium **shrub**; young growth sticky; **branches** densely covered in greyish hairs, rough; **leaves** to 6 cm x up to 0.7 cm, alternate, linear to narrow-lanceolate, petiolate, densely hairy, greyish, margins entire, apex pointed; **flowers** tubular, to about 3 cm long, cream, pink or orange, exterior glabrous, lobes pointed, solitary or rarely 2 per axil, on long, slender, hairy pedicels; **stamens** not exserted; **calyx lobes** broad, pointed, about 1 cm long, pinkish, hairy; **fruit** ovoid, about 0.6 cm long.

An inhabitant of low-rainfall regions such as the Austin, Coolgardie and Irwin Districts, where it occurs on or around granite outcrops. Poorly known in cultivation but showing signs of being adaptable, this ornamental species has potential for use in arid and semi-arid areas and possibly temperate regions. Needs good drainage and plenty of sunshine. Tolerant of most frosts. Worthy of trial as a container plant in temperate zones. Propagate from seed or cuttings.

The var. *lancifolia* Kraenzlin has much broader leaves, to 2 cm wide, and it may become a separate species in the forthcoming revision. It occurs from the Warburton region near the NT border, and extends westward to the Austin District. The calyx is often more showy than the flowers.

Eremophila platythamnos Diels
(broad or spreading bush)
SA, WA, NT   Desert Foxglove
1.5-2 m x 1.5-2 m   Aug-Oct; also sporadic

Small **shrub**; young growth sticky; **branches** glabrous, shiny; **leaves** to 1 cm x about 0.6 cm, alternate, obovate, sessile, glabrous, sticky, margins entire, apex pointed; **flowers** tubular, about 2 cm long, violet, exterior hairy, constricted near base, lobes rounded, solitary, axillary, near ends of branchlets; **stamens** not exserted; **calyx lobes** to 1 cm long, blunt, lemon yellow, hairy or glabrous; **fruit** ovoid, to 0.7 cm x 0.5 cm.

This spreading, showy species occurs in arid and semi-arid regions, where it usually grows on red sands. Cultivated mainly by enthusiasts to date, but commonly grown in SA. Requires very well drained soils, and is best suited to low-rainfall areas which experience high temperatures over extended periods. May succeed as a container plant in temperate regions. Propagate from seed or cuttings.

Previously known as *E. exotrachys* Kraenzlin.

Eremophila polyclada (F. Muell.) F. Muell.
(many branches)
Qld, NSW, Vic, SA

Flowering Lignum;
Lignum Fuchsia-bush;
Twiggy Emu-bush

1-2.5 m x 1.5-4 m   Sept-May; also sporadic

Small to medium, spreading **shrub**; **branches** rigid, glabrous, smooth, often entangled; **branchlets** short; **leaves** 2-6 cm x up to 0.4 cm, alternate, linear, glabrous, sparse, bright green, apex pointed; **flowers** tubular, to about 3 cm long, white or purplish white, with green or brownish spots, exterior glabrous, lobes blunt and spreading, solitary or rarely 2 per axil, on slender, glabrous pedicels, profuse; **stamens** not exserted; **calyx lobes** to 0.6 cm long, pointed, overlapping, mainly glabrous; **fruit** oblong, to 1.5 cm long, glabrous.

Relatively well known in cultivation. This is an inland species which usually occurs in heavy soils which are subjected to flooding. It has proved very adaptable in cultivation, growing in temperate and semi-arid zones, on a wide range of soils. It develops best on soils with a heavy texture. Withstands some shade but prefers a situation with full sun. Hardy to extended dry periods and most frosts. Can grow into a wide, spreading mound, but is readily restricted by pruning. Propagate from seed or cuttings which produce roots readily. Sometimes young shoots on cuttings may be attacked by grey mould or dieback. When preparing cuttings, retain as many nodes as possible to allow for shoot development.

Eremophila pterocarpa W. Fitzg.
(winged fruits)
WA
2-3 m x 1.5-3 m   April-Nov; also sporadic

Medium to tall **shrub**; **branches** rough; **branchlets** with silver-white hairs; **leaves** 2-2.5 cm x up to 0.4 cm, alternate or opposite, lanceolate, crowded, densely minutely hairy, margins entire, apex pointed, yellowish green; **flowers** tubular, to 2 cm long, pink to dull red, exterior hairy, lobes blunt, solitary, axillary, on slender

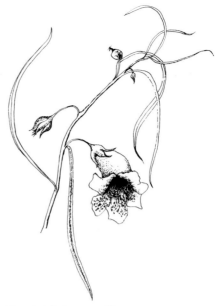

Eremophila polyclada form, × ·55

pedicels; **stamens** slightly exserted; **calyx lobes** short, pointed, slightly hairy; **fruit** prominently winged, about 1 cm long.

An outstanding silver-foliaged species from the Austin District, where it usually occurs on clay flats. It is commonly cultivated in SA, but not as well known in other states. An adaptable species which does best in relatively well drained soils in a sunny situation. Tolerates alkaline soils. Hardy to most frosts and extended dry periods. Propagate from seed or cuttings which may be slow to produce roots. Cuttings may rot if propagating medium is kept too moist.

Eremophila punctata Chinn.
(dotted)
WA
0.5-1.5 m x 0.3-1 m   June-Sept

Dwarf to small **shrub**; **branches** slightly ribbed, densely hairy; **leaves** 0.5-2.6 cm x 0.2-0.5 cm, alternate, oblanceolate, crowded, glabrous, glandular, margins toothed near the pointed apex; **flowers** tubular, to 3 cm long, pale lilac to purple, exterior hairy, lobes pointed, solitary, axillary, on hairy pedicels to 2 cm long, near ends of branchlets; **stamens** not exserted; **calyx lobes** to 1.5 cm long, green to blackish purple, overlapping, pointed, hairy; **fruit** ovoid, to 0.7 cm long, hairy.

A most ornamental recently described species. It extends from the Great Victoria and southern Gibson Deserts, to near Meekatharra in the west. Occurs on clay-loams or rocky hillsides. Little-known in cultivation, but has potential for use in areas of low rainfall and high temperatures. Must have excellent drainage and maximum sunshine for good growth. Hardy to most frosts. Worth trying as a container plant in temperate zones. Propagate from seed or cuttings.

Eremophila platythamnos, × ·75

Eremophila punicea

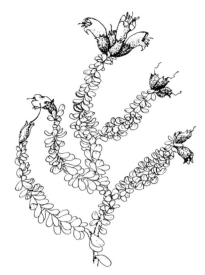

Eremophila punicea, × ·6

Eremophila punicea S. Moore
(scarlet)
WA   Crimson Eremophila
0.15-1 m x 0.3-1.5 m   July-Aug; also sporadic
    Dwarf, woolly **shrub**; **branches** densely hairy; **leaves**
about 1 cm long x 0.5 cm, alternate, oblong to spathulate, crowded, densely hairy, margins thickened or
revolute, apex blunt, greenish grey; **flowers** tubular, to
about 2 cm, deep pink to red or mauve, exterior hairy,
lobes solitary, axillary, on short pedicels, near ends of
branchlets; **stamens** exserted; **calyx lobes** to 1.2 cm
long, pointed, woolly; **fruit** ovoid, about 0.6 cm x
0.4 cm, ribbed, glabrous.
    This densely hairy species usually occurs on red
loam, and is found in the Austin, Coolgardie and Irwin
Districts. It is rare in cultivation, as it is mainly grown
by enthusiasts. Best suited to areas with low rainfall
and high temperatures. Tolerant of most frosts, but
bark may suffer damage from heavy frost. Could make
a decorative container plant in temperate regions.
Propagate from seed or cuttings which can be slow to
form roots.

Eremophila purpurascens Chinn.
(becoming purple)
WA
1-1.5 m x 1-1.5 m   Aug-Oct
    Small **shrub**; **branches** erect, rigid, forked, warty,
glabrous; **leaves** 0.5-1.4 cm x 0.2-0.7 cm, alternate,
obovate to spathulate, dark green, spreading, thick,
fleshy, warty, glabrous, margins entire, apex blunt;
flowers tubular, about 2.5 cm long, light purple, exterior hairy, lobes pointed, single, lower one reflexed,
solitary, axillary, on slender S-shaped pedicels to
2.5 cm long, on branchlets; **stamens** exserted; **calyx
lobes** broad, about 1 cm long, overlapping, purplish,
hairy; **fruit** conical, about 0.4 cm long.

456

A recently described and highly ornamental species
that is becoming well known in cultivation. In nature it
is restricted to the granite hills around Norseman.
Seems adaptable to a wide range of soils, provided
drainage is adequate. It needs plenty of sunshine. Ideal
for use in areas of low rainfall and high temperatures,
and worthy of trial in temperate zones. Propagate from
seed or cuttings.
    Allied to *E. alternifolia*, which has narrower and
longer leaves, and densely warty branches.

Eremophila pustulata S. Moore
(blistered)
WA
0.5-1 m x 0.5-1 m   Sept
    Dwarf **shrub**; **branches** ascending, many, warty,
glabrous; **leaves** to about 0.8 cm x 0.2 cm, alternate,
linear-oblanceolate, glabrous, succulent, smooth
above, warty below, margins entire, apex blunt;
flowers tubular, to 1.2 cm long, pale lilac to lilac, exterior glabrous, lobes pointed, constricted near base,
solitary, axillary, on slender pedicels, near ends of
branchlets; **stamens** not exserted; **calyx lobes** about
0.5 cm long, overlapping, hairy on inside; **fruit** ovoid,
about 0.6 cm x 0.4 cm, hairy.
    From the Austin and Coolgardie Districts, where it
grows on sand-dunes. It is rare in cultivation. Should
do best in well drained soils, in a situation that receives
plenty of sunshine. Tolerates calcareous soils, most
frosts and extended dry periods. Propagate from seed
or cuttings which produce roots readily. Also successfully grafted on to *Myoporum parvifolium* rootstock.
    Closely related to *E. caerulea* which has hairy flowers
and narrower leaves which are warty on both surfaces,
and *E. weldii* which has purplish flowers and blunt
leaves.

Eremophila racemosa, × ·65

Eremophila racemosa Endl.
(flowers in racemes)
WA
0.3-2 m x 0.3-1.5 m   Sept-Nov; also sporadic
Dwarf to small **shrub**; **branches** usually erect, glabrous, slightly warty on mature parts; **leaves** 1.3-5 cm x 0.2-0.8 cm, alternate, narrow or oblanceolate, pale green, glabrous, margins entire, apex pointed; **flowers** tubular, to 2.5 cm long, initially orange and yellowish orange, mature to reddish purple, exterior glabrous, constricted near base, lobes pointed and reflexed, solitary, axillary, on straight or S-shaped pedicels, profuse; **stamens** exserted; **calyx lobes** to 0.7 cm long, pointed, glabrous exterior; **fruit** globular, to 1.5 cm diameter, beaked, spotted before maturity.
A very showy upright species which is becoming well known in cultivation. During flowering, plants are most colourful, with the various shades of yellow, orange, pink, red and purple. It has a very limited distribution in the Lake Cronin-Forrestiana region north of Ravensthorpe, where it grows in lateritic soils. Has adapted well and is grown successfully from NSW to WA. Soils need to be relatively well drained, and a sunny situation is best. Tolerant of extended dry periods and most frosts. Needs protection from wind, as plants are prone to blow over. Tip pruning from planting is recommended as it can help to promote bushy growth; otherwise plants often become very leggy. Successfully cultivated as a container plant. Propagate from seed or cuttings which readily produce roots.
E. bicolor Chinn. is considered conspecific.

Eremophila ramiflora B. Dell
(flowers from branches)
WA
1-3 m x 1.5-4 m   July-Sept; also sporadic
Small to medium **shrub**; young growth sticky; **branches** hairy, sticky; **branchlets** thick, rigid, rough, resinous; **leaves** 1.2-2.5 cm x up to 1 cm, alternate, lanceolate, crowded, spreading, reflexed, keeled and folded lengthwise, glabrous, sticky, margins usually entire, wavy, apex pointed; **flowers** tubular, to 3 cm long, magenta with dark spots, exterior hairy, lobes pointed, lower one strongly reflexed, 1-3 on pedicels to 1.2 cm long, in upper axils, can be profuse; **stamens** exserted; **calyx lobes** enlarge after flowering to 2 cm long, broad, green becoming red, shiny; **fruit** ovoid, to 1 cm long.
A recently described ornamental species from the Eremaean Botanical Province, where it occurs in the south-eastern Ashburton and eastern Austin Districts. It grows on red loams that often have a high stone content, overlying laterite. Evidently not well known in cultivation, it should be best suited to arid or semi-arid zones. Probably tolerant of most frosts. Propagate from seed or cuttings.
It is closely allied to *E. fraseri*, which has reddish brown flowers and larger calyx lobes.

Eremophila resinosa F. Muell.
(resinous)
WA
0.5-1 m x 0.5-1 m   Oct-Jan
Dwarf, spreading **shrub**; young growth sticky; **branches** densely covered in white hairs, warty; **leaves** to 1 cm x 0.3 cm, alternate, cuneate-obovate, erect to spreading, greyish green, sparse to dense covering of short hairs, margins entire, apex pointed or blunt; **flowers** tubular, about 1.5 cm long, violet, exterior hairy, lobes pointed, constricted near base, solitary, axillary, nearly sessile, near ends of branchlets; **stamens** not exserted; **calyx lobes** narrow, pointed, hairy; **fruit** ovoid, about 0.6 cm long, hairy.
A free-flowering species from the south-western corner of WA, where it grows in a range of soil types. It is rare in cultivation and mainly grown by enthusiasts. Has potential for wider use. Needs soils that are well drained, and likes plenty of sunshine. Hardy to frosts and extended dry periods. Will grow well in containers. Tip pruning helps to promote bushy growth. Propagate from seed or cuttings which strike readily.
E. kochii Ewart is included in this species.

Eremophila rotundifolia F. Muell.
(round leaves)
SA, WA, NT
1-1.5 m x 1-1.5 m   Aug-Nov
Small **shrub**; **branches** densely covered in greyish hairs, warts obscured by hairs; **leaves** 0.6-1.2 cm x 0.5-1.2 cm, alternate, ovate to orbicular, thick, rigid, crowded, usually glabrous above and slightly hairy below, margins entire, apex recurved; **flowers** tubular, to 2.5 cm long, lilac to blue, exterior hairy, lobes blunt, solitary, axillary, on greyish pedicels to 1 cm long, near ends of branchlets; **stamens** not exserted; **calyx lobes** to 1 cm long, pinkish, with long hairs; **fruit** ovoid, to 0.7 cm long, hairy.
This species, from arid regions, has greyish foliage and is grown mainly by enthusiasts. It is most ornamental and has potential for greater use in areas with low rainfall and high temperatures. Needs soils with good to excellent drainage, and a situation with

Eremophila resinosa, × ·65

457

Eremophila saligna

maximum sunshine. It may succeed as a container plant in temperate regions. The grey foliage is useful in defining edges of paths or traffic-ways at night. Propagate from seed or cuttings which may be slow to produce roots.

Eremophila saligna (S. Moore) C. Gardner
(willow-like)
WA   Willowy Eremophila
1-2 m x 0.6-1.5 m   Sept-Oct; rarely sporadic
Small erect **shrub**; **branches** erect, glabrous, glands in line; **leaves** to 4 cm x 0.6 cm, alternate, lanceolate, glabrous, distantly toothed margins, apex pointed and hooked; **flowers** tubular, to 1.5 cm long, white, exterior glabrous, lobes more or less pointed, 1-4 slender pedicels per axil; **stamens** not exserted; **calyx lobes** pointed, glabrous; **fruit** cylindrical, to 1 cm long, glabrous.

This eremophila is very similar to *Myoporum platycarpum*, but is readily distinguished during flowering. It occurs in the Coolgardie District, where it grows on rocky slopes and red-brown earths in *Eucalyptus* woodland. Not commonly cultivated. It requires well drained soils, with partial or full sunshine. Hardy to most frosts and extended dry periods. Propagate from seed or cuttings which can be slow to form roots.

Eremophila santalina (F. Muell.) F. Muell.
(resembling the genus *Santalum*)
SA
2-6 m x 1.5-5 m   Aug-Oct; also sporadic
Small to tall **shrub**; **branches** glabrous; **branchlets** pendulous; **leaves** 3-8 cm x up to 1 cm, alternate, lanceolate, glabrous, dark green, margins entire, apex pointed; **flowers** tubular, to 2 cm long, whitish, exterior glabrous, lobes blunt, 1-2 slender pedicels per axil, profuse; **stamens** not exserted; **calyx lobes** broad, overlapping, pointed, glabrous; **fruit** globular, to 0.9 cm across, slightly sticky.

A graceful species from the Flinders Ranges. Very suitable for use in regions with high temperatures and low rainfall. It will grow in most well drained soils, and does best in a sunny situation. Will tolerate some shade. A useful screening or windbreak plant. Worth trying in large containers in temperate regions. Propagate from seed or cuttings which can be slow to form roots. Grafting can be successful, using *E. bignoniiflora*. *Myoporum parvifolium* has also been used as rootstock, but it can be short-lived. *M. insulare* and *M. montanum* are long-lived and more suitable.

Eremophila sargentii (S. Moore) Chinn.
(after Oswald H. Sargent, 19-20th-century WA pharmacist and botanical collector)
WA
0.8-1.5 m x 1-1.5 m   Sept-Nov
Dwarf to small **shrub**; young growth sticky; **branches** many, upright, glabrous, grooved, sticky; **leaves** about 0.4 cm x 0.1 cm, alternate, narrow-linear, sessile, crowded, glabrous, apex recurved or hooked; **flowers** tubular, about 1.5 cm long, blue, exterior glabrous, lobes blunt, solitary, axillary, on slender pedicels, along branchlets; **stamens** not exserted; **calyx lobes**

glabrous, tips recurved; **fruit** ovoid, about 0.35 cm x 0.2 cm, glabrous.

A small heath-like species from the Avon District, where it is confined to the Wongan Hills area. It is cultivated mainly by enthusiasts, but should prove adaptable to a wide range of soil types in arid to temperate regions. Hardy to most frosts, and withstands extended dry periods. Worthy of trial as a container plant because of its compact foliage and blue flowers. Propagate from seed or cuttings.

It is suspected that bees collect propolis from the leaves (see *E. exilifolia*).

Eremophila scaberula W. Fitzg.
(minutely rough)
WA
0.6-1.5 m x 1-1.5 m   Oct-Nov
Dwarf to small **shrub**; young growth slightly sticky; **branches** with raised, decurrent lines; **leaves** to 1 cm x about 0.1 cm, alternate, linear to subulate, thick, crowded, rough, slightly sticky; **flowers** tubular, small, violet or purple, exterior glabrous, solitary, axillary; **stamens** not exserted; **calyx lobes** glabrous, with pointed, recurved apex; **fruit** narrow-ovoid, glabrous.

A little-known species from the Avon District. It is known only from the original collection at the beginning of this century. Evidently not in cultivation, but should be best suited to a situation with plenty of sunshine, in soils that have good drainage. Tolerant of most frosts and dryness. Propagate from seed or cuttings.

Eremophila scoparia (R. Br.) F. Muell.
(broom-like)
NSW, Vic, SA, WA   Scotia Bush; Silver Emu-bush
1-3 m x 0.6-2 m   June-Nov; also sporadic
Small to medium, broom-like **shrub**; **branches** erect, slender, 4-ribbed, grey to yellowish, smooth or warty; **leaves** 0.7-1.5 cm x about 0.1 cm, mostly opposite, linear to narrow-linear or somewhat terete, silvery, apex pointed and hooked; **flowers** tubular, to 2 cm long, white or pale blue to violet, exterior scaly, lobes blunt, 1-2 on short pedicels in upper axils; **stamens** not exserted; **calyx lobes** small, pointed; **fruit** ovoid, to 0.6 cm long, beaked.

This ornamental, erect species has silvery foliage. It usually occurs on red sandy loams in low-rainfall regions, where it commonly grows as an undershrub in woodland. Plants can be narrow and upright or more bushy. At present it is not widely cultivated, but has potential for greater use in arid and semi-arid areas. Hardy to most frosts. Tolerant of alkaline soils. Tip pruning from an early age may help to induce lateral growth. Propagate from seed or cuttings which usually produce roots readily. If bushy plants are desired it is best to select cuttings from dense shrubs.

Eremophila serpens Chinn.
(creeping)
WA   Creeping Eremophila
prostrate x 1.5-3 m   most of year
Prostrate, spreading **shrub**; young growth often purplish; **branches** creeping, self layering, purplish, fleshy,

Eremophila santalina   W.R. Elliot

warty; **leaves** 3-5 cm x 0.5-1.5 cm, alternate, oblance-olate, erect to slightly spreading, glabrous, glandular, toothed upper half, apex pointed; **flowers** tubular, to about 2.5 cm long, lime green with purple on upper lobe, exterior warty, 1-2 per axil, on pedicels to 1 cm long; **stamens** exserted, bright purple; **calyx lobes** pointed, enlarging after flowering to about 1 cm long; **fruit** globular, to about 1 cm diameter, papery exterior, grey.

A vigorous groundcovering species from the Cool-gardie District, where it occurs only in the Hyden-Lake King-Newdegate region. It has adapted very well to cultivation in semi-arid and temperate zones. Prefers soil with good drainage, but will tolerate extended wet periods and even short periods of inundation. Will grow well in semi-shade to full sun. Resents highly alkaline soils. Tolerant of most frosts. Especially useful for heavy soil embankments. An excellent living mulch. The flowers are very attractive to native honey-eating birds. Propagate from seed, cuttings which strike readily or by separation of layered branches which can be grown in containers until established prior to planting.

Eremophila serrulata (A. Cunn. ex A.DC.) Druce
(finely toothed)
NSW, SA, WA, NT   Green Fuchsia-bush
1-2 m x 1-2 m   June-Sept; also sporadic
Small **shrub**; young growth sticky; **branches** often leggy, slightly hairy; **branchlets** sticky; **leaves** 2-8.5 cm x 1-2 cm, alternate, ovate to lanceolate, slightly hairy, very sticky, margins irregularly toothed or entire, apex blunt or pointed; **flowers** tubular, to about 2.5 cm long, green to brownish green, exterior hairy, lobes pointed or with lower one reflexed, solitary, axillary, on slender S-shaped pedicels to 2 cm long; **stamens** exserted; **calyx lobes** overlapping, enlarging and spreading after flowering to 1.5 cm long; **fruit** globular, to 0.8 cm diameter, glabrous.

This variable species grows in differing habitats over its wide distribution in the arid and semi-arid regions. It is not widely cultivated, which is probably due to its commonly open, leggy growth habit and insignificant flowers. It has potential, as tip pruning of young plants helps to promote bushy growth, and the flowers are regularly visited by honey-eating birds seeking nectar. Plants need relatively good drainage and should grow in most soil types. They will grow in dappled shade but prefer partial or full sun. Hardy to frost. Propagate from seed or cuttings which may be slow to form roots.

Allied to *E. undulata*, which is smaller and compact and has white, hairy branches. See also *E. gibbosa*.

Eremophila spathulata W. Fitzg.
(spoon-shaped)
WA
0.5-1 m x 0.8-1.5 m   Aug-Sept
Dwarf **shrub**; young growth slightly sticky; **branches** rough; **leaves** 1-1.5 cm x up to 1 cm, alternate, spath-ulate, dense covering of hairs, margins entire, apex rounded, often ending in small point; **flowers** tubular, to about 2.5 cm, deep violet with yellow base, exterior hairy, lobes blunt, solitary, on hairy pedicels, near ends of branchlets; **stamens** not exserted; **calyx lobes** broad, hairy; **fruit** ovoid, 0.5-0.8 cm x about 0.5 cm, beaked, hairy.

An ornamental species from the Austin District, where it often occurs on stony soils. As yet it is rare in cultivation, but it has potential for greater use in arid and semi-arid regions. Needs a situation that receives maximum sunshine and has good drainage. Tolerant of most frosts. May succeed as a container plant in temperate climes. Propagate from seed or cuttings which are difficult to strike.

Eremophila spectabilis C. Gardner
(remarkable; spectacular)
WA   Showy Eremophila
1-2 m x 1-2 m   July
Small **shrub**; young growth sticky; **branches** sticky, with short white hairs; **leaves** to 7 cm x about 0.3 cm, alternate, linear to linear-lanceolate, sessile, hairy, sticky, margins faintly toothed, apex pointed; **flowers** tubular, to 3.5 cm long, pale to deep violet or purple, exterior glabrous, lobes pointed, solitary, axillary, on slender pedicels to 2 cm long, near ends of branchlets; **stamens** not exserted; **calyx lobes** shiny, green, enlarging after flowering to 2.5 cm long; **fruit** narrow-ovoid, about 1 cm long, hairy.

459

Eremophila spinescens

An attractive eremophila which occurs on stony soils amongst the mulga scrublands of the Austin District, near Meekatharra. At this stage it is poorly known in cultivation, but warrants wider use. Probably best suited to regions with low rainfall and high temperatures. Hardy to most frosts. Suited to large containers. Propagate from seed or cuttings which may be slow to form roots.

This species is closely related to *E. freelingii*, which has broader, entire leaves and shorter pedicels.

Eremophila spinescens Chinn.
(somewhat spiny)
WA
0.3-0.5 m x 0.5-1.5 m   July-Oct
   Dwarf **shrub**; **branches** many, ending in sharp point, hairy; **leaves** 0.5-2.5 cm x 0.1-0.4 cm, alternate, linear to oblanceolate, sessile, glabrous or slightly hairy, sometimes in clusters, margins usually entire or with scattered teeth, apex pointed, green or green with purplish tonings; **flowers** tubular, to 2 cm long, pale blue to dark purple, exterior hairy to glabrous, lobes blunt, solitary, axillary, on short, hairy pedicels, on branchlets; **stamens** not exserted; **calyx lobes** overlapping, to 1 cm long, green to blackish purple, usually hairy; **fruit** ovoid, to 1 cm long, densely hairy.
   A recently described species from the arid region of central WA, where it grows on light brown to red-brown clay-loams, in depressions or near salt lakes. Not well known in cultivation. It displays its flowers well amongst the spiny branches, and should be useful in arid and semi-arid regions. Probably best suited to hot, well drained situations. Worth trying as a container plant in temperate zones. Propagate from seed or cuttings.
   Allied to *E. battii*, which has soft branches and cuneate leaves.

Eremophila strehlowii E. Pritzel =
                              E. macdonnellii F. Muell.

Eremophila strongylophylla F. Muell. =
                              E. mackinlayi F. Muell.

Eremophila sturtii R. Br.
(after Charles Sturt, 19th-century explorer and botanist)
Qld, NSW, Vic, SA, NT   Turpentine-bush
1-4 m x 1-3 m   Sept-Nov; also sporadic
   Small to medium **shrub**; **bark** dark grey; new growth sticky; **branches** slender, glabrous; **leaves** 2-4 cm x up to 0.2 cm, alternate, linear or somewhat terete, bright green, highly aromatic, margins entire, apex pointed; **flowers** tubular, to 1.7 cm long, lilac to pale mauve, sweetly scented, exterior hairy, lobes blunt, solitary, axillary, on short pedicels to 1.2 cm long, profuse, near ends of branchlets; **stamens** not exserted; **calyx lobes** whitish green, to about 1 cm long, broad; **fruit** ovoid, to 0.7 cm long, beaked, hairy.
   A widespread species that occurs on a range of soil types, but is not usually found on flood plains. Although decorative and cultivated to a limited extent, it is considered a weed of some pastoral areas because it can form large colonies by suckering if roots are damaged, plants are burnt or if areas are denuded by overgrazing. It will grow in most well drained soils, and appreciates plenty of sunshine. Hardy to frost and tolerant of extended dry periods. Propagate from seed which is difficult to germinate, from cuttings which may be difficult to strike or from root suckers.

Eremophila subfloccosa Benth.
(somewhat woolly)
SA, WA
0.5-1.5 m x 1-2 m   Aug-Feb; also sporadic
   Dwarf, spreading **shrub**; young growth woolly; **branches** many, densely hairy; **leaves** 2-4 cm x up to 0.8 cm, alternate, ovate-oblong to elliptical-oblong, initially densely hairy, becoming glabrous on upper surface, margins entire, apex blunt; **flowers** tubular, to 2.5 cm long, greenish yellow, exterior glabrous or slightly hairy, lobes pointed, solitary, axillary, more or less sessile, profuse, along branches; **stamens** exserted; **calyx lobes** narrow, hairy; **fruit** subglobular, 0.5-1 cm x 0.4-0.9 cm, glabrous.
   This species is widely distributed over the semi-arid region of WA, and also occurs in the Eyre Peninsula in SA. The WA form is not as hairy as the one from SA which will become a subspecies in the revision. Is mainly grown by enthusiasts, but has potential for greater use in areas with low rainfall and high temperatures. Has proved difficult to maintain as a garden plant in temperate regions, but may succeed in containers. Responds well to light pruning. Propagate from seed or cuttings which may produce roots quickly but can be hard to maintain in the humid atmosphere of the propagation structure, due to attack by grey mould.
   E. glabra var. *viridiflora* will become a subspecies of *E. subfloccosa* in the revision.

Eremophila ternifolia Chinn.
(leaves in 3s)
WA
0.3-0.5 m x 0.5-1 m   Oct-Feb
   Dwarf, spreading **shrub**; **branches** flattened, hairy;

Eremophila subfloccosa, × ·7

leaves 0.6-1.1 cm x up to 0.4 cm, in alternate whorls of 3, lanceolate to elliptic, glabrous, green above, often reddish brown below, margins entire, apex blunt; **flowers** tubular, to 1 cm long, lilac with white below, exterior hairy, lobes blunt, solitary, sessile, near ends of branchlets; **stamens** not exserted; **calyx lobes** narrow, pointed, hairy; **fruit** ovoid, to about 0.3 cm long, beaked, hairy.

A recently described species from the Wongan Hills region of the Avon District, where it is very rare. Occurs on red clays between low, stony hills. This species should be conserved in nature, and it is also important that it be maintained in cultivation so that there is a plant bank available for future use. It has been grown successfully for some years at Adelaide Botanic Gardens. Should prove adaptable to a wide range of soil types, and will probably do best in a situation with plenty of sunshine. Propagate from seed or cuttings which strike readily.

Eremophila tetraptera C. White
(4-winged)
Qld
1-3 m x 1-2.5 m   June-July; also sporadic

Small to medium **shrub**; young growth slightly sticky; **branches** initially hairy, becoming glabrous; **leaves** 1.5-6 cm x 0.2-0.7 cm, alternate, linear-lanceolate, flat, fleshy, initially hairy, becoming glabrous, often clustered on short branchlets, margins entire, apex blunt or pointed; **flowers** tubular, to about 2.5 cm long, red, exterior glabrous, lobes blunt, solitary, axillary, on slender, glabrous S-shaped pedicels to 2 cm long; **stamens** exserted; **calyx lobes** to 1 cm long, pointed, glabrous; **fruit** ovoid, to 1.2 cm long, with 4 longitudinal wings about 0.5 cm wide.

From the Gregory North District, where it grows in proximity to the Diamantina River, in the channel country. It is cultivated mainly by enthusiasts, and needs well drained soils, with plenty of sunshine. Has potential for greater use in areas of low rainfall and high temperatures, where it could be used as a low windbreak and screening plant. Tolerant of most frosts. Responds well to summer watering. Propagate from seed or cuttings which can be very slow to form roots. Some growers find semi-hardwood cuttings the most successful. *Myoporum insulare* is a suitable grafting rootstock for this species.

Eremophila undulata Chinn.
(wavy)
WA
0.3-0.5 m x 0.5-1.5 m   May-Sept

Dwarf, spreading **shrub**; **branches** terete, densely covered in long white hairs; **leaves** 2-3.5 cm x 0.5-1 cm, alternate, oblong to elliptic, petiolate, hairy, margins entire, wavy, apex blunt; **flowers** tubular, 1.5-2 cm, greenish brown to yellowish brown, exterior hairy, lobes pointed, lower lobe reflexed, solitary, axillary, on S-shaped hairy pedicels to 1.7 cm long, near ends of branchlets; **stamens** exserted; **calyx lobes** overlapping, broad, pointed, slightly hairy; **fruit** globular, to 0.7 cm long, exterior papery.

This recently described species has a restricted distribution in an area of the Great Victoria Desert, where it is common. It grows on red-brown clay-loams, in mulga and amongst spinifex. At this stage it is very rare in cultivation. Needs are for high temperatures and low rainfall, although it will tolerate most frosts. Will grow in shaded situations in semi-arid or arid zones. May succeed as a container plant in temperate regions. Propagate from seed or cuttings.

Allied to *E. serrulata* which is a larger shrub with slightly hairy branches.

Eremophila veronica (S. Moore) C. Gardner
(after the genus *Veronica*)
WA
0.6-1.3 m x 0.6-1.5 m   Sept-Jan

Dwarf to small, dense **shrub**; **branches** many; **leaves** to 0.5 cm x about 0.1 cm, alternate, but usually in whorls of 3, more or less terete, falcate, erect to spreading, crowded, glabrous, apex blunt; **flowers** tubular, to about 1 cm, blue or lilac, exterior glabrous, lobes blunt, solitary, axillary, on short pedicels, near ends of branchlets; **stamens** scarcely exserted; **calyx lobes** short, pointed, hairy margins; **fruit** globular, to 0.3 cm x about 0.2 cm, hairy on upper part.

This small heath-like species occurs in the Coolgardie District. It is not well known in cultivation and is probably best suited to relatively well drained soils, with a warm to hot situation. Should be hardy to most frosts and extended dry periods. Propagate from seed or cuttings.

Eremophila verrucosa Chinn.
(covered with warts)
WA
1-2 m x 0.6-1.5 m   May-Nov; also sporadic

Small erect **shrub**; **branches** many, erect, silvery, warty; **leaves** 0.5-1.6 cm x about 0.2 cm, opposite, decussate, linear-lanceolate, somewhat terete, spreading, silvery, crowded, often reflexed, margins entire, apex pointed and often hooked; **flowers** tubular, 1.5-2.2 cm long, pale blue to mauve, exterior with long stellate hairs, lobes blunt, 1-2 on very short pedicels, in upper axils, profuse; **stamens** not exserted; **calyx lobes** pointed, short, silvery; **fruit** ovoid, to about 0.5 cm long, very hairy.

This recently described species occurs in the northern and Nullarbor regions of SA, where it usually grows in skeletal soils on slopes. Poorly known in cultivation, but the silvery appearance is appealing and the species will probably become better known. Will need very well drained soils and maximum sunshine. Should be tolerant of most frosts and extended dry periods. Pruning from an early stage should help to promote bushy growth. Worth trying as a container plant. The ssp. *brevistellata* Chinn. differs with the shorter hairs on the flowers. It is known only from near Ooldea, where it occurs on red sandy loams overlying limestone.

Propagate both forms from seed or cuttings.

Eremophila species (Kalgoorlie)
see *E. drummondii*

Eremophila 'versicolor'
see under *E. biserrata* Chinn.

Eremophila virens

Eremophila virens, × ·5

Eremophila virens C. Gardner
(green)
WA
3-5 m x 1.5-3.5 m   June-Sept

Medium to tall **shrub**; young growth sticky; **branches** erect, sparse, slender, sticky; **leaves** to 8 cm x 3.5 cm, alternate, oblong-lanceolate to ovate-lanceolate, folded lengthwise, petiolate, glabrous, shiny, sticky, margins entire, apex pointed; **flowers** tubular, to 2 cm long, green, exterior very hairy, lobes pointed, upper ones short, lower long and reflexed, 1-2 slender pedicels to 2 cm long, in upper axils, profuse; **stamens** exserted; **calyx lobes** to 1 cm long, margins hairy, shiny; **fruit** ovoid to globular, 0.3-0.5 cm x about 0.3 cm, glabrous.

In nature this sticky, shrubby species occurs in isolated locations on granitic soils, north of Merredin in the Coolgardie District. It is gazetted as a rare species. It is relatively rare in cultivation, but could prove useful as a screening plant (if pruned while young) in areas that have low rainfall and high temperatures. Will need fairly well drained soils. Young plants may benefit from tip pruning to promote bushy growth. Propagate from seed or cuttings which may be extremely slow to form roots. Successfully grafted on to *Myoporum insulare* rootstock. Endangered species such as this one need to be propagated and grown in many locations to ensure that they do not become extinct. Plants also need to be re-established in their native habitat.

Eremophila virgata W. Fitzg.
see under *E. interstans* (S. Moore) Diels

Eremophila viscida Endl.
(sticky)
WA   Varnish Bush
2-6 m x 1.5-5 m   Sept-Dec

Medium to tall **shrub**; **branches** glabrous, sticky, shiny brown; **leaves** 5-10 cm x up to about 1 cm, alternate, lanceolate to elliptic, folded lengthwise, shiny, sticky, glabrous, margins entire, apex pointed; **flowers** tubular, about 2 cm long, white to pale yellow or reddish, with purple-spotted interior, exterior hairy, on slender pedicels, in upper axils; **stamens** exserted; **calyx lobes** to about 0.7 cm long, grey-blue or reddish, blunt, veined; **fruit** ovoid, 0.5-0.7 cm x about 0.4 cm, compressed, hairy on the upper part.

A fast growing, showy eremophila that has a wide distribution covering the Austin, Avon, Coolgardie, Eyre and Roe Districts. It is usually found in open woodland in association with *Eucalyptus loxophleba* (York Gum) on low hills and plains. At present it is not well known in cultivation, but has potential for greater use in semi-arid areas as a low windbreak and screening plant. Soils need to be relatively well drained and it prefers a warm to hot, sunny situation. Propagate from seed or cuttings. Can be grafted successfully using *Myoporum insulare* as the rootstock.

Eremophila weldii F. Muell.
(after Sir Frederick A. Weld, 19th-century New Zealand parliamentarian and Governor of WA and Tas)
SA, WA
0.2-1.3 m x 0.4-1.5 m   Aug-Dec

Dwarf to small **shrub**; **branches** glabrous, slightly warty; **leaves** 0.4-1.2 cm x about 0.2 cm, alternate, oblanceolate to obovate, slightly concave, warty below, glabrous, margins entire, apex pointed; **flowers** tubular, to 1.3 cm long, whitish, violet or purplish, with spotted interior, exterior glabrous, constricted at base, lobes pointed, pedicels usually solitary, in axils along branchlets; **stamens** not exserted; **calyx lobes** to 0.5 cm, pointed, margins hairy; **fruit** ovoid, to 0.5 cm long, beaked, hairy.

A widespread species occurring on Kangaroo Island, SA, and extending westwards to the Coolgardie District, WA. It is found mainly in coastal or slightly

Eremophila viscida, × ·45

Eremophila willsii   W.R. Elliot

inland situations. Has proved adaptable in cultivation, growing in semi-arid to temperate climates. Does best in soils that have good drainage, and is useful as a container plant because of its long flowering period. Will tolerate semi-shade but prefers plenty of sunshine. Hardy to most frosts, and withstands extended dry periods. Responds well to pruning. A dwarf form from Balladonia, WA, that reaches 0.2 m x about 0.5 m is becoming popular in cultivation. Propagate from seed or cuttings which usually strike readily. The dwarf form is more difficult.

Eremophila willsii F. Muell.
(after William J. Wills, 2 I.C. of the Burke and Wills Expedition, 1860-61)
Qld, SA, WA, NT
0.5-2 m x 0.5-2 m   July-Oct
   Dwarf to small **shrub**; young growth sticky; **branches** hairy, erect; **leaves** 1-4.5 cm x 0.4-2 cm, alternate, obovate to elliptic, sessile, with rusty hairs, margins serrate, apex pointed; **flowers** tubular, to 2.5 cm long, reddish purple, exterior hairy, purple patches in throat, prominent white hairs on upper lobe, lobes pointed, 1-3 per axil, near ends of branchlets; **stamens** not exserted; **calyx lobes** overlapping, to 1.8 cm long, hairy; **fruit** ovoid, to about 1 cm long, beaked.
   An ornamental eremophila that occurs in arid regions, where it usually grows on red sands and sandy loams. It is mainly cultivated by enthusiasts, but has potential for wider use. Needs very well drained soils and maximum sunshine. Seems difficult to establish unless on red sand-hills. Tolerant of most frosts. May need pruning to maintain bushy growth. Propagate from seed or cuttings.
   The var. *integrifolia* Ewart from NT has entire leaves but at this stage its botanical status is uncertain.
   E. leonhardiana E. Pritzel is included in this species.

Eremophila woollsiana F. Muell. and
Eremophila woollsiana var. **dentata** Ewart & J. White
         = *E. lehmanniana* (Sonder ex Lehm.) Chinn.

Eremophila youngii F. Muell.
(after Jess Young, member of 1875 Giles expedition, and botanical collector)
WA
1-3 m x 0.6-2 m   June-Jan
   Small to medium, upright **shrub**; young growth velvety; **branches** erect, greyish, warty; **leaves** 1.5-6 cm x

Eremophila weldii, × ·65

Eremophila youngii, × ·55

463

up to 0.5 cm, opposite, linear-lanceolate, slightly warty, greyish green, margin entire, apex pointed and hooked; **flowers** tubular, to 2.5 cm long, pink to red, exterior hairy, lobes pointed, solitary, axillary, on pedicels to 0.5 cm long, along branchlets; **stamens** exserted; **calyx lobes** narrow, pointed, to 0.5 cm long; **fruit** ovoid, to about 1 cm long, beaked, ribbed, hairy.

This ornamental species occurs in the Gibson and Great Victoria Deserts. It is usually found on alluvial soils. Although from arid regions, it has proved hardy under a wide range of conditions, provided it has excellent drainage and plenty of sunshine. Hardy to frost. Responds well to pruning. Suitable for narrow spaces and for cultivation as a container plant. Propagate from seed or cuttings.

ERIA Lindley
(from the Greek *erion*, wool; in reference to parts of the inflorescence and flowers which in some species bear woolly hairs)
Orchidaceae

Epiphytic **orchids**; **rhizome** short or long; **pseudobulbs** erect, thin and wiry or fleshy, crowded or widely spaced; **leaves** terminal on fleshy pseudobulbs or scattered along thin pseudobulbs, simple, entire; **inflorescence** a crowded raceme; **flowers** small to large, dull coloured to colourful, frequently not opening widely, often hairy on the outside; **labellum** lobed or entire; **fruit** a capsule.

A large genus of about 375 species of orchids mostly found in tropical Asia and extending through the Pacific region to New Guinea and north-eastern Australia. They mostly grow on trees or rocks. The genus is extremely variable in features of growth habit, and size and colour of the flowers. They are not highly popular with orchid growers. In Australia there are 6 species, all confined to the rainforests of north-eastern Qld, where they grow on trees or rocks, usually in well lit situations. Only a couple of the Australian species have horticultural merit with the others having small dull coloured flowers which may not open if the weather is adverse. All species can be grown readily on slabs or in pots of coarse material. They have a very fine, much-branched root system which can rot readily, so they must have a very coarse mixture and should not be overpotted. All Australian species are cold sensitive and require a heated glasshouse. Propagate by division of the clumps into 4-6 bulb sections or from seed sown on sterilized media.

Eria dischorensis Schltr.
(from the area of Dischor in New Guinea)
Qld

Oct-Dec

Epiphytic **orchid** forming small clumps; **pseudobulbs** 2-4 cm x 1.3-1.7 cm, ovoid, fleshy, not crowded, when young covered with brown, papery bracts; **leaf** 7-13.5 cm x 2-3.5 cm, solitary, oblong, dark green, unequally notched at the apex; **racemes** 6-8 cm long, bearing 4-8 flowers on pedicels about 1 cm long; **flowers** about 0.8 cm across, cream with a few red spots; **labellum** 3-lobed.

Eria eriaeoides   E.R. Rotherham

This orchid, which is also found in New Guinea, is uncommon in Australia and is rarely collected. It is found in north-eastern Qld, growing on trees in rainforests. Plants grow readily in a pot of coarse mixture or mounted on a slab of weathered hardwood, cork or tree-fern. In cold districts they need the protection of a heated glasshouse, while in the tropics a bushhouse is suitable. Humid conditions with some air movement are excellent for the growth of this species. Propagate by division of the clumps into 4-5 bulb sections or from seed.

Eria eriaeoides (Bailey) Rolfe
(like the genus *Eria*)
Qld

Aug-Oct; also sporadic

Epiphytic **orchid** forming medium to large clumps; **pseudobulbs** 4-10 cm x 1-1.5 cm, cylindrical, fleshy, green, widely spaced, when young covered by white, papery bracts; **leaves** 8-20 cm x 2-3 cm, 2 at the end of each pseudobulb, lanceolate to narrow-ovate, bright green, thin-textured, unequally notched at the apex; **racemes** 3-5 cm long, bearing 3-12 flowers on pedicels about 0.5 cm long; **flowers** about 0.4 cm across, dull white to mauve, not opening widely; **labellum** 3-lobed.

A common orchid found on trees, frequently near watercourses in the rainforests of north-eastern Qld. It

Eria inornata   D.L. Jones

is not a showy orchid, and frequently the flowers may pollinate without opening. It grows readily in a pot of coarse material or on a slab of weathered hardwood, cork or tree-fern. In temperate areas it needs the protection of a heated glasshouse. Plants prefer fairly humid conditions throughout the year, but should be kept fairly dry over winter. Propagate by division of the clumps into 4-5 bulb sections or from seed.

Eria fitzalanii F. Muell.
(after E. Fitzalan, original collector)
Qld

Aug-Oct

Epiphytic **orchid** forming medium to large clumps; **pseudobulbs** 10-20 cm x 2-4 cm, ovoid, fleshy, crowded, usually brownish; **leaves** 10-30 cm x 3-5 cm, 3-4 from near the apex of the pseudobulb, obovate, dark green; **racemes** 10-30 cm long, bearing numerous flowers on pedicels about 1 cm long; **flowers** 1-1.5 cm across, cream; **labellum** entire or 3-lobed.

A common orchid of north-eastern Qld, found on trees or rocks in rainforests, usually in fairly open, well lit situations. Plants are most common in lowland areas but also extend to about 800 m alt. The species grows readily in cultivation but is extremely cold sensitive, and in temperate regions must be grown in a heated glasshouse with a minimum temperature of 10°C. In

these circumstances plants should be kept dry over winter. In the tropics this orchid can be grown in a bushhouse or even on a suitable garden tree. It can be grown in a pot of a suitable coarse material or attached to a slab of weathered hardwood or tree-fern. Propagate by division of the clumps into 4-5 bulb sections or from seed.

Eria inornata Hunt
(without adornment)
Qld

Aug-Oct

Epiphytic **orchid** forming medium to large clumps; **pseudobulbs** 10-25 cm x 2.5-5 cm, ovoid, fleshy, crowded, greenish; **leaves** 20-35 cm x 3-5 cm, 3-4 from near the apex of the pseudobulb, linear, dark green; **racemes** 15-30 cm long, bearing numerous flowers on pedicels about 0.8 cm long; **flowers** about 1 cm across, white, cream or yellowish, not opening widely; **labellum** concave, deeply 3-lobed.

This orchid is widely distributed in the rainforests of north-eastern Qld, usually growing on rocks or trees in humid, fairly well lit situations near water. Plants can be quite showy when in flower. They grow easily in cultivation, with requirements similar to those of *E. fitzalanii*. Propagate by division of the clumps into 4-6 bulb sections or from seed.

Eria irukandjiana St Cloud
(after a tribe of Aborigines that inhabited the rainforest area where the orchid grows)
Qld

Nov-Dec

Epiphytic **orchid** forming small clumps; **pseudobulbs** 5-12 cm x 0.5-1 cm, conical, fleshy, crowded, bright green; **leaves** 5-12 cm x 0.5-1 cm, 2-3 at the apex of the pseudobulb, linear-lanceolate, erect, bright green, the apex unequally notched; **racemes** 0.5-1.5 cm long, erect, fleshy, bearing 7-12 crowded flowers on pedicels 0.2-0.3 cm long; **flowers** about 0.3 cm across, pinkish to dull white, opening widely; **labellum** not lobed.

An uncommon orchid of north-eastern Qld, found on trees in rainforest and occasionally on casuarinas in open forest. It is of compact growth habit but the flowers are small and dull coloured, and the species is of interest mainly to the orchid enthusiast. It can be easily grown in a small pot of coarse material or mounted on a slab of weathered hardwood, cork or tree-fern fibre. Plants are very cold sensitive and in temperate regions must be over-wintered in a heated glasshouse with a minimum temperature of 10°C. They also like fairly humid conditions. Propagate by division of the clumps into 4-6 bulb sections or from seed.

Eria queenslandica Hunt
(from Queensland)
Qld

Aug-Oct

Epiphytic **orchid** forming medium to large clumps; **pseudobulbs** 3-6 cm x 1-1.2 cm, ovoid to cylindrical, fleshy, green, crowded; **leaves** 6-12 cm x 1-2 cm, 2 at the apex of each pseudobulb, lanceolate, dark green, fairly thick-textured; **racemes** 1-4 cm long, hairy, bearing 3-12 flowers on pedicels 0.2-0.3 cm long; **flowers** about 0.3 cm across, white, cream or pink, not opening widely; **labellum** 3-lobed.

A common orchid found on rainforest trees in north-eastern Qld. The flowers are small and dull coloured, and the species has little appeal except to the orchid enthusiast. Plants grow readily in cultivation, requiring conditions similar to those of *E. fitzalanii*. Propagate by division of the clumps into 4-6 bulb sections or from seed.

ERIACHNE R. Br.
(from the Greek *erion*, wool; *achne*, a glume; in reference to the densely hairy florets within the glumes)
Poaceae

Annual or perennial **grasses** forming clumps or tussocks; **leaves** flat or inrolled, sometimes rigid and spine-like; **culms** simple or branched; **inflorescence** a contracted panicle; **spikelets** solitary, flattened; **lemmas** awned or awnless.

A genus of about 35 species, all found in Australia, and 3 of which extend to Asia. Most are endemic. They are grasses characteristic of inland regions, although a few species are found on the coast. They grow mostly on poor soils and are of limited importance as pasture grasses. A few species have horticultural appeal, but they are rarely grown. Propagate from seed which may need storage for up to 12 months before sowing or by division of the clumps.

Eriachne benthamii Hartley
(after George Bentham, 19th-century English botanist)
Qld, SA, WA, NT   Swamp Wanderrie Grass
40-90 cm tall   Aug-Jan

Perennial **grass** forming dense, coarse glaucous tussocks; **leaves** 8-15 cm x 0.2-0.4 cm, flat or inrolled, leathery, glaucous, with thickened white margins; **culms** 40-90 cm tall, smooth, covered with white powdery blooms; **panicle** 8-13 cm x 2-3.5 cm, fairly dense; **spikelets** 0.7-0.9 cm long; **lemmas** with a recurved awn.

A common grass of low-lying areas subject to partial inundation. It frequently grows in heavy soils. In cultivation it develops into a dense tussock with an overall bluish green appearance. Best grown in moist soil or where it can be watered regularly. Suited to tropical, subtropical and warm inland regions. Propagate by division of the rhizome or from seed.

Eriachne helmsii (Domin) Hartley
(after Richard Helms, 19-20th-century naturalist)
Qld, NSW, SA, WA, NT   Woollybutt Wanderrie
Grass
45-90 cm tall   Aug-Dec

Perennial **grass** forming coarse, erect, straggly tussocks; **leaves** 2.5-10 cm x 0.2-0.4 cm, flat or inrolled, stiff and spreading; **culms** 45-90 cm tall, wiry, bluish green, the nodes thickened; **panicle** 5-11 cm x 1-2 cm, loose and open; **spikelets** 0.5 cm long, straw coloured; **lemmas** awnless, with silky hairs.

A tough grass of inland regions, found on stony, shallow soils. Plants have limited horticultural appeal but could be grown in a rockery or amongst shrubs. Propagate by division or from seed.

Eriachne mucronata R. Br.
(ending in a sharp point)
Qld, NSW, SA, WA, NT   Mountain Wanderrie
Grass
30-60 cm tall   Aug-Dec

Perennial **grass** forming loose or dense, compact tussocks; **leaves** 2-4 cm x 0.1-0.2 cm, linear, stiffly spreading, flat or folded, sharply pointed; **culms** 30-60 cm tall, thin, wiry, usually branched; **panicle** 2.5-4 cm, narrow, purplish; **spikelets** 0.5-0.6 cm long.

A tough grass of semi-arid and arid regions. It consists of 2 forms, a slender type forming loose tussocks and a dense type forming compact tussocks reminiscent of spinifex. The latter type has some appeal as a rockery plant, although the tussocks are spiny. It requires a sunny aspect, in well drained soil. Clumps can be rejuvenated by burning. Best suited to inland tropical and subtropical regions. Propagate from seed or by division.

Eriachne muelleri Domin = *E. pallescens* R. Br.

Eriachne nervosa Ewart & Cookson
(prominent nerves or veins)
Qld, NT   Plains Wanderrie Grass
60-90 cm tall   Aug-Dec

Perennial **grass** forming dense, coarse glaucous tussocks; **leaves** 10-17 cm x 0.2-0.4 cm, flat, leathery, glaucous, the sheathing base shiny; **culms** 60-90 cm tall, glabrous, purplish, shiny; **panicle** 8-12 cm x 2-3 cm, fairly dense, purplish; **spikelets** up to 1.2 cm long, hairy, with long, terminal awn-like points.

This grass is locally common on heavy clay soils subject to partial inundation, and around the margins of drainage channels and permanent water-holes. New growth is a fresh bluish green, and the species could have potential for cultivation. Would require conditions similar to those of *E. benthamii*. Propagate by division of the rhizome or from seed.

Eriachne pallescens R. Br.
(becoming paler)
Qld, NSW, NT   Wanderrie Grass
30-80 cm tall   Oct-Dec

Perennial **grass** forming erect tussocks; **leaves** 5-15 cm x 0.1-0.3 cm, linear, inrolled, basal or scattered on the culms; **culms** 30-80 cm tall, slender, glabrous, branched in the lower parts; **panicle** 6-15 cm long, very sparse; **spikelets** about 0.6 cm long, purplish, borne on long, slender stalks; **lemmas** with awns up to 0.6 cm long.

A common grass of coastal areas, usually growing in sandy soils. It has limited ornamental appeal but can be grown in a sunny garden position. Requires well drained soil. May naturalize but is not a problem. Propagate from seed or by division.

Previously known as *E. muelleri* Domin.

Erigeron stellatus   T.L. Blake

Ericaceae Juss.

A large family of dicotyledons consisting of about 100 genera and 3000 species. It is cosmopolitan in its distribution but is rather poorly developed in Australia, where it is perhaps replaced by members of the Epacridaceae family. In some countries, species of Ericaceae form important and extensive communities, frequently on wet or peaty soils. They are small woody shrubs, rarely trees, and some have an epiphytic or lithophytic growth habit. The flowers are tubular, often with flared petals, and are frequently colourful and showy. Some groups are very popular in cultivation, and exotic species and hybrids are widely grown in parks and public and private gardens throughout Australia, eg. rhododendrons. A few species are important for commercial fruit production and are exemplified by cranberries and blueberries in the genus *Vaccinium*. Australian representation in the family numbers about 5 genera and 6 species. Some of these are very popular plants, grown for their ornamental flowers. The placement of the genus *Wittsteinia* is subject to controversy and may be placed here or in Epacridaceae. Australian genera: *Agapetes, Gaultheria, Pernettya, Rhododendron* and *?Wittsteinia.*

ERICHSENIA Hemsley

(after Frederick O. Erichsen, an engineer with the Goldfields Water Scheme)
Fabaceae

A monotypic endemic genus restricted to southern WA.

Erichsenia uncinata Hemsley

(hooked)
WA
0.5-1 m x 0.6-1 m   Sept-Dec

Dwarf **shrub**; **branches** terete, glabrous, slender, green; **leaves** to 1.2 cm long, simple, terete, apex prominently hooked; **flowers** pea-shaped, about 1.2 cm across, yellow with red stripes, in axillary and terminal racemes; **calyx** densely hairy; **pods** about 0.4 cm x 0.2 cm.

This showy pea-flowered species is uncommon in cultivation. It occurs in the Avon and Coolgardie Districts, where it grows on yellow sand. Will need very well drained, light to medium soils in a situation that receives partial to full sun. In nature it withstands fire by reshooting from its rootstock, so it should tolerate heavy pruning. Has potential as a container plant. Propagate from seed which will need pre-sowing treatment (see Volume 1, page 206). Cuttings of firm young growth are worthy of trial.

ERIGERON L.

(from the Greek *erigeron*, the name used by Theophrastus to describe *Senecio vulgaris* because of its hairy appearance)
Asteraceae

Annual or perennial **herbs**; **leaves** alternate, entire or toothed; **inflorescence** daisy-like, terminating simple scape or short branches; **bracts** in 2-4 rows; **receptacle** naked; **ray florets** female; **disc florets** bisexual; **pappus** of a few bristles; **anthers** obtuse at base; **style branches** dilated above the stigmatic lines; **achenes** flat, not beaked, hairy.

A cosmopolitan genus of over 200 species. Australia has a representation of about 7 species, with 4 endemic. In general the Australian species do not have much appeal for cultivation. *E. stellatus*, a Tas endemic, is grown to a limited extent.

Propagate from seed or cuttings or by division of creeping rhizomes.

Erigeron stellatus (Hook.f.) W. M. Curtis

(starry)
Tas
prostrate x 0.5-1 m   Nov-Feb

Spreading perennial **herb**; **rhizomes** slender, branched; **branches** develop from ends of rhizomes and terminate in dense rosettes of leaves; **leaves** 1-2 cm x 0.3-0.5 cm, linear-oblong to narrow-spathulate, slightly concave, thick, margins entire or with some hairs, apex rounded; **flower-heads** daisy-like, about 1.5 cm across, solitary, on simple, short scapes; **ray florets** white or cream; **disc florets** yellow.

This Tas endemic is common on the subalpine herb-fields. An uncommon species in cultivation, it is grown mainly by botanical institutions in UK. For cultivation it will need moist but well drained, acid to neutral soils containing plenty of organic matter. It will also require a sunny situation. Has proved successful as a container plant. Flowers well during summer when the pots are placed on capillary matting to maintain a constant moisture supply, or when pots are buried in moist, coarse sand. Propagate from seed or cuttings or by division of the rhizomes.

Eriocaulaceae Desv.

A large family of monocotyledons comprising about 13 genera and 1200 species, mainly developed in tropical and subtropical regions, and especially significant in South America. They are perennial herbs with grass-like leaves and small unisexual flowers in dense heads. Male and female flowers occur in each head but may be variously arranged. Many species are aquatic, or grow in areas subject to partial inundation. The family is represented in Australia by about 19 species in the solitary genus *Eriocaulon*.

ERIOCAULON L.

(from the Greek *erion*, wool; *caulon*, a stalk; in reference to the woolly stalks of some species)

Eriocaulaceae

Annual or perennial, tufted **herbs**; **leaves** in a basal tuft, linear; **flower-heads** globular, terminating an erect peduncle; **flowers** small, unisexual, crowded in heads, each flower subtended by a bract, the outer flowers mostly female, inner mostly male; **fruit** a capsule.

A large genus of about 400 species mainly found in tropical regions. There are about 20 species named from Australia and a few others undescribed. They are small annual or perennial herbs found in temporary or permanent shallow water. They are interesting plants but are rarely grown. They are suitable for bog gardens or moist depressions etc. Propagate by transplants or sowing the very fine seed on the surface of moist soil under glass.

Eriocaulon australe R. Br.

(southern)

Qld, NSW, NT

30-45 cm tall   Nov-Feb

Tufted **herb**; **leaves** including peduncles 30-45 cm long, grass-like, 0.2-0.6 cm across, flat or concave, bright green, hairy below the middle; **scapes** hairy below the middle; **flower-heads** 0.6-0.8 cm across, white, the bracts closely overlapping and mealy; **perianth segments** joined.

A widely distributed species found on inundated ground, often in coastal districts. Tufts are large and attractive. Plants can be grown in a pot with the base submerged in water, or in a bog garden. Plants are annuals if the area where they grow dries out. Propagate by transplants or perhaps from seed sown on moist soil under glass.

Eriocaulon carsonii F. Muell.

(derivation unknown)

NSW, SA   Salt Pipewort

3-8 cm tall   Nov-Feb

Small tufted **herb**; **leaves** 2.5-7 cm x 0.4-1 cm, slender, grass-like, broad at the base; **peduncles** 3-8 cm long, slender; **flower-heads** about 0.2 cm across, globular, yellowish; **bracts** elliptic.

A small, rare species that forms dense colonies in shallow, permanent water of inland regions. Could have potential as an aquarium plant. Also reported to grow in saline soils. Propagate by transplants or from seed.

Eriocaulon cinereum R. Br.

(ash coloured)

Qld, WA, NT

2.5-12 cm tall   Nov-Jan

Small tufted **herb**; **leaves** 1-5 cm long, linear or

thread-like; **peduncles** 2.5-12 cm tall, thread-like; **flower-heads** 0.2-0.3 cm across, hemispherical, white to ash coloured; **bracts** lanceolate.

Found on damp and inundated ground of tropical regions. A dainty species suitable for bog gardens or a pot with the base standing in water. Probably very cold sensitive. Propagate from seed or by transplants.

Eriocaulon scariosum Smith

(dry and membranous)

Qld, NSW, Vic   Common Pipewort

7-13 cm tall   Oct-Dec

Small tufted **herb**; **leaves** 2-3 cm long, very thick at the base, tapering to a fine point; **peduncles** 7-10 cm tall, slender; **flower-heads** about 0.4 cm across, globular, light brown, the perianth segments copiously ciliate; **bracts** orbicular, dry and membranous.

A widespread species found on muddy ground and low-lying areas subject to periodic inundation. It is of interest to enthusiasts, and can be grown in bog gardens or in a pot with the base submerged in water. Propagate from seed or by transplants. On suitable sites the species may naturalize itself.

Eriocaulon setaceum L.

(bristly)

Qld, WA, NT

aquatic   Nov-Dec

Submerged **aquatic** forming slender strands; **stem** up to 90 cm long, simple and unbranched, brittle; **leaves** 2.5-12 cm long, filiform, numerous on the stems, each stem bearing at its apex a cluster of flower peduncles, each 4-35 cm long; **flower-heads** 0.2-0.5 cm across, greyish, dense, globular, each consisting of den-

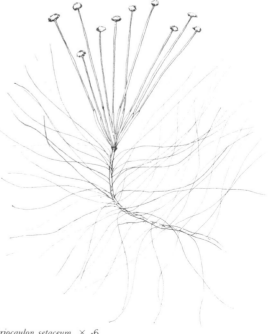

Eriocaulon setaceum, × ·6

Eriochilus cucullatus   J. Fanning

sely crowded, small male and female flowers; **capsule** 3-celled, splitting when ripe.

An interesting aquatic confined to warm waters of tropical Australia. The leafy stems are relatively inconspicuous but the emergent flower-heads are much more noticeable. The species can be grown in tropical aquaria with heated water, and may need fairly strong light to flower. Propagate by inserting the bases of fairly long pieces of stem bearing healthy leaves in a pot of sand in the aquarium (see Aquarium Plants and Aquatics, Volume 2, pages 216-18).

Eriocaulon smithii R. Br.
(after Sir J. E. Smith)
Qld
7-13 cm tall   Oct-Jan
Small tufted **herb**; **leaves** 2-3 cm long, linear, sometimes broader at the base; **peduncles** 7-13 cm tall, slender; **flower-heads** 0.3-0.5 cm across, white, globose; **bracts** obovate, bearing a few hairs.

A common species found in shallow water in central and southern Qld. Makes an attractive bog plant. Can also be grown in a pot with the base standing in water. Suitable for subtropical regions. Propagate by transplants or from seed.

Eriocaulon spectabile F. Muell.
(very showy)
Qld, NT
6-14 cm tall   Dec-March
Small tufted **herb**; **leaves** 4-8 cm long, broad at the base, tapering to a linear point; **peduncles** 6-14 cm long, finely striate; **flower-heads** 0.4-0.6 cm across, straw coloured to silvery white, globular; **bracts** ovate, white.

Eriochilus cucullatus

An annual or perennial herb found in damp situations subject to periodic inundation. Attractive when in flower. A useful plant for bog gardens or the margins of dams and ponds. Can also be grown in a pot with the base submerged in water. Plants are very cold sensitive. Propagate by transplants or from seed.

Eriocaulon tortuosum F. Muell.
(twisted or entangled)
NT
10-20 cm   Dec-March
Small tufted **herb**; **leaves** 1-2 cm long, linear-lanceolate; **peduncles** 10-20 cm tall, irregularly twisted; **flower-heads** 0.2-0.4 cm across, white, globular; **bracts** wedge-shaped or spathulate, slightly hairy.

A little-known species found in the Top End of NT and originally collected from the Victoria River. Its twisted flower-stems are unusual, and the species could have potential for shallow ponds or bog gardens in the tropics. Can also be grown in a pot kept wet by submerging the base in water. Propagate by transplants or from seed.

ERIOCHILUS R. Br.
(from the Greek *erion*, wool; *cheilos*, a lip; in reference to the hairy labella of these orchids)
Orchidaceae
Small terrestrial **orchids**; **tuber** rounded, subterranean; **leaf** solitary, basal, glabrous or hairy, usually ovate; **flower-stem** slender to filiform; **flowers** small, the lateral sepals conspicuous and spreading, the petals much reduced and dull coloured; **labellum** strongly recurved, hairy; **fruit** a capsule.

A small genus of 3 species of terrestrial orchids all endemic to Australia. They are slender orchids with dainty, interesting flowers which are easily passed over. Flowers appear in the autumn-winter, followed by the leaves, and the plants die back to the subterranean tuber over summer. They are mainly of interest to enthusiasts specializing in terrestrial orchids. Results in cultivation are mixed, with the plants tending to die out following drastic disturbance or any setback. If left undisturbed they can be grown for many years. Best results have been obtained using mixtures based on sandy loams derived from coastal heathlands in combination with coarse sand and broken down leaf mould or eucalypt shavings. Watering should be regular while the plants have above-ground growth, and the soil mixture should be allowed to dry out while they are dormant. At least one species can be propagated successfully by sprinkling seed around the bases of established plants.

Eriochilus autumnalis R. Br. =
                    E. cucullatus (Labill.) H. G. Reichb.

Eriochilus cucullatus (Labill.) H. G. Reichb.
(hooded)
Qld, NSW, Vic, Tas, SA   Parsons Bands
5-25 cm tall   Dec-May
Small, slender terrestrial **orchid**; **leaf** 0.5-3.5 cm x 0.3-0.8 cm, ovate, ground hugging, dull green; **flower-stem** 5-25 cm tall, filiform, bearing 1-5 flowers; **flowers**

469

about 1.5 cm across, most commonly white but occasionally pale to bright pink, the lateral sepals conspicuous and spreading or slightly deflexed; **labellum** strongly recurved, hairy.

A widely distributed orchid, common in a variety of habitats from coastal heathlands through open forest to subalpine grasslands. In Qld it is only found in the south-east. The dainty flowers appear in the autumn, and the leaf emerges and becomes conspicuous well after flowering has finished. Plants in subalpine communities frequently have bright rosy pink flowers. This species can be successfully grown in a pot of sandy mixture containing eucalypt sawdust or shavings. The mixture should be kept dry over summer, and watering should be commenced with the emergence of the flower-spikes in the autumn. Seedlings can sometimes be induced to grow by sprinkling seed around the bases of established plants.

Eriochilus dilatatus Lindley
(enlarged; widened)
WA   White Bunny Orchid
10-40 cm tall   March-June
Small, slender terrestrial **orchid**; **leaf** 2-6 cm x 0.5-1 cm, ovate to ovate-lanceolate, dark green, often carried well above the ground; **flower-stem** 10-40 cm tall, very slender, bearing 1-13 flowers; **flowers** about 1.5 cm across, white, the lateral sepals conspicuous and spreading; **labellum** strongly recurved, hairy.

Although restricted to the South-West Botanical Province, this orchid is widely distributed and often locally common. It is somewhat variable, especially in such features as vigour, and number of flowers. It can be successfully grown, and has requirements similar to those of *E. cucullatus*, although plants may die quickly following a setback.

Eriochilus scaber Lindley
(rough to the touch)
WA   Pink Bunny Orchid
5-15 cm tall   July-Sept
Small, slender terrestrial **orchid**; **leaf** 0.5-3 cm x 0.6-1.2 cm, ovate to cordate, dark green, hairy; **flower-stem** 5-15 cm tall, filiform, hairy, bearing 1-3 flowers; **flowers** about 1.5 cm across, bright pink, the lateral sepals conspicuous and spreading; **labellum** strongly recurved, densely hairy.

This species is endemic to the South-West Botanical Province, growing in open forest and heathland, and flowering well only after bushfires. In cultivation it has proved to be rather difficult to maintain, and seems to prefer a peaty soil mixture and abundance of moisture while above-ground.

Eriochiton sclerolaenoides (R. H. Anders.) A. J. Scott
   = *Maireana sclerolaenoides* (F. Muell.) P. G. Wilson

ERIOCHLAMYS Sonder & F. Muell. ex Sonder
(from the Greek *erion*, wool; *chlamys*, a cloak; in reference to the woolly hairs around the flowers)
Asteraceae
A monotypic endemic genus distributed throughout

the mainland states in semi-arid to arid regions. Until recently it was thought that this genus contained 2 distinct species.

Eriochlamys behrii Sonder & F. Muell. ex Sonder
(after Dr. Hermann Behr, early botanical collector, SA)
Qld, NSW, Vic, SA, WA, NT   Woolly Mantle
0.05-0.15 m x 0.1-3 m   Sept-Nov; also sporadic
Dwarf, annual **herb**; **branches** ascending to erect, densely woolly-hairy; **leaves** 0.3-1 cm long, alternate, linear to oblanceolate, sessile, margins revolute, woolly or becoming glabrous; **flower-heads** solitary or compound, globular, to about 1 cm diameter, whitish with green woolly bracts, terminal; **fruit** an achene.

This small annual is a fairly common component of inland grassland, growing in a wide range of soils. If growing vigorously it can have an extended flowering period, beginning in April-May and continuing until March the next year. It is poorly known in cultivation, but should have application in inland regions and could prove to be a decorative pot plant if grown in clumps. Propagate from seed or cuttings.

ERIOSTEMON Smith
(from the Greek *erion*, wool; *stemon*, stamen)
Rutaceae   Wax Flowers
Shrubs or small **trees**; **branchlets** angular or terete, often warty, glabrous or slightly hairy, sometimes glaucous; **leaves** alternate, simple, flat to terete, often warty, glabrous to hairy, often strongly aromatic; **inflorescence** axillary or terminal, in umbels, cymes, racemes, or a solitary flower; **flowers** star-like, petals 5 (rarely 4), white, pink, red, mauve or blue; **stamens** 10 (rarely 8); **style** solitary; **fruit** composed of up to 5 cocci.

An endemic genus of 33 species occurring in Qld, NSW, Vic, Tas, SA and WA. Habitats vary from subalpine regions in south-eastern Australia to semi-arid regions of WA, with many species found between this range. Eriostemons nearly always grow in very well drained, slightly acid to neutral soils. Plants rarely occur on alkaline soils. Often the soils contain a high proportion of gravel, and some are skeletal in nature.

Eriostemons are usually undershrubs, as part of open forest or woodland communities, but they may also occur in exposed, rocky situations. Most species grow in areas where there is a great variation in temperatures. Minimum temperatures can be below freezing during winter, while maximums are often above 35°C in summer. Most species are hardy to frost.

Some species in eastern Australia receive rainfall in excess of 1500 mm, whereas those in WA may receive less than 300 mm. In most areas the main rainfall occurs during the cool winter months.
Cultivation
Some *Eriostemon* species are popular in cultivation, but not many are grown. *E. myoporoides* and *E. verrucosus*, along with their various selections and cultivars, are the most commonly cultivated. They seem to be the most adaptable species in the genus. *E. myoporoides* was

470

for many years (dating back to the beginning of this century) one of the most frequently grown Australian plants, because of its hardiness to a wide range of conditions. Along with *Grevillea rosmarinifolia* it was one of the few species readily available from nurseries prior to 1930.

Other eastern species such as *E. australasius* and *E. buxifolius* are now finding their way into gardens, after not being readily available due to propagation difficulties. Of the western species, *E. nodiflorus* and *E. spicatus* are the best-known.

Some eriostemons were introduced into England during the early 19th century, and were much admired by gardeners. Amongst those grown (with the introduction date in brackets) was *E. buxifolius* (1822), regarded as amongst the best due to its strong growth and wonderful display of flowers. Others were *E. australasius* (1824), *E. myoporoides* (1824) and *E. scaber* (1840). Some hybrids were also grown. *E.* 'intermedius', with parents *E. buxifolia* and *E. myoporoides*, was renowned for its compact growth. Another was *E.* 'pulchellus', a hybrid between *E. hispidulus* and *E. myoporoides*.

In cultivation most species adapt well, provided the soils have adequate drainage. Generally they grow well in semi-shade to full sun. The eastern species are more tolerant of shade than those from the west. Most species withstand extended dry periods once they are established, and they are not usually prone to damage from frost.

Eriostemons are commonly called Wax Flowers, because of their smooth, thickish petals. Their main ornamental feature is the massed display of starry flowers, produced over a long period. Flower colour is usually white to pale pink, but before the flowers open the buds of some species are highly decorative, with pink to red tonings. Some flowers are mauve to bluish, eg. *E. nodiflorus* and *E. spicatus*. A couple of species have red flowers, eg. *E. coccineus*, but these are little-known in cultivation.

Quite a few species have sweetly aromatic foliage. Some people find this fragrance pleasant, and grow plants near paths to enjoy the aroma as branches are brushed past, or as leaves are crushed. Others find the aroma displeasing, and place the plants in a position where contact with the foliage will be avoided.

Eriostemons are suitable for many uses. *E. myoporoides* is useful as a low screening or windbreak shrub. The white-flowered species are excellent for lighting up a dull position with their bright, starry flowers. Quite a few species grow well on embankments, eg. *E. verrucosus*. Some grow well in containers, eg. *E. australasius* and *E. buxifolius*, and many species have potential for wider use in this way, especially those that occur naturally in semi-arid regions, eg. *E. thryptomenoides* and *E. tomentellus*.

Most species respond well to slow-release fertilizers. Some plants can become chlorotic and this can be remedied by application of a nitrogenous fertilizer if they are deficient in nitrogen, or iron chelates if the soil is neutral to alkaline and causing iron deficiency.

Some plants can become leggy, eg. *E. australasius* and *E. nodiflorus*, and most species respond well to pruning. *E. verrucosus* is one species which will withstand severe

pruning, and some of its forms will produce sucker growth close to the main stems.

Eriostemons are not subject to many pests or diseases. The most common pest is scale, which can very quickly cover the branches and undersides of leaves if prevailing conditions are satisfactory. Usually the resultant sooty mould which develops on the exudate of the scale is evidence of a large build-up of the pest. Spraying with white oil on a cool day should provide adequate control. Follow-up spraying may be necessary.

Collar rot can occur if plants are placed too deeply at planting, or if soil or mulch is mounded up around the stem. Some species are susceptible to *Phytophthora cinnamomi* attack. This is more likely to occur if soils are poorly drained.

Propagation

Most species are propagated from cuttings, due to the problem of germination inhibitors in seed. Grafting of some species is practised and may become more widely used in the future.

Germination from seed can be sporadic. In general, western species germinate more readily than those from the east. A successful method that has had limited use is to break the hard testa by nicking it, and then place the seed in running water for 10-20 days. By then the inhibitors have been removed and this allows germination to occur (see Volume 1, page 207).

Propagation from cuttings can be a slow process with some species, eg. *E. australasius*, *E. buxifolius*, *E. myoporoides* and *E. verrucosus*. Others such as *E. angustifolius*, *E. nodiflorus* and *E. spicatus* usually form roots quickly.

Best results are gained from the use of firm young material that is still very whippy. Very soft material is also suitable, provided it can be prevented from wilting when placed in the propagation area.

The use of rooting hormones and bottom heat is usually beneficial in promoting better and quicker rooting.

It has been found that cuttings from certain plants produce roots more readily than those from others, eg. *E. australasius*.

Grafting is not widely used at present, although in England during the 19th century *E. buxifolius* and *Correa alba* were successfully used as rootstocks. There is need for more trials with a wider range of species before grafting can be recommended as a viable propagation method.

Eriostemon angustifolius Paul G. Wilson
(narrow leaves)
Qld, Vic, SA   Narrow-leaf Wax Flower
1-1.5 m x 1-1.5 m   June-Nov
   Small **shrub**; **branchlets** slender, warty; **leaves** to 1 cm x about 0.1 cm, subterete to clavate, subsessile, warty, glabrous, shiny, apex acute to blunt; **flowers** to 2 cm across, white with pink midrib, pink in bud, 1-4 in terminal sessile clusters; **petals** with glabrous exterior.

   Generally occurs in low-rainfall regions. The main distribution is from north-central Vic, westwards to south-eastern SA. There are isolated occurrences in the Darling Downs District of Qld and the Flinders

Ranges, SA, where there is a form with large flowers and leaves. Needs very well drained, light to medium soils, with partial or full sun. Tolerant of extended dry periods and most frosts.

This species was known as *E. difformis* var. *teretifolius* Benth. The ssp. *montanus* Paul G. Wilson is confined to the Grampians in western Vic. It differs from ssp. *angustifolius* in its shorter, broader, thick, flat leaves, and the interior of the petals is usually without hairs. It is more common in cultivation than the typical form. Makes an excellent container plant. Can be slow-growing. A selection of this variety with red flower-buds that open to pale pink is a most ornamental form.

The ssp. *montanus* has affinities to *E. difformis*, the petals of which are hairy inside and outside.

Propagate all forms from cuttings.

Eriostemon apiculatus Paul G. Wilson
(short point)
WA
0.6-1.2 m x 0.5-1 m   Sept-Oct
Dwarf to small **shrub**; **branchlets** terete, glabrous or slightly hairy; **leaves** to 1 cm long, narrow-clavate, thick, slightly falcate, convex above, somewhat convex below, slightly hairy or glabrous, apex with prominent point; **flowers** to 1.5 cm across, white to pink, 1-4 in terminal clusters, surrounded by leaves and often branchlets; **petals** exterior glabrous.

This species has a very limited distribution. Presently it is only known from north-east of Norseman, where it grows amongst outcrops of basalt rocks. In cultivation it will need very well drained soil, with a warm to hot situation that receives plenty of sun. Should tolerate most frosts and extended dry periods. Worthy of trial as a container plant in temperate regions. Propagate from cuttings and possibly seed.

Eriostemon australasius Pers.
(Australian)
Qld, NSW   Pink Wax Flower
1-2.5 m x 0.6-1.5 m   Aug-Nov
Small to medium, erect **shrub**; **branches** usually erect; **branchlets** angular, smooth, slightly hairy; **leaves** 3-7 cm x about 0.8 cm, narrow-oblong to elliptic, tapering to base, flat, leathery, faintly nerved, entire, glandular, becoming glabrous, apex ending in very short mucro; **flowers** to 4 cm across, pink to mauve-pink, rarely white, borne in upper axils, profuse.

The typical form of this outstanding ornamental species occurs in coastal regions from Fraser Island in south-eastern Qld, southwards to Sydney, where it is common on sandstone. It is also found in the Stanthorpe region in the border ranges. A very desirable species which has been tried for many years but has continually proved difficult to maintain. For successful cultivation it requires very well drained soils. Has adapted well to clay-loams. Will grow well in dappled shade through to full sun. Prefers some protection for its root system. Plants seem to prefer no artificial watering during summer. Tolerant of most frosts and will withstand limited periods of dryness. Responds well to pruning and light application of slow-release fertilizer. Grows well in a pot and is an excellent con-

tainer plant. Flowers are long-lasting, and most useful as a cut flower for indoor decoration. Propagation has proved to be difficult. Seed needs pre-sowing treatment, such as nicking the hard coat and then washing in flowing water for 10-14 days, but even then the germination percentage is very low. Plants can be raised from cuttings but generally results are slow. Some forms of this species produce roots more readily than others. Selection of such forms is enabling this decorative species to become more readily available.

Previously known as *E. lanceolata* Gaertner f.

The ssp. *banksii* (Cunn. ex Endl.) Paul G. Wilson differs from the typical form in its broadly elliptic leaves of up to 3.5 cm x 1.2 cm. It occurs in coastal north-eastern Qld, where it can grow to a small tree. Is virtually unknown in cultivation.

Eriostemon banksii Cunn. ex Endl. =
E. australasius ssp. *banksii* (Cunn. ex Endl.) Paul G. Wilson

Eriostemon brevifolius Cunn. ex Endl.
(short leaves)
NSW
1-1.5 m x 1-1.5 m   Aug-Oct
Small **shrub**; **branchlets** warty, slightly hairy; **leaves** to 0.5 cm long, more or less terete, flattish above, rounded below, warty, becoming glabrous, apex rounded, ending in a small black point; **flowers** about 1 cm across, white to pink, 1-4 in terminal sessile clusters; **petals** with glabrous exterior, slightly hairy interior.

An ornamental species with white to pink flowers. It has a restricted distribution, and occurs in the South-Western Plains, mainly near Griffith. Has had only limited cultivation. Suited to well drained soils, with partial or full sun. Withstands extended dry periods and most frosts. Well worth growing as a container plant. Propagate from cuttings.

The NSW form of what was known as *E. difformis* var. *teretifolius* Benth. is now referable to *E. brevifolius*. *E. angustifolius* is similar, but differs in not having leaves ending in a short black point.

Eriostemon buxifolius ssp. *buxifolius*. × ·65

Eriostemon australasius ssp. *australasius*

W.R. Elliot

Eriostemon brucei F. Muell.
(after Lt. Col. John Bruce, 19th-century Acting Governor of SA)
WA
1-2 m x 0.6-1.5 m   July-Nov
Small **shrub**; **branches** glaucous; **branchlets** erect, very warty; **leaves** 0.5-2 cm, nearly terete, sessile, grooved above, glabrous at maturity, acute apex; **flowers** to 1 cm across, white to pink, glabrous, solitary, axillary, profuse along branchlets.

A widespread species from north of Perth to north of Mullewa, extending inland into the arid regions. It occurs on a wide range of soils including clays, granite, laterite or red sand, and is sometimes found on salt affected sands. Not commonly cultivated. Should adapt to well drained soils and probably needs a warm to hot situation. Will tolerate shade in semi-arid and arid regions. Tolerant of most frosts and extended dry periods. Worth trying as a container plant in temperate zones.

There are a further 2 subspecies:
ssp. *brevifolius* Paul G. Wilson has densely hairy branches and short, broad leaves. Flowers have mauve buds which open to white during Aug-Sept. It is not well known. It has been collected only in the western Eremaean Botanical Province, from Paynes Find to Sandstone, where it occurs on granite.
ssp. *cinereus* Paul G. Wilson has densely hairy branches, branchlets and leaves. The leaves are concave. The mauve flowers produced from June-Aug have a lavender perfume. It occurs on lateritic soils in the upper Murchison River area.

Propagate all forms from cuttings. Some success has been achieved with seed.

Eriostemon buxifolius Smith
(leaves similar to those of the genus *Buxus*)
NSW
0.6-1.5 m x 0.8-2 m   Aug-Nov
Dwarf to small **shrub**; **branches** not warty, slightly hairy; **leaves** 0.6-1.2 cm long, broadly elliptical to orbicular, sessile, crowded, leathery, flat to concave above, keeled and glandular below, margins entire, base cordate, apex rounded, ending in short point; **flowers** to 2.5 cm across, pink in bud, opening white to pale pink, axillary, solitary, forming leafy spikes.

A most ornamental species that is becoming popular in cultivation. It is found mainly in areas around Sydney, but extends southward to near Jervis Bay. This species is confined to sandstone, usually growing in heathland. It needs well drained soils and seems to adapt to clay-loams. Will withstand full sun but prefers dappled shade to partial sun. Frost tolerant. Usually has compact growth, but responds well to tip pruning to promote lateral branching. An excellent container plant. This species was highly regarded in England

Eriostemon 'Clearview Apple-blossom'

during the 19th century. A double-flowered cultivar known as *E. buxifolius* 'Parry's Double' is in cultivation.

The ssp. *obovatus* (G. Don) Paul G. Wilson has obovate leaves with narrow, wedge-shaped bases and warty margins. It is restricted to the Broken Bay region. Propagate all forms from cuttings which may be slow to form roots.

E. buxifolius also grades into *E. scaber*, resulting in considerable variation and showing characteristics of both species.

E. buxifolius hybridizes readily in nature with *E. myoporoides*, and resultant plants are in cultivation.

Eriostemon 'Clearview Apple-blossom'
see under *E. myoporoides*

Eriostemon 'Clearview Bouquet'
see under *E. myoporoides*

Eriostemon 'Clearview Double'
see under *E. verrucosus*

Eriostemon 'Clearview Pink'
see under *E. myoporoides*

Eriostemon coccineus C. Gardner
(scarlet)
WA
0.5-1 m x 0.5-0.8 m   Aug-Oct

Dwarf, erect **shrub**; **branchlets** warty, becoming corky; **leaves** to 1.5 cm x about 0.1 cm, linear, flat above, 2 rows of prominent glands, apex blunt; **flowers** 1 cm x 0.5 cm, appear cylindrical, pendent, red, solitary, axillary, on long glabrous pedicel.

This unusual eriostemon, with its pendent red flowers, is known only from the sandheaths west of Coolgardie. Evidently not very well known in cultivation. It requires very well drained, light to medium soils and plenty of sunshine. Frost and drought tolerant. Could need regular pruning to promote bushy growth. Should do well as a container plant in temperate regions. It is recorded that honey-eaters visit the flowers. Propagate from cuttings and possibly seed.

This species is closely allied to *E. pachyphyllus*, which differs in its erect white flowers.

Eriostemon deserti E. Pritzel
(of the desert)
WA
1-3 m x 0.8-1.5 m   July-Oct

Small to medium **shrub**; **branches** erect; **branchlets** initially green, becoming brown and slightly corky; **leaves** 0.6-3 cm, needle-like, very warty, narrow groove above, glabrous; **flowers** about 0.5 cm across, white, axillary.

An inland species that occurs from Yalgo in the north, extending south-east to near Kalgoorlie. It grows on lateritic soils, red loam or sand. This eriostemon has small flowers, and is evidently not cultivated. It has potential for use in arid and semi-arid zones. Will need very well drained soils and a warm to hot situation. Propagate from cuttings and possibly seed.

Eriostemon difformis Cunn. ex Endl.
(unevenly shaped)
Qld, NSW, Vic
1-3 m x 1-2 m   March-July; also sporadic

Small to medium, compact **shrub**; **branchlets** slightly warty, lightly hairy; **leaves** to 1 cm x 0.1 cm, subterete, linear-oblong, fleshy, tuberculate, convex above, concave or channelled below, glabrous; **flowers** about 1 cm across, white, 1-4 in terminal sessile clusters; **petals** hairy.

This is a dweller of the slopes and plains of the Great Dividing Range. It has generally proved to be difficult to maintain in cultivation, although one form (ssp. *smithianus*) has proved more reliable in a range of well drained soils. To date the species has been grown only to a limited extent. Should be best suited to well drained soils in a warm location that receives partial or full sun. Tolerant of most frosts and extended dry periods. Propagate from cuttings.

This species consists of 2 subspecies. The typical form is found to the west of the Great Dividing Range.

The ssp. *smithianus* (Benth.) Paul G. Wilson inhabits the eastern regions of the Great Dividing Range and the coastal areas of south-eastern Qld. A further occurrence is on the Macleay River, on the coast of north-eastern NSW. It can be distinguished by its broad, flat, leathery leaves. Grows to about 1 m x 1 m. This ssp. is more common in cultivation than the typical form, and is more reliable.

E. difformis is similar to *E. angustifolius* which differs in its pink to red buds and the exterior of the petals, which are glabrous.

Eriostemon difformis var. **teretifolius** Benth.
see under *E. angustifolius* Paul G. Wilson
E. brevifolius Cunn. ex Endl.
E. gardneri Paul G. Wilson

Eriostemon difformis ssp. *smithianus*. × ·65

474

Eriostemon gardneri   W.R. Elliot

Eriostemon ericifolius Cunn. ex Benth.
(leaves similar to those of the genus *Erica*)
NSW

1-2 m x 1-2 m   May-Oct

Small **shrub**; **branchlets** slightly warty, slightly hairy; **leaves** to 1 cm x 0.1 cm, terete, slightly warty, glabrous or slightly hairy, apex blunt; **flowers** about 1.5 cm across, lilac, 1-6 in terminal sessile clusters; **petals** hairy, glandular; **stamens** woolly near base, with long hairs at apex; **sepals** narrow-triangular.

Recognized as a rare species in nature and confined to central-eastern NSW, between Peak Hill and Denman. It is an ornamental species with well displayed flowers. Has had only limited cultivation, but should do best in well drained soils, with partial or full sun. Hardy to most frosts, and withstands extended dry periods. Has potential as a container plant. Propagate from cuttings.

Some forms of *Philotheca salsolifolia* are similar. They can be distinguished by the very hairy stamens and the broader sepals.

Eriostemon falcatus Paul G. Wilson
(sickle-shaped)
WA

0.3-0.8 m x 0.5-1 m   Aug-Nov

Dwarf, multi-stemmed **shrub**; **branches** many; **branchlets** densely warty, with lines of short hairs; **leaves** about 0.6 cm long, narrow, terete, falcate, warty, glabrous, apex blunt; **flowers** about 1.5 cm across, white, terminal, solitary, profuse; **petals** exterior glabrous.

An uncommon species, presently known only from the original collection near Southern Cross. Effort must be made to conserve this species in nature and cultivation if plants have survived the onslaught of clearing. Should be suited to very well drained soils, with partial or full sun. Probably tolerant of most frosts and extended dry periods. Propagate from cuttings and possibly seed.

Eriostemon fitzgeraldii C. R. P. Andrews
(after W. V. Fitzgerald, early 20th-century WA botanist)
WA

0.3-0.6 m x 0.5-1 m   Sept-Oct

Dwarf, multi-stemmed **shrub**; **branchlets** slightly rough, becoming corky with maturity; **leaves** to 0.5 cm x 0.1 cm, semi-terete, smooth and sometimes grooved above, rounded and warty below, glabrous, apex blunt; **flowers** about 0.7 cm across, white, terminal or axillary, pedicels very short.

A compact species that occurs on sandy soils between Norseman and Esperance. Little-known in cultivation, but its compact growth habit should have wide appeal. Will probably grow well in light to medium soils that have good drainage. Hardy to most frosts and extended dry periods. Has potential as a container plant. Propagate from cuttings.

Eriostemon gardneri Paul G. Wilson
(after Charles Gardner, 20th-century WA botanist)
WA

0.6-1.5 m x 0.8-1.5 m   June-Oct

Dwarf to small **shrub**; **branches** many; **branchlets** slightly hairy; **leaves** 0.5-1 cm long, semi-terete, glandular, glabrous, flat above, convex below, apex blunt, ending in small gland; **flowers** about 1.2 cm across, white, pink in bud, terminal and usually solitary, profuse; **petals** exterior glabrous.

A most ornamental species with pink flower-buds opening to white. It occurs over a wide area, and on many soil types from sand through to clay-loam. On the coast it is found from east of Esperance to Cheyne Beach in the west, and it extends inland towards Lake King, north of Ravensthorpe. Not commonly cultivated, it is best suited to well drained soils, with partial or full sun. Coastal forms could be susceptible to some frost damage in cold districts. Has potential as a container plant. Propagate from cuttings.

Prior to 1970 this species was included under *E. difformis* var. *teretifolius* Benth.

Eriostemon glaber Paul G. Wilson
(glabrous)
WA

0.5-1 m x 0.5-1 m   July-Oct

Dwarf **shrub**; **branchlets** warty, becoming corky, glabrous; **leaves** to 0.5 cm long, nearly sessile, semi-terete to narrow-obovoid, flat above, convex below, warty, glabrous; **flowers** to 1.5 cm across, white, often tinged with pink, 1-3 in terminal clusters; **petals** exterior glabrous.

A decorative dwarf shrub with pink-tinged white

475

Eriostemon halmaturorum

flowers. Its natural habitat extends from east of Geraldton in a south-eastern direction to north-west of Merredin, covering a wide range of soil types. The species should prove to be adaptable to most well drained soils. It prefers partial or full sun. Is tolerant of most frosts, and withstands extended dry periods. Suited to container planting. Propagate from cuttings or seed which has been successful on a limited scale.

Eriostemon halmaturorum F. Muell. =
E. *linearis* Cunn. ex Endl.

Eriostemon 'Heyfield Double'
see under E. *verrucosus*

Eriostemon 'Heyfield Formal'
see under E. *verrucosus*

Eriostemon 'Heyfield Semi'
see under E. *verrucosus*

Eriostemon hispidulus Sieber ex Sprengel
(somewhat rough or with small bristles)
NSW
0.6-1.5 m x 0.8-1.5 m   Aug-Nov
Dwarf to small **shrub**; **branchlets** terete, slightly warty and hairy; **leaves** 1-3.5 cm x about 0.4 cm, narrow-obovate to narrow-spathulate, sessile, smooth and hairy above, sparsely hairy below, margins revolute, prominent midrib below, apex blunt, ending in small point; **flowers** about 1.5 cm across, pink in bud, opening white to pink, 1-3 per axil, profuse.

This species is not well known in cultivation, but it has potential for greater use. It occurs from Wisemans Ferry on the Hawkesbury River in the north to Cole Vale, south-west of Sydney. As it grows on sandstone it needs very well drained soils, and does best in dappled shade or partial sun. It will tolerate full sun. Withstands frost and extended dry periods. Responds well

Eriostemon myoporoides ssp. *leichhardtii.* × ·6

476

to pruning and is suitable for use as a container plant. Propagate from cuttings which can be slow to form roots.

E. *hispidulus* hybridizes with E. *myoporoides* in the Blue Mountains, and hybrid forms are possibly in cultivation.

Eriostemon hybrid 'Decumbent'
see under E. *obovalis*

Eriostemon lanceolatus Gaertner f. =
E. *australasius* Pers.

Eriostemon linearis Cunn. ex Endl.
(linear)
NSW, SA
1-2 m   Aug-Nov; also sporadic
Small **shrub**; **branches** glandular, slightly hairy to glabrous; **leaves** 0.5-2 cm x about 0.1 cm, terete, glandular, glabrous, apex acute to blunt, crowded; **flowers** about 1 cm across, white, borne in axils of branchlets; **petals** with hairy interior and margins.

A dweller of rocky hills from the Gawler Ranges in SA to Wilcannia in NSW. It is not well known in cultivation. Should be best suited to very well drained soils that receive plenty of sunshine. Frost and drought tolerant. It has proved difficult to maintain in cultivation. Worth trying as a container plant in temperate regions. Propagate from cuttings and possibly from seed after treatment.

The form from the Flinders Ranges has been referred to as E. *halmaturorum* F. Muell.

E. *difformis* has affinities to E. *linearis*, but is distinguished by its hairy stems and fleshy leaves.

Eriostemon 'Mountain Giant'
see under E. *myoporoides*

Eriostemon myoporoides DC.
(resembling the genus *Myoporum*)
Qld, NSW, Vic   Long-leaf Wax Flower
1.5-5 m x 1.5-3 m   April-Dec
Small to tall **shrub**; **branchlets** terete, usually warty, sometimes glaucous, glabrous; **leaves** 2-12 cm x 0.7-2.5 cm, oblong to broad-obovate, sessile, often leathery, entire, darker green above than below, apex rounded to pointed, glabrous, aromatic; **flowers** to 2.5 cm across, white, in axillary, stalked clusters of 2-8, pink buds, profuse.

A widely distributed species found in a variety of habitats and soil types, but always where drainage is good. It is most common in mountainous regions, but is also known from coastal heathlands. It is a very familiar and popular species in cultivation, and it has proved to be extremely hardy and adaptable. The species was introduced into England during 1824. It prefers neutral to acid soils to those which are alkaline. Needs good drainage and, in wet soils, may be sensitive to root diseases. Does best in dappled shade or partial sun, but will tolerate semi-shade and full sun. Hardy to frosts and extended dry periods. Responds well to light or heavy pruning, and appreciates mulching.

The species is extremely variable, and 5 subspecies and many cultivars are recognized.

Eriostemon myoporoides 'Profusion'   W.R. Elliot

All forms are generally propagated from cuttings of firm new growth, which can be slow to produce roots. Bottom heat and overhead misting usually help to develop roots more quickly. Only limited success has been achieved in germinating seed.

Cultivars and Hybrids

Over the years many forms have been selected for their specific merits in growth habit or flower arrangement and display. Some of these cultivars are highly ornamental.

'Clearview Apple-blossom' — not well known in cultivation. It produces red buds and semi-double flowers which open to pure white. Originally collected north of Briagolong in East Gippsland, Vic.

'Clearview Bouquet' — a form with very profuse flowering. The pink buds which open to white flowers are borne in dense clusters near the ends of branchlets, and in leaf axils. Also from the Briagolong area.

'Clearview Pink' — a selection that has red buds, and as the flowers mature the petal tips gain reddish tonings.

'Mountain Giant' — there is some confusion regarding this selection, as similar forms grow in East Gippsland and to the north-east of Melbourne in areas such as Lake Mountain and Toolangi. Further examination of plants from both areas is needed to verify whether they are significantly different. Leaves are large and bluish green. Dense clusters of up to 13 pink-and-white flowers, each of about 2 cm across, provide a most decorative display. Plants can reach 3-4 m in height. Initially they may be slow-growing, but once established will grow quickly. Best suited to a sheltered, moist situation. Cuttings can be very difficult to strike.

'Poorinda Patience' — this selection originated from W Tree Creek in East Gippsland, Vic. It grows as a bushy shrub, to about 2 m tall. Has large leaves and produces dense clusters of up to 20 flowers in the axils.

'Profusion' — a compact shrub of about 1.5 m x

E. myoporoides ssp. **acutus** (Blakeley) Paul G. Wilson. Occurs in central NSW. **Leaves** concave, oblong-elliptic, to 4 cm x 0.4-0.6 cm, apex acute to blunt, ending in a small point; **flowers** solitary or up to 4 per cluster, on short stalks. It is common in cultivation, and has proved adaptable. Has been sold as the 'short-leaf form'.

E. myoporoides ssp. **conduplicatus** Paul G. Wilson. Occurs in south-eastern Qld near Stanthorpe, and in north-eastern NSW. Differs with its smooth **branches** and **leaves** which are 4-7 cm long, and folded together lengthwise; 1-4 white **flowers** per axil; **sepals** and **petals** are tinged with red. This subspecies is poorly known in cultivation.

E. myoporoides ssp. **epilosus** Paul G. Wilson. Its distribution is in the border mountains between Qld and NSW, near Stanthorpe. It has warty stems and obovate **leaves** to 3 cm x 0.7 cm, with a rounded apex; usually there is 1 **flower** per axil. Has been found adaptable in cultivation.

E. myoporoides ssp. **leichhardtii** (Benth.) Paul G. Wilson. Restricted to the summits of Glasshouse Mountains, north of Brisbane, Qld. It is a spreading **shrub** of 1-1.5 m x 1.5-2.5 m; **branchlets** are very warty and the oblong-cuneate **leaves** of 2-3.5 cm x 0.6-0.8 cm have a rounded apex which can be notched, ending in a short point. Deep rosy pink buds open to pale pink **flowers**. There is only 1 per axil. Poorly known in cultivation but grows well in subtropical regions.

E. myoporoides ssp. **queenslandicus** (C. White) Paul G. Wilson. Known from the sandy wallum flats near Moreton Bay and Wide Bay, Qld. It grows to about 0.5 m x 1-1.5 m, and has many minutely warty **branchlets**. The **leaves** are narrow-elliptic to narrow-obovate, 1.5-2.5 cm x up to 0.3 cm, concave above, smooth or slightly warty, with the apex blunt or pointed; **flowers** 1 per axil, profuse along the branches. Buds are pink, opening to white with the backs of the petals tinged with pink. Has had only limited cultivation.

Eriostemon myoporoides ssp. *queenslandicus*, × ·7

477

Eriostemon myoporoides 'Mountain Giant'

W.R. Elliot

1.5 m. A most floriferous selection, producing white flowers during June-Nov.

A limited number of hybrids, with *E. myoporoides* as one of the parents, have been produced. The following are the best known of those in cultivation:

'Poorinda' — parents are an East Gippsland form of *E. myoporoides* and a form of *E. verrucosus* from Mount Arapiles in western Vic. It grows to 1.5 m x 0.75 m. Pink buds open to white. Flowering is during Sept-Jan.

'Swanson' — also a hybrid with *E. verrucosus* as the other parent. It has slightly warty branches and broad, rounded leaves. Produces large white flowers. Known also as *E.* 'Swanson's Hybrid'.

'Stardust' — very similar to *E.* 'Swanson', having the same parentage. This adaptable hybrid has very warty branches, and is a most profuse flowering shrub of about 1.5-2 m tall. It was popular during the late 1960s and early 1970s.

Eriostemon nodiflorus Lindley
(flowers from the nodes)
WA
0.5-1.5 m x 0.8-1.5 m   July-Nov
   Dwarf to small **shrub**; **branches** more or less erect, slender, not dense, slightly hairy when young; **branchlets** short; **leaves** to 1.5 cm x 0.2 cm, sessile, semiterete, slender, glabrous or slightly hairy, glandular, apex pointed or blunt; **flowers** about 1 cm across, blue to pink, initially in terminal clusters of about 3 cm

diameter, or becoming spike-like with new growth emerging.

A showy species from the Darling Range, where it grows in lateritic soils, peaty sand or sandy soils. Well known in cultivation, although it can be short-lived. Suited to most types of soil, provided drainage is adequate. Grows well in dappled shade and partial sun. Will tolerate full sun. Often develops an open growth habit, which is rectified by pruning after flowering. Regular tip pruning is also beneficial. Hardy to most frosts but does not like extended dry periods. Does well in pots.

The ssp. *lasiocalyx* (Domin) Paul G. Wilson has blue flowers in small heads (to 1.5 cm diameter) and is not as well known in cultivation. It occurs from near Cape Arid on the south coast and westward to near Manjimup.

Propagate both forms from seed which should germinate within 5-14 weeks. Cuttings produce roots readily.

E. spicatus can be confused with this species. It is distinguished by its terminal racemes of flowers.

Eriostemon nutans Paul G. Wilson
(nodding)
WA
0.5-1 m x 0.5-1 m   July-Aug
   Dwarf **shrub**; **branchlets** terete, warty, hairy; **leaves** about 1 cm long, linear, thick, flattened above, roun-

ded and very warty below, glabrous, apex blunt; **flowers** about 1 cm x 0.7 cm, appearing cylindrical, pale red, pendent, solitary, axillary, borne on long recurved pedicels.

A rare species, known only from an isolated population in the Ninghan region, where it grows on the sandy flats. Evidently not in cultivation. Will need very well drained, light to medium soils, with partial or full sun. Should tolerate most frosts and extended dry periods. Propagate from cuttings and possibly seed.

Allied to *E. coccineus*, which has thinner leaves, and flowers of a deeper red.

Eriostemon obovalis Cunn.
(inversely egg-shaped)
NSW
0.5-1 m x 0.6-1.5 m   June-Nov
Dwarf **shrub**; **branchlets** terete, hairy, not or only slightly warty; **leaves** about 0.8 cm x 0.4-0.6 cm, broadly cuneate, notched apex, leathery, slightly warty below, glabrous or with slightly hairy margins; **flowers** to about 1.2 cm across, white tinged with pink, glabrous, axillary, solitary, profuse.

A most ornamental species confined to the Blue Mountains, where it grows on sandstone. It is not cultivated very much, probably due to confusion with *E. verrucosus* which differs in its warty, glabrous branchlets. Has proved difficult to maintain in cultivation. Best grown in well drained, light to medium soils. Tolerates dappled shade, partial sun or full sun. Withstands most frosts, and extended dry periods. Plants tend to become leggy but respond well to regular pruning. Suitable as a container plant. Propagate from cuttings.

A hardy and showy cultivar known as *E.* hybrid 'Decumbent' or *E.* species 'Decumbent' is thought to have *E. obovalis* as one parent. Plants of this cultivar develop to about 0.6 m x 1.5 m. The leaves are about 2 cm x 1 cm, obovate and concave, and margins on new growth are often purple. Apex can be notched. Flowers are up to 2 cm across, white to pale pink, and exterior of petals and buds is pink.

Eriostemon obovalis, × ·7

Eriostemon pachyphyllus Paul G. Wilson
(thick leaves)
WA
0.6-1.5 m x 0.6-1 m   May-Oct
Dwarf to small **shrub**; **branchlets** slightly glandular, becoming corky; **leaves** to 0.5 cm x 0.2 cm, oblong, petiolate, fleshy, slightly convex above, blistery below, glabrous, margin wavy, apex blunt; **flowers** about 0.6 cm long, white, erect, solitary, axillary.

An uncommon species which is known from only one population, west of Coolgardie. It grows in sand-heath on sand or red loam over limestone. Evidently not in cultivation. Will need very well drained, light to medium soils, with a warm to hot situation that receives plenty of sunshine. Frost and drought tolerant. Propagate from cuttings and possibly seed.

Is allied to *E. coccineus*, which has pendent red flowers.

Eriostemon pinoides Paul G. Wilson
(similar to the genus *Pinus*)
WA
0.3-0.8 m x 0.3-1 m   Aug-Oct; also sporadic
Dwarf **shrub**; **branchlets** minutely warty near nodes; **leaves** about 1 cm long, needle-shaped, slightly twisted, not warty, glabrous, grooved above, triangular below, prickly apex; **flowers** about 1 cm across, dark red to reddish pink, 1-3 terminating branchlets; **pedicel** to 0.7 cm long, glabrous.

An unusual species with dark red to reddish pink flowers. It inhabits lateritic sands on or near the west coast, between the Hill River and Coorow. Apparently not in cultivation. It needs well drained soils but should tolerate clay-loams. Is suited to a sunny situation. Withstands most frosts and extended dry periods. In temperate regions it may grow best in containers. Propagate from cuttings.

Eriostemon 'Poorinda'
                              see under *E. myoporoides*

Eriostemon 'Poorinda Patience'
                              see under *E. myoporoides*

Eriostemon 'Profusion' = *E. myoporoides* 'Profusion'

Eriostemon pungens Lindley
(ending in a sharp hard point)
Vic, SA   Prickly Wax Flower
0.2-0.6 m x 0.5-1 m   Sept-Nov
Dwarf **shrub**; **branches** spreading, not dense; **branchlets** short, slender, glabrous or slightly hairy; **leaves** 0.8-2.5 cm x up to 0.2 cm, linear to linear-lanceolate, glabrous or slightly rough, ending in a pungent point, initially erect, becoming reflexed; **flowers** about 1 cm across, white with mauve to pink exterior, axillary, along branchlets; **pedicels** about 0.5 cm long.

In nature this eriostemon inhabits sandy soils and rocky places in central and western Vic, and extends to Eyre Peninsula in SA. It is not common in cultivation. Is most likely to succeed in very well drained, light to medium soils. Needs a warm to hot position. Tolerant of frost and extended dry periods. Grows well in con-

Eriostemon rhomboideus

Eriostemon pungens, × 1

tainers in temperate zones. Responds well to pruning.

A prostrate form from the Golton Gorge in the Grampians, Vic, is a recent introduction to cultivation. It forms a dense bright green mat, and flowers profusely.

Propagate from cuttings which produce roots readily.

Eriostemon rhomboideus Paul G. Wilson
(rhomboid)
WA
0.5-1 m x 0.5-1 m   Aug-Oct
Dwarf, multi-stemmed **shrub**; **branches** many; **branchlets** slightly glandular, glabrous; **leaves** to 0.5 cm long, broad-elliptic to ovoid or rhomboidal, thick, flat or sometimes rounded and glandular below, glabrous, margins wavy, apex blunt; **flowers** about 1.2 cm across, white to pale pink, 1-3 in sessile terminal clusters; **petals** glabrous.

The natural distribution of this species is mainly confined to the inland region of the south-western corner. It occurs on a wide range of soils including red sand, clays and granitic loam. Evidently not well known in cultivation. Is best suited to well drained soils, in a warm to hot situation receiving plenty of sunshine. Frost and drought tolerant. Has potential as a container plant. Propagate from cuttings.

Eriostemon 'Semmens Double Wax Flower'
see under *E. verrucosus*

Eriostemon scaber Paxton
(rough)
NSW
0.5-1 m x 0.6-1.5 m   Aug-Nov
Dwarf **shrub**; **branchlets** terete, smooth, hairy; **leaves** to 2.5 cm long, semi-terete, sessile, often falcate, deeply concave above, warty below, apex acute, mucronate; **flowers** to 1.4 cm across, white to pink, axillary, solitary, profuse; **petals** glabrous.

This most attractive species is restricted to sandstone, mainly in the area to the west and south of Sydney. It must have well drained soils, but is adaptable to various amounts of light. A position with semi-shade, dappled shade or full sun will suffice. Dense growth can be promoted by tip pruning from planting time, but it also responds to heavier pruning if required.

Tolerant of most frosts and dry periods. Grows well as a container plant. The species was introduced to England during 1840.

The ssp. *latifolius* Paul G. Wilson differs in its very warty stems, and oblong-elliptic leaves of about 1.2 cm x 0.3 cm that are only slightly concave on the upper surface. It occurs in south-eastern NSW.

Propagate both forms from cuttings.

The two subspecies hybridize in nature with *E. buxifolius*. *E. myoporoides* apparently grades into *E. scaber* ssp. *latifolius* in the Pigeon House Range. Some of these hybrids may be in cultivation.

Eriostemon sericeus Paul G. Wilson
(silky)
WA
1-2 m x 0.6-1.5 m   Aug-Oct; also sporadic
Small **shrub**; **branchlets** smooth, hairy; **leaves** to 0.4 cm long, elliptic to obovate, fleshy, flat above, rounded below, glabrous or with hairy margins, apex rounded; **flowers** about 2 cm across, white to pink, terminal, usually solitary, profuse; **petals** silky on both sides.

An ornamental and unusual eriostemon with silky petals. It is not well known in cultivation. It occurs near Geraldton, and extends eastward to Sandstone in the Austin District. This species grows on various sands that may contain lateritic gravel. In cultivation it will need very well drained, light to medium soils, with partial or full sun. Best suited to arid and semi-arid zones. Could be difficult to establish in temperate and tropical regions, although it is worthy of trial as a container plant. Should tolerate most frosts. Propagate from cuttings. There has been limited success with propagation from seed.

Eriostemon scaber, × ·7

480

Eriostemon tomentellus   W.R. Elliot

Eriostemon species 'Decumbent'

see under *E. obovalis*

Eriostemon spicatus A. Rich
(flowers in spikes)
WA
0.3-1 m x 0.5-1 m   June-Dec
Dwarf **shrub**; **branches** slender; **branchlets** slightly hairy or glabrous; **leaves** 0.6-2 cm long, sessile, linear to narrow-elliptic, glandular, concave above, slightly hairy to glabrous, apex pointed; **flowers** about 0.8 cm across, white, mauve, pink or bluish, in a terminal raceme to about 20 cm long.

An outstanding ornamental eriostemon that occurs near the coast from the Geraldton area in the north, and extends south to near Albany. It grows on a wide range of gravels and soils including some which are calcareous. Has adapted well to cultivation. It prefers well drained soils, with plenty of sunshine, and a location where there is considerable air movement. Tolerates most frosts and limited dry periods. Responds well to pruning after flowering, which helps to promote bushy growth. Can suffer damage from grey mould as flowering finishes, particularly during humid weather. An excellent container plant. Propagate from seed which usually germinates within 4-10 weeks. Cuttings of firm young growth form roots readily.

There has been some confusion between this species and *E. nodiflorus*. They are readily distinguished as *E. nodiflorus* has tight clusters of flowers, while *E. spicatus* has its flowers in long terminal racemes.

Eriostemon 'Stardust'

see under *E. myoporoides*

Eriostemon 'Swanson'

see under *E. myoporoides*

Eriostemon thryptomenoides S. Moore
(resembling the genus *Thryptomene*)
WA
0.5-1 m x 0.5-1 m   July-Oct
Dwarf **shrub**; **branchlets** smooth, slightly hairy; **leaves** to about 0.3 cm long, clavate, shortly petiolate, flattened above, glandular, glabrous, apex rounded; **flowers** about 1.2 cm across, white with reddish brown strip on exterior of petals, terminal, solitary, profuse.

This species is restricted to the area from around Merredin and extending north-west to Wubin in the Avon District. It occurs on sandy clay, and lateritic and granitic soils. Should adapt to a wide range of well drained soils in cultivation, in a position with partial or full sun. It is not well known to date, but its interesting white and reddish brown flowers should ensure that it becomes more widely planted. Withstands most frosts and is hardy to extended dry periods. Well suited as a container plant. Propagate from cuttings or seed. There has been only limited success with seed germination.

Has some affinities to *E. tomentellus*, which is distinguished by its hairy branchlets, and leaves that have a black-pointed apex.

Eriostemon tomentellus Diels
(minutely tomentose)
WA
0.6-1 m x 0.5-1 m   June-Oct
Dwarf **shrub**; **branches** many; **branchlets** minutely hairy, slightly warty, can be glaucous; **leaves** about 0.4 cm long, clavate to somewhat terete, flat above, convex below, hairy, apex blunt, with prominent black point; **flowers** about 1 cm across, white with pink to red rib on exterior of petals, 1-4 in terminal clusters, profuse.

A widespread species from the Austin, Avon and Coolgardie Districts, where it grows on sandy clays and red or yellow sands. It is not well known in cultivation. Should be best suited to well drained soils and a position where it receives plenty of sunshine. Certainly has potential as it is decorative in bud as well as in flower. Is worth growing as a container plant in temperate regions. May prove difficult to establish in tropical areas. Propagate from cuttings and possibly seed.

Eriostemon trachyphyllus F. Muell.
(rough leaves)
NSW, Vic   Rock Wax Flower
2-7 m x 1-4 m   Aug-Nov
Medium **shrub** to small **tree**; **branchlets** terete, warty, willowy, glaucous, glabrous; **leaves** to 6 cm x

Eriostemon verrucosus

1 cm, oblong-elliptic to narrow-obovate, narrowing to
base, slightly warty, glabrous, margins slightly recurved, apex acute to rounded, with small point; **flowers**
to 1.5 cm across, white, axillary, solitary or in clusters
of 3, on slender pedicels, profuse.

This large species usually grows in cool locations on
rocky hills, from near Singleton in the north and extending to East Gippsland. The rounded bushes can be
literally covered with white blossoms. Not commonly
cultivated to date. Has proved adaptable to most well
drained soils, and grows well in dappled shade or partial sun. Hardy to frost and extended dry periods. Propagate from cuttings which can be very slow to form
roots. It is generally regarded as extremely difficult to
strike cuttings of this species.

Eriostemon verrucosus A. Rich.
(covered with warts)
Vic, Tas, SA   Fairy Wax Flower
0.3-4 m x 0.3-2 m   June-Nov; also sporadic

Dwarf to medium **shrub**; **branchlets** terete, many
glandular warts, glabrous; **leaves** to 2 cm x 1 cm,
narrow to broad-obcordate, sessile, flat to folded
lengthwise, smooth above, very warty below, aromatic;
flowers to 2 cm across, pink in bud, opening white to
pale pink, axillary, along branchlets, profuse.

An extremely ornamental and variable species. In
nature it grows in a wide range of soils, and under
various climatic conditions. There are many forms in
cultivation, selected from different locations. It grows
as a low shrub in the goldfields area of Central Victoria, and as an upright shrub of about 4 m in height in
Gippsland and at Mount Arapiles near the Grampians
in western Vic. Has proved hardy in most soils that are
not prone to waterlogging, but does not tolerate alkalinity. Will grow successfully in dappled shade to full
sun. Frost and drought tolerant. Highly suited to container planting. Can be pruned very hard to promote
bushy growth. Propagate from cuttings which can be
slow to form roots.

The following are some of the selected forms known
to be in cultivation:

'Arapiles' — has an open growth habit, and can
reach 4 m in height. Has pinkish flowers. From Mount
Arapiles near Horsham in western Vic.

'Clearview Double' — also known as E. 'Heyfield
Double'. Has flowers of about 2.5 cm across, with 3
rows of pink-tinged petals. Originated near Heyfield in
eastern Vic.

'Heyfield Formal' — produces pure white, formal
flowers of about 2 cm across. Flowering time is Oct-
Nov. Plants grow to about 0.6 m tall.

'Heyfield Semi' — very similar to E. 'Heyfield Formal', but has erect petals which are not as numerous as
in that cultivar.

'Semmens Double Wax Flower' — this ornamental
and popular form from near Bendigo, Vic, has multi-
petalled white flowers of about 2.5 cm across. They are
displayed on slender, arching branchlets. Plants will
grow to 0.5-1 m x 0.5-1.5 m. Requires excellent
drainage and prefers a semi-shaded situation. They
often develop by suckering lightly. Hard pruning can
further induce suckering. Makes a very attractive container subject. For many years this form has been

Eriostemon verrucosus   W.R. Elliot

Eriostemon verrucosus (pale pink-flowered form)   T.L. Blake

482

known by a number of names including 'Bendigo Double Wax Flower', 'J. Semmens Wax Flower' and 'Semmens Double Wax Flower'. Propagation from cuttings can be slow. Best results are gained when firm young growth is used. Grafting on to *E. myoporoides* has been successful.

Eriostemon virgatus Hook.f.
(twiggy)

Vic, Tas

Tasmanian Wax Flower;
Twiggy Wax Flower

1-4 m x 0.6-2 m   Sept-Dec

Small to medium, erect **shrub**; **branches** long, slender; **branchlets** small, often in clusters, triangular, becoming terete, warty, glabrous; **leaves** 1-2.5 cm x up to 0.5 cm, narrow, oblong-cuneate to narrow-obovate, warty, margins slightly recurved, prominent midrib below, apex blunt or notched; **flowers** to 1.5 cm across, usually 4-petalled, sometimes 5-petalled, white, sometimes tinged with pink, axillary, near ends of branches and branchlets, profuse.

Originally thought to be a Tas endemic species; however, it also occurs in East Gippsland, Vic. In Tas it is found near the coast in the east, south-west and west. It usually grows in moist to wet forests. In cultivation it requires well drained soils in a situation with dappled shade or partial sun. Hardy to frost. Regular tip pruning of young plants can help to promote bushy growth. Propagate from cuttings. This species is easily distinguished from other eriostemons by its 4-petalled flowers (sometimes 5-petalled).

Eriostemon wonganensis Paul G. Wilson
(in reference to it occurring in the Wongan Hills)

WA

0.8-1.5 m x 0.5-1 m   Aug-Oct

Dwarf to small, multi-stemmed **shrub**; **branches** slender, erect, glandular, with corky strips, glabrous; **leaves** 0.5-1 cm x about 0.1 cm, slender, somewhat terete, erect, very warty, glabrous, apex pointed; **flowers** to 0.7 cm across, white with pink strip on exterior of petals, axillary, profuse, near ends of branchlets.

A recently described endangered species. It occurs only in the Wongan Hills, about 150 km north-east of Perth, where it grows in red soil overlying fractured greenstone. In cultivation it will require very well drained soils, with plenty of sunshine. Tolerant of most frosts and drought. This eriostemon needs to be cultivated as it is in danger of extinction in the wild. There are only a small number of plants on private land. Propagate from seed or cuttings.

ERQDIOPHYLLUM F. Muell.
(from the Greek *erodios*, a heron; *phyllon*, leaf; in reference to the similarity of the leaves to a heron's bill)
Asteraceae

Perennial **herbs**; **leaves** alternate, deeply lobed; **flower-heads** terminal, solitary, many-flowered; **receptacle** convex, with scales; **ray florets** female; **disc florets** tubular, outer female, inner male; **fruit** an achene.

An endemic genus with 2 species, of semi-arid

Eriostemon verrucosus 'Semmen's Double Wax Flower'   W.R. Elliot

regions in SA and WA. *E. elderi*, the Koonamoore Daisy, is a most ornamental species and is cultivated to a limited extent.

Erodiophyllum acanthocephalum Stapf
(prickly head)

WA

0.2-0.5 m x 0.3-0.8 m   Aug-Nov

Dwarf **herb**; **branches** spreading; **leaves** 6-12 cm long, alternate, thin, deeply lobed, with narrow segments, hairy below; **flower-heads** daisy-like, about 5 cm across, solitary, terminal on stalks 10 cm or more in length; **ray florets** blue to pale violet; **disc florets** yellow; **fruiting head** about 2 cm across, with prominent prickly bracts; **achene** about 0.3 cm long, ribbed.

A showy member of the genus, but evidently not in cultivation. It is recorded as growing in the red soils of the Austin District, between Sandstone and Laverton. Undoubtedly it will require a warm to hot location. Well drained soils that retain moisture but do not become waterlogged should be to its liking. Experimentation with this species is needed to learn more about its cultivation requirements. Should withstand most frosts and extended dry periods.

Erodiophyllum elderi F. Muell.
(after Sir Thomas Elder, 19th-century sponsor of exploring expeditions)

NSW, SA, WA   Koonamoore Daisy

0.15-0.4 m x 0.2-0.4 m   Aug-Dec

Dwarf annual or perennial **herb** with rough hairs; **stems** erect; **leaves** 5-10 cm long, alternate, deeply lobed, usually with further segments, hairy; **flower-heads** daisy-like, to about 5 cm across, terminal,

483

solitary; **ray florets** purple; **disc florets** yellow; **fruiting head** about 2 cm across; **achene** woody, about 0.3 cm long, ribbed.

An outstanding ornamental daisy from semi-arid and arid zones, where it inhabits heavy soils that are periodically inundated. These can be calcareous. Occasionally it is found in sand. Has been successfully cultivated, but not to a great extent. It seems that the main requirements are for warmth and plenty of sunshine for most of the year. Should be tolerant of a wide range of soils, although probably prefers those of heavy texture. Although occurring on heavy soils in nature, good drainage should provide satisfactory results. Frost and drought tolerant. Has potential as a container plant in temperate regions. Propagate from seed which can be hard to remove from the fruit. Some growers cut the fruit with a knife to release the seed. Seed does not need pre-sowing treatment, and usually germinates within 3-8 weeks. Cuttings produce roots readily.

ERODIUM L'Her. ex Ait.

(from the Greek *erodios*, a heron; in reference to the similarity of the carpels to a heron's head and beak)
Geraniaceae

Annual or perennial **herbs**; **leaves** petiolate, dissected or deeply lobed; **flowers** in umbel-like cymes, axillary; **sepals** 5; **petals** 5, white, pink, purplish to blue; **stamens** 5, alternating with 5 staminodes; **fruit** splits into 5 mericarps, each with a spirally coiled awn.

A cosmopolitan genus of about 60 species. In Australia it has a representation of 8 species. Three are endemic and the others are introduced weeds. The native species have small purplish to blue, or rarely pinkish flowers. They are as yet poorly known in cultivation. Could be useful container plants. Propagate from seed which does not need any pre-sowing treatment or from cuttings which strike readily.

Erodium angustilobum Carolin
(narrow lobes)
SA, WA
prostrate-0.2 m x 0.3-1 m   July-Sept

Dwarf annual or perennial **herb**; **stems** spreading or ascending, hairy; **leaves** 5-20 cm x 2-4 cm, petiolate, deeply lobed or dissected, hairy, lobes 4-6, oblong to narrow-oblong, toothed; **flowers** to 1.5 cm across, blue or pinkish, on hairy pedicels; **calyx** with short, appressed hairs.

A dweller of sandy soils in low-rainfall regions. This species will need very well drained soils, in a position that receives plenty of sun. Tolerant of most frosts and extended dry periods. Worth trying in pots. Propagate from seed or cuttings.

Erodium crinitum Carolin
(long hair)
Qld, NSW, Vic, SA, WA, NT   Blue Heron's Bill;
                                 Native Crow-foot
0.1-0.4 m x 0.3-1 m   July-Oct

Dwarf annual or perennial **herb**; **stems** decumbent to ascending, hairy; **leaves** 1.5-4 cm x 1.5-3 cm, petiolate, with 3 main lobes, toothed; **flowers** usually in umbels of 2-6, about 1.5 cm across, blue with yellowish to white veins near the base, on hairy pedicels; **calyx hairs** spreading.

Widespread throughout the mainland. This attractive species occurs on sandy or clay soils, generally in open situations. Plants are not difficult to cultivate. Will need well drained soils, with partial or full sun. It may naturalize in some regions. Hardy to frost and extended dry periods. Has potential as a container plant. Propagate from seed or cuttings.

Erodium cygnorum Nees
(from the Latin *cygnus*, swan; after the Swan River)
SA, WA, NT
0.01-0.5 m x 0.3-1 m   July-Oct

Dwarf perennial **herb** with a fleshy rootstock; **stems** often many, with stiff, simple hairs; **leaves** 1.5-4 cm x up to 3 cm wide, with 3 main lobes; **petioles** slender, about 2 cm long, with scattered simple hairs; **flowers** usually in umbels of 2-5, about 1.5 cm across, bluish purple with yellowish to white veins near the base; **pedicels** almost glabrous; **calyx hairs** short, appressed.

Generally this species has its distribution in south-western WA, with other populations occurring in more arid regions and in the tropical zones. It is usually found in open, sandy depressions. Should prove adaptable in cultivation. Will need very well drained soils, and partial or full sun. Tolerant of frost and drought. Has potential as a container plant.

The ssp. *glandulosum* Carolin, from NSW, SA, WA and NT, grows taller, to 0.7 m. It has glandular hairs on stems and leaves. The bluish purple petals have red veins near the base.

Propagate both forms from seed or cuttings.

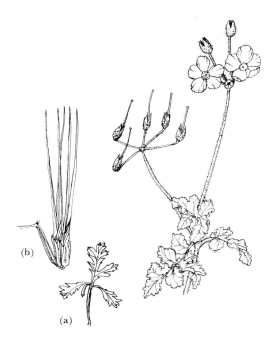

Erodium crinitum, × ·6   (a) leaf variant   (b) seeds

ERVATAMIA Stapf
(from the southern Indian plant name of nandi-ervatam)
Apocynaceae

Shrubs or small **trees**, branching dichotomously; **sap** milky; **leaves** simple, alternate, entire; **inflorescence** a terminal or leaf-opposed cyme; **flowers** tubular at the base, the lobes conspicuously rotate, white or yellow; **calyx** of 5 lobes; **corolla lobes** 5; **fruit** an elongated capsule.

A large genus of about 80 species widely distributed in tropical regions with 3 or 4 species in Australia. They are attractive shrubs for a shady garden situation and are deserving of wider recognition. Propagate from seed which, for best results, should be sown fresh or from cuttings.

Ervatamia angustisepala (Benth.) Domin
(narrow sepals)
Qld, NSW   Banana Bush
1-2.5 m x 1-3 m   Sept-Nov

Small to medium, spreading **shrub**; **sap** milky, bleeding readily from wounds; **leaves** 8-15 cm x 1.5-3.5 cm, oblanceolate or elliptical, thin-textured, bright green and shiny above, paler beneath, the apex acute to acuminate, gradually tapered to the petiole; **cymes** few-flowered, terminal on branchlets; **flowers** 1.5-2 cm across, white, sweetly fragrant, the corolla tube about 1 cm long; **fruit** 2.5-4 cm long, carried in pairs, banana or clog-shaped, yellow, mature Feb-March.

A widely distributed shrub extending from north-eastern Qld to north-eastern NSW. It grows on rainforest margins, tracks and clearings in rainforest, and prefers the drier types of rainforest. Flowers are very ornamental and the fruit is interesting and decorative. Plants form an open, spreading bush that is never dense. The species has excellent horticultural potential and is grown to some extent in tropical and subtropical regions. It will grow in a sunny situation, but the plants have a much better appearance if they are in a situation where the sun is filtered rather than direct. Needs well drained soil. Plants can be quite fast growing in good conditions, and they respond to watering and small amounts of fertilizer. Propagate from seed which is best sown fresh or from cuttings.

Ervatamia orientalis (R. Br.) Domin
(from Asia)
Qld, NT
1-4 m x 1-3 m   Aug-Nov

Small to tall **shrub** with a spreading habit; young shoots glabrous; **leaves** 5-16 cm x 2-4 cm, elliptical to oblong, bluntly pointed, dark green and shiny, narrowed gradually to the petiole; **cymes** few-flowered, terminal on branchlets; **flowers** 1.5-2 cm across, white, sweetly fragrant, the corolla tube 1-1.2 cm long; **fruit** 1.5-2.5 cm long, ovoid, falcate, 3-angled, yellow.

An attractive shrub of north-eastern Qld and NT, growing along stream banks and in dry rainforests. The flowers are sweetly perfumed and showy. An ideal garden shrub for the tropics, and could also be grown in warmer parts of the subtropics. Best appearance is attained in a semi-shady situation. Requires soil of free

Ervatamia angustisepala 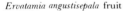   D.L. Jones

Ervatamia angustisepala fruit   D.L. Jones

drainage, and responds to liberal applications of mulch on the soil surface. Propagate from seed and perhaps cuttings.

Ervatamia pubescens (R. Br.) Domin
(softly hairy)
Qld, NT
1-3 m x 1-3 m   Aug-Dec

Small to medium **shrub** with a spreading habit; young shoots softly hairy; **leaves** 10-16 cm x 2-5 cm, elliptical to oblong, dark green and glabrous above, softly hairy beneath, bluntly pointed, narrowed gradually to the petiole; **cymes** few-flowered, hairy, terminal on branchlets; **flowers** 1.5-2.2 cm across, white, sweetly fragrant, the corolla tube about 1.2 cm long; **fruit** 1.5-2.5 cm long, ovoid, falcate, angular, bright orange.

A shrub of north-eastern Qld and parts of NT, usually growing in shaded situations such as along stream banks and rainforest margins. It can be distinguished from the other species by its softly hairy foliage. Plants bear delightfully fragrant, large white flowers in small clusters, and are followed by colourful fruit. An ideal shrub for tropical and perhaps warm subtropical gardens. Needs a shady, semi-protected situation in well drained soil. Propagate from fresh seed.

ERYCIBE Roxb.
(from an old Malabar name)
Convolvulaceae

Tall woody **climbers**; **leaves** alternate, simple, entire, petiolate; **inflorescence** of racemes or cymes or terminal panicles; **flowers** small, bisexual, funnel-shaped, the basal part a short tube, the upper part flared, consisting of 5 deeply bi-lobed sections; **stigma** with 5-10 rays; **fruit** a fleshy berry containing a single seed.

A genus of about 66 species distributed through India, China and Japan with a solitary species in north-eastern Qld. It is a vigorous climber rarely encountered in cultivation. Propagate from seed which has a limited viability and must be sown fresh.

Erycibe coccinea (Bailey) Hoogl.
(scarlet)
Qld
5-15 m tall   April-June

Tall woody **climber**; young shoots covered with rusty hairs; **leaves** 7-10 cm x 3-5 cm, ovate to elliptical, leathery, dark green and glabrous above, dull and somewhat hairy beneath, the margins entire, the apex pointed; **racemes** axillary, shorter than the leaves, the upper ones forming a terminal panicle; **flowers** 0.8-1 cm across, yellow, funnel-shaped, hairy outside, dense in the panicles; **berries** about 1.2 cm across, ovoid, red.

A climber restricted to north-eastern Qld, where it grows in rainforests. It is a strong climber which has some ornamental appeal, but is probably too vigorous for a home garden. Could be useful in large municipal gardens. Best suited to tropical and perhaps sub-tropical regions. Requires well drained soil. Propagate from seed which must be sown fresh. The species was previously known as *E. paniculata* Roxb. var. *coccinea* Bailey.

Erycibe paniculata Roxb. var. coccinea Bailey =
E. coccinea (Bailey) Hoogl.

ERYNGIUM L.
(from the Greek *eryggion*, a name used by Dioscorides for a similar thistle-like plant which was believed to cure flatulence)
Apiaceae

Annual or perennial **herbs** with prickly leaves and bracts, perennial species with a carrot-like underground stem; **leaves** radical and scattered along the stems, entire or deeply lobed, the lobes pungent; **flowers** small, in compact thistle-like heads, usually bluish, each surmounted by an involucre of stiff, spreading, pungent bracts.

A genus of about 230 species distributed in tropical and temperate parts of the world with 6 species indigenous to Australia and a further 2 species naturalized as weeds. A couple of exotic species are also grown as garden plants. The native species usually grow in low-lying, heavy soils. They are extremely prickly but are of interest because of the delicate blue and purple colourations of their flower-heads and bracts. A couple of species are cultivated on a limited scale. Propagate by transplants and in some cases from stem or root cuttings. Seed can be difficult to germinate and may improve with storage.

Eryngium expansum F. Muell.
(spreading)
Qld, NSW
prostrate x 0.6-1.2 m   Aug-Jan

Perennial prostrate **herb**; **stems** 30-60 cm long; **radical leaves** 5-8 cm x 1-1.5 cm, oblong to obovate, leathery, the margins lobed or with coarse prickly teeth, spreading in a rosette; **stem-leaves** shorter, divided into 3-5 prickly lobes; **flower-heads** about 1 cm long, globular, blue; **bracts** pungent, spreading.

An interesting perennial found along the flood plains of coastal rivers from central Qld to north-eastern NSW. It has smaller flower-heads than other species and is not quite as colourful. It can be grown in a container or open, sunny garden position. Propagate from seed or by transplants.

Eryngium pinnatifidum Bunge
(lobed leaves)
WA   Blue Devil
0.3-0.6 m x 0.5-1 m   Aug-Jan

Perennial **herb**; **stems** stiff, semi-erect or spreading; **leaves** 10-15 cm long, linear, with a few scattered, long prickly lobes, in a basal clump and scattered along the stems; **flower-heads** about 2 cm long, ovoid, whitish, becoming blue or purple; **bracts** 1.5-3 cm long, rigid, pungent, purple.

This species is restricted to WA, and is widespread north and south of Perth, usually growing in heavy clay soils. Its responses in cultivation are unknown, but it is an interesting decorative species worthy of trial. Propagate by transplants or from seed.

Eryngium plantagineum F. Muell.
(like the genus *Plantago*)
Qld, NSW, Vic   Eryngo
0.2-0.6 m x 0.2-0.5 m   Oct-Jan

Annual or biennial **herb**; **stems** 20-60 cm tall, erect, usually branching only near the apex; **leaves** 10-25 cm x 0.2-0.4 cm, ovate to lanceolate, in a radical rosette, usually deeply divided into spiny, spreading segments, but sometimes linear and with a few marginal teeth, marked with transverse partitions; **flower-heads** 1-2.5 cm long, cylindrical, blue to purplish; **bracts** 1-2.5 cm long, 5-10, spreading, sharply pointed.

A widely distributed plant found on heavy clay soils subject to periodic flooding. Plants are very prickly but are appealing and colourful when in flower. Can be grown in a pot or as a garden plant in inland regions. Propagate from seed which may germinate best after a period of storage.

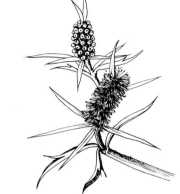

Eryngium plantagineum, × ·3

Eryngium rostratum, × ·01

Eryngium rostratum Cav.
(beaked)
Qld, NSW, Vic, Tas, SA   Blue Devil
0.1-0.6 m x 0.3-0.5 m   Aug-Feb

Low-growing annual or biennial **herb** of a metallic blue colouration; **stems** 10-60 cm tall, erect, stiff and rigid, much-branched; **leaves** 10-25 cm long, stiff, finely divided and longer than the stem-leaves; **flower-heads** 1.5-2 cm long, cylindrical to ovoid, blue; **bracts** about 2 cm long, 8-20 per head, rigid, purplish.

A widely distributed species found on moist, clayey soils, usually inland and often growing in colonies. The plants take on a very attractive bluish colour which intensifies as they mature. Blue devils are very attractive but are extremely prickly. They make a decorative container plant and can also be grown in a rockery or interspersed amongst shrubs in an open, sunny situation. They are very hardy to frost and dryness. Propagate from seed or by transplants.

Eryngium supinum J. Black
(prostrate)
Qld, NSW, SA
prostrate x 0.3-0.6 m   Oct-Feb

Prostrate or low-growing perennial **herb**; **stems** 10-40 cm long, stiff, cylindrical; **leaves** 10-20 cm x 0.6-1.8 cm, the basal ones largest and entire or with a few teeth near the apex, the stem-leaves shorter and broader and with 3 very sharp lobes; **flower-heads** 1.5-2.3 cm long, cylindrical, blue; **bracts** 0.5-0.7 cm long, 5-8 per head, spreading, sharply pointed.

A perennial herb found on low-lying clay soils subject to flooding for part of the year. Plants form a sprawling prickly clump but are interesting and colourful in flower. Can be grown in gardens and are attractive when planted amongst rocks. Needs to be flooded at intervals or plants may die out. Can be propagated by transplants or from cuttings or seed.

Eryngium vesiculosum Labill.
(bladder-like)
Qld, NSW, Vic, SA, Tas   Prickfoot
prostrate x 0.3-0.6 m   Oct-Feb

Perennial **herb** with short, prostrate **branches**; **radical leaves** 5-10 cm x 1-1.4 cm, oblanceolate, on long petioles, bright green, the margins with numerous coarse spiny teeth, in a rosette; **stem-leaves** shorter, divided into 3-5 prickly lobes; **flower-heads** about 1 cm long, pale blue, globular; **bracts** 1-2 cm long, 8-10 per head, pungent.

A widespread perennial herb forming small colonies in marshy situations. Plants develop a conspicuous rosette before sending out the flowering stems each season. Although the leaves are prickly, the plants make an excellent container specimen and are most decorative when in flower. For best results the base of the container should be submerged in water. Plants can also be established in an open, sunny garden situation but they need plenty of moisture and do well in a bog garden or around the margins of springs etc. They are very frost hardy. Propagate from seed, root cuttings or by transplants.

487

ERYTHRINA L.

(from the Greek *erythros*, red; in reference to the predominant flower colour of the genus)
Fabaceae

Shrubs or **trees**; trunk, branches and petioles often
armed with thorns; **timber** light but tough; **leaves**
alternate, trifoliolate; **leaflets** 2 or 3, broad, entire or
lobed; **inflorescence** an axillary or terminal raceme;
flowers pea-shaped, large, usually red, clustered; **calyx**
campanulate, entire or toothed; **standard** erect, recurved or incurved; **wings** short; **keel** short; **stamens**
united at the base; **fruit** a woody pod.

A large genus of about 100 species, well developed in
the tropics and subtropics with 3-6 species native to
Australia. Many exotic species, especially those from
Africa, are widely grown and prized for their massed
flowering during the warmer months of the year. These
are hardy plants, well suited to Australia's climate. The
native species are decorative but have not achieved the
recognition they deserve. Propagate from seed which
has a very hard coat and requires treatment prior to
sowing (see Volume 1, page 205). Because the seeds
are large, nicking or slicing the coat with a knife is
usually the easiest method. Some species can also be
propagated from cuttings taken while the plants are
leafless.

Erythrina phlebocarpa Bailey
(veined pods)

Qld   Coral Tree
6-12 m x 3-8 m   Aug-Dec

Small to medium **tree**; **branchlets** bearing numerous
small black prickles; **leaves** 25-40 cm long, trifoliolate;
leaflets 8-17 cm x 10-20 cm, rhomboidal, thin-
textured, light green, the terminal one much larger
than the lateral ones; **racemes** 15-35 cm long, terminal
on the branches; **flowers** 4-5 cm long, bright red,
usually in whorls of 4 in the raceme; **standard** about
4.5 cm x 1.4-1.6 cm, the margins infolded; **keel** about
1.2 cm long; **pods** 8-12 cm x 2-2.5 cm, constricted between the seeds, strongly veined, containing 1-2 seeds;
seeds about 1.4 cm long, red.

A tree found in vine thickets and along rainforest
margins of Cape York Peninsula. Plants are deciduous
during the dry season, and flower while leafless. The
flowers attract birds. The wood is soft and may have
been used by the Aborigines for shields, canoes etc.
This species is similar to a few exotic coral trees which
are cultivated in tropical and subtropical regions. It
grows readily in parks or gardens in an open, sunny
situation and well drained soil. Propagate from scarified seed or leafless cuttings which strike easily.

Erythrina variegata L. var. orientalis (L.) Merr.
(variegated; from the east)

Qld, NT   Coral Tree
5-10 m x 4-8 m   July-Nov

Tall **shrub** or small **tree**; **bark** brown, furrowed and
corky with age; **leaves** 10-15 cm long, trifoliolate;
leaflets 3-12 cm x 5-13 cm, variable in shape but
usually broadly 3-lobed, the lobes at right angles to the
main vein, the terminal leaflet larger than the others,
the leaves becoming yellow before falling; **racemes** 10-
23 cm long, erect, terminal on the branchlets; **flowers**

3-4 cm long, red, with a broad standard; **pods** 8-12 cm
x 0.8-1 cm, tapering to a fine point, containing 2-4
seeds; **seeds** about 1.2 cm long, yellow to dark red,
shiny.

A widespread Asian species which extends to parts
of northern Australia, where it grows in open woodland. The trees are deciduous for a long period in the
late dry season, usually flowering while leafless. Young
plants are spiny but older plants are usually unarmed,
although strong flushes of growth may bear spines.
The area where this species grows is subject to regular
fires but the trees are protected by the thick, corky
bark. The leaves are palatable to stock. Aborigines
hollowed the trunks of the trees for use as dugout
canoes. The flowers are attractive to birds. Horticulturally this species is an excellent hardy plant for
tropical, subtropical and drier inland regions. It will
grow in relatively poor soils but requires free drainage.
A sunny situation is essential. Propagate from seed
which needs nicking or scarification prior to sowing or
from cuttings which strike readily if taken while the
plants are leafless. Seedlings are reported to be susceptible to damping off fungus. Young trees transplant
readily while leafless. The typical form has variegated
leaves and, although not native to Australia, it is grown
here.

Erythrina verspertilio Benth.
(leaflets shaped like a bat's wings)

Qld, NSW, SA, WA, NT   Bat's Wing Coral Tree;
                               Bean Tree
6-12 m x 3-5 m   Aug-Nov

Medium **shrub** to small straggly **tree**; **bark** grey,
thick, furrowed and corky; **branches** bearing stout
black thorns; **leaves** 10-15 cm long, bifoliolate to trifoliolate; **leaflets** 7-12 cm x 5-12 cm, broadly wedge-
shaped, divided into 2 divergent linear lobes up to 6 cm
long, the opposite ones resembling a bat's open wings,
the terminal leaflet often longer and narrower than the
others but sometimes completely absent; **racemes** 5-
25 cm long, erect, terminal on the branchlets; **flowers**
3-4 cm long, scarlet to orange-red; **standard** about
3 cm long; **keel** about 1.2 cm long; **pods** 6-12 cm x 1.5-
1.8 cm, often uneven, black, bearing 2-8 seeds; **seeds**
about 1.2 cm long, scarlet, orange or yellow, glossy.

A very widespread species which grows in a variety
of situations from rainforest margins to harsh open
country. In very inhospitable conditions it may grow as
a low, stunted shrub, whereas in high-rainfall areas it
can form a substantial tree with a spreading canopy.
Trees flower mainly while they are leafless but sometimes a second flush of flowers occurs with the new
growth. The flowers vary in colour and there is scope
for selection of different forms. Aborigines used the
light but tough timber to make shields and troughs for
carrying food and water. They also collected and ate
the roots raw and used the flowers for decoration. The
flowers attract birds. The species is an excellent horticultural subject and deserves to be much more widely
grown. It demands free drainage around its roots but
will grow in a tremendous range of poor soils. Once
established, plants are very hardy to dryness but they
respond to watering by developing a denser canopy and
retaining their leaves longer. Plants require a sunny

Erythrina verspertilio foliage   D.L. Jones

situation and resent cool, misty or humid conditions. They succeed well in drier tropical, subtropical and inland regions. Propagate from seed which needs nicking or scarification prior to sowing. Cuttings may prove difficult but selected forms can be readily grafted on to seedlings. Young trees transplant readily while leafless. The species is somewhat variable in a number of features. Plants from Central Australia have bi-lobed leaves and may represent a distinct species.

Erythrina species
Qld   Pine Mountain Coral Tree
6-10 m x 3-6 m   Oct-Nov

Tall **shrub** or small **tree**; **bark** grey, corky; **trunk** with short thorny outgrowths; **branchlets** thorny; **leaves** 12-22 cm long, trifoliolate, on a long petiole; **leaflets** 6-12 cm x 4-10 cm, light green, the margins entire; **racemes** 8-15 cm long, terminal or in the upper axils, densely flowered; **flowers** 2.5-3.5 cm long, reddish, with a prominent white standard; **standard** reddish outside, white inside, curved; **keel** reddish, curved; **pods** 15-20 cm x 0.6-1 cm, black, constricted between the seeds; **seeds** 6-10 per pod, shiny red, mature Jan-Feb.

An undescribed species restricted to the Pine Mountain-Mount French area near Ipswich. It grows in free-draining, stony soils in stunted, dry rainforest, where it may be an emergent tree. Plants are deciduous for a period prior to and during flowering. Birds are attracted to the flowers. This is an extremely decorative coral tree and is well worthy of culture in subtropical and adjacent inland regions. Has been introduced into cultivation by enthusiasts, and prefers a sunny situation in very freely draining soils. Propagate from seed which is hard and needs treatment to germinate.

ERYTHROPHLEUM Afzel. ex G. Don
(from the Greek *erythros*, red; *phloios*, bark; in reference to red sap prominent in an African species)
Caesalpiniaceae

Trees; **sap** often red; **bark** hard; **leaves** bipinnate; **pinnae** in opposite pairs; **pinnules** alternate, simple, entire; **inflorescence** a terminal panicle; **calyx lobes** 5; **petals** 5; **stamens** 10, free; **fruit** a leathery pod.

A genus of about 17 species found in the drier tropical regions of Africa and Asia with a solitary species endemic in northern Australia. Propagate from seed which is hard and needs treatment to germinate or from root cuttings.

Erythrophleum chlorostachys (F. Muell.) Baillon
(lime green spikes)
Qld, WA, NT   Cooktown Ironwood;
                 Northern Ironwood
8-15 m x 3-8 m   Nov-March

Small to medium **tree** with a wide, spreading canopy; **bark** dark grey to blackish, deeply furrowed to flaky; **timber** dark brown, heavy, very hard; **leaves** bipinnate; **pinnae** 4-8 cm long, in 2-3 opposite pairs; **leaflets** 3-5 cm x 2-3 cm, 4-9 on each pinna, oval to rounded, dark green and leathery, the apex blunt or slightly notched; **panicles** 2.5-7.5 cm long, terminal; **flowers** about 0.6 cm long, lime green to pale yellow, crowded on the racemes; **pods** 10-15 cm x 2.5-4 cm, dark brown to black, brittle; **seeds** dark brown, 3-6 per pod.

A tough plant widely distributed through northern Australia in open forest country. It usually grows in sandy soils, and the plants shed their leaves during dry periods but quickly produce fresh new growth following rains. The timber is extremely hard and difficult to cut, hence the common name. It is highly valued for fence-posts and structural purposes since it is resistant to termite attack. It also makes an excellent fire-wood. The leaves are extremely poisonous to animals and there are many cases of heavy mortalities caused by this species. Even dry leaves are poisonous, and the sucker growth is extremely toxic. Ironwood is an outstanding hardy tree for dry inland and tropical regions, producing excellent shade and withstanding adversity and neglect. Plants retain their leaves longer if watered, and respond to fertilizers. Because of the extreme toxicity to grazing animals, the species cannot be

Erythrophleum chlorostachys   D.L. Jones

planted anywhere near where they feed. Plants tolerate a fair degree of coastal exposure. They require a sunny situation and well drained soil. Propagate from scarified seed, by transplanting suckers or seedlings or from root cuttings.

Erythroxylaceae Kunth.

A family of dicotyledons consisting of 2 genera and about 250 species found mainly in tropical and subtropical zones with a concentration of development in South America and Madagascar. They are trees or shrubs with alternate, stipulate leaves. Many species produce valuable timber, and the leaves are used for extraction of essential oils and the alkaloid cocaine which is valued in medicine for its narcotic properties. The family is represented in Australia by 3 species in the genus *Erythroxylum*.

ERYTHROXYLUM P. Browne

(from the Greek *erythros*, red; *xylon*, wood; the wood of some species is red)
Erythroxylaceae

Shrubs or **trees**; **leaves** entire, alternate, stipulate; **stipules** united at the base, encircling the petiole, deciduous or persistent; **flowers** small, whitish, solitary or clustered in axils; **sepals** 5 or 6, united or free; **petals** 5 or 6, with a 2-lobed appendage; **stamens** 10 or 12; **fruit** a drupe.

A genus of about 250 species of tropical trees and shrubs with a preponderance in South America. Three species are found in northern and eastern Australia. They are rare in cultivation and have limited horticultural appeal. Propagate from seed which has limited viability and must be sown fresh.

Erythroxylum australe F. Muell.

(southern)
Qld, NSW
6-12 m x 3-6 m   Aug-Oct

Tall **shrub** or small **tree** with slender, spreading branches; **leaves** 0.5-2.5 cm x 0.3-0.8 cm, elliptical to oblong, rounded at the apex, tapering to the stalk; **flowers** about 0.2 cm across, greenish, solitary or clustered in the leaf axils; **drupe** 0.5-0.7 cm long, oblong, yellowish, 3-celled.

A widely distributed species extending from northeastern Qld to north-eastern NSW, usually in drier scrubs and rainforests. The leaves contain a yellow dye and the red wood is tough and very attractive when polished. The species has limited horticultural appeal. It can be easily grown in well drained soil and is adaptable to a variety of conditions. Hardy once established, although young plants may need some protection. Propagate from seed which is best sown fresh.

Erythroxylum ecarinatum Burck

(without a keel)
Qld
8-15 m x 3-5 m   July-Oct

Small **tree**; **trunk** slender; **branchlets** with prominent lenticels, the bases of leaf-stalks ringed by scars from fallen stipules; **leaves** 5-10 cm x 1.5-4 cm, lanceolate, bright green, the apex drawn out into a point; **flowers** about 0.2 cm across, bell-shaped, white, in groups of 3-7 in the axils; **drupe** 1-1.5 cm long, oval, yellowish, mature Feb-March.

A fairly uncommon tree in Australia, restricted to highland rainforests of the Atherton Tableland. It is also found in New Guinea. The timber is durable in the ground, and the trunks are used for tank stands, fences etc. The tree has an attractive habit and could be usefully planted in subtropical and perhaps temperate gardens. Its cultural requirements are largely unknown but young plants need some protection from direct sun. Propagate from seed which germinates best if sown fresh.

Erythroxylum ellipticum R. Br. ex Benth.

(elliptical)
Qld, NT
6-10 m x 3-5 m   Oct-Dec

Tall **shrub** or small **tree**; **bark** corky; **branchlets** flattened; **leaves** 3-6 cm x 1-2 cm, ovate to obovate or elliptical, blunt, thin-textured; **flowers** about 0.4 cm across, white, in axillary clusters of 3-6, on pedicels up to 0.8 cm long; **drupe** 0.6-0.8 cm long, oblong, reddish, with sweet pulp.

This species is restricted to dry scrubs of the Burke and Cook Districts of northern Qld and the Top End of NT. Plants may become deciduous during long dry periods. The timber is very durable and useful for fence-posts, and is also excellent for cabinet work. The species is virtually unknown in cultivation but could be useful for drier tropical and adjacent inland regions. Requires well drained soil. Propagate from seed which must be sown fresh.

Escalloniaceae Dum.

A small family of dicotyledons consisting of 150 species in 7-8 genera found in tropical and temperate zones with a proliferation in South America. They are woody shrubs with opposite or alternate, often toothed leaves, and racemes of flowers. The Australian representation comprises 7 genera and 15 species, most of which occur in rainforests. A number of exotic species of *Escallonia* are common garden shrubs in temperate Australia but few native species of the family are grown. Members of this family may be included in Saxifragaceae by some botanists. Australian genera: *Abrophyllum, Anopterus, Argophyllum, Corokia, Cuttsia, Polyosma* and *Quintinia*.

Darwin

Gardner

BOTANICAL PROVINCE

Fitzgerald

Victoria River District

Dampier

NORTHERN

Hall

NORTHER

Mueller

Canning

Fortescue

WESTERN

AUSTRALIA

Central Austra

Carnegie

Keartland

EREMAEAN BOTANICAL PROVINCE

Giles

Carnarvon

Ashburton

North Western

Helms

Austin

SOL

Irwin

Eucla

Nullarbor

SOUTH-WEST BOTANICAL

SOUTH-WESTERN
INTERZONE

Avon

Coolgardie

Roe

PROVINCE

Darling

Eyre

Botanical Regions of Australia

Glossary of Technical Terms

abaxial On the side of a lateral organ away from the axis; the lower side of a leaf or petiole.

aberrant Unusual or atypical; differing from the normal form.

abrupt Changing suddenly rather than gradually.

abscise To shed or throw off.

abscission Shedding of plant parts, eg. leaves. This may be natural resulting from old age or premature as a result of stress.

acaulescent Without a trunk.

accessory buds Lateral buds associated with a main bud such as in a leaf axil. They usually develop only if the main bud is damaged.

accessory roots Lateral roots developed from the base of the trunk (as in palms) as opposed to those arising from the seed root system.

acerose Very slender or needle-shaped.

achene A small dry one-seeded fruit which does not split at maturity, eg. *Clematis, Senecio, Helichrysum.*

acicular Needle-shaped.

actinomorphic Symmetrical and regular; usually applied to flowers, eg. *Wahlenbergia.*

aculeate Bearing short sharp prickles.

acuminate Tapering into a long drawn out point.

acute Bearing a short sharp point.

adaxial On the side of a lateral organ next to the axis; the upper side of a leaf or petiole.

adnate Fused tightly together so that separation without damage is impossible.

adventitious Arising in irregular position, eg. adventitious roots, adventitious buds.

aerial roots Adventitious roots arising on stems and growing in the air.

aff., affinity A botanical reference used to denote an undescribed species closely related to an already described species.

after-ripening The changes that occur in a dormant seed and render it capable of germinating.

aggregate fruit A fruit formed by the coherence of ovules that were distinct while in the flower, eg. *Rubus.*

albumen An old term used for the endosperm of seeds.

alternate Borne at different levels in a straight line or in a spiral.

amino acid An organic compound which is a structural unit of protein.

androecium Collectively the male parts of a flower, ie. the stamens.

angiosperm A major group of plants which bear seeds within an ovary.

annual A plant completing its life cycle within 12 months.

annular Prominent ring scars left on the trunk of certain palms after leaf fall, eg. *Archontophoenix.*

annulate Bearing annular rings on the trunk.

anomalous An abnormal or freak form.

anther The pollen-bearing part of a stamen.

anthesis The process of flowering.

apetalous Without petals.

apical dominance The dominance of the apical growing shoot which produces hormones and prevents lateral buds developing while it is still growing actively.

apiculate With a short pointed tip.

apomixis Seed development without the benefit of sexual fusion.

appendage A small growth attached to an organ.

appressed Pressed flat against something.

aquatic A plant growing wholly or partially submerged in water.

arborescent With a tree-like growth habit.

arboretum A collection of planted trees.

arcuate Arched.

aril A fleshy or papery appendage produced as an outgrowth from the outer coat of a seed.

aristate Bearing a small bristle.

articulate Jointed or having swollen nodes, eg. the culms of grasses.

asexual reproduction Reproduction by vegetative means without the fusion of sexual cells.

attenuated Drawn out.

auricle An ear-like appendage, eg. surrounding the bases of the fronds of *Angiopteris.*

auriculate Bearing auricles.

auxin A growth-regulating compound controlling many growth processes such as bud-break, root development, seed germination etc.

awn A bristle-like appendage, eg. on the seeds of many grasses.

axil Angle formed between adjacent organs in contact; commonly applied to the angle between a leaf and the stem.

axillary Borne within the axil.

axis The main stem of a plant or part of a plant.

barbed Bearing sharp backward-sloping hooks as on *Calamus* or *Rubus.*

bearded Bearing a tuft of hairs.

berry A simple, fleshy, many-seeded fruit with 2 or more compartments which do not split open when ripe.

biennial A plant completing its life cycle in 2 years, usually growing vegetatively in the first year.

495

Glossary of Technical Terms

bifid Deeply notched for more than half its length.

bifoliolate With 2 leaflets to a leaf.

bifurcate Forked into 2 parts.

bi-lobed Two-lobed.

bilocular With two cavities.

bipinnate Twice pinnately divided.

bisexual Both male and female sexes present.

blade The expanded part of a leaf.

bole The trunk of a tree.

bottom heat A propagation term used to denote the application of artificial heat in the basal region of the cutting.

bract Leaf-like structure which subtends a flower-stem or inflorescence.

bracteole A small leaf-like structure found on a flower-stem.

branchlet A small slender branch.

bristle A short stiff hair.

bulb An enlarged thickened stem containing a bud surrounded by thickened leaf scales.

bulbil A small bulb produced in a leaf axil; a specialized bud produced at the junction of main veins on the fronds of certain ferns.

bulbous Bulb-shaped.

burr A prickly fruit.

bush A low, thick shrub, usually without a distinct trunk.

caducous Falling off early.

caespitose Growing in a tuft or tussock.

calcareous An excess of lime, as in soil.

callus Growth of undifferentiated cells; in orchids an organelle developed on the labellum.

calyx All of the sepals.

cambium The growing tissue lying just beneath the bark.

campanulate Bell-shaped.

cane A reed-like plant stem.

canopy The cover of foliage of a tree or community.

capitate Enlarged and head-like.

capitulate An inflorescence consisting of sessile flowers in a head, as in the family Asteraceae.

capsule A dehiscent dry fruit containing many seeds.

carinate Bearing a keel.

carpel Female reproductive organ.

catkin A pendent spike-like inflorescence composed of unisexual apetalous flowers; often used loosely for any rod-like inflorescence.

caudex A trunk-like axis in monocotyledons and some ferns.

cauliflory The production of flowers from the trunk and larger branches.

cauline Attached to the stem.

chlorophyll The green pigment of leaves and other organs, important as a light-absorbing agent in photosynthesis.

ciliate With a fringe of hairs.

cirrus A whip-like organ bearing recurved hooks used as an aid for climbing; it arises as an extension of the leaf rhachis and is present in some species of *Calamus*.

cladode A stem modified to serve as a leaf.

clavate Club-shaped.

claw The narrowed base of a sepal or petal; the staminal bundles of the flowers of the genus *Melaleuca*.

cleistogamous A term applied to self pollinating flowers which do not open, eg. *Viola*.

clone A group of vegetatively propagated plants with a common ancestry.

coccus A single unit of a multiple fruit which splits at maturity.

column A fleshy growth in the flowers of orchids formed by the union of the stigmas and stamens.

compound leaf A leaf with two or more separate leaflets.

compressed Flattened laterally.

cone A woody fruit in gymnosperms formed by sporophylls arranged spirally on an axis; other woody fruits such as those of the genus *Banksia* and *Casuarina* are also called cones.

confluent Running together; said of compound leaves where the leaflets remain united and do not separate.

congested Crowded closely together.

conifer A cone-bearing tree with needle-shaped or scale-like leaves; a gymnosperm.

connate Fused or joined together.

contorted Twisted.

contracted Narrowed.

contractile root A specialized root developed by bulbs to maintain the bulb at a suitable level in the soil.

convoluted Rolled around and overlapping as in the leaves of a young shoot.

coppice shoot A shoot developing from a dormant bud in the trunk or larger branches of a tree; a very common feature of eucalypts.

cordate Heart-shaped.

coriaceous Leathery in texture.

corolla All of the petals.

corymb An inflorescence where the branches start at different points but reach about the same height to give a flat-topped effect.

costa The rib of a costapalmate leaf.

costapalmate Palmate leaves with a well developed rib which is an extension of the petiole into the blade and is the equivalent of the rhachis, eg. *Corypha elata*.

cotyledon The seed leaf of a plant.

crenate The margin cut regularly into rounded teeth.

crenulate The margin cut into fine, rounded teeth.

crisped The margins very wavy or crumpled.

cross Offspring or hybrid.

cross-fertilization Fertilization by pollen from another flower.

cross-pollination Transfer of pollen from flower to flower.

crown That part of a shrub or tree above the first branch on the trunk.

crownshaft A series of tightly packed specialized tubular leaf-bases which terminate the trunk of some pinnate-leaved palms.

crozier Young coiled fern fronds.

culm Flowering stem of grasses or sedges.

cultivar A horticultural variety of a plant or crop.

cuneate Wedge-shaped.

cupule A bowl or cup-shaped calyx developed at the base of some fruit.

cymbiform Boat-shaped.

cyme An inflorescence where the branches are opposite

and the flowers open sequentially downwards starting from the terminal of each branch.

cymose Divided like a cyme.

damping off A condition in which young seedlings are attacked and killed by soil-borne fungi.

deciduous Falling or shedding of any plant part.

decumbent Reclining on the ground but with the tips ascending, as of branches.

decurrent Running downwards beyond the point of junction, as of leaves, phyllodes, leaflets, lobes etc.

decussate Opposite leaves in four rows along the stem.

deflexed Abruptly turning downwards.

dehiscent Splitting or opening when mature.

dentate Toothed.

denticulate Finely toothed.

depauperate A weak plant or one imperfectly developed.

depressed Flattened at the end.

determinate With a definite cessation of growth in the main axis.

dichotomous Regularly forking into equal branches.

dicotyledons A section of the Angiosperms bearing two seed leaves in the seedling stage.

diffuse Widely spreading and much branched; of open growth.

digitate Spreading like the fingers of a hand from one point.

dimorphic Existing in two different forms.

dioecious Bearing male and female flowers on separate plants.

diploid Having two sets of chromosomes.

disc floret The tubular flowers in the centre of heads of the family Asteraceae.

dissected Deeply divided into segments.

distichous Alternate leaves arranged along the axis in two opposite rows.

divaricate Widely spreading and straggling.

divided Separated to the base.

domatia Small structures in the leaves of some rainforest plants; they occur in the axils of the midrib and main lateral veins and may be either little tufts of hair or small, sunken, hooded enclosures (see *foveole*).

dormancy A physical or physiological condition that prevents growth or germination even though external factors are favourable.

dorsal sepal Sepal subtending the column of orchid flowers.

drupaceous Drupe-like.

drupe A fleshy indehiscent fruit with seed(s) enclosed in a stony endocarp.

drupelet One drupe of an aggregate fruit made up of drupes, eg. *Rubus*.

dune A mound formed from wind-blown sand.

ebracteate Without bracts.

ecology Study of the interaction of plants and animals within their natural environment.

effuse Very open and loosely spreading.

elliptic Oval and flat and narrowed to each end which is rounded.

elongate Drawn out in length.

emarginate Having a notch at the apex.

embryo Dormant plant contained within a seed.

endemic Restricted to a particular country, region or area.

endocarp A woody layer surrounding a seed in a fleshy fruit.

endosperm Tissue rich in nutrients surrounding the embryo in seeds.

ensiform Sword-shaped.

entire Whole; not toothed or divided in any way.

enzyme A specialized protein capable of promoting a chemical reaction.

ephemeral A plant completing its life cycle within a very short period, ie. 3-6 months.

epicarp The outermost layer of fruit.

epidermis The outer layer of cells which protects against drying and injury.

epigeal A term used for roots which grow above-ground.

epiphyte A plant growing on or attached to another plant but not parasitic.

equable A term used to describe the endosperm of seed when it is smooth and uniform.

erect Upright.

evergreen Remaining green and retaining leaves throughout the year.

exocarp Outermost layer of the fruit wall.

exotic A plant introduced from overseas.

exserted Protruding beyond the surrounding parts.

exstipulate Without stipules.

falcate Sickle-shaped.

family A taxonomic group of related genera.

farinaceous Containing starch; appearing as if covered by flour.

fasciculate Arranged in clusters.

ferruginous Rusty brown colour.

fertile bract Bract on an inflorescence which subtends the flowering branches (cf. *sterile bract*).

fertilization The act of union of the pollen gametes and egg cells in the ovule.

fetid Having an offensive odour.

fibrillose Bearing fine fibres or threads.

fibrose Containing fibres.

filament The stalk of the stamen supporting the anther.

filiform Long and very slender; thread-like.

fimbria The fringe; often applied to the fine hair-like fringes of a scale.

fimbriate Fringed with fine hairs.

flabellate Fan-shaped.

flabellum A term sometimes applied to the united pair of terminal leaflets of a pinnate leaf, eg. *Hydriastele*.

flaccid Soft, limp, lax.

flagellum A whip-like organ that bears curved hooks and is used as an aid to climbing; it is a modified inflorescence and arises in a leaf axil, eg. *Calamus*.

flexuose Having a zigzag form.

floccose Having tufts of woolly hairs.

flora The plant population of a given region; also a book detailing the plant species of an area.

floral leaf A specialized leaf subtending flowers or an inflorescence and differing from normal foliage leaves.

floret The smallest unit of a compound flower.

floriferous Bearing numerous flowers.

foliaceous Leaf-like.

foliolate Bearing leaflets.

497

Glossary of Technical Terms

follicle A dry fruit formed from a single carpel and which splits along one line when ripe.

forest A plant community dominated by trees.

forked Divided into nearly equal parts.

form A botanical division below a species.

foveole A sunken, often hooded structure found on the leaves of some rainforest plants (see *domatia*).

free Not joined to any other part.

frond Leaf of a fern or palm.

fruit The seed-bearing organ developed after fertilization.

fruitlet Small fruits forming part of an aggregate fruit, eg. *Rubus*.

fugacious Falling or withering away very early.

fungicide A chemical used to control fungus diseases.

fused Joined or growing together.

fusiform Spindle-shaped; narrowed to both ends from a swollen middle.

galea A hood or helmet-shaped structure formed by fusion of petals and sepals, eg. *Pterostylis*.

gamete One of the sex cells, either male or female.

gemma A vegetative bud by which a plant propagates.

gene A hereditary factor located in linear order on a chromosome.

geniculate Bent like a knee.

genus A taxonomic group of closely related species.

germination The active growth of an embryo resulting in the development of a young plant.

gibbous Humped.

glabrous Without hairs; smooth.

gland A fluid-secreting organ.

glandular Bearing glands.

glaucous Covered with a bloom giving a bluish lustre.

globoid Globe-like, globular, spherical.

globose Globular; almost spherical.

glume The bract subtending the spikelets of grasses and sedges.

glutinous Covered with a sticky exudation.

granular Covered with small grains.

growth regulator A synthetic compound which can control growth and flowering responses in plants and seeds.

growth split A vertical crack or split that develops in the trunk of a fast growing tree.

gymnosperm A major group of plants which bear seeds not enclosed within an ovary.

gynoecium Collectively the female parts of a flower.

habit The general appearance of a plant.

habitat The environment in which a plant grows.

halophyte A plant which grows in saline soils.

hapaxanthic A term describing clumping palms, individual stems of which die after flowering (cf. *monocarpic, pleonanthic*).

haploid Having a single set of chromosomes.

hastate Shaped like an arrow-head and with spreading basal lobes.

haustorial Said of a parasite which is able to absorb water and nutrients directly from a host by a specialized attachment.

head A composite cluster of flowers, as in the family Asteraceae.

herb A plant which produces a fleshy rather than woody stem.

herbaceous A perennial plant which dies down each year after flowering.

herbicide A chemical used to control weeds.

hermaphrodite Bearing both male and female sex organs in the same flower.

hilum The scar left on the seed at its point of detachment from the seed stalk.

hirsute Covered with long, spreading coarse hairs.

hispid Covered with stiff bristles or hairs.

hispidulous Minutely hispid.

hoary Covered with short white hairs giving the surface a greyish appearance.

hormone A chemical substance produced in one part of a plant and inducing a growth response when transferred to another part.

hybrid Progeny resulting from the cross-fertilization of parents.

hybrid swarm Variable population resulting from complex crossing such as between the hybrids themselves or between the hybrids and the parents.

hybrid vigour The increase in vigour of hybrids over their inbred parents.

hypocotyl Part of the embryo between the cotyledons and primary root.

imbricate Overlapping.

imparipinnate Pinnate leaves bearing a single terminal leaflet which extends from the end of the rhachis.

incised Cut sharply and deeply.

incurved Curved inwards.

indehiscent Not splitting open at maturity.

indeterminate Growing on without termination.

indigenous Native to a country, region or area.

indumentum The hairy covering on plant parts.

induplicate Leaflets folded longitudinally with the Vs (leaf-margins) opened upwards.

indusium A membrane covering a fern sorus; a cup-shaped structure protecting the stigma in Goodeniaceae.

inferior Below some other part; often used in reference to an ovary when held below other layers of the perianth.

inflorescence The flowering structure of a plant.

infructescence A term used to describe a fruiting inflorescence.

inhibitor A chemical substance which prevents a growth process.

insecticide A chemical used to control insect pests.

internode The part of a stem between two nodes.

involucre A cluster of overlapping bracts surrounding the base of the flower-heads.

involute Rolled inwards.

jointed Bearing joints or nodes.

juvenile The young stage of growth before the plant is capable of flowering.

keel A ridge like the base of a boat; in pea-shaped flowers the basal part formed by the union of two petals.

labellum A lip; in orchids the petal in front of the column.

498

lacerate Irregularly cut or torn into narrow segments.

laciniate Cut into narrow slender segments.

lamina The expanded part of a leaf.

lanceolate Lance-shaped; narrow and tapering at each end, especially the apex.

lateral Arising from the main axis; arising at the side of.

lax Open and loose.

leaf-base Specialized expanded and sheathing part of the petiole where it joins the trunk, as in palms.

leaflet A segment of a compound leaf.

legume A dry fruit formed from one carpel and splitting along two lines.

lemma The lower bract enclosing the flower of grasses.

lepidote Dotted with persistent, small, scurfy, peltate scales, as on some palm leaves.

liana, liane A large woody climber.

lignotuber A woody swelling containing dormant buds at the base of a trunk, eg. mallee eucalypts.

ligule A strap-shaped organ; in grasses a growth at the junction of leaf-sheath and blade.

ligulate Strap-shaped.

linear Long and narrow with parallel sides.

littoral Growing in communities near the sea.

lobe A segment of an organ as the result of a division.

loculus A compartment within an ovary.

loricate Covered with overlapping scales, eg. *Calamus* fruits.

mallee A shrub or tree with many stems arising from at or below ground level.

mangrove A specialized plant growing in salt water and gathering oxygen through specialized roots (pneumatophores).

marcescent Short-lived and withering while still attached to the plant.

marginal Attached to or near the edge.

maritime Belonging to or growing near the sea.

marlock A shrub or tree with waxy stems arising from a point on the trunk above ground level.

marsh A swamp.

mealy Covered with flour-like powder.

membranous Thin-textured.

meristem A growing point or an area of active cell division.

merous The number of parts of a flower that makes up any whorl, eg. petals, stamens; usually written 4-merous etc.

mesocarp Middle layer of a fruit wall.

midrib The principal vein that runs the full length.

miticide (also *acaracide*) A chemical used to control mites.

moniliform Constricted at regular intervals and appearing bead-like.

monocarpic A term describing plants which flower once, then die (cf. *hapaxanthic, pleonanthic*).

monocotyledons A section of the Angiosperms bearing a single seed leaf at the seedling stage.

monoecious Bearing male and female flowers on the same plant.

monopodial A stem with a single main axis which grows forward at the tip.

monotypic A genus with a single species.

morphology The form and structure of a plant.

mucronate With a short sharp point.

mucronulate With a very small point.

mutation A change in the genetic constitution of a plant or part of a plant.

myccorrhiza A beneficial relationship between the roots of a plant and fungi or bacteria resulting in a nutrient exchange system. Some plants cannot grow without such a relationship.

nectar A sweet fluid secreted from a nectary.

nectary A specialized gland which secretes nectar.

nematicide A chemical used to control nematodes.

nematode A minute worm-like animal, some species of which attack plants.

nerves The fine veins which traverse the leaf-blade.

node A point on the stem where leaves or bracts arise.

nodule A small swollen lump on roots; in legumes the nodules contain symbiotic bacteria of the genus *Rhizobium*.

nut A dry indehiscent one-seeded fruit.

nutlet A small nut enclosing a single seed.

obcordate Cordate with the broadest part above the middle.

oblanceolate Lanceolate with the broadest part above the middle.

oblate Nearly spherical, but noticeably broader than long; flattened from above.

oblique Slanting; unequal-sided.

obovate Ovate with the broadest part above the middle.

obtuse Blunt or rounded at the apex.

offset A growth arising from the base of another plant.

olivaceous Dark olive green.

operculum A structure formed from the fusion of petals and sepals and protecting the stamens and style when in bud, eg. *Eucalyptus*.

opposite Arising on opposite sides but at the same level.

orbicular Nearly circular.

order A taxonomic group of related families.

organelle A small plant organ.

osmosis Diffusion of water through a membrane caused by different concentrations of salts on either side of the membrane.

oval Rounded but longer than wide.

ovary The part of the gynoecium which encloses the ovules.

ovate Egg-shaped in longitudinal section.

ovoid Egg-shaped.

ovule The structure within the ovary which becomes seed after fertilization.

palea The upper bract enclosing the flower of grasses.

paleaceous Clothed with papery scales.

palmate Divided like a hand.

palmatifid Lobed like a hand.

palmet A dwarf-growing palm, eg. *Linospadix* spp.

panicle A much-branched racemose inflorescence.

paniculate Arranged in a panicle.

pappus A tuft of feathery bristles representing a modified calyx, on the seeds of Asteraceae.

parasite A plant growing or living on or in another plant, eg. mistletoe.

paripinnate Compound pinnate leaves lacking a terminal leaflet.

parthenocarpy Development of fruit without fertilization and seed formation.

Glossary of Technical Terms

patent Spreading out.

pectinate Shaped like a comb.

pedicel The stalk of a flower in a compound inflorescence.

pedicellate Growing on a pedicel.

peduncle The main axis of a compound inflorescence or the stalk of a solitary flower.

pedunculate Growing on a peduncle.

peltate Circular with the stalk attached in the middle on the undersurface.

pendent Hanging downwards.

penninerved The veins branching pinnately.

perennial A plant living for more than two years.

perfoliate United around the stem, as in leaves.

perianth A collective term for all of the petals and sepals of a flower.

pericarp The hardened ovary wall that surrounds a seed.

persistent Remaining attached until mature; not falling prematurely.

pesticide A chemical used to control pests. The word pests in this case refers to a range of plant enemies including fungi, bacteria, nematodes, insects, mites etc.

petal A segment of the inner perianth whorl or corolla.

petaloid Petal-like; resembling a petal.

petiole The stem or stalk of a leaf.

petiolate Bearing a petiole.

phloem Part of the vascular system of plants concerned with the movement and storage of nutrients and hormones.

photosynthesis The conversion of carbon dioxide from the atmosphere to sugars within green parts of the plant, using chlorophyll and energy from the sun's rays.

phylloclade A stem acting in the capacity of a leaf (cladode).

phyllode A modified petiole acting as a leaf.

pilose With scattered long simple hairs.

pinna A primary segment of a divided leaf.

pinnate Once-divided with the divisions extending to the midrib.

pinnatifid Once-divided with the divisions not extending to the midrib.

pinnule The segment of a compound leaf divided more than once.

pistil The ovule and seed-bearing organ of the gynoecium.

pistillate flowers Female flowers.

pistillode A sterile pistil, often found in male flowers.

pleonanthic Plants which flower regularly each year after reaching maturity (cf. *hapaxanthic, monocarpic*).

plicate Folded longitudinally.

plumose Feather-like from fine feathery hairs.

pneumatophore Specialized roots of mangroves carrying oxygen to the plant.

pod A dry non-fleshy fruit that splits when ripe to release its seeds.

pollen Haploid male cells produced by the anthers.

pollination The transference of the pollen from the anther to the stigma of a flower.

pollinium An aggregated mass of pollen grains found in the Orchidaceae and Asclepiadaceae.

polyembryony The condition where a seed produces more than one embryo.

polygamous Having mixed unisexual and bisexual flowers together.

polymorphic Consisting of many forms; a variable species, eg. *Grevillea glabella*.

praemorse As though bitten off, as in the leaflet tips of *Ptychosperma*.

prickle A small spine borne irregularly on the bark or epidermis.

procumbent Spreading on the ground without rooting, as of branches.

proliferous Bearing offshoots and other processes of vegetative propagation.

prophyll The first (or outer) sheathing bract of a palm inflorescence.

prostrate Lying flat on the ground.

proteoid roots A specialized root development found in species of the family Proteaceae. Numerous short roots develop in a compact mop-like clump.

protuberance A swelling or bump.

pseudobulb Thickened bulb-like stems of orchids bearing nodes.

pubescent Covered with short soft downy hairs.

pulvinus A cushion-like growth of inflated cells on the leaf or leaflet-stalk at its junction with the blade, or on spines or inflorescence branches, usually in pairs.

punctate Marked with spots or glands.

pungent Very sharply pointed; also smelling strongly.

pyriform Pear-shaped.

raceme A simple unbranched inflorescence with stalked flowers.

racemose In the form of a raceme.

radical Arranged in a basal rosette.

radicle The undeveloped root of the embryo.

ray floret The outermost, flattened florets of the inflorescence in the family Asteraceae.

recurved Curved backwards.

reduplicate Leaflets folded with the Vs (leaf-margins) opened downwards.

reflexed Bent backwards and downwards.

regular Symmetrical, especially of flowers.

reniform Kidney-shaped.

repent With a creeping growth habit.

reticulate A network, as of veins.

retuse With a slight notch at the apex.

revolute With the margins rolled backwards.

rhachilla A small rhachis; the secondary and lesser ones of a compound inflorescence.

rhachis The main axis of a compound leaf or an inflorescence.

rhizome An underground stem.

rib The section of the petiole of a costapalmate leaf that extends into the blade.

rosette A group of leaves radiating from a centre.

rostrate With a beak.

rotate The lobes of the corolla spreading horizontally like the spokes of a wheel.

rugose Wrinkled.

rugulose Finely wrinkled.

ruminate Folded like a stomach-lining; used to describe the folds in the endosperm of some seeds.

runner A slender trailing shoot forming roots at the nodes.

saccate Pouch or sac-like.

sagittate Shaped like an arrow-head, with the basal lobes pointing downwards.

samara An indehiscent winged seed.

saprophyte A plant which derives its food from dead or decaying organic matter.

scabrous Rough to the touch.

scale A dry, flattened, papery body; sometimes also used as a term for rudimentary leaf.

scandent Climbing.

scape Leafless peduncle arising near the ground; it may bear scales or bracts and the foliage leaves are radical.

scarious Thin, dry and membranous.

sclerophyll A plant (or forest) with hard stiff leaves.

scrub Strictly a plant community dominated by shrubs; often used loosely for rainforests.

scurfy Bearing small, flattened, papery scales.

secondary thickening The increase in trunk diameter as the result of cambial growth.

section A taxonomic subgroup of a genus containing closely related species.

secund With all parts directed to one side.

seed A mature ovule consisting of an embryo, endosperm and protective coat.

seed-coat The protective covering of a seed; also called testa.

seedling A young plant raised from seed.

segment A subdivision or part of an organ, eg. sepal is a segment of the calyx.

self pollination Transfer of pollen from stamen to stigma of same flower.

sepal A segment of the calyx or outer whorl of the perianth.

sepaloid Sepal-like.

septate Divided by partitions.

serrate Toothed with sharp, forward-pointing teeth.

serrulate Finely serrate.

sessile Without a stalk, pedicle or petiole.

seta A bristle.

setaceous Shaped like a bristle.

setose Bristly.

shrub A woody plant that remains low (less than 6 m) and usually with many stems or trunks.

simple Undivided; of one piece.

sinuate With a wavy margin.

sinus A junction; a specialized term for flowers of the orchid genus *Pterostylis* describing the point of junction of the lateral sepals.

slip A cutting.

soboliferous Bearing creeping, rooting stems; sometimes interpreted as bearing suckers, and applied to clumping plants.

sorus A cluster of sporangia on the fronds of ferns.

spadix An inflorescence which is a fleshy spike with flowers more or less sunken in the axis, and usually enclosed by a spathe.

spathe A large sheathing bract which encloses a spadix.

spathulate Spatula-shaped; with a broad top and tapering base.

spear-leaf The erect, unopened young leaf of a palm.

species A taxonomic group of closely related plants, all possessing a common set of characters which set them apart from another species.

spicate Arranged like or resembling a spike.

spike A simple unbranched inflorescence with sessile flowers.

spikelet A small spike bearing one or more flowers; in grasses a small unit composed of glumes and florets.

spine A sharp rigid structure.

spinescent Bearing spines or ending in spines.

spinule A weak spine.

spinulose With small spines.

sporangium A case that bears spore.

spore A reproductive unit which does not contain an embryo.

sporeling A young fern plant.

spur A tubular sac-like projection of a flower, often containing nectar.

stamen The male part of a flower producing pollen, consisting of an anther and a filament.

staminate flowers Male flowers.

staminode A sterile stamen; often of different form, eg. petaloid.

standard The dorsal petal, usually confined to flowers of the family Fabaceae.

stellate Star-shaped or of star-like form.

stem-clasping Enfolding a stem.

sterile Unable to reproduce.

sterile bracts Bracts on an inflorescence which do not subtend a flower or flowering branch (cf. *fertile bracts*).

stigma The usually enlarged area of the style receptive to pollen.

stipe A stalk or leaf-stalk.

stipitate Stalked.

stipule Small bract-like appendage borne in pairs at the base of the petiole.

stolon A basal stem growing just below the ground surface and rooting at intervals.

stoloniferous Bearing stolons; spreading by stolons.

strain An improved selection within a variety; also cultivar.

strand plant A plant growing near the sea.

stratification The technique of burying seed in coarse sand so as to expose it to periods of cold temperatures or to soften the seed-coat.

striate Marked with narrow lines or ridges.

strobilus A cone.

strophiolate Bearing a strophiole.

strophiole An appendage arising from the seed-coat near the hilum.

style Part of the gynoecium connecting the stigma with the ovary.

sub Beneath, nearly, approximately.

subclimber Shrubby plant that can produce long shoots which need support of surrounding plants.

subcordate Nearly cordate.

subfamily A taxonomic group of closely related genera within a family.

subspecies A taxonomic subgroup within a species used to differentiate geographically isolated variants.

subulate Narrow and drawn out to a fine point.

succulent Fleshy or juicy.

sucker A shoot arising from the roots or the trunk below ground level.

sulcate Grooved or furrowed.

Glossary of Technical Terms

superior Above some other part; often used in reference to an ovary when held above other layers of the perianth.

symbiosis A beneficial association of different organisms.

taproot The perpendicular main root of a plant.

taxon A term used to describe any taxonomic group, eg. genus, species.

taxonomy The classification of plants or animals.

tendril A plant organ modified to support stems used in climbing.

tepal A term used for the perianth segments when the sepals and petals are alike, eg. Liliaceae.

teratology The study of abnormal or aberrant forms.

terete Slender and cylindrical.

terminal The apex or end.

ternate Divided or arranged in threes.

terrestrial Growing in the ground.

testa The outer covering of the seed; the seed-coat.

tetragonous With four angles.

tetrahedral With four sides.

thorn A reduced branch ending in a hard sharp point.

tomentose Densely covered with short, matted, soft hairs.

tomentum A covering of matted, soft hairs.

tortuous Twisted; with irregular bending.

transpiration The loss of water vapour to the atmosphere through openings in the leaves of plants.

tree A woody plant that produces a single trunk and a distinct elevated head.

triad A group of three.

tribe A taxonomic group of related genera within a family or subfamily.

trichomes A term used to describe outgrowths of the epidermis such as scales or hairs.

trifid Divided into three to about the middle.

trifoliolate A compound leaf with three leaflets.

trigonous With three angles.

trilobed, trilobate With three lobes.

tripinnate Divided three times.

triquetrous With three angles or ridges.

truncate Ending abruptly as if cut off.

trunk The main stem of a tree.

tuber The swollen end of an underground stem or stolon.

tubercle Small tubers produced on aerial stems, eg. *Dioscorea*.

tuberculate With knobby or warty projections.

tuberoid The swollen end of an underground root.

tuberous Swollen and fleshy; resembling a tuber.

tufted Growing in small, erect clumps.

turgid Swollen or bloated.

twiner Climbing plants which ascend by twining of their stems or rhachises.

type form The form of a variable species from which the species was originally described.

umbel An inflorescence in which all the stems arise at the same point and the flowers lie at the same level.

umbellate Like an umbel.

undulate Wavy.

unequal Of different sizes.

unilateral One-sided.

unilocular With one cavity.

unisexual Of one sex only: staminate or pistillate.

united Joined together, wholly or partially.

urceolate Urn-shaped.

valvate Opening by valves; with perianth segments overlapping in the bud.

valve A segment of a woody fruit.

variegated Where the basic colour of a leaf or petal is broken by areas of another colour, usually white, pale green or yellow.

variety A taxonomic subgroup within a species used to differentiate variable populations.

vascular bundle The internal conducting system of plants.

vascular plant A plant bearing water conducting tissue in its organs.

vegetation The whole plant communities of an area.

vegetative Asexual development or propagation.

vein The conducting tissue of leaves.

veinlet A small or slender vein.

velamen A veil; spongy epidermis of epiphytic roots.

venation The pattern formed by veins.

vernalization The promotion of flowering.

verrucose Rough and warty.

verticillate Arranged in whorls.

viable Alive and able to germinate, as of seeds.

villous Covered with long, soft, shaggy hairs.

virgate Twiggy.

viscid Coated with a sticky or glutinous substance.

viviparous Germinating while still attached to the parent plant, eg. mangroves.

watershoot A strong rapid growing shoot arising from the trunk or main stem.

whorl Three or more segments (leaves, flowers etc.) in a circle at a node.

wing A thin, dry, membranous expansion of an organ; the side petals of flowers of the family Fabaceae.

woolly Bearing long, soft matted hairs.

xeromorph A plant with drought resistant features.

xerophyte A drought resistant plant.

zygomorphic Asymmetrical and irregular; usually applied to flowers, eg. *Anigozanthos*.

zygote A fertilized egg.

Further Reading

Adams, G. M. (1980) *Birdscaping Your Garden*, Rigby, Adelaide.

Allen, H. H. (1961) *Flora of New Zealand*, Volume 1, Government Printer, Wellington, NZ.

Anderson, R. H. (1967) *The Trees of New South Wales*, Government Printer, NSW.

Armitage, I. (1978) *Acacias of New South Wales*, New South Wales Region, Society for Growing Australian Plants, Sydney.

Aston, H. I. (1973) *Aquatic Plants of Australia*, Melbourne University Press.

Audas, J. W. (193?) *Native Trees of Australia*, Whitcombe & Tombs Ltd., Melbourne.

Australian Plant Study Group (1980) *Grow What Where*, Thomas Nelson (Australia) Ltd., Melbourne.

Australian Plant Study Group (1982) *Grow What Wet*, Thomas Nelson (Australia) Ltd., Melbourne.

Australian Systematic Botany Society (1981) *Flora of Central Australia*, (Ed J. Jessop), A. H. & A. W. Reed, Sydney.

Austrobaileya (1977-) Periodical Journal of Queensland Herbarium, Department of Primary Industries, Brisbane, Government Printer, Qld.

Bailey, F. M. (1899-1902) *The Queensland Flora*, Queensland Government, H. J. Diddams, Qld.

Bailey, F. M. (1909) *A Comprehensive Catalogue of Queensland Plants*, Government Printer, Qld.

Baines, J. A. (1981) *Australian Plant Genera*, Society for Growing Australian Plants, Sydney.

Baker, K. F. (Ed) (1957) *The U.C. System for Producing Healthy Container-Grown Plants*, University of California.

Beadle, N. C. W. (1971-80) *Students Flora of North-Eastern New South Wales*, Parts 1-4, University of New England, Armidale.

Beadle, N. C. W., Evans, O. D. and Carolin, R. C. (1972) *Flora of the Sydney Region*, A. H. & A. W. Reed, Sydney.

Beard, J. S. (Ed) (1965) *Descriptive Catalogue of West Australian Plants*, Society for Growing Australian Plants, Sydney.

Beauglehole, A. C. (1980) *Victorian Vascular Plant Checklist*, Western Victorian Field Naturalists Clubs Association, Portland, Vic.

Bentham, G. and Mueller, F. (1863-78) *Flora Australiensis*, Parts 1-7, Lovell Reeve & Co., London.

Black, J. M. (1943-57) *Flora of South Australia*, Parts 1-4, Government Printer, Adelaide.

Black, J. M. (1978) *Flora of South Australia*, Part 1 (3rd Edition), Government Printer, SA.

Blackall, W. E. (1954) *How to Know Western Australian Wildflowers*, Part 1, University of Western Australia Press, Perth.

Blackall, W. E. and Grieve, B. J. (1956-80) *How to Know Western Australian Wildflowers*, Parts 2-4, University of Western Australia Press, Perth.

Blake, S. T. and Roff, C. (1972) *The Honey Flora of Queensland*, Government Printer, Qld.

Blake, T. L. (1981) *A Guide to Darwinia and Homoranthus*, Society for Growing Australian Plants, Maroondah Group, Ringwood, Vic.

Blombery, A. M. (1967) *A Guide to Native Australian Plants*, Angus & Robertson, Sydney.

Blombery, A. M. (1972) *What Wildflower is That?*, Paul Hamlyn, Sydney.

Blombery, A. and Rodd, A. N. (1982) *Palms*, Angus & Robertson, Sydney.

Bonney, N. B. (1977) *An Introduction to the Identification of Native Flora in the Lower South-East of South Australia*, South-East Community College, Mt. Gambier.

Boomsma, C. D. (1972) *Native Trees of South Australia*, Woods & Forests Department, SA.

Bridgeman, P. H. (1976) *Tree Surgery*, David & Charles Ltd., London.

Brooks, A. E. (1973) *Australian Native Plants for Home Gardens*, 6th Edition, Lothian Publishing Co., Melbourne.

Brown, A. and Hall, N. (1968) *Growing Trees on Australian Farms*, Forestry & Timber Bureau, Canberra.

Brunonia (1978-) Periodical Journal of the Herbarium Australiense, CSIRO, Melbourne.

Burbridge, N. T. (1963) *Dictionary of Australian Plant Genera*, Angus & Robertson, Sydney.

Burbridge, N. T. and Gray, M. (1970) *Flora of the Australian Capital Territory*, Australian National University Press, Canberra.

Bureau of Flora and Fauna (1981-) *Flora of Australia*, (Ed A. S. George), Australian Government Publishing Service, Canberra.

Burgman, M. A. and Hopper, S. D. (1982) *The Western Australian Wildflower Industry*, Department of Fisheries & Wildlife, Perth.

Canberra Botanic Gardens (1971-) *Growing Native Plants* (series), Australian Government Publishing Service, Canberra.

Carman, J. K. (1978) *Dyemaking with Eucalypts*, Rigby, Adelaide.

Chippendale, G. M. (Ed) (1968) *Eucalyptus Buds and Fruits*, Forestry & Timber Bureau, Canberra.

Further Reading

Chippendale, G. M. (1973) *Eucalypts of the Western Australian Goldfields*, Australian Government Publishing Service, Canberra.

Clements, M. A. (1982) *Preliminary Checklist of Australian Orchidaceae*, National Botanic Gardens, Canberra.

Clifford, H. T. and Constantine, J. (1980) *Ferns, Fern Allies and Conifers of Australia*, University of Queensland Press, St. Lucia, Qld.

Cochrane, R. G., Fuhrer, B.A., Rotherham, E. R. and Willis, J. H. (1968) *Flowers and Plants of Victoria*, A. H. & A. W. Reed, Sydney.

Conabere, B. and Garnet, J. Ros (1974) *Wildflowers of South-Eastern Australia*, Volumes 1-2, Thomas Nelson (Australia) Ltd., Melbourne.

Contributions from the Queensland Herbarium (1968-1975), Department of Primary Industries, Brisbane, Government Printer, Qld.

Costermans, L. (1981) *Native Trees and Shrubs of South-Eastern Australia*, Rigby, Adelaide.

Costin, A. B., Gray, M., Totterdell, C. J. and Wimbush, D. J. (1979) *Kosciusko Alpine Flora*, CSIRO, Melbourne and William Collins, Sydney.

Cribb, A. B. and Cribb, J. W. (1974) *Wild Food in Australia*, William Collins, Sydney.

Cribb, A. B. and Cribb, J. W. (1981) *Useful Wild Plants in Australia*, William Collins, Sydney.

Cribb, A. B. and Cribb, J. W. (1981) *Wild Medicine in Australia*, William Collins, Sydney.

Cunningham, G. M., Mulham, W. E., Milthorpe, P. L. and Leigh, J. H. (1981) *Plants of Western New South Wales*, Soil Conservation Service of New South Wales.

Curtis, W. M. (1956-79) *The Students Flora of Tasmania*, Parts 1-4a, Government Printer, Tas.

Curtis, W. M. and I. Morris, D. K. (1975) *The Students Flora of Tasmania*, Part 1, 2nd Edition, Government Printer, Tas.

Dockrill, A. W. (1969) *Australian Indigenous Orchids*, Volume 1, Society for Growing Australian Plants, Sydney.

Eichler, H. (1965) *Supplement to J. M. Black's Flora of South Australia*, 2nd Edition, Government Printer, Adelaide.

Elliot, G. M. (1979) *Australian Plants for Small Gardens and Containers*, Hyland House, Melbourne.

Elliot, G. M. (1981) *Fun With Australian Plants*, Hyland House, Melbourne.

Elliot, R. (1975) *An Introduction to the Grampians Flora*, Algona Guides, Melbourne.

Erickson, R. (1958) *Triggerplants*, Paterson Brokensha Pty. Ltd., Perth.

Erickson, R. (1968) *Plants of Prey in Australia*, Lamb Publications, Osborne Park, WA.

Erickson, R., George, A. S., Marchant, N. G. and Morcombe, M. K. (1973) *Flowers and Plants of Western Australia*, A. H. & A. W. Reed, Sydney.

Everist, S. K. (1974) *Poisonous Plants of Australia*, Angus & Robertson, Sydney.

Ewart, A. J. (1930) *Flora of Victoria*, Government Printer, Vic.

Fairhall, A. R. (1970) *West Australian Native Plants in Cultivation*, Pergamon Press, Sydney.

Floyd, A. G. (1960-) *NSW Rainforest Trees* (series),

Research Notes, Forestry Commission of New South Wales, Sydney.

Francis, W. D. (1970) *Australian Rainforest Trees*, 3rd Edition, Forestry & Timber Bureau, Canberra.

Fuller, L. and Badans, R. (1980) *Wollongong's Native Trees*, Fuller, Wollongong.

Galbraith, J. (1977) *Collins Field Guide to the Wild Flowers of South-East Australia*, William Collins, Sydney.

Gardner, C. A. (1975) *Wildflowers of Western Australia*, West Australian Newspapers Ltd., Perth.

Gardner, C. A. and Bennetts, H. W. (1956) *The Toxic Plants of Western Australia*, West Australian Newspapers Ltd., Perth.

Gemmell, N. (1980) *Trees and Places, a South Australian Study*, Investigator Press Pty. Ltd., SA.

Goodman, R. D. (1973) *Honey Flora of Victoria*, Department of Agriculture, Vic.

Green, J. W. (1981) *Census of the Vascular Plants of Western Australia*, Western Australian Herbarium, Department of Agriculture, Perth.

Groves, R. H. (1981) *Australian Vegetation*, Cambridge University Press, Cambridge, England.

Guilfoyle, W. R. (about 1927) *Australian Plants Suitable for Gardens, Parks, Timber Reserves etc.*, Whitcombe & Tombs Ltd., Melbourne.

Hadlington, P. W. and Johnston, J. A. (1977) *A Guide to the Care and Cure of Australian Trees*, New South Wales University Press Ltd.

Hall, N. (1972) *The Use of Trees & Shrubs in the Dry Country of Australia*, Forestry & Timber Bureau, Canberra.

Hall, N., Johnston, R. D. and Chippendale, G. M. (1970) *Forest Trees of Australia*, Australian Government Publishing Service, Canberra.

Handweavers and Spinners Guild of Victoria (1974) *Dyemaking With Australian Flora*, Rigby, Adelaide.

Harmer, J. (1975) *North Australian Plants*, Part 1, Society for Growing Australian Plants, Sydney.

Harris, T. Y. (1977) *Gardening with Australian Plants — Shrubs*, Thomas Nelson (Australia) Ltd., Melbourne.

Harris, T. Y. (1979) *Gardening with Australian Plants — Small Plants and Climbers*, Thomas Nelson (Australia) Ltd., Melbourne.

Harris, T. Y. (1980) *Gardening with Australian Plants — Trees*, Thomas Nelson (Australia) Ltd., Melbourne.

Hartley, W. (1979) *A Checklist of Economic Plants in Australia*, CSIRO, Melbourne.

Hartmann, H. T. and Kester, D. E. (1975) *Plant Propagation Principles and Practices*, Third Edition, Prentice Hall, USA.

Hearne, D. A. (1975) *Trees for Darwin & Northern Australia*, Australian Government Publishing Service, Canberra.

Heywood, V. H. (1978) *Flowering Plants of the World*, Mayflower Books Inc., New York, USA.

Hockings, F. D. (1980) *Friends and Foes of Australian Gardens*, A. H. & A. W. Reed, Sydney.

Holliday, I., and Watton, G. (1975) *A Field Guide to Banksias*, Rigby, Adelaide.

Hyland, B. P. M. (1982) *A Revised Card Key to the Rain-*

forest Trees of North Queensland, CSIRO, Melbourne.

Jacobs, S. W. L. and Pickard, J. (1980) *Plants of New South Wales,* Government Printer, Sydney.

Jones, D. L. and Clemesha, S. C. (1976) *Australian Ferns and Fern Allies,* A. H. & A. W. Reed, Sydney.

Jones, D. L. and Gray, B. (1977) *Australian Climbing Plants,* A. H. & A. W. Reed, Sydney.

Journal of the Adelaide Botanic Gardens (1976-) Adelaide Botanic Gardens, SA.

Keighery, G. (1979 unpublished) *Notes on the Biology and Phytogeography of Western Australian Plants,* Kings Park and Botanic Gardens, West Perth.

Kelly, S. (1969) *Eucalypts,* Volume 1, Thomas Nelson (Australia) Ltd., Melbourne.

Kelly, S. (1978) *Eucalypts* Volume II, Thomas Nelson (Australia) Ltd., Melbourne.

Kings Park Research Notes (1973-) Kings Park Board, Kings Park and Botanic Gardens, West Perth.

Kleinschmidt, H. E. and Johnson, R. W. (1979) *Weeds of Queensland,* Queensland Department of Primary Industries, Brisbane.

Landscape Australia, (1979-) Periodical Journal of the Australian Institute of Landscape Architects, Landscape Publications, Melbourne.

Launceston Field Naturalists Club (1981) *Guide to Flowers and Plants of Tasmania,* M. Cameron (Ed), A. H. & A. W. Reed, Sydney.

Lazarides, M. (1970) *Grasses of Central Australia,* Australian National University Press, Canberra.

Lear, R. and Turner, T. (1977) *Mangroves of Australia,* University of Queensland Press, St. Lucia, Qld.

Leeper, G. W. (Ed) (1970) *The Australian Environment,* CSIRO and Melbourne University Press.

Leigh, J. and Boden, R. (1979) *Australian Flora in the Endangered Species Convention* — CITES, Australian National Parks & Wildlife Service, Canberra.

Leigh, J., Briggs, J. and Hartley, W. (1981) *Rare or Threatened Australian Plants,* Australian National Parks & Wildlife Service, Canberra.

Lord, E. E. and Willis, J. H. (1982) *Shrubs and Trees for Australian Gardens,* 5th Edition, Lothian Publishing Co., Melbourne.

Loudon, W. (Ed Mrs Loudon) (1855) *Loudon's Encyclopaedia of Plants,* Longman, Brown, Green and Longmans, London.

Maiden, J. H. (1889) *The Useful Native Plants of Australia,* Facsimile Edition, Alexander Bros Pty. Ltd., Melbourne.

Mathias, M. E. (1976) *Colour For the Landscape,* Brooke House Publishers, California, USA.

McCubbin, C. (1971) *Australian Butterflies,* Thomas Nelson (Australia) Ltd., Melbourne.

McLuckie, J. and McKee, H. S. (1962) *Australian & New Zealand Botany,* Horwitz-Graeme, Sydney.

Molyneux, B. (1980) *Grow Native,* Anne O'Donovan, Melbourne.

Moore, L. B. and Edgar, E. (1980) *Flora of New Zealand,* Volume 2, Government Printer, Wellington, NZ.

Muelleria (1965-) Periodical Journal of the National Herbarium of Victoria, Government Printer, Vic.

Newbey, K. (1968-72) *West Australian Plants for Hor-*

ticulture, Parts 1-2, Society for Growing Australian Plants, Sydney.

Newbey, K. (1982) *Growing Trees on Western Australian Farms,* Farm Management Foundation of Australia (Inc.), Mosman Park, WA.

Nicholls, W. A. (1964) *Orchids of Australia,* Complete Edition (Ed D. L. Jones and T. B. Muir), Thomas Nelson (Australia) Ltd., Melbourne.

Nuytsia (1970-) Bulletin of the Western Australian Herbarium, Department of Agriculture, Western Australia, Government Printer, WA.

Penfold, A. R. and Willis, J. L. (1961) *The Eucalypts,* Leonard Hill, London.

Plumridge, J. (1976) *How to Propagate Plants,* Lothian Publishing Co., Melbourne.

Pryor, L. D. and Johnson, L. A. S. (1971) *Classification of the Eucalypts,* Australian National University, Canberra.

Rogers, F. J. C. (1971) *Growing Australian Native Plants,* Thomas Nelson (Australia) Ltd., Melbourne.

Rogers, F. J. C. (1975) *Growing More Australian Native Plants,* Thomas Nelson (Australia) Ltd., Melbourne.

Rogers, F. J. C. (1978) *A Field Guide to Victorian Wattles,* Revised Edition, F. J. C. Rogers.

Rotherham, E. R., Briggs, B. C., Blaxell, D. F. and Carolin, R. C. (1975) *Flowers & Plants of New South Wales and Southern Queensland,* A. H. & A. W. Reed, Sydney.

Royal Horticultural Society (1956) *Dictionary of Gardening,* 4 Volumes plus Supplement, 2nd Edition with corrections, Oxford University Press, England.

Rye, B. L. and Hopper, S. D. (1981) *A Guide to the Gazetted Rare Flora of Western Australia,* Department of Fisheries & Wildlife, Perth.

Rye, B. L., Hopper, S. D. and Watson, L. E. (1980) *Commercially Exploited Vascular Plants Native in Western Australia: Census, Atlas and Preliminary Assessment of Conservation Status,* Department of Fisheries & Wildlife, Perth.

Salter, B. (1977) *Australian Native Gardens and Birds,* Ure Smith, Sydney.

Sharr, F. A. (1978) *Western Australian Plant Names and their Meanings,* University of Western Australia Press, WA.

Simon, B. K. (1978) *A Preliminary Check-list of Australian Grasses,* Botany Branch, Department of Primary Industry, Brisbane.

Simpfendorfer, K. J. (1975) *An Introduction to Trees for South-Eastern Australia,* Inkata Press, Victoria.

Smith, A. W. and Stearn, W. T. (1972) *A Gardener's Dictionary of Plant Names,* Enlarged Edition, Cassell & Company Ltd., London.

Society for Growing Australian Plants, *Australian Plants,* Quarterly, Society for Growing Australian Plants, Sydney.

Society for Growing Australian Plants, Canberra Region (1976) *Australian Plants For Canberra Gardens,* 2nd Edition, Society for Growing Australian Plants, Canberra.

Society for Growing Australian Plants, Maroondah Group, Vic., Miscellaneous publications, Society

Further Reading

for Growing Australian Plants, Maroondah Group, Ringwood, Vic.

Society for Growing Australian Plants, Western Suburbs Branch, Queensland (1979) *Perfumed and Aromatic Australian Plants*, Society for Growing Australian Plants, Western Suburbs Branch, Indooroopilly, Qld.

Sparnon, N. (1967) *The Beauty of Australia's Wildflowers*, Ure Smith, Sydney.

Spooner, P. (Ed) (1973) *Practical Guide to Home Landscaping*, Readers Digest, Sydney.

Stead Memorial Wildlife Research Foundation, Miscellaneous publications, Stead Memorial Wildlife Research Foundation, Sydney.

Stearn, W. T. (1973) *Botanical Latin*, 2nd Edition, David and Charles, Newton Abbot, England.

Stones, M. and Curtis, W. (1967-78) *The Endemic Flora of Tasmania*, Parts 1-6, The Ariel Press, London.

Telopea (1975-) Contributions from the National Herbarium of New South Wales, Government Printer, NSW.

Tothill, J. C. and Hacker, J. B. (1973) *The Grasses of Southeast Queensland*, University of Queensland Press, St. Lucia, Qld.

Wells, J. S. (1955) *Plant Propagation Practices*, Macmillan, New York, USA.

Western Australian Herbarium Research Notes (1978-) Western Australian Department of Agriculture, Perth.

Wheeler, D. J. B., Jacobs, S. W. L. and Norton, B. E. (1982) *Grasses of New South Wales*, The University of New England, Armidale, NSW.

Whibley, D. J. E. (1980) *Acacias of South Australia*, Government Printer, Adelaide.

Williams, K. A. W. (1979) *Native Plants Queensland*, Volume 1, K. A. W. Williams, North Ipswich, Qld.

Willis, J. C. (1973) *A Dictionary of the Flowering Plants and Ferns*, 8th Edition, Revised by H. K. Airy-Shaw, Cambridge University Press, London.

Willis, J. H. (1962-72) *A Handbook to Plants in Victoria*, Volumes 1 and 2, Melbourne University Press.

Wilson, G. (1975) *Landscaping with Australian Plants*, Thomas Nelson (Australia) Ltd., Melbourne.

Wilson, G. (1980) *Amenity Planting in Arid Zones*, Canberra College of Advanced Education, Canberra.

Womersley, J. S. (Ed) (1978) *Handbooks of the Flora of Papua New Guinea* Volume 1, Melbourne University Press.

Wrigley, J. W. and Fagg, M. (1979) *Australian Native Plants*, William Collins, Sydney.

Zimmer, G. F. (1956) *A Popular Dictionary of Botanical Names and Terms*, Routledge, Kegan & Paul Ltd., London.

506

Common Name Index

Index

Index

Index

Index